Data Mining Algorithms

Data Mining Algorithms: Explained Using R

Paweł Cichosz

Department of Electronics and Information Technology
Warsaw University of Technology
Poland

WILEY

Library of Congress Cataloging-in-Publication Data

Cichosz, Pawel, author.
  Data mining algorithms : explained using R / Pawel Cichosz.
       pages cm
  Summary: "This book narrows down the scope of data mining by adopting a heavily modeling-oriented perspective" – Provided by publisher.
  Includes bibliographical references and index.
  ISBN 978-1-118-33258-0 (hardback)
  1. Data mining. 2.   Computer algorithms. 3.   R (Computer program language)   I. Title.
  QA76.9.D343C472 2015
  006.3′12–dc23

                                                                                          2014036992

A catalogue record for this book is available from the British Library.

ISBN: 9781118332580

Typeset in 10/12pt Times by Laserwords Private Limited, Chennai, India

To my wife, Joanna, and my sons,
Grzegorz and Łukasz

Contents

Acknowledgements

With the rise and rapidly growing popularity of online idea sharing methods, such as blogs and wikis, traditional books are no longer the only way of making large portions of text available to a wide audience. The former are particularly suitable for collaborative or social writing and readings undertakings, often with mixed reader–writer roles of particular participants. For individual writing and reading efforts the traditional book form (although not necessarily tied to the paper media) still remains the best approach. On one hand, it clearly assigns full and exclusive responsibility for the contents to the author, with no easy excuses for errors and other deficiencies. On the other hand, there are several other people engaged in the publishing process who help to give the book its final shape and protect the audience against a totally flawed work.

As the author of this book, I feel indeed totally responsible for all its imperfections, only some which I am aware of, but I have no doubts that there are many more of them. With that being said, several people from the editorial and production team worked hard to make the imperfect outcome of my work worth publishing. My thanks go, in particular, to Richard Davies, Prachi Sinha Sahay, Debbie Jupe, and Kay Heather from Wiley for their encouragement, support, understanding, and reassuring professionalism at all stages of writing and production. Radhika Sivalingam, Lincy Priya, and Yogesh Kukshal did their best to transform my manuscript into a real book, meeting publication standards. I believe there are others who contributed to this book's production that I am not even aware of and I am grateful to them all, also.

I was thoughtless enough to share my intention to write this book with my colleagues from the Artificial Intelligence Applications Research Group at the Warsaw University of Technology. While their warm reception of this idea and constant words of encouragement were extremely helpful, I wished I had not done that many times. It would have been so much easier to give up if I had kept this in secret. Perhaps the ultimate reason why I continued to work despite hesitations is that I knew they would keep asking and I would be unable to find a good excuse. Several thoughts expressed in this book were shaped by discussions during our group's seminar meetings. Interacting with my colleagues from the analytics teams at Netezza Poland, IBM Poland, and iQor Poland, with which I had an opportunity to work on some data mining projects at different stages of writing the book, was also extremely stimulating, although the contents of the book have no relationships with the projects I was involved with.

I owe special thanks to my wife and two sons, who did not directly contribute to the contents of this book, but made it possible by allowing me to spend much of my time that should be normally devoted to them on this work and providing constant encouragement. If you guys can read these thanks in a published copy of the book, then it means it is all over at last and we will hopefully get back to normal life.

Preface

Data mining

Data mining has been a rapidly growing field of research and practical applications during the last two decades. From a somewhat niche academic area at the intersection of machine learning and statistics it has developed into an established scientific discipline and a highly valued branch of the computing industry. This is reflected by data mining becoming an essential part of computer science education as well as the increasing overall awareness of the term "data mining" among the general (not just computing-related) academic and business audience.

Scope

Various definitions of data mining may be found in the literature. Some of them are broad enough to include all types of data analysis, regardless of the representation and applicability of their results. This book narrows down the scope of data mining by adopting a heavily modeling-oriented perspective. According to this perspective the ultimate goal of data mining is delivering *predictive models*. The latter can be thought of as computationally represented chunks of knowledge about some domain of interest, described by the analyzed data, that are capable of providing answers to queries *transcending the data*, i.e., such that cannot be answered by just extracting and aggregating values from the data. Such knowledge is discovered from data by capturing and generalizing useful relationship patterns that occur therein.

Activities needed for creating predictive models based on data and making sure that they meet the application's requirements fall in the scope of data mining as understood in this book. Analytical activities which do not contribute to model creation – although they may still deliver extremely useful results – remain therefore beyond the scope of our interest. This still leaves a lot of potential contents to be covered, including not only modeling algorithms, but also techniques for evaluating the quality of predictive models, transforming data to make modeling algorithms easier to apply or more likely to succeed, selecting attributes most useful for model creation, and combining multiple models for better predictions.

Modeling view

The modeling view of data mining is by no means unique for this book. It is actually the most natural and probably the most wide-spread view of data mining. Nevertheless, it deserves some more attention in this introductory discussion, which is supposed to let the reader know what this book is about. In particular, it is essential to underline – and it will be repeatedly underlined on several other occasions throughout the book – that a useful data mining model is not merely a description of some patterns discovered in the data. In other words, it does not only and not mainly represent knowledge about the data, but also – and much more importantly – knowledge about the domain from which the data originates.

The *domain* can be considered a set of entities from the real world about which knowledge is supposed to be delivered by data mining. These can be people (such as customers, employees, patients), machines and devices (such as car engines, computers, or ATMs), events (such as car failures, purchases, or bank transactions), industrial processes (such as manufacturing electronic components, energy production, or natural resources exploitation), business units (such as stores or corporate departments), to name only a few typical possibilities. Such real-world entities – in this book referred to as *instances* – are, usually incompletely and imperfectly, described by a set of features – in this book referred to as *attributes*. A *dataset* is a subset of the domain, described by the set of available attributes, usually – assuming a tabular data representation – with rows corresponding to instances and columns corresponding to attributes. Data mining can then be viewed as an analytic process that uses one or more available datasets from the same domain to create one or more *models* for the domain, i.e., models that can be used to answer queries not just about instances from the data used for model creation, but also about any other instances from the same domain. More directly and technically, speaking, if some attributes are generally available (observable) and some attributes are only available on a limited dataset (hidden), then models can often be viewed as delivering predictions of hidden attributes wherever their true values are unavailable. The unavailable attribute values to be predicted usually represent properties or quantities that are hard and costly to determine, or (more typically) that become known later than are needed. The latter justifies the term "prediction" used when referring to a model's output. The attribute to be predicted is referred to as the *target attribute*, and the observable attributes that can be used for prediction are referred to as the *input attributes*.

Tasks

The most common types of predictive models – or queries they can be used to answer – correspond to the following three major data mining tasks.

Classification. Predicting a discrete target attribute (representing the assignment of instances to a fixed set of possible classes). This could be distinguishing between good and poor customers or products, legitimate and fraudulent credit card transactions or other events, assigning failure types and recommended repair actions to faulty technical devices, etc.

Regression. Predicting a numeric target attribute which represents some quantity of interest. This could be an outcome or a parameter of an industrial process, an amount of money earned or spent, a cost or gain due to a business decision, etc.

Clustering. Predicting the assignment of instances to a set of similarity-based clusters. Clusters are not predetermined, but discovered as part of the modeling process, to achieve possibly high intracluster similarity and possibly low intercluster similarity.

Most real-world data mining projects include one or more instantiations of these three generic tasks. Similarly, most of data mining research contributes, modifies, or evaluates algorithms for these three tasks. These are also the tasks on which this book is focused.

Origin

Data mining techniques have their roots in two fields: machine learning and statistics. With the former traditionally addressing the issue of acquiring knowledge or skill from supplied training information and the latter the issue of describing the data as well as identifying and approximating relationships occurring therein, they both have contributed modeling algorithms. They have also become increasingly closely related, which makes it difficult and actually unnecessary to put hard separating boundaries between them. With that being said, their common terminological and notational conventions remain partially different, and so do background profiles of researchers and practitioners in these fields. Wherever this difference matters, this book is much closer to machine learning than statistics, to the extent that the description of "strictly statistical" techniques – appearing rather sparingly – may be found oversimplified by statisticians. In particular, the formulations of major data mining tasks in Chapter 1 assume the inductive learning perspective.

The brief discussion of the modeling view of data mining presented in the previous section makes it possible to encounter this book's bias toward machine learning for the first time. The terms "domain," "instance," "attribute," and "dataset," in particular, have their counterparts that are more common in statistics, such as "population," "observation," "variable," and "sample."

Motivation

The book is intended to be a practical, technically oriented guide to data mining algorithms, focused on clearly explaining their internal operation and properties as well as major principles of their application. According to the general perspective of data mining adopted by the book, it encompasses all analytic processes performed to produce predictive models from available data and verify whether and to what extent they meet the application's requirements. The book will cover the most important algorithms for building classification, regression, and clustering models, as well as techniques used for attribute selection and transformation, model quality evaluation, and creating model ensembles.

The book will hopefully appeal to the reader, either already familiar with data mining to some extent or just approaching the field, by its practical and technical, utility-driven perspective, making it possible to quickly start gaining his or her own hands-on experience. The reader will be given an opportunity to become familiar with a number of data mining algorithms, presented in a systematic, coherent, and relatively easy to follow way. By studying their description and examples the reader will learn how they work, what properties they exhibit, and how they can be used.

The book is not intended to be a "data mining bible" providing a complete coverage of the area, but rather to selectively focus on a number of algorithms that:

- are known to work well for the most common data mining tasks,
- are good representatives of typical data mining techniques,

- can be well explained to the general technically educated audience without an excessive required mathematical and computing background,

- can be used to illustrate good practices as well as caveats of data mining.

It is not supposed to be a business-oriented manager's guide to data mining or a bird's eye-perspective overview of the field, either. Topics covered by the book are discussed in a technical way, with a level of detail believed to be adequate for most practical needs, albeit not overwhelming. This includes the presentation of the internal mechanisms, properties, and usage scenarios of algorithms that are not extremely complex mathematically or implementationally and offer the potential of excellent results in many applications, but may need expertise and experience to be used fruitfully. The ambition of the book is to help the reader develop that expertise and experience.

The book's technical and practical orientation, with very limited theoretical background, but a relatively high level of detail on algorithm internal operation and application principles, makes it appropriate for a mixed audience consisting of

- students of computer science and related fields,

- researchers working on experimental or applied research projects in any area where data analysis capabilities are used,

- analysts and engineers working with data and creating or using predictive models.

The book should be particularly appealing to computer scientists and programmers due to its extensive use of R code examples, as explained below.

While the little-background assumption makes the book suitable as an introductory text, the level of detail and precision puts it actually on an advanced or semi-advanced level, since on many occasions – whenever it is justified by practical utility – it discusses issues that tend too be overlooked or taken lightly in typical introductions to data mining.

Organization

The book is divided into the following parts:

Part I. Preliminaries.
Part II. Classification.
Part III. Regression.
Part IV. Clustering.
Part V. Getting better models.

Part I contains two chapters, as summarized below.

Chapter 1: Tasks. This chapter introduces the major data mining tasks algorithms for which are presented in the book: classification, regression, and clustering.

Chapter 2: Basic statistics. This chapter is be devoted to simple techniques for performing data exploration tasks, usually referred to as basic statistics, that are often applied before any modeling algorithms or used internally by some modeling algorithms.

Part II contains five chapters listed below.

Chapter 3: Decision trees. This chapter presents algorithms for creating decision tree classification models, shortly referred to as decision tree algorithms.

Chapter 4: Naïve Bayes classifier. This chapter presents arguably the simplest useful classification algorithm, the naïve Bayes classifier.

Chapter 5: Linear classification. This chapter is devoted to classification algorithms that adopt a linear model representation. Since they are largely based on linear regression algorithms, forward references to Chapter 8 are unavoidable.

Chapter 6: Misclassification costs. This chapter systematically discusses the issue of nonuniform misclassification costs in the classification task and techniques that can be used to create cost-sensitive classification models.

Chapter 7: Classification model evaluation. This chapter is devoted to techniques used to evaluate classification models: performance measures serving as model quality indicators on a particular dataset, and evaluation procedures used to reliably estimate their expected values on new data. Since the presented evaluation procedures are applicable to regression and clustering models as well, the extensive discussion presented in this chapter makes it possible to keep the chapters on regression and clustering model evaluation considerably shorter.

The contents of Part III is summarized below.

Chapter 8: Linear regression. This chapter presents regression algorithms that employ a linear model representation and parameter estimation techniques used to find model parameters.

Chapter 9: Regression trees. This chapter presents regression tree algorithms, which are decision trees adapted for the regression task. This makes it possible to refer to Chapter 3 extensively and focus mostly on regression-specific issues.

Chapter 10: Regression model evaluation. This chapter is devoted to techniques used to assess the quality of regression models. It will focus mostly on regression model performance measures, since evaluation procedures applicable to regression models are the same as those for classification models presented in Chapter 7.

Part IV contains four chapters listed below.

Chapter 11: Dissimilarity measures. This chapter presents several dissimilarity and similarity measures used for clustering.

Chapter 12: k-Centers clustering. This chapter is devoted to the popular *k*-centers family of clustering algorithms.

Chapter 13: Hierarchical clustering. This chapter presents algorithms for creating hierarchical clustering models.

Chapter 14: Clustering model evaluation. This chapter discusses several quality measures used for clustering model evaluation.

Part V is more heterogenic than the preceding parts, as it covers a set of diverse techniques that can be used to improve the quality of classification, regression, or clustering models. The corresponding list of chapters is presented below.

Chapter 15: Model ensembles. This chapter reviews ensemble modeling algorithms that combine multiple models for the same task classification or regression task for better predictive power.

Chapter 16: Kernel methods. This chapter discusses the possibility of improving the capabilities of liner classification and regression models by employing kernel functions to perform implicit nonlinear data transformation into a higher-dimensional space. This is also an opportunity to discuss the support vector machines and support vector regression algorithms, with which kernel functions are most often combined.

Chapter 17: Attribute transformation. This chapter presents selected techniques used for transforming data prior to applying modeling algorithms, to make them easier to apply and more likely to deliver good models.

Chapter 18: Discretization. This chapter addresses one specific type of data transformation, the discretization of continuous attributes, that is particularly often applied and for which several different algorithms have been developed.

Chapter 19: Attribute selection. This chapter is devoted to attribute selection algorithms, used to select a subset of available attributes with the highest predictive utility.

The last chapter in the book, Chapter 20, is placed outside the division into parts, since it contains data mining case studies that put the algorithms covered in the previous chapters at real work. Existing R implementations and publicly available datasets will be used to demonstrate the process of searching for the possibly best model step by step. This will include data transformation whenever necessary or useful, attribute selection, modeling algorithm application with parameter tuning, and model evaluation.

Notation

Different families of data mining algorithms are often presented in the literature using different notational conventions. Even for different descriptions of the same or closely related algorithms, it is not uncommon to adopt different notation. This is hardly acceptable in a single book, though, even if it covers a variety of techniques, sometimes with completely different origin. To keep the notation consistent throughout the book, it sometimes departs considerably from notational standards typical for particular algorithms and subareas of data mining. The adopted unified notation may therefore sometimes appear nonstandard or even awkward to readers accustomed to other conventions. It should be still much less confusing than changing notation from chapter to chapter.

This book's notational conventions are governed by the modeling view of data mining presented above and similarly biased toward machine learning rather than statistics. They include, in particular, explicit references to instances as elements of the domain, attributes as functions assigning values to instances, and datasets as subsets of the domain. For datasets there are also subscripting conventions used to refer to their subsets satisfying equality or inequality conditions for a particular attribute. Most of them are introduced early in the book, in Chapters 1 and 2, giving the reader enough time to get used to them before passing to modeling algorithms. They are all collected and explained in Appendix A for easy reference.

R code examples

One of primary features of this book is the extensive use of examples, which is necessary to make the desired combination of depth, precision, and readability possible. All of these examples contain R code snippets, in most cases sufficiently simple to be at least roughly comprehensible without prior knowledge of the language. There are two types of these examples.

Algorithm operation illustrations. These are numbered examples included in book sections devoted to particular algorithms or single major steps of more complex algorithms, supposed to help explain the details of internal algorithm calculations. Most of them present R implementations of either complete algorithms (for simple ones) or single steps thereof (for more complex ones). This supplements the natural language, pseudocode, or math formula algorithm description, adding some more clarification, specificity, and appeal. While serving the illustrative purpose mostly, usually inefficient and suited to single simple usage scenarios only, some portions of this example code may actually be useful for practical applications as well. This is at least likely for functions which have no direct counterparts in existing R packages. The insufficient efficiency and flexibility of these illustrative implementations will hopefully encourage the readers to develop more useful modified versions thereof.

Case studies. These bigger examples grouped in the book's final chapter are more realistic demonstrations of performing data mining tasks using standard R implementations of the algorithms described in the book and publicly available datasets. Their purpose is not to explain how particular algorithms work – which is the responsibility of the other chapters and their examples – or how particular functions should be called – which is easy to learn from their R help pages. Instead, their primary focus is on how to solve a given task using one or more available algorithms. They present the process of searching for good quality predictive models that may include data transformation, attribute selection, model building, and model evaluation. They will hopefully encourage the readers to approach similar tasks by their own.

Of those, the first category occupies much more space and corresponds to about 10 times more code lines. It can also be considered a distinctive feature of this book. This is because, while it is a relatively common practice for contemporary books on data mining or statistics to use R (or another analytic toolbox) to provide algorithm usage demonstrations or case studies, the idea of R code snippets for algorithm operation illustrations is probably relatively uncommon.

R is an increasingly popular programming language and environment for data analysis, sometimes referred to as the "lingua franca" of this domain, with a huge set of contributed packages available from the CRAN repository,[1] providing implementations of various analytic algorithms and utility functions. The book uses the R language extensively as a pedagogical tool, but does not teach it nor requires the readers to learn it. This is because the example code can be run and the results can looked up with barely any knowledge of R. Elementary knowledge of any general-purpose programming language augmented by a small set of R-specific

---

[1] *Comprehensive R Archive Network* (http://cran.r-project.org/web/packages).

features, such as its ubiquitous vectorization and logical indexing, should be sufficient to get at least some rough understanding of most of the presented example code snippets.

However, investing some effort into learning the basics of R will definitely pay off, making it possible to fully exploit the illustrative value of this book's examples. They will hopefully encourage at least some readers to study readily available tutorials and provide useful starting points for such self-study. Making the reader familiar with R is not therefore the purpose of the book, but may become its beneficial side effect.

Needless to say, R code algorithm illustrations and case studies need data. In some examples tiny and totally unrealistic datasets are used to make it possible to manually verify the results. In some other examples slightly larger artificial datasets are generated. On several occasions, however, publicly available real datasets are used, available in CRAN packages and originating from the *UCI Machine Learning Repository*.[2] These are listed in Appendix C.

It is not uncommon for some R functions defined in examples to be reused by examples in other chapters. This is due to the natural relationships among data mining algorithms, some having common operations and some being typically used in combinations. To make it easier to run the example code with such dependences, all potentially reusable functions defined in particular chapters are grouped into corresponding R packages. They all share the same dmr name prefix and are available from the book's website.

These packages, referred to as DMR packages thereafter, should be thought of as simple containers for example functions and in many respects they do not meet commonly adopted R package standards. The documentation is particularly lacking, limited to references to the books section and example numbers, but this is forgivable given the fact that they are not distributed as standalone software tools, but as a supplementary material for the book. Some frequently reused utility functions that have no illustrative value on their own and therefore are not included in any of this book's examples, are grouped in the dmr.util package. In the first example of each chapter all packages used in subsequent examples – both DMR and CRAN ones – are explicitly loaded. Additionally, whenever a function from another chapter is mentioned, the corresponding example and package containing its definition are indicated in a margin note, as demonstrated here for the err function. Appendix B contains the list of all DMR and CRAN packages used in this book.

> EX. 7.2.1
> dmr.claseval

Since the primary role of the code is didactic and illustrative, it is written without any care for efficiency and error handling, and it may not always demonstrate a good R programming style. Testing was mostly limited to the presented example calls. Since the book's chapters were created over an extended period of time, there are some noticeable inconsistencies between their R code illustrations. With all those fully justified disclaimers, the R code snippets are believed to deserve the space they occupy in the book and – while some readers may choose to skip them while reading – the definitely recommended way to use the book is to stop at each example not only to run the code and inspect the results, but ideally also to understand how it matches the preceding text or equations.

---

[2] http://archive.ics.uci.edu/ml

Website

A companion website is maintained for the book at:

http://www.wiley.com/go/data_mining_algorithms

which contains:

R code snippets.
- code from numbered examples: a single separate file for each example,
- code from case studies: a single separate file for each section of Chapter 20.

DMR packages. A package file for download for each DMR package listed in Appendix B.

Links to CRAN packages. A link to the package home page at the CRAN repository for each CRAN package, as listed in Appendix B.

Links to datasets. A link to the dataset home page at the UCI repository for each dataset listed in Appendix C.

Further readings

To prevent this section from outweighing the rest of this short chapter, it will not attempt to review or evaluate other books on related topics. Putting aside space constraints, it would be hardly possible for such a review to at least approach completeness – given the rapidly growing number of published titles – and objectivity – given the fact the book may compete with some of them. A much more realistic to achieve and also more useful role of this section is therefore to refer the reader to a subjectively selected and incomplete set of other books that may best serve the purpose of complementing this book. This will be particularly focused on areas where it is lacking, either in form or in content, either by deliberate design or due to imperfect execution.

Several data mining books give a considerably broader picture of data mining than will be presented in the foregoing chapters (e.g., Han *et al.* 2011; Hand *et al.* 2001; Tan *et al.* 2013). They contain a greater selection of algorithms for the major data mining tasks, some discussion of more specific tasks or application domains, as a well as a more adequate representation of data mining techniques that do not directly originate from machine learning. In particular, Hand *et al.* (2001) provide a much better representation of statistical foundations of data mining and present techniques not necessarily directly related to predictive modeling. These include, in particular, techniques for data exploration and visualization and association discovery algorithms. The latter are also extensively discussed by Tan *et al.* (2013). They also give much more attention to the clustering task and present a rich set of clustering algorithms. The book by Han *et al.* (2011) has a particularly rich table of contents, covering an impressive range diverse of techniques with both statistical and machine learning roots, as well as presenting a database perspective on data analysis, although with regression deliberately omitted. It additionally includes the discussion of several specific data mining tasks and application domains. Several popular algorithms for classification and regression are presented in the classic book by Duda and Hart (1973).

The machine learning perspective adopted by this book makes it naturally related to books on machine learning. Mitchell (1997) gives an excellent, even if now somewhat dated, coverage of the essential machine learning algorithms, some of which happen to be also presented in this book. This is also the case for Cichosz (2007), mostly focusing on the same three major inductive learning tasks that are presented in this book. These books make it clear, however, that there are several interesting and useful algorithms that address learning tasks not related to data mining, such as the reinforcement learning or regular language learning tasks. Bishop (2007) addresses machine learning from a more statistical viewpoint, particularly focusing on algorithms based on Bayesian inference. A similar statistics-oriented view of machine learning with a broader selection of algorithms is presented by Hastie *et al.* (2011). Rather than reviewing many different machine learning algorithms, Abu-Mostafa *et al.* (2012) insightfully discuss some basic issues of inductive learning, including overfitting and model evaluation. An extensive representation of machine learning algorithms is described by Russell and Norvig (2011) in a general artificial intelligence perspective, highlighting their links to other algorithmic tasks related to achieving intelligent behavior.

Books that present several different modeling algorithms may not always be able to allocate sufficient space to the discussion of data preparation and preprocessing techniques. This is in striking contrast with the practice of data mining projects, where they take a substantial, if not dominating, part of the overall effort. These are by no means limited to data transformation and attribute selection, covered in Chapters 17 and 19, respectively, but additionally include assembling data from different sources and fixing, alleviating, or bypassing data quality issues, to name only the two most important other activities. These and several others are comprehensively reviewed by Pyle (1999). Weiss and Indurkhya (1997) also pay high attention to data preparation and preprocessing.

It is not uncommon for more practically oriented data mining books to describe specific software tools apart from algorithms and present their application examples (e.g., James *et al.* 2013; Janert 2010; Nisbet *et al.* 2009; North 2012). They differently share space allocated to discussing general algorithm properties and demonstrating specific implementation usage. The books by Witten *et al.* (2011) andWilliams (2011) also belong to the above category, but they not only recommend or describe, but also contribute software tools: the *Weka* library of Java data mining algorithm implementations and the *Rattle* graphical front-end for selected R analytic functions. The former provides an extensive set of functionally rich modeling algorithms and related techniques for data transformation and attribute selection and the latter makes it possible to use the analytic capabilities of R without actually writing any R code. Witten *et al.* (2011) do not actually spend much time discussing code and software usage in their book, presenting instead the operation principles and properties of the implemented algorithms. With respect to data mining task and algorithm selection, it may be the most similar other book to this one, although the presentation method differs substantially. Torgo (2010) uses the R language to teach data mining, but in a different way than adopted by this book: by demonstrating complete, extensive, and realistic case studies for specific application domains. General data mining methodology and algorithm discussion is presented in the case study context, interleaved with R code examples that demonstrate how existing or newly created R functions can be used to apply the discussed techniques. The book does not leave the reader alone when learning R, with some introductory description of the language included and illustrated.

There are several other useful books for learning R, mostly combining the presentation of language constructs with the demonstrations of its application to analytic tasks (e.g., Kabacoff 2011; Teetor 2011). Unlike those, Matloff (2011) is entirely or almost entirely focused on R as a programming language, extensively discussing also the advanced and unobvious language features that deserve much more attention than they usually receive. The official R language definition (R Development Core Team 2010) is not necessarily a good self-learning material, but remains invaluable for reference.

References

Abu-Mostafa YS, Magdon-Ismail M and Lin HT 2012 *Learning from Data*. AMLBook.

Bishop CM 2007 *Pattern Recognition and Machine Learning*. Springer.

Cichosz P 2007 *Learning Systems (in Polish)* 2nd edn. WNT.

Duda RO and Hart PE 1973 *Pattern Classification and Scene Analysis*. Wiley.

Han J, Kamber M and Pei J 2011 *Data Mining: Concepts and Techniques* 3rd edn. Morgan Kaufmann.

Hand DJ, Mannila H and Smyth P 2001 *Principles of Data Mining*. MIT Press.

Hastie T, Tibshirani R and Friedman J 2011 *The Elements of Statistical Learning: Data Mining, Inference, and Prediction* 2nd edn. Springer.

James G, Witten D, Hastie T and Tibshirani R 2013 *An Introduction to Statistical Learning with Applications in R*. Springer.

Janert PK 2010 *Data Analysis with Open Source Tools*. O'Reilly.

Kabacoff R 2011 *R in Action*. Manning Publications.

Matloff N 2011 *The Art of R Programming: A Tour of Statistical Software Design*. No Starch Press.

Mitchell T 1997 *Machine Learning*. McGraw-Hill.

Nisbet R, Elder J and Miner G 2009 *Handbook of Statistical Analysis and Data Mining Applications*. Academic Press.

North MA 2012 *Data Mining for the Masses*. Global Text Project.

Pyle D 1999 *Data Preparation for Data Mining*. Morgan Kaufmann.

R Development Core Team 2010 *R: A Language and Environment for Statistical Computing* R Foundation for Statistical Computing.

Russell S and Norvig P 2011 *Artificial Intelligence: A Modern Approach* 3rd edn. Prentice Hall.

Tan PN, Steinbach M and Kumar V 2013 *Introduction to Data Mining* 2nd edn. Addison-Wesley.

Teetor P 2011 *R Cookbook*. O'Reilly.

Torgo L 2010 *Data Mining with R: Learning with Case Studies*. Chapman and Hall.

Weiss SM and Indurkhya N 1997 *Predictive Data Mining: A Practical Guide*. Morgan Kaufmann.

Williams G 2011 *Data Mining with Rattle and R: The Art of Excavating Data for Knowledge Discovery*. Springer.

Witten IH, Frank E and Hall MA 2011 *Data Mining: Practical Machine Learning Tools and Techniques* 3rd edn. Morgan Kaufmann.

Part I

PRELIMINARIES

1

Tasks

1.1 Introduction

This chapter discusses the assumptions and requirements of the three major data mining tasks this book focuses on: classification, regression, and clustering. It adopts a machine learning perspective, according to which they are all instantiations of *inductive learning*, which consists in generalizing patterns discovered in the data to create useful *knowledge*. This perfectly matches the predictive modeling view of data mining adopted by this book, according to which the ultimate goal of data mining is delivering models applicable to new data. While the book also discusses tasks that are not directly related to model creation, their only purpose is to make the latter easier, more reliable, and more effective. These auxiliary tasks – attribute transformation, discretization, and attribute selection – are not discussed here. Their definitions are presented in the corresponding chapters.

Inductive learning is definitely the most commonly studied learning scenario in the field of machine learning. It assumes that the learner is provided with *training information* (usually – but not necessarily – in the form of examples) from which it has to derive knowledge via inductive inference. The latter is based on discovering patterns in training information and generalizing them appropriately. The learner is not informed and has no possibility to verify with certainty which of the possible generalizations are correct and can only be hoped, but never guaranteed to succeed.

Inductive learning is the source of many data mining algorithms as well as of their theoretical justifications. This is the area where the domains of machine learning and data mining intersect. But even data mining algorithms that do not originate from machine learning can be usually seen as some explicit or implicit forms of inductive learning. The analyzed data plays the role of training information, and the models derived therefrom represent the induced knowledge. In particular, the three most widely studied and practically exercised data mining tasks, classification, regression, and clustering, can be considered inductive learning tasks. This chapter provides some basic background, terminology, and notation that is common for all of them.

Data Mining Algorithms: Explained Using R, First Edition. Paweł Cichosz.
© 2015 John Wiley & Sons, Ltd. Published 2015 by John Wiley & Sons, Ltd.

1.1.1 Knowledge

Whenever discussing any form of learning, including inductive learning, the term "knowledge" is frequently used to refer to the expected result of the learning process. It is unfortunately difficult to provide a satisfactory definition of knowledge, consistent with the common understanding as well as technically useful, without an extensive discussion of psychological and philosophical theories of mind and reasoning, which are undoubtedly beyond the scope of our interest here. It makes sense therefore to adopt a simple indirect surrogate definition which does not try to explain what knowledge is, but explains what purpose it can serve. This purpose of knowledge is *inference*.

1.1.2 Inference

Inference can be considered the process of using some available knowledge to derive some new knowledge. Given the fact that knowledge is both the input and output of inference, the above idea of defining knowledge as something used for inference may appear pretty useless and creating an infinite definition loop. It is not necessarily quite that bad since, in the context of inductive learning, different types of inference are employed when using training information to derive knowledge and when using this derived knowledge. These are *inductive inference* and *deductive inference*, and their role in inductive learning is schematically illustrated below.

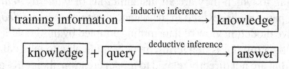

1.1.2.1 Inductive inference

It is common to describe inductive inference as a "specific-to-general" reasoning process that uses a number of individual observations to generate laws that match these known observations and can be used to predict currently unknown future observations. This simplified view does not encompass all possible variations of inductive inference, but is perfectly sufficient for our needs. In the diagram presented above, the role of inductive inference is to derive (general) knowledge from the available training information (which can be considered specific knowledge).

Clearly, inductive inference is fallible and there is no guarantee that it would yield true conclusions when applied to true premises. Appropriately designed inductive inference mechanisms can reduce but never eliminate the risk of arriving at wrong conclusions. Actually, in the case of inductive learning, it makes more sense to speak about different quality levels of the induced knowledge rather than of its true or false (correct or incorrect) status. The main effort in inductive learning research is devoted to maximizing the quality of knowledge derived from not necessarily reliable training information via inductive reasoning.

1.1.2.2 Deductive inference

In contrast, deductive inference is infallible and its conclusions are guaranteed to be satisfied whenever its premises are. It does not necessarily have to be a "general-to-specific" reasoning process, although tends to be presented as such when opposed to inductive inference. Actually, it can be used with both general and specific premises, and general and specific conclusions, although performing deductive inference based on knowledge derived via inductive learning does indeed conform to the popular "general-to-specific" pattern. It is noteworthy, by the way, that the well-known and useful number-theoretical theorem proving scheme called mathematical induction, which does follow a hybrid "specific-and-general-to-general" reasoning path, is in fact a form of deductive inference.

As shown in the above diagram, the induced knowledge is used to deduce an answer to a specified query. If the training information contained a number of known "historical" cases or observations, then typically the query presents one or more new cases or observations some interesting aspects of which remain unknown. The deductive inference process is supposed to supply these missing interesting aspects as an answer.

The infallibility of deductive inference by no means guarantees receiving correct answers to all queries. This would be the case only if the knowledge on which the deduction is based were perfectly correct, which cannot be expected in practice.

1.2   Inductive learning tasks

The three key data mining tasks, classification, regression, and clustering, are all based on the inductive learning paradigm. The essence of each of them is to inductively derive from *data* (representing training information), a *model* (representing knowledge) that has predictive utility, i.e., can be deductively applied to new data. Whereas these tasks, also called *predictive modeling* tasks, by no means exhaust the scope of inductive learning tasks studied in the field of machine learning, they represent the most widely applicable and useful variations thereof from a data mining perspective. They make a number of common assumptions about their input and output and have some common issues that deserve particular attention.

1.2.1   Domain

The *domain*, designated by X, is the set of all entities that are considered in a given inductive learning task. These can be customers, transactions, devices, or whatever is the subject of our interest.

1.2.2   Instances

Any single element of the domain, $x \in X$, is an *instance*. Instances constitute both training information for model creation and queries for model application.

1.2.3   Attributes

Instances, which may be some entities of the real world, are not directly observable. Their observable representation is provided by *attributes*. An attribute is a function $a : X \to A$ that assigns an attribute value to each instance from the domain. Unless discussing a specific example domain, we will assume that there are n attributes defined on the domain X, $a_1 : X \to A_1, a_2 : X \to A_2, \ldots, a_n : X \to A_n$.

Depending on the codomain A, attributes can be divided into different types, which may be treated differently by data mining algorithms. For most algorithms, it is sufficient to distinguish the following three major attribute types:

Nominal. Having a finite number of discrete values with no total order relation.

Ordinal. Having a finite number of discrete values with a total order relation.

Continuous (aka numerical, linear). Having numerical values.

These attribute types are best characterized by the basic relational and arithmetic operations that can be reasonably performed on their values rather than the "physical" representation of their values. Nominal attribute values can only be tested for equality. Ordinal attributes can be tested for both equality and inequality. For continuous attributes, we can perform inequality tests and all arithmetic operations defined for real numbers. It is not untypical to find nominal and ordinal attribute values represented by integer numbers assigned according to some encoding, but this representation does not make them liable for any arithmetics. On the other hand, continuous attributes can actually take only a small number of discrete numerical values, but these values can be treated as real numbers and used for whatever calculation or transformation that can be applied to real numbers.

In many cases, the distinction between attribute types is not quite crisp and the same attribute could be reasonably considered both nominal and ordinal, or both ordinal and continuous. In such a situation, the data miner has to judge or experimentally verify whether adopting a meaningful order relation for a nominal attribute could be helpful or harmful, or whether permitting some arithmetics on numerically represented values of an ordinal attributes might lead to some improvement.

Since instances can only be observable via their attribute values, it is common to identify them with the corresponding attribute value vectors. When speaking of an instance x, we will usually mean the vector of values $a_1(x), a_2(x), \ldots, a_n(x)$.

1.2.4   Target attribute

For the classification and regression tasks, a single attribute is distinguished as the *target attribute*. It represents the property of instances that the created model should be able to predict, based on the other attributes. Inductive learning tasks with a designated target attribute are referred to as *supervised learning* tasks, whereas those with no target attribute are referred to as *unsupervised learning* tasks. The same terms are also used when referring to algorithms for these two types of tasks. The values of the target attribute are assumed to be generally unavailable except for a subset of the domain used for model creation and evaluation.

1.2.5   Input attributes

Some or all of nontarget attributes are considered *input attributes*, the values of which are assumed to be generally available for the whole domain, so that the model can use them for generating its predictions. This general availability does not exclude the possibility of missing values, which is one of the common practical data quality issues.

1.2.6   Training set

Training information is represented by a *training set*, which is a subset of the domain. For any inductive learning task, a set of instances from the domain has to be available. Then the training set $T \subseteq D$ is the set of instances actually used for model creation, where $D \subset X$ denotes the set of all available instances. If there is a distinguished target attribute for a given inductive learning tasks, its values are assumed to be known on D, and the available instances are called labeled. Selecting a subset of the whole available set of instances D for model creation can be motivated by various reasons, such as computational savings and intention to leave out some data for other purposes, including model evaluation.

There is some ambiguity about the term "training set" that has to be clarified. It is used here in a broader sense, as the whole subset of the domain is used for inductive learning (model creation). When discussing particular data mining algorithms, we will also be using this term in a narrower sense, referring to the subset of instances used for a single algorithm run. Several runs may be required before the final model is obtained (e.g., for algorithm selection, parameter tuning, attribute selection). These runs cannot use the whole available set of instances, since some have to be held out for model evaluation. This is necessary to provide basis for making decisions about the utility of particular algorithms, parameter setups, or attribute subsets, for which these runs are performed.

Since instances are not observable directly, but through their attribute values, when solving inductive learning tasks, we deal actually not with sets of instances, but with *datasets* – which are sets of attribute value vectors describing particular instances. A dataset can be usually thought of as a table with rows corresponding to instances, and columns corresponding to attributes (similarly to a table in a relational database or a spreadsheet). In practice, it is common to simply identify sets of instances with the corresponding datasets.

1.2.7   Model

Inductive learning tasks consist in finding (or generating), based on the provided training set, a *model h* representing knowledge that can be applied to all instances $x \in X$ in an automated way. This is a function that takes an instance as its argument (query) and returns a *prediction* as its value (answer). It can be therefore considered a new attribute, inductively derived from the training data, but defined (i.e., computable) for the whole domain.

What actually has to be predicted depends on the task. It is, in particular, the target attribute for the classification and regression tasks (the class and the target function value, respectively), for which the new attribute represented by the model is an approximation of an existing attribute that is only provided for a limited dataset, but remains unknown for the rest of the domain. For the clustering task, on the other hand, it is an entirely new attribute that represents the similarity structure discovered from the data, i.e., cluster membership assigning arbitrary instances from the domain to one of the similarity-based clusters.

1.2.8   Performance

The quality of predictions provided by a model is called the model's *performance*. It is not a big challenge to achieve good *training performance*, i.e., the quality of predictions generated on the training set. Of much greater interest is the *true performance*, i.e., the expected quality of predictions on the whole domain, including (mostly or entirely) previously unseen instances. It can be estimated in the process of model evaluation, using appropriate performance measures (task-specific) and evaluation procedures (mostly task-independent). Performance measures and evaluation procedures for inductive learning tasks are discussed in Chapters 7, 10, and 14.

1.2.9   Generalization

The true performance is also called the generalization performance, since to predict well on new, previously unseen data, the model has to encompass appropriate generalizations of patterns detected in the training set. Generalization is the essence of inductive learning. The exact generalization mechanisms are largely task- and algorithm-specific, but the common effect can be described in a simplistic way as making predictions for new instances based on their similarity to known training instances. For most algorithms, the simplicity is not determined based on any explicit similarity measure, though, but rather implied by their internal operating mechanisms and model representation.

1.2.10   Overfitting

Poor generalization leads to *overfitting*, which is a nightmare of inductive learning. A model h is considered overfitted to a training set T if there exists another model h' for the same task that performs worse than h on the training set, but performs better on the whole domain (including unseen data). The essence of overfitting is therefore a discrepancy between a model's good training performance (performance on the training set) and its poor true performance (expected performance on the whole domain).

1.2.11   Algorithms

Algorithms that solve an inductive learning task, i.e., generate models based on a given training set, are called *inductive learning algorithms* or *modeling algorithms*. Although an algorithm producing an arbitrarily poor model is formally a learning algorithm, it is natural to restrict one's interest to algorithms that attempt to optimize some explicitly specified or (more typically) implicitly assumed performance measure.

1.2.11.1   Weight-sensitive algorithms

Weight-sensitive modeling algorithms accept a vector of weights w containing a numerical weight $w_x \geq 0$ for each training instance $x \in T$. When a weight vector is specified for a weight-sensitive algorithm, it attempts to optimize the correspondingly weighted version of the performance measure normally assumed. For integer weights, this is roughly equivalent to using a modified training set T_w in which each instance $x \in T$ is replicated w_x times.

1.2.11.2   Inductive bias

Unlike in deductive inference, where all possible conclusions that can be derived are strictly determined by the premises, the training information used for inductive learning only narrows down the space of possible models, but does not strictly determine the model that will be obtained. There may be many models fitting the same set of training instances, and different possible generalizations of the patterns discovered therein. The criteria used by an inductive learning algorithm to select one of them for a given training set, which may be stated explicitly or implied by its operating mechanisms and model representation, are called the *inductive bias*. The inductive bias is not a deficiency of inductive learning algorithms, it is in fact a necessity: it guides the inductive inference process toward (hopefully) the most promising generalizations.

The inductive bias usually takes one or both of the following two forms:

Representation bias. A model representation method is adopted that makes it possible to represent only a small subset of possible models fitting a given dataset.

Preference bias. A preference measure is adopted that favors some models against the others based on their properties.

It is particularly common to see the preference for model simplicity as the inductive bias, which is believed to reduce the risk of overfitting according to Ockham's razor principle.

1.2.12 Inductive learning as search

It is a very useful and insightful perspective to view the inductive learning process as search through the space of possible models, directed by the training information and the inductive bias. The goal of this search is to find a model that fits the training set and can be expected to generalize well. It might appear that the model space should be as rich as possible, as it increases the chance that it actually contains a sufficiently good model. Unfortunately, a rich model space is likely to contain a number of poor models that fit the training set by a mere chance and would not perform well on the whole domain, and the inductive bias used might not be able to avoid choosing one of them. Such situation is referred to as *oversearching* and is one of the most important factors that increases the risk of overfitting.

The search perspective is by most means a conceptual framework that provides useful insights making it easier to understand the inductive learning process and some possible related caveats, but some learning algorithms actually do perform search not only from a conceptual, but also from a technical viewpoint, by applying some general purpose or tailored search techniques to identify the best model.

1.3 Classification

Classification is one of the fundamental cognitive processes used to organize and apply our knowledge about the world. It is common both in everyday life and in business, where we might want to classify customers, employees, transactions, stores, factories, devices, documents, or any other types of instances into a set of predefined meaningful classes or categories. It is therefore not surprising that building classification models by analyzing available data is one of the central data mining tasks that attracted more research interest and found more applications than any other task studied in the field.

The classification task consists in assigning instances from a given domain, described by a set of discrete- or continuous-valued attributes, into a set of classes, which can be considered values of a selected discrete target attribute, also called the target *concept*. Correct class labels are generally unknown, but are provided for a subset of the domain. It can be used to create the classification model, which is a machine-friendly representation of the knowledge needed to classify any possible instance from the same domain, described by the same set of attributes. This follows the general assumptions of inductive learning, of which the classification task is the most common instantiation.

The assumed general unavailability of class labels, but their availability for a given subset of the domain, may seem at first inconsistent, but it is essential for the idea of inductive inference on which all data mining methods are based. It also perfectly corresponds to the requirements of most practical applications of classification, where the class represents some property of classified instances that is either hard and costly to determine, or (more typically) that becomes known later than is needed. This is why applying a classification model to assign class labels to instances is commonly referred to as *prediction*.

To see more precisely how the classification task instantiates the general inductive learning task, we only need to discuss those aspects of the latter where some classification-specific complements or comments can be added.

1.3.1   Concept

The term "concept" comes from the traditional machine learning terminology and is used to refer to a classification function $c : X \to C$, representing the true assignment of all instances from the domain to a finite set of classes (or categories) C. It can be considered simply a selected target nominal attribute. Concept values will be referred to as *class labels* or *classes*.

A particularly simple, but interesting kind of concepts is that with just a two-element set of classes, which can be assumed to be $C = \{0, 1\}$ for convenience. Such concepts are sometimes called *single concepts*, opposed to *multiconcepts* with $|C| > 2$. Single concepts best correspond to the original notion of concepts, borrowed by machine learning from cognitive psychology. An instance x is said to "belong to" or "be an example of" concept c when $c(x) = 1$. When $c(x) = 0$, the instance is said "not to belong to" or "to be a negative example of" concept c. Classification tasks with single concepts will be referred to as two-class classification tasks.

1.3.2   Training set

The target concept is assumed to be unknown in general, except for some set of instances $D \subset X$ (otherwise no data mining would be possible). Some or all of these available *labeled instances* constitute the *training set* $T \subseteq D$.

Example 1.3.1 As a simple example of a training set for the classification task, consider the classic *weather* data with 14 instances, four input attributes, and one target attribute. The following R code reads this dataset to an R dataframe [dmr.data] and summarizes the distribution of attributes (outlook, temperature, humidity, windy) and the target concept (play).

```
weather <- read.table(text="
        outlook temperature humidity    wind play
    1     sunny         hot     high  normal   no
    2     sunny         hot     high    high   no
    3  overcast         hot     high  normal  yes
    4     rainy        mild     high  normal  yes
    5     rainy        cold   normal  normal  yes
    6     rainy        cold   normal    high   no
    7  overcast        cold   normal    high  yes
    8     sunny        mild     high  normal   no
    9     sunny        cold   normal  normal  yes
   10     rainy        mild   normal  normal  yes
   11     sunny        mild   normal    high  yes
   12  overcast        mild     high    high  yes
```

```
  13 overcast         hot   normal normal  yes
  14 rainy           mild    high   high   no")

summary(weather)
```

The first four attributes describe weather conditions and the last attribute is the target concept that classifies them as appropriate or inappropriate for playing sports. The extremely small size of this dataset makes it totally unrealistic and unsuitable for any experiments evaluating the performance of classification algorithms, but absolutely perfect for illustrating the calculations needed for their operation. All such calculations, specified by mathematical equations and implemented by illustrative R code, can be easily verified "manually." This is why the *weather* dataset is frequently used in examples presented in chapters on classification in this book.

Example 1.3.2 A modified version of the dataset from the previous example, in which the `temperature` and `humidity` attributes are continuous, will also be used occasionally. This will be referred to as the *weatherc* data. The following R code reads this dataset to an R dataframe and summarizes the attribute distributions.

`dmr.data`

```
weatherc <- read.table(text="
     outlook temperature humidity   wind play
  1     sunny        27        80 normal   no
  2     sunny        28        65   high   no
  3  overcast        29        90 normal  yes
  4     rainy        21        75 normal  yes
  5     rainy        17        40 normal  yes
  6     rainy        15        25   high   no
  7  overcast        19        50   high  yes
  8     sunny        22        95 normal   no
  9     sunny        18        45 normal  yes
 10     rainy        23        30 normal  yes
 11     sunny        24        55   high  yes
 12  overcast        25        70   high  yes
 13  overcast        30        35 normal  yes
 14     rainy        26        85   high  no")

summary(weatherc)
```

1.3.3 Model

A *classification model* $h : X \rightarrow C$ produces class predictions for all instances $x \in X$ and is supposed to be a good approximation of the target concept c on the whole domain. Classification models are briefly called *classifiers*, although the latter term sometimes also refers to classification algorithms, used to create classification models.

1.3.3.1 Scoring classifiers

For two-class classification tasks (single concepts), a particular kind of *scoring* classification models deserves special interest. These are the classification models that predict class labels in a two-step process: they first map instances into real numbers called scores and then they assign one class label (1, by convention) to instances with sufficiently high scores and the other class label (0) to the remaining instances.

More precisely, a scoring model is represented by a scoring function $\pi : X \to \mathcal{R}$ and a labeling function $\lambda : \mathcal{R} \to \{0, 1\}$. The former assigns real-valued scores to all instances from the domain, and the latter converts these scores to class labels using a cutoff rule, such as

$$\lambda(r) = \begin{cases} 1 & \text{if } r \geq \theta \\ 0 & \text{otherwise} \end{cases} \tag{1.1}$$

where θ is a cutoff value. The model is then the composition of its scoring and labeling functions, $h(x) = \lambda(\pi(x))$.

It is a common convention to consider scoring classification models sharing the same scoring function and differing only in the labeling function (i.e., using different cutoff values) as the same single model, working in different *operating points*. Classification algorithms capable of generating scoring classification models typically create a scoring function and a cutoff value for one *default* operating point, but a number of other operating points can be obtained by using different cutoff values.

Classification models that generate class labels directly, without scoring and labeling functions, are sometimes called *discrete* classifiers.

1.3.3.2   Probabilistic classifiers

A related interesting and useful special kind of classification models are *probabilistic* classifiers, which estimate class probabilities for instances being classified, and then make predictions based on these probabilities. A probabilistic classifier assigns to each instance $x \in X$ and class $d \in C$ a probability estimate $P(d|x)$ of instance x belonging to class d of the target concept c. The estimated class probabilities can be used to generate class labels using the obvious *maximum-probability* rule:

$$h(x) = \arg\max_{d \in C} P(d|x) \tag{1.2}$$

or – under nonuniform misclassification costs – the less obvious minimum-cost rule, as discussed in Section 6.3.3.

For two-class tasks, probabilistic classifiers constitute a particularly common subclass of scoring classifiers, with the estimated probabilities of class 1 for particular instances considered scores, i.e., $\pi(x) = P(1|x)$.

1.3.4   Performance

The exact meaning of "good approximation" of the target concept that is expected from a classification model is established by classification model performance measures, but – informally – we want the model to usually provide correct class labels, as far as possible. Even a model that is wrong in most cases remains a (poor) model, but –needless to say – poor models are not of particular interest in the classification task. The most commonly adopted performance measure is the *misclassification error* which is the fraction of instances from a dataset or the whole domain misclassified by the model. This and other performance measures for classification models are discussed in Section 7.2.

1.3.5 Generalization

As any inductive model, a classification model should be judged "good" or "poor" not just based on its performance on the training set, but on its (expected) performance on the whole domain. In other words, we care not only and not mainly for the classification accuracy on the training set, but also on new previously unseen instances to which the model could be applied. This requires classification algorithms to not only *discover* relationships between class labels and attribute values in the training set, but also to *generalize* them so that they can be expected to hold on new data.

1.3.6 Overfitting

A classification model is overfitted if another model predicts class labels for the whole domain better despite yielding worse training set predictions. This is typically defined with respect to the misclassification error, but arbitrary performance measures can be used as well. Many classification algorithms include mechanisms supposed to reduce the risk of overfitting.

1.3.7 Algorithms

Algorithms that solve the classification task, i.e., generate classification models based on a supplied training set, are called *classification algorithms*. Two special types of classification algorithms deserve particular interest: weight-sensitive algorithms and cost-sensitive algorithms. They are capable of creating models with particular properties that are sometimes desirable.

1.3.7.1 Weight-sensitive algorithms

A weight-sensitive classification algorithm – like other weight-sensitive modeling algorithms – accepts a vector of weights w containing a numerical weight $w_x \geq 0$ for each training instance $x \in T$. It is not uncommon, though, to have weights assigned solely based on classes, with ω_d for each $d \in C$ being the weight of all training instances of class d, i.e., $w_x = \omega_{c(x)}$. When a weight vector is specified for a weight-sensitive algorithm, it attempts to optimize the correspondingly weighted version of the performance measure normally assumed, where each instance's weight is applied to its contribution to the performance measure. Typically, this is the weighted misclassification error instead of the usual one. For integer weights, this is roughly equivalent to using a modified training set T_w in which each instance $x \in T$ is replicated w_x times.

1.3.7.2 Cost-sensitive algorithms

Cost-sensitive classification algorithms take into account that the severity of misclassifying instances may vary across different true and predicted class combinations, with some being more acceptable than others. Such algorithms accept a misclassification cost specification on input and adopt the mean misclassification cost as the performance measure to minimize during model creation. Several approaches to achieving cost sensitivity are discussed in Chapter 6.

1.4   Regression

Similar to classification, regression is an inductive learning task that has been extensively studied and can be widely encountered in practical applications. It can be informally characterized as "classification with continuous classes," which means that regression models predict numerical values rather than discrete class labels. This relationship to the classification task makes it possible to describe the regression task by referring to the latter where appropriate and highlighting the differences where necessary.

The term "regression" tends to be sometimes used in a narrow technical meaning referring to statistical algorithms for fitting parametric regression models. We adopt here a broader view in which regression is presented in a completely algorithm-independent way as one of the major data mining tasks, and any algorithm that solves this task will be considered a regression algorithm. This makes regression equivalent to *numerical prediction*, practical instantiations of which are nearly as common as those of classification. In particular, we might want to predict prices, demand, production or sales volumes, resource consumption, physical parameters, etc.

The regression task consists in assigning numerical values to instances from a given domain, described by a set of discrete or continuous-valued attributes. This assignment is supposed to approximate some target function, generally unknown, except for a subset of the domain. This subset can be used to create the regression model, which is a machine-friendly representation of the relationships between the target function and the attributes that makes it possible to predict unknown target function values for any possible instance from the same domain. The regression task adopts therefore the same general scenario of inductive learning that has been presented above for the classification task. In practical applications, the target function represents some interesting property of instances from the domain that is either difficult and costly to determine, or (more typically) that becomes known later than is needed. Subsections below add regression-specific highlights to what has been presented above for inductive learning in general.

1.4.1   Target function

The *target function* $f : X \to \mathcal{R}$ represents the true assignment of numerical values to all instances from the domain. Target function values will briefly be called *target values* or *target labels*.

1.4.2   Training set

The *training set* $T \subseteq D \subset X$ for regression consists of some or all labeled instances for which target function values are available, despite being unknown in general.

Example 1.4.1 As a simple example of a training set for the regression task, consider a modified version of the *weatherc* data, in which the play attribute originally representing the target concept is replaced by a new continuous playability attribute, which now represents the target function. This will be referred to as the *weatherr* data. The following R code reads this dataset to an R dataframe and summarizes the attribute distributions.

dmr.data

```
weatherr <- read.table(text="
        outlook temperature humidity    wind playability
    1     sunny          27       80  normal        0.48
    2     sunny          28       65    high        0.46
    3  overcast          29       90  normal        0.68
    4     rainy          21       75  normal        0.52
    5     rainy          17       40  normal        0.54
    6     rainy          15       25    high        0.47
    7  overcast          19       50    high        0.74
    8     sunny          22       95  normal        0.49
    9     sunny          18       45  normal        0.64
   10     rainy          23       30  normal        0.55
   11     sunny          24       55    high        0.57
   12  overcast          25       70    high        0.68
   13  overcast          30       35  normal        0.79
   14     rainy          26       85    high       0.33")

summary(weatherr)
```

1.4.3 Model

A *regression model* $h : X \to \mathcal{R}$ can be used to generate numerical predictions for all instances $x \in X$ and supposed to provide a good approximation of the target function f on the whole domain.

1.4.4 Performance

The exact meaning of "good approximation" is established by regression model performance measures, but – informally – we want the model to usually provide predictions that are not far away from the true target values. One commonly adopted performance measure is the mean sum of squared differences between the true and predicted values, referred to as the *mean square error*. This and other regression performance measures are discussed in Chapter 10.

1.4.5 Generalization

Generalization is no less crucial for regression than for classification. Regression algorithms have to not only *discover* relationships between the target function and attribute values in the training set, but also to *generalize* them so that they can be expected to hold on new data.

1.4.6 Overfitting

Poor generalization leads to overfitting, which is the same serious problem for regression as for the classification and can be defined in the same way. Many regression algorithms include mechanisms supposed to reduce the risk of overfitting.

1.4.7 Algorithms

A *regression algorithm* generates a regression model based on a given training set.

1.5 Clustering

Clustering is an inductive learning task that differs from the classification and regression tasks from the same family by the lack of a predetermined target attribute to be predicted. It can be thought of as classification with autonomously discovered rather than predefined classes, which are based on similarity patterns identified in the data.

The clustering task consists in dividing a set of instances from a given domain, described by a number of discrete or continuous-valued attributes, into a set of clusters based on their similarity, and creating a model that can map arbitrary instances from the same domain to these clusters. This can be considered a superposition of two subtasks:

Cluster formation. The identification of similarity-based groups in the analyzed data.

Cluster modeling. Creating a model for cluster membership prediction.

The latter is clearly a classification task, with clusters identified within the first subtask used as classes. This could be performed, in principle, using any available classification algorithm. It is usually more convenient not to separate these two subtasks, though, and most clustering algorithms handle both cluster formation and cluster modeling. It makes it possible for the criteria used to identify clusters to be subsequently reused for cluster membership prediction.

1.5.1 Motivation

The utility of the clustering task may not be as self-evident as for the classification and regression tasks and deserves some more explanation. Some typical reasons to perform clustering are listed below, along with example applications where they are likely to appear.

- Clustering can provide useful insights about the similarity patterns present in the data, and a clustering model can be considered as knowledge *per se*. Some example applications where this is the case include

  — customer segmentation,

  — point of sale segmentation,

  — document catalog creation.

- Clustering can be performed on a selected subset of *observable* attributes that are easily available for all instances, and used to predict *hidden* attributes that are impossible or difficult to determine for some instances based on cluster membership. This is similar to classification or regression with multiple target attributes with sparingly available values. Such situation occurs in the following example applications:

  — customer clustering based on socio-demographic attributes used to predict attributes describing purchase behavior,

  — point-of-sale segmentation based on location, building, and local population features, used to predict attributes describing selling performance.

- Clustering performed on a set of "normal" instances can be used for anomaly detection, by issuing alerts for new instances that do not fit any existing cluster. This is a possible approach to various diagnostics applications, such as

  — network traffic clustering, used for intrusion detection,

  — credit card transaction clustering, used for fraud detection,

  — sensor signal clustering, used for device fault detection.

- Clustering can be used as a domain decomposition method for some further data mining tasks, which may be easier to perform within homogeneous clusters. Example applications include

  — customer clustering based on socio-demographic and purchase history attributes, and classification with respect to loyalty within clusters,

  — customer clustering based on socio-demographic and purchase history attributes, and predicting reaction to incentives within clusters,

  — credit card account clustering based on cardholder socio-demographic attributes and transaction history attributes, and classification with respect to fraud likelihood within clusters,

  — product clustering based on technical specification and usage attributes, and demand forecasting within clusters,

  — used vehicle clustering and price prediction within clusters.

To more formally define the clustering task, we only need to slightly modify the classification task definition wherever these two differ.

1.5.2   Training set

The *training set* $T \subseteq D$ is a subset of the available dataset $D \subset X$ used to create a clustering model.

Unlike for the classification and regression tasks, training instances are not normally assumed to be labeled by any target attribute values, since no target attribute is considered for the clustering task. Particular instantiations of the clustering task may adopt some other assumptions, though. In particular, one of the typical clustering usage scenarios assumes that the set of attributes is divided into subsets of *observable* and *hidden* attributes. If this is the case, only the former are assumed to be available for the whole domain, but the dataset D consists of instances for which the latter are known as well.

Example 1.5.1 As a simple example of a training set for the clustering task, consider a modified version of the *weatherc* data, in which the `play` attribute originally representing the target function is dropped, as demonstrated by the following R code. This will be referred to as the *weathercl* data.

`dmr.data`

```
weathercl <- weatherc[,-5]
```

1.5.3   Model

The clustering task consists in creating, based on the provided training set, a model $h : X \to C_h$ that is computable for all $x \in X$ and maps them into a model-specific set of clusters C_h. This is formally nearly the same requirement as for classification models, except for an apparently small but substantial difference consisting in the set of "classes" C_h not being predetermined, but identified as part of model creation. It is therefore more instructive to think of the clustering model creation process as of cluster identification (i.e., determining C_h) rather than cluster modeling (i.e., determining h for given C_h), since once the set of clusters is determined, their representation usually makes the actual mapping function straightforward to obtain.

Without a predetermined target attribute, the clustering model is not required to approximate any kind of target concept or function. This deprives the model creation process from any explicit and objective guidance, which is so essential for the classification and regression tasks, making clustering an unsupervised inductive learning task. The only requirement adopted for the clustering task is to identify clusters based on similarity patterns observed in the set of training instances. This can only be stated informally when discussing the general task formulation and is explicitly or implicitly formalized by specific clustering algorithms.

1.5.4   Crisp vs. soft clustering

As presented above, a clustering model represents the so-called *crisp* clustering in which all clusters are disjoint, i.e., every instance is assigned to exactly one cluster. This is the same as with classes in the classification task. Departures from this view of clustering, however, are not quite uncommon. Unlike "objectively" existing, predefined classes, which serve the purpose of separating distinct types of entities, clusters formed by a clustering algorithm extract similarity patterns that may not be strong enough to justify definite distinctions. This is why sometimes *soft* or *fuzzy* clustering models are considered that may assign a single instance to multiple clusters at some membership level.

1.5.5   Hierarchical clustering

A variation of the clustering task receiving special attention requires that the clustering model be *hierarchical*. A hierarchical clustering model can be thought of as a set of ordinary (flat) clustering models organized in a tree structure. Each internal tree node represents both a flat clustering model and a cluster of the flat clustering model from its parent node. The root node represents a special level-0 cluster covering the whole domain. Leaves represent just clusters, with no further clustering models assigned to them. The model in the root node is applied to the whole domain and maps it to level-1 clusters. These clusters correspond to descendant nodes with subsequent models that partition them into subclusters, etc.

1.5.6   Performance

Given the unsupervised nature of the clustering task, clustering model performance can be hardly evaluated in a truly objective and application-independent way. Still there is a number of clustering model performance measures that may be helpful in judging the suitability of a given model for a given application. These are presented in Chapter 14.

1.5.7   Generalization

As for other inductive learning models, a clustering model is expected, in principle, to generalize relationships discovered in the training set and make them applicable to the whole domain. In other words, we care not only and not mainly for fitting all the similarity patterns in the training set, but also for capturing those that would hold for new previously unseen instances to which the model could be applied. On the other hand, the importance of this form of generalization required for the clustering task tends to be underestimated sometimes, with all or most of the attention paid to achieving the best possible match between the identified set of clusters and the training set. This is sufficient for applications where similarity patterns captured by the clustering model are not supposed to be applied for prediction.

1.5.8   Algorithms

Algorithms that generate clustering models based on a given training set are called *clustering algorithms*. Since the clustering task in its general formulation does not specify any strict requirements for the exact way of capturing the similarity patterns in the data by the clustering algorithm, different algorithms take substantially different approaches. They may be roughly categorized as follows:

(Dis)similarity-based clustering. Using a predefined or user-specified explicit measure of instance similarity to drive the cluster formation and modeling processes.

Probabilistic clustering. Using probability distributions and probabilistic inference to drive the cluster formation and modeling processes.

Conceptual clustering. Using a (usually symbolic) conceptual cluster representation to drive the cluster formation and modeling processes.

The scope of clustering algorithms presented in this book is limited to (dis)similarity-based clustering.

1.5.9   Descriptive vs. predictive clustering

The definition of clustering presented in this section and assumed later in this book's clustering chapters adopts a predictive modeling perspective. It is not uncommon to see practical applications of clustering that focus on descriptive modeling only, though. The capability of predicting cluster membership for new instances is neither needed nor used whenever the purpose of clustering is just to discover and present similarity patterns in the data. Several practical implementations of clustering algorithms do not provide the prediction functionality, making it possible to determine cluster membership for training instances only.

1.6   Practical issues

The definitions of inductive learning tasks presented above are somewhat idealized. For practical tasks some compromises are often necessary. They are mostly related to imperfect data, which may not provide full or reliable instance descriptions.

1.6.1   Incomplete data

The above descriptions of attributes as functions mapping instances to attribute values might suggest that the values of all attributes are available for all instances. This is usually not the case in practice, where for some instances some attribute values may be missing. Such an incomplete dataset can either be "repaired" in a preprocessing phase, or handled in some special way by modeling algorithms.

1.6.2   Noisy data

Similarly, having described attributes (including the target attribute) as functions mapping instances to attribute values or classes, we might expect the available attribute values to be always perfectly reliable. It is often not the case in practice, where attribute values (including instance target labels) can be corrupted by some noise. Sometimes incorrect attribute values can be corrected or unreliable instances filtered out during a preprocessing phase, but usually the presence of noise has to be accepted as unavoidable. Moreover, some noise not only cannot be usually eliminated, but also in many cases it cannot be even detected. To take the simplest example, do two instances with exactly the same attribute values but different class labels result from noise or rather from an insufficient set of attributes which cannot fully differentiate instances from different classes? Such questions can be often asked, but rarely answered, unless we accept a somewhat evasive answer that both hypotheses represent simply two different views on the same phenomenon. The fact is that all useful data mining algorithms have to assume the risk of data being affected by noise and not blindly trust any apparent patterns. This limited confidence in data is actual at the heart of good generalization.

1.7   Conclusion

Many data mining algorithms, both those originating from machine learning and those developed in the field of statistics, are based on the inductive learning paradigm. The three most common data mining tasks, classification, regression, and clustering, follow this paradigm particularly directly and can be therefore called inductive learning tasks. This chapter has provided some background information, assumptions, and terminology that apply to all of them. The entirety of this book is devoted to algorithms solving these tasks and to closely related techniques used to improve model quality.

The formulations of the inductive learning tasks presented in this chapter and subsequently adopted throughout the book are simple and generic. Their more specific or enhanced versions are sometimes studied in the literature and employed in applications where necessary. These may adopt special assumptions about the properties of the target concept or the target function for the classification and regression tasks (such as the number of classes, relationships among classes, or target function value distribution), fulfill special requirements for cluster formation and representation methods for clustering (such as soft clusters), or adjust to special domain and training set properties (such as the number and types of attributes, or the number and availability of training instances). Leaving such interesting and useful extensions beyond the scope of this book is a regrettable necessity, dictated by the adopted level of detail, precision, and R code illustration coverage. Extending the scope of the book would require either compromising on the former, or making its size unmanageable for a single author.

This is also one additional reason why other data mining tasks and the corresponding algorithms that definitely deserve attention are not included in the book. This applies, in particular, to association and temporal pattern discovery, geospatial data analysis, time series analysis, or survival analysis. While some of these tasks can be viewed as forms of classification or regression, they are much better handled by dedicated algorithms. In some cases, the latter tend to be more mathematically refined than those presented in the book and do not fit the "maximum usefulness at minimum complexity" principle adopted here. Some of these tasks also make substantially different assumptions about data representation, making the instance-attribute scheme introduced in this chapter inapplicable or awkward. This is why such tasks and algorithms would require a considerably different form of presentation and including them would make the book not only overly large or overly superficial, but also inconsistent.

1.8   Further readings

The three major data mining tasks introduced in this chapter are discussed in most data mining books which cover predictive modeling (e.g., Abu-Mostafa *et al.*, 2012, Cios *et al.*, 2007, Han *et al.*, 2011, Hand *et al.*, 2001, Tan *et al.*, 2013, Witten *et al.*, 2011). While they may use partially different terminology and notation, emphasize different aspects of these tasks, or present different motivation and application examples, they ultimately arrive at the same basic assumptions and requirements. Kohavi and Provost (1998) collect concise definitions for some of the most commonly used terms.

While all the three tasks can be presented as instantiations of inductive learning, it is classification learning that has received most attention in the machine learning literature (Cichosz, 2007; Mitchell, 1997). This learning task, also referred to as concept learning, is also one of the major topics of machine learning theoretical work. Most of this work has been done within the scope of the computational learning theory, which focuses on the learnability of concept classes and characterizing their hardness, deriving requirements for the number and quality of training instances, and establishing the properties and performance bounds of specific learning algorithms (e.g., Kearns and Vazirani, 1994, Valiant, 1984, Vapnik, 1998).

However, a complementary approach that highlights different types of inductive inference used to derive models from training information, possibly augmented with background knowledge, also received some attention and brought insightful results (Michalski, 1983). Viewing inductive learning as searching the model space for the most justified generalizations of training data, as proposed by Mitchell (1982) for classification learning, remains a valid view of other data mining tasks. This is also the case for the idea of bias as a necessary component of any inductive learning process (Haussler, 1988; Mitchell, 1980).

The classification and clustering tasks in text domains, where instances are text documents or messages of any kind, become text mining tasks. While some of general-purpose classification and clustering algorithms handle text data quite well, after transforming it to an appropriate representation, there are also several dedicated text mining algorithms as well as more specific text mining tasks that do not have their general data mining counterparts (Aggarwal and Zhai, 2012; Weiss, 2010).

The *weather* data first presented in this chapter and then used several times throughout the book comes from Quinlan (1986) and is quite popular in the machine learning and data mining literature (e.g., Witten *et al.* 2011).

References

Abu-Mostafa YS, Magdon-Ismail M and Lin HT 2012 *Learning from Data*. AMLBook.

Cichosz P 2007 *Learning Systems (in Polish)* 2nd edn. WNT.

Cios KJ, Pedrycz W, Swiniarski RW and Kurgan L 2007 *Data Mining: A Knowledge Discovery Approach*. Springer.

(eds. Aggarwal CC and Zhai CX) 2012 *Mining Text Data*. Springer.

Han J, Kamber M and Pei J 2011 *Data Mining: Concepts and Techniques* 3rd edn. Morgan Kaufmann.

Hand DJ, Mannila H and Smyth P 2001 *Principles of Data Mining*. MIT Press.

Haussler D 1988 Quantifying inductive bias: AI learning algorithms and Valiant's learning framework. *Artificial Intelligence* **36**, 177–221.

Kearns M and Vazirani U 1994 *An Introduction to Computational Learning Theory*. MIT Press.

Kohavi R and Provost F 1998 Glossary of terms: Editorial for the special issue on applications of machine learning and the knowledge discovery process. *Machine Learning* **30**, 271–274.

Michalski RS 1983 A theory and methodology of inductive learning In *Machine Learning: An Artificial Intelligence Approach* (ed. Michalski RS, Carbonell JG and Mitchell TM) Tioga.

Mitchell T 1980 The need for biases in learning generalizations. Technical Report CBM-TR-117, Rutgers University, New Brunswick, NJ.

Mitchell T 1997 *Machine Learning*. McGraw-Hill.

Mitchell TM 1982 Generalization as search. *Artificial Intelligence* **18**, 203–226.

Quinlan JR 1986 Induction of decision trees. *Machine Learning* **1**, 81–106.

Tan PN, Steinbach M and Kumar V 2013 *Introduction to Data Mining* 2nd edn. Addison-Wesley.

Valiant L 1984 A theory of the learnable. *Communications of the ACM* **27**, pp. 1134–1142.

Vapnik VN 1998 *Statistical Learning Theory*. Wiley.

Weiss SM 2010 *Fundamentals of Predictive Text Mining*. Springer.

Witten IH, Frank E and Hall MA 2011 *Data Mining: Practical Machine Learning Tools and Techniques* 3rd edn. Morgan Kaufmann.

2

Basic statistics

2.1 Introduction

This chapter presents a very limited subset of basic statistical techniques used for data mining. They can be divided into two categories:

Distribution description. Techniques used to describe the observed distribution of a single attribute in a dataset,

Relationship detection. Techniques used to detect relationships between two attributes in a dataset.

Techniques of these two categories can be used to explore the data, assess their quality, and identify the most promising directions for further, more refined analysis, usually involving model creation. They can also be internally employed by modeling algorithms as auxiliary operations (most typically, criteria used to make some decisions).

The latter is the primary way by which other chapters, mostly devoted to modeling algorithms, refer to basic statistics. Making such references possible without redirecting the reader to external sources is also the primary motivation for this chapter. It makes no attempt to replace a proper basic statistics tutorial. The scope of presented techniques and – more importantly – the depth of their discussion are far too limited for this purpose. What it achieves, however, is presenting a small set of the most commonly used basic statistical techniques using a perspective, terminology, and notation fully consistent with the remaining chapters in this book.

Example 2.1.1 For most statistical techniques presented in this chapter, implementations are available in the standard `stats` R package. Despite that, simple illustrative implementations are presented in a series of examples, to make the provided mathematical formulae easier to understand and verify. For completeness and consistency, this applies even to the simplest statistics, such as the mean or variance. The illustrative reimplementations of functions available in R have the `bs.` prefix prepended to their names to avoid

---

Data Mining Algorithms: Explained Using R, First Edition. Paweł Cichosz.
© 2015 John Wiley & Sons, Ltd. Published 2015 by John Wiley & Sons, Ltd.

shadowing their standard counterparts. They are usually considerably limited compared to the latter, but in the most basic case – equivalent to them. This is demonstrated by example calls, using the *weather*, *weatherc*, and *weatherr* datasets introduced in Examples 1.3.1, 1.3.2, and 1.4.1. They are loaded by the R code presented below. `dmr.data`

```
data(weather, package="dmr.data")
data(weatherc, package="dmr.data")
data(weatherr, package="dmr.data")
```

Some basic statistics functions defined in subsequent examples have no direct standard R counterparts, although the same functionality can be achieved by appropriately calling one or more other functions. They can be considered simple convenience wrappers around the latter. Most of them are actually used by example code presented in other chapters.

2.2   Notational conventions

Notational conventions used by this chapter are consistent with those adopted throughout this book, but not necessarily with those commonly used in basic statistics textbooks and tutorials. The general inductive learning perspective and the corresponding terminology are assumed. We speak therefore about domain (rather than population), datasets (rather than samples), attributes (rather than variables), etc.

This chapter presents definitions of a number of statistics calculated based on a dataset and concerning a single attribute or an attribute pair. This is written as $\text{stat}_S(a)$ for single attributes and $\text{stat}_S(a_1, a_2)$ for attribute pairs. Additional subscripts or superscripts are included whenever the statistic depends on other parameters.

Dataset-calculated values of some statistics are used as estimators of the corresponding unknown values on the whole domain. If that is the case, the former are designated by Latin letters and the latter – by the corresponding Greek letters (and without dataset subscripts). For example, $m_S(a)$ designates the mean of attribute a calculated on dataset S and $\mu(a)$ its mean on the whole domain.

Statistic definitions typically refer to data subsets satisfying certain conditions, based on attribute values. This book's standard data subset notation is used for this purpose. Whenever referring to data-estimated attribute value probabilities, this book's standard probability notation is employed. These notational conventions are summarized in Appendix A.

2.3   Basic statistics as modeling

Techniques presented in this chapter can be actually reduced to mathematical formulae that calculate certain quantities based on a dataset. A *statistic* (in a narrow sense) is just that: the value of a formula calculated on a dataset. What such basic statistics produce on output are essentially single numbers. This is indeed very far from knowledge representations delivered by modeling algorithms.

Despite that gap in complexity, basic statistics and models have something important in common. They both are created or calculated based on a limited dataset, but intended to adequately represent the properties of the whole domain. For a model, this usually means that it would be capable of delivering good quality predictions for arbitrary, possibly previously unseen, instances from the domain. For basic statistics, this usually means one of the following:

> For *distribution description statistics*. The calculated statistic value reasonably approximates the unknown value of the same statistic on the whole domain; speaking in statistical terms, a statistic value calculated on a dataset is used as an *estimator* of the corresponding value on the domain,
>
> For *relationship detection statistics*. The relationship detected on a dataset is likely to hold on the whole domain.

This makes it possible and justified to treat basic statistics as a particularly simple, degenerate form of modeling, which uses a dataset to discover the knowledge expected to be applicable to the whole domain. It is worthwhile to keep this perspective in mind.

2.4   Distribution description

Statistics used to describe the distribution of attributes, often called *descriptive statistics*, are particularly simple. Most of them are also widely and well known. Clearly, continuous and discrete attribute distributions need different descriptions and will be discussed separately.

2.4.1   Continuous attributes

The two most important properties of continuous attributes are *location* (indicating the most typical or representative observed values) and *dispersion* (indicating how much individual observed values differ).

2.4.1.1   Location measures

Location measures provide a quick and simple means of identifying where the values of a continuous attribute lie on the number axis. This can be thought of as characterizing the whole set of attribute values observed in a dataset with a single value, which is of course extremely imperfect, but still extremely useful.

Mean   The simplest and most commonly used location measure is the *mean*. The mean of attribute a on dataset S is calculated as follows:

$$m_S(a) = \frac{1}{|S|} \sum_{x \in S} a(x) \qquad (2.1)$$

It is easy to interpret and efficient to calculate, and serves well as a location measure in many situations, but is susceptible to outliers. Even a single considerably outlying value may distort the mean to a degree making it totally misleading. This is why care is needed when using the mean.

The sum of squared differences between attribute values and the mean is less than the sum of squared differences between attribute values and any other number:

$$m_S(a) = \arg\min_{v \in \mathcal{R}} \sum_{x \in S} (a(x) - v)^2 \qquad (2.2)$$

This can be easily verified by taking the derivative of the above sum with respect to v and equating it to 0.

Example 2.4.1 The following R code implements and demonstrates mean calculation.

```
bs.mean <- function(v) { sum(v)/length(v) }

  # demonstration
bs.mean(weatherc$temperature)
mean(weatherc$temperature)
```

Weighted mean In some situations, it is desirable to assign nonuniform weights to instances when calculating statistics. This is necessary when they are used within weight-sensitive modeling algorithms. The mean, as the most often used statistic, is also the one the weighted form of which is definitely the most often needed. It is defined as follows:

$$m_{S,w}(a) = \frac{1}{\sum_{x \in S} w_x} \sum_{x \in S} w_x a(x) \tag{2.3}$$

where w_x is the weight assigned to instance x.

Example 2.4.2 The following R code implements and demonstrates weighted mean calculation.

```
bs.weighted.mean <- function(v, w=rep(1, length(v))) { sum(w*v)/sum(w) }

  # demonstration
bs.weighted.mean(weatherc$temperature, ifelse(weatherc$play=="yes", 5, 1))
weighted.mean(weatherc$temperature, ifelse(weatherc$play=="yes", 5, 1))
```

Median A location measure more computationally demanding to calculate, but robust with respect to outliers, is the *median*. For attribute a and dataset S it is a value $\text{med}_S(a)$ that partitions the dataset into two equally sized subsets. This may not be always exactly possible, so a more rigorous definition is given by the following two conditions:

$$\frac{|S_{a \leq \text{med}_S(a)}|}{|S|} \geq \frac{1}{2} \tag{2.4}$$

$$\frac{|S_{a \geq \text{med}_S(a)}|}{|S|} \geq \frac{1}{2} \tag{2.5}$$

where here and thereafter the $|S_{\text{condition}}|$ notation is used to designate the subset of S satisfying the condition. Informally, it is a middle value of a, with about a half of the dataset having values below and above the median. If dataset size is an odd number, the middle value can be identified exactly as the kth consecutive value of attribute a after ordering, where $k = (|S| + 1)/2$. Otherwise the two middle-most values are averaged. These are the k_1th and k_2th values after ordering, for $k_1 = |S|/2$ and $k_2 = |S|/2 + 1$.

The sum of absolute differences between attribute values and the median is less than the sum of absolute differences between attribute values and any other number:

$$\text{med}_S(a) = \arg\min_{v \in R} \sum_{x \in S} |a(x) - v| \tag{2.6}$$

This property is somewhat less straightforward to verify than the corresponding total squared difference minimization property of the mean.

Example 2.4.3 Median calculation is implemented and demonstrated by the following R code. Notice that `k1` and `k2` are equal if `m` (the dataset size) is odd.

```
bs.median <- function(v)
{
  k1 <- (m <- length(v))%/%2+1
  k2 <- (m+1)%/%2
  ((v <- sort(v))[k1]+v[k2])/2
}


# demonstration
bs.median(weatherc$temperature)
bs.median(weatherc$temperature[weatherc$play=="yes"])
median(weatherc$temperature)
median(weatherc$temperature[weatherc$play=="yes"])
```

Weighted median For weighted data, the subset size conditions appearing in the definition of the median have to be replaced by the corresponding conditions based on the subset weight sums:

$$\frac{\sum_{x \in S_{a \leq \text{med}_{S,w}(a)}} w_x}{\sum_{x \in S} w_x} \geq \frac{1}{2} \tag{2.7}$$

$$\frac{\sum_{x \in S_{a \geq \text{med}_{S,w}(a)}} w_x}{\sum_{x \in S} w_x} \geq \frac{1}{2} \tag{2.8}$$

Unlike for the ordinary unweighted median, simple middle index calculation is not sufficient to determine the weighted median and the weight sums need to actually calculated and examined.

Example 2.4.4 The R code presented below implements and demonstrates weighted median calculation. Since there is no equivalent standard R function, the results are verified by applying the `median` function to appropriately resampled data, simulating the effect of weighting. The `shift.right` utility function is used to shift the cumulative weight sum to the right. $\boxed{\texttt{dmr.util}}$

```
weighted.median <- function(v, w=rep(1, length(v)))
{
  v <- v[ord <- order(v)]
  w <- w[ord]
  tw <- (sw <- cumsum(w))[length(sw)]
```

```
    mean(v[which(sw>=0.5*tw & tw-shift.right(sw, 0)>=0.5*tw)])
}

    # demonstration
weighted.median(weatherc$temperature, ifelse(weatherc$play=="yes", 5, 1))
median(c(weatherc$temperature[weatherc$play=="no"],
        rep(weatherc$temperature[weatherc$play=="yes"], 5)))
weighted.median(weatherc$temperature, ifelse(weatherc$play=="yes", 0.2, 1))
median(c(weatherc$temperature[weatherc$play=="yes"],
        rep(weatherc$temperature[weatherc$play=="no"], 5)))
```

2.4.1.2 Rank and order statistics

Rank and order statistics are based on attribute value ordering in a dataset. They are not necessarily very useful *per se*, but they are used to calculate other distribution description and relationship detection statistics.

Rank The *rank* of instance x with respect to attribute a on dataset S, designated by $r_{S,a}(x)$, is the ordinal number of x after sorting S nondecreasingly by a. There are several ranking schemes, differing in the way of handling ties, i.e., assigning ranks to instances with equal attribute values. These include:

Competition ranking (1 2 2 4). Instances with equal attribute values all receive the same rank and then a gap is left to adjust for the number of those instances.

Dense ranking (1 2 2 3). Instances with equal attribute values all receive the same rank and then no gap is left.

Ordinal ranking (1 2 3 4). Instances with equal attribute values receive different consecutive ranks in an arbitrary order.

Fractional ranking (1 2.5 2.5 4). Instances with equal attribute values receive the same rank, equal to the mean of ranks they would receive under ordinal ranking.

The latter is the most common ranking strategy for basic statistics. Whenever referring to ranks below, fractional ranking is assumed (unless explicitly otherwise noted – one such exception is the next subsection devoted to order statistics).

Example 2.4.5 Rank calculation is implemented and demonstrated by the following R code. The fractional ranking (1 2.5 2.5 4) technique is used, which is the default for the standard rank function in R. Notice, by the way, that it is the competition ranking (1 2 2 4) that is calculated and assigned the r.min variable in the first line.

```
bs.rank <- function(v)
{
  r.min <- match(v, sort(v))
  r.max <- length(v)+1-match(v, rev(sort(v)))
  (r.min+r.max)/2
}
```

```
 # demonstration
bs.rank(weatherr$playability)
rank(weatherr$playability)
```

Order   Order statistics can be though of as an inverse of ranks. Informally, the kth order statistic is the kth least attribute value. More precisely, the kth *order statistic* of attribute a is the attribute's value for the instance that has rank k with respect to a under *ordinal ranking*:

$$o_S^{(k)}(a) = a(r_{S,a}^{-1}(k)) \qquad (2.9)$$

where $r_{S,a}^{-1}$ denotes the inverse rank, satisfying

$$r_{S,a}^{-1}(r_{S,a}(x)) = x \qquad (2.10)$$

Example 2.4.6 The R code presented below implements and demonstrates order statistic calculation. There is no built-in R direct equivalent (the `order` function serves a different purpose), but the correctness can be verified using the `rank` function (with ordinal ranking, as requested via the `ties.method="first"` argument).

```
ord <- function(v, k=1:length(v))
{
    sort(v)[k]
}

 # demonstration
ord(weatherr$playability, 11)
weatherr$playability[rank(weatherr$playability, ties.method="first")==11]
ord(weatherr$playability, 10:13)
weatherr$playability[rank(weatherr$playability, ties.method="first") %in% 10:13]
```

2.4.1.3   Quantiles

If the median splits the set of attribute values into halves, then *quantiles* can be used to achieve arbitrary other uniform splitting. Roughly speaking, the order-p quantile of attribute a on dataset S cuts out the lower $p \cdot 100\%$ values of a occurring in S. It is more strictly defined as a number such that for at least $p|S|$ instances the values of a are less than or equal to the quantile and for at least $(1-p)|S|$ instances the values of a are greater than or equal to the quantile:

$$|S_{a \leq q_S^{(p)}(a)}| \geq p|S| \qquad (2.11)$$

$$|S_{a \geq q_S^{(p)}(a)}| \geq (1-p)|S| \qquad (2.12)$$

This definition is not unambiguous and may be satisfied by several values. In the "perfect" case when $p(|S|-1)$ is an integer, it is natural to use $o_S^{(p(|S|-1)+1)}(a)$ as the order-p quantile of a. If $p|S|$ is an integer, then any number from the $[o_S^{(p|S|)}(a), o_S^{(p|S|+1)}(a))$ interval can be used as $q_S^{(p)}(a)$ and a natural convention is to take the average of the interval's boundaries. In general, one of several existing quantile estimation techniques has to be used that interpolate between two consecutive order statistic values:

$$q_S^{(p)}(a) = (1 - \beta)o_S^{(k)}(a) + \beta o_S^{(k+1)}(a) \tag{2.13}$$

where $k = \lfloor p|S| + b \rfloor$ for some $-1 < b < 1$ and $0 \le \beta \le 1$. Particular quantile estimation techniques differ in b and β values used. The quantile function in R provides nine quantile estimation techniques, specified via the type parameter, and the R *type* is a common way to refer to them. Noteworthy approaches include:

R type 3.

$$b = -\frac{1}{2} \tag{2.14}$$

$$\beta = \begin{cases} 0 & \text{if } p|S| + b - k = 0 \\ 1 & \text{otherwise} \end{cases} \tag{2.15}$$

R type 7.

$$b = 1 - p \tag{2.16}$$

$$\beta = p|S| + b - k \tag{2.17}$$

R type 8.

$$b = (p + 1)/3 \tag{2.18}$$

$$\beta = p|S| + b - k \tag{2.19}$$

R type 9.

$$b = \frac{p}{4} + \frac{3}{8} \tag{2.20}$$

$$\beta = p|S| + b - k \tag{2.21}$$

Quantiles of orders $\frac{1}{m}, \frac{2}{m}, \dots, \frac{m-1}{m}$ are usually used together and referred to as *m*-quantiles:

$$q_S^{m,i}(a) = q_S^{(i/m)}(a) \tag{2.22}$$

for $i = 1, 2, \dots, m - 1$. For $m = 4$ (which is the most popular choice) we receive *quartiles*:

$$Q_S^1(a) = q_S^{4,1}(a) = q_S^{(0.25)}(a) \tag{2.23}$$

$$Q_S^2(a) = q_S^{4,2}(a) = q_S^{(0.5)}(a) = \text{med}_S(a) \tag{2.24}$$

$$Q_S^3(a) = q_S^{4,3}(a) = q_S^{(0.75)}(a) \tag{2.25}$$

referred to as the *first*, *second*, and *third quartile*, respectively. Quartiles are an extremely useful and commonly used means of quickly characterizing the distribution of a continuous attribute in an easily comprehensible way. With a middle half of attribute values falling between the first and the second quartile, bottom 25% below and top 25% above, quartiles make it possible to notice the major properties of the distribution. These three numbers (or five numbers, adding the minimum and maximum) provide a very good concise distribution description that is readable at a glance. This matters a lot whenever hundreds of attributes have to be analyzed.

Example 2.4.7 The R code presented below presents an illustrative simplified reimplementation of the standard `quantile` function and demonstrates its usage. The implemented quantile estimation technique is R type 7, corresponding to the default `type=7` argument of the latter. The function can be easily modified or enhanced for other estimation techniques.

```
bs.quantile <- function(v, p=c(0, 0.25, 0.5, 0.75, 1))
{
  b <- 1-p
  k <- floor((ps <- p*length(v))+b)
  beta <- ps+b-k
  `names<-`((1-beta)*(v <- sort(v))[k]+beta*(ifelse(k<length(v), v[k+1], v[k])), p)
}

 # demonstration
bs.quantile(weatherc$temperature)
quantile(weatherc$temperature)
bs.quantile(weatherc$temperature[weatherc$play=="yes"])
quantile(weatherc$temperature[weatherc$play=="yes"])
```

2.4.1.4 Dispersion measures

Dispersion or *spread* measures assess the level of variability observed in the set of values of a continuous attribute. Location measures indicate where the values typically lie, whereas dispersion measures indicate how often and how much they depart from this typical location.

Variance The most commonly used dispersion measure is the *variance*, defined as the mean squared difference between attribute values and the mean:

$$s_S^2(a) = \frac{1}{|S| - 1} \sum_{x \in S} (a(x) - m_S(a))^2 \tag{2.26}$$

Dividing by $|S| - 1$ rather than by $|S|$ – which matters for small datasets only – is necessary to make it an *unbiased* estimator of the variance on the whole domain, $\sigma^2(a)$. The variance calculated without this correction:

$$s'^2_S(a) = \frac{1}{|S|} \sum_{x \in S} (a(x) - m_S(a))^2 = m_S(a^2) - m_S^2(a) \tag{2.27}$$

underestimates the true domain variance if calculated on a small dataset.

Example 2.4.8 Variance calculation is implemented and demonstrated by the R code presented below.

```
bs.var <- function(v) { sum((v-mean(v))^2)/(length(v)-1) }

 # demonstration
bs.var(weatherr$playability)
var(weatherr$playability)
```

Example 2.4.9 It is sometimes technically more convenient to receive the variance of 0 rather than a missing value for one-element datasets. This is accomplished by the wrapper around the standard `var` function defined by the following R code. The `var1` function returns 0 rather than NA for one-element value vectors and returns NaN rather than fails for empty vectors.

```
## variance that returns 0 for 1-element vectors and NaN for empty vectors
var1 <- function(v) { switch(min(length(v), 2)+1, NaN, 0, var(v)) }

# demonstration
var1(1:2)
var1(1)
var1(weatherr$temperature[weatherr$playability<0.75])
var1(weatherr$temperature[weatherr$playability>=0.75])
var1(weatherr$temperature[weatherr$playability>=0.8])
```

Weighted variance   Similarly as for location measures, it may be sometimes necessary to incorporate instance weights to dispersion measure calculation. This is the case, in particular, whenever they are used within weight-sensitive modeling algorithms. Not surprisingly, the most popular weighted dispersion measure is the weighted variance. The following formula

$$s'^2_{S,w}(a) = \frac{1}{\sum_{x \in S} w_x} \sum_{x \in S} w_x(a(x) - m_S(a))^2 \tag{2.28}$$

defines the most straightforward weighted variance estimator, which is unfortunately biased. The unbiased weighted variance estimator is calculated in a somewhat more complex way as follows:

$$s^2_{S,w}(a) = \frac{\sum_{x \in S} w_x}{\left(\sum_{x \in S} w_x\right)^2 - \sum_{x \in S} w_x^2} \sum_{x \in S} w_x(a(x) - m_S(a))^2 \tag{2.29}$$

Example 2.4.10 The following code defines a function that calculates (the unbiased estimator of) the weighted variance.

```
weighted.var <- function(v, w=rep(1, length(v)))
{
  sw <- sum(w)
  ssw <- sum(w^2)
  wm <- weighted.mean(v, w)
  sw/(sw^2-ssw)*sum(w*(v-wm)^2)
}

# demonstration
weighted.var(weatherr$playability)
weighted.var(weatherr$playability, ifelse(weatherr$outlook=="rainy", 2, 1))
```

Example 2.4.11 For technical convenience, similarly as before for the ordinary variance, the following R code defines the `weighted.var1` function which is a wrapper around the `weighted.var` function, returning 0 rather than NA for one-element vectors and returning NaN rather than failing for empty vectors.

```
## weighted variance that returns 0 for 1-element vectors and NaN for empty vectors
weighted.var1 <- function(v, w=rep(1, length(w)))
{ switch(min(length(v), 2)+1, NaN, 0, weighted.var(v, w)) }

  # demonstration
weighted.var1(1:2, 1:2)
weighted.var1(1, 2)
weighted.var1(weatherr$temperature[weatherr$playability<0.75],
              weatherr$playability[weatherr$playability<0.75])
weighted.var1(weatherr$temperature[weatherr$playability>=0.75],
              weatherr$playability[weatherr$playability>=0.75])
weighted.var1(weatherr$temperature[weatherr$playability>=0.8],
              weatherr$playability[weatherr$playability>=0.8])
```

Standard deviation   The square root of the variance – bringing back the dispersion to the attribute's original scale of values for convenience – is the *standard deviation*. Assuming the unbiased estimator, it is calculated as

$$s_S(a) = \sqrt{\frac{1}{|S| - 1} \sum_{x \in S} (a(x) - m_S(a))^2} \tag{2.30}$$

Example 2.4.12 Standard deviation calculation is implemented and demonstrated by the following R code:

```
bs.sd <- function(v) { sqrt(sum((v-mean(v))^2)/(length(v)-1)) }

  # demonstration
bs.sd(weatherr$playability)
sd(weatherr$playability)
```

Coefficient of variation   A further attempt to measure the dispersion in a directly comprehensible way is the *coefficient of variation*, calculated as the quotient of the standard deviation and the mean:

$$v_S(a) = \frac{s_S(a)}{m_S(a)} \tag{2.31}$$

This makes it immediately clear how much attribute values are spread out around the mean relatively to the mean itself. Notice that the coefficient of variation is a signed dispersion measure, with the sign inherited from the mean.

Example 2.4.13 The following R code shows how to calculate and use the coefficient of variation. There is no standard R function for this purpose.

```
varcoef <- function(v) { sqrt(sum((v-(m <- mean(v)))^2)/(length(v)-1))/m }

  # demonstration
varcoef(weatherr$playability)
varcoef(-weatherr$playability)
```

Relative standard deviation An unsigned version of the coefficient of variation is the *relative standard deviation*:

$$|v_S(a)| = \frac{s_S(a)}{|m_S(a)|} \tag{2.32}$$

Example 2.4.14 The R code presented below implements relative standard deviation calculation and demonstrates its usage. There is no standard R function for this purpose.

```
relsd <- function(v) { abs(varcoef(v)) }

  # demonstration
relsd(weatherr$playability)
relsd(-weatherr$playability)
```

Median absolute deviation With the variance being the mean squared difference between attribute values and its mean, it is clearly prone to being distorted by outliers. The same applies to the other related dispersion measures discussed above. This is why they have to be used with care when making judgments about the distribution of an attribute. Sometimes much less popular, but more robust dispersion measures may be employed. Of those, the *median absolute deviation* – defined as the median of the differences between attribute values and its median – is the most natural and straightforward choice:

$$\text{mad}_S(a) = \text{med}_S(|a - \text{med}_S(a)|) \tag{2.33}$$

Interestingly, the median absolute deviation can serve as an estimator of the true domain standard deviation for a normally distributed continuous attribute, if using a sufficiently large dataset. For this purpose, it has to be scaled by a factor of

$$\frac{1}{\Phi^{-1}\left(\frac{3}{4}\right)} \approx 1.482602 \tag{2.34}$$

where Φ^{-1} is the inverse cumulative distribution function of the standard normal distribution. It is not uncommon to incorporate this factor when calculating the median absolute deviation by default. This approach to estimating the standard deviation eliminates the impact of outliers.

Example 2.4.15 The implementation and demonstration of the median absolute deviation is provided by the following R code:

```
bs.mad <- function(v, scale=1/qnorm(0.75)) { scale*median(abs(v-median(v))) }

  # demonstration
bs.mad(weatherr$playability, scale=1)
mad(weatherr$playability, constant=1)
bs.mad(weatherr$playability)
mad(weatherr$playability)
sd(weatherr$playability)
```

Interquartile range The difference between the third and the first quartiles, referred to as the *interquartile range*, can also be considered a simple, but very natural and intuitive measure of dispersion:

$$\text{iqr}_S(a) = Q_S^3(a) - Q_S^1(a) \tag{2.35}$$

It represents the range of attribute values that covers the middle half of the dataset.

Example 2.4.16 The calculation of the interquartile range is implemented and demonstrated by the following R code. There is no standard R function available for this purpose.

```
iqr <- function(v)   { unname(diff(quantile(v, c(0.25, 0.75)))) }

  # demonstration
iqr(weatherc$temperature)
```

Quartile dispersion coefficient The *quartile dispersion coefficient* measures the dispersion by relating the interquartile range to the sum of the first and third quartiles:

$$\text{qd}_S(a) = \frac{Q_S^3(a) - Q_S^1(a)}{Q_S^3(a) + Q_S^1(a)} \tag{2.36}$$

Example 2.4.17 The following R code implements and demonstrates the quartile coefficient of dispersion. There is no standard R function available for this purpose.

```
qd <- function(v) { unname(diff(q <- quantile(v, c(0.25, 0.75)))/sum(q)) }

  # demonstration
qd(weatherc$temperature)
```

2.4.1.5   Outlier detection

A simple but commonly used quartile-based outlier detection technique filters out the attribute value that are below the first quartile or above the third quartile by more than a margin proportional to the interquartile range:

$$a(x) < Q_S^1(a) - \beta(Q_S^3(a) - Q_S^1(a)) \tag{2.37}$$

$$a(x) > Q_S^3(a) + \beta(Q_S^3(a) - Q_S^1(a)) \tag{2.38}$$

where usually $\beta >= 1$, with 1.5 being a popular default.

Example 2.4.18 Quartile-based outlier detection is implemented by the R code presented below. Essentially the same technique is provided in R by the `boxplot` function. Its primary purpose is to produce a simple, but very popular and useful visual summary of a continuous attribute distribution that will be discussed later in Section 2.6.1, but it can be requested not to actually produce a plot, but only calculate the underlying statistics. However, due to the latter using a slightly different quartile calculation formula than the default of the quantile function, the results may sometimes differ. This is demonstrated by the example calls. They use a very restrictive value of the interquartile range multiplier to identify at least one outlying value.

```
is.outlier <- function(v, b=1.5)
{ v<(q <- quantile(v, c(0.25, 0.75)))[1]-b*(r <- diff(q)) | v>q[2]+b*r }

weatherc$temperature[is.outlier(weatherc$temperature, 0.5)]
boxplot(weatherc$temperature, range=0.5, plot=FALSE)
boxplot(weatherc$temperature, range=0.49, plot=FALSE)
```

2.4.2   Discrete attributes

The variety of measures describing different properties of continuous attribute distribution is in striking contrast with discrete attribute distribution description. The latter is basically limited to identifying the most frequently occurring values and estimating value probabilities.

2.4.2.1   Mode

The *mode* or *modal value* is the most frequently occurring attribute value in the dataset:

$$\mod_S(a) = \arg \max_{v \in A} |S_{a=v}| \tag{2.39}$$

While the mode can be identified for both continuous and discrete attributes, it is much more useful for the latter than for the former and hence listed here. It is commonly used within modeling algorithms whenever a single representative discrete attribute value has to be selected for a data subset.

Example 2.4.19 The following R code defines the modal function for finding the mode of a given vector of discrete or continuous values, identified as the (first, in case of ties) most frequently occurring value in the vector. There is no equivalent standard R function. The name modal is used to avoid clashing with the mode function, which serves an entirely different purpose in R. The flevels utility function is used to retrieve factor levels as a factor rather than as a character vector.

`dmr.util`

```
modal <- function(v)
{
  m <- which.max(table(v))
  if (is.factor(v))
    flevels(v)[m]
  else
    sort(unique(v))[m]
}

  # demonstration
modal(weather$outlook)
modal(weatherr$temperature)
```

2.4.2.2 Weighted mode

Since the mode is often needed to make decisions within modeling algorithms, those of them that are weight sensitive actually have to use the weighted mode:

$$\text{mod}_{S,w}(a) = \arg\max_{v \in A} \sum_{x \in S_{a=v}} w_x \tag{2.40}$$

This simply replaces subset counts with the corresponding instance weight sums.

Example 2.4.20 The following R code implements and demonstrates weighted mode calculation. The `weighted.table` auxiliary function is used for creating the weighted contingency table.

`dmr.util`

```
weighted.modal <- function(v, w=rep(1, length(v)))
{
  m <- which.max(weighted.table(v, w=w))
  if (is.factor(v))
    factor(levels(v)[m], levels=levels(v))
  else
    sort(unique(v))[m]
}

# demonstration
weighted.modal(weather$outlook)
weighted.modal(weather$outlook, w=ifelse(weather$play=="yes", 2, 1))
```

2.4.2.3 Probability

The standard approach in estimating discrete attribute value probabilities is to use relative frequencies. The probability of attribute a taking value v estimated on dataset S is then

$$P_S(a = v) = \frac{|S_{a=v}|}{|S|} \tag{2.41}$$

Such a frequency estimation (also referred to as the empirical probability) is generally a perfectly reasonable approach whenever the dataset is sufficiently large. This can be safely assumed if probabilities are estimated as part of data exploration rather than model creation.

One can similarly estimate conditional attribute value probabilities:

$$P_S(a_1 = v_1 | a_2 = v_2) = \frac{|S_{a_1=v_1, a_2=v_2}|}{|S_{a_2=v_2}|} \tag{2.42}$$

or joint probabilities for attribute value pairs:

$$P_S(a_1 = v_1, a_2 = v_2) = \frac{|S_{a_1=v_1, a_2=v_2}|}{|S|} \tag{2.43}$$

It is rather problematic, though, to similarly proceed with a larger number of attributes, as it would require very large datasets for the estimates to be reliable.

Example 2.4.21 The following R code implements and demonstrates single attribute value probability estimation.

```
prob <- function(v, v1) { sum(v==v1)/length(v) }

  # demonstration
prob(weather$outlook, "rainy")
```

Example 2.4.22 The following R code implements and demonstrates full discrete probability distribution estimation for one or more attributes. This is usually more convenient than estimating probabilities for a single value or value combination at a time.

```
pdisc <- function(v, ...) { (count <- table(v, ..., dnn=NULL))/sum(count) }

  # demonstration
pdisc(weather$outlook)
pdisc(weather$outlook, weather$temperature)
```

Example 2.4.23 The R code presented below implements and demonstrates conditional attribute value probability estimation for a pair of discrete attributes.

```
## conditional probability distribution P(v1|v2)
pcond <- function(v1, v2)
{
  t(apply(count <- table(v1, v2, dnn=NULL), 1, "/", colSums(count)))
}

  # demonstration
pcond(weather$outlook, weather$play)
```

Weighted probability It is sometimes desirable to take into account instance weights when using data for probability estimation. The probability of attribute a taking value v estimated on dataset S using per-instance weights w can be calculated as

$$P_{S,w}(a = v) = \frac{\sum_{x \in S_{a=v}} w_x}{\sum_{x \in S} w_x} \tag{2.44}$$

The same approach can be applied when estimating conditional or joint attribute value probabilities.

Example 2.4.24 The following R code implements and demonstrates single attribute value weighted probability estimation.

```
weighted.prob <- function(v, v1, w=rep(1, length(v))) { sum(w[v==v1])/sum(w) }

  # demonstration
weighted.prob(weather$outlook, "rainy")
weighted.prob(weather$outlook, "rainy", w=ifelse(weather$play=="yes", 2, 1))
```

Example 2.4.25 The following R code implements and demonstrates full discrete weighted probability distribution estimation for one or more attributes, using the `weighted.table` function.

dmr.util

```
## weighted discrete probability distribution
weighted.pdisc <- function(v, ..., w=rep(1:length(v)))
{
  (count <- weighted.table(v, ..., w=w))/sum(count)
}

  # demonstration
weighted.pdisc(weather$outlook, w=ifelse(weather$play=="yes", 2, 1))
weighted.pdisc(weather$outlook, weather$temperature,
               w=ifelse(weather$play=="yes", 2, 1))
```

2.4.2.4 Impurity

Impurity for discrete attributes can be thought of as a rough counterpart of dispersion for continuous attributes. It represents that part to which one or a small number of the most frequent values dominate over other less frequent values, with low impurity meaning high domination. The two most widely known and used impurity measures are the *entropy* adopted from the information theory, where it is used to measure the expected information contents of a message, and the *Gini index* adopted from economy, where it is used to measure the inequality of wealth. Both can be used to characterize the impurity of discrete attribute distribution in a set of instances.

Entropy   The *entropy* of attribute $a : X \to V$ on dataset S is defined as follows:

$$E_S(a) = \sum_{v \in A} -P_S(a = v) \log P_S(a = v) \tag{2.45}$$

The base 2 logarithm is traditionally most often used, but it is actually irrelevant, as the only effect of changing the base is scaling the calculated value up or down.

The entropy reaches its maximum value when attribute values are distributed uniformly (maximum impurity) and minimum values when only one value is represented (maximum purity). In the latter case, $0 \log 0$ should be assumed to be 0, which is numerically convenient and mathematically justified by the fact that $p \log p$ is 0 in the limit for p approaching 0.

Example 2.4.26 The R code presented below defines the `entropy` function for entropy calculation based on a vector of discrete values. The `entropy.p` function that it internally uses operates directly on a probability distribution (i.e., a vector of probabilities) and can also come handy. It uses the `plogp` utility function for calculating $p \log (p)$ for arbitrary $p \geq 0$ that returns 0 for $p = 0$. Simple usage demonstrations are provided for the two entropy functions.

dmr.util

```
## entropy for discrete probability distributions
entropy.p <- function(p) { sum(-plogp(p)) }

entropy <- function(v) { entropy.p(pdisc(v)) }

  # demonstration
```

```
entropy.p(c(1/5, 2/5, 3/5))
entropy(weather$outlook)
entropy(weather$play)
entropy(weather$play[weather$outlook=="overcast"])
entropy(weather$play[weather$outlook!="overcast"])
```

Gini index   The *Gini index* of attribute $a : X \to V$ on dataset S is defined as follows:

$$\text{GI}_S(a) = \sum_{v \in A} P_S(a = v)(1 - P_S(a = v)) = 1 - \sum_{v \in A} P_S^2(a = v) \tag{2.46}$$

Similarly for the entropy, its maximum and minimum values correspond to the maximum and minimum impurity, respectively.

Example 2.4.27 The following R code defines the `gini` function that calculates the Gini index for a vector of discrete values. The internally used `gini.p` function operates directly on a probability distribution.

```
## Gini index for discrete probability distributions
gini.p <- function(p) { 1-sum(p^2) }

gini <- function(v) { gini.p(pdisc(v)) }

  # demonstration
gini.p(c(1/5, 2/5, 3/5))
gini(weather$outlook)
gini(weather$play)
gini(weather$play[weather$outlook=="overcast"])
gini(weather$play[weather$outlook!="overcast"])
```

Example 2.4.28 To get some more insight into the two impurity measures presented above, consider the simplest binary attribute case, where taking $P_S(1) = p$ and $P_S(0) = 1 - p$, we can rewrite the entropy as $-p \log p - (1 - p) \log (1 - p)$ and the Gini index as $1 - p^2 - (1 - p)^2$. The following R code will plot them as functions of p in the $(0, 1)$ interval.

```
  # plot the entropy
curve(-x*log(x)-(1-x)*log(1-x), from=0, to=1,
      xlab="p", ylab="", ylim=c(-0.02, 0.7), lty=1)
  # and add the plot of the Gini index
curve(1-x^2-(1-x)^2, from=0, to=1, add=TRUE, lty=2)
legend("topright", legend=c("entropy", "gini"), lty=1:2)
```

The resulting plot is shown in Figure 2.1.

2.4.3   Confidence intervals

As mentioned above, descriptive statistics calculated on a dataset can be considered estimators of their unknown values on the whole domain. The quality of such estimators can be assessed using a *confidence interval*.

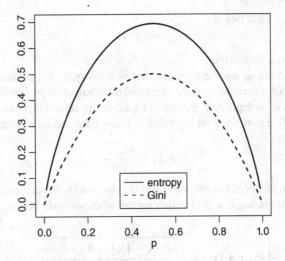

Figure 2.1 The plots of the entropy and the Gini index impurity measures for a binary attribute.

Let ζ be the unknown value of a descriptive statistic on the whole domain X and z_S – its value calculated on dataset $S \subset X$. Different values $z_{S_1}, z_{S_2}, \ldots$ will be obtained for different datasets $S_1, S_2, \ldots$. They may be all used as *estimates* of ζ and considered *realizations* of the corresponding random variable z. The latter is referred to as *estimator* of ζ.

A *confidence interval* for ζ at the *confidence level* $1 - \delta$ is any interval to which ζ belongs with the probability $1 - \delta$, where $0 < \delta < 1$. Confidence intervals may be determined by appropriate interval estimation techniques using a single estimate z_S, based on a single dataset S, or multiple estimates $z_{S_1}, z_{S_2}, \ldots$ based on several datasets $S_1, S_2, \ldots$.

2.4.3.1   Parametric interval estimation

Parametric interval estimation techniques use a single estimate z_S to determine a confidence interval that is centered at z_S and has width adjusted according to the specified confidence level based on some known (or assumed) properties of the distribution of z. More specifically, the estimator's probability density function is used to determine a margin $\Delta_{S,\delta}^{(z)}$ such that the probability of z being between $z_S - \Delta_{S,\delta}^{(z)}$ and $z_S + \Delta_{S,\delta}^{(z)}$ is $1 - \delta$.

To see how the confidence interval margin can be derived, consider the following standardized form of the estimate:

$$\frac{z_S - \zeta}{\sigma(z)} \tag{2.47}$$

assuming the estimator's standard deviation $\sigma(z)$ is known (or can be estimated). Assuming additionally that the distribution function of z is known, then the corresponding inverse cumulative distribution function Ψ^{-1} can be used to identify $u_\delta = \Psi^{-1}(\frac{\delta}{2})$ such that

$$-u_\delta < \frac{z_S - \zeta}{\sigma(z)} < u_\delta \tag{2.48}$$

with probability $1 - \delta$. This yields

$$\Delta_{S,\delta}^{(z)} = u_\delta \sigma(z) \tag{2.49}$$

as the confidence interval's margin.

One specific example of such estimation that is directly useful in data mining tasks is applied to the probability parameter of the binomial distribution, representing the probability of success in a series of independent Bernoulli trials. In our terms, this can be interpreted as the probability of a binary attribute taking value 1. Using dataset S to estimate this probability we have

$$P_S(a = 1) = \frac{|S_{a=1}|}{|S|} \tag{2.50}$$

If the dataset is drawn uniformly at random from the domain (with each instance drawn independently), the corresponding true probability on the whole domain belongs to the following interval:

$$\left(P_S(a = 1) - u_\delta \sqrt{\frac{P_S(a = 1)(1 - P_S(a - 1))}{|S|}} \right., \tag{2.51}$$

$$\left. P_S(a = 1) + u_\delta \sqrt{\frac{P_S(a = 1)(1 - P_S(a - 1))}{|S|}} \right) \tag{2.52}$$

with probability $1 - \delta$, where

$$u_\delta = -\Phi^{-1}\left(\frac{\delta}{2}\right) = \Phi^{-1}\left(1 - \frac{\delta}{2}\right) \tag{2.53}$$

and Φ^{-1} is the inverse cumulative distribution function of the standard normal distribution. This form of the confidence interval for binary attribute value probability estimation relies on some approximations that do not hold for very small datasets.

It can be easily verified that $\Phi(u_\delta) - \Phi(-u_\delta) = 1 - \delta$. This means that the probability of a random value drawn from the standard normal distribution falling to the $(-u_\delta, u_\delta)$ interval is equal $1 - \delta$. For the particularly popular confidence level $1 - \delta = 0.95$, we have $u_\delta \approx 1.96$, which appears to be quite easy to remember (add 1 to the first and last digit of 0.95).

Example 2.4.29 Parametric interval estimation for the probability parameter of the binomial distribution (i.e., binary attribute value probability estimation) is implemented and illustrated by the following R code. The *weather* dataset used for this illustration is actually too small for the estimated interval to be reliable.

```
prob.ci.par <- function(v, v1=1, delta=0.05)
{
  list(p=(p <- prob(v, v1)),
       low=p-(u <- qnorm(1-delta/2))*(sp <- sqrt(p*(1-p)/length(v))),
       high=p+u*sp)
}

  # demonstration
prob.ci.par(weather$play, "yes")
prob.ci.par(weather$play, "yes", delta=0.01)
prob.ci.par(weather$play, "yes", delta=0.1)
```

2.4.3.2 Bootstrapping interval estimation

Bootstrapping interval estimation does not rely on any knowledge or assumptions about the distribution of the estimator. Instead, it "simulates" the availability of multiple datasets from the same domain and uses the observed distribution of the estimated statistic on these "simulated" multiple datasets to derive confidence interval bounds.

"Simulating" multiple datasets is accomplished using *bootstrap samples* of the original single datasets –samples drawn uniformly at random with replacement, typically of the same size as the original dataset. This yields a number of estimates, $z_{S_1}, z_{S_2}, \ldots, z_{S_m}$, where $S_1, S_2, \ldots, S_m$ are the bootstrap samples of S and m is typically several hundred or more. We can treat these estimates as values of an attribute z' defined as $z'(x) = z_{S_x}$ for $x \in M = \{1, 2, \ldots, m\}$ and determine the confidence interval based on the distribution of z' on M. The most straightforward way to do this is to use the appropriate quantiles as interval bounds, yielding the following confidence interval:

$$\left(q_M^{\frac{\delta}{2}}(z'), q_M^{1-\frac{\delta}{2}}(z') \right) \tag{2.54}$$

for the confidence level $1 - \delta$.

Bootstrapping interval estimation is an attractive alternative to parametric interval estimation if the properties of the distribution of the estimated statistic required for the latter are unknown, hard to determine, or rely on unsatisfied assumptions (such as sufficiently large data). If this is not the case, though, parametric techniques should be preferred, as yielding more precise interval bounds. Bootstrapping techniques are approximate by nature. Even a large number of bootstrap samples drawn from the original dataset do not guarantee adequate "simulation" of multiple datasets from the same domain, particularly if the dataset is small.

Example 2.4.30 The following R code illustrates the bootstrapping-based approach in estimating binary attribute value probability. The results differ noticeably from those obtained using the parametric approach demonstrated in the previous example, which is to be expected for such small data.

```
prob.ci.boot <- function(v, v1=1, delta=0.05, m=1000)
{
  q <- unname(quantile(sapply(1:m, function(i) prob(sample(v, replace=TRUE), v1)),
                    probs=c(delta/2, 1-delta/2)))
  list(p=prob(v, v1), low=q[1], high=q[2])
}

# demonstration
prob.ci.boot(weather$play, "yes")
prob.ci.boot(weather$play, "yes", delta=0.01)
prob.ci.boot(weather$play, "yes", delta=0.1)
```

2.4.4 *m*-Estimation

As already mentioned before, statistics describing the attribute distribution are not only used to examine the data prior to running data mining algorithms, but also within data mining algorithms. It is not uncommon for the latter to partition the data into small subsets during model creation, typically by one or more attribute-value conditions that may be satisfied for few instances only. Whenever this is the case, distribution description statistics calculated on such subsets are no longer reliable estimators of the corresponding values on the whole domain. This issue is particularly severe for probability estimation, but sometimes may also be important for other commonly used statistics, including the mean and the variance.

The issue of unreliable estimation from small data subsets can be resolved by combining estimates calculated on such subsets with some prior estimates, which can be based on the background knowledge, adopted assumptions, or calculated from larger data. The combination is performed by introducing a number of fictitious instances to the subset, with some prior value of the estimated parameter.

2.4.4.1   Probability m-estimation

Simple frequency-based probability estimates based on small data subsets may be unreliable. In particular, estimating near-zero or near-one probabilities requires sufficiently large datasets and their estimates obtained on small data are likely to be exact 0s or 1s. This is usually undesirable. It would essentially mean assuming unobserved is equivalent to impossible, whereas considering unobserved unlikely is a much safer choice for inductive learning. In many cases, we may actually know for sure that some attribute values, not observed in a small data subset on which probability estimation is performed, have nonzero occurrence probabilities. This may be implied by domain knowledge or by having observed these values in a superset of the dataset.

The technique of probability m-estimation incorporates m fictitious instances with an *a priori* assumed probability estimate. The resulting *m-estimate*, also called the *Cestnik estimate*, of the probability of attribute a taking the value v is then calculated using dataset S as follows:

$$P_{S,m,p_{0,a=v}}(a=v) = \frac{|S_{a=v}| + mp_0}{|S| + m} \tag{2.55}$$

where $p_{0,a=v}$ denotes the prior probability estimate of attribute a taking value v. In the simplest case when all attribute values are considered equally likely *a priori*, we have

$$p_{0,a=v} = \frac{1}{|A|} \tag{2.56}$$

Example 2.4.31 The following R code defines the `mest` function that calculates the m-estimate of probability given instance counts and the `mprob` function that uses the former to m-estimate the probability of a single attribute value. It is straightforward to similarly implement other wrappers around `mest` for more complex usage scenarios (e.g., conditional probabilities or joint probability distributions).

```
## m-estimate of probability of an event occurring n1 out of n times
## incorporating m fictitious instances
mest <- function(n1, n, m=2, p0=1/m) { (n1+m*p0)/(n+m) }

mprob <- function(v, v1, m=nlevels(v), p0=1/nlevels(v))
{ mest(sum(v==v1), length(v), m, p0) }

   # demonstration
mest(0, 10, 1, 0.5)
```

```
mest(0, 10, 2, 0.5)
mest(10, 10, 1, 0.5)
mest(10, 10, 2, 0.5)

mprob(weather$outlook, "rainy", m=0)
mprob(weather$outlook, "rainy")
mprob(weather$play[weather$outlook=="overcast"], "no", m=0)
mprob(weather$play[weather$outlook=="overcast"], "no")
mprob(weather$play[weather$outlook=="overcast"], "no", m=3, p0=0.5)
```

If we assume $p_{0,a=v} = \frac{1}{|A|}$ and additionally $m = |A|$ for probability m-estimation, we receive a special case known as the *Laplace estimate*:

$$P_{S,1}(a = v) = \frac{|S_{a=v}| + 1}{|S| + |A|} \tag{2.57}$$

also known as *add-one* or *Laplace smoothing*. It assumes that there is always one unseen occurrence of the value the probability of which is being estimated. A parameterized version of this estimator is sometimes used as

$$P_{S,l}(a = v) = \frac{|S_{a=v}| + l}{|S| + l|A|} \tag{2.58}$$

where $l > 0$ controls the intensity of smoothing.

Example 2.4.32 The mprob function from the previous example with default parameter settings actually performs Laplace probability estimation, as demonstrated below.

```
## Laplace estimate of probability of an event occurring n1 out of n times
## with m possible outcomes
laest <- function(n1, n, m=2) { mest(n1, n, m)   }

laprob <- function(v, v1) { mprob(v, v1) }

   # demonstration
laest(0, 10, 2)
mest(0, 10, 2)
laest(10, 10, 2)
mest(10, 10, 2)

laprob(weather$outlook, "rainy")
mprob(weather$outlook, "rainy", m=3, p0=1/3)
laprob(weather$play[weather$outlook=="overcast"], "no")
mprob(weather$play[weather$outlook=="overcast"], "no", m=2, p0=0.5)
```

2.4.4.2 Mean *m*-estimation

The technique of m-estimation can be similarly applied to estimating the mean. Introducing a prior mean estimate in this case would prevent one overly trusting a possibly unreliable mean estimate obtained on a small data subset. As before, this is accomplished by incorporating m

fictitious instances with an assumed prior mean value m_0. The m-estimated mean of attribute a on data S is then calculated as follows:

$$m_{S,m,m_0}(a) = \frac{\sum_{x \in S} a(x) + mm_0}{|S| + m}$$

(2.59)

Example 2.4.33 The following R code implements and demonstrates mean m-estimation. The default prior mean value, equal to the actual mean of the input vector, makes the m-estimated mean equal to the ordinary mean.

```
## m-mean that incorporates m fictitious values with a specified mean m0
mmean <- function(v, m=2, m0=mean(v)) { (sum(v)+m*m0)/(length(v)+m) }

# demonstration
mmean(weatherr$playability)
mmean(weatherr$playability, m=0)
mmean(weatherr$playability, m0=0.5)
mmean(weatherr$playability, 5, 0.5)
mmean(weatherr$playability[weatherr$temperature<25], m=0)
mmean(weatherr$playability[weatherr$temperature<25], m0=mean(weatherr$playability))
```

2.4.4.3 Variance m-estimation

When the variance is used within modeling algorithms to make some model-building decisions based on data subsets, it may be also reasonable to apply m-estimation and combine the actually observed variance with a prior estimate. This can be accomplished by incorporating m fictitious instances with an assumed *a priori* variance value s_0^2. If using a small data subset, these instances would bias the estimated variance towards the prior value. The m-*variance* of attribute a on dataset S can be calculated using the following formula:

$$s_{S,m,s_0^2}^2(a) = \frac{(|S| - 1)s_S^2(a) + (m - 1)s_0^2}{|S| + m - 2}$$

(2.60)

This combines the observed variance of the actual attribute values in S and the assumed variance of the m fictitious instances in the same way as during pooled variance calculation.

While the above m-variance estimator may be sufficient for simple usage scenarios, it is actually oversimplified since it assumes that the mean value used for variance calculation is estimated in the usual way rather than m-estimated. A better variance m-estimate is obtained by using an m-estimate of the mean

$$s_{S,m,m_0,s_0^2}^2(a) = \frac{\sum_{x \in S} (a(x) - m_{S,m,m_0}(a))^2 + (m - 1)s_0^2}{|S| + m - 2}$$

(2.61)

where m_0 and s_0^2 are the prior mean and variance estimates, respectively, and $m_{S,m,m_0}(a)$ is the m-estimated mean.

Example 2.4.34 The following code defines a function that incorporates m fictitious values with a specified prior mean and variance values. The default priors, equal to the actual mean and variance of the input vector, makes the m-estimated variance equal to the ordinary variance.

```
## m-variance that incorporates m fictitious values with a specified variance s02
mvar <- function(v, m=2, m0=mean(v), s02=var(v))
{ (sum((v-mmean(v, m, m0))^2)+max(m-1, 0)*s02)/max(length(v)+m-2, 1)  }

    # demonstration
mvar(weatherr$playability)
mvar(weatherr$playability, m=0)
mvar(weatherr$playability, s02=0.05)
mvar(weatherr$playability, m=5, s02=0.05)
mvar(weatherr$playability[weatherr$temperature<25], m=0)
mvar(weatherr$playability[weatherr$temperature<25],
      m0=mean(weatherr$playability), s02=var(weatherr$playability))
```

2.4.4.4 Obtaining priors

There are three major possible sources of priors for m-estimation, already mentioned above:

- background knowledge,
- more or less arbitrary assumptions adopted in the lack of background knowledge (such as the assumption of all attribute values being equiprobable),
- estimates from a superset of S, $S_0 \supset S$.

The last approach yields the following priors for the m-estimated probability, mean, and variance:

$$p_0 = P_{S_0}(a = v) \tag{2.62}$$

$$m_0 = m_{S_0}(a) \tag{2.63}$$

$$s_0^2 = s_{S_0}^2(a) \tag{2.64}$$

It makes most sense when, within a modeling algorithm, $S = T_{\text{condition}}$ is a subset of the training set satisfying a particular condition, and $S_0 = T$ is the full training set.

2.5 Relationship detection

Techniques for detecting relationships between attributes are particularly interesting and useful for data mining. They are used

- during initial data exploration, to get the assessment of the overall data quality and predictive utility,
- during attribute selection, within attribute selection filters, and
- during model creation, within modeling algorithms.

Generally speaking, a relationship between two attributes is a property that makes it possible to better-than-randomly predict one attribute based on the other. This is whenever knowing the value of one attribute "narrows down" the distribution of the other.

2.5.1   Significance tests

Most statistical relationship detection techniques can be viewed as statistical *significance tests*. As usual, statistics calculated on a dataset are used to infer about the properties of the domain. The purpose of relationship detection techniques is often not just to detect or measure a relationship between attributes in a particular dataset, but also – and more importantly – to identify relationships that also hold on the domain. A relationship is said to be *significant* if it is likely to hold on the whole domain. Statistical significance tests are used to detect such significant relationships.

2.5.1.1   Null and alternative hypotheses

A statistical significance test is a decision procedure making it possible to choose between the following two *statistical hypotheses* about a possible relationship (or other phenomenon of interest) based on a dataset:

Null hypothesis. The relationship does not hold on the domain (even if it is observed on the dataset).

Alternative hypothesis. The relationship holds on the domain (and this is why it is observed on the dataset).

The null hypothesis is the "dull" one – it states that there is nothing to be discovered, and if the data suggest otherwise, it is entirely due to chance. A different dataset from the same domain would not probably confirm any such observations. The alternative hypothesis is the "interesting" one – it states the observations made using the dataset are not due to chance and are likely to hold on the domain. The specific formulations of the null and alternative hypotheses depend on the particular type of relationship being examined.

2.5.1.2   Statistic

One of the two statistical hypotheses is accepted and the other is rejected based on the value of a *test statistic*. It is calculated on the dataset using an appropriate formula or algorithm. More precisely, the statistic is a random variable taking different values for different datasets from the same domain, and a particular value calculated for one particular dataset is a realization of that random variable.

Clearly, if the test statistic is supposed to make it possible to judge whether the relationship being examined does or does not hold on the domain, it has to incorporate some measure of its observed strength on the dataset. The particular way of achieving this is specific to particular tests. Another requirement for the test statistic is that it must have a known probability distribution under the null hypothesis. Some properties of the distribution of the attributes under consideration may have to be known or assumed to satisfy this requirement. Once determined, the distribution makes it possible to assess how a particular value calculated on the analyzed dataset is likely or unlikely assuming the null hypothesis. A significance test is actually comprised of a test statistic and a statistical inference procedure used to reject or accept the null hypothesis.

2.5.1.3   *p*-Value

The known test statistic distribution under the null hypothesis is used to calculate the probability of achieving a value equal to or more extreme than that calculated on the dataset if the null hypothesis were true. This probability, referred to as *p-value*, compared against a threshold called the *significance level*, yields the tests's decision criterion. If the *p*-value is below the significance level (i.e., the probability of the statistic value obtained for the dataset is sufficiently low under the null hypothesis assumption), the null hypothesis is rejected. Otherwise the alternative hypothesis is rejected. Popular significance level values are 0.001, 0.01, and 0.05.

2.5.1.4   False positives and false negatives

Using the *p*-value to reject one of the two hypotheses may result in the following two types of errors:

> *False positive (aka type I error).* The incorrect rejection of the null hypothesis that is actually true (an unexisting relationship is detected that does not hold on the domain).
>
> *False negative (aka type II error).* The incorrect rejection of the alternative hypothesis that is actually true (there is a relationship on the domain that remains undetected).

With a large significance level the risk of false positives is increased and a small significant level increases the risk of false negatives.

The probabilities of these two types of errors, also referred to as the false positive rate and the false negative rate, are the most important quality criteria for statistical tests (related, by the way, to the analogous quality criteria for classification models discussed in Section 7.2.4).

2.5.1.5   Relationship significance vs. relationship strength

The significance of a discovered relationship between attributes, represented by the *p*-value of the applied statistical test, should not be confused with the relationship's strength. While the latter is usually captured by the test statistic, it is not the only factor that contributes to the significance. As it will become clear after some popular tests are presented, a low *p*-value, indicating statistical significance, may also be obtained for weak relationships on sufficiently large datasets. While such relationships are likely to hold on the whole domain, they are not necessarily interesting, as their predictive utility is low. A reasonable approach to detect useful attribute relationships may be therefore to consider both their significance, represented by the *p*-value, and strength, often represented by the test statistic, and focus on those that are both significant and sufficiently strong.

In some applications of relationship detection, it is actually only the strength of the relationship that matters and its significance is immaterial. This is the case, in particular, whenever one has to identify one or more attributes most closely related to the target attribute of a classification or regression task, regardless of the significance of these relationships. This is a typical situation when making model construction decisions within modeling algorithms, sometimes based on small attribute subsets. Some relationship detection statistics that are only or predominately used for this purpose will be presented without the accompanying statistical inference procedure, leading to p-value calculation. On the other hand, when exploring the properties of a large dataset, all strong relationships are practically guaranteed to be significant.

2.5.2   Continuous attributes

The most common approach to detecting relationships between continuous attributes is by measuring their correlation. Attributes are said to be correlated if they exhibit a similar pattern of high and low values over the dataset, i.e., the tendency to simultaneously increase or decrease. The strength of such a pattern can be measured using the widely known linear correlation coefficient and the not so widely known rank correlation coefficient. They can be both considered test statistics, accompanied by inference procedures using them within significance tests.

2.5.2.1   Pearson's linear correlation

The linear correlation coefficient, also known as Pearson's correlation coefficient, measures the strength of linear relationship between two continuous attributes, i.e., the degree to which their relationship approaches a linear function. It is calculated using the following formula:

$$\rho_S(a_1, a_2) = \frac{\sum_{x \in S}(a_1(x) - m_S(a_1))(a_2(x) - m_S(a_2))}{\sqrt{\sum_{x \in S}(a_1(x) - m_S(a_1))^2 \sum_{x \in S}(a_2(x) - m_S(a_2))^2}} \tag{2.65}$$

If $a_1(x) = \alpha a_2(x) + \beta$ for some $\alpha \neq 0$ and all $x \in X$ then $\rho_S(a_1, a_2) = \text{sgn}(\alpha)$ and the two attributes exhibit a perfectly linear relationship on the dataset. Otherwise the correlation coefficient is a number between -1 and 1 the absolute value of which indicates the strength of the relationship and the sign of which indicates its direction (positive if the values of one attribute tend to increase with increasing values of the other attribute). Of course, Pearson's correlation coefficient and the corresponding test may fail to detect nonlinear relationships or underestimate their strength.

The most typical inference procedure using the linear correlation coefficient checks the null hypothesis that its value on the domain is actually 0. One simple way to determine the p-value, applicable if the two attributes can be assumed to have normal distributions, is to consider an auxiliary statistic defined as follows:

$$t_S^{(\rho)}(a_1, a_2) = \rho_S(a_1, a_2) \sqrt{\frac{|S| - 2}{1 - \rho_S^2(a_1, a_2)}} \tag{2.66}$$

which has Student's standard t-distribution with $|S| - 2$ degrees of freedom under the null hypothesis. This is a reasonably good approximation even if the assumption is not satisfied, except for very small datasets, for which other approaches may be used. The p-value is determined as

$$p_S^{(\rho)}(a_1, a_2) = 2 \left(1 - \Phi_{|S|-2}^{(t)} \left(|t_S^{(\rho)}(a_1, a_2)| \right) \right) \tag{2.67}$$

where $\Phi_k^{(t)}$ denotes the standard cumulative t-distribution function with k degrees of freedom. This is the probability of a value equal or above $|t_S^{(\rho)}(a_1, a_2)|$, or equal or below $-|t_S^{(\rho)}(a_1, a_2)|$.

Example 2.5.1 Pearson's correlation is implemented and demonstrated by the R code presented below. The `corl` function calculates the correlation coefficient value and the *p*-value. The `cor.test` function with `method="pearson"` is used for comparison.

```
corl.test <- function(v1, v2)
{
  rho <- sum((v1-(m1 <- mean(v1)))*(v2-(m2 <- mean(v2))))/
          sqrt(sum((v1-m1)^2)*sum((v2-m2)^2))
  ts <- rho*sqrt((df <- length(v1)-2)/(1-rho^2))
  list(rho=rho, statistic=ts, p.value=2*(1-pt(abs(ts), df)))
}

  # demonstration
corl.test(weatherr$temperature, weatherr$playability)
cor.test(weatherr$temperature, weatherr$playability, method="pearson")
corl.test(weatherr$temperature, -weatherr$playability)
cor.test(weatherr$temperature, -weatherr$playability, method="pearson")
```

2.5.2.2 Spearman's rank correlation

To make it possible to discover arbitrary monotonic relationships, regardless of their linearity, one can consider the linear correlation of instance ranks with respect to the two attributes rather than the linear correlation of the attributes themselves. Such a linear correlation of ranks is called the rank correlation or Spearman's correlation:

$$\varrho_S(a_1, a_2) = \rho_S(r_{S,a_1}, r_{S,a_2}) \tag{2.68}$$

assuming fractional ranking. If ties (duplicate attribute values) are absent and no fractional ranks appear, this is equivalent to the following formula:

$$\varrho_S(a_1, a_2) = 1 - \frac{6 \sum_{x \in S} (r_{S,a_1}(x) - r_{S,a_2}(x))^2}{|S|(|S|^2 - 1)} \tag{2.69}$$

The rank correlation coefficient takes the value 1 or −1 if the attributes a_1 and a_2 exhibit any strictly increasing or decreasing (respectively) relationship on S. Intermediate values indicate value levels of monotonic relationship.

Inference using Spearman's correlation coefficient is possible by testing the null hypothesis that its value is 0 on the domain. There are multiple possible approaches to determining the *p*-value, including that the presented above for Pearson's correlation, which is also applicable here. For smaller data more refined approaches are used.

Example 2.5.2 The following R code implements and demonstrates rank correlation calculation. The `corr.test` function simply calls the `corl.test` function from the previous examples for attribute ranks.

```
corr.test <- function(v1, v2) { corl.test(rank(v1), rank(v2)) }

  # demonstration
corr.test(weatherr$temperature, weatherr$playability)
```

```
cor.test(weatherr$temperature, weatherr$playability, method="spearman")
corr.test(weatherr$temperature, -weatherr$playability)
cor.test(weatherr$temperature, -weatherr$playability, method="spearman")
```

2.5.3 Discrete attributes

Discovering relationships between two discrete attributes is typically based on observing how their joint distribution differs from their marginal distributions. In each case, the distributions are estimated using the same dataset on which relationship detection is performed. There are multiple specific ways of measuring the degree to which the distributions differ, using absolute frequencies or probabilities (relative frequencies) of single attribute values and attribute value pairs.

2.5.3.1 χ^2 Test

The most widely known technique used for detecting relationships between discrete attributes employs the following χ^2 statistic:

$$\chi_S^2(a_1, a_2) = \sum_{v_1 \in A_1} \sum_{v_2 \in A_2} \frac{\left(|S_{a_1=v_1, a_2=v_2}| - e_{a_1=v_1, a_2=v_2}\right)^2}{e_{a_1=v_1, a_2=v_2}} \tag{2.70}$$

where

$$e_{a_1=v_1, a_2=v_2} = \frac{|S_{a_1=v_1}| \cdot |S_{a_2=v_2}|}{|S|} \tag{2.71}$$

is the expected absolute frequency of the values v_1 and v_2 for the attributes a_1 and a_2, respectively, under the null hypothesis of the two attributes being unrelated. The latter means that knowing the value of one of them does not alter the distribution of the other. The χ^2 statistic compares the actual count of each value combination observed on the dataset with the expected one. High values, obtained if the actual and observed counts differ considerably, indicate a strong relationship.

The χ^2 statistic takes discrete numerical values, since it depends on discrete attribute value count frequencies on a finite set of instances. Its distribution under the null hypothesis can be approximated, however, by the continuous χ^2 distribution with $(|A_1| - 1)(|A_2| - 1)$ degrees of freedom. The p-value is the probability of a χ^2 value equal or greater than that calculated on the dataset:

$$p_S^{(\chi^2)}(a_1, a_2) = 1 - \Phi_{(|A_1|-1)(|A_2|-1)}^{(\chi^2)}(\chi_S^2(a_1, a_2)) \tag{2.72}$$

where $\Phi_k^{(\chi^2)}$ denotes the cumulative χ^2 distribution function with k degrees of freedom. The approximation holds for sufficiently large data. More specifically, it is considered safe to use if at least 80% of expected frequencies are greater or equal than 5 and none of them are 0. For small datasets not satisfying these criteria, discontinuity corrections may be applied or other tests used.

Example 2.5.3 The χ^2 test is implemented and demonstrated by the following R code, with the standard `chisq.test` function used for comparison.

```
bs.chisq.test <- function(v1, v2)
{
  o12 <- table(v1, v2)
  e12 <- table(v1)%*%t(table(v2))/sum(o12)
  chi2 <- sum((o12-e12)^2/e12)
  list(statistic=chi2, p.value=1-pchisq(chi2, (nrow(o12)-1)*(ncol(o12)-1)))
}

  # demonstration
bs.chisq.test(weather$outlook, weather$play)
chisq.test(weather$outlook, weather$play)
```

2.5.3.2   Loglikelihood ratio test

The loglikelihood ratio test, also known as the G-test, is an increasingly popular alternative to the χ^2 test. It is based on the following statistic:

$$G_S(a_1, a_2) = 2 \sum_{v_1 \in A_1} \sum_{v_2 \in A_2} |S_{a_1=v_1, a_2=v_2}| \ln \frac{|S_{a_1=v_1, a_2=v_2}|}{e_{a_1=v_1, a_2=v_2}} \qquad (2.73)$$

which takes a different approach to comparing the actual and expected counts of all attribute value pairs. As for the χ^2 statistic, the distribution of the G statistic under the null hypothesis of the two attributes being unrelated can be approximated by the χ^2 distribution and the approximation is actually better in this case (although it may still need discontinuity corrections for small datasets). This is why the loglikelihood ratio test may be usually preferred (unless performing the calculations "manually," for which the χ^2 test is more convenient). The p-value is the probability of a G value equal to or greater than that calculated on the dataset:

$$p_S^{(G)}(a_1, a_2) = 1 - \Phi_{(|A_1|-1)(|A_2|-1)}^{(\chi^2)} \left(G_S(a_1, a_2) \right) \qquad (2.74)$$

Example 2.5.4 The following R code defines the g.test function implementing the loglikelihood ratio test and demonstrates its application. There is no standard equivalent R function that could be used for comparison.

```
g.test <- function(v1, v2)
{
  o12 <- table(v1, v2)
  e12 <- table(v1)%*%t(table(v2))/sum(o12)
  g <- 2*sum(o12*log(o12/e12), na.rm=TRUE)
  list(statistic=g, p.value=1-pchisq(g, (nrow(o12)-1)*(ncol(o12)-1)))
}

  # demonstration
g.test(weather$outlook, weather$play)
```

2.5.3.3   Conditional entropy

The entropy, presented above as an impurity measure, can also serve as a relationship measure for discrete attributes. This is possible by calculating the impurities of one attribute in subsets to which the dataset is partitioned by the other attribute and averaging them, weighted by subset sizes. The resulting quantity is called the *conditional entropy* and defined as follows:

$$E_S(a_1|a_2) = \sum_{v \in A_2} P_S(a_2 = v)E_S(a_1) \tag{2.75}$$

The conditional entropy of attribute a_1 and given attribute a_2 is therefore an indicator of the level to which knowing the values of a_2 makes the values of a_1 predictable (if low impurity is interpreted as predictability).

Notice that, strictly speaking, a small conditional entropy value does not necessarily indicate a relationship between attributes if the (unconditional) entropy of a_1 is the same or similarly small. The conditional entropy is also asymmetric, as it measures the utility of a_2 for predicting a_1 only and not the other way round.

Example 2.5.5 The following R code defines and demonstrates the `entropy.cond` function which calculates the conditional entropy of the first argument given the second argument.

```
entropy.cond <- function(v1, v2)
{
  p12 <- pdisc(v1, v2)
  p2 <- colSums(p12)
  sum(p2*mapply(function(i, p2i) entropy.p(p12[,i]/p2i), 1:ncol(p12), p2))
}

# demonstration
entropy.cond(weather$play, weather$outlook)
entropy.cond(weather$play, weather$outlook=="rainy")
```

2.5.3.4 Mutual information

Another information-theoretic approach to measuring the strength of the attribute relationship is *mutual information*, calculated as follows:

$$I_S(a_1, a_2) = \sum_{v_1 \in A_1} \sum_{v_2 \in A_2} P(a_1 = v_1, a_2 = v_2) \log \frac{P(a_1 = v_1, a_2 = v_2)}{P(a_1 = v_1) \cdot P(a_2 = v_2)} \tag{2.76}$$

The statistic takes high values if the joint probability distribution of a_1 and a_2 differs considerably from their marginal distribution.

Unlike the conditional entropy, the mutual information is symmetric and measures the strength of the actual relationship between attributes. It can be easily verified that it is actually closely related to the conditional entropy in the following way:

$$I_S(a_1, a_2) = E_S(a_1) - E_S(a_1|a_2) = E_S(a_2) - E_S(a_2|a_1) \tag{2.77}$$

This makes it possible to view the mutual information as the reduction of impurity of one attribute due to the other attribute.

The mutual information is also directly related to the loglikelihood ratio statistic:

$$G_S(a_1, a_2) = 2 \ln b |S| I_S(a_1, a_2) \tag{2.78}$$

where b is the base of the logarithm used for calculating $I_S(a_1, a_2)$. Unlike the G statistic, the mutual information itself is not used for assessing the significance of the relationship between attributes, but just measuring its strength. In this application, it is much more convenient than the χ^2 or G statistic, since it does not depend on the data size.

Example 2.5.6 The following R code implements mutual information calculation, using the `pdisc` function from Example 2.4.22 for discrete probability distribution estimation and the `logp` function for calculating the base 2 logarithm of probabilities. Then its application is demonstrated.

`dmr.util`

```
mutinfo <- function(v1, v2)
{
  p12 <- pdisc(v1, v2)
  p1 <- rowSums(p12)
  p2 <- colSums(p12)
  sum(p12*logp(p12/(p1%o%p2)), na.rm=TRUE)
}

  # demonstration
mutinfo(weather$outlook, weather$play)
  # this should be the same
entropy(weather$play)-entropy.cond(weather$play, weather$outlook)
entropy(weather$outlook)-entropy.cond(weather$outlook, weather$play)
g.test(weather$outlook, weather$play)$statistic/(2*log(2)*nrow(weather))
```

2.5.3.5 Symmetric uncertainty

Another closely related convenient symmetric measure of discrete attribute relationship strength is the *symmetric uncertainty*, defined as follows:

$$U_S(a_1, a_2) = \frac{2I_S(a_1, a_2)}{E_S(a_1) + E_S(a_2)} \qquad (2.79)$$

This applies a kind of normalization to the mutual information. The resulting normalized quantity is usually a more reliable relationship strength measure in applications where relationships for multiple attribute pairs have to be evaluated and compared. In particular, it may give superior results when applied in attribute selection filters.

Example 2.5.7 The following R code defines the `symunc` function for symmetric uncertainty calculation, based on the `mutinfo` function from the previous example and the `entropy` function defined in Example 2.4.26.

```
## symmetric uncertainty for discrete vectors
symunc <- function(v1, v2)
{
  2*mutinfo(v1, v2)/(entropy(v1)+entropy(v2))
}

  # demonstration
symunc(weather$outlook, weather$temperature)
symunc(weather$outlook, weather$play)
```

2.5.4   Mixed attributes

The most straightforward approach to detecting relationships between two attributes, one of which is discrete and the other is continuous, is based on observing whether and how the distribution of the continuous attribute differs in data subsets determined by the values of the discrete attribute. This is relatively easy given the fact that the continuous attribute distribution may be sufficiently well characterized by a small number of descriptive statistics that are easy to calculate. In fact, commonly used mixed attribute relationship detection techniques focus on the location of the continuous attribute only, measuring the degree of location differences in subsets corresponding to the values of the discrete attributes.

2.5.4.1   *t*-Test

The simplest and most widely known technique applicable to detecting relationships between discrete and continuous attributes is the *t-test*. In its most popular form, it assumes that there are two data subsets S_0 and S_1 coming from two subdomains X_0 and X_1 of the same domain for which the same continuous attribute a is available. The test objective is to choose between the null hypothesis that the means of a in X_0 and X_1 are equal and the alternative hypothesis that they differ. It uses the following *t*-statistic:

$$t_{S_0,S_1}(a) = \frac{m_{S_0}(a) - m_{S_1}(a)}{s_{S_0,S_1}(a)\sqrt{\frac{1}{|S_0|} + \frac{1}{|S_1|}}} \tag{2.80}$$

The standard deviation estimator $s_{S_0,S_1}(a)$ used above is based on the so-called *pooled variance*, i.e., combining the variance estimators obtained for the two subsets:

$$s_{S_0,S_1}^2(a) = \frac{(|S_0| - 1)s_{S_0}^2(a) + (|S_1| - 1)s_{S_1}^2(a)}{|S_0| + |S_1| - 2} \tag{2.81}$$

$$s_{S_0,S_1}(a) = \sqrt{s_{S_0,S_1}^2(a)} \tag{2.82}$$

This only makes sense under the assumption that the variance of attribute a in the two subdomains is equal.

Assuming additionally that the attribute has normal distribution, the *t*-statistic has the standard Student's *t*-distribution with $|S_0| + |S_1| - 2$ degrees of freedom and the *p*-value can be determined as

$$p_{S_0,S_1}^{(t)}(a) = 2\left(1 - \Phi_{|S_0|+|S_1|-2}^{(t)}\left(|t_{S_0,S_1}(a)|\right)\right) \tag{2.83}$$

where $\Phi_k^{(t)}$ is the cumulative distribution function of the standard *t*-distribution with k degrees of freedom. The combination of the above statistic and inference procedure is known as Student's unpaired two-sample *t*-test. A modified version thereof that does not assume variance equality in subdomains is known as Welch's test.

The *t*-test is applicable to discrete and continuous attribute relationship detection only in the most basic case when the discrete attribute is binary. Its two values determine two subsets of the available dataset and, correspondingly, two subdomains, making it possible to use the

t-test. More specifically, we have

$$S_0 = S_{a_{01}=0} \tag{2.84}$$

$$S_1 = S_{a_{01}=1} \tag{2.85}$$

where a_{01} is the binary attribute and S is the dataset used for relationship detection.

Example 2.5.8 The following R code implements and demonstrates Student's t-test. The standard `t.test` function used for comparison is called with the `var.equal=TRUE` argument since it does not assume variance equality by default.

```
bs.t.test <- function(v, v01)
{
  m <- unname(tapply(v, v01, mean))
  s2 <- unname(tapply(v, v01, var))
  cn <- unname(tapply(v, v01, length))
  sp <- sqrt((s2[1]*(cn[1]-1)+s2[2]*(cn[2]-1))/(sum(cn)-2))

  ts <- (m[1]-m[2])/(sp*sqrt(sum(1/cn)))
  list(statistic=ts, p.value=2*(1-pt(abs(ts), sum(cn)-2)))
}


# demonstration
bs.t.test(weatherc$temperature, weatherc$play)
t.test(temperature~play, weatherc, var.equal=TRUE)
```

2.5.4.2 One-way ANOVA (F-test)

The counterpart of the t-test that can be used to compare the means of a continuous attribute in more than two subsets is the F-test. More precisely, there is a family of F-tests serving different purposes and the one referred to here – for the means of several subsets – is the best-known member of this family. This particular F-test is also referred to as one-way ANOVA (analysis of variance) with a completely randomized design. This test is applicable to detecting a relationship between a continuous attribute and an arbitrary discrete attribute.

The test uses the F-statistic defined as follows:

$$F_S(a_1, a_2) = \frac{\sum_{v_2 \in A_2} |S_{a_2=v_2}| (m_{S_{a_2=v_2}}(a_1) - m_S(a_1))^2 / (|A_2| - 1)}{\sum_{v_2 \in A_2} \sum_{x \in S_{a_2=v_2}} (a_1(x) - m_{S_{a_2=v_2}}(a_1))^2 / (|S| - |A_2|)} \tag{2.86}$$

This definition adopts the relationship detection perspective, with a continuous attribute a_1 and a discrete attribute a_2 observed on dataset S. The values of attribute a_2 are used to partition dataset S into subsets. The numerator contains the sum of the squared differences between the per-subset means and the overall mean of a_1. The denominator contains the sum of the squared differences between a_1 values and the corresponding subset mean. For the special case of a_2 being binary, for which the ordinary t-test is applicable, the F-statistic can be easily verified to be equal the square of the corresponding t-statistic.

The F-statistic has the F-distribution with $|A_2| - 1, |S| - |A_2|$ degrees of freedom under the null hypothesis of the means of a_1 in all the subdomains to which the domain is partitioned by a_2 being equal. This is based on the assumption that the continuous attribute has a normal distribution with equal variance in the subdomains. The p-value can then be obtained as follows:

$$p_S^{(F)}(a_1, a_2) = 1 - \Phi_{|A_2|-1,|S|-|A_2|}^{(F)} \left(F_S(a_1, a_2) \right) \tag{2.87}$$

where $\Phi_{k_1,k_2}^{(F)}$ is the cumulative distribution function of the F-distribution with k_1, k_2 degrees of freedom.

Example 2.5.9 The F-test for one-way ANOVA is implemented and demonstrated by the following R code. The standard `anova` function applied to a simple single-attribute linear model created with `lm` can be used to achieve the same effect. It is also demonstrated that in the binary case the F-statistic is equal to the square of the t-statistic.

```
f.test <- function(v1, v2)
{
  subsets <- split(v1, v2)
  m <- unname(sapply(subsets, mean))
  cn <- unname(sapply(subsets, length))
  m.a <- mean(v1)
  cn.a <- length(v1)

  f <- (sum(cn*(m-m.a)^2)/((k <- length(subsets))-1))/
        (sum(sapply(1:length(subsets),
                    function(i) sum((subsets[[i]]-m[i])^2)))/((cn.a-k)))
  list(statistic=f, p.value=1-pf(f, k-1, cn.a-k))
}

# demonstration
f.test(weatherc$temperature, weatherc$outlook)
f.test(weatherc$temperature, weatherc$play)
anova(lm(temperature~outlook, weatherc))
anova(lm(temperature~play, weatherc))
abs(sqrt(f.test(weatherc$temperature, weatherc$play)$statistic)-
  abs(t.test(temperature~play, weatherc, var.equal=TRUE)$statistic))
```

2.5.4.3 Mann–Whitney–Wilcoxon test

The t-test and its extensions are *parametric tests* that assume a known distribution of the continuous attribute of interest (normal in the case of the t-test) and estimate the parameters of this distribution (mean and variance in the case of the t-test). Whenever the distribution is unknown or known to be different than assumed by the test, the results may be unreliable. In such situations *nonparametric* tests are a noteworthy alternative, relying on no assumptions about the distribution.

The Mann–Whitney–Wilcoxon test (also known as the Mann–Whitney test or the Wilcoxon test) can be considered a nonparametric counterpart of the t-test. It is applicable to detecting the relationship between a discrete binary attribute and a continuous attribute, which can be seen as comparing the location of the latter in the two subsets determined by the former. Contrary to the t-test, however, the Mann–Whitney–Wilcoxon test does not compare the means, but rather counts the number of values in one subset that are below each value in the other subset. The test is based on the following statistic:

$$U_{S_0,S_1}(a) = \sum_{x_1 \in S_1} \left(\left| \{x_0 \in S_0 \mid a(x_0) < a(x_1)\} \right| \right.$$

$$\left. + \frac{1}{2} \left| \{x_0 \in S_0 \mid a(x_0) = a(x_1)\} \right| \right) \tag{2.88}$$

where a is the continuous attribute of interest, and S_0 and S_1 are two data subsets of the dataset S corresponding to the two binary attribute values:

$$S_0 = S_{a_{01}=0} \tag{2.89}$$

$$S_1 = S_{a_{01}=1} \tag{2.90}$$

The following complementary definition:

$$U'_{S_0,S_1}(a) = \sum_{x_0 \in S_0} \left(\left| \{x_1 \in S_1 \mid a(x_1) < a(x_0)\} \right| \right.$$

$$\left. + \frac{1}{2} \left| \{x_1 \in S_1 \mid a(x_1) = a(x_0)\} \right| \right) \tag{2.91}$$

in which the roles of S_0 and S_1 are swapped can also be used, clearly yielding

$$U'_{S_0,S_1}(a) = |S_0| \cdot |S_1| - U_{S_0,S_1}(a) \tag{2.92}$$

The above definitions make the statistic easy to understand, but in practical implementation the following equivalent definitions may be more convenient:

$$U_{S_0,S_1}(a) = \sum_{x \in S_0} r_{S_0 \cup S_1, a}(x) - \frac{|S_0|(|S_0| + 1)}{|S_0|} \tag{2.93}$$

$$U'_{S_0,S_1}(a) = \sum_{x \in S_1} r_{S_0 \cup S_1, a}(x) - \frac{|S_1|(|S_1| + 1)}{|S_1|} \tag{2.94}$$

where $r_{S_0 \cup S_1, a}(x)$ is the (fractional) rank of instance x with respect to attribute a on the combined dataset. The U statistic takes values between 0 and $|S_0| \cdot |S_1|$, and approaching either of these extremes indicates a strong relationship.

To perform statistical inference based on the U statistic, one may use a tabularized distribution for very small datasets and a normal approximation for larger datasets. The latter uses the following standardized version thereof:

$$z^{(U)}_{S_0,S_1}(a) = \frac{U_{S_0,S_1}(a) - m^{(U)}_{S_0,S_1}}{s^{(U)}_{S_0,S_1}} \tag{2.95}$$

where

$$m^{(U)}_{S_0,S_1} = \frac{|S_0| \cdot |S_1|}{2} \tag{2.96}$$

$$s^{(U)}_{S_0,S_1} = \sqrt{\frac{|S_0| \cdot |S_1|(|S_0| + |S_1| + 1)}{12}} \tag{2.97}$$

where the latter is valid under the assumption of no ties occurring in rank calculation (a more complex formula is necessary if it is not satisfied). The standardized statistic has approximately the standard normal distribution under the null hypothesis of no relationship and hence the p-value can be determined in the usual way:

$$p_{S_0,S_1}^{(U)}(a) = 2\left(1 - \Phi\left(|z_{S_0,S_1}^{(U)}(a)|\right)\right) \tag{2.98}$$

where Φ is the standard normal cumulative distribution function.

Example 2.5.10 The Mann–Whitney–Wilcoxon test is implemented and demonstrated by the following R code. The standard `wilcox.test` function is used for comparison, with the `correct=FALSE` and `exact=FALSE` arguments that disable the discontinuity correction used by default and exact distribution calculation for small data. This makes it produce the same results as the simple reimplementation `bs.wilcox.test`. The lines that are commented out provide an equivalent alternative implementation of the U statistic calculation.

```
bs.wilcox.test <- function(v, v01)
{
  subsets <- split(v, v01)
  ranks <- unname(split(rank(v), v01))
  cn <- unname(sapply(subsets, length))
  mu <- cn[1]*cn[2]/2
  su <- sqrt(cn[1]*cn[2]*(cn[1]+cn[2]+1)/12)

  u <- sum(ranks[[1]])-cn[1]*(cn[1]+1)/2
# u <- sum(sapply(subsets[[2]],
#                 function(v2) sum(v2<subsets[[1]])+sum(v2==subsets[[1]])/2))
  list(statistic=u, p.value=2*(1-pnorm(abs(u-mu)/su)))
}

  # demonstration
bs.wilcox.test(weatherc$temperature, weatherc$play)
wilcox.test(temperature~play, weatherc, exact=FALSE, correct=FALSE)
```

2.5.4.4 Kruskal–Wallis test

The Kruskal–Wallis test is the nonparametric counterpart of one-way ANOVA that generalizes the Mann–Whitney–Wilcoxon test to multivalued discrete attributes. The underlying statistic is defined for continuous attribute a_1, discrete attribute a_2, and dataset S as follows:

$$K_S(a_1, a_2) = (|S| - 1)\frac{\sum_{v_2 \in A_2} |S_{a_2=v_2}|(\bar{r}_{S_{a_2=v_2}}(a_1) - \bar{r}_S(a_1))^2}{\sum_{v_2 \in A_2} \sum_{x \in S_{a_2=v_2}} (r_{S_{a_2=v_2},a_1}(x) - \bar{r}_S(a_1))^2} \tag{2.99}$$

where

$$\bar{r}_{S_{a_2=v_2}}(a_1) = \frac{1}{|S_{a_2=v_2}|} \sum_{x \in S_{a_2=v_2}} r_{S,a_1}(x) \tag{2.100}$$

$$\bar{r}_S(a_1) = \frac{1}{|S|} \sum_{x \in S} r_{S,a_1}(x) = \frac{|S| + 1}{2} \tag{2.101}$$

are the average ranks with respect to a_1 on subset $S_{a_2=v_2}$ and on the complete dataset S, respectively. For sufficiently large datasets, the distribution of the K statistic is approximated by the χ^2 distribution with $|A_2| - 1$ degrees of freedom. This makes it possible to calculate the p-value:

$$p_S^{(K)}(a_1, a_2) = 1 - \Phi_{|A_2|-1}^{(\chi^2)} \left(K_S(a_1, a_2) \right) \tag{2.102}$$

where $\Phi_k^{(\chi^2)}$ denotes the cumulative χ^2 distribution function with k degrees of freedom.

Example 2.5.11 The following R code defines the bs.kruskal.test which is a simple reimplementation of the standard kruskal.test function. The two implementations of the Kruskal test are then applied to two attribute pairs of the *weatherc* data, one with the binary discrete attribute play and the other with the three-valued discrete attribute outlook. Notice that for the former the resulting p-value agrees with that obtained using the Mann–Whitney–Wilcoxon test in the previous example.

```
bs.kruskal.test <- function(v1, v2)
{
  subsets <- split(v1, v2)
  ranks <- split((rank.all <- rank(v1)), v2)
  cn <- unname(sapply(subsets, length))
  mr <- sapply(ranks, mean)
  mr.a <- mean(rank.all)

  k <- (length(v1)-1)*sum(cn*(mr-mr.a)^2)/
          sum(sapply(ranks, function(r) sum((r-mr.a)^2)))
  list(statistic=k, p.value=1-pchisq(k, length(subsets)-1))
}

# demonstration
bs.kruskal.test(weatherc$temperature, weatherc$play)
kruskal.test(temperature~play, weatherc)
bs.kruskal.test(weatherc$temperature, weatherc$outlook)
kruskal.test(temperature~outlook, weatherc)
```

2.5.5 Relationship detection caveats

Relationship detection techniques should be used with care to avoid drawing unjustified conclusions. Two particularly common and severe caveats are related to unsatisfied assumptions and multiple tests.

2.5.5.1 Unsatisfied assumptions

Most significance tests used for relationship detection rely on some assumptions. Parametric tests may assume the attributes of interest being distributed normally, as in the case of the t-test. Such assumptions may often be unsatisfied. Both parametric and nonparametric tests may use some approximations valid for sufficiently large datasets only. Even though the sufficient data size is typically as little as a few dozen, sometimes the tests may be applied to smaller datasets, requiring exact calculations.

Whenever test assumptions are unsatisfied, the distribution of the test statistic (under the null hypothesis) may differ from that assumed by the test's inference procedure and used to determine the p-value. This may lead to an incorrect p-value and, possibly, making a wrong decision.

This is not to say that statistical tests should never be used if their assumptions are unknown to be satisfied or known to be unsatisfied. What is necessary, though, is being aware of the possible consequences. Unsatisfied assumptions do not necessarily make statistical tests useless, but they make them less reliable. One may use them to get insights about the data (and the domain) rather that make definite statements. The underlying test statistics remain useful measures of relationship strength.

2.5.5.2 Multiple tests

Even if all necessary assumptions are satisfied and the calculated p-value is correct, the decision to accept or reject the null hypothesis may be wrong. This is because the p-value is the probability of the observed statistic value under the null hypothesis and, even if it is small, the null hypothesis may still be true. The rejection of the null hypothesis is justified, though, because the observed result would be very unlikely if it were true.

The above justification to reject the null hypothesis is no longer valid if we consider a series of multiple significance tests performed on the same dataset (e.g., when considering relationships for several attribute pairs). What is unlikely in a single experiment becomes more likely in a series of experiments. This means, roughly speaking, a statistic value suggesting a statistically significant relationship may occur by chance if we try several times. To avoid an increased false positive rate one should therefore adjust the significance level at which null hypotheses are rejected accordingly. One very simple and commonly used approach, known as the Bonferroni correction, is to divide the significance level used by the number of hypotheses tested. With each individual null hypothesis out of a family of m hypotheses being tested rejected if the corresponding p-value is below $\frac{\delta}{m}$, then the "effective significance level" for the whole family does not exceed δ. This is actually an overly conservative approach that keeps the false positive rate low at the cost of an increased false negative rate and more refined correction techniques exist.

2.6   Visualization

This book, in general, and this chapter in particular, definitely do not give justice to data visualization techniques. The role of these extremely powerful tools in data exploration can be hardly overestimated, but their use in the book is only marginal, limited to the presentation of model predictive performance measures. This a direct consequence of its major focus being not on understanding data but on understanding modeling algorithms. This is by no means to say the latter is more important than the former (actually, the opposite is more likely to be true), but just to acknowledge the consciously limited scope of data mining techniques that are covered in the book. It affects this chapter particularly strongly. However, this section will at least briefly introduce the absolutely simplest visualization techniques that are actually used in some of the subsequent chapters.

2.6.1   Boxplot

One of the simplest, yet extremely popular and useful visualization technique, is the *box-plot*. It is basically a graphical summary of a continuous attribute's distribution, including the first and third quartiles – represented by the top and bottom edges of a rectangular box, the median – represented by a horizontal line (or, sometimes, a dot) inside the box, and the minimum and maximum values – represented by the so-called whiskers extending above and below the box. The latter are usually determined with outlying values filtered out using the quartile-based rules given by Equations 2.37 and 2.38 . Some or all of those may be included in the plot as additional dots above or below the whiskers.

A boxplot makes it possible to notice some major properties of a continuous attribute at a glance: where it is located, how dispersed it is, whether or not it is distributed symmetrically, and possibly how many outlying values it has. While mostly the same information can be seen from quartile values directly, the graphical form is much easier to interpret quickly. This is particularly important when comparing the distribution of the same continuous attribute in different data subsets, as all major differences can be immediately spotted out if boxplots for all these subsets are presented side by side.

Example 2.6.1 The examples included in this section, unlike those presented above, will not implement the visualization techniques being discussed, but just use existing R functions to produce actual graphical illustrations. The following R code produces boxplots for the `playability` attribute in the *weatherr* data: one with outlying values identified (using an untypically small interquartile range multiplier for the sake of illustration) and the other in subsets determined by the values of the `outlook` attribute. The plots are presented in Figure 2.2.

```
par(mfrow=c(1, 2))
boxplot(weatherr$playability, range=0.5, col="grey", main="playability")
boxplot(playability~outlook, weatherr, col="grey", main="playability")
```

2.6.2   Histogram

Not so immediately readable as the boxplot, but capable of providing a more detailed picture of a continuous attribute's distribution, is the *histogram*. It represents occurrence counts or

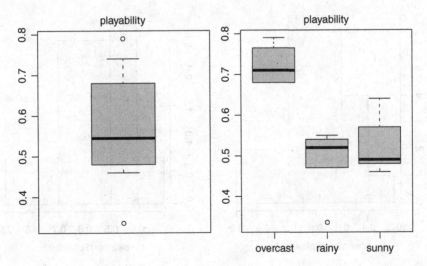

Figure 2.2   The boxplots for the `temperature` attribute in the *weathercl* data.

relative frequencies for a set of intervals by the heights or areas of vertical bars "standing" on these intervals. More precisely, it is usually the area that is used to represent the relative frequency (i.e., estimated probability) and the height that is used to represent the occurrence count, but conventions vary. For equal-width intervals the difference obviously disappears. In any case, a histogram is basically a graphical – and therefore more friendly to the human eye – presentation of an interval frequency table. It makes it easy to identify the overall shape of a distribution, assess its asymmetry, peakedness, and modality (the number of peaks). Of those, only asymmetry can be revealed by a boxplot. On the other hand, histograms are not so easy to quickly compare as boxplots and therefore less useful in detecting attribute distribution differences across multiple data subsets.

Example 2.6.2 Two histograms for the `playability` attribute in the *weatherr* data are produced by the following R code: one displaying occurrence counts and the other displaying relative frequencies for the same set of nonequal-weight intervals. The plots are presented in Figure 2.3. Notice the difference in bar heights.

```
par(mfrow=c(1, 2))
hist(weatherr$playability, breaks=c(0.3, 0.4, 0.5, 0.7, 0.9), probability=FALSE,
    col="grey", main="")
hist(weatherr$playability, breaks=c(0.3, 0.4, 0.5, 0.7, 0.9), probability=TRUE,
    col="grey", main="")
```

2.6.3 Barplot

The *barplot* is the "least statistical" of the simple visualization techniques presented in this section. It is just a graphical representation of a set of numbers by the lengths of the set corresponding horizontal or vertical bars. This is not particularly useful to examine the distributions of continuous attributes, but makes it easy to quickly compare a small set of values, e.g., the predictive performance of several algorithms, the predictive utility of several attributes, etc.

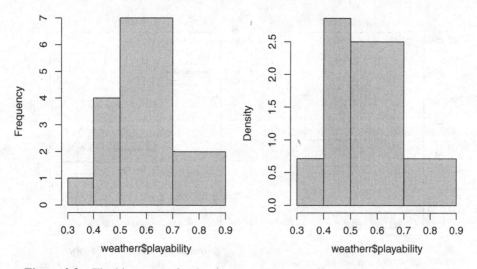

Figure 2.3 The histograms for the `playability` attribute in the *weatherr* data.

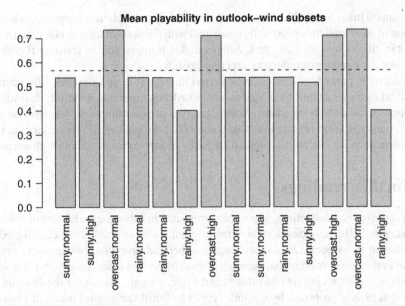

Figure 2.4 The barplot of the mean `playability` values in subsets determined by the `outlook` and `wind` attributes in the *weatherr* data.

Example 2.6.3 The R code shown below produces a somewhat artificial barplot illustration using the *weatherr* data. The resulting barplot is displayed in Figure 2.4. It calculates the mean values of the `playability` attributes within subsets corresponding to all the combinations of the values of the `outlook` and `wind` attributes using the `ave` function and plots them using the `barplot` function. A horizontal line representing the whole-dataset mean is added.

```
par(mar=c(7, 4, 4, 2))
barplot(`names<-`(ave(weatherr$playability, weatherr$outlook, weatherr$wind),
          interaction(weatherr$outlook, weatherr$wind)),
     las=2, main="Mean playability in outlook-wind subsets")
lines(c(0, 17), rep(mean(weatherr$playability), 2), lty=2)
```

2.7 Conclusion

The primary purpose of this chapter is to provide a quick source of information on basic statistical techniques that can be referred to by other chapters in this book. Many of them are mentioned in the description of data mining algorithms and used in their illustrative implementations. This chapter makes it possible to get some brief summary of their purpose and operation principles without having to resort to external sources. The scope of presented techniques and the presentation depth are subordinated to this goal. Only techniques that are actually referred to and some directly related techniques serving the same purpose are included. No extensive statistical background is provided and simplifications are adopted to make the description both concise and readable. It is noteworthy that the notation (and, to some extent, terminology) used in this chapter considerably departs from that typically used by statisticians. This is a possibly controversial, but a deliberate choice, supposed to keep the presentation of basic statistics maximally consistent with the presentation of other data mining techniques in this book's other chapters. All this makes it impossible to consider this chapter a replacement for a proper introductory text on statistics.

A considerable part of the content is delivered through R code examples, often containing simplified reimplementations of existing standard functions that hopefully facilitate the interpretation of the underlying maths. Some examples provide functions that have no direct existing R counterparts (at least in the most commonly used packages). Those serve not only the illustration purpose, but are also actually used by example code from other chapters.

2.8   Further readings

Unlike for more refined data mining algorithms presented in subsequent chapters of this book, it is not necessarily the best idea to refer to the original books or articles that contributed particular statistical techniques. Due to their large number and diversity it makes more sense to restrict one's attention to sources that cover all or most of them in a consistent form. It is worthwhile to allow some exceptions from this general rule, though. One is for the *m*-estimation technique specifically proposed by Cestnik (1990) to fulfill the special needs of classification algorithms that often have to estimate class or attribute value probabilities on small data subsets. Karalič and Cestnik (1991) applied the same idea of combining empirical and prior estimates to the *m*-estimation of the mean and the variance. The second is for bootstrap estimation methods introduced by Efron (1979) and more extensively described by Efron and Tibshirani (1994). Finally, the third of those specific references is the discussion of statistical tests that can be used to compare the performance of classification algorithms by Dietterich (1998), including also a high-level summary of different types of questions arising in inductive learning that can be answered using statistics.

Basic statistics remain a standard part of most data mining courses and are often – although not always – at least partially and superficially covered by data mining books (e.g., Han *et al.*, 2011, Tan *et al.*, 2013). Some of them go actually much deeper into statistics (e.g., Hand *et al.*, 2001) than "proper" introductory statistics sources. The latter may be still worthwhile to consult for a better feel of traditional statistical thinking and methodology. There is a lot to choose from, since techniques presented in this chapter are sufficiently elementary on one hand and sufficiently popular on the other hand to be widely covered by nearly all introductory statistics textbooks. Many of them that not only present a greater variety of related techniques, but also – and much more importantly – provide much more in-depth discussions of their underlying assumptions, strengths, and limitations. Books that adopt an informal and intuitive rather that math-loaded and rigorous presentation style may be more preferred for gentle introduction to statistics. Freedman *et al.* (2007), Urdan (2010), and Witte and Witte (2009) all excel in presenting not always straightforward and intuitive statistical techniques in an astonishingly straightforward and intuitive way so that they read nearly like novels. For those that feel more comfortable with going into mathematical symbols and equations, not necessarily preceded by several paragraphs of plain language explanations, Wilcox (2009) or Kiemele *et al.* (1997) may appear more useful. The latter is more oriented toward industrial applications and may appeal to practitioners, whereas the former enriches the presentation of classical techniques by a fresh perspective inspired by contemporary statistical research.

References

Cestnik B 1990 Estimating probabilities: A crucial task in machine learning *Proceedings of the Ninth European Conference on Artificial Intelligence (ECAI-90)*. Pitman.

Dietterich TG 1998 Approximate statistical tests for comparing supervised classification learning algorithms. *Neural Computation* **10**, 1895–1924.

Efron B 1979 Bootstrap methods: Another look at the jackknife. *The Annals of Statistics* **7**, 1–26.

Efron B and Tibshirani R 1994 *An Introduction to the Bootstrap*. Chapman and Hall.

Freedman DA, Pisani R and Purves R 2007 *Statistics* 4th edn. Norton.

Han J, Kamber M and Pei J 2011 *Data Mining: Concepts and Techniques* 3rd edn. Morgan Kaufmann.

Hand DJ, Mannila H and Smyth P 2001 *Principles of Data Mining*. MIT Press.

Karalič A and Cestnik B 1991 The Bayesian approach to tree-structured regression *Proceedings of the Thirteenth International Conference on Information Technology Interfaces (ITI-91)*. University Computing Centre, University of Zagreb, Croatia.

Kiemele MJ, Schmidt SR and Berdine RJ 1997 *Basic Statistics: Tools for Continuous Improvement* 4th edn. Air Academy Press.

Tan PN, Steinbach M and Kumar V 2013 *Introduction to Data Mining* 2nd edn. Addison-Wesley.

Urdan TC 2010 *Statistics in Plain English* 3rd edn. Routlege.

Wilcox RR 2009 *Basic Statistics: Understanding Conventional Methods and Modern Insights*. Oxford University Press.

Witte RS and Witte JS 2009 *Statistics* 10th edn. Wiley.

Part II
CLASSIFICATION

3

Decision trees

3.1 Introduction

In many applications, we not only want to just *use* the created classification model to accurately classify instances, but we may also want to *inspect* the model. This makes it possible to explain its predictions, modify it, or combine with some existing background knowledge. In such applications, where both high classification accuracy and human readability of the model are required, the obvious method of choice for most data miners will be *decision trees*.

Decision tree algorithms have been studied for many years and belong to those data mining algorithms for which particularly numerous refinements and variations have been proposed. One can therefore speak about a family of algorithms that share the same model representation and algorithm operation schemes, but may differ in several details. The space for this diversity is increased by the two-phase process usually performed to create decision tree models, consisting of decision tree growing and pruning. It is hardly possible to describe all these algorithm variations with the level of detail adopted by this book without some substantial omissions and compromises. Only the most common ones will be discussed and not all of them will be illustrated with R examples.

Example 3.1.1 This chapter contains examples that illustrate the major algorithmic operations related to decision trees using the *weather* data, the small size of which makes it easy to manually verify the results. To illustrate continuous attribute handling, the *weatherc* data will be used. The following code prepares the environment for subsequent examples by loading the datasets as well as DMR and CRAN packages that will be needed.

> Ex. 1.3.1
> dmr.data

> Ex. 1.3.2
> dmr.data

```
library(dmr.claseval)
library(dmr.stats)
library(dmr.util)
```

Data Mining Algorithms: Explained Using R, First Edition. Paweł Cichosz.
© 2015 John Wiley & Sons, Ltd. Published 2015 by John Wiley & Sons, Ltd.

```
library(rpart)
library(rpart.plot)
library(lattice)

data(weather, package="dmr.data")
data(weatherc, package="dmr.data")
```

3.2   Decision tree model

A decision tree is a hierarchical structure that represents a classification model. Internal tree nodes correspond to *splits* applied to decompose the domain into regions, and terminal nodes assign class labels to regions believed to be sufficiently small or sufficiently uniform. For convenience, we will reserve the term *node* to internal nodes only and refer to terminal nodes as *leaves*.

Example 3.2.1 A graphical illustration of domain decomposition represented by a decision tree is created by the R code presented below. It generates a simple artificial training set for decision tree growing, which is performed using the rpart function – the R implementation of decision trees used in this book. Then the structure of the decision tree is plotted using the prp function provided by the rpart.plot package. The corresponding domain decomposition is visualized by a level plot produced by the levelplot function from the lattice package, with different shades of gray corresponding to class labels, which can be directly matched to the corresponding tree leaves. The produced illustration is presented in Figure 3.1.

```
dtdat <- expand.grid(a1=seq(1, 10, 3), a2=seq(1, 10, 3))
dtdat$c <- as.factor(ifelse(dtdat$a1<=7 & dtdat$a2<=1, 1,
                     ifelse(dtdat$a1<=7 & dtdat$a2<=7, 2,
                     ifelse(dtdat$a1<=7, 3,
                     ifelse(dtdat$a2<=4, 4, 5)))))
  # decision tree structure
prp(rpart(c~., dtdat, minsplit=2, cp=0))
  # the corresponding domain decomposition
levelplot(c~a1*a2, dtdat, at=0.5+0:5, col.regions=gray(seq(0.1, 0.9, 0.1)),
          colorkey=list(at=0.5+0:5))
```

3.2.1   Nodes and branches

Splits are specified by some relational conditions based on selected attributes that may have two or more outcomes. Formally, a split can be represented by a test function $t : X \to R_t$ that maps instances into split outcomes. A separate outgoing branch is associated with each possible outcome of a node's split. The relationship between the parent node and its descendant nodes, conceptually represented by the branches linking the former to the latter, does not always have to be explicitly represented in the decision tree data structure. In particular, when binary splits are used, the relationship can be implicitly represented by an appropriate node numbering scheme, e.g., the descendants of node numbered k can be numbered $2k$ and $2k + 1$.

If a split's outcome can be unambiguously determined for any possible instance, then it does partition the domain into disjoint subsets, corresponding to the outgoing branches. It is

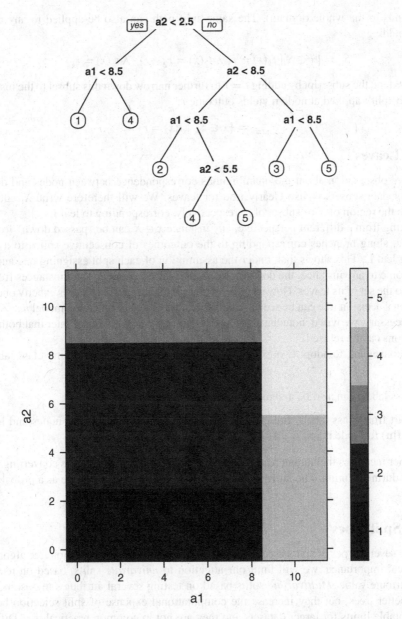

Figure 3.1 Domain decomposition by a decision tree.

therefore easy to see that each node **n** of a decision tree corresponds to a region (subset) of the domain

$$X_{\mathbf{n}} = \{x \in X \mid t_1(x) = r_1 \wedge t_2(x) = r_2 \wedge \cdots \wedge t_k(x) = r_k\} \tag{3.1}$$

determined by the sequence of splits $t_1, t_2, \ldots, t_k$ and their outcomes $r_1, r_2, \ldots, r_k$ occurring on the path from the root to the node (and the root, which has the empty path of splits, corresponds to the whole domain). The same split path can also be applied to any dataset $S \subset X$, yielding

$$S_n = \{x \in S \mid t_1(x) = r_1 \wedge t_2(x) = r_2 \wedge \cdots \wedge t_k(x) = r_k\} \tag{3.2}$$

We will extend the subscript by adding $t = r$ to further narrow down this subset to the instances for which split t applied at node $\mathbf{n}$ yields outcome r:

$$S_{\mathbf{n},t=r} = \{x \in S_n \mid t(x) = r\} \tag{3.3}$$

3.2.2   Leaves

The above observation about the unambiguous correspondence between nodes and domain regions or data subsets is also clearly true for leaves. We will therefore write X_l and S_l to designate the region of X or subset of S, respectively, corresponding to leaf l.

Looking from a different perspective, any instance $x \in X$ can be "passed down" from the root node, along branches corresponding to the outcomes of consecutive splits, to a corresponding leaf l_x. This shows that, under the assumption of each split assigning one and only one outcome to any instance, the decision tree represents a mapping of all instances from the domain to the set of its leaves. Now if we further assume that each leaf stores exactly one class label, then a decision tree can be seen as a representation of a classification model $h : X \rightarrow C$. When necessary, we will denote the class label of leaf l by d_l. We will see later that both these assumptions can be relaxed.

It is convenient to adopt two extensions to the assumption about storing class labels in leaves:

1. class labels can also be assigned to nodes: $d_{\mathbf{n}}$ for node $\mathbf{n}$,

2. apart from class labels, full class distribution can be stored in both nodes and leaves: $P(d|\mathbf{n})$ for node $\mathbf{n}$ and $P(d|l)$ for leaf l, in both cases for all $d \in C$.

The former increases the human readability of decision trees and facilitates converting nodes to leaves during pruning. The latter makes it possible to use a decision tree as a probabilistic classifier.

3.2.3   Split types

There are several types of splits used for decision trees. Following all decision tree algorithms of practical importance, we will limit our attention to *univariate* splits, based on testing a single attribute value. *Multivariate* splits based on testing several attributes may sometimes lead to better trees, but they increase the computational expense of split selection beyond the acceptable limits for larger datasets and they are not in common practical use. Different types of splits are characterized by the form of the test function used. Defining a univariate split on attribute a with the test function t requires specifying how $t(x)$ is determined based on $a(x)$ for all $x \in X$.

3.2.3.1   Nominal attributes

For nominal attributes, it is common to use one of the following two split types:

Value-based. With a test function defined as $t(x) = a(x)$.

Equality-based. With a test function defined as

$$t(x) = \begin{cases} 1 & \text{if } a(x) = v \\ 0 & \text{otherwise} \end{cases} \tag{3.4}$$

where $v \in A$.

A value-based split is just an attribute, and each outcome corresponds to one attribute value. An equality-based split is binary and assigns one outcome to instances with a specified attribute value and the other to the remaining instances. There is little sense to use both value-based or equality-based splits in the same tree, so it is a standard practice for decision tree algorithm implementations to consider either one or the other.

Both split types can be made more flexible (but also considerably more costly to select) by replacing single values by subsets of an attributes's codomain. This yields the following split types:

Subset-based. With a test function defined as

$$t(x) = \begin{cases} 1 & \text{if } a(x) \in V_1 \\ 2 & \text{if } a(x) \in V_2 \\ \ldots \\ k & \text{if } a(x) \in V_k \end{cases} \tag{3.5}$$

where $V_1, V_2, \ldots, V_k \subset A$ constitute a disjoint partition of the codomain of A.

Membership-based. With a test function defined as

$$t(x) = \begin{cases} 1 & \text{if } a(x) \in V \\ 0 & \text{otherwise} \end{cases} \tag{3.6}$$

where $V \subset A$. Again, only one of these two split types is considered by particular implementations of decision tree algorithms.

3.2.3.2 Continuous attributes

The single most common split type for continuous attributes uses the inequality relation:

Inequality-based. With a test function defined as

$$t(x) = \begin{cases} 1 & \text{if } a(x) \leq v \\ 0 & \text{otherwise} \end{cases} \tag{3.7}$$

where $v \in A$.

A more flexible, but more costly to select split type assigns different outcomes to several intervals of an attribute's co-domain.

Interval-based. With a test function defined as

$$t(x) = \begin{cases} 1 & \text{if } a(x) \in I_1 \\ 2 & \text{if } a(x) \in I_2 \\ \ldots \\ k & \text{if } a(x) \in I_k \end{cases} \tag{3.8}$$

where $I_1, I_2, \ldots, I_k \subset A$ are intervals constituting a disjoint partition of the codomain of a. This is clearly the same as a subset-based split applied to a continuous attribute.

3.2.3.3 Ordinal attributes

Ordinal attributes, which share some properties with both nominal and continuous ones, can be treated as either of these types, depending on whether one wants to exploit the available order relation or not.

3.2.3.4 Binary splits

Binary splits – equality- or membership-based for nominal attributes and inequality-based for continuous attributes – are particularly popular due to their implementational convenience. As already noted above, with each node having the same number of descendants a simple node numbering scheme is sufficient to represent the tree structure. It is a fairly common, but not universally adopted convention that for binary splits the left branch corresponds to the true outcome and the right branch to the false outcome of the underlying condition. It only matters for decision tree printing or drawing.

3.3 Growing

The most important process needed to create a decision tree model from a given training set is called *growing*. As this term borrowed from "real" (biological) trees suggests, it is usually a sequential process during which new nodes or leaves are added step by step. Adding new nodes or leaves is performed in a top-down fashion, starting from a single root node. This paradigm, followed by all decision tree algorithms in wide practical use, is called *top-down decision tree induction* (TDIDT).

3.3.1 Algorithm outline

Since a decision tree is a recursive structure, it is no surprise that top-down growing algorithms are usually formulated as recursive procedures which create the first root node based on the whole training set and then call themselves for the subsets obtained by applying the split selected for that node. While there are no major problems with this "obvious" recursive algorithm formulation, it may be more instructive and practically useful to consider a not-so-common alternative iterative formulation presented below.

```
 1: create the root node and mark it as open;
 2: assign all training instances from T to the root node;
 3: while there are open nodes do
 4:     select an open node n;
 5:     calculate class distribution P(d|n) for d ∈ C based on Tₙ;
 6:     assign class label dₙ;
 7:     if stop criteria are satisfied for n then
 8:         mark n as a closed leaf;
 9:     else
10:         select a split t : X → Rₜ for n;
11:         for all split outcomes r ∈ Rₜ do
12:             create a descendant node nᵣ corresponding to r and mark it as open;
```

```
13:        assign all instances from T_{n,t=r} to n_r;
14:    end for
15:    mark n as a closed node;
16:  end if
17: end while
```

This algorithm operates by maintaining a set of *open* nodes and processing them until they become *closed* nodes or leafs. The latter are fixed elements of the decision tree being grown and require no further processing. The former can be considered node stubs. They only have the corresponding subset of instances assigned, but neither class distribution estimated nor class label or split selected. At the beginning the first and only open node is the root node of the tree. Then whenever an open node is to be converted to a closed node, its descendants (corresponding to the outcomes of the selected split) are created as new open nodes. Since no new nodes are created when an open node is converted to a closed leaf, the algorithm is guaranteed to terminate provided that a stop criterion will be eventually satisfied on all paths in the tree.

The processing required to turn an open node into a closed node or leaf always includes:

- *class distribution calculation* based on the corresponding subset of training instances,

- *class label assignment*, strictly necessary only for leaves, but useful for nodes as well,

- checking the *stop criteria*, which determine whether the open node will become a closed node or a closed leaf.

Whenever none of the stop criteria is satisfied, the following additional operations are performed:

- *split selection* based on the corresponding subset of training instances,

- *split application*, i.e., creating descendant (open) nodes, corresponding to every possible split outcome r of the selected split t and partitioning the current subset of training instances T_n into subsets $T_{n,t=r}$ assigned to the newly created descendant nodes.

These operations, required to fully specify a decision tree growing algorithm, will be reviewed below.

Example 3.3.1 The *weather* data will be used in a series of R code examples illustrating the basic operations performed during tree growing. To facilitate the adaptation of these examples to other datasets, the following R code sets variables used to refer to the dataset, attribute names, and the class attribute name.

> Ex. 1.3.1
> dmr.data

```
data <- weather
attributes <- names(weather)[1:4]
class <- names(weather)[5]
```

The following R code initializes the tree by creating the root node and assigning to it all training instances. These operations, for greater reusage convenience, are grouped into a function which, when called with the data variable set as shown above, initializes tree growing for the *weather* data.

```
init <- function()
{
  clabs <<- factor(levels(data[[class]]),
                   levels=levels(data[[class]]))      # class labels
  tree <<- data.frame(node=1, attribute=NA, value=NA, class=NA, count=NA,
                      'names<-'(rep(list(NA), length(clabs)),
                               paste("p", clabs, sep=".")))
  cprobs <<- (ncol(tree)-length(clabs)+1):ncol(tree)  # class probability columns
  nodemap <<- rep(1, nrow(data))
  n <<- 1
}

init()
```

In this and subsequent examples, a decision tree is represented by a data frame with rows corresponding to nodes and the following columns:

node: node number (starting from 1 for the root node),
attribute: the attribute used for the split in the node,
value: the value used for the split in the node,
class: the class label assigned to the node,
count: the number of instances in the node,
p.*: the estimated class probabilities for each class in the node.

This is sufficient to represent a binary decision tree with equality-based or inequality-based splits, to which the examples are limited. The assignment of instances to decision tree nodes is represented by a vector containing the numbers of nodes to which the corresponding instances are assigned. Notice that the init function also sets two auxiliary variables: clabs storing class labels and cprobs storing the indices of class probability columns in the tree data frame. It also initializes the n variable, indicating the currently processed node.

Subsequent examples will demonstrate the remaining steps of the decision tree growing process using the tree initialized in this example. They will assume the data, attributes, and class variable assignments presented above and follow the same convention of grouping R expressions performing modifications of the tree structure into functions referring to these variables for easier reusage. It is worthwhile to inspect the tree data frame and the nodemap vector whenever they are modified.

3.3.2   Class distribution calculation

This operation determines, for an open node $\mathbf{n}$, the corresponding class probabilities $P(d|\mathbf{n})$ for all classes $d \in C$ based on the set of training instances $T_{\mathbf{n}}$ assigned to the node:

$$P(d|\mathbf{n}) = P_{T_{\mathbf{n}}}(d) = \frac{|T_{\mathbf{n}}^d|}{|T_{\mathbf{n}}|} \tag{3.9}$$

where d in the superscript is used to designate the selection of instances of class d. Class probabilities can be stored in the node regardless of whether it becomes a closed node or a closed leaf later. They will be used for class label assignment and split selection.

Example 3.3.2 The following R code uses the `pdisc` function for estimating discrete probability distributions to implement class probability calculation for a decision tree node. The resulting `class.distribution` function, which also takes care of setting the instance count, is then applied to the root node of the previously initialized decision tree for the *weather* data.

Ex. 2.4.22
dmr.stats

```
class.distribution <- function(n)
{
   tree$count[tree$node==n] <<- sum(nodemap==n)
   tree[tree$node==n,cprobs] <<- pdisc(data[nodemap==n,class])
}

class.distribution(n)
```

3.3.3   Class label assignment

As mentioned above, the first direct application of class probabilities calculated for an open node is to select a class label for it. Except for a special situation of nonuniform misclassification costs, discussed extensively in Chapter 6, assigning the most probable class label is always the most reasonable choice:

$$d_{\mathbf{n}} = \arg\max_{d\in C} P(d|\mathbf{n}) \qquad (3.10)$$

It minimizes the probability of misclassification of any instance if the node became a closed leaf.

In the case of more than one class having exactly the same maximum probability, it makes most sense to break ties based on which of them is more probable in the parent node, unless **n** is the root node, for which ties can be broken arbitrarily or based on some background knowledge. One particularly important situation when this parent-based tie breaking policy will have to be applied is when the set of training instances assigned to the node is empty, resulting in $P(d|\mathbf{n}) = 0$ for all $d \in C$. It can happen if the corresponding outcome of the split used in the parent node is not obtained for any of its training instances. This does not have to be taken into account only if all available splits are binary, since a binary split yielding the same outcome for all training instances to which it is applied would be useless and should not be selected in the parent node anyway.

Example 3.3.3 The following R code continues the decision tree growing demonstration for the *weather* data by performing class label assignment, based on the class probability estimates calculated in the previous example. This actually assigns the numerical representation of the maximum-probability class, but it may be easily converted to the original representation using class labels stored in the `clabs` variable upon the completion of decision tree growing.

```
class.label <- function(n)
{
   tree$class[tree$node==n] <<- which.max(tree[tree$node==n,cprobs])
}

class.label(n)
```

3.3.4   Stop criteria

The stop criteria are used to decide whether a given open node requires no further split and should become a closed leaf. It is most instructive to consider first the *most definite* stop criteria which, when satisfied, would make applying any further splits impossible or undoubtedly pointless. They will be referred to as *strict* stop criteria. Their *relaxed* forms may be used to prevent growing overly large trees.

3.3.4.1   Strict criteria

The following three strict stop criteria for decision tree growing can be applied to each node **n**:

Uniform class. All training instances in the node are of the same class $d_\mathbf{n}$, i.e., $P(d|\mathbf{n}) = 0$ for $d \neq d_\mathbf{n}$; this makes any further splits totally useless, since any possible descendants of **n** would receive the very same class label $d_\mathbf{n}$ and any possible subtree starting in **n** would classify any instances in exactly the same way as **n** alone.

No instances left. The set of training instances assigned to the node is empty, i.e., $T_\mathbf{n} = \emptyset$, and splitting the empty set clearly cannot lead to any further improvement; similarly as above, any possible descendants of **n** would receive the very same class label $d_\mathbf{n}$ and any possible subtree starting in **n** would classify any instances in exactly the same way as **n** alone.

No splits left. There is no split that can be applied to further partition the current subset of training instances $T_\mathbf{n}$, either because all available splits have been already used up on the path from the root to **n**, or every split not yet used gives the same outcome for all instances from $T_\mathbf{n}$, which would put them to the same branch; this means, again, that any possible descendants of **n** would receive the very same class label $d_\mathbf{n}$ and any possible subtree starting in **n** would classify all instances in exactly the same way as **n** alone.

This first of these stop criteria is the "regular" and most desired one. It corresponds to the situation when the sequence of splits applied on the path from the root to the current node has successfully identified a subset of training instances of the same class. When this node becomes a leaf, it will therefore accurately classifies all of these instances (and hopefully achieves good accuracy for new instances as well). The second stop criterion, which may have to be employed only if nonbinary splits are used (otherwise we would have already stopped at the parent node due to the third criterion), does not give a similar comfort of creating a leaf that accurately classifies some portion of the training set, but it will not contribute to any training set inaccuracy, either. The leaf created in this situation does not participate in the classification of the training set at all, but it has to be there in case there are some new instances to which the tree will be eventually applied arriving at that point. The class label inherited from the parent node provides the best possible way of classifying them. The last stop criterion can be rather considered an "emergency" one: we are forced to stop without reaching a uniform class since there is no possibility of growing further. This is the only situation in which the created leaf will inaccurately classify some of the training instances. It is important to notice, however, that it can only happen if there are instances in the training set that cannot be separated by the available set of splits despite belonging to different classes. Since it is reasonable to assume that any complete set of splits should allow one to separate any two instances which differ on least at one attribute value, those must be instances from different classes with the same attribute value vectors. It is not at all uncommon for realistic datasets and can result from either data noise or an insufficient set of attributes. Since these indistinguishable instances will be assigned the most probable class, the resulting inaccuracy on the training set will be as small as possible (and completely unavoidable).

3.3.4.2 Relaxed criteria

It is easy to see that the first two of the above strict stop criteria can be easily relaxed. Instead of requiring a single uniform class, we may choose to stop when the probability of the dominating class is sufficiently high. Similarly, instead of waiting for the empty set of instances we may choose to stop when the current subset of training instances is sufficiently small. It is common to use such relaxed stop criteria. They result in creating smaller trees that do not achieve the highest possible training set accuracy, but may actually generalize better and turn out to be more accurate on new data. This corresponds to the belief that simplicity reduces the risk of overfitting, known as Ockham's razor. Another (and usually more reliable) way of creating smaller trees with better generalization properties is pruning, and such stop criteria can be considered attempts to make computational savings by preventing growing subtrees that would likely get pruned anyway.

A more technical, but sometimes useful stop criterion for decision tree growing is based on the maximum tree depth. No further splits are performed after a sufficiently long path has been created (i.e., for nodes created on the specified maximum tree level). In a sense, this can be thought of as a relaxed version of the "no splits left" criterion, as it sets the limit on the number of splits used along each tree path.

Example 3.3.4 The following R code demonstrates simple stop criteria checking for the root node of the decision tree for the *weather* data. Two of the three criteria discussed above are checked, in their relaxed versions: "uniform class" (by comparing the probability of the dominating class against the maxprob threshold) and "no instances left" (by comparing the number of instances corresponding to the node against the minsplit threshold). Additionally, the more technical maximum tree depth criterion is checked, by comparing the node's number to 2^maxdepth. The three stop criteria parameters:

- maxprob – the maximum dominating class probability allowed for a split,
- minsplit – the minimum number of instances required for a split,
- maxdepth – the maximum tree depth,

are specified via variables for the sake of this demonstration. All these criteria are obviously found to be unsatisfied for the root node of the decision tree for the *weather* data.

```
maxprob <- 0.999
minsplit <- 2
maxdepth <- 8

stop.criteria <- function(n)
{
  n>=2^maxdepth || tree$count[tree$node==n]<minsplit ||
    max(tree[tree$node==n,cprobs])>maxprob
}

stop.criteria(n)
```

3.3.5   Split selection

If the stop criteria presented above are used in the basic strict form, the resulting decision tree will achieve the maximum training set accuracy possible with the available set of splits, regardless of the split selection method. As we have seen, only leaves created due to the lack of splits that could separate training instances of different classes contribute to misclassification, and only a richer set of available splits (corresponding to an enhanced set of attributes) could overcome the problem. If we were only concerned with the training set performance, splits could be selected completely arbitrarily, e.g., at random or in the lexicographic order of attributes and values. Of course, we are concerned with generalization, i.e., the expected performance on new data, and this is why split selection matters.

3.3.5.1   Preference for simplicity

Most, if not all, practically used decision tree algorithms use split selection as an opportunity (or one of opportunities) to introduce the preference for simplicity, which is one of the most common types of inductive bias. Following the principle of Ockham's razor, smaller trees are believed to be less prone to overfitting. Despite some controversies about the validity of this principle in general, it lies behind the standard approach to split selection, which can be phrased as follows: try to select splits that, under given stop criteria, will lead to growing small trees. Whenever a split for a node is selected, it should preferably make the subtrees connected to this node as small as possible.

There are several possible formal and informal interpretations of decision tree complexity or decision tree size. The one that is particularly simple and easy to adopt for split selection is that based on the average path length (i.e., the average number of splits between the root node and leaves). To minimize the average path length, each node's split should minimize the average length of subpaths starting in the node. Although this minimization cannot be achieved exactly without actually building all complete subtrees for each candidate split (which would turn the growing process into an exhaustive search over all possible trees), reasonably good heuristics can be used for approximate minimization. The idea is to select splits that decrease the impurity of class distribution in the resulting subsets, i.e., increase the domination of one or more classes over the others. This is motivated by the hope that a subset containing only or mostly instances of one class will be reached after a small number of splits, and a leaf will be created.

3.3.5.2   Impurity measures

Several different measures can be used to characterize the impurity of class distribution in a set of instances. The two particularly popular ones are the entropy and the Gini index, defined in Section 2.4.2.

3.3.5.3   Split evaluation

To see how impurity measures such as the entropy or Gini index can be used for split evaluation, consider a node $\mathbf{n}$ and a candidate split $t : X \rightarrow R_t$. By applying an impurity measure to the set of estimated class probabilities in this node, i.e., calculating $E_{T_\mathbf{n}}(c)$ or $\mathrm{GI}_{T_\mathbf{n}}(c)$, we would only see how good or bad this node is at identifying one or more dominating classes, without finding out anything about the candidate split. To achieve the latter, we have to consider subsets $T_{\mathbf{n},t=r}$ for all $r \in R_t$ to which the current set of training instances $T_{\mathbf{n}}$ is partitioned by the candidate split, and measure the class distribution impurity for them. This requires that the class probability estimates be calculated for each of these subsets:

$$P_{T_{\mathbf{n},t=r}}(d) = P_{T_{\mathbf{n}}}(c = d | t = r) \tag{3.11}$$

Using the entropy as an example, we would then measure the class distribution impurity in the subset corresponding to outcome r of the candidate split t as $E_{T_{\mathbf{n},t=r}}(c)$. The split can be evaluated by the weighted average of such per-outcome entropies:

$$E_{T_{\mathbf{n}}}(c|t) = \sum_{r \in R_t} \frac{|T_{\mathbf{n},t=r}|}{|T_{\mathbf{n}}|} E_{T_{\mathbf{n},t=r}}(c) \tag{3.12}$$

The weights in this average are based on the proportions in which the current set of training instances is partitioned into subsets corresponding to the possible outcomes of t, i.e., subset entropies are weighted proportionally to subset sizes. Notice that, since $\frac{|T_{\mathbf{n},t=r}|}{|T_{\mathbf{n}}|}$ can be interpreted as the estimated probability of split t taking outcome r, $P_{T_{\mathbf{n}}}(t = r)$, the latter is in fact the conditional entropy of the target concept given the split, as it matches directly the definition of the latter presented in Section 2.5.3. The same averaging should be applied to the Gini index or any other impurity measure used. Split selection can be then performed by *minimizing* any of these weighted average impurity measures over all available splits (assuming they all assign smaller values to smaller impurity).

It is not uncommon to use slightly modified split evaluation functions, measuring not just the class distribution impurity obtained after the split, but rather the *decrease* of the impurity. It is defined as the difference between the impurity for the subset of instances corresponding to the node in which the split is to be applied and the weighted average impurity of the subsets corresponding to split outcomes. For the entropy, the difference can be written as follows:

$$\Delta E_{T_{\mathbf{n}}}(c|t) = E_{T_{\mathbf{n}}}(c) - E_{T_{\mathbf{n}}}(c|t) \tag{3.13}$$

and is called the *information gain*, although, by comparing to the definition presented in Section 2.5.3, it can also be easily seen to be the same as the mutual information for the target concept c and split t, $I_{T_{\mathbf{n}}}(c,t)$. Since the first term does not depend on the evaluated split, the information gain will lead to selecting exactly the same splits as the entropy (as long as it is maximized rather than minimized). Its advantage is that it not only indicates which split is the best, but also how much improvement it gives. This is sometimes used to define an additional stop criterion that converts a node to a leaf when the improvement due to the best available split is too small. The corresponding difference for the Gini index can be defined similarly and used in the same way.

Example 3.3.5 The following R code snippet defines the `weighted.impurity` function that calculates the weighted impurity based on the supplied class probability distributions and instance counts for the true and false split condition outcomes. The impurity measure passed via the `imp` parameter is assumed to accept a probability distribution on input. The `entropy.p` or `gini.p` functions for calculating the entropy and the Gini index can be used for this purpose. The application of the `weighted.impurity` function to evaluating the split based on the `outlook==overcast` condition is demonstrated.

Ex. 2.4.26, 2.4.27
dmr.stats

```
weighted.impurity <- function(pd1, n1, pd0, n0, imp=entropy.p)
{
  weighted.mean(c(imp(pd1), imp(pd0)), c(n1, n0))
}

  # weighted impurity of play for outlook=overcast and outlook!=overcast
weighted.impurity(pdisc(weather$play[weather$outlook=="overcast"]),
                  sum(weather$outlook=="overcast"),
                  pdisc(weather$play[weather$outlook!="overcast"]),
                  sum(weather$outlook!="overcast"))
```

Example 3.3.6 The following R code defines functions that can be used to evaluate and select splits using the entropy and the Gini index as impurity measures, and then demonstrates their application to the root node of the tree for the *weather* data. The `entropy.p` and `gini.p` functions are used for calculating the entropy and the Gini index based on supplied discrete probability distributions.

> Ex. 2.4.26,
> 2.4.27
> `dmr.stats`

Notice that the `split.eval` function detects useless splits that give the same outcome for all instances and makes sure that they will not be selected (by returning `Inf`). Its return value is checked by the `split.select` function which actually makes no split selection if the best available split is useless. This in fact indirectly implements the "no splits left" stop criterion.

It is also worthwhile to notice that the `split.select` and `split.eval` functions include support for inequality-based splits for continuous attributes. The corresponding parts of the code are easily located by the references to the `is.numeric` function. The `midbrk` utility function is used to determine inequality thresholds for continuous attribute splits. This capability is not demonstrated at this point, though, since there are no continuous attributes in the *weather* data.

> `dmr.util`

```
split.eval <- function(av, sv, cl)
{
  cond <- !is.na(av) & (if (is.numeric(av)) av<=as.numeric(sv) else av==sv)

  pd1 <- pdisc(cl[cond])
  n1 <- sum(cond)
  pd0 <- pdisc(cl[!cond])
  n0 <- sum(!cond)

  if (n1>0 && n0>0)
    weighted.impurity(pd1, n1, pd0, n0, imp)
  else
    Inf
}

split.select <- function(n)
{
  splits <- data.frame()
  for (attribute in attributes)
  {
    uav <- sort(unique(data[nodemap==n,attribute]))
    if (length(uav)>1)
      splits <- rbind(splits,
                      data.frame(attribute=attribute,
```

```
                        value=if (is.numeric(uav))
                                 midbrk(uav)
                               else as.character(uav),
                        stringsAsFactors=FALSE))
  }

  if (nrow(splits)>0)
    splits$eval <- sapply(1:nrow(splits),
                          function(s)
                          split.eval(data[nodemap==n,splits$attribute[s]],
                                     splits$value[s],
                                     data[nodemap==n,class]))
  if ((best.eval <- min(splits$eval))<Inf)
    tree[tree$node==n,2:3] <<- splits[which.min(splits$eval),1:2]
  best.eval
}

  # entropy-based split selection
imp <- entropy.p
split.select(n)

  # Gini index-based split selection
imp <- gini.p
split.select(n)
```

As we can see, the split `outlook=overcast` is ranked the best with respect to both the entropy and the Gini index.

3.3.5.4 Reducing split selection complexity

While the issues of computational complexity and implementation efficiency are generally not addressed by this book, two possibilities of computational savings for membership-based splits and inequality-based splits are worthwhile to mention, since they are based on simple properties of these split types. For these two split types, the complexity of split selection can be reduced by narrowing down the set of candidate splits.

Consider candidate membership-based splits for attribute $a : X \to A$ at node **n**. Assuming a two-class classification task, all possible values of a can be ordered monotonically with respect to the conditional probabilities of class 1 given attribute values in this node:

$$P_{T_\mathbf{n}}(c = 1|a = v_1) \le P_{T_\mathbf{n}}(c = 1|a = v_2) \le \cdots \le P_{T_\mathbf{n}}(c = 1|a = v_{|A|}) \qquad (3.14)$$

It is easy to verify that class impurity can be only minimized by one of the following splits:

$$t(x) = \begin{cases} 1 & \text{if } a(x) \in \{v_1, \dots, v_k\} \\ 0 & \text{otherwise} \end{cases} \qquad (3.15)$$

for $k = 1, 2, \dots, |A| - 1$. This observation – valid both for the entropy and the Gini index used as impurity measures – reduces the number of splits to consider from $2^{|A|} - 2$ to $|A| - 1$.

Reducing the number of candidate splits for continuous attributes is also based on attribute value ordering, but it does not require the two-class assumption. In this case, all values of a continuous attribute a in the subset of training instances corresponding to node $\mathbf{n}$ are ordered monotonically $v_1 \leq v_2 \leq \cdots \leq v_k$. An inequality-based split can separate each pair of adjacent values, but it can be verified that splits separating values v_j, v_{j+1} such that for all $d \in C$ the corresponding conditional class probabilities are the same:

$$P_{T_\mathbf{n}}(c = d | a = v_j) = P_{T_\mathbf{n}}(c = d | a = v_{j+1}) \tag{3.16}$$

cannot minimize class impurity and therefore do not have to be considered as candidates.

3.3.6   Split application

When a split $t : X \rightarrow R_t$ for a node $\mathbf{n}$ has been selected, new descendant nodes $\mathbf{n}_r$ for all $r \in R_t$ can be created, corresponding to each of its possible outcomes. These new nodes are marked as open, to get processed in subsequent iterations of the algorithm. The set of training instances $T_\mathbf{n}$ corresponding to the parent node has to be partitioned into subsets corresponding to the newly created descendant nodes, $T_{\mathbf{n}, t=r}$ for $r \in R_t$. This completes the current iteration of the algorithm and node $\mathbf{n}$ can be marked as closed.

Example 3.3.7 The following R code shows how to apply the previously selected split to partition the training set into subsets. This results in creating two new nodes, numbered 2 and 3, of the decision tree for the *weather* data. The partitioning is accomplished by altering the assignment of instances to nodes represented by the `nodemap` vector. Notice that both equality-based splits for discrete attributes and inequality-based splits for continuous attributes are supported by the `split.apply` function, although it is only the former that are needed for the *weather* data.

```
split.apply <- function(n)
{
    tree <<- rbind(tree,
                  data.frame(node=(2*n):(2*n+1),
                             attribute=NA, value=NA, class=NA, count=NA,
                             'names<-'(rep(list(NA), length(clabs)),
                                       paste("p", clabs, sep="."))))

    av <- data[[tree$attribute[tree$node==n]]]
    cond <- !is.na(av) & (if (is.numeric(av))
                             av<=as.numeric(tree$value[tree$node==n])
                          else av==tree$value[tree$node==n])
    nodemap[nodemap==n & cond] <<- 2*n
    nodemap[nodemap==n & !cond] <<- 2*n+1
}

split.apply(n)
```

3.3.7   Complete process

All operations performed during a single iteration of top-down decision tree growing have been presented above. After a number of iterations, when there are no more open nodes left, the growing process terminates, yielding a completely grown tree.

Example 3.3.8 It is worthwhile to note that the code from the previous examples illustrating the operations of class distribution calculation, class label assignment, split selection, and split application can be applied again to process the newly created nodes, by passing consecutive node numbers via the n argument of the corresponding functions. Of course, split selection and split application only makes sense for nodes not satisfying the stop criteria. It can be verified that after 13 iterations the growing process will complete, yielding the decision tree represented by the following data frame (with class labels converted to the original representation):

	node	attribute	value	class	count	p.no	p.yes
1	1	outlook	overcast	yes	14	0.3571429	0.6428571
2	2	<NA>	<NA>	yes	4	0.0000000	1.0000000
3	3	humidity	high	no	10	0.5000000	0.5000000
4	6	outlook	rainy	no	5	0.8000000	0.2000000
5	7	wind	high	yes	5	0.2000000	0.8000000
6	12	wind	high	no	2	0.5000000	0.5000000
7	13	<NA>	<NA>	no	3	1.0000000	0.0000000
8	14	outlook	rainy	no	2	0.5000000	0.5000000
9	15	<NA>	<NA>	yes	3	0.0000000	1.0000000
10	24	<NA>	<NA>	no	1	1.0000000	0.0000000
11	25	<NA>	<NA>	yes	1	0.0000000	1.0000000
12	28	<NA>	<NA>	no	1	1.0000000	0.0000000
13	29	<NA>	<NA>	yes	1	0.0000000	1.0000000

It is also quite straightforward to use the code from the above series of *weather* examples to create a very simple and inefficient, but easy to understand and working implementation of decision tree growing. The only missing piece is the main loop. Such an implementation is presented below. The grow.dectree function implemented by the following R code organizes the iterative processing of decision tree nodes, using the init, class.distribution, class.label, stop.criteria, split.eval, split.select, and split.apply functions from the previous examples as its internal functions. Two additional utility functions, x.vars and y.var, are used to extract the attribute and class names from the input formula, to make the implementation applicable to other datasets. Apart from the formula and dataset, the grow.dectree function receives the impurity evaluation function and stop criteria parameters as its optional arguments. The class attribute of the resulting tree object is set to dectree to enable the appropriate prediction method dispatching and a method for conversion to a data frame is provided.

`dmr.util`

```
## a simple decision tree growing implementation
grow.dectree <- function(formula, data,
                         imp=entropy.p, maxprob=0.999, minsplit=2, maxdepth=8)
{
  init <- function()
  {
    clabs <<- factor(levels(data[[class]]),
                     levels=levels(data[[class]]))     # class labels
    tree <<- data.frame(node=1, attribute=NA, value=NA, class=NA, count=NA,
                        `names<-`(rep(list(NA), length(clabs)),
                                  paste("p", clabs, sep=".")))
    cprobs <<- (ncol(tree)-length(clabs)+1):ncol(tree)  # class probability columns
    nodemap <<- rep(1, nrow(data))
```

```
    n <<- 1
}

next.node <- function(n)
{
  if (any(opn <- tree$node>n))
    min(tree$node[opn])
  else Inf
}

class.distribution <- function(n)
{
  tree$count[tree$node==n] <<- sum(nodemap==n)
  tree[tree$node==n,cprobs] <<- pdisc(data[nodemap==n,class])
}

class.label <- function(n)
{
  tree$class[tree$node==n] <<- which.max(tree[tree$node==n,cprobs])
}

stop.criteria <- function(n)
{
  n>=2^maxdepth || tree$count[tree$node==n]<minsplit ||
    max(tree[tree$node==n,cprobs])>maxprob
}

split.eval <- function(av, sv, cl)
{
  cond <- !is.na(av) & (if (is.numeric(av)) av<=as.numeric(sv) else av==sv)
  pd1 <- pdisc(cl[cond])
  n1 <- sum(cond)
  pd0 <- pdisc(cl[!cond])
  n0 <- sum(!cond)

  if (n1>0 && n0>0)
    weighted.impurity(pd1, n1, pd0, n0, imp)
  else
    Inf
}

split.select <- function(n)
{
  splits <- data.frame()
  for (attribute in attributes)
  {
    uav <- sort(unique(data[nodemap==n,attribute]))
    if (length(uav)>1)
      splits <- rbind(splits,
                      data.frame(attribute=attribute,
                                 value=if (is.numeric(uav))
                                         midbrk(uav)
                                       else as.character(uav),
                                 stringsAsFactors=FALSE))
  }

  if (nrow(splits)>0)
    splits$eval <- sapply(1:nrow(splits),
                          function(s)
                            split.eval(data[nodemap==n,splits$attribute[s]],
                                       splits$value[s],
```

```
                                        data[nodemap==n,class]))
    if ((best.eval <- min(splits$eval))<Inf)
      tree[tree$node==n,2:3] <<- splits[which.min(splits$eval),1:2]
    best.eval
  }

  split.apply <- function(n)
  {
    tree <<- rbind(tree,
                   data.frame(node=(2*n):(2*n+1),
                              attribute=NA, value=NA, class=NA, count=NA,
                              'names<-'(rep(list(NA), length(clabs)),
                                        paste("p", clabs, sep="."))))

    av <- data[[tree$attribute[tree$node==n]]]
    cond <- !is.na(av) & (if (is.numeric(av))
                            av<=as.numeric(tree$value[tree$node==n])
                          else av==tree$value[tree$node==n])
    nodemap[nodemap==n & cond] <<- 2*n
    nodemap[nodemap==n & !cond] <<- 2*n+1
  }

  tree <- nodemap <- n <- NULL
  clabs <- cprobs <- NULL
  class <- y.var(formula)
  attributes <- x.vars(formula, data)

  init()
  while (is.finite(n))
  {
    class.distribution(n)
    class.label(n)
    if (!stop.criteria(n))
      if (split.select(n)<Inf)
        split.apply(n)
    n <- next.node(n)
  }
  tree$class <- clabs[tree$class]
  'class<-'(tree, "dectree")
}

## convert a dectree object to a data frame
as.data.frame.dectree <- function(x, row.names=NULL, optional=FALSE, ...)
{ as.data.frame(unclass(x), row.names=row.names, optional=optional) }

  # grow a decision tree for the weather data
tree <- grow.dectree(play~., weather)

  # grow a decision tree for the weatherc data
treec <- grow.dectree(play~., weatherc)

  # data frame conversion
as.data.frame(tree)
as.data.frame(treec)
```

The `grow.dectree` function is called above for the *weather* data, yielding the tree presented before, and additionally for the *weatherc* data. The latter gives an opportunity to illustrate the continuous attribute support capability of the `split.eval`, `split.select`, and `split.apply` functions, and yields the following tree:

```
   node   attribute    value class count      p.no     p.yes
 1    1      outlook overcast   yes    14 0.3571429 0.6428571
 2    2         <NA>     <NA>   yes     4 0.0000000 1.0000000
 3    3  temperature       25    no    10 0.5000000 0.5000000
 4    6  temperature       16   yes     7 0.2857143 0.7142857
 5    7         <NA>     <NA>    no     3 1.0000000 0.0000000
 6   12         <NA>     <NA>    no     1 1.0000000 0.0000000
 7   13     humidity       85   yes     6 0.1666667 0.8333333
 8   26         <NA>     <NA>   yes     5 0.0000000 1.0000000
 9   27         <NA>     <NA>    no     1 1.0000000 0.0000000
```

3.4   Pruning

Decision tree pruning is an insurance policy against overfitting motivated by Ockham's razor. It can be considered an inverse of growing that results in cutting off some overgrown subtrees and replacing them by leaves with the intention to improve the tree's generalization capability. It might seem totally unreasonable to waste computational resources on growing a tree and then pruning it by just throwing away several nodes with so carefully selected splits. One might be tempted to avoid this waste by using more refined stop criteria that would prevent creating poor nodes and growing useless subtrees. This approach is known as *prepruning*.

After some more thought, however, one can easily conclude that it is not possible, in general, to reliably evaluate the utility of any subtree without actually growing it. While we might try to use relaxed stop criteria to prevent growing subtrees that would likely get pruned later, there is some inherent risk associated with such computational savings, which should be always taken with care and adopted only when the computation time really matters. In most cases, pruning appears to be the best way of achieving good generalization with decision trees.

To fully describe a pruning algorithm, we need to specify the following major components thereof:

Pruning operators. Which determine how the operation of cutting off nodes from the tree is exactly performed.

Pruning criterion. Which determines how to judge whether a pruning operator should be applied to a given node.

Pruning control strategy. Which determines the order in which candidate nodes for pruning are considered.

These are discussed more extensively below, with particular emphasis put on pruning criteria which are the essential components of pruning algorithms, and to which the other components are usually adjusted appropriately.

Example 3.4.1 In subsequent examples illustrating decision tree pruning, we will continue to use the *weather* data, but instead of the tree created by the simple decision tree growing implementation presented previously, we will use a tree grown by the much more refined implementation provided by the `rpart` package, calling the `rpart` function as follows:

```
rptree <- rpart(play~., weather, minsplit=2)
```

The resulting tree can be easily verified to be essentially identical to the one presented in the previous example, subject to slightly different assignment of split outcomes to left/right branches and of course a different representation. Using `rpart` for pruning examples provides an opportunity to demonstrate how to extract split information from an `rpart` object. The reader may consider using the code presented in the forthcoming examples to actually implement pruning for trees created either using `rpart` or `grow.dectree` as an exercise.

3.4.1   Pruning operators

The basic and most common pruning operator, employed by all known decision tree pruning algorithms, is the *subtree cutoff* operator, which replaces a selected node by a leaf. The leaf should be assigned the majority class label of the corresponding subset of training instances. This is one reason why it is convenient to assign class labels to nodes during tree growing – when it comes to pruning, the labels are already there. This operator can be thought of as "ungrowing" an overgrown subtree.

A somewhat less obvious and much less commonly used pruning operator is the *node removal* operator, which consists in replacing a selected node by one of its descendants. More precisely, the descendant node with the largest corresponding subset of training instances takes place of its parent node, and the other descendants are dropped (with any subtrees rooted at them completely removed). This operator is a relatively gentle way of "withdrawing" from a split that turned out not to be very useful, without regrowing the whole subtree. It would usually make sense only if the descendant nodes being dropped correspond to a small subset of instances.

Example 3.4.2 A graphical illustration of the most common subtree cutoff operator is generated by the following R code. It plots both the original `rpart` tree for the *weather* data created in the previous example and a modified version thereof, in which cutting off the node with the `outlook=sunny` split is simulated by selecting an appropriate data subset. Figure 3.2 shows the resulting trees, before and after subtree cutoff.

```
  # simulate cutoff of the subtree starting from the outlook=sunny split
rptree.stc <- rpart(play~., weather, subset=!(outlook %in% c("rainy", "sunny") &
                                    humidity=="high" & outlook!="sunny"),
              minsplit=2, cp=0)

prp(rptree, varlen=0, faclen=0, main="Before subtree cutoff")
prp(rptree.stc, varlen=0, faclen=0,  main="After subtree cutoff")
```

3.4.2   Pruning criterion

The pruning criterion is used to make decisions whether a given pruning operator should be applied to a given node or not. This is done by comparing the original subtree rooted at the node and a new subtree (in particular, a single leaf for the most common subtree cutoff operator) with respect to some quality measures. The number of pruning criteria described in the literature and used by practical implementations is quite large and still likely to increase. We will briefly describe just a few representative examples. For simplicity, the discussion will be limited to the first and most common subtree cutoff pruning operator, but some of the presented criteria can be adapted to the node removal operator as well.

Before subtree cutoff

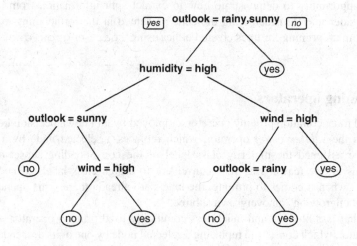

After subtree cutoff

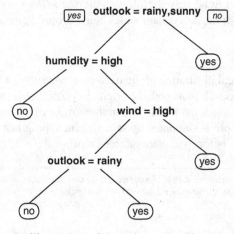

Figure 3.2   An illustration of the subtree cutoff pruning operator.

3.4.2.1   Reduced error pruning

The idea of reduced error pruning comes directly from the goal of pruning, which is to improve the generalization capability of a decision tree. Assuming the misclassification error is used as the performance measure, the criterion is based on comparing the error of the original subtree and the leaf that would be replacing it. In fact, the original subtree would (nearly) always win if the error were calculated on the training set, but this clearly would not say anything about generalization. The reduced error pruning assumes instead, according to standard model evaluation principles, that a separate labeled dataset R is maintained, called the *pruning set*, and used to calculate the errors to be compared.

Let $e_R(\mathbf{n})$ and $e_R(\mathbf{l})$ denote the errors of the original node $\mathbf{n}$ and the replacing leaf $\mathbf{l}$, respectively, measured on the pruning set R. Both these errors are calculated as the fractions of instances in $R_\mathbf{n}$ – the subset of R corresponding to $\mathbf{n}$ – misclassified by the subtree rooted at node $\mathbf{n}$ and by leaf $\mathbf{n}$, respectively. The reduced error pruning criterion would recommend to replace node $\mathbf{n}$ by leaf $\mathbf{l}$ if

$$e_R(\mathbf{l}) \leq e_R(\mathbf{n}) \tag{3.17}$$

If the leaf performs no worse than the original subtree on a separate dataset, then using it to replace the subtree may improve the tree's generalization capability, since it is likely to *reduce* the true misclassification error (which justifies the name of this criterion, by the way).

It could be easily considered the perfect pruning criterion unless its application were so often prevented by "data economy" issues. To provide sufficiently reliable error estimates even for low-level nodes (where pruning is most likely to be needed), a considerable number of pruning instances may be required, at least comparable to the number of training instances used for growing. Unless we have a plethora of labeled data which have to be sampled anyway due to computational constraints, it may be unreasonable to leave out such a substantial portion of the available data from the growing process, since this may considerably impact the quality of the grown tree. Even a perfectly pruned, but poorly grown tree may turn out inferior to a better grown, but not so perfectly pruned tree. This justifies the existence and common use of other pruning criteria, which assume that the same set of training instances must be used for both growing and pruning.

Example 3.4.3 The following R code defines a set of simple functions that calculate the error of a given leaf of an `rpart` decision tree on a given dataset and demonstrates how they can be applied to the decision tree grown for the *weather* data. The subset of instances corresponding to the leaf is determined by using a rule extracted from the tree, represented by the conjunction of split conditions occurring on the path from the root to the leaf. This is accomplished by the `rewrite.splits` and `extract.rule` functions.

```
## transform rpart split conditions to a convenient form
rewrite.splits <- function(cond)
{
  ss <- strsplit(cond, "=")[[1]]
  attribute=ss[1]
  values <- sapply(strsplit(ss[2], ",")[[1]], deparse)
  newcond <- paste(attribute, "==", values, sep="", collapse="|")
  return(paste("(", newcond, ")", sep=""))
}

## extract a rule from an rpart tree corresponding to the path from the root
## to a given node
extract.rule <- function(rp, node)
{
  path <- path.rpart(rp, node, print.it=FALSE)[[1]][-1]
  ifelse(length(path)>0, paste(sapply(path, rewrite.splits), collapse="&"), "TRUE")
}

## calculate the error of a given node, if treated as a leaf
leaf.error <- function(rp, node, data, class)
{
  rule <- extract.rule(rp, node)
  dsub <- eval(parse(text=rule), data)
  lab <- levels(class)[rp$frame$yval[row.names(rp$frame)==node]]
  sum(lab!=class[dsub])/nrow(data[dsub,])
}
```

```
   # error of node 1, if treated as a leaf
leaf.error(rptree, 1, weather, weather$play)
   # error of node 3, which is actually a leaf
leaf.error(rptree, 3, weather, weather$play)
   # error of node 4, if treated as a leaf
leaf.error(rptree, 4, weather, weather$play)
```

The following R code implements node error calculation, as the error of the complete subtree rooted at a given node, and again demonstrates its application to the decision tree grown for the *weather* data.

```
## check whether a given node is a leaf of an rpart tree
rp.leaf <- function(rp, node)
{
  rp$frame$var[row.names(rp$frame)==node]=="<leaf>"
}

## calculate the number of instances corresponding to a node
node.card <- function(rp, node, data)
{
  rule <- extract.rule(rp, node)
  dsub <- eval(parse(text=rule), data)
  nrow(data[dsub,])
}

## calculate the error of the subtree rooted at a given node
node.error <- function(rp, node, data, class)
{
  if (rp.leaf(rp, node))
    leaf.error(rp, node, data, class)
  else
  {
    el <- node.error(rp, 2*node, data, class)
    nl <- node.card(rp, 2*node, data)
    er <- node.error(rp, 2*node+1, data, class)
    nr <- node.card(rp, 2*node+1, data)
    weighted.mean(c(el, er), c(nl, nr))
  }
}

   # error of node 1
node.error(rptree, 1, weather, weather$play)
   # error of node 3, which is actually a leaf
node.error(rptree, 3, weather, weather$play)
   # error of node 4
node.error(rptree, 1, weather, weather$play)
```

Although all illustrative R code presented in this book is written without any care for efficiency, with the readability only kept in mind, this one stands out as particularly inefficient (which, luckily, is no problem for the small dataset used). This is because it requires independently extracting multiple tree paths and applying them as rules to identify the corresponding subsets of instances. The error of an internal tree node is calculated as the weighted average of its descendants' errors. It is easy to see that this approach is exactly equivalent to directly using the subtree rooted at the node to classify the given set of instances and calculating the misclassification rate, as long as the leaf.error function returns the misclassification rate of a given leaf.

In this example, node error calculation is demonstrated using the same dataset on which the tree was grown, which violates the key assumption of reduced error pruning and serves the illustration purpose only. Since the tree perfectly fits the training set which is also used to calculate errors, we are guaranteed to obtain an error value of 0 for any node. It is therefore not at all surprising that the following R code, which checks the reduced error pruning criterion, finds no candidates for pruning.

```
# check which nodes satisfy the REP criterion
sapply(as.integer(row.names(rptree$frame)),
      function(node)
      {
         !rp.leaf(rptree, node) &&
           leaf.error(rptree, node, weather, weather$play) <=
             node.error(rptree, node, weather, weather$play)
      })
```

3.4.2.2 Pessimistic pruning

The idea of pessimistic pruning is based on the obvious observation that the error of a subtree estimated on the same set of instances on which the subtree was grown has to be optimistically biased and therefore needs a pessimistic correction. The criterion can be written as

$$e_T(\mathbf{l}) \leq \tilde{e}_T(\mathbf{n}) \tag{3.18}$$

which is essentially the same inequality as for reduced error pruning with the exception that the training set T is used to estimate errors and the error of the original subtree rooted at node $\mathbf{n}$ is pessimistically corrected, which is designated by using a tilde. The correction is achieved by adding a correction term as follows:

$$\tilde{e}_T(\mathbf{n}) = e_T(\mathbf{n}) + \sqrt{\frac{\hat{e}_T(\mathbf{n})(1 - \hat{e}_T(\mathbf{n}))}{|T_{\mathbf{n}}|}} \tag{3.19}$$

The correction term can be thought of as an estimate of the error's standard deviation, and adding it to the original error is equivalent to taking an upper boundary of the corresponding confidence interval, but this should be considered rather an intuitive explanation of this approach than a strict and formal justification. Since the training set misclassification error of unpruned subtrees may often be 0, the standard deviation estimate uses a modified error estimate $\hat{e}_T(\mathbf{n})$. The simplest way of calculating it that works reasonably well in practice is to always assume that "a half of an instance" is misclassified:

$$\hat{e}_T(\mathbf{n}) = \frac{|\{x \in T_{\mathbf{n}} \mid h_{\mathbf{n}}(x) \neq c(x)\}| + 0.5}{|T_{\mathbf{n}}|} \tag{3.20}$$

where $h_{\mathbf{n}}$ denotes the classification model represented by the subtree rooted at $\mathbf{n}$. A somewhat more modern and more refined approach would be to use an m-estimate instead:

$$\hat{e}_T(\mathbf{n}) = \frac{|\{x \in T_{\mathbf{n}} \mid h_{\mathbf{n}}(x) \neq c(x)\}| + mp}{|T_{\mathbf{n}}| + m} \tag{3.21}$$

or its Laplace version obtained when assuming $p = 0.5$ and $m = 2$:

$$\hat{e}_T(\mathbf{n}) = \frac{|\{x \in T_{\mathbf{n}} \mid h_{\mathbf{n}}(x) \neq c(x)\}| + 1}{|T_{\mathbf{n}}| + 2} \tag{3.22}$$

These two probability estimation techniques are presented in Section 2.4.4.

Example 3.4.4 The following R code defines a function for calculating the pessimistic estimate of node error (in the last presented version, using the Laplace estimator) and demonstrates how it can be applied to the decision tree grown for the *weather* data.

```
## calculate the PEP error of the subtree rooted at a given node
node.pep.error <- function(rp, node, data, class)
{
  e <- node.error(rp, node, data, class)
  n <- node.card(rp, node, data)
  e1 <- (e*n+1)/(n+2)
  e + sqrt(e1*(1-e1)/n)
}

  # PEP error of node 1
node.pep.error(rptree, 1, weather, weather$play)
  # PEP error of node 4
node.pep.error(rptree, 1, weather, weather$play)
```

It turns out that the pessimistic error pruning criterion finds no candidates for pruning in the tree, which can be verified using this R code:

```
# check which nodes would get pruned under the PEP criterion
sapply(as.integer(row.names(rptree$frame)),
    function(node)
    {
      !rp.leaf(rptree, node) &&
        leaf.error(rptree, node, weather, weather$play)<=
          node.pep.error(rptree, node, weather, weather$play)
    })
```

3.4.2.3   Minimum error pruning

The minimum error pruning (MEP) criterion is a more elegant and better controllable way of achieving the same objectives as pessimistic error pruning, i.e., comparing training set errors of the node under consideration and the replacing leaf with some compensation for the inherent optimistic bias. It estimates the error of any leaf **l** as **1**'s complement of its m-estimated accuracy. It is calculated using the technique of m-estimation presented in Section 2.4.4 as follows:

$$\hat{e}_T(\mathbf{l}) = 1 - \frac{|\{x \in T_{\mathbf{l}} \mid c(x) = d_{\mathbf{l}}\}| + mp}{|T_{\mathbf{l}}| + m} \tag{3.23}$$

where in the lack of better knowledge the *a priori* probability of class $d_{\mathbf{l}}$ is taken to be $p = \frac{1}{|C|}$ or estimated on the complete training set, and m remains an adjustable parameter. For any node **n**, the error estimate $\hat{e}_T(\mathbf{n})$ is obtained as the average of error estimates for its descendants, weighted by the numbers of training instances corresponding to them:

$$\hat{e}_T(\mathbf{n}) = \sum_{\mathbf{n}' \in N(\mathbf{n})} \frac{|T_{\mathbf{n}'}|}{|T_{\mathbf{n}}|} \hat{e}_T(\mathbf{n}') \tag{3.24}$$

where $N(\mathbf{n})$ is the set of all the descendants (nodes and leaves) of node $\mathbf{n}$.

In effect, the error estimates for leaves calculated as shown above are propagated upward to nodes. Such error up-propagation would clearly result in exactly the same node error as calculated in the usual direct way if leaf errors were plain misclassification rates, but since they are based on m-estimated leaf accuracies to avoid optimistic bias, the resulting node error will also be pessimistically corrected, to an extent depending on the value of m.

Now if we consider the particular candidate node for pruning $\mathbf{n}$ and the leaf $\mathbf{l}$ that could replace it, the MEP criterion is the obvious inequality for the not-so-obvious error estimates:

$$\hat{e}_T(\mathbf{l}) \le \hat{e}_T(\mathbf{n}) \tag{3.25}$$

Since the errors of leaves are calculated based on their m-estimated accuracy, they are the more affected by the m "fictitious" instances the smaller subset of their "real" instances. Even perfectly accurate leaves of the original subtree may therefore achieve quite poor error estimates for sufficiently large m, and these estimates are propagated upward to obtain the error estimate of the whole original subtree. The leaf considered as a replacement will have the number of instances equal to the number of all training instances corresponding to the original subtree, which will be usually substantially larger than the number of instances corresponding to each of its individual leaves. Its error estimate will therefore be much less affected by the m "fictitious" instances and it can win in the comparison. As this interpretation clearly shows, the m parameter can be used to adjust the aggressiveness of pruning (the larger, the more nodes are likely to get pruned) and it should be tuned for a particular dataset. Noisy datasets, for which there is a higher risk of overfitting, will usually require larger values of m.

Example 3.4.5 The following R code defines a function for calculating the minimum error pruning error estimate for leaves, and demonstrates its application to the decision tree grown for the *weather* data.

```
## calculate the MEP error of a given node, if treated as a leaf
leaf.mep.error <- function(rp, node, data, class, m)
{
  e <- leaf.error(rp, node, data, class)
  n <- node.card(rp, node, data)
  nc <- (1-e)*n
  p <- as.double(pdisc(class)[rp$frame$yval[row.names(rp$frame)==node]])
  1-(nc+m*p)/(n+m)
}


  # MEP error of node 1, if treated as a leaf, for m=0, 2, 5
leaf.mep.error(rptree, 1, weather, weather$play, m=0)
leaf.mep.error(rptree, 1, weather, weather$play, m=2)
leaf.mep.error(rptree, 1, weather, weather$play, m=5)
  # MEP error of node 3, which is actually a leaf, for m=0, 2, 5
leaf.mep.error(rptree, 3, weather, weather$play, m=0)
leaf.mep.error(rptree, 3, weather, weather$play, m=2)
leaf.mep.error(rptree, 1, weather, weather$play, m=5)
```

```
  # MEP error of node 4, if treated as a leaf, for m=0, 2, 5
leaf.mep.error(rptree, 4, weather, weather$play, m=0)
leaf.mep.error(rptree, 4, weather, weather$play, m=2)
leaf.mep.error(rptree, 4, weather, weather$play, m=5)
```

For $m = 0$, we obtain the regular leaf error, and for larger m the amount of pessimistic compensation increases. The following R code shows how the corresponding node error estimate can be calculated through error up-propagation.

```
## calculate the MEP error of the subtree rooted at a given node
node.mep.error <- function(rp, node, data, class, m)
{
  if (rp.leaf(rp, node))
    leaf.mep.error(rp, node, data, class, m)
  else
  {
    el <- node.mep.error(rp, 2*node, data, class, m)
    nl <- node.card(rp, 2*node, data)
    er <- node.mep.error(rp, 2*node+1, data, class, m)
    nr <- node.card(rp, 2*node+1, data)
    weighted.mean(c(el, er), c(nl, nr))
  }
}

  # MEP error of node 1 for m=0, 2, 5
node.mep.error(rptree, 1, weather, weather$play, m=0)
node.mep.error(rptree, 1, weather, weather$play, m=2)
node.mep.error(rptree, 1, weather, weather$play, m=5)
  # MEP error of node 3, which is actually a leaf, for m=0, 2, 5
node.mep.error(rptree, 3, weather, weather$play, m=0)
node.mep.error(rptree, 3, weather, weather$play, m=2)
node.mep.error(rptree, 3, weather, weather$play, m=5)
  # MEP error of node 4 for m=0, 2, 5
node.mep.error(rptree, 4, weather, weather$play, m=0)
node.mep.error(rptree, 4, weather, weather$play, m=2)
node.mep.error(rptree, 4, weather, weather$play, m=5)
```

For $m = 0$, we obtain the regular node error, which is equal to 0 for a tree perfectly fitted to the training set, and for larger m, the amount of pessimistic compensation increases. Finally, the following R code verifies whether there are any candidates for MEP for a few different values of m:

```
  # check which nodes would get pruned under the MEP criterion
  # for m=2
sapply(as.integer(row.names(rptree$frame)),
       function(node)
       {
         !rp.leaf(rptree, node) &&
           leaf.mep.error(rptree, node, weather, weather$play, m=2)<=
             node.mep.error(rptree, node, weather, weather$play, m=2)
       })

  # for m=5
sapply(as.integer(row.names(rptree$frame)),
       function(node)
```

```
{
    !rp.leaf(rptree, node) &&
        leaf.mep.error(rptree, node, weather, weather$play, m=5)<=
            node.mep.error(rptree, node, weather, weather$play, m=5)
})

# for m=10
sapply(as.integer(row.names(rptree$frame)),
    function(node)
    {
        !rp.leaf(rptree, node) &&
            leaf.mep.error(rptree, node, weather, weather$play, m=10)<=
                node.mep.error(rptree, node, weather, weather$play, m=10)
    })
```

As we can see, no node would get pruned for $m = 2$, for $m = 5$ the MEP is satisfied for exactly one node (node 5), and for $m = 10$ it is additionally satisfied for the root node.

3.4.2.4 Cost-complexity pruning

Instead of trying to reliably estimate the expected error on new data using the training set, which is hard to achieve, cost-complexity pruning makes an explicit use of Ockham's razor. If smaller trees should be indeed expected to generalize better, then the original subtree has to be punished for its complexity before its error can be fairly compared with the error of the replacement leave. This yields the following criterion:

$$e_T(\mathbf{l}) \leq e_T(\mathbf{n}) + \alpha \mathbf{C}(\mathbf{n}) \tag{3.26}$$

where the $\mathbf{C}(\mathbf{n})$ represents the complexity of the subtree rooted at $\mathbf{n}$, measured by the number of nodes. Adding the error to the number of nodes is of course as arbitrary as it goes, and there is no universally good way of setting the α coefficient, called the *complexity parameter*, other than tuning. Although, the method is simple and (in a sense) elegant, and given the fact that parameter tuning may be required for some other pruning criteria as well, it may not be a bad choice. The complexity parameter represents the required amount of error reduction per single node. All nodes that do not yield at least this amount of error reduction should be pruned off the tree.

Whereas the inequality presented above describes the essential idea of cost-complexity pruning simply and consistently with the description of other pruning criteria, it does not accurately represent the standard way of applying this approach to pruning, which is somewhat more refined. It consists in identifying, for a given value of α, the smallest pruned tree that minimizes the sum

$$e_T(\mathbf{n}_1) + \alpha \mathbf{C}(\mathbf{n}_1) \tag{3.27}$$

where $\mathbf{n}_1$ is the root node. Such a tree is called the optimally pruned tree with respect to α. It can be shown that for any two values of the complexity parameter α_1 and α_2, the corresponding optimally pruned trees are either identical or can be made identical by pruning off some subtrees of the larger of them (which would be clearly the one corresponding to the smaller value of the complexity parameter). The number of different optimally pruned trees corresponding to all possible values of the complexity parameter is at most equal to the number of nodes in the original, unpruned tree. Each of those trees corresponds to an interval of α values for which it remains optimally pruned. As α is increased above the upper boundary of this interval, one or more bottom-level nodes (i.e., one-level subtrees) will get pruned off. This observation suggests that cost-complexity pruning can be applied in a bottom-up order and, for a fixed value of the complexity parameter, only one-level subtrees (consisting of a node with its descendant leaves) have to be considered (if a node does not get pruned, then its parent node does not have to be considered for pruning, since it would not get pruned either).

An important consequence of the last observation is that the maximum value of the complexity parameter for which a node would be retained (not pruned off) can be determined immediately after its split has been selected, by assuming its descendants would be leaves. It makes it possible to efficiently identify the sequence of the optimally pruned trees and the corresponding complexity parameter intervals after completing the growing process. Finding the right value of α does not therefore require explicit parameter tuning by separately pruning the tree with respect to a number of complexity parameter values and comparing the quality of the obtained pruned trees. Instead, all possible pruned trees can be considered as soon as the original tree has been grown. Selecting one of them still requires a quality measure, which is typically based on the cross-validated misclassification error (i.e., the k-fold cross-validation technique, discussed in Section 7.3.4, is employed internally to estimate the expected effects of pruning). The most obvious approach would be to prune the tree at the complexity level corresponding to the minimum cross-validated error. This may be referred to as *minimum-error cost-complexity pruning*. There may be, however, several optimally pruned trees, corresponding to different complexity parameter intervals, achieving very similar error levels. This would make the choice of the ultimate pruned tree somewhat arbitrary. A reasonable rule of thumb is to estimate the standard deviation of error during cross-validation and then select the smallest tree of those with the error within the distance of one standard deviation from the minimum error. This may be referred to as *one-standard-deviation cost-complexity pruning*. Cost-complexity pruning is the pruning method implemented in the rpart package in R.

The implicit complexity parameter tuning by internal k-fold cross-validation makes cost-complexity pruning computationally costly and nondeterministic. Different cross-validated performance estimates can be obtained on multiple independent runs, resulting in different pruning complexity level selections. A more modern reinterpretation of cost-complexity pruning is based on the minimum encoding principle, where both the error and the complexity are expressed by information-theoretic measures and therefore can be added quite naturally without any complexity parameter used as an "exchange rate" between them. This saves the computations needed to tune the complexity parameter by cross-validation and avoids the associated nondeterminism, but is arbitrary to some extent by assuming the direct relationship between complexity and predictive performance.

3.4.3   Pruning control strategy

The pruning control strategy is responsible for deciding, whenever the pruning criterion is satisfied for more than one node and pruning operator, which operator and for which node should be applied first. The order may have considerable impact on the overall effect of pruning, and different control strategies are likely to produce different final trees of different quality. The control strategy is usually one of the following:

Bottom-up. This considers nodes for pruning starting from the last level and going upward.

Top-down. This considers nodes for pruning starting from the root node and going downward.

Best-first. This considers nodes and operators for pruning in the order implied by the possible resulting improvement indicated by the pruning criterion.

The control strategy can also be some mix of them. The first two strategies only resolve the order of nodes, which is nearly always sufficient (especially given the fact that it is rather uncommon to use more than one pruning operator), but otherwise they would have to fall back to the best-first strategy. The latter exploits the nature of pruning criteria specified by inequalities which, when satisfied, can also be used to measure the expected improvement as the difference between the compared quantities (e.g., error estimates before and after pruning).

A successful pruning algorithm needs a good combination of the pruning criterion and the pruning control strategy. Some pruning criteria tend to work well with one control strategy and poorly with another. When a pruning algorithm is designed, it is rather the pruning criterion that comes first and the control strategy is chosen to best match the criterion. The bottom-up and best-first strategies (pure or in a combined form) appear to be much more frequent than the top-down strategy, which usually turns out too aggressive. In particular, all of the most common pruning criteria presented above are typically coupled with the bottom-up strategy.

3.4.4 Conversion to rule sets

An interesting alternative to decision tree pruning is changing the model representation by converting the tree to a rule set and then pruning in this new representation. A decision tree can be converted to a set of rules in a straightforward way, with exactly one rule corresponding to every path from the root node to a leaf. Such a rule is created by writing down a conjunction of conditions corresponding to all splits on a given path, followed by the class label of the leaf the path leads to. For a path represented by a sequence of splits $t_1, t_2, \ldots, t_k$ and their outcomes $r_1, r_2, \ldots, r_k$ leading to a leaf l with the class label d_l the following rule would be created, presented in a simple logic-like notation:

$$t_1(x) = r_1 \wedge t_2(x) = r_2 \wedge \cdots \wedge t_k(x) = r_k \rightarrow d_l \tag{3.28}$$

3.4.4.1 Pruning rule sets

Pruning rule sets are similar to pruning decision trees with the following important differences:

Different operators. The main pruning operator is *condition removal* which removes a selected single condition from the conjunctive antecedent of a selected rule, and an additional pruning operator is *rule removal* which eliminates a selected rule from the rule set completely.

Different control strategies. The bottom-up and top-down strategies are not applicable to rule sets for obvious reasons, and either the best-first or last-first strategies (with the order of conditions corresponding to the order of splits on the path from which the rule originated) can be applied.

The pruning criteria remain mostly the same, except for those specifically designed for trees and dependent of their hierarchical structure, like MEP, and cost-complexity pruning, although some criteria specifically designed for rule sets can also be employed.

More important than the plain pruning criteria is, however, the proper way of their application. Whatever error estimates (or, more generally, quality measures) we would like to employ to evaluate the effect of the condition removal operator, they can be applied either on a per-rule basis or to the whole rule set. The former means that we only observe how removing a condition from a rule affects the estimated quality of this single rule, and in the latter case we are concerned with the resulting change of the quality of the whole rule set. Taking the reduced error pruning as an example, we might compare the before and after pruning error of the rule or of the whole rule set.

It may not be quite self-evident why the above distinction actually matters. When we alter one rule at a time, without touching any other rules, the resulting change of error for the whole rule set might appear to be necessarily the same as for this single rule. Some more thought is needed to see that it indeed may be the same, but not quite necessarily. The original rule set obtained from a decision tree has a nice disjoin property, which means that there is exactly one rule matching any possible instance from the whole domain (as long as all split outcomes used in rule conditions can be unambiguously determined). This is because every instance always reaches a single leaf of the original tree (under the same assumption). When a condition of a rule gets removed, there may be more possible instances satisfying its conjunctive conditions, possibly including some instances already covered by other rules. It opens the possibility of some instances being covered by several *overlapping* rules with the same or different class labels, which means that the observed change of error for a single pruned rule might not be the same as the observed change of error for the whole rule set. Since the ultimate objective of pruning is to improve the generalization capability of the rule set as a whole, it should be used to evaluate the effects of condition removals for single rules.

The possibility of overlapping rules altered by pruning implies the risk of conflicts during rule set application, if two or more rules matching the same instance do not predict the same unique class label. This requires some *conflict resolution* techniques be employed to choose the most appropriate class label. Discussing them extensively is beyond the scope of this chapter, but one common choice is a weighted voting scheme, with rule weights based on the number of covered training instances.

Note that in decision tree pruning when we consider replacing a node with a leaf, they both have the same sets of corresponding (pruning or training) instances, used for calculating errors or other quality estimates. It is not the case for rule pruning, when the set of instances covered by a rule will likely increase after condition removal and only because of this its quality may change. Luckily, it is easy to see that whenever a single rule appears to be improved due to pruning, the whole rule set will also be usually improved. This is because the rule, to appear better, must correctly classify most of the additional instances it started to cover after pruning, which usually prevents it from degrading the rule set's quality (under a reasonable conflict resolution technique). Only the reverse is not always true, i.e., even if a single rule appears to get worse after pruning, the whole set does not necessarily get worse. This means that the simple approach of judging the effects of condition removal in a single rule based only on the change of the error of this rule is also acceptable and should be safe in most cases.

3.4.4.2   Flexibility of rule set pruning

Whenever a node with its subtree gets pruned in a decision tree and replaced by a leaf, all paths going through the node are cut and terminated in the leaf. To achieve an equivalent effect in a rule-set representation one would have to replace all rules with some common prefix of their condition conjunctions (corresponding to the path from the root to the pruned node) and different remainders (corresponding to the paths from the pruned node to leaves) with a single rule, with the conjunction of conditions cut to this common prefix. It appears quite a radical operation for a rule set, showing how "gentle" the condition removal operator is in comparison. This "gentleness" of pruning is the major advantage of conversion to rule sets, as it allows one to perform much more fine-grained model modifications, possibly leading to better generalization.

Not only the operation of removing a single condition from a single rule is much more delicate than cutting off a complete subtree, but it is also capable of detecting the utility of conditions or the lack thereof depending on the context provided by other conditions. Even the condition corresponding to the split from the root node of the original decision tree, although it will be usually perfectly useful in most rules, may turn out to be unnecessary or harmful in some rules and get removed from them. In a decision tree every node either stays in place or goes away from all paths it belongs to, although – even assuming perfect split selection – it may not be necessary for all of them. By replacing a hierarchical tree structure by a flat rule-set representation, we release every node from its fixed position in the hierarchy and make it possible to independently either sustain or drop any condition in any rule. Again, this possibility to evaluate the utility of rule conditions in a particular context provided by other conditions allows one to perform much more fine-grained model modifications, possibly leading to better generalization.

It is not uncommon to give yet another argument supporting the conversion of decision trees to rule sets: the human readability of models. This argument should not be taken without some skepticism, though. While both decision trees and rule sets are symbolic representations of classification models and are thus undoubtedly human readable, it is hard to give unquestionable reasons to find rule sets superior to decision trees in that respect. This may be a matter of subjective personal preference rather than any objective advantages. Actually, it may be even easier to argue about some disadvantages of the rule-set representation. While rule sets might indeed appear more readable on a very superficial "syntactic" level, their correct interpretation is often more difficult, given the possible overlapping of rules (i.e., multiple rules covering the same instances). Whereas appropriate conflict-resolution strategies can successfully deal with this issue during model application, they hardly help a human comprehend the model and explain its predictions.

3.5   Prediction

Applying a decision tree model to an arbitrary dataset from the same domain which the training set used during modeling came from makes it possible to generate predicted class labels. A decision tree which stores class distribution information for its leaves can also be used as a probabilistic classifier, i.e., to predict class probabilities.

3.5.1 Class label prediction

The application of a previously grown and possibly pruned decision tree to generate class label predictions for a dataset boils down to identifying leaves where particular instances from the dataset land when propagated through the tree. This requires for each instance sequentially applying splits and descending along branches corresponding to their outcomes, starting from the root node, until a leaf is reached, which provides the predicted class label for the instance. Clearly, the split application operation previously discussed for decision tree growing is used here multiple times, until all instances arrive at the corresponding leaves.

Example 3.5.1 The R code presented below implements the decision tree prediction operation, assuming the tree representation as created in the growing examples. Notice that the internal recursive `descend` function closely follows the pattern of the previously presented `split.apply` function, and – as the latter – supports both equality-based splits for discrete attributes and inequality-based splits for continuous attributes. Predictions are then generated for the *weather* data, assuming that the previously grown decision tree is stored in the `tree` variable, and for the *weatherc* data, assuming that the previously grown decision tree is stored in the `treec` variable, as shown in Example 3.3.8.

```
## decision tree prediction
predict.dectree <- function(tree, data)
{
  descend <- function(n)
  {
    if (!is.na(tree$attribute[tree$node==n]))  # unless reached a leaf
    {
      av <- data[[tree$attribute[tree$node==n]]]
      cond <- !is.na(av) & (if (is.numeric(av))
                                av<=tree$value[tree$node==n]
                            else
                                av==tree$value[tree$node==n])

      nodemap[nodemap==n & cond] <<- 2*n
      nodemap[nodemap==n & !cond] <<- 2*n+1
      descend(2*n)
      descend(2*n+1)
    }
  }

  nodemap <- rep(1, nrow(data))
  descend(1)
  tree$class[match(nodemap, tree$node)]
}

  # decision tree prediction for the weather data
predict(tree, weather)
  # decision tree prediction for the weatherc data
predict(treec, weatherc)
```

3.5.2 Class probability prediction

As described above, class label assignment is usually the straightforward maximization of the node's or the leaf's class probability estimate, with the only exception being the case of nonuniform misclassification costs, discussed in Chapter 6. Sometimes it may be more convenient, however, not to assign class labels during tree growing (or not use them later, even if assigned), but to postpone the class label assignment process till the prediction time. The decision tree as a probabilistic classification model would output the estimated class probability distribution for each instance it is applied to. When an instance x to be classified reaches a single leaf $\mathbf{l}$, the tree's output can be obtained as

$$P(d|x) = P(d|\mathbf{l}) \qquad (3.29)$$

which gives the estimated probability of instance x belonging to class d, for all $d \in C$. A separate decision-making mechanism may ultimately use the probability distribution to classify the instance to whatever class appears to be most appropriate. This gives us the flexibility to change the misclassification costs anytime without altering the decision tree. It also transforms a decision tree model into a scoring classifier, which can be analyzed and optimized via the receiver operating characteristic curve, as presented in Section 7.2.5.

3.6   Weighted instances

Notice that the use of data instances for decision tree growing and pruning is limited to counting. Indeed, all operations performed in both these phases of creating decision tree models that directly use the data only need the sizes of subsets of instances satisfying appropriate conditions. These conditions, in general, can be conjunctions including:

- the selection of instances associated with a particular node,

- the selection of instances with a particular split outcome,

- the selection of instances of a particular class.

Specifically, when growing a tree from a training set T, to apply stop and split selection criteria at node $\mathbf{n}$, one just needs $|T_{\mathbf{n}}^d|$ and $|T_{\mathbf{n},t=r}^d|$ for all classes d, for all splits t and their outcomes r. Similarly, when pruning a tree based on pruning set R, to apply pruning criteria to node n, it is sufficient to determine $R_{\mathbf{n}}^d$ for all classes d. It makes it possible for decision tree induction algorithms to use *weighted* instances, i.e., to act as weight-sensitive algorithms.

Achieving the weight sensitivity is nearly effortless, boiling down to appropriately redefining counts as weight sums. For any subset of instances the size of which is normally used, it should be replaced by the sum of weights for the instances in this set. In particular, for the counts mentioned above, one should assume

$$|T_{\mathbf{n},t=r}^d| = \sum_{x \in T_{\mathbf{n},t=r}^d} w_x \qquad (3.30)$$

$$|T_{\mathbf{n}}^d| = \sum_{x \in T_{\mathbf{n}}^d} w_x \qquad (3.31)$$

$$|R_{\mathbf{n}}^d| = \sum_{x \in R_{\mathbf{n}}^d} w_x \qquad (3.32)$$

where w_x denotes the weight assigned to instance x.

The most important benefit resulting from the capability of using weighted instances is the possibility of easily incorporating nonuniform misclassification costs. The instance weighting technique is directly applicable to decision trees and can be used to create cost-sensitive decision tree models whenever misclassification costs are specified on a per-class basis (i.e., via a cost vector). This possibility is extensively discussed and demonstrated in Section 6.3.1.

3.7 Missing value handling

The problem of missing attribute values is common for all data mining algorithms. Many of them can benefit from some general-purpose imputation techniques applied in a data preprocessing phase, as discussed in Section 17.3.4. Some algorithms may just degrade gracefully when applied to incomplete data and some others may employ special techniques to handle missing values during model building and application. Decision trees fall into this last category, although such "internal" missing value handling induces considerable extra computational cost and the preprocessing approach may be sometimes a viable alternative.

There are two major approaches to internal missing value handling in decision tree induction and prediction:

Fractional instances. Whenever a split on an attribute with missing value is considered or applied, each incomplete instance is virtually replaced by several instances corresponding to all possible split outcomes, with a fractional "copy count,"

Surrogate splits. A number additional splits are stored for each node, apart from the ordinary main split, and used to dispatch instances for which the outcome of the main split cannot be determined due to missing attribute values.

Some more details about these two approaches are provided below.

3.7.1 Fractional instances

The idea of fractional instances is to always consider all possible split outcomes for an instance with a missing value of the split attribute and assign them appropriate weights or probabilities, based on the observed distribution of known outcomes. This is equivalent to replacing the original instance with several virtual instances, one per split outcome, each having a fractional number of "copies," equal to the corresponding outcome probability.

Consider a node $\mathbf{n}$ with a split $t : X \to R_t$ which has to be evaluated or applied. If the outcome of split t cannot be determined due to a missing attribute value, each possible outcome $r \in R_t$ is considered with the probability

$$P(t = r|\mathbf{n}) = \frac{|T_{\mathbf{n},t=r}|}{|T_{\mathbf{n}}| - |T_{\mathbf{n},t=?}|} \tag{3.33}$$

where $T_{\mathbf{n},t=?}$ denotes the subset of $T_{\mathbf{n}}$ for which the outcome of test t is unknown due to missing values of the tested attribute. Subtracting the number of instances in this set in the denominator above, it is necessary to obtain split outcome probabilities based on instances for which the outcome is known only. Instead of assigning an instance with a missing value to a single subset, corresponding to a single split outcome, its fractions are then assigned to the subsets corresponding to all possible outcomes. Such virtual fractional instances can be further processed in exactly the same way as real regular instances with just one exception: Whenever the size of a subset of instances has to be determined, each regular instance counts as 1 whereas each fractional instance counts as the corresponding probability weight. This is the same redefinition of subset size as weight sum that was presented in Section 3.6 when discussing the use of weighted instances.

3.7.1.1 Growing with fractional instances

During tree growing fractional instances make it possible to evaluate and use splits on attributes with missing values. Of all the major growing steps discussed above, only split selection (or, more precisely, split evaluation) and split application need to be substantially modified.

Split evaluation with fractional instances   Recall that split evaluation by a selected impurity measure is based on the class probability distribution for each possible split outcome, i.e., $P_{T_{\mathbf{n},t=r}}(d)$ for all $d \in C$ and $r \in R_t$. If the split outcome is unknown for some attribute values, these should be estimated as follows:

$$P_{T_{\mathbf{n},t=r}}(d) = \frac{|T^d_{\mathbf{n},t=r}| + P(t = r|\mathbf{n})|T^d_{\mathbf{n},t=?}|}{|T_{\mathbf{n},t=r}| + P(t = r|\mathbf{n})|T_{\mathbf{n},t=?}|} \quad (3.34)$$

This is equivalent to assigning fractions of instances with missing values of the tested attribute to all possible split outcomes, with "copy counts," equal to the corresponding outcome probabilities. Notice that no explicit assignment of fractional "copy counts" to individual instances is actually needed for split evaluation. Instead, a fraction of the number of instances of class d with an unknown split outcome is added to the number of instances of class d and split outcome r.

Other than this modified way of class probability estimation for split outcomes, no other changes are needed in the split evaluation process. In particular, the calculation of the entropy, Gini index, or any other impurity measure goes exactly the usual way, since they are all based on class probability distributions.

Split application with fractional instances   Unlike for split evaluation, during split application individual instances have their fractional "copy counts" assigned, also referred to as weights. This is necessary to dispatch an instance with a missing value of the split attribute along all branches, corresponding to all split outcomes. For a node $\mathbf{n}$ with split $t : X \to R_t$ and instance x for which $t(x)$ is unknown the "copy count" or weight of x at descendant node $\mathbf{n}_r$, corresponding to outcome r of split t, will be calculated as

$$w_{x,\mathbf{n}_r} = P(t = r|\mathbf{n})w_{x,\mathbf{n}} \quad (3.35)$$

where $w_{x,\mathbf{n}}$ is the "copy count" of the instance x at the node $\mathbf{n}$. For the root node the "copy counts" of all instances are set to 1 or user-specified instance weights. The latter makes it possible to use fractional instances for missing value handling with weight-sensitive decision trees.

It is not unlikely that before reaching a leaf a fractional instance will arrive at another node where a split on the same or another attribute with a missing value will have to be evaluated or applied. This will lead to creating "fractions of fractions": the already fractional instances will be replaced by new fractional instances, with probability weights obtained by multiplying the previous weight by split outcome probabilities. In effect, when an instance arrives to a leaf, it may have gone through several "fractionization" operations and its probability weight may be the product of the corresponding split outcome probabilities.

As mentioned above, the direct consequence of "copy counts" being assigned to instances at nodes is that all instance counts need to be calculated as "copy count" sums. In other words, subset size needs to be reinterpreted as "copy count" sum over subset members, i.e.

$$|S| = \sum_{x \in S} w_{x,\mathbf{n}} \tag{3.36}$$

for any subset $S \subseteq T_{\mathbf{n}}$.

Example 3.7.1 The previously presented decision tree growing implementation took a very primitive approach to missing value handling, considering the split condition unsatisfied for instances with missing values (i.e., always assigning such instances to the branch correspond- ing to the false split outcome). The following R code contains a modified implementation that includes missing value support by the technique of fractional instances. Notice the following major changes:

- the mapping of instances to nodes is no longer represented by a simple vector, but by a matrix containing the `instance`, `node`, and `weight` columns, to make it possible to have (fractions of) a single instance assigned to multiple nodes,

- class distribution calculation takes instance weights into account and is per- formed using the `weighted.pdisc` function,

> Ex. 2.4.25
> dmr.stats

- the `split.eval` and `split.apply` functions perform instance fractionization.

The `grow.dectree.frac` function is applied to modified versions of the *weather* and *weatherc* datasets, with some attribute values removed.

```
## a simple decision tree growing implementation
## with missing value support using fractional instances
grow.dectree.frac <- function(formula, data,
                        imp=entropy.p, maxprob=0.999, minsplit=2, maxdepth=8)
{
  nmn <- function(n) { nodemap[,"node"]==n }             # nodemap entries for node n
  inn <- function(n)
  { nodemap[nodemap[,"node"]==n,"instance"] }            # instances at node n
  wgn <- function(n) { nodemap[nodemap[,"node"]==n,"weight"] }   # weights at node n

  init <- function()
  {
    clabs <<- factor(levels(data[[class]]),
                  levels=levels(data[[class]]))          # class labels
    tree <<- data.frame(node=1, attribute=NA, value=NA, class=NA, count=NA,
                    `names<-`(rep(list(NA), length(clabs)),
                         paste("p", clabs, sep=".")))
```

```
    cprobs <<- (ncol(tree)-length(clabs)+1):ncol(tree)   # class probability columns
    nodemap <<- cbind(instance=1:nrow(data), node=rep(1, nrow(data)),
                      weight=rep(1, nrow(data)))
    n <<- 1
}

next.node <- function(n)
{
  if (any(opn <- tree$node>n))
    min(tree$node[opn])
  else Inf
}

class.distribution <- function(n)
{
  tree[tree$node==n,"count"] <<- sum(wgn(n))
  tree[tree$node==n,cprobs] <<- weighted.pdisc(data[inn(n),class], w=wgn(n))
}

class.label <- function(n)
{
  tree$class[tree$node==n] <<- which.max(tree[tree$node==n,cprobs])
}

stop.criteria <- function(n)
{
  n>=2^maxdepth || tree[tree$node==n,"count"]<minsplit ||
    max(tree[tree$node==n,cprobs])>maxprob
}

split.eval <- function(av, sv, cl, w)
{
  cond <- if (is.numeric(av)) av<=as.numeric(sv) else av==sv
  cond1 <- !is.na(av) & cond    # true split outcome
  cond0 <- !is.na(av) & !cond   # false split outcome

  pd1 <- weighted.pdisc(cl[cond1], w=w[cond1])
  n1 <- sum(w[cond1])
  pd0 <- weighted.pdisc(cl[cond0], w=w[cond0])
  n0 <- sum(w[cond0])
  pdm <- weighted.pdisc(cl[is.na(av)], w=w[is.na(av)])
  nm <- sum(w[is.na(av)])

  if (nm>0)
  {
    p1 <- if (n1+n0>0) n1/(n1+n0) else 0.5
    p0 <- 1-p1
    pd1 <- (n1*pd1 + p1*nm*pdm)/(n1+p1*nm)
    n1 <- n1 + p1*nm
    pd0 <- (n0*pd0 + p0*nm*pdm)/(n0+p0*nm)
    n0 <- n0 + nm*p0
  }

  if (n1>0 && n0>0)
    weighted.impurity(pd1, n1, pd0, n0, imp)
  else
    Inf
}

split.select <- function(n)
{
```

```
    splits <- data.frame()
    for (attribute in attributes)
    {
      uav <- sort(unique(data[inn(n),attribute]))
      if (length(uav)>1)
        splits <- rbind(splits,
                       data.frame(attribute=attribute,
                                 value=if (is.numeric(uav))
                                         midbrk(uav)
                                     else as.character(uav),
                                 stringsAsFactors=FALSE))
    }

    if (nrow(splits)>0)
      splits$eval <- sapply(1:nrow(splits),
                            function(s)
                            split.eval(data[inn(n),splits$attribute[s]],
                                      splits$value[s],
                                      data[inn(n),class], wgn(n)))
    if ((best.eval <- min(splits$eval))<Inf)
      tree[tree$node==n,2:3] <<- splits[which.min(splits$eval),1:2]
    return(best.eval)
}

split.apply <- function(n)
{
  tree <<- rbind(tree,
                 data.frame(node=(2*n):(2*n+1),
                           attribute=NA, value=NA, class=NA, count=NA,
                           `names<-`(rep(list(NA), length(clabs)),
                                     paste("p", clabs, sep="."))))

  av <- data[nodemap[,"instance"],tree$attribute[tree$node==n]]
  cond <- if (is.numeric(av)) av<=as.numeric(tree$value[tree$node==n])
          else av==tree$value[tree$node==n]
  cond1 <- !is.na(av) & cond   # true split outcome
  cond0 <- !is.na(av) & !cond  # false split outcome

  n1 <- sum(nodemap[nmn(n) & cond1,"weight"])
  n0 <- sum(nodemap[nmn(n) & cond0,"weight"])
  nm <- sum(nodemap[nmn(n) & is.na(av),"weight"])

  nodemap[nmn(n) & cond1,"node"] <<- 2*n
  nodemap[nmn(n) & cond0,"node"] <<- 2*n+1

  if (nm>0)
  {
    p1 <- if (n1+n0>0) n1/(n1+n0) else 0.5
    p0 <- 1-p1
    newnn <- nodemap[nmn(n) & is.na(av),,drop=FALSE]
    nodemap[nmn(n) & is.na(av),"weight"] <<-
      p1*nodemap[nmn(n) & is.na(av),"weight"]
    nodemap[nmn(n) & is.na(av),"node"] <<- 2*n
    newnn[,"weight"] <- p0*newnn[,"weight"]
    newnn[,"node"] <- 2*n+1
    nodemap <<- rbind(nodemap, newnn)
  }
}

tree <- cprobs <- nodemap <- n <- NULL
clabs <- cprobs <- NULL
```

```
    class <- y.var(formula)
    attributes <- x.vars(formula, data)

    init()
    while (is.finite(n))
    {
      class.distribution(n)
      class.label(n)
      if (!stop.criteria(n))
        if (split.select(n)>Inf)
          split.apply(n)
      n <- next.node(n)
    }
    tree$class <- clabs[tree$class]
    'class<-'(tree, "dectree.frac")
}

## convert a dectree.frac object to a data frame
as.data.frame.dectree.frac <- function(x, row.names=NULL, optional=FALSE, ...)
{ as.data.frame(unclass(x), row.names=row.names, optional=optional) }

  # grow a decision tree for the weather data with missing attribute values
weatherm <- weather
weatherm$outlook[1] <- NA
weatherm$humidity[1:2] <- NA
treem <- grow.dectree.frac(play~., weatherm)

  # grow a decision tree for the weatherc data with missing attribute values
weathercm <- weather
weathercm$temperature[1:2] <- NA
weathercm$humidity[1] <- NA
treecm <- grow.dectree.frac(play~., weathercm)

  # data frame conversion
as.data.frame(treem)
as.data.frame(treecm)
```

3.7.1.2 Prediction with fractional instances

When a decision tree is applied to an instance with missing attribute values, things get only a little more complicated. Split application is performed as during growing until all instances reach leaves. It is important to underline that split outcome probabilities required for instance fractionization are the same as those estimated during decision tree growing, based on the training set.

Each of fractional instances replacing an original instance with missing values will clearly end up in a different leaf, with a possibly different class label. To make the final classification decision, a weighted voting mechanism may be used, with weights assigned based on instance fractions. More exactly, we can interpret the fraction of x arriving to l (i.e., the probability weight of the corresponding fractional instance) as the probability of leaf l for instance x, taking $P(l|x) = w_{x,l}$, and classify x to the class with the largest sum of leaf probabilities:

$$h(x) = \arg \max_{d \in C} \sum_{l} P(l|x) \mathbb{I}_{d_l=d} \tag{3.37}$$

where the summation runs over all decision tree leaves and the $\mathbb{I}_{condition}$ notation is used to denote an indicator function that yields 1 when the *condition* is satisfied and 0 otherwise.

A more refined approach that should be usually preferred uses class distribution information for decision tree leaves, as during probabilistic classification. Class label voting is replaced by probability voting (or, actually, averaging), which is basically the application of the law of total probability:

$$P(d|x) = \sum_{l} P(l|x)P(d|l) \tag{3.38}$$

Again, the summation runs over all leaves. This approach combines class probabilities stored in decision tree leaves to obtain predicted class probabilities for the instance being classified.

Example 3.7.2 The R code presented below implements decision tree prediction with fractional instances. The internal `descend` function is a minor modification of the `split.apply` function used for growing. The probability voting approach is used to calculate class probabilities and output the highest probability classes.

```
## decision tree prediction
## with missing value support using fractional instances
predict.dectree.frac <- function(tree, data)
{
  nmn <- function(n) { nodemap[,"node"]==n }  # nodemap entries for node n

  descend <- function(n)
  {
    if (!is.na(tree$attribute[tree$node==n]))  # unless reached a leaf
    {
      av <- data[nodemap[,"instance"],tree$attribute[tree$node==n]]
      cond <- if (is.numeric(av)) av<=as.numeric(tree$value[tree$node==n])
              else av==tree$value[tree$node==n]
      cond1 <- !is.na(av) & cond    # true split outcome
      cond0 <- !is.na(av) & !cond   # false split outcome

      nodemap[nmn(n) & cond1, "node"] <<- 2*n
      nodemap[nmn(n) & cond0, "node"] <<- 2*n+1

      if (sum(nodemap[nmn(n) & is.na(av), "weight"])>0)
      {
        n1 <- tree$count[tree$node==2*n]
        n0 <- tree$count[tree$node==2*n+1]
        p1 <- if (n1+n0>0) n1/(n1+n0) else 0.5
        p0 <- 1-p1

        newnn <- nodemap[nmn(n) & is.na(av),,drop=FALSE]
        nodemap[nmn(n) & is.na(av),"weight"] <<-
          p1*nodemap[nmn(n) & is.na(av),"weight"]
        nodemap[nmn(n) & is.na(av), "node"] <<- 2*n
        newnn[,"weight"] <- p0*newnn[,"weight"]
        newnn[,"node"] <- 2*n+1
        nodemap <<- rbind(nodemap, newnn)
      }

      descend(2*n)
      descend(2*n+1)
    }
  }
```

```
  nodemap <- cbind(instance=1:nrow(data)*, node=rep(1, nrow(data)),
                      weight=rep(1, nrow(data)))
  descend(1)

  clabs <- factor(levels(tree$class), levels=levels(tree$class))
  votes <- merge(nodemap, as.data.frame(tree)[,c("node", "class",
                                          paste("p", clabs, sep="."))])
  cprobs <- (ncol(votes)-length(clabs)+1):ncol(votes)
  clabs[by(votes, votes$instance,
           function(v) which.max(colSums(v$weight*v[,cprobs])))]
}

  # decision tree prediction for the weather data with missing attribute values
predict(treem, weatherm)

  # decision tree prediction for the weatherc data with missing attribute values
predict(treecm, weathercm)
```

3.7.2 Surrogate splits

This approach to handling missing values is based on multiplicating splits rather than frac-
tionizing instances. If the outcome of a split cannot be determined for an instance, the latter
is leaved unchanged in its original form, but a surrogate split (based on another attribute) is
used instead of the former. Surrogate splits have to be selected during decision tree growing
and stored at nodes, just as primary splits. It is only the split selection criterion that differs. A good surrogate split should give a partitioning of the training instances corresponding
to the node as similar as possible to that produced by the primary split, but using another
attribute. This means that as many instances as possible should go to the same branches as
by the primary split. Whenever the outcome of the primary split cannot be determined, a
good surrogate split will most likely be dispatched to the same branch as it would be by the
primary split.

A necessary condition for a split to be considered as a surrogate split for a primary split
selected before is to have the same number of possible outcomes. The technique of surrogate
splits is usually used with binary splits, where this is not a problem. A natural evaluation function of a binary split $t' : X \rightarrow \{0, 1\}$ as a candidate surrogate for a binary split $t : X \rightarrow \{0, 1\}$
at node $\mathbf{n}$ can be defined as

$$\delta_{t,t'} = \frac{|T_{\mathbf{n},t=0} \cap T_{\mathbf{n},t'=0}| + |T_{\mathbf{n},t=1} \cap T_{\mathbf{n},t'=1}|}{|T_{\mathbf{n}}|} \tag{3.39}$$

One should also make sure that for each binary split t_1 a "complementary" split t_2 is available
such that $t_2(x) = 1 - t_1(x)$ for all $x \in X$. In other words, for any possible split condition we
should have two splits, one where the "true" logical value corresponds to the first outcome
and the other where the "false" logical value corresponds to the first outcome. This is to make
sure that the best possible surrogate split is available (otherwise we may not be able to find
good surrogate splits even with very strongly related attributes).

Typically, an ordered list of several surrogate splits are selected for a single primary split,
to minimize the chance that the values of the attributes tested by the surrogate splits will

also be missing. If no surrogate split can dispatch an instance with an excessive number of missing values, it may be either stopped at the current node or dispatched to the most frequent branch.

Surrogate splits solve the problem of determining the most outgoing branch if the primary split cannot be applied. They do not help at all in split selection (actually, they add quite a bit of extra split selection work), which is performed in the usual way based on the instances with known attribute values only. Since surrogate splits are only inherently imperfect approximations of the primary splits, it is reasonable to avoid splits on attributes with many missing values, which is easily accomplished by introducing a penalty coefficient to the split evaluation function.

3.8   Conclusion

Decision trees have been intensively investigated and applied for nearly 30 years and they still remain both an interesting research topic and an extremely useful practical tool. Although most of the key ideas, covered by this chapter, come from quite early algorithms (CART, ID3, and C4.5) and are quite simple, they are apparently brilliant enough to be easily applicable to many diverse domains and often produce about the most accurate classifiers one can create from data, which are also human readable. They can still be compared favorably with some not-so-readable classifiers created by some more recent not-so-easy-to-use approaches. Their human readability can be even exploited to permit human participation in the model creation process, when necessary or desirable, by making all or selected growing and pruning steps subject to human verification and possibly modification (e.g., an apparently slightly worse split could be selected if it has a better interpretation based on the available domain knowledge or the analyst is more confident about its true predictive utility). When the human readability can be sacrificed, decision trees can be used as components of model ensembles to reach even higher accuracy levels, as we will see in Chapter 15. This unquestionable success of decision trees is one good reason they deserve special attention.

Another reason is the high instructive value of decision trees which allow one to encounter, understand, and appreciate some major issues common for the whole data mining domain, such as noise, overfitting, Ockham's razor, or missing value handling. Studying decision tree algorithms and practicing their application is a good way to learn what is the essence of data mining, and then move forward to other algorithms in a much more comforted way.

3.9   Further readings

Decision tree growing and pruning algorithms are discussed to some extent by most data mining and machine learning books (e.g., Cios *et al.* 2007; Han *et al.* 2011; Mitchell 1997; Tan *et al.* 2013; Witten *et al.* 2011). Similar to this chapter, these presentations are most often based on the two best known specific decision tree induction algorithms, *CART* (Breiman *et al.* 1984) and C4.5 (Quinlan 1993). The former served as prototype of the algorithm implemented by the `rpart` package in R (Therneau and Atkinson 1997), extensively used throughout this book. The latter was preceded by the ID3 algorithm (Quinlan 1986), which marked the beginning of the rapidly growing interest in decision trees and their applications. The idea of decision tree model representation is actually much older and dates back to the early work of Hunt (1962) in the field of machine learning. There were also some pioneering decision tree studies in statistics (Friedman 1977; Henrichon and Fu 1969; Meisel and Michalopoulos 1973).

Other noteworthy specific decision tree algorithms include *CHAID* (Kass 1980) and *QUEST* (Loh and Shih 1997). Murthy (1997) provides a comprehensive survey of the decision tree literature and Rokach and Maimon (2007) give an extensive and relatively contemporary account of decision tree algorithms and their applications. The number of research articles presenting various enhancements to the basic tree growing and pruning schemes is enormous and only a tiny subset of them can be mentioned here.

The CART, C4.5, and most other decision tree algorithms share a number of common design principles. These include using univariate splits (based on single attributes) and assuming the simultaneous availability of the complete training set. Brodley and Utgoff (1995) depart from the former, by presenting a tree growing algorithm with multivariate splits, and Utgoff (1989) departs from the latter, by presenting an incremental tree growing algorithm that can modify the tree structure based on sequentially arriving data portions. Incremental decision tree induction was later revisited by Domingos and Hulten (2000), who presented an algorithm that asymptotically approaches the result of standard batch algorithms using constant memory and time per instance.

Univariate and nonincremental decision tree growing algorithms may still differ in several details. One of them is split selection and several different choices in this regard were reviewed and experimentally investigated by Mingers (1989b) and Buntine and Niblett (1992). It is decision tree pruning, though, that creates the most space for diversity, has attracted the most research attention, and is likely to have the biggest model quality impact. Reduced error pruning and a basic version of pessimistic pruning were introduced by Quinlan (1987). The MEP algorithm was proposed by Niblett and Bratko (1986) and enhanced by Cestnik and Bratko (1991). Cost-complexity pruning was presented as a part of the *CART* algorithm by Breiman *et al.* (1984). These and other tree pruning techniques are surveyed by Breslow and Aha (1997) and experimentally compared by Mingers (1989a) and Esposito *et al.* (1997). Extending the idea of cost-complexity pruning, Bohanec and Bratko (1994) proposed a more refined dynamic programming-based pruning algorithm that identifies the smallest sufficiently accurate tree by generating the sequence of most accurate trees within given size constraints. Decision tree pruning is also a natural application area for minimum description length inference, not covered in this book (Kononenko 1998; Mehta *et al.* 1995; Quinlan 1989).

Preference for simplicity during split selection and pruning are two ways of introducing the idea of Occam's razor (Blumer *et al.* 1987) to decision tree induction. As argued by Webb (1996), simpler trees do not have to be universally superior to more complex ones. The technique of decision tree grafting may sometimes improve true performance by increasing complexity, i.e., adding nodes (Webb 1997).

The two major techniques for missing value handling in decision trees, surrogate splits, and fractional instances, come from the *CART* (Breiman *et al.* 1984) and *C4.5* (Quinlan 1993) algorithms, respectively. Algorithm-independent missing value imputation techniques (Little and Rubin 2002; Liu *et al.* 1997; Pyle 1999) may be employed instead if these are unavailable in a particular implementation used or turn out too computationally demanding.

With few minor exceptions, computational efficiency issues for data mining algorithms are not addressed by this book. They may be crucial, however, for successful applications to very large datasets, not necessarily fitting into main memory. This motivates tree growing algorithm variations specifically designed to handle large data (e.g., Gehrke *et al.* 2000; Mehta *et al.* 1996; Shafer *et al.* 1996).

References

Blumer A, Ehrenfeucht A, Haussler D and Warmuth MK 1987 Occam's razor. *Information Processing Letters* **24**, 377–380.

Bohanec M and Bratko I 1994 Trading accuracy for simplicity in decision trees. *Machine Learning* **15**, 223–250.

Breiman L, Friedman JH, Olshen RA and Stone CJ 1984 *Classification and Regression Trees*. Chapman and Hall.

Breslow LA and Aha DW 1997 Simplifying decision trees: A survey. *The Knowledge Engineering Review* **12**, 1–40.

Brodley CE and Utgoff PE 1995 Multivariate decision trees. *Machine Learning* **19**, 45–77.

Buntine WL and Niblett T 1992 A further comparison of splitting rules for decision-tree induction. *Machine Learning* **8**, 75–86.

Cestnik B and Bratko I 1991 On estimating probabilities in tree pruning *Proceedings of the European Working Session on Learning (EWSL-91)*. Springer.

Cios KJ, Pedrycz W, Swiniarski RW and Kurgan L 2007 *Data Mining: A Knowledge Discovery Approach*. Springer.

Domingos P and Hulten G 2000 Mining high-speed data streams *Proceedings of the Sixth ACM SIGKDD International Conference on Knowledge Discovery and Data Mining*. ACM Press.

Esposito F, Malerba D and Semeraro G 1997 A comparative analysis of methods for pruning decision trees. *IEEE Transactions on Pattern Analysis and Machine Intelligence* **19**, 476–491.

Friedman JH 1977 A recursive partitioning decision rule for non-parametric classification. *IEEE Transactions on Computers* **26**, 404–408.

Gehrke J, Ramakrishnan R and Ganti V 2000 RainForest – a framework for fast decision tree construction of large datasets. *Data Mining and Knowledge Discovery* **4**, 127–162.

Han J, Kamber M and Pei J 2011 *Data Mining: Concepts and Techniques* 3rd edn. Morgan Kaufmann.

Henrichon EG and Fu KS 1969 A nonparametric partitioning procedure for pattern classification. *IEEE Transactions on Computers* **18**, 614–624.

Hunt EB 1962 *Concept Learning: An Information Processing Problem*. Wiley.

Kass GV 1980 An exploratory technique for investigating large quantities of categorical data. *Applied Statistics* **29**, 119–127.

Kononenko I 1998 The minimum description length-based decision tree pruning *Proceedings of the Fifth Pacific Rim Internation Conference on Artificial Intelligence (PRICAI-98)*. Springer.

Little RJA and Rubin DB 2002 *Statistical Analysis with Missing Data* 2nd edn. Wiley.

Liu WZ, White AP, Thompson SG and Bramer MA 1997 Techniques for dealing with missing values in classification *Proceedings of the Second International Symposium on Intelligent Data Analysis (IDA-97)*. Springer.

Loh WY and Shih YS 1997 Split selection methods for classification trees. *Statistica Sinica* **7**, 815–840.

Mehta M, Agrawal R and Rissanen J 1996 SLIQ: A fast scalable classifier for data mining *Proceedings of the Fifth Intl Conference on Extending Database Technology (EDBT-96)*. Springer.

Mehta M, Rissanen J and Agrawal R 1995 MDL-based decision tree pruning *Proceedings of the First International Conference on Knowledge Discovery and Data Mining (KDD-95)*. AAAI Press.

Meisel WS and Michalopoulos DA 1973 A partitioning algorithm with application in pattern classification and the optimization of decision trees. *IEEE Transactions on Computers* **22**, 93–103.

Mingers J 1989a An empirical comparison of pruning methods for decision tree induction. *Machine Learning* **4**, 227–243.

Mingers J 1989b An empirical comparison of selection measures for decision tree induction. *Machine Learning* **3**, 319–342.

Mitchell T 1997 *Machine Learning*. McGraw-Hill.

Murthy SK 1997 Automatic construction of decision trees from data: A multi-disciplinary survey. *Data Mining and Knowledge Discovery* **2**, 345–389.

Niblett T and Bratko I 1986 Learning decision rules in noisy domains *Proceedings of the Sixth Annual Technical Conference on Research and Development in Expert Systems*. Cambridge University Press.

Pyle D 1999 *Data Preparation for Data Mining*. Morgan Kaufmann.

Quinlan JR 1986 Induction of decision trees. *Machine Learning* **1**, 81–106.

Quinlan JR 1987 Simplifying decision trees. *International Journal of Man-Machine Studies* **27**, 221–234.

Quinlan JR 1989 Inferring decision trees using the minimum description length principle. *Information and Computation* **80**, 227–248.

Quinlan JR 1993 *C4.5: Programs for Machine Learning*. Morgan Kaufmann.

Rokach L and Maimon O 2007 *Data Mining with Decision Trees: Theory and Applications*. World Scientific.

Shafer J, Agrawal R, and Mehta M 1996 SPRINT: A scalable parallel classifier for data mining *Proceedings of the Second International Conference on Very Large Databases*. Morgan Kaufmann.

Tan PN, Steinbach M and Kumar V 2013 *Introduction to Data Mining* 2nd edn. Addison-Wesley.

Therneau TM and Atkinson EJ 1997 An introduction to recursive partitioning using the RPART routines. Technical Report 61, Mayo Clinic.

Utgoff PE 1989 Incremental induction of decision trees. *Machine Learning* **4**, 161–186.

Webb GI 1996 Further experimental evidence against the utility of Occam's razor. *Journal of Artificial Intelligence Research* **4**, 397–417.

Webb GI 1997 Decision tree grafting *Proceedings of the Fifteenth International Joint Conference on Artificial Intelligence (IJCAI-97)*. Morgan Kaufmann.

Witten IH, Frank E and Hall MA 2011 *Data Mining: Practical Machine Learning Tools and Techniques* 3rd edn. Morgan Kaufmann.

4

Naïve Bayes classifier

4.1 Introduction

The naïve Bayes classifier is one of the simplest approaches to the classification task that is still capable of providing reasonable accuracy. Whereas in many cases it cannot compete with much more refined algorithms, such as decision trees, it sometimes does not stay far behind, and it may be even superior for certain specific application domains, with text classification being the most prominent example. Its simplicity – conceptual, implementational, and computational – makes it easy and inexpensive to try besides or before more sophisticated classifiers.

Example 4.1.1 Examples illustrating the naïve Bayes classifier will use the ultra small *weather* and *weatherc* datasets from Examples 1.3.1 and 1.3.2, respectively. These datasets, as well as DMR packages required to run some of example code snippets, are loaded by the following R code.

```
library(dmr.stats)
library(dmr.util)

data(weather, package="dmr.data")
data(weatherc, package="dmr.data")
```

`dmr.data`

4.2 Bayes rule

Bayesian inference, of which the naïve Bayes classifier is a particularly simple example, is based on the Bayes rule that relates conditional and marginal probabilities. More exactly, it shows how the conditional (posterior) probability of an event can be calculated based on its

---

Data Mining Algorithms: Explained Using R, First Edition. Paweł Cichosz.
© 2015 John Wiley & Sons, Ltd. Published 2015 by John Wiley & Sons, Ltd.

marginal (prior) probability and the inverse conditional probability. For two events A and B, the rule can be written as

$$P(A|B) = \frac{P(A)P(B|A)}{P(B)} \qquad (4.1)$$

where

- $P(A)$ is the *prior probability* of A,

- $P(A|B)$ is the conditional probability of A given B, also called the *posterior probability* of A,

- $P(B|A)$ is the conditional probability of B given A, and

- $P(B)$ is the probability of B.

In the most common setting, the rule is applied to inference about a set of mutually exclusive events $A_1, A_2, \ldots, A_k$ that exhaust the probability space, i.e.,

$$P(A_i \cap A_j) = 0 \quad \text{for } i \neq j \qquad (4.2)$$

$$\sum_{i=1}^{k} P(A_i) = 1 \qquad (4.3)$$

Then from the law of total probability

$$P(B) = \sum_{j=1}^{k} P(A_j)P(B|A_j) \qquad (4.4)$$

which allows one to rewrite the Bayes rule as

$$P(A_i|B) = \frac{P(A_i)P(B|A_i)}{\sum_{j=1}^{k} P(A_j)P(B|A_j)} \qquad (4.5)$$

This form shows that $P(B)$ acts as a normalizing constant, ensuring that

$$\sum_{i=1}^{k} P(A_i|B) = 1 \qquad (4.6)$$

Note that, unlike the posterior probabilities $P(A_i|B)$, the inverse conditional probabilities $P(B|A_i)$ do not have to and usually do not sum up to 1.

Typically, $A_1, A_2, \ldots, A_k$ represent a set of alternative *hypotheses*, and B represents some available *evidence* that may affect their probability. Without taking the evidence into account, the hypotheses have their prior probabilities assigned. The Bayes rule shows how to incorporate the evidence and obtain posterior hypothesis probabilities. Each of the inverse conditional probabilities $P(B|A_i)$ can be considered a measure of the extent to which the evidence supports (or refutes) the corresponding hypothesis A_i.

Example 4.2.1 The following R code implements the Bayes rule (combined with the total probability law) for an exhaustive set of mutually exclusive events and demonstrates its application. The `bayes.rule` function takes vectors of prior and inverse conditional probabilities as well as an event number as arguments and returns the posterior probability for the selected event. If the event number argument is a vector, the corresponding vector of posterior probabilities is returned.

```
## calculate the posterior probability given prior and inverse probabilities
bayes.rule <- function(prior, inv)
{
  prior*inv/sum(prior*inv)
}

  # posterior·probabilities
bayes.rule(c(0.2, 0.3, 0.5), c(0.9, 0.9, 0.5))

  # let P(burglery)=0.001,
  # P(alarm|burglery)=0.95,
  # P(alarm|not burglery)=0.005
  # calculate P(burglery|alarm)
bayes.rule(c(0.001, 0.999), c(0.95, 0.005))[1]
```

4.3   Classification by Bayesian inference

There are two major approaches in applying Bayesian inference to the classification task:

Model-probability inference. Based on calculating posterior model probabilities given a dataset.

Class-probability inference. Based on calculating posterior class probabilities given attribute values.

The first approach appears attractive as it could allow one to identify the most probable model. It is practical only for a limited set of candidate models, however, which have to be preselected either using background knowledge or some other algorithms. Moreover, assigning prior probabilities to such candidate models is nontrivial. This can be indirectly achieved using information theoretic approaches, such as the minimum description length principle, but discussing them is beyond the scope of this book.

The naïve Bayes classifier follows the second approach, which does not promise that much, but also does not imply so many difficulties. Without explicitly considering any candidate models and their probabilities, it actually does create a probabilistic model that estimates class probabilities for an instance based on its attribute values.

4.3.1   Conditional class probability

The class-probability approach to Bayesian inference applies the Bayes rule to calculate probabilities of the following form:

$$P(c = d \mid a_1 = v_1, a_2 = v_2, \dots, a_n = v_n) \tag{4.7}$$

which should be read as the probability of an instance belonging to class d if its attribute values are $v_1, v_2, \dots, v_n$, respectively. The capability of calculating such probabilities for arbitrary

$d \in C$ and $v_1 \in A_1, v_2 \in A_2, \ldots, v_n \in A_n$ immediately permits the probabilistic classification of any instance x by substituting its attribute values $a_1(x), a_2(x), \ldots, a_n(x)$ for $v_1, v_2, \ldots, v_n$:

$$P(d|x) = P(c = d \,|\, a_1 = a_1(x), a_2 = a_2(x), \ldots, a_n = a_n(x)) \qquad (4.8)$$

This is the probability of an instance belonging to class d if its all attribute values are the same as for x.

By applying the Bayes rule to the conditional class probability we get

$$
\begin{aligned}
&P(c = d \,|\, a_1 = v_1, \ldots, a_n = v_n) \\
&= \frac{P(c = d)P(a_1 = v_1, \ldots, a_n = v_n \,|\, c = d)}{P(a_1 = v_1, \ldots, a_n = v_n)}
\end{aligned}
\qquad (4.9)
$$

The probability from the denominator is actually a normalizing constant that does not depend on the class label. It can be simply ignored if instances are ultimately classified to the most probable classes, as the class maximizing the Bayes numerator will also maximize the probability. Otherwise, if the actual probabilities are required (e.g., if nonuniform misclassification costs have to be handled as discussed in Section 6.3.3 or ROC analysis performed as discussed in Section 7.2.5), it can be calculated using the total probability law as the sum of the Bayes numerators for all classes:

$$P(a_1 = v_1, \ldots, a_n = v_n) = \sum_{d \in C} P(c = d)P(a_1 = v_1, \ldots, a_n = v_n \,|\, c = d) \qquad (4.10)$$

since the classes in C are obviously assumed to be mutually exclusive and exhaust all possibilities. The probabilities occurring in the Bayes numerator – the class prior probability and the conditional joint probability of attribute values given the class – need more attention and will be discussed separately below.

4.3.2   Prior class probability

The prior class probability, $P(c = d)$, can be directly estimated from the training set T, assuming its representativeness, as the relative frequency of instances of class d:

$$P(c = d) = P_T(c = d) = \frac{|T^d|}{|T|} \qquad (4.11)$$

where T^d denotes the subset of T containing instances of class d. If the training set is not representative, the correct class priors must be provided by the available domain knowledge.

Example 4.3.1 The following R code applies the `pdisc` function for estimating discrete probability distributions to estimate prior class probabilities for the *weather* data.

Ex. 2.4.22
dmr.stats

```
# prior class probabilities for the weather data
pdisc(weather$play)
```

4.3.3   Independence assumption

The conditional joint probability of attribute values given the class cannot be directly estimated from even a perfectly representative training set of any realistic size. For most practical
datasets, there are numerous attribute value combinations not appearing at all, and most of
those that do appear, appear exactly once. This would lead as to estimating probabilities for
most attribute value combinations as 0 or $1/|T|$. Such estimates would be clearly useless
for classification, because they do not allow one to differentiate probabilities for instances
with different attribute values. Therefore, the conditional joint probability of attribute
values given the class is calculated as the product of per-attribute marginal conditional
probabilities:

$$P(a_1 = v_1, \ldots, a_n = v_n \mid c = d) = \prod_{i=1}^{n} P(a_i = v_i | c = d) \qquad (4.12)$$

Equation 4.12 holds only if attributes are conditionally independent given the class, which is
unfortunately usually not true. The naïve Bayes classifier adopts this *independence assumption*, ignoring the fact that it is rarely satisfied. This is what it owes the term "naïve" in
its name to, although there is actually more pragmatism than naïvety in pretending something obviously false is true just to achieve some benefit. This is exactly what happens here.
Due to the independence assumption, the naïve Bayes classifier avoids directly estimating
the joint conditional probability of attribute values, reducing it to a much simpler problem
of estimating marginal conditional attribute value probabilities. The price for this simplification is that the probabilities calculated based on the unsatisfied assumption may be incorrect (or, speaking more openly, are guaranteed to be incorrect). The reason why this makes
sense is that incorrect class probabilities still may and quite frequently do permit correct
classification.

4.3.4   Conditional attribute value probabilities

The probabilities of attribute values given the class can be directly estimated from the training
data as follows:

$$P(a_i = v_i | c = d) = P_{T^d}(a_i = v_i) = \frac{|T^d_{a_i=v_i}|}{|T^d|} \qquad (4.13)$$

where $T^d_{a_i=v_i}$ denotes the subset of T^d consisting of instances for which the value of attribute a_i
is v_i. Such frequency-based estimation is perfectly sufficient, assuming the representativeness
of the training set.

Example 4.3.2 The following R code demonstrates estimating the conditional
attribute value probabilities given the class for the *weather* data using the
pcond function.

| Ex. 2.4.23 |
| dmr.stats |

```
  # conditional attribute value probabilities given the class
pcond(weather$outlook, weather$play)
pcond(weather$temperature, weather$play)
pcond(weather$humidity, weather$play)
pcond(weather$wind, weather$play)
```

4.3.5   Model construction

As demonstrated above, to achieve the desired capability of calculating the conditional class probability given attribute values it is sufficient to estimate the following probabilities from the training set:

- $P(c = d)$ for each class $d \in C$,

- $P(a_i = v_i | c = d)$ for each class $d \in C$, each attribute a_i, and each value $v_i \in A_i$.

This set of probabilities constitutes model representation for the naïve Bayes classifier. To create such a model, one needs to estimate all these probabilities from the training set, which reduces to simple counting required to obtain $|T^d|$ and $|T^d_{a_i = v_i}|$. It does not take much effort to see that these counts can be calculated by a single iteration through the training set.

Example 4.3.3 Contrary to the previous examples, the R code presented below does not use any built-in counting facilities available in R, but shows how a naïve Bayes classification model can be obtained with a single iteration over the training set. This "naïve" implementation makes it more explicit what data operations are actually necessary to create a naïve Bayes model and how they would be implemented in conventional programming languages. Unfortunately, it is much less efficient than a more R-style implementation could be, and the reader may find it worthwhile to exercise rewriting it as such. The x.vars and y.var functions are used for extracting input and target attribute names from the supplied formula. The class attribute of the created model representation is set to nbc to enable appropriate prediction method dispatching. As before, the implementation is illustrated using the *weather* data.

```
dmr.util
```

```
## create a naive Bayes classifier
nbc <- function(formula, data)
{
  class <- y.var(formula)
  attributes <- x.vars(formula, data)

  cc <- integer(nlevels(data[[class]]))  # initialize class counts
  names(cc) <- levels(data[[class]])
  avc <- sapply(attributes,              # initialize attribute-value-class counts
                function(a)
                matrix(0, nrow=nlevels(data[[a]]), ncol=nlevels(data[[class]]),
                        dimnames=list(levels(data[[a]]), levels(data[[class]]))))

  for (i in (1:nrow(data)))              # iterate through training instances
  {
    cc[data[[class]][i]] <- cc[data[[class]][i]]+1  # increment class count
    for (a in attributes)                # increment attribute-value-class counts
      avc[[a]][data[[a]][i],data[[class]][i]] <-
        avc[[a]][data[[a]][i],data[[class]][i]]+1
  }

  # calculate probability estimates based on counts
  `class<-`(list(prior=cc/sum(cc),
              cond=sapply(avc, function(avc1)
```

```
                                    t(apply(avc1, 1, "/", colSums(avc1))))),
            "nbc")
}

  # naive Bayes classifier for the weather data
nbw <- nbc(play~., weather)
```

4.3.6 Prediction

Applying the naïve Bayes classifier to predict class probabilities for a given instance x is even more straightforward. We just need to multiply the prior class probability and the conditional probabilities of the instance's attribute values given the class, i.e., use $P(a_i = v_i | c = d)$ with $v_i = a_i(x)$:

$$P(d|x) = \frac{1}{b} P(c = d) \prod_{i=1}^{n} P(a_i = a_i(x)|c = d) \tag{4.14}$$

where the normalizing constant b is obtained as

$$b = \sum_{d' \in C} P(c = d') \prod_{i=1}^{n} P(a_i = a_i(x)|c = d') \tag{4.15}$$

This is a simple rewrite of Equation 4.8 which incorporates the application of the Bayes rule according to Equation 4.9 and the independence assumption according to Equation 4.12, and additionally applies the total probability law according to Equation 4.10. Note that all probabilities needed to classify an arbitrary instance are estimated during model construction, and prediction requires just selecting and multiplying an appropriate subset of them, corresponding to the attribute values of the classified instance x.

Example 4.3.4 The following R code implements prediction using the naïve Bayes classification model, as created in the previous example, and demonstrates its application to the *weather* data.

```
## naive Bayes prediction for a single instance
predict1.nbc <- function(model, x)
{
  aind <- names(x) %in% names(model$cond)
  bnum <- model$prior*apply(mapply(function(a, v)
                                  model$cond[[a]][v,], names(model$cond), x[aind]),
                           1, prod)
  bnum/sum(bnum)
}

## naive Bayes prediction for a dataset
predict.nbc <- function(model, data)
{
  t(sapply(1:nrow(data), function(i) predict1.nbc(model, data[i,])))
}

  # make predictions for the weather data
predict(nbw, weather)
```

4.4   Practical issues

The naïve Bayes algorithm may need some minor enhancements before it is ready to work using real-world data. This section reviews the most important practical issues that need to be taken care of.

4.4.1   Zero and small probabilities

The plain frequency approach to estimating conditional attribute value probabilities presented above leads to a problem when a particular attribute value does not occur in training instances of a particular class. Such a situation is very common for realistic datasets and cannot be avoided. If this happened for value v_j of attribute a_j in class d, we would estimate $P(a_j = v_j | c = d) = 0$. If such a zero probability appears during the classification of an instance x:

$$P(a_j = a_j(x) | c = d) = 0 \tag{4.16}$$

the calculated product of probabilities will also be 0:

$$P(c = d) \prod_{i=1}^{n} P(a_i = a_i(x) | c = d) = 0 \tag{4.17}$$

As long as this does not happen for all classes, it would just imply $P(d|x) = 0$, which, although "radical," still permits the classification of x. Things get much worse if we have one or more zero probabilities for all classes, e.g.,

$$P(a_{j_1} = a_{j_1}(x) | c = d_1) = 0 \tag{4.18}$$

$$P(a_{j_2} = a_{j_2}(x) | c = d_2) = 0 \tag{4.19}$$

$$\dots$$

Such a situation, which is not at all unlikely for practical datasets, makes prediction for instance x impossible, because we obtain $P(d|x) = 0$ for all $d \in C$.

One way to avoid the problem of zero probabilities is to simply replace any conditional attribute value probabilities estimated as 0 by a small positive number:

$$P(a_i = v_i | c = d) = \begin{cases} \dfrac{|T^d_{a_i = v_i}|}{|T^d|} & \text{if } T^d_{a_i = v_i} \neq \emptyset \\ \epsilon & \text{otherwise} \end{cases} \tag{4.20}$$

For this approach to make sense, ϵ should be considerably less than $1/|T^d|$, which is the probability estimate that would be obtained if there were one instance of class d with the value of a_i equal to v_i.

Whereas the above should work well in most cases, a more elegant and safer solution is to use the technique of m-estimation described in Section 2.4.4:

$$P(a_i = v_i | c = d) = \frac{|T^d_{a_i = v_i}| + mp}{|T^d| + m} \tag{4.21}$$

where – in the lack of domain knowledge suggesting otherwise – p is set to $1/|A_i|$. This corresponds to incorporating m fictitious instances to the estimation, for which all attribute values are equally probable. When additionally $m = |A_i|$, we obtain the Laplace estimator of the following form:

$$P(a_i = v_i | c = d) = \frac{|T^d_{a_i=v_i}| + 1}{|T^d| + |A_i|}$$
(4.22)

A less severe, but sometimes also important issue is associated with probabilities that are just close to 0. The multiplication of several small numbers may lead to numeric underflow. This will also result in inability to classify an instance, whenever the underflow appears for all classes. One common trick that helps to reduce the risk of this problem is to calculate probability logarithms instead of plain probabilities:

$$\log\left(P(c = d) \prod_{i=1}^{n} P(a_i = a_i(x) | c = d) \right)$$

$$= \log P(c = d) + \sum_{i=1}^{n} \log P(a_i = a_i(x) | c = d)$$
(4.23)

This transforms multiplication into addition, which is less prone to numerical underflow.

Example 4.4.1 The following R code defines a modified version of the function for estimating conditional probabilities, using m-estimation to avoid zero probabilities. The latter is performed by the mest function. The probability estimates produced by the original function defined previously and the modified version are compared using the *weather* data.

Ex. 2.4.31
dmr.stats

```
## m-estimated conditional probability distribution P(v1|v2)
mpcond <- function(v1, v2, p=1/nlevels(v1), m=nlevels(v1))
{
  count <- table(v1, v2, dnn=NULL)
  t(apply(count, 1, function(cnt, sumcnt) mest(cnt, sumcnt, p, m), colSums(count)))
}

  # conditional attribute value probabilities given the class
pcond(weather$outlook, weather$play)
mpcond(weather$outlook, weather$play)
mpcond(weather$outlook, weather$play, m=0)
mpcond(weather$outlook, weather$play, m=1)
```

4.4.2 Linear classification

There is one noteworthy consequence of working with probability logarithms rather than probabilities, as suggested above to bypass numerical problems with probability multiplication. This is not actually a practical issue, but this is probably the best opportunity to mention it. Consider the ratio of class probabilities in the two-class case:

$$\frac{P(1|x)}{1 - P(1|x)} = \frac{P(1|x)}{P(0|x)} = \frac{P(c = 1) \prod_{i=1}^{n} P(a_i = a_i(x) | c = 1)}{P(c = 0) \prod_{i=1}^{n} P(a_i = a_i(x) | c = 0)}$$
(4.24)

After taking the logarithm and rearranging terms, we receive

$$\log \frac{P(1|x)}{1 - P(1|x)} = \log P(c = 1) - \log P(c = 0)$$

$$+ \sum_{i=1}^{n} (\log P(a_i = a_i(x)|c = 1) - \log P(a_i = a_i(x)|c = 0)) \tag{4.25}$$

It can be rewritten as

$$\log \frac{P(1|x)}{1 - P(1|x)} = \sum_{i=1}^{n} \sum_{v \in A_i} a_{i,v}(x) w_{i,v} + w_{n+1} = g(x) \tag{4.26}$$

where

$$w_{n+1} = \log P(c = 1) - \log P(c = 0) \tag{4.27}$$

$$w_{i,v} = \log P(a_i = v|c = 1) - P(a_i = v|c = 0) \tag{4.28}$$

$$a_{i,v}(x) = \mathbb{I}_{a_i(x)=v} \tag{4.29}$$

and the indicator function $\mathbb{I}_{\text{condition}}$ takes value 1 if condition is satisfied and 0 otherwise. This reveals that the naïve Bayes classifier can be viewed as a linear classifier in the modified space of binary attributes $a_{i,v}$ (for each $i = 1, 2, \ldots, n$ and $v \in A_i$), i.e., it belongs to the family of classification algorithms discussed extensively in Chapter 5. More precisely, if the natural logarithm is used in the equations above, they directly correspond the linear logit representation, described in Section 5.2.4, since it can be easily verified that

$$P(1|x) = \frac{e^{g(x)}}{e^{g(x)} + 1} \tag{4.30}$$

The naïve Bayes classifier can be therefore considered a particularly simple and imperfect method of estimating linear logit classifier parameters.

4.4.3   Continuous attributes

All the above references to conditional attribute value probabilities of the form $P(a_i = v_i|c = d)$ are obviously valid for discrete attributes only. If some or all attributes are continuous, a simple, but perfectly reasonable workaround would be to discretize them. This is usually the best way to proceed unless this extra data preprocessing step constitutes a problem. This section shows how continuous attributes can be handled by the naïve Bayes classifier by itself in such a situation.

The idea is to replace probabilities by the appropriate probability density function values for continuous attributes. If a_i is a continuous attribute, we would therefore use $g_i^d(v_i)$ instead of $P(a_i = v_i|c = d)$, where g_i^d denotes the probability density function of attribute a_i within class d. If this function is available, we just need to take its value for v_i, i.e., when making prediction for instance x, for $a_i(x)$.

The remaining issue of identifying the required probability density functions for continuous attributes within particular classes is typically solved by assuming that they are all distributed normally and estimating the mean and variance parameters from the training set.

For attribute a_i and class d, we would estimate these parameters as

$$m_i^d = \frac{1}{|T^d|} \sum_{x \in T^d} a_i(x) \tag{4.31}$$

and

$$(s_i^d)^2 = \frac{1}{|T^d| - 1} \sum_{x \in T^d} (a_i(x) - m_i^d)^2 \tag{4.32}$$

respectively.

Example 4.4.2 The following R code defines functions for estimating conditional means and variances of continuous attributes and demonstrates their application to the *weatherc* data, containing two continuous attributes.

> Ex. 1.3.2
> dmr.data

```
## conditional mean
mcond <- function(v1, v2)
{
  tapply(v1, v2, mean)
}

## conditional variance
vcond <- function(v1, v2)
{
  tapply(v1, v2, var)
}

  # conditional mean and variance of attribute values given the class
mcond(weatherc$temperature, weatherc$play)
vcond(weatherc$temperature, weatherc$play)
mcond(weatherc$humidity, weatherc$play)
vcond(weatherc$humidity, weatherc$play)
```

4.4.4 Missing attribute values

Missing attribute values are likely to decrease model quality for any modeling algorithm, when occurring for training instances, or classification accuracy, when occurring for classified instances. The naïve Bayes classifier is no exception here, but it is noteworthy that at least missing values constitute no algorithmic problem and do not increase computational complexity in any way. They can be handled easily and naturally, and the resulting deterioration of classification accuracy will be as graceful as possible.

In the model construction phase, the naïve Bayes classifier uses attribute values to estimate their conditional probabilities within particular classes. Whenever an attribute's value is missing for an instance, it is simply not used in this estimation. More precisely, the counts needed to estimate $P(a_i = v_i | c = d)$ should only include instances with *known values* of a_i. We could rewrite the simple formula for frequency-based estimation as

$$P(a_i = v_i | c = d) = \frac{|T_{a_i = v_i}^d|}{|T^d| - |T_{a_i = ?}^d|} \tag{4.33}$$

where $T^d_{a_i=?}$ denotes the subset of instances of class d for which the values of attribute a_i are missing (and obviously $T^d_{a_i=v_i}$ only includes instances for which the values of attribute a_i are available and equal to v_i). Corrections to avoid zero probabilities can be applied in the same way as presented before.

When a naïve Bayes model is applied to make prediction for an instance with missing values of some attributes, the corresponding factors from the conditional probability product are unavailable. More precisely, if $a_i(x)$ is missing for the instance x being classified, the $P(a_i = a_i(x)|c = d)$ factor is unavailable, since one cannot choose the appropriate conditional attribute value probability $P(a_i = v_i|c = d)$ without knowing what to substitute for v_i. To accept the unavoidable fact that attributes with missing values provide no information that could impact the prediction, it is then sufficient just to skip the unavailable factors or (equivalently) take them as equal to 1.

The above solution perfectly fits most common practical needs in situations where missing attribute values indeed carry no information that could affect prediction. In some domains, however, missing attribute values can be actually meaningful, i.e., the unavailability of an attribute's value for an instance may be somehow related to the properties of this instances relevant for its classification. In such cases, when $P(a_i = ?|c = d)$ differs substantially across classes, it may be more reasonable to treat "missing" as an additional possible attribute value.

Example 4.4.3 The `pcond` function for estimating conditional attribute value probabilities given the class used in Example 4.3.2 remains correct for attributes with missing values, which is illustrated by the following R code.

```
  # weather data with missing values
weatherm <- weather
weatherm$outlook[1] <- NA
weatherm$humidity[1:2] <- NA

  # conditional attribute value probabilities given the class
  # with and without missing values
pcond(weather$outlook, weather$play)
pcond(weatherm$outlook, weatherm$play)
pcond(weather$humidity, weather$play)
pcond(weatherm$humidity, weatherm$play)
```

4.4.5   Reducing naïvety

Since the conditional attribute value independence assumption is the source of the naïvety of the naïve Bayes classifier, it can be made *not-so-naïve* by relaxing the assumption. In general, this would lead to calculating the probability of class given attribute values as follows:

$$P(c = d \mid a_1 = v_1, \dots, a_n = v_n) = P(c = d) \prod_{i=1}^{n} P(a_i = v_i|c = d, a_{U_i} = v_{U_i}) \qquad (4.34)$$

where U_i is the set of the numbers of attributes on which a_i is assumed to directly depend given the class and $a_{U_i} = v_{U_i}$ is a notational shortcut for the sequence of conditions $a_j = v_j$

for all $j \in U_i$. The attributes with numbers in U_i are called the *immediate predecessors* of a_i. This term relates to *Bayesian networks* of which the naïve Bayes classifier is a particularly simple representative. They represent links between attributes by directed edges in an acyclic graph, with the plain naïve Bayes model containing only edges from the target concept to all attributes. Permitting attributes to have some other attributes as their immediate predecessors corresponds to inserting additional edges to the naïve Bayes network structure. They are referred to as *augmenting edges*, and the resulting classification model – as the *augmented naïve Bayes model*.

As written above, the relaxed independence assumption states that each attribute is conditionally independent of all other attributes given the class and the values of its immediate predecessors. This makes it possible to calculate the conditional probability of attribute values given the class as the product of per-attribute probabilities, each conditional on the class and the values of its immediate predecessors. Since the latter have to be estimated from the data, the numbers of the immediate predecessors of each attribute should be small – otherwise the estimates would be unreliable (except for very large data). To keep things simple, it is common to restrict augmented naïve Bayes models to a single immediate predecessor attribute:

$$P(a_1 = v_1, \ldots, a_n = v_n \mid c = d) = \prod_{i=1}^{n} P(a_i = v_i | c = d, a_{u_i} = v_{u_i}) \qquad (4.35)$$

where u_i is the number of the single immediate predecessor of a_i. This simplification makes it possible to determine augmenting edges to add using dedicated algorithms, more efficient than general-purpose algorithms for deriving Bayesian network structures from data. One noteworthy example is the *tree-augmented naïve Bayes classifier* (TAN) which identifies edges by finding the maximum weighted spanning tree with respect to the conditional mutual information for attribute pairs given the class:

$$I_T(a_1, a_2 | c) = \sum_{d \in C} P_T(c = d) I_{T^d}(a_1, a_2) \qquad (4.36)$$

where $I_{T^d}(a_1, a_2)$ is the mutual information for attribute pair a_1, a_2 calculated according to Equation 2.76 on the subset of the training set limited to class d. This yields the smallest subset of edges that connect all attributes and indicate maximally strong relationships. All edges are assumed to be directed outward from an arbitrarily chosen "root" attribute, which guarantees no cycles.

An interesting simpler alternative approach that avoids pairwise mutual information calculation considers each attribute as a single common predecessor for all others. More specifically, if we assume a_j to be the single immediate predecessor of all other attributes, then

$$P_j(a_1 = v_1, \ldots, a_n = v_n \mid c = d) =$$

$$P(a_j = v_j | c = d) \prod_{i=1}^{n} P(a_i = v_i | c = d, a_j = v_j) \qquad (4.37)$$

where $P(a_i = v_i | c = d, a_j = v_j) = 1$ for $i = j$. Probabilities obtained with each attribute used as the single common immediate predecessor (i.e., for $j = 1, 2, \ldots, n$) are then averaged, yielding

$$P(a_1 = v_1, \dots, a_n = v_n \mid c = d) =$$

$$\frac{1}{n} \sum_{j=1}^{n} P(a_j = v_j \mid c = d) \prod_{i=1}^{n} P(a_i = v_i \mid c = d, a_j = v_j) \qquad (4.38)$$

Some computation may be saved by skipping attribute a_j from consideration as an immediate predecessor if $P(a_j = v_j)$ is sufficiently small. The resulting prediction scheme is known as the *averaged one-dependence estimators* algorithm (AODE).

The not-so-naïve versions of the naïve Bayes classifier substantially increase the computational complexity of model creation and prediction. When this is acceptable, though, they may improve the predictive performance considerably.

4.5   Conclusion

Despite – but also due to – its simplicity, the naïve Bayes classifier holds an unrivaled position in most contemporary data mining toolboxes. Whereas often beaten accuracy-wise by more refined classification algorithms, it has unquestionable advantages with respect to computational efficiency of both model creation and prediction. The former is also easily parallelizable. This becomes particularly important for very large datasets, for which other algorithms may require subsampling. It is also extremely easy to apply, with no parameters to adjust (in the basic version).

But leaving apart the computational cost and convenience, the naïve Bayes classifier still has some attractive properties. Two of them are particularly worthwhile to mention. First, its inherent inability to accurately fit the training data make it extraordinarily resistant to overfitting. Whereas underfitting is not necessarily better, it is definitely much more apparent and easy to notice. Eliminating the risk of model deficiency that is not so straightforward to detect is hard to overestimate. Second, the naïve Bayes classifier can handle numerous attributes without needing to internally make any selections. This is useful when there are no outstandingly strong relationships between attributes and the target concept, but rather a large number of attributes have some small impact on class membership, and the contributions of all of them need to be incorporated during classification. One important domain where it tends to be true is text classification, for which the naïve Bayes classifier belongs to most successful algorithms.

4.6   Further readings

The naïve Bayes classifier is covered by most data mining and machine learning books (e.g., Cios *et al.*, 2007; Han *et al.*, 2011; Mitchell, 1997; Tan *et al.*, 2013; Witten *et al.*, 2011). Some of them also discuss other, more advanced forms of Bayesian inference, used for model creation, prediction, and model selection (e.g., Bishop 2007; Hand *et al.*, 2001; Theodoridis and Koutroumbas 2008).

The Bayes rule was derived by Bayes (1763) more than 250 years ago and still remains the foundation of many probabilistic inference techniques in statistics, machine learning, and artificial intelligence. Probably the first to present the basic naïve Bayes algorithm were Duda and Hart (1973). Despite its simplicity, it keeps attracting not only practical, but also research interest. Several experimental studies have demonstrated its sometimes surprisingly good classification performance, despite the oversimplified conditional independence assumption (e.g., Cestnik 1990; Clark and Niblett 1989; Langley *et al.*, 1992). Subsequent investigations have confirmed that correct class label predictions can be obtained even if the latter is unsatisfied, leading to incorrect probability calculations (Domingos and Pazzani 1996, 1997; Hand and Yu 2001; Rish 2001). This definitely brought more appreciation to the algorithm.

A basic approach to applying the naïve Bayes classifier to text classification is described by Mitchell (1997) and several text-oriented variations of the algorithm are reviewed by Lewis (1998). McCallum and Nigam (1998) highlighted relationships between text representation – with binary word presence/absence or numerical word occurrence count attributes – and probability calculations, using the Bernoulli and multinomial distributions. Some other issues related to naïve Bayes text classifications and possible workarounds are discussed by Rennie *et al.* (2003). The popularity of the Bayesian approach to unsolicited mail detection is largely due to Graham (2002, 2003), who presented successful spam filters using a modified form of the naïve Bayes classifier.

Bayesian networks, of which the naïve Bayes classifier can be considered a particularly simple representative, were introduced by Pearl (1988) as knowledge representation and inference methods. Algorithms for creating Bayesian network models from data are reviewed by Russell and Norvig (2011). Augmented naïve Bayes models – with edges between selected attribute nodes added – reduce the degree of naïvety while avoiding the complexity of full Bayesian network learning (Friedman *et al.*, 1997; Keogh and Pazzani 1999). The similarly effective but computationally simpler averaged one-dependence estimators (AODE) approach was presented by Webb *et al.* (2005).

References

Bayes T 1763 An essay towards solving a problem in the doctrine of chances. *Philosophical Transactions of the Royal Society of London* **53**, 370–418.

Bishop CM 2007 *Pattern Recognition and Machine Learning*. Springer

Cestnik B 1990 Estimating probabilities: A crucial task in machine learning *Proceedings of the Ninth European Conference on Artificial Intelligence (ECAI-90)*. Pitman.

Cios KJ, Pedrycz W, Swiniarski RW and Kurgan L 2007 *Data Mining: A Knowledge Discovery Approach*. Springer.

Clark P and Niblett T 1989 The CN2 induction algorithm. *Machine Learning* **3**, 261–283.

Domingos P and Pazzani M 1996 Beyond independence: Conditions for the optimality of the simple Bayesian classifier *Proceedings of the Thirteenth International Conference on Machine Learning (ICML-96)*. Morgan Kaufmann.

Domingos P and Pazzani M 1997 On the optimality of the simple Bayesian classifier under zero-one loss. *Machine Learning* **29**, 103–137.

Duda RO and Hart PE 1973 *Pattern Classification and Scene Analysis*. Wiley.

Friedman N, Geiger D and Goldszmidt M 1997 Bayesian network classifiers. *Machine Learning* **29**, 131–163.

Graham P 2002 A plan for spam. Reprinted in Graham, P., *Hackers and Painters: Big Ideas from the Computer Age*, 2004, O'Reilly.

Graham P 2003 Better Bayesian filtering. Presented at the *2003 Spam Conference*.

Han J, Kamber M and Pei J 2011 *Data Mining: Concepts and Techniques* 3rd edn. Morgan Kaufmann.

Hand DJ, Mannila H and Smyth P 2001 *Principles of Data Mining*. MIT Press.

Hand DJ and Yu K 2001 Idiot's Bayes – not so stupid after all?. *International Statistical Review* **69**, 385–399.

Keogh E and Pazzani M 1999 Learning augmented Bayesian classifiers: A comparison of distribution-based and classification-based approaches *Proceedings of the Seventh International Workshop on Artificial Intelligence and Statistics*. Morgan Kaufmann.

Langley P, Iba W and Thompson K 1992 An analysis of Bayesian classifiers *Proceedings of the Tenth National Conference on Artificial Intelligence (AAAI-92)*. AAAI Press.

Lewis DD 1998 Naive (Bayes) at forty: The independence assumption in information retrieval *Proceedings of the Tenth European Conference on Machine Learning (ECML-98)*. Springer.

McCallum A and Nigam K 1998 A comparison of event models for naive Bayes text classification *Proceedings of the AAAI/ICML-98 Workshop on Learning for Text Categorization*. AAAI Press.

Mitchell T 1997 *Machine Learning*. McGraw-Hill.

Pearl J 1988 *Probabilistic Reasoning in Intelligent Systems*. Morgan Kaufmann.

Rennie JDM, Shih L, Teevan J and Karger DR 2003 Tackling the poor assumptions of naive Bayes classifiers *Proceedings of the Twentieth International Conference on Machine Learning (ICML-03)*. AAAI Press.

Rish I 2001 An empirical study of the naive Bayes classifier *Proceedings of the IJCAI-2001 Workshop on Empirical Methods in Artificial Intelligence*.

Russell S and Norvig P 2011 *Artificial Intelligence: A Modern Approach* 3rd edn. Prentice Hall.

Tan PN, Steinbach M and Kumar V 2013 *Introduction to Data Mining* 2nd edn. Addison-Wesley.

Theodoridis S and Koutroumbas K 2008 *Pattern Recognition* 4th edn. Academic Press.

Webb GI, Boughton J and Wang Z 2005 Not so naive Bayes: Aggregating one-dependence estimators. *Machine Learning* **58**, 5–24.

Witten IH, Frank E and Hall MA 2011 *Data Mining: Practical Machine Learning Tools and Techniques* 3rd edn. Morgan Kaufmann.

5

Linear classification

5.1 Introduction

The linear model representation is a special case of the parametric representation which assumes that the model's predictions are calculated by applying a representation function to attribute values and a set of real-valued parameters. This is particularly natural and extremely common for regression models which make real-valued predictions. The same approach can also be adopted to represent classification models, though. Moreover, such models can be created by the same or nearly the same algorithms as those that normally deliver regression models. This can be achieved in several ways, some of which are discussed in this chapter. The chapter will focus on issues related to adopting parametric regression methods to the classification task. This is essentially based on using a composite model representation function, consisting of a real-valued inner representation function and a discrete outer representation function that assigns class labels based on the former.

According to this book's task-oriented organization, chapters devoted to classification algorithms precede those covering regression algorithms. A book must have a linear structure and of different possible presentation orders; this one is believed to provide the best didactic value. For consistency, this linear classification chapter appears before Chapter 8 devoted to linear regression. However, the linear model representation and the corresponding modeling algorithms for the classification task can be most clearly viewed as modifications of those developed for the regression task. This imposes a considerable number of forward references which are likely to make it rather uncomfortable to read this chapter without at least briefly skimming Chapter 8.

Example 5.1.1 Selected approaches to parametric classification model representation discussed in this chapter will be illustrated by simple R code examples. Some examples visually explain particular representations using plots, and some demonstrate the

---

Data Mining Algorithms: Explained Using R, First Edition. Paweł Cichosz.
© 2015 John Wiley & Sons, Ltd. Published 2015 by John Wiley & Sons, Ltd.

parameter estimation process. The plots will be produced by functions from the `lattice` package. Example code snippets will use auxiliary functions from several DMR packages. Parameter estimation for linear classifications models will be demonstrated using the *Pima Indians Diabetes* data, available in the `mlbench` package. Additionally, the *weatherc* data will be used to illustrate discrete attribute processing. The following R code loads the required packages and the datasets. The larger of those is partitioned into training and test subsets.

Ex. 1.3.2
dmr.data

```
library(dmr.claseval)
library(dmr.linreg)
library(dmr.regeval)
library(dmr.trans)
library(dmr.util)

library(lattice)

data(weatherc, package="dmr.data")
data(PimaIndiansDiabetes, package="mlbench")

set.seed(12)
rpid <- runif(nrow(PimaIndiansDiabetes))
pid.train <- PimaIndiansDiabetes[rpid>=0.33,]
pid.test <- PimaIndiansDiabetes[rpid<0.33,]
```

The code snippet presented below defines a linear function of two attributes, named `lcg.plot`, and uses it to generate the `lcdat.plot` dataset that will be used to produce plots illustrating the linear classification model representation. Additional artificial training and test datasets for parameter estimation examples are also generated, using the `ustep` function for unit step calculation. Notice that the `lcg` function of four attributes used for class label generation is actually quadratic rather than linear, so it may be impossible to ideally fit linear classification models. The `lcg.plot` function (which is linear) is plotted as a plane in a three-dimensional space in Figure 5.1.

dmr.util

```
set.seed(12)

  # dataset for surface plots
lcg.plot <- function(a1, a2) { 2*a1-3*a2+4 }
lcdat.plot <- 'names<-'(expand.grid(seq(1, 5, 0.05), seq(1, 5, 0.05)), c("a1", "a2"))
lcdat.plot$g <- lcg.plot(lcdat.plot$a1, lcdat.plot$a2)
lcdat.plot$c <- as.factor(ustep(lcdat.plot$g))

  # datasets for parameter estimation examples
lcg <- function(a1, a2, a3, a4) { a1^2+2*a2^2-a3^2-2*a4^2+2*a1-3*a2+2*a3-3*a4+1 }
lcdat <- data.frame(a1=runif(400, min=1, max=5), a2=runif(400, min=1, max=5),
                    a3=runif(400, min=1, max=5), a4=runif(400, min=1, max=5))
lcdat$c <- as.factor(ustep(lcg(lcdat$a1, lcdat$a2, lcdat$a3, lcdat$a4)))
lcdat.train <- lcdat[1:200,]
lcdat.test <- lcdat[201:400,]

print(wf.g <- wireframe(g~a1+a2, lcdat.plot, col="grey50", zoom=0.8))
```

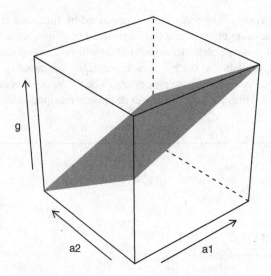

Figure 5.1   A linear representation function.

5.2   Linear representation

Linear representation is the most common instantiation of the parametric representation family that will be more thoroughly discussed in Section 8.2 in the regression context, but can be summarized as follows:

- A fixed model *representation function* is adopted that determines the model's predicted value for an instance based on the instance's attribute values and a vector of model parameters.

- Creating a model based on a training set consists in estimating its parameters.

This is in contrast to nonparametric representation, where both the representation function and parameters have to be derived from the data as part of the model creation process. We have actually already encountered these two types of model representation. Decision trees discussed in Chapter 3 can be viewed as instantiations of nonparametric representation, with the tree structure playing the role of the representation function, and per-leaf class distributions serving as model parameters. The naïve Bayes classifier from Chapter 4 adopts a parametric representation, on the other hand, using prior class probabilities and conditional attribute value probabilities as parameters to the fixed representation function that calculates conditional class probabilities given attribute values based on the Bayes rule and the independence assumption.

In principle, parametric model representation is applicable to both classification and regression models, since the employed representation function can be real valued or discrete valued. However, a model representation function is useful only if reasonably efficient and effective parameter estimation algorithms are available. This unquestionably favors real-valued representation functions. Several approaches to parametric classification are therefore based on wrapping the latter so that they can be used for class label prediction. This is most natural and easiest to achieve with two-class classification tasks. Since multiclass tasks can be transformed to two-class tasks, the discussion in this chapter is limited to the latter.

Example 5.2.1 Because of linear classification being an adaptation of linear regression to the classification task, R code presented in this chapter's examples will use several functions defined in linear regression examples presented in Chapter 8 that implement the linear representation and model parameter estimation. To make the forthcoming examples easier to follow, the code snippet presented below demonstrates the application of the most essential of those: `repf.linear` from Example 8.2.2 that implements the linear representation function, `grad.linear` from Example 8.3.3 that implements its gradient used for parameter estimation, and `predict.par` from Example 8.2.1 which is the prediction $\boxed{\texttt{dmr.linreg}}$ method for parametric models. The three functions are applied to the first 10 instances from the `lcdat.plot` dataset, using a simple linear regression model with the parameter vector exactly matching the `lcg.plot` function. The model is assumed to be represented by a list containing two components, named `repf` (the representation function) and `w` (the parameter vector).

```
 # parameter vector for the lcg.plot function
w.plot <- c(2, -3, 4)
repf.linear(lcdat.plot[1:10,1:2], w.plot)
grad.linear(lcdat.plot[1:10,1:2], w.plot)
 # parametric model for the lcg.plot function
m.plot <- `class<-`(list(repf=repf.linear, w=w.plot), "par")
predict(m.plot, lcdat.plot[1:10,1:2])
```

5.2.1   Inner representation function

The representation function for parametric classification is the composite of a real-valued *inner representation function* and another (outer) function that assigns binary class labels (as always in this book, assumed to be from the $\{0, 1\}$ set) based on its values. The inner representation function is calculated based on attribute values and model parameters:

$$g(x) = F(\mathbf{a}(x), \mathbf{w}) \tag{5.1}$$

More specifically, for the linear representation we have

$$g(x) = \sum_{i=1}^{n} w_i a_i(x) + w_{n+1} \tag{5.2}$$

or, assuming $a_{n+1}(x) = 1$ for all x to include the *intercept* term w_{n+1} in the summation

$$g(x) = \sum_{i=1}^{n+1} w_i a_i(x) = \mathbf{w} \circ \mathbf{a}(x) \tag{5.3}$$

where $\circ$ denotes the dot product operator, $\mathbf{w}$ is the parameter vector, and $\mathbf{a}(x)$ is the vector of attribute values $a_1(x), a_2(x), \ldots, a_{n+1}(x)$. Whenever referring to $\mathbf{w}$ or $\mathbf{a}(x)$ in this chapter, they will be assumed to contain $n + 1$ elements, i.e., include the intercept term w_{n+1} and the corresponding fictitious attribute value $a_{n+1}(x)$. When the last elements of these vectors have to be omitted, this will be explicitly indicated by adding the $1 : n$ subscripts.

Example 5.2.2 The following R code identifies the parameter vector that yields the best linear approximation of the quadratic `lcg` function used to generate class labels for the example training dataset created in Example 5.1.1. To achieve this, the standard `lm` function for linear regression is applied to a modified copy of the dataset, containing real-valued `lcg` function labels instead of the original class labels. The resulting parameter vector (rearranged to match the different position of the intercept term assumed in our example code), referred to as the "perfect" parameter vector, may be then used to assess the potential of predictive performance that would be possible to obtain in the totally unrealistic case in which the representation function underlying class labels in the data were directly available. This assumption is only adopted to illustrate the linear representation of classification models and the parameter estimation process. The `mse` function is used to calculate the mean square error for the linear approximation of `lcg` using the "perfect" parameter vector. The `predict.par` function is applied to obtain parametric model predictions.

> Ex. 10.2.3
> dmr.regeval

> Ex. 8.2.1
> dmr.linreg

```
lcdat.train.lr <- lcdat.train[,1:4]
lcdat.train.lr$g <- lcg(lcdat.train$a1, lcdat.train$a2,
                        lcdat.train$a3, lcdat.train$a4)

  # "perfect" parameter vector
w.perf <- lm(g~., lcdat.train.lr)$coef[c(2:5, 1)]

  # "perfect" predictions
mse(predict.par(list(repf=repf.linear, w=w.perf), lcdat[,1:4]),
    lcg(lcdat$a1, lcdat$a2, lcdat$a3, lcdat$a4))
```

5.2.2 Outer representation function

There are two major approaches to assigning binary class labels based on linearly represented real-valued inner predictions:

Boundary modeling. Assuming that the inner representation function represents a boundary between regions of different classes,

Probability modeling. Assuming that the inner representation function represents, possibly indirectly, class probabilities.

In boundary modeling, hypersurfaces (in the attribute value space) separating positive and negative instances, called *decision boundaries*, are represented parametrically. They partition the domain into regions, with each region assigned a class label.

Probability modeling is a family of approaches that use a parametric representation of class probabilities. For two-class tasks this reduces to representing the probability of class 1. The latter may be then used to predict class labels as with any probabilistic classifiers, i.e., by using the maximum probability rule, the minimum cost rule presented in Section 6.3.3, or adjusting operating points by the ROC analysis or similar methods, as discussed in Section 7.2.5.

These two approaches lead to the following two most commonly used types of outer representation functions for linear classification:

- threshold representation, which is a standard way to perform boundary modeling,

- logit representation, which is is the most popular instantiation of probability modeling.

We will see that, while they differ in important details, they have actually a lot in common.

5.2.3   Threshold representation

For two-class classification tasks partitioning the domain into the positive and negative regions can be easily achieved by comparing a parametric representation function against a threshold. Without loss of generality, the latter may be assumed to be 0, which yields the following model representation:

$$h(x) = H(g(x)) = \begin{cases} 1 & \text{if } g(x) \geq 0 \\ 0 & \text{otherwise} \end{cases} \tag{5.4}$$

For a threshold parametric classification model defined as above predictions are obtained by applying the unit step function H to the inner representation function g. The latter determines a hypersurface in the $(n + 1)$-dimensional space (with dimensions corresponding to $a_1, a_2, \dots, a_n$, and g). By comparing against 0 the projection of this hypersurface to n dimensions (corresponding to $a_1, a_2, \dots, a_n$) is determined. In general, it may yield one or more n-dimensional hypersurfaces where g crosses the $a_1, a_2, \dots, a_n$ hyperplane. The model function h, which is a binary-valued function in an n-dimensional space, assigns 0 or 1 to regions separated by a number of n-dimensional surfaces, obtained by the projection of an $(n + 1)$-dimensional surface.

It is common to use the sign rather than the unit step function for threshold parametric classification models, assuming class labels are from the $\{-1, 1\}$ set rather than the $\{0, 1\}$ set. This chapter sticks with the latter, to preserve consistency with conventions used for presenting other classification algorithms in this book. However, on several occasions the binary true or predicted class labels will be used as numbers in equations (and, correspondingly, code examples), whereas the discussion of classification in other chapters usually does not rely on the numeric interpretation of class labels.

The threshold representation instantiated for linear classification takes the following form:

$$h(x) = H(\mathbf{w} \circ \mathbf{a}(x)) = \begin{cases} 1 & \text{if } \mathbf{w} \circ \mathbf{a}(x) \geq 0 \\ 0 & \text{otherwise} \end{cases} \tag{5.5}$$

In this case, the decision boundary separating the domain regions assigned the 0 and 1 class labels, represented by the parameter vector, is a hyperplane in n dimensions. The target concept is said to be *linearly separable* on a dataset if there exists a hyperplane that separates all instances of different classes in the dataset (i.e., there exists a parameter vector that yields correct predictions for all instances in the dataset). The dataset is then also said to be *linearly separable* with respect to the target concept.

Example 5.2.3 The linear threshold representation is illustrated by the R code presented below, using the `lcg.plot` function from the previous example as the linear inner representation function. The comparison of its value against 0 determines two-dimensional regions that are assigned the 1 and 0 class labels using the `ustep` function for unit step calculation.

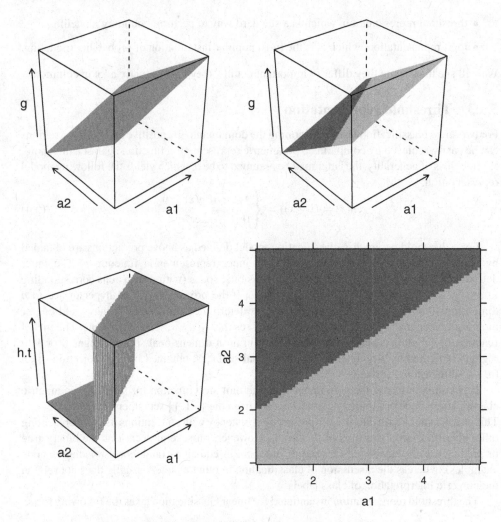

Figure 5.2   The linear threshold classification model representation.

Plots illustrating this representation are presented in Figure 5.2. The top left plot presents the plane determined by the linear inner representation function. The top right plot presents the same plane using different shades, darker when the inner representation function is below 0 and lighter elsewhere. The bottom left plot presents the effect of applying the unit step function to the values of the linear inner representation function. The bottom right plot shows the corresponding projection to two dimensions, using a darker shade for regions assigned class 0 and a lighter shade for regions assigned class 1.

```
h.t <- function(a1, a2) { ustep(lcg.plot(a1, a2)) }

lcdat.plot$h.t <- h.t(lcdat.plot$a1, lcdat.plot$a2)

wf.g.t <- wireframe(g~a1+a2, lcdat.plot, drape=TRUE, at=c(-100, 0, 100),
```

```
                    col="transparent", col.regions=c("grey30", "grey70"),
                    colorkey=FALSE, zoom=0.8)
wf.h.t <- wireframe(h.t~a1+a2, lcdat.plot, col="grey50", zoom=0.8)
l.h.t <- levelplot(h.t~a1+a2, lcdat.plot, at=c(0, 0.5, 1),
                    col.regions=c("grey30", "grey70"), colorkey=FALSE)


print(wf.g, split=c(1, 1, 2, 2), more=TRUE)
print(wf.g.t, split=c(2, 1, 2, 2), more=TRUE)
print(wf.h.t, split=c(1, 2, 2, 2), more=TRUE)
print(l.h.t, split=c(2, 2, 2, 2))
```

Example 5.2.4 The following code defines the `repf.threshold` function which takes an inner representation function on input and returns a composite representation function, with a threshold outer representation function applied. Used with the `repf.linear` function, it implements the linear threshold representation. To demonstrate the representation function, it is used in combination with the "perfect" parameter vector determined in Example 5.2.2 to create the perfect threshold model for the example dataset. The quality of its predictions, generated using the `predict.par` function, is evaluated using the misclassification error. The latter is calculated using the `err` function.

> Ex. 8.2.2
> dmr.linreg

> Ex. 8.2.1
> dmr.linreg

> Ex. 7.2.1
> dmr.claseval

```
## threshold representation function
repf.threshold <- function(repf) { function(data, w) ustep(repf(data, w)) }

  # "perfect" threshold model
perf.threshold <- `class<-`(list(repf=repf.threshold(repf.linear), w=w.perf), "par")
  # test set error
err(predict(perf.threshold, lcdat.test[,1:4]), lcdat.test$c)
```

Example 5.2.5 The R code presented below defines the `linsep.sub` function that identifies the linearly separable subset of a given dataset, following a similar approach as demonstrated in Example 5.2.2. It applies the `lm` function to perform linear regression, with class labels converted to numeric target function values from the $\{-1, 1\}$ set, and then verifies for which instances the resulting model yields correct threshold predictions. The function is applied to determine the linearly separable subsets of the example training and test sets.

```
## identify a linearly separable subset of data
linsep.sub <- function(formula, data)
{
  class <- y.var(formula)
  attributes <- x.vars(formula, data)
  aind <- names(data) %in% attributes

  wlm <- lm(make.formula(paste("(2*as.num0(", class, ")-1)", sep=""), attributes),
          data)$coef
  wpar <- c(wlm[-1], wlm[1])  # rearrange for predict.par
  predict.par(list(repf=repf.threshold(repf.linear), w=wpar), data[,aind])==
    data[[class]]
}
```

```
# linearly separable training and test subsets
lcdat.ls <- linsep.sub(c~., lcdat)
lcdat.train.ls <- lcdat[1:200,][lcdat.ls[1:200],]
lcdat.test.ls <- lcdat[201:400,][lcdat.ls[201:400],]
```

5.2.4 Logit representation

The most popular approach to parametric class probability modeling uses the *logit* representation

$$P(1|x) = \frac{e^{g(x)}}{e^{g(x)} + 1} \tag{5.6}$$

The inner representation function g does not therefore represent directly $P(1|x)$, but rather the *logit* or *log-odds* thereof:

$$g(x) = \text{logit}(P(1|x)) \tag{5.7}$$

where for $p \in [0, 1]$:

$$\text{logit}(p) = \ln \frac{p}{1 - p} \tag{5.8}$$

and therefore:

$$g(x) = \ln \frac{P(1|x)}{1 - P(1|x)} = \ln \frac{P(1|x)}{P(0|x)} \tag{5.9}$$

$$P(1|x) = \text{logit}^{-1}(g(x)) \tag{5.10}$$

The effective representation function for $P(1|x)$ is then the composite of the inverse logit function logit^{-1} (also called the logistic function) and the inner representation function g.

The logit representation combined with a linear inner representation function

$$P(1|x) = \text{logit}^{-1}(\mathbf{w} \circ \mathbf{a}(x)) = \frac{e^{\mathbf{w} \circ \mathbf{a}(x)}}{e^{\mathbf{w} \circ \mathbf{a}(x)} + 1} \tag{5.11}$$

is used to represent *linear logit* classification models, more commonly referred to as *logistic regression* models.

Notice that under the maximum-probability prediction rule, the condition for instance x to be assigned class 1

$$P(1|x) \geq P(0|x) \tag{5.12}$$

entails (as long as $P(1|x) < 1$):

$$\frac{P(1|x)}{P(0|x)} > 0 \tag{5.13}$$

and therefore

$$g(x) = \ln \frac{P(1|x)}{P(0|x)} > 0 \tag{5.14}$$

This shows that the maximum-probability class predictions for the logit representation are identical to those produced by the threshold representation if the underlying inner

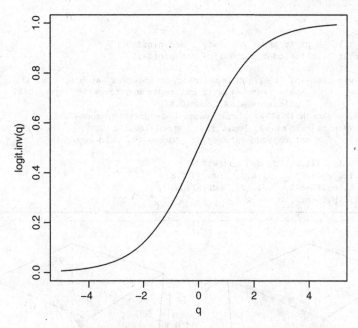

Figure 5.3   The inverse logit function.

representation function and model parameters are the same. The advantage of the logit representation, though, is the capability of predicting class probabilities.

Example 5.2.6 The following R code defines the `logit.inv` function and plots the inverse logit (logistic) curve. It is a sigmoid curve that can be seen as a smoothed and differentiable counterpart of the unit step function. The plot is presented in Figure 5.3.

The inverse logit function is then applied to the `lcg.plot` inner representation function to obtain class probabilities and class label assignments based on the logit representation. The representation is visualized by a series of subsequently generated plots, presented in Figure 5.4. The top left plot illustrates the inner representation function. The top right plot presents the corresponding probabilities of class 1, plotted in a darker shade where below 0.5 and in a lighter shade elsewhere. The bottom left plot presents the effect of applying the maximum-probability rule to assign 0 or 1 class labels. The bottom right plot shows the corresponding projection to two dimensions, a darker shade for regions assigned class 0 and a lighter shade for regions assigned class 1. As expected, the two bottom plots that show class label assignments are identical to those produced in Example 5.2.3.

```
logit.inv <- function(q) { (e <- exp(q))/(e+1) }

curve(logit.inv(x), from=-5, to=5, xlab="q", ylab="logit.inv(q)")

p1.lt <- function(a1, a2) { logit.inv(lcg.plot(a1, a2)) }
h.lt <- function(a1, a2) { ustep(p1.lt(a1, a2), 0.5) }
```

```
lcdat.plot$p1.lt <- p1.lt(lcdat.plot$a1, lcdat.plot$a2)
lcdat.plot$h.lt <- h.lt(lcdat.plot$a1, lcdat.plot$a2)

wf.p1.lt <- wireframe(p1.lt~a1+a2, lcdat.plot, drape=TRUE, at=c(0, 0.5, 1),
                      col="transparent", col.regions=c("grey30", "grey70"),
                      colorkey=FALSE, zoom=0.8)
wf.h.lt <- wireframe(h.lt~a1+a2, lcdat.plot, col="grey50", zoom=0.8)
l.h.lt <- levelplot(h.lt~a1+a2, lcdat.plot, at=c(-100, 0, 100),
                    col.regions=c("grey30", "grey70"), colorkey=FALSE)

print(wf.g, split=c(1, 1, 2, 2), more=TRUE)
print(wf.p1.lt, split=c(2, 1, 2, 2), more=TRUE)
print(wf.h.lt, split=c(1, 2, 2, 2), more=TRUE)
print(l.h.lt, split=c(2, 2, 2, 2))
```

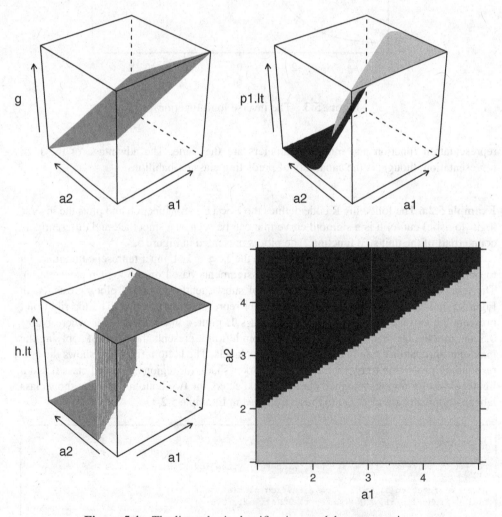

Figure 5.4   The linear logit classification model representation.

Example 5.2.7 The code presented below implements wrapper generation around an inner representation function, combining it with the logit outer representation to produce a composite logit representation function. When applied to the `repf.linear` function, this yields the linear logit representation. To provide a demonstration, the prediction quality of a linear logit model using the "perfect" parameter vector from Example 5.2.2 is evaluated. Model quality is measured by the misclassification error for class label prediction and by the loglikelihood for class probability predictions. The latter is calculated using the `loglik01` function.

> Ex. 8.2.2
> `dmr.linreg`

> Ex. 7.2.17
> `dmr.claseval`

```
## logit representation function
repf.logit <- function(repf) { function(data, w) logit.inv(repf(data, w)) }

  # "perfect" logit model
perf.logit <- `class<-`(list(repf=repf.logit(repf.linear), w=w.perf), "par")
  # test set error
err(ustep(predict(perf.logit, lcdat.test[,1:4]), 0.5), lcdat.test$c)
  # test set loglikelihood
loglik01(predict(perf.logit, lcdat.test[,1:4]), lcdat.test$c)
```

5.3   Parameter estimation

Parameter estimation is the process of identifying model parameters based on a given training set that are likely to yield good performance. This can be viewed as an optimization process in which the space of possible parameter vectors is searched for the one that optimizes an adopted performance measure. In general, several different performance measures and optimization methods could be used for this purpose.

One strategy that is applicable to the composite parametric classification model representations presented above is to adopt the delta rule and the gradient descent algorithm for linear regression, presented in Section 8.3, to the estimation of linear classification model parameters. This is a particularly simple approach to the underlying optimization problem that may suffer from slow convergence, and the resulting algorithms can be explained and understood using elementary maths, as well as illustrated by plain-vanilla implementations.

5.3.1   Delta rule

Unfortunately, applying the delta rule directly to parametric classification may take more than just replacing the real-valued target function f with a discrete (binary) target concept c. Even though the 0 and 1 class labels can be used as numbers, the resulting update rule may not be valid with respect to a reasonable classification performance measure. It is not even directly applicable to the parametric classification representations presented above. This is because h is nondifferentiable due to using the step function for the threshold representation and probability-based class label assignment for the logit representation.

Rather than trying to directly apply the delta rule to a classification model h just by replacing f with c, we will consider therefore its application to the underlying inner representation function g. The correct way of doing this clearly depends on the particular type of classification representation and will be discussed separately for the three representations presented before.

5.3.1.1   Threshold representation

For the threshold representation, applying the delta rule is vastly simplified by the fact that it is only the sign of the inner representation function that really matters and its values can be arbitrary as long as they are on the right side of 0. This makes it easier to overcome the obstacles encountered when trying to directly rewrite the delta rule for this representation as follows:

$$\mathbf{w} := \mathbf{w} + \beta(c(x) - H(g(x)))\nabla_{\mathbf{w}}H(g(x)) \tag{5.15}$$

temporarily ignoring the nondifferentiability of the unit step function H. If the latter were approximated by a differentiable sigmoid counterpart, its derivative would take a maximum value at 0 (corresponding to the slope of the corresponding sigmoid curve when crossing 0) and gradually decay when moving away from 0 either side. This has the effect of performing the largest parameter updates near the decision boundary, and smaller or negligible updates farther away. Notice, however, that the latter is not necessary if the actual values of g are irrelevant and only its sign matters, and therefore $\nabla_{\mathbf{w}}H(g(x))$ can be safely replaced with $\nabla_{\mathbf{w}}g(x)$. For the same reason, the step-size β can be omitted (assumed to be 1). This makes it possible to write down the delta rule for the threshold representation in the following simplified form:

$$\mathbf{w} := \mathbf{w} + (c(x) - h(x))\nabla_{\mathbf{w}}g(x)$$

$$= \begin{cases} \mathbf{w} + \nabla_{\mathbf{w}}g(x) & \text{if } c(x) = 1 \text{ and } h(x) = 0 \\ \mathbf{w} - \nabla_{\mathbf{w}}g(x) & \text{if } c(x) = 0 \text{ and } h(x) = 1 \\ \mathbf{w} & \text{otherwise} \end{cases} \tag{5.16}$$

which can be further simplified to

$$\mathbf{w} := \begin{cases} \mathbf{w} + c_-(x)\nabla_{\mathbf{w}}g(x) & \text{if } h(x) \neq c(x) \\ \mathbf{w} & \text{otherwise} \end{cases} \tag{5.17}$$

where $c_-(x) = 2c(x) - 1$ is the counterpart of c with class labels converted from $\{0, 1\}$ to $\{-1, 1\}$ for arithmetic convenience.

This leaves model parameters unchanged if instance x is classified correctly and modifies them by adding or subtracting $\nabla_{\mathbf{w}}g(x)$ otherwise. If a parameter vector that yields no misclassifications on the training set is eventually arrived at, no further parameter updates will occur, i.e., the parameter estimation process will converge. If such a parameter vector does not exist (i.e., separating classes perfectly on the training set is impossible using the adopted inner representation function), parameter updates will continue to take place indefinitely long (unless a gradually decaying step size value is used), although a reasonably good parameter vector may be sometimes obtainable by stopping the parameter estimation process at some point when no further improvement seems to be possible.

The above rule is easily instantiated for linear threshold classification by substituting the inner representation function gradient. Since $\nabla_{\mathbf{w}}g(x) = \mathbf{a}(x)$, the error-minimization delta rule for the linear threshold representation takes the following form:

$$\mathbf{w} := \begin{cases} \mathbf{w} + c_-(x)\mathbf{a}(x) & \text{if } h(x) \neq c(x) \\ \mathbf{w} & \text{otherwise} \end{cases} \tag{5.18}$$

Iteratively applied, this yields a parameter-estimation algorithm for the linear threshold classifier, often referred to as the *perceptron* algorithm.

The above discussion is by no means a proper derivation of the parameter update rule for linear threshold classification and serves only as an intuitive explanation of its relationships to the linear regression delta rule. Section 5.3.3 will explain why it works by more rigorous arguments.

Example 5.3.1 The following R code defines the `grad.threshold` function that implements gradient calculation for the threshold representation function, using the specified inner representation function and its gradient. This is necessary to use the `gradient.descent` function, as will be demonstrated later. As discussed above, the derivative of the nondifferentiable unit step function is taken to be 1, and therefore simply the supplied inner representation function gradient is returned. Applied to the `grad.linear` function, implementing the linear representation gradient, it yields the linear threshold gradient.

| Ex. 8.3.3 |
| dmr.linreg |

| Ex. 8.3.1 |
| dmr.linreg |

```
## threshold representation gradient
grad.threshold <- function(grad) { grad }

  # linear threshold gradient for the "perfect" parameter vector
grad.threshold(grad.linear)(lcdat.train[1:10,1:4], w.perf)
```

5.3.1.2 Logit representation

Since in the logit representation it is the class 1 probability that is represented parametrically, an appropriate probabilistic performance measure is needed to guide the parameter estimation process. This is necessary to ensure that the resulting model parameter vector will indeed yield probability estimates that fit the training set. Two such directly related measures, the likelihood and the loglikelihood, are presented in Section 7.2.6. It may be unclear how to modify the delta rule to suite the adopted performance measure, but it can be derived following the pattern of its derivation for regression from Section 8.3.1.

Consider the likelihood of training set T with respect to the target concept c given model probability estimates π:

$$P(T, c|\pi) = \prod_{x \in T} P(x|\pi) = \prod_{x \in T} P(c(x)|x) \tag{5.19}$$

where

$$P(1|x) = \pi(x) \tag{5.20}$$
$$P(0|x) = 1 - \pi(x) \tag{5.21}$$

Notice that

$$P(c(x)|x) = \pi(x)^{c(x)}(1 - \pi(x))^{1-c(x)} \tag{5.22}$$

and therefore

$$P(T, c|\pi) = \prod_{x \in T} \left(\pi(x)^{c(x)}(1 - \pi(x))^{1-c(x)} \right) \tag{5.23}$$

Let us now take the natural logarithm of the above expression as the loglikelihood of the training set T with respect to the target concept c given model probability estimates π:

$$L_{T,c}(\pi) = \sum_{x \in T} \Big(c(x) \ln \pi(x) + (1 - c(x)) \ln (1 - \pi(x)) \Big) \tag{5.24}$$

Of course, unlike the misclassification error, both the likelihood and the loglikelihood need to maximized by parameter estimation, which would therefore lead to *gradient ascent* rather than gradient descent parameter estimation. Equivalently, one may consider minimizing the logarithmic loss (log-loss) defined as the negated loglikelihood.

Using the loglikelihood as the underlying performance measure for the delta rule

$$\mathbf{w} := \mathbf{w} + \beta \nabla_{\mathbf{w}} L_{T,c}(\pi) \tag{5.25}$$

requires determining its gradient with respect to the parameter vector. We can proceed with this as follows:

$$\nabla_{\mathbf{w}} L_{T,c}(\pi) = \sum_{x \in T} \left(c(x) \frac{1}{\pi(x)} + (1 - c(x)) \frac{-1}{1 - \pi(x)} \right) \nabla_{\mathbf{w}} \pi(x)$$

$$= \sum_{x \in T} \frac{c(x) - \pi(x)}{\pi(x)(1 - \pi(x))} \nabla_{\mathbf{w}} \pi(x) \tag{5.26}$$

Since for the logit representation

$$\pi(x) = \frac{e^{g(x)}}{e^{g(x)} + 1} = \frac{1}{1 + e^{-g(x)}} \tag{5.27}$$

we have

$$\nabla_{\mathbf{w}} \pi(x) = \pi(x)(1 - \pi(x)) \nabla_{\mathbf{w}} g(x) \tag{5.28}$$

Notice, by the way, that the possibility of calculating the derivative value based on the function value for the same argument is a nice property of the logistic function that is often exploited by parameter estimation algorithms that use the function in their model representation. Finally, we arrive at the following form of the loglikelihood gradient:

$$\nabla_{\mathbf{w}} L_{T,c} = \sum_{x \in T} (c(x) - \pi(x)) \nabla_{\mathbf{w}} g(x) \tag{5.29}$$

which surprisingly yields the same delta rule form as given by Equation 8.15 for parametric regression, with just $c(x)$ and $\pi(x)$ appearing in place of $f(x)$ and $h(x)$:

$$\mathbf{w} := \mathbf{w} + \beta \sum_{x \in T} (c(x) - \pi(x)) \nabla_{\mathbf{w}} g(x) \tag{5.30}$$

The corresponding incremental delta rule, describing the per-instance parameter update, is obtained by dropping the summation

$$\mathbf{w} := \mathbf{w} + \beta (c(x) - \pi(x)) \nabla_{\mathbf{w}} g(x) \tag{5.31}$$

Given $\nabla_{\mathbf{w}} g(x) = \mathbf{a}(x)$, this further becomes

$$\mathbf{w} := \mathbf{w} + \beta (c(x) - \pi(x)) \mathbf{a}(x) \tag{5.32}$$

for linear logit classification.

It turns out that modifying parameters as if one were attempting to minimize the mean square error of class 1 probabilities compared to true class labels on the training set results in maximizing the loglikelihood of the latter when using the logit representation. This is definitely a nice property of this representation.

Example 5.3.2 The following R code defines the `grad.logit` function that generates the logit representation function gradient wrapper around the specified inner representation function and its gradient. Applied to the linear representation function and gradient, it yields the linear logit gradient. It can be seen to implement Equation 5.28, using the `rmm` utility function to multiply all rows of the gradient matrix obtained for the inner representation function by the corresponding $\pi(x)(1 - \pi(x))$ values.

`dmr.util`

```
## logit representation gradient
grad.logit <- function(repf, grad)
{ function(data, w) rmm(grad(data, w), (p <- repf.logit(repf)(data, w))*(1-p)) }

    # linear logit gradient for the "perfect" parameter vector
grad.logit(repf.linear, grad.linear)(lcdat.train[1:10,1:4], w.perf)
```

5.3.2   Gradient descent

Either the batch or incremental (stochastic) gradient descent algorithm can be used in combination with the delta rules derived above for the threshold and logit parametric classification. For the latter, it would be actually *gradient ascent*, given the fact the delta rule is derived to maximize the loglikelihood rather than minimize the error. No changes in the gradient-based algorithms for parametric regression presented in Section 8.3 are required other than substituting the appropriate delta rules.

The simple form of the inner linear representation function makes the optimization process relatively fast, computationally efficient, and resistant to local optima, which is a considerable advantage over nonlinear representations.

Example 5.3.3 The R code presented below implements the delta rule for the threshold and logit representations, providing the last missing piece necessary to apply the gradient descent parameter estimation algorithm.. For the threshold representation the delta rule, assuming the objective of misclassification error minimization, is implemented by the `delta.err` function, which is actually the same as `delta.mse` from Example 8.3.2 for regression. This is justified by Equation 5.16. For the logit representation the delta rule is implemented by the `delta.loglik` function, based on Equation 5.26 which assumes loglikelihood maximization. The two delta rule implementations are then used in example calls that apply the `gradient.descent` function to estimate parameters of linear threshold and logit models for the example artificial data (both the original dataset and its linearly separable subset). Similar example calls are presented for the *Pima Indians Diabetes* dataset, which is not linearly separable. For the latter, the binary 0/1 predictions have to be converted to the original class labels. Notice that the negated loglikelihood is used as the performance measure specified via the `perf` argument for the logit representation, so that the stop criterion of the gradient descent algorithm works correctly.

Ex. 8.3.3
`dmr.linreg`

The performance of the obtained models is evaluated by calculating their training and test set error and loglikelihood values.

```
## calculate parameter update based on given true and predicted values,
## gradient, and step-size using the delta rule for loglikelihood minimization
delta.loglik <- function(true.y, pred.y, gr, beta)
{
  d <- ifelse(is.finite(d <- rmm(gr, 1/(pred.y*(1-pred.y)))), d, 1)
  colSums(beta*rmm(d, true.y-pred.y))
}

## calculate parameter update based on given true and predicted values,
## gradient, and step-size using the delta rule for error minimization
delta.err <- delta.mse

  # linear threshold for the artificial data
gdl.th <- gradient.descent(c~., lcdat.train, w=rep(0, 5),
                           repf=repf.threshold(repf.linear),
                           grad=grad.threshold(grad.linear),
                           delta=delta.err, perf=err,
                           beta=1, batch=TRUE, eps=0.03)
gdl.th.ls <- gradient.descent(c~., lcdat.train.ls, w=rep(0, 5),
                           repf=repf.threshold(repf.linear),
                           grad=grad.threshold(grad.linear),
                           delta=delta.err, perf=err,
                           beta=1, batch=TRUE, eps=0.001)

  # linear logit for the artificial data
gdl.lt <- gradient.descent(c~., lcdat.train, w=rep(0, 5),
                           repf=repf.logit(repf.linear),
                           grad=grad.logit(repf.linear, grad.linear),
                           delta=delta.loglik, perf=function(p, y) -loglik01(p, y),
                           beta=0.01, batch=TRUE, eps=15.4)
gdl.lt.ls <- gradient.descent(c~., lcdat.train.ls, w=rep(0, 5),
                           repf=repf.logit(repf.linear),
                           grad=grad.logit(repf.linear, grad.linear),
                           delta=delta.loglik,
                           perf=function(p, y) -loglik01(p, y),
                           beta=0.1, batch=TRUE, eps=3)

  # linear threshold for the Pima Indians Diabetes data
pid.gdl.th <- gradient.descent(diabetes~., pid.train, w=rep(0, ncol(pid.train)),
                           repf=repf.threshold(repf.linear),
                           grad=grad.threshold(grad.linear),
                           delta=delta.err, perf=err,
                           beta=1, batch=TRUE, eps=0.28, niter=10000)

  # linear logit for the Pima Indians Diabetes data
pid.gdl.lt <- gradient.descent(diabetes~., pid.train, w=rep(0, ncol(pid.train)),
                           repf=repf.logit(repf.linear),
                           grad=grad.logit(repf.linear, grad.linear),
                           delta=delta.loglik,
                           perf=function(p, y) -loglik01(p, y),
                           beta=1e-7, batch=TRUE, eps=250, niter=1e6)

  # training set error
err(predict(gdl.th$model, lcdat.train[,1:4]), lcdat.train$c)
err(ustep(predict(gdl.lt$model, lcdat.train[,1:4]), 0.5), lcdat.train$c)
```

```
err(predict(gdl.th.ls$model, lcdat.train.ls[,1:4]), lcdat.train.ls$c)
err(ustep(predict(gdl.lt.ls$model, lcdat.train.ls[,1:4]), 0.5), lcdat.train.ls$c)

err(factor(predict(pid.gdl.th$model, pid.train[,-9]),
           levels=0:1, labels=levels(pid.train$diabetes)),
           pid.train$diabetes)
err(factor(ustep(predict(pid.gdl.lt$model, pid.train[,-9]), 0.5),
           levels=0:1, labels=levels(pid.train$diabetes)),
     pid.train$diabetes)

  # training set loglikelihood
loglik01(predict(gdl.th$model, lcdat.train[,1:4]), lcdat.train$c)
loglik01(predict(gdl.lt$model, lcdat.train[,1:4]), lcdat.train$c)

loglik01(predict(gdl.th.ls$model, lcdat.train.ls[,1:4]), lcdat.train.ls$c)
loglik01(predict(gdl.lt.ls$model, lcdat.train.ls[,1:4]), lcdat.train.ls$c)

loglik01(predict(pid.gdl.th$model, pid.train[,-9]), pid.train$diabetes)
loglik01(predict(pid.gdl.lt$model, pid.train[,-9]), pid.train$diabetes)

  # test set error
err(predict(gdl.th$model, lcdat.test[,1:4]), lcdat.test$c)
err(ustep(predict(gdl.lt$model, lcdat.test[,1:4]), 0.5), lcdat.test$c)

err(predict(gdl.th.ls$model, lcdat.test.ls[,1:4]), lcdat.test.ls$c)
err(ustep(predict(gdl.lt.ls$model, lcdat.test.ls[,1:4]), 0.5), lcdat.test.ls$c)

err(factor(predict(pid.gdl.th$model, pid.test[,-9]),
           levels=0:1, labels=levels(pid.train$diabetes)),
     pid.test$diabetes)
err(factor(ustep(predict(pid.gdl.lt$model, pid.test[,-9]), 0.5),
           levels=0:1, labels=levels(pid.train$diabetes)),
     pid.test$diabetes)

  # test set loglikelihood
loglik01(predict(gdl.th$model, lcdat.test[,1:4]), lcdat.test$c)
loglik01(predict(gdl.lt$model, lcdat.test[,1:4]), lcdat.test$c)

loglik01(predict(gdl.th.ls$model, lcdat.test.ls[,1:4]), lcdat.test.ls$c)
loglik01(predict(gdl.lt.ls$model, lcdat.test.ls[,1:4]), lcdat.test.ls$c)

loglik01(predict(pid.gdl.th$model, pid.test[,-9]), pid.test$diabetes)
loglik01(predict(pid.gdl.lt$model, pid.test[,-9]), pid.test$diabetes)
```

When applied to the artificial dataset, the gradient descent algorithm (used in batch mode due to the computational efficiency advantage of all-data prediction and gradient calculations compared to per-instance incremental calculations in R) successfully identifies model parameters that yield reasonably good classification performance for each of the three representations used, despite the nonlinearity of the "true" representation function used to generate the data. On the linearly separable data subset they all yield perfectly accurate classification. Not surprisingly, the threshold model is inferior to the other two with respect to the loglikelihood, as it does not predict probabilities (it is evaluated using the loglikelihood for the sake of illustration only).

For the real linearly inseparable *Pima Indians Diabetes* data, the gradient descent algorithms fails to converge with a linear threshold representation. While it apparently achieves the training set misclassification error level specified as a stop criterion, the returned parameter vector – which incorporates one additional update (due to the particular implementation of the gradient.descent function) – exhibits an extremely poor and totally useless classification performance. Much more successful results are obtained for the linear logit representation, for which the gradient descent algorithm converges – albeit very slowly – to a quite reasonable performance level.

5.3.3   Distance to decision boundary

For a linear threshold model using the parameter vector $\mathbf{w}$ the signed distance between instance x and the decision boundary (i.e., the hyperplane separating instances with positive and negative representation function values) can be calculated as follows:

$$\delta_{\mathbf{w}}(x) = \frac{\sum_{i=1}^{n} w_i a_i(x) + w_{n+1}}{\sqrt{\sum_{i=1}^{n} w_i^2}} = \frac{\mathbf{w} \circ \mathbf{a}(x)}{\|\mathbf{w}_{1:n}\|} \tag{5.33}$$

where $\mathbf{w}_{1:n}$ denotes the parameter vector with the w_{n+1} (intercept) parameter omitted and $\|\mathbf{w}_{1:n}\|$ denotes its L^2 (Euclidean) norm. This is basically a signed version of the Euclidean distance between the $\langle a_1(x), a_2(x), \ldots, a_n(x) \rangle$ point, representing the instance in the n-dimensional space of attribute values, and its projection on the $w_1 a_1(x) + w_2 a_2(x) + \cdots + w_n a_n(x) + w_{n+1} = 0$ hyperplane. The sign indicates whether the point lies on the positive or negative side of the hyperplane. This may be referred to as the signed distance between instance x and the decision boundary for the parameter vector w.

If instance x is misclassified and $c(x) = 1$, we have $\delta_{\mathbf{w}}(x) < 0$. If instance x is misclassified and $c(x) = 0$, we have $\delta_{\mathbf{w}}(x) > 0$. Therefore, the absolute distance between the *misclassified* instance x and the decision boundary is $(1 - 2c(x))\delta_{\mathbf{w}}(x) = -c_-(x)\delta_{\mathbf{w}}(x)$. Minimizing the sum of such distances *over all misclassified instances* in the training set may be considered a reasonable approach to parameter estimation for linear threshold models. Simplifying this further to the minimization of

$$\sum_{x \in T_{h \neq c}} -c_-(x)\mathbf{w} \circ \mathbf{a}(x) \tag{5.34}$$

leads to the following per-instance parameter update rule:

$$\mathbf{w} := \begin{cases} \mathbf{w} + \beta \nabla_{\mathbf{w}} c_-(x)\mathbf{w} \circ \mathbf{a}(x) & \text{if } h(x) \neq c(x) \\ \mathbf{w} & \text{otherwise} \end{cases} \tag{5.35}$$

This can be immediately seen to reduce to the very same linear threshold delta rule presented above. This better explains its effects by showing that it actually minimizes the distance between misclassified instances and the separating hyperplane. This implies, in particular, that the gradient descent algorithm using the delta rule for linear threshold models is guaranteed to find the hyperplane perfectly separating the 0 and 1 classes on the training set, if it only exists (i.e., the target concept is linearly separable on the training set), for which the quantity to be minimized is equal to 0. Otherwise, the gradient descent algorithm will not converge (i.e., parameter updates will not cease to occur, unless a gradually decaying step size value is used), although it may arrive at a reasonably good parameter vector after a number of iterations.

5.3.4   Least squares

Unlike for linear regression, with the ordinary least-squares (OLS) algorithm presented in Section 8.3.3 available as an usually preferred alternative to the iterative parameter estimation based on the delta rule, there is no closed-form solution for the minimum-error or maximum-likelihood parameter vector of linear classification models. A workaround is available, however, for linear threshold models, for which it is only the sign of the inner representation function that matters. This makes it possible to estimate model parameters as if solving a regression task with a two-valued target function, taking a negative value (e.g., -1) for instances of class 0 and a positive value, e.g., 1, for instances of class 0.

Consider the mean square error of a regression model h_- with respect to the target function $c_- : X \rightarrow \{-1, 1\}$ defined as $c_-(x) = 2c(x) - 1$, assuming the class labels from the $\{0, 1\}$ set are used as numbers:

$$\mathrm{mse}_{c_-, T}(h_-) = \frac{1}{|T|} \sum_{x \in T} (c_-(x) - h_-(x))^2 \tag{5.36}$$

While minimizing the above error does not guarantee minimizing the misclassification error of the classification model using the same parameter vector (which can be defined as $h(x) = H(h_-(x))$, it is likely to yield good classification performance, with h_- usually positive for instances of class 1 and negative for instances of class 0. Therefore using the least-squares method to estimate the parameters of regression model h_- may be considered a reasonable approximate approach to parameter estimation for the corresponding classification model h. It is not guaranteed to identify the separating hyperplane even if it exists, contrary to the delta rule with the gradient descent algorithm, but is usually much faster and easier to use due to no convergence problems.

Example 5.3.4 The following R code defines a modified version of the `ols` function adjusted for linear threshold classification. The function relabels training instances with values from the $\{-1, 1\}$ set, using the `as.num0` function, to obtain the numerical representation of the original class labels, performs the usual ordinary least-squares calculation, and returns the obtained parameter vector with the linear threshold representation function. The `x.vars` and `y.var` functions are used for extracting input and target attribute names from the supplied formula. The `ols.threshold` function is then applied to create a linear threshold model for the example artificial dataset and the linearly separable subset thereof, as well as the *Pima Indians Diabetes*.

Ex. 8.3.4
dmr.linreg

dmr.util

dmr.util

```
## estimate linear threshold model parameters using the OLS method
ols.threshold <- function(formula, data)
{
  class <- y.var(formula)
  attributes <- x.vars(formula, data)
  aind <- names(data) %in% attributes

  amat <- cbind(as.matrix(data[,aind]), intercept=rep(1, nrow(data)))
  cvec <- 2*as.num0(data[[class]])-1
  `class<-`(list(repf=repf.threshold(repf.linear),
            w=solve(t(amat)%*%amat, t(amat)%*%cvec)),
         "par")
}
```

```
# least-squares linear threshold for the artificial data
ols.th <- ols.threshold(c~., lcdat.train)
ols.th.ls <- ols.threshold(c~., lcdat.train.ls)

# least-squares linear threshold for the Pima Indians Diabetes data
pid.ols.th <- ols.threshold(diabetes~., pid.train)

# training set error
err(predict(ols.th, lcdat.train[,1:4]), lcdat.train$c)
err(predict(ols.th.ls, lcdat.train.ls[,1:4]), lcdat.train.ls$c)
err(factor(predict(pid.ols.th, pid.train[,-9]),
           levels=0:1, labels=levels(pid.train$diabetes)),
    pid.train$diabetes)

# test set error
err(predict(ols.th, lcdat.test[,1:4]), lcdat.test$c)
err(predict(ols.th.ls, lcdat.test.ls[,1:4]), lcdat.test.ls$c)
err(factor(predict(pid.ols.th, pid.test[,-9]),
           levels=0:1, labels=levels(pid.train$diabetes)),
    pid.test$diabetes)
```

Notice that the OLS method does not yield a perfectly accurate model for the linearly separable subset of the artificial data. It works reasonably well for the more realistic and linearly inseparable *Pima Indians Diabetes*, though, for which the gradient descent algorithm failed to converge.

5.4   Discrete attributes

It is common for classification tasks to involve discrete attributes, either alone or along with continuous attributes. For parametric classification to be applicable to such tasks the inner representation function and its gradient must be capable of handling discrete attributes. This can be achieved using the binary encoding attribute transformation described in Section 17.3.5 that is also used for linear regression. It is based on replacing (explicitly, in the dataset, or implicitly, in internal algorithm calculations only) a discrete k-valued attribute $a : X \rightarrow \{v_1, v_2, \dots , v_k\}$ with k or – in the preferred nonredundant form – $k - 1$ binary attributes that can then be treated as continuous (i.e., numerical) for all calculations. These binary attributes are usually assumed to take values from the $\{0, 1\}$ or $\{-1, 1\}$ sets.

Example 5.4.1 The R code presented below creates discrete attribute-enabled wrappers around the linear logit representation function and its gradient using the repf.disc and grad.disc functions. This is sufficient to apply the gradient.descent function with discrete attributes Actually, all functions for linear classification parameter estimation could be applied to datasets with discrete attributes after transforming them using the discode function. For the purpose of illustration, however, a modified discrete attribute-enabled version of the ols.threshold function is defined below that applies this data transformation internally. It also takes care of setting the representation function of the returned model appropriately using repf.disc, so that no explicit data transformation is needed for prediction.

> Ex. 8.4.1
> dmr.linreg

> Ex. 17.3.5
> dmr.trans

Linear classification parameter estimation with discrete attributes is then demonstrated for the *weatherc* data, containing two discrete and two continuous attributes. The original no and yes class labels are assigned to the obtained 0 and 1 predictions.

```
## estimate linear threshold model parameters using the OLS method
## for data with discrete attributes
ols.threshold.disc <- function(formula, data)
{
  class <- y.var(formula)
  attributes <- x.vars(formula, data)
  aind <- names(data) %in% attributes

  amat <- cbind(as.matrix(discode(~., data[,aind])), intercept=rep(1, nrow(data)))
  cvec <- 2*as.num0(data[[class]])-1
  `class<-`(list(repf=repf.disc(repf.threshold(repf.linear)),
              w=solve(t(amat)%*%amat, t(amat)%*%cvec)),
          "par")
}

  # gradient descent for the weatherc data
w.gdl <- gradient.descent(play~., weatherc, w=rep(0, 6),
                          repf=repf.disc(repf.threshold(repf.linear)),
                          grad=grad.disc(grad.threshold(grad.linear)),
                          delta=delta.mse, perf=err,
                          beta=1, batch=TRUE, eps=0.2)

  # OLS for the weatherc data
w.ols <- ols.threshold.disc(play~., weatherc)

  # training set error
err(factor(ustep(predict(w.gdl$model, weatherc[,1:4]), 0.5),
          levels=0:1, labels=c("no", "yes")),
    weatherc$play)
err(factor(ustep(predict(w.ols, weatherc[,1:4]), 0.5),
          levels=0:1, labels=c("no", "yes")),
    weatherc$play)
```

As we can see, the gradient descent algorithm does not converge due to the linear insepa-rability of the *weatherc* data and the obtained error level is rather unimpressive.

5.5 Conclusion

Linear classifiers belong to the first inductive learning algorithms studied, dating back at least to the 1950s. This early work focused on linear threshold classification and was limited by the linear separability requirement. The interest in linear classification models then decayed for several decades due to this limitation, particularly given the increasing popularity of more refined algorithms, such as decision trees, that can not only cope with linearly inseparable data, but also produce human-readable models.

Recent years have seen the renewal of interest in linear classification. While still suffering from the linearity limitation and lack of human readability, they have strengths that make them an attractive choice for some applications. This is particularly the case for linear logit models, capable of class probability prediction, and explicit loglikelihood maximization. This makes them ideal probabilistic classifiers as long as the inner representation function is sufficient, and as such, well suited to applications where probabilistic prediction is needed (e.g., involving nonuniform misclassification costs, as discussed in Chapter 6). Another reason of linear classification regaining attention is the possibility of overcoming the linearity limitation by effectively transforming the task being solved from its original representation to an enhanced representation using kernel methods. In combination with an alternative approach to parameter estimation that increases the resistance to overfitting, they deliver high quality predictions in many applications. This is discussed in Chapter 16.

Unlike for regression, the parametric approach to classification is not necessarily the most common and not always the most successful. Its disadvantages include:

- very limited (if any) human readability of models (only with the simplest inner representation functions the impact of particular attributes on the predicted classes can be assessed,

- prediction quality directly dependent on the choice of the inner representation function, with the most popular linear classification insufficient in many cases,

- parameter estimation by numerical optimization that may be time-consuming and prone to failures for nonlinear inner representation functions.

These are counterbalanced, however, by the simplicity of model representation, the capability to predict class probabilities (with loglikelihood maximization) by logit classifiers.

5.6   Further readings

Linear classification models are heavily grounded in both machine learning and statistics traditions. More precisely, it is the former that mostly influenced the development of linear threshold classifiers and the latter that contributed generalized linear models on which the logit classifiers are based. This mixed origin of linear classification models is to some extent reflected by their coverage in the literature, with some books representing both the traditions (Bishop 2007; Duda and Hart 1973; Hand et al. 2001; Hastie et al. 2011; Theodoridis and Koutroumbas 2008; Witten et al. 2011) and some more focused on approaches with machine learning roots (e.g., Abu-Mostafa et al. 2012; Tan et al. 2013). Of the former, Bishop (2007) and Theodoridis and Koutroumbas (2008) are particularly noteworthy for the completeness, consistency, and depth of presentation.

Linear threshold classifiers, often referred to as perceptrons, appeared in the early romantic period of artificial intelligence and machine learning research as computational models of neurons (Rosenblatt 1958), extending even earlier related work (McCulloch and Pitts 1943). This is, by the way, how artificial neural networks originated, which form a somewhat separate specific subdomain of machine learning, contributing neurophysiologically inspired algorithms for classification, regression, and clustering (Hertz et al. 1991), including in particular nonlinear classification and regression algorithms. Some of them are also covered by general data mining and machine learning books (Bishop 2007; Hand et al. 2001; Mitchell 1997; Tan et al. 2013). Minsky and Papert (1969) in their seminal book examined the properties of perceptrons and identified their linear separability limitation. Nonlinear perceptrons with parameters estimated using the backpropagation algorithm (Rumelhart et al. 1986) remain the most popular type of neural networks.

The logit classification algorithm, often referred to as logit (or logistic) regression, is a specific instantiation of generalized linear models (GLM) introduced by Nelder and Wedderburn (1972) and later extensively discussed by McCullagh and Nelder (1989). This discussion covers, in particular, a general approach to parameter estimation and a variety of different link functions and target distributions that yield specific GLM instantiations. An in-depth presentation of several variations of logit models is given by Hilbe (2009). Ng and Jordan (2001) compared logit models and naïve Bayes classifiers. Perlich *et al.* (2003) evaluated the predictive performance of logit models and decision trees for datasets of varying size. Bishop (2007) and Agresti (2013) can also be referred to for more details about logit classification.

There are more approaches to linear classification than presented in this chapter, originating both from machine learning and statistics. The most noteworthy of the latter is linear discriminant analysis, first presented by Fisher (1936) and then extensively studied in different versions, overviewed by McLachlan (2004). The former include modified versions of the original perceptron algorithm, capable of handling linearly inseparable data (Gallant 1990) and the Winnow algorithm that applies multiplicative rather than additive parameter updates (Littlestone 1988), to faster identify a separating hyperplane, particularly if there are many irrelevant attributes. Freund and Schapire (1999) extended the perceptron algorithm to ensure a possibly large classification margin (the distance between the separating hyperplane and the nearest correctly separated positive and negative instances), achieving properties partially similar to the SVM algorithm presented in Chapter 16 in a much simpler way.

References

Abu-Mostafa YS, Magdon-Ismail M and Lin HT 2012 *Learning from Data*. AMLBook.

Agresti A 2013 *Categorical Data Analysis* 3rd edn. Wiley.

Bishop CM 2007 Pattern recognition and machine learning.

Duda RO and Hart PE 1973 *Pattern Classification and Scene Analysis*. Wiley.

Fisher RA 1936 The use of multiple measurements in taxonomic problems. *Annals of Eugenics* **7**, 179–188.

Freund Y and Schapire RE 1999 Large margin classification using the perceptron algorithm. *Machine Learning* **37**, 277–296.

Gallant SI 1990 Perceptron-based learning algorithms. *IEEE Transactions on Neural Networks* **1**, 179–191.

Hand DJ, Mannila H and Smyth P 2001 *Principles of Data Mining*. MIT Press.

Hastie T, Tibshirani R and Friedman J 2011 *The Elements of Statistical Learning: Data Mining, Inference, and Prediction* 2nd edn. Springer.

Hertz J, Krogh A and Palmer RG 1991 *Introduction to the Theory of Neural Computation*. Addison-Wesley.

Hilbe JM 2009 *Logistic Regression Models*. Chapman and Hall.

Littlestone N 1988 Learning quickly when irrelevant attributes abound: A new linear-threshold algorithm. *Machine Learning* **2**, 285–318.

McCullagh P and Nelder JA 1989 *Generalized Linear Models* 2nd edn. Chapman and Hall.

McCulloch W and Pitts W 1943 A logical calculus of the ideas immanent in nervous activity. *Bulletin of Mathematical Biophysics* **7**, 115–133.

McLachlan GJ 2004 *Discriminant Analysis and Statistical Pattern Recognition*. Wiley.

Minsky ML and Papert SA 1969 *Perceptrons*. MIT Press.

Mitchell T 1997 *Machine Learning*. McGraw-Hill.

Nelder J and Wedderburn R 1972 Generalized linear models. *Journal of the Royal Statistical Society A* **135**, 370–384.

Ng AY and Jordan MI 2001 On discriminative vs. generative classifiers: A comparison of logistic regression and naive Bayes *Advances in Neural Information Processing Systems 14 (NIPS-01)*. A Bradford Book.

Perlich C, Provost F and Simonoff JS 2003 Tree induction vs. logistic regression: A learning-curve analysis. *Journal of Machine Learning Research* **4**, 211–255.

Rosenblatt F 1958 The perceptron: A probabilistic model for information storage and organization in the brain. *Psychological Review* **65**, 386–408.

Rumelhart DE, Hinton GE and Williams RJ 1986 Learning internal representations by error propagation In *Parallel Distributed Processing: Explorations in the Microstructure of Cognition* (eds Rumelhart DE and McClelland JL) vol. 1 MIT Press.

Tan PN, Steinbach M and Kumar V 2013 *Introduction to Data Mining* 2nd edn. Addison-Wesley.

Theodoridis S and Koutroumbas K 2008 *Pattern Recognition* 4th edn. Academic Press.

Witten IH, Frank E and Hall MA 2011 *Data Mining: Practical Machine Learning Tools and Techniques* 3rd edn. Morgan Kaufmann.

6

Misclassification costs

6.1 Introduction

In several practical classification tasks, it is not devoid of significance how the performance of a classification model differs for particular classes of the target concept. Models exhibiting seemingly the same performance in terms of overall misclassification rate may vastly differ in actual utility depending on which classes they predict successfully and for which they fail. This is especially true for all kinds of diagnostic or anomaly detection tasks where some model mistakes may be more severe or more tolerable than the others.

To adequately describe the requirements for classification models in such situations, real-valued *misclassification costs* are used, assigned to particular pairs of predicted and true classes. They can not only provide additional performance criteria for model evaluation, but also – and more importantly – get incorporated into model construction, to make it cost-sensitive.

When discussing misclassification cost incorporation techniques, we will have to refer to classification model quality measures, defined in Chapter 7: the misclassification error, the weighted misclassification error, the mean misclassification cost, and the confusion matrix. This forward reference is unavoidable, because – while it is natural and logical to present model evaluation techniques *after* cost-sensitive model creation techniques – the former can be only fully justified and understood by referring to the latter.

Example 6.1.1 The discussion of techniques for incorporating misclassification costs will be illustrated by a series of examples in R, using the implementations of the decision tree and naïve Bayes classification algorithms provided by the `rpart` and `e1071` packages, applied to the *Vehicle Silhouettes* dataset, available in the `mlbench` package. The `ipred` package, providing a bagging ensemble modeling implementation, will be employed by some

---

Data Mining Algorithms: Explained Using R, First Edition. Paweł Cichosz.
© 2015 John Wiley & Sons, Ltd. Published 2015 by John Wiley & Sons, Ltd.

demonstrations. Functions from the `dmr.claseval` and `dmr.util` packages will also be used. The following code sets up the environment for these examples by loading the required packages and the dataset, as well as creating baseline (cost-insensitive) models. Apart from the original dataset, a modified two-class version is used with the `opel` and `saab` classes aggregated to a new `car` class, and the `bus` and `van` classes aggregated to a new `other` class. The original and modified datasets are randomly split into training and test subsets, for simple hold-out evaluation. A fixed initial seed of the random number generator is used for this partitioning to make the results exactly reproducible.

```
library(dmr.claseval)
library(dmr.util)

library(rpart)
library(e1071)
library(ipred)

data(Vehicle, package="mlbench")

set.seed(12)
rv <- runif(nrow(Vehicle))
v.train <- Vehicle[rv>=0.33,]
v.test <- Vehicle[rv<0.33,]

  # two-class version
Vehicle01 <- Vehicle
Vehicle01$Class <- factor(ifelse(Vehicle$Class %in% c("opel", "saab"),
                                 "car", "other"))
v01.train <- Vehicle01[rv>=0.33,]
v01.test <- Vehicle01[rv<0.33,]

  # cost-insensitive decision trees
v.tree <- rpart(Class~., v.train)
v01.tree <- rpart(Class~., v01.train)

  # cost-insensitive naive Bayes classifiers
v.nb <- naiveBayes(Class~., v.train)
v01.nb <- naiveBayes(Class~., v01.train)

  # misclassification error for cost-insensitive models
v.err.b <- list(tree=err(predict(v.tree, v.test, type="c"), v.test$Class),
                nb=err(predict(v.nb, v.test), v.test$Class))

v01.err.b <- list(tree=err(predict(v01.tree, v01.test, type="c"), v01.test$Class),
                  nb=err(predict(v01.nb, v01.test), v01.test$Class))
```

To evaluate the baseline models created in this example as well as cost-sensitive models created in subsequent examples the simple `err`, `mean.cost`, and `confmat` functions will be used, defined in Examples 7.2.1, 7.2.3, and 7.2.4 in Chapter 7, which calculate the misclassification error, mean misclassification cost, and confusion matrix, respectively. The above code uses only the first of those to calculate the test set misclassification error of the created baseline, cost-insensitive models.

```
dmr.claseval
```

6.2 Cost representation

Misclassification costs can be based on the actual *objective* cost of decisions made using a model's predictions. In some applications, the corresponding domain knowledge is available that makes it possible to precisely determine or roughly estimate the actual cost of wrong predictions, expressed in money, power consumption, human effort, or any other meaningful units. The usage of misclassification costs is not limited to such cases, though. Even if no domain knowledge-based costs can be specified, *subjective* costs can be used during model construction to make the model more sensitive to some classes that are considered more interesting or harder to predict.

6.2.1 Cost matrix

Misclassification costs can be specified as a $|C| \times |C|$ matrix ρ where $\rho[d_1, d_2]$ is the misclassification cost of predicting class d_1 for an instance of a true class d_2. The matrix is usually assumed to contain 0s on the main diagonal (i.e., $\rho[d, d] = 0$ for all $d \in C$). Typically, positive integer numbers are used for the remaining entries, with 1 corresponding to the least expensive misclassification. This is usually the most intuitive way of specifying costs, although in general arbitrary nonnegative real numbers are permitted. Such noninteger costs may be necessary if objective misclassification costs based on domain knowledge are used.

Example 6.2.1 Cost matrices for the original and two-class versions of the *Vehicle Silhouettes* dataset, to be used in subsequent examples, are created by the following R code. The cost matrices are used to calculate the mean misclassification cost achieved by the baseline cost-insensitive models on the test sets.

```
v.rm <- matrix(0, nrow=nlevels(Vehicle$Class), ncol=nlevels(Vehicle$Class),
               dimnames=list(predicted=levels(Vehicle$Class),
                             true=levels(Vehicle$Class)))

v.rm["bus","opel"] <- 7
v.rm["bus","van"] <- 0.2
v.rm["bus","saab"] <- 7
v.rm["opel","bus"] <- 1.4
v.rm["opel","saab"] <- 1
v.rm["opel","van"] <- 1.4
v.rm["saab","bus"] <- 1.4
v.rm["saab","opel"] <- 1
v.rm["saab","van"] <- 1.4
v.rm["van","bus"] <- 0.2
v.rm["van","opel"] <- 7
v.rm["van","saab"] <- 7

 # two-class version
v01.rm <- matrix(0, nrow=nlevels(Vehicle01$Class), ncol=nlevels(Vehicle01$Class),
               dimnames=list(predicted=levels(Vehicle01$Class),
                             true=levels(Vehicle01$Class)))
v01.rm["other","car"] <- 5
v01.rm["car","other"] <- 1
```

```
# mean  misclassification cost for cost-insensitive models
v.mc.b <- list(tree=mean.cost(predict(v.tree, v.test, type="c"), v.test$Class, v.rm),
            nb=mean.cost(predict(v.nb, v.test), v.test$Class, v.rm))

v01.mc.b <- list(tree=mean.cost(predict(v01.tree, v01.test, type="c"),
                      v01.test$Class, v01.rm),
              nb=mean.cost(predict(v01.nb, v01.test), v01.test$Class, v01.rm))
```

The cost matrices assumed for the original and two-class data versions are presented in Table 6.1. For the former, the most expensive misclassification is to predict bus or van for instances of class opel and saab and the least expensive is to predict bus for van or van for bus. The cost matrix includes fractional costs and costs below 1, which is rarely encountered in practice. For the two-class version of the *Vehicle Silhouettes* data, the cost of misclassifying car as other is equal to 5 and the cost of misclassifying other as car is equal to 1.

6.2.2   Per-class cost vector

Sometimes a simplified misclassification cost representation is considered, with the cost matrix replaced by a per-class cost vector ρ of length $|C|$, where $\rho[d]$ is the misclassification cost of predicting an incorrect class for an instance of class d. Misclassification costs in this cost representation may be easier to incorporate to the modeling process, as we will see later. This is why it is not uncommon to reduce the original cost matrix representation to a simpler cost vector form by averaging or summing on a per-class basis.

Note that for two-class classification tasks, for which it is particularly common to analyze nonuniform misclassification costs, the matrix and vector cost representations are exactly equivalent. This is because the cost matrix with zeros on the main diagonal (no cost of correct predictions) contains only two nonzero entries, $\rho[0, 1] = \rho[1]$ and $\rho[1, 0] = \rho[0]$.

Incorporating per-class misclassification costs is closely related to the issue of unbalanced classes, which is another problem commonly encountered in classification tasks. When the proportion of particular classes in the training set substantially differs, error-minimizing classification algorithms may deliver models that fail to correctly detect instances of the less frequent class or classes. Resampling the training set to alter the class distribution, commonly performed to overcome the issue, may be then seen to implicitly incorporate per-class misclassification costs – with mistakes for less frequent classes considered more costly. This

Table 6.1 Misclassification cost matrices for the original and two-class versions of the *Vehicle Silhouettes* data.

	c					c	
h	bus	opel	saab	van	*h*	car	other
bus	0	7	7	0.2	car	0	1
opel	1.4	0	1	1.4	other	5	0
saab	1.4	1	0	1.4			
van	0.2	7	7	0			

makes the model more sensitive to less frequent classes as they are more costly if not recognized properly. Handling unbalanced classes is therefore one particular use-case of techniques for incorporating per-class misclassification costs.

Example 6.2.2 The following R code creates per-class cost vectors for the original and two-class versions of the *Vehicle Silhouettes* dataset by simply averaging the columns of the cost matrices presented before. The average misclassification cost for a given true class is therefore used to define per-class costs. This is performed using the `rhom2c` function. The `rhoc2m` function is also defined that handles the inverse transformation, i.e., creating the matrix representation of a given per-class cost vector, which will be useful to calculate the mean misclassification cost using the `mean.cost` function, which is also demonstrated below. Of course for the two-class data version the per-class cost vector is exactly equivalent to the original cost matrix. This is why it is not used for model evaluation here and thereafter.

```
rhom2c <- function(rhom)
{
  apply(rhom, 2, sum)/(nrow(rhom)-1)
}

rhoc2m <- function(rhoc)
{
  rhom <- matrix(rhoc, nrow=length(rhoc), ncol=length(rhoc), byrow=TRUE,
                 dimnames=list(predicted=names(rhoc), true=names(rhoc)))
  `diag<-`(rhom, 0)
}

  # per-class cost vectors
v.rc <- rhom2c(v.rm)
v01.rc <- rhom2c(v01.rm)

  # per-class cost vectors in a matrix representation
v.rcm <- rhoc2m(v.rc)
v01.rcm <- rhoc2m(v01.rc)

  # misclassification cost for cost-insensitive models
  # with respect to the per-class cost vector
v.mcc.b <- list(tree=mean.cost(predict(v.tree, v.test, type="c"),
                               v.test$Class, v.rcm),
            nb=mean.cost(predict(v.nb, v.test), v.test$Class, v.rcm))
```

6.2.3   Instance-specific costs

Neither the simplified per-class cost vector nor the full cost matrix can adequately represent misclassification costs that depend not only on the classes of instances, but also on the instances themselves. Such instance-specific misclassification costs cannot be handled by typical cost-sensitive classification algorithms, although the need for such cost representation may naturally arise in some applications. This may be the case, in particular, for some anomaly or fraud detection tasks, where the cost of undetected anomalies or frauds may depend on their specific features (e.g., credit card transaction amount).

Both the misclassification cost representations discussed above, the pairwise cost matrix and the per-class cost vector, have their natural instance-specific extensions. They consist in assuming a cost function ρ that assigns to each instance $x \in X$ the corresponding instance-specific cost matrix or cost vector. For the instance-specific cost matrix representation, $\rho(x)[d_1, d_2]$ is the cost of predicting $h(x) = d_1$ if $c(x) = d_2$. For the instance-specific cost vector representation $\rho(x)[d]$ is the cost of predicting $h(x) \neq d$ if $c(x) = d$. In both cases ρ needs to be specified based on domain-specific knowledge.

This chapter focuses on the basic instance-independent cost matrix and cost vector representations, but possible extensions of the presented techniques to instance-specific costs will be suggested where appropriate.

6.3   Incorporating misclassification costs

Whereas some classification algorithms may be designed as naturally cost-sensitive, many commonly used algorithms are not. Still most (if not all) of them can be used to create classification models that try to minimize misclassification costs. There are several general techniques that can be used to make classification algorithms cost-sensitive, reviewed below. They could be considered misclassification cost wrappers that turn a cost-insensitive algorithm into a cost-sensitive algorithm without altering its internal operation.

6.3.1   Instance weighting

One approach to incorporating misclassification costs that is not algorithm specific, but applicable to a wider class of classification algorithms, is based on *instance weighting*. It consists in assigning appropriately selected weights to training instances, which will turn a weight-sensitive classification algorithm into a cost-sensitive algorithm.

A cost-insensitive classification algorithm, to be considered successful, should attempt to minimize the expected misclassification error of the generated model on new data. Whereas it cannot be usually achieved by strictly minimizing the training set error, due to the risk of overfitting, many classification algorithms do in fact perform a constrained form of such minimization – i.e., they minimize the training set error subject to some constraints supposed to prevent overfitting. These constraints are algorithm specific and directly related to the adopted inductive bias.

The same can be said about weight-sensitive algorithms, already mentioned in Section 1.3.7, with just one complement: they accept numerical weights assigned to training instances and minimize (subject to overfitting-prevention constraints) the corresponding weighted misclassification error, as defined in Section 7.2.2.

Now assume that weights are assigned to training instances based on their classes, i.e., $w_x = \omega_{c(x)}$, where ω_d for $d \in C$ is a weight value corresponding to class d. Then the weighted training set error of model h can be written as

$$e_{c,T,\omega}(h) = \frac{\sum_{x \in T_{h \neq c}} \omega_{c(x)}}{\sum_{x \in T} \omega_{c(x)}} = \beta \frac{\sum_{x \in T} \rho[h(x), c(x)]}{|T|} \qquad (6.1)$$

where

$$\beta = \frac{|T|}{\sum_{x \in T} \omega_{c(x)}} \qquad (6.2)$$

serves as a scaling factor and

$$\rho[d_1, d_2] = \begin{cases} 0 & \text{if } d_1 = d_2 \\ \omega_{d_2} & \text{otherwise} \end{cases} \tag{6.3}$$

The cost matrix ρ defined as above actually assigns misclassification costs to particular classes rather than class pairs and is therefore equivalent to the simple per-class cost vector representation. The cost matrix notation is used only for consistency with the mean misclassification cost definition presented in Section 7.2.3.

Indeed, the expression we have arrived at can be easily verified to represent the mean misclassification cost of model h with respect to concept c and cost matrix ρ on the training set, scaled by a model-independent factor. This demonstrates that, whenever all misclassifications for the same true class are associated with the same cost, using this cost as an instance weight for training instances of this class effectively turns a weight-sensitive algorithm into a cost-sensitive algorithm, since minimizing the weighted training set error is equivalent to minimizing the mean training set misclassification cost.

For the instance-specific cost vector representation, similarly setting the weight of each training instance to the corresponding misclassification cost, $w_x = \rho(x)[c(x)]$, we can see the weighted misclassification error to represent the mean misclassification cost with respect to the instance-specific cost function ρ:

$$\begin{aligned} e_{c,T,w}(h) &= \frac{\sum_{x \in T_{h \neq c}} w_x}{\sum_{x \in T} w_x} \\ &= \beta \frac{\sum_{x \in T_{h \neq c}} \rho(x)[c(x)]}{|T|} \end{aligned} \tag{6.4}$$

It is worthwhile to underline, though, that minimizing the training set mean misclassification cost with instance-specific costs may be only hoped to lead to the minimization of the corresponding expected mean misclassification cost on new data if there is a relationship between the cost function and attributes. This relationship needs to be (implicitly) discovered and generalized by the classification algorithm, similarly as the relationship between the target concept and attributes is (explicitly) discovered. This naturally happens as a part of the inductive learning process used for classification model creation. Assigning greater weights to instances that are costly to misclassify makes the model not only more sensitive to these particular instances, but also to other similar instances from the domain, implicitly assumed to be costly to misclassify, too. This implicit learning of the relationship between the cost function and attributes is therefore a side effect of the explicit learning of the relationship between the target concept and attributes.

Although the presented approach works fully only with the simplified per-class cost vector representation (or with its extended instance-specific version), it suits sufficiently well with many practical applications, which do not require the power of a full cost matrix. In particular, it is perfectly sufficient for two-class classification tasks, which occur in many kinds of diagnostic and anomaly detection applications. These are actually the most typical examples of applications that require cost-sensitivity.

Example 6.3.1 Both decision trees and the naïve Bayes classifier could easily use weighted training instances, but of the R implementations of these algorithms used for the examples in this chapter only `rpart` supports this functionality, which will be demonstrated here. The following R code defines the `mc.weight` function that is a cost-sensitive wrapper generator using the weighting approach. For an algorithm specified as `alg` – assumed to take the `weights` argument – it returns a list of two functions, for model creation and for prediction. The former uses the per-class misclassification cost vector specified via the `rho` argument to assign weights to training instances and then calls the original algorithm with these weights. The latter is actually the unchanged prediction function specified via the `predf` argument, defaulting to `predict`, since the weighting approach modifies model creation only. The `y.var` utility function is used to extract the class attribute name from the supplied R formula. The `mc.weight` function is applied   `dmr.util` to generate a weighting cost-sensitive wrapper around the `rpart` decision tree algorithm. The mean misclassification costs, errors, and confusion matrices for the trees created by this wrapper for the original and two-class versions of the *Vehicle Silhouettes* data are calculated. For mean misclassification cost calculation on the original dataset both the simplified per-class cost representation (cost vector) and the full cost matrix representation is used. It makes no sense for the two-class version of the dataset, for which these two cost representations are equivalent.

```
## generate an instance-weighting cost-sensitive wrapper
mc.weight <- function(alg, predf=predict)
{
  wrapped.alg <- function(formula, data, rho, ...)
  {
    class <- y.var(formula)
    w <- rho[data[[class]]]
    do.call(alg, list(formula, data, weights=w, ...))
  }

  list(alg=wrapped.alg, predict=predf)
}

  # weighting wrapper around rpart
rpart.w <- mc.weight(rpart, predf=function(...) predict(..., type="c"))

  # decision trees with instance weighting
v.tree.w <- rpart.w$alg(Class~., v.train, v.rc)
v01.tree.w <- rpart.w$alg(Class~., v01.train, v01.rc)

  # mean misclassification cost with respect to the cost matrix
v.mc.w <- list(tree=mean.cost(rpart.w$predict(v.tree.w, v.test), v.test$Class, v.rm))
v01.mc.w <- list(tree=mean.cost(rpart.w$predict(v01.tree.w, v01.test),
                     v01.test$Class, v01.rm))

# mean misclassification cost with respect to the per-class cost vector
v.mcc.w <- list(tree=mean.cost(rpart.w$predict(v.tree.w, v.test),
                     v.test$Class, v.rcm))

  # misclassification error
v.err.w <- list(tree=err(rpart.w$predict(v.tree.w, v.test), v.test$Class))
v01.err.w <- list(tree=err(rpart.w$predict(v01.tree.w, v01.test), v01.test$Class))
```

```
# confusion matrix
confmat(rpart.w$predict(v.tree.w, v.test), v.test$Class)
confmat(rpart.w$predict(v01.tree.w, v01.test), v01.test$Class)
```

Not surprisingly, the cost-sensitive decision trees achieve a less mean misclassification cost and a greater error than their cost-insensitive counterparts. For the original four-class data, the cost reduction occurs not only with respect to the simplified per-class costs, but also – and even more substantially – with respect to the full cost matrix. This is because weighting the `opel` and `saab` classes fives times more than the `bus` and `van` classes reduced the number of misclassifications that are the most costly under both these cost representations, increasing the number of other misclassifications instead. Some of these new misclassifications are less expensive according to the full cost matrix than according to the per-class cost vector (in particular, misclassifying `opel` as `saab` or `saab` as `opel`). This can be easily observed by comparing the confusion matrices. For the two-class data version, the improvement is achieved simply by reducing the number of the more costly `car` as `other` misclassifications and increasing the number of the less costly `other` as `car` misclassifications, which makes the decision tree biased toward the `car` class.

6.3.2   Instance resampling

The approach presented above, applicable to weight-sensitive algorithms only, can be easily adapted to arbitrary algorithms in an approximate way. Consider the effect of giving a positive integer weight w_x to a training instance x used by a weight-sensitive algorithm. The effect is, for most algorithms, equivalent to replicating the instance w_x times in the training set. This is what one can do for any algorithm. The resulting *replication* approach is therefore another algorithm-independent technique of misclassification cost incorporation.

Instance replication can exactly "simulate" weighting as long as all weights are positive integers. This is in fact quite common in practice, since misclassification costs are often specified as integers for convenience (and the least "expensive" mistake usually costs 1), but in principle it does not have to be always the case. For noninteger weights (corresponding to noninteger misclassification costs) that cannot be transformed to integers by scaling up appropriately one should therefore use a more general *instance resampling* technique in which a resampled training set is drawn from the original set at random with replacement, using the following probability distribution based on instance weights:

$$p_x = \frac{w_x}{\sum_{x' \in T} w_{x'}} \tag{6.5}$$

Such direct resampling, due to its nondeterminism, will yield different resampled training set on each application, leading to different cost-sensitive models for the same original training set, which may be undesirable. This variance can be partially reduced by a different, indirect version of resampling, using a mix of replication (for integer weight parts) and *undersampling* (for fractional weight parts). The latter consists in selecting another copy of an instance to be included in the sample with the probability equal to the fractional part of the corresponding weight. For example, x with weight w_x, we will therefore have $\lfloor w_x \rfloor$ copies and another copy with the probability $w_x - \lfloor w_x \rfloor$. This makes the expected number of copies of instance x equal

w_x, as in direct resampling, but additionally all instances are guaranteed to have at least as many copies as the integer parts of their weights.

If using indirect resampling to handle noninteger weights, it makes sense to scale them up so that they are all above 1, i.e., at least one copy of each instance will be included in the resulting data sample, with no instances entirely omitted due to undersampling. This prevents data loss, but may require enlarging the training set considerably, particularly if misclassification costs differ vastly, which may be undesirable for efficiency reasons. In such cases, weights may have to be left unchanged or even scaled down, so that their sum (which is the expected data sample size) stays within reasonable limits.

Example 6.3.2 The following R code defines the `mc.resample` function, which is an indirect resampling cost-sensitive wrapper generator for arbitrary classification algorithms. Following the pattern of the previous example, it returns a two-element list of functions, for model creation and prediction. The former replicates training instances according to the integer part of the supplied per-class cost vector `rho` and undersamples them according to the its fractional part. The latter is again the unmodified prediction function specified by the `predf` argument. Resampling wrappers around decision trees and the naïve Bayes classifier are then created and evaluated using the original and two-class versions of the *Vehicle Silhouettes* data, as in the previous example. For the original four-class dataset, the mean cost calculation is performed using both the simplified per-class costs and the full cost matrix.

```
## generate an instance-resampling cost-sensitive wrapper
mc.resample <- function(alg, predf=predict)
{
  wrapped.alg <- function(formula, data, rho, ...)
  {
    class <- y.var(formula)
    w <- rho[data[[class]]]
    rs <- na.omit(c(rep(1:nrow(data), floor(w)),
                    ifelse(runif(nrow(data))<=w-floor(w), 1:nrow(data), NA)))
    do.call(alg, list(formula, data[rs,], ...))
  }

  list(alg=wrapped.alg, predict=predf)
}

  # resampling wrapper around rpart
rpart.s <- mc.resample(rpart, predf=function(...) predict(..., type="c"))

  # resampling wrapper around naiveBayes
naiveBayes.s <- mc.resample(naiveBayes)

  # decision trees with instance resampling
v.tree.s <- rpart.s$alg(Class~., v.train, v.rc)
v01.tree.s <- rpart.s$alg(Class~., v01.train, v01.rc)

  # naive Bayes with instance resampling
v.nb.s <- naiveBayes.s$alg(Class~., v.train, v.rc)
v01.nb.s <- naiveBayes.s$alg(Class~., v01.train, v01.rc)

  # mean misclassification cost with respect to the cost matrix
v.mc.s <- list(tree=mean.cost(rpart.s$predict(v.tree.s, v.test), v.test$Class, v.rm),
```

```
            nb=mean.cost(naiveBayes.s$predict(v.nb.s, v.test),
                         v.test$Class, v.rm))
v01.mc.s <- list(tree=mean.cost(rpart.s$predict(v01.tree.s, v01.test),
                         v01.test$Class, v01.rm),
            nb=mean.cost(naiveBayes.s$predict(v01.nb.s, v01.test),
                         v01.test$Class, v01.rm))

# mean misclassification cost with respect to the per-class cost vector
v.mcc.s <- list(tree=mean.cost(rpart.s$predict(v.tree.s, v.test),
                         v.test$Class, v.rcm),
            nb=mean.cost(naiveBayes.s$predict(v.nb.s, v.test),
                         v.test$Class, v.rcm))

 # misclassification error
v.err.s <- list(tree=err(rpart.s$predict(v.tree.s, v.test), v.test$Class),
            nb=err(naiveBayes.s$predict(v.nb.s, v.test), v.test$Class))
v01.err.s <- list(tree=err(rpart.s$predict(v01.tree.s, v01.test), v01.test$Class),
            nb=err(naiveBayes.s$predict(v01.nb.s, v01.test), v01.test$Class))

 # confusion matrix
confmat(rpart.s$predict(v.tree.s, v.test), v.test$Class)
confmat(naiveBayes.s$predict(v.nb.s, v.test), v.test$Class)

confmat(rpart.s$predict(v01.tree.s, v01.test), v01.test$Class)
confmat(naiveBayes.s$predict(v01.nb.s, v01.test), v01.test$Class)
```

Instance resampling can be seen to increase the error, but reduce the mean misclassification cost. The confusion matrices show how this is obtained by reducing the number of the most costly misclassifications and increasing the number of the less expensive ones. This can be observed both for the decision tree and naïve Bayes models. For the former, one might have expected exactly the same effect as seen previously with instance weighting, but it turns out not to be the case. It is easy to verify, by inspecting the v.tree1 and v.tree2 (or v01.tree1 and v01.tree2) objects, that their splits are the same, but the trees using the replicated training sets are grown to a greater depth, as if different stop criteria were applied. The stop criteria are actually the same, specified by the default parameter setup of rpart, but with more training instances they tend to be satisfied later (on lower tree levels). The trees built with instance weighting could be made identical to those built with the equivalent instance resampling by pruning at an appropriately selected complexity level, but this is not within the scope of this example.

6.3.3   Minimum-cost rule

Instance weighting or instance resampling can deal with simplified cost matrices only (per-class cost vectors). A more powerful technique is possible, incorporating arbitrary cost matrices, for probabilistic classifiers, which are capable of predicting estimated class probabilities $P(d|x)$ for each instance $x \in X$ and class $d \in C$. Whereas, without considering misclassification costs, such probabilities are used to predict class labels using the maximum-probability rule, a different *minimum-cost rule* can be applied to incorporate misclassification costs.

Consider the expected misclassification cost associated with predicting class d for instance x which can be expressed as follows:

$$\sum_{d' \in C} \rho[d, d']P(d'|x) \tag{6.6}$$

This calculates the expected cost by considering all possible mistakes that can be made when predicting class d for instance x and summing up their cost, multiplied by their probabilities. The minimum-cost rule states that one should predict the class label that minimizes the expected misclassification cost, i.e.,

$$h(x) = \arg\min_{d \in C} \sum_{d' \in C} \rho[d, d']P(d'|x) \tag{6.7}$$

It can be easily verified that for uniform misclassification costs the rule is equivalent to the maximum-probability rule.

The minimum-cost rule can be easily extended to the instance-specific cost matrix representation of misclassification costs just by using the specific cost matrix for the instance being classified:

$$h(x) = \arg\min_{d \in C} \sum_{d' \in C} \rho(x)[d, d']P(d'|x) \tag{6.8}$$

This makes it applicable to the most powerful misclassification cost representation.

Notice that the calculation of the expected misclassification cost of predicting each possible class for instance x can also be formulated as a matrix multiplication operation $\rho\pi_x$, where π_x is the column vector of class probabilities for instance x (i.e., $P(d'|x)$ for all $d' \in C$). The result produced by this multiplication is the vector of expected misclassification cost of predicting each possible class for x. This observation makes it possible to rewrite the minimum-cost rule as

$$h(x) = \arg\min_{d \in C}(\rho\pi_x)[d] \tag{6.9}$$

In the general form presented above, the minimum-cost rule can be applied to arbitrary classification tasks and arbitrary cost matrices, as long as a probabilistic classification model is used. The particular case of two-class classification tasks with $C = \{0, 1\}$ still deserves some more attention due to its wide popularity in practical applications. Notice that for such tasks, the expected misclassification cost of predicting class 1 for instance x can be expressed as

$$P(0|x)\rho[1, 0] + P(1|x)\rho[1, 1] \tag{6.10}$$

and, assuming a zero cost of correct predictions and writing $\rho[0]$ instead of $\rho[1, 0]$, further simplified to $P(0|x)\rho[0]$. Similarly, the expected cost of predicting class 0 for instance x is $P(1|x)\rho[1]$. Now the condition for class 1 to be the minimum-cost class for instance x is that its expected cost is no greater than that associated with class 0, which can be written as

$$P(0|x)\rho[0] \leq P(1|x)\rho[1] \tag{6.11}$$

After substituting $1 - P(1|x)$ for $P(0|x)$, the inequality can be easily solved yielding

$$P(1|x) \geq \frac{\rho[0]}{\rho[0] + \rho[1]} \tag{6.12}$$

This makes it possible to write the special form of the minimum-cost rule for two-class tasks:

$$h(x) = \begin{cases} 1 & \text{if } P(1|x) \geq \frac{\rho[0]}{\rho[0]+\rho[1]} \\ 0 & \text{otherwise} \end{cases} \tag{6.13}$$

with $\rho[0]$ and $\rho[1]$ denoting the cost of misclassifying an instance of true class 0 and 1, respectively.

The rule determines a probability cutoff value that leads to the minimization of misclassification costs. Clearly the cutoff is equal to 0.5, which corresponds to the maximum-probability rule, if both types of mistakes have the same cost.

The minimum-cost rule is definitely an elegant and versatile technique for misclassification cost incorporation. Its two distinctive properties deserve particular appreciation: the capability to handle a full cost matrix, with possibly different misclassification cost for each possible pair of classes, and the possibility of using a cost-insensitive model for cost-sensitive classification. The latter is particularly useful when misclassification costs are likely to change during model exploitation or cannot be incorporated during model creation for whatever other reasons. The price for this adaptability is that no explicit representation of a cost-sensitive model is available, and a cost-insensitive model is just used in a cost-sensitive way.

Another downside is that the actual classification performance of the minimum-cost rule heavily depends on the quality of class probability predictions generated by the model to which it is applied. Even classification algorithms that are known to usually produce highly accurate models with respect to class label prediction, are not necessarily guaranteed and even likely to deliver models capable of reliable class probability prediction. This may be the case, in particular, with algorithms for which the minimization of the misclassification error is the primary objective rather than accurate class probability estimation. Many practically used classification algorithms, including in particular decision trees, can be argued to fall in this category. For such algorithms, predicted class probabilities often take only a small number of distinct values, which may come close to the 0 and 1 extremes. Even the inherently probabilistic naïve Bayes classifier may deliver poor class probability predictions due to its unsatisfied independence assumption. This is why the minimum-cost rule does not always perform in practice up to expectations, yielding only some minor mean misclassification cost reduction. It can be expected to work best with algorithms that are purposely designed to maximize the quality of class probability predictions, such as logit classification.

One possible way to improve the effectiveness of misclassification cost minimization with the minimum-cost rule is to alter the classification algorithm known to be good at predicting class labels to make it deliver better probability predictions. In particular, for decision trees possible modifications include different stop, split selection, and pruning criteria, as well as probability smoothing at leaves. A simpler approach that may be expected to work is to adopt the technique of bootstrapping, following the pattern of ensemble modeling, discussed in Chapter 15. In particular, a probabilistic version of the bagging technique from Section 15.5.1 may be employed, in which the probability predictions of multiple base models, created using the same probabilistic classification algorithm and different training set bootstrap samples, are averaged. An additional advantage is that a deterministic classification algorithm can be used for base model creation as well, and the distribution of class label predictions of the multiple obtained models yields estimated class probabilities.

Example 6.3.3 The following R code implements the minimum-cost rule in the general multiclass form, using the matrix multiplication formulation from Equation 6.9. The `mincostclas1` function classifies a single instance based on its class probability vector and the supplied cost matrix. The `mincostclas` function wraps an `apply` call to the former, making it applicable to multiple instances. It takes predicted class probabilities (with rows representing instances and named columns representing classes, exactly as returned by `predict` for most probabilistic classification models in R) and a cost matrix on input, and generates predicted class labels on output. The `mc.mincost` function is a minimum-cost wrapper generator that leaves the original model creation algorithm unchanged, but appends the (optionally) supplied misclassification cost matrix to the created model object. This keeps the interface of the wrapped modeling algorithm consistent with that used by the wrappers presented in the previous examples, although the cost matrix is actually used at the prediction time only, when class probabilities are generated and the `mincostclas` function is applied to them. The cost matrix supplied for model creation may be overridden by specifying a different cost matrix for prediction. Minimum-cost wrappers for decision trees and the naïve Bayes classifier are then created and demonstrated, similarly as in the previous examples: their class label predictions are evaluated using the misclassification error, mean misclassification cost, and confusion matrix. For decision trees, the complexity parameter `cp` is raised to 0.025 from the default value of 0.01 to prevent excessive tree depth that would push class probabilities toward 0 and 1. Although the minimum-cost rule is applied with the full cost matrix, for consistency with the previous examples the mean misclassification cost is calculated both with respect to the full cost matrix and with respect to the simplified cost matrix (per-class cost vector), which was the only cost representation handled by the techniques demonstrated previously. This clearly matters only for the original four-class dataset, as for the two-class version the cost matrix and cost vector representations are equivalent.

```
## minimum-cost rule for a single instance
mincostclas1 <- function(p, rho)
{ factor(which.min(rho%*%p), levels=1:length(p), labels=names(p)) }

## minimum-cost rule for multiple instances
mincostclas <- function(p, rho) { apply(p, 1, mincostclas1, rho) }

## generate a minimum-cost wrapper
mc.mincost <- function(alg, ppredf=predict)
{
  wrapped.alg <- function(formula, data, rho=NULL, ...)
  {
    list(model=alg(formula, data, ...), rho=rho)
  }

  wrapped.predict <- function(model, data, rho=NULL, ...)
  {
    mincostclas(ppredf(model$model, data, ...),
                if (is.null(rho)) model$rho else rho)
  }

  list(alg=wrapped.alg, predict=wrapped.predict)
}
```

```
  # minimum-cost wrapper around rpart
rpart.m <- mc.mincost(rpart)

  # minimum-cost wrapper around naiveBayes
naiveBayes.m <- mc.mincost(naiveBayes, ppredf=function(...) predict(..., type="r"))

  # decision trees with minimum-cost prediction
v.tree.m <- rpart.m$alg(Class~., v.train, v.rm, cp=0.025)
v01.tree.m <- rpart.m$alg(Class~., v01.train, v01.rm, cp=0.025)

  # naive Bayes with minimum-cost prediction
v.nb.m <- naiveBayes.m$alg(Class~., v.train, v.rm)
v01.nb.m <- naiveBayes.m$alg(Class~., v01.train, v01.rm)

  # mean misclassification cost with respect to the cost matrix
v.mc.m <- list(tree=mean.cost(rpart.m$predict(v.tree.m, v.test), v.test$Class, v.rm),
              nb=mean.cost(naiveBayes.m$predict(v.nb.m, v.test),
                           v.test$Class, v.rm))
v01.mc.m <- list(tree=mean.cost(rpart.m$predict(v01.tree.m, v01.test),
                               v01.test$Class, v01.rm),
                nb=mean.cost(naiveBayes.m$predict(v01.nb.m, v01.test),
                            v01.test$Class, v01.rm))
# mean misclassification cost with respect to the per-class cost vector
v.mcc.m <- list(tree=mean.cost(rpart.m$predict(v.tree.m, v.test),
                              v.test$Class, v.rcm),
               nb=mean.cost(naiveBayes.m$predict(v.nb.m, v.test),
                           v.test$Class, v.rcm))

  # misclassification error
v.err.m <- list(tree=err(rpart.m$predict(v.tree.m, v.test), v.test$Class),
               nb=err(naiveBayes.m$predict(v.nb.m, v.test), v.test$Class))

v01.err.m <- list(tree=err(rpart.m$predict(v01.tree.m, v01.test), v01.test$Class),
                 nb=err(naiveBayes.m$predict(v01.nb.m, v01.test), v01.test$Class))

  # confusion matrix
confmat(rpart.m$predict(v.tree.m, v.test), v.test$Class)
confmat(naiveBayes.m$predict(v.nb.m, v.test), v.test$Class)

confmat(rpart.m$predict(v01.tree.m, v01.test), v01.test$Class)
confmat(naiveBayes.m$predict(v01.nb.m, v01.test), v01.test$Class)
```

Rather surprisingly, the mean cost improvement due to the minimum-cost rule observed for the decision tree models on the original four-class data is noticeably less than previously obtained with instance weighting or replication, despite those technique's inability to handle the full cost matrix. At the same time, the misclassification error increase is greater. This may be because the minimum-cost rule does not alter the model structure in any way; it only affects class label assignment based on class probabilities. One could fully benefit from its cost minimization capability only for a perfect probabilistic model, providing reliable class probability predictions. Decision trees are built with the purpose of discriminating between classes rather than estimating probabilities, and techniques that adjust the tree structure to the misclassification costs may indeed perform better. Using relaxed stop criteria to avoid extreme probabilities at leaves may have actually prevented reaching a sufficient class discriminative power.

For the naïve Bayes models, the improvement observed for the two-class version of the data is exactly the same as for the instance resampling technique. This is because for two-class tasks, the minimum-cost rule accounts to using a probability cutoff value other than the default 0.5, which is equivalent to the corresponding modification of class prior probabilities than can be obtained by instance replication (or weighting). For the original four-class data, there is some more improvement than previously obtained using instance resampling.

6.3.4   Instance relabeling

A more refined misclassification cost handling technique based on the minimum-cost rule consists in training *instance relabeling*, and then creating an ordinary, cost-insensitive model based on the relabeled training set. The relabeling operation assigns each training instance the minimum-cost class label, determined using the minimum-cost rule. Unlike the previously described application of the minimum-cost rule, this does not change the way of model application, but – by altering the training set – the model itself. This may be an advantage particularly if a comprehensible model representation is used, such as a decision tree, because then the cost-sensitive model is explicitly represented and available for human inspection.

Since the minimum-cost rule, as shown above, is applicable both to the standard instance-independent cost matrix representation and to its instance-specific extension, so is the instance relabeling technique. Training instances may be therefore relabeled based on their specific cost matrices. Of course, the cost function that assigns cost matrices to instances must depend on instance attributes for this approach to actually reduce the expected mean misclassification cost on new data, and the relationship has to be implicitly captured by the classification algorithm applied.

Class probabilities needed to determine minimum-cost class labels for training instances do not have to be obtained using the same algorithm as the one subsequently used to create the final model based on the relabeled training set. This is a potential advantage of this technique over the minimum-cost rule alone: an algorithm optimized toward probability prediction may be applied in the relabeling phase and another algorithm (or the same algorithm with a different parameter setup) in the modeling phase. This makes it possible, in particular, to use ensemble modeling-based class probability prediction with multiple models created based on bootstrap training set samples, as already suggested above for the minimum-cost rule, and then create a single cost-sensitive model based on the relabeled training set.

Example 6.3.4 As for the techniques described before, the instance relabeling technique is implemented by the following R code as a wrapper generator. It can be applied to an arbitrary classification algorithm, specified via the `alg` argument, with a possibly different algorithm, capable of delivering probabilistic predictions, used to generate class probability estimates for training instances. The latter is specified via the `palg` argument and defaults to the same algorithm as the one specified for model creation. The modified modeling function first creates a probabilistic model used for class probability prediction for training instances and assigns them their minimum-cost class labels using the `mincostclas` function from the previous example. Then it creates the final cost-sensitive model on the relabeled data. The prediction function is left unchanged. The relabeling wrappers for decision trees and the naïve Bayes classifier are then generated and evaluated on the original and two-class versions of the *Vehicle Silhouettes* data, just like in the previous examples. For decision trees, the cp parameter is again set to 0.025 to reduce tree complexity and avoid near-0 or near-1 class probabilities. Two different relabeling decision tree wrappers are created: one with the same algorithm used both in the relabeling and modeling phases and the other which uses bootstrapping class probability estimation for relabeling. The latter is performed with the implementation of the bagging algorithm provided by the ipred package. Notice the aggregation="a" argument specified for bagging prediction, which requests that class probability averaging be used instead of class label voting.

```
## generate an instance-relabeling cost-sensitive wrapper
mc.relabel <- function(alg, palg=alg, pargs=NULL, predf=predict, ppredf=predict)
{
  wrapped.alg <- function(formula, data, rho, ...)
  {
    class <- y.var(formula)
    model <- do.call(palg, c(list(formula, data), pargs))
    prob <- ppredf(model, data)
    data[[class]] <- mincostclas(prob, rho)
    alg(formula, data, ...)
  }

  list(alg=wrapped.alg, predict=predf)
}

  # relabeling wrapper around rpart
rpart.l <- mc.relabel(rpart, pargs=list(cp=0.025),
                      predf=function(...) predict(..., type="c"))

  # relabeling wrapper around rpart using bagging for probability estimation
rpart.bagg.l <- mc.relabel(rpart, bagging,
                           pargs=list(control=rpart.control(cp=0.025)),
                           predf=function(...) predict(..., type="c"),
                           ppredf=function(...) predict(..., type="p",
                                                        aggregation="a"))

  # relabeling wrapper around naiveBayes
naiveBayes.l <- mc.relabel(naiveBayes, ppredf=function(...) predict(..., type="r"))

  # decision trees with instance relabeling
v.tree.l <- rpart.l$alg(Class~., v.train, v.rm)
v.tree.bagg.l <- rpart.bagg.l$alg(Class~., v.train, v.rm)
v01.tree.l <- rpart.l$alg(Class~., v01.train, v01.rm)
v01.tree.bagg.l <- rpart.bagg.l$alg(Class~., v01.train, v01.rm)

  # naive Bayes with instance relabeling
v.nb.l <- naiveBayes.l$alg(Class~., v.train, v.rm)
v01.nb.l <- naiveBayes.l$alg(Class~., v01.train, v01.rm)

  # mean misclassification cost with respect to the cost matrix
v.mc.l <- list(tree=mean.cost(rpart.l$predict(v.tree.l, v.test), v.test$Class, v.rm),
               tree.bagg=mean.cost(rpart.bagg.l$predict(v.tree.bagg.l, v.test),
                         v.test$Class, v.rm),
               nb=mean.cost(naiveBayes.l$predict(v.nb.l, v.test),
                            v.test$Class, v.rm))
v01.mc.l <- list(tree=mean.cost(rpart.l$predict(v01.tree.l, v01.test),
                      v01.test$Class, v01.rm),
```

```
        tree.bagg=mean.cost(rpart.bagg.l$predict(v01.tree.bagg.l, v01.test),
                            v01.test$Class, v01.rm),
        nb=mean.cost(naiveBayes.l$predict(v01.nb.l, v01.test),
                            v01.test$Class, v01.rm))

# mean misclassification cost with respect to the per-class cost vector
v.mcc.l <- list(tree=mean.cost(rpart.l$predict(v.tree.l, v.test),
                            v.test$Class, v.rcm),
            tree.bagg=mean.cost(rpart.bagg.l$predict(v.tree.bagg.l, v.test),
                            v.test$Class, v.rcm),
            nb=mean.cost(naiveBayes.l$predict(v.nb.l, v.test),
                            v.test$Class, v.rcm))

  # misclassification error
v.err.l <- list(tree=err(rpart.l$predict(v.tree.l, v.test), v.test$Class),
                tree.bagg=err(rpart.bagg.l$predict(v.tree.bagg.l, v.test),
                            v.test$Class),
                nb=err(naiveBayes.l$predict(v.nb.l, v.test), v.test$Class))
v01.err.l <- list(tree=err(rpart.l$predict(v01.tree.l, v01.test), v01.test$Class),
                tree.bagg=err(rpart.bagg.l$predict(v01.tree.bagg.l, v01.test),
                            v01.test$Class),
                nb=err(naiveBayes.l$predict(v01.nb.l, v01.test), v01.test$Class))

  # confusion matrix
confmat(rpart.l$predict(v.tree.l, v.test), v.test$Class)
confmat(rpart.bagg.l$predict(v.tree.bagg.l, v.test), v.test$Class)
confmat(naiveBayes.l$predict(v.nb.l, v.test), v.test$Class)

confmat(rpart.l$predict(v01.tree.l, v01.test), v01.test$Class)
confmat(naiveBayes.l$predict(v01.nb.l, v01.test), v01.test$Class)
```

The instance relabeling technique turns out to work better than the minimum-cost rule for decision trees on the original four-class dataset only if class probabilities are estimated with bagging. This cost-sensitive wrapper yields a substantial misclassification cost reduction (still less than with simple instance weighting or resampling, but considerably greater than observed in the previous example). Interestingly, the effect is obtained without significantly increasing the misclassification error. Unfortunately, instance relabeling fails to bring any improvement for the naïve Bayes classifier. This may result from the algorithm's inability to sufficiently well fit to the relabeled training set.

6.4   Effects of cost incorporation

There are several reasons why cost-sensitive algorithms obtained by using the wrapper techniques reviewed in this chapter may fail to deliver satisfactory results.

1. Techniques that only work with the per-class cost vector representation (instance weighting and instance resampling) may be insufficient to bring any improvement for more complex cost matrices. This is particularly likely if per-class costs obtained after flattening the latter into a vector form are not very diverse, making the effect of weighting or resampling negligible.

2. Techniques based on predicted class probabilities (minimum-cost, instance relabeling) may not reveal their potential if the underlying probabilistic classifier estimates class probabilities poorly. This is particularly likely for models created by algorithms focusing on error minimization that fit to the training set well and tend to deliver near-0 or near-1 class probabilities.

3. A classification algorithm, tweaked to minimize the mean misclassification cost instead of the misclassification error, cannot be guaranteed to succeed at the former, just as – in its original cost-insensitive form – it could not be guaranteed to succeed at the latter.

The risk of failure is inevitable in inductive learning, but sufficiently good models are obtained for many practical tasks. Similarly, techniques for misclassification cost incorporation may not minimize the mean misclassification cost, but are likely to usually reduce it compared to models produced by the original cost-insensitive algorithms.

Example 6.4.1 This example collects and graphically presents the misclassification cost and error levels achieved by all the cost incorporation techniques, with baseline models included for comparison. The following code can be used to produce barplots displaying these results. Horizontal lines show the performance levels for baseline cost-insensitive models. Of course, they depend on the particular random partitioning of the datasets into training and test subsets and, given the variance of the hold-out procedure, different partitionings are likely to yield somewhat different results.

```
 # mean misclassification cost with respect to the cost matrix
v.mc <- c(tree=v.mc.b$tree,
            tree.w=v.mc.w$tree,
            tree.s=v.mc.s$tree,
            tree.m=v.mc.m$tree,
            tree.l=v.mc.l$tree,
            tree.bagg.l=v.mc.l$tree.bagg,
            nb=v.mc.b$nb,
            nb.s=v.mc.s$nb,
            nb.m=v.mc.m$nb,
            nb.l=v.mc.l$nb)


v01.mc <- c(tree=v01.mc.b$tree,
            tree.w=v01.mc.w$tree,
            tree.s=v01.mc.s$tree,
            tree.m=v01.mc.m$tree,
            tree.l=v01.mc.l$tree,
            tree.bagg.l=v01.mc.l$tree.bagg,
            nb=v01.mc.b$nb,
            nb.s=v01.mc.s$nb,
            nb.m=v01.mc.m$nb,
            nb.l=v01.mc.l$nb)

 # mean misclassification cost with respect to the per-class cost vector
v.mcc <- c(tree=v.mcc.b$tree,
            tree.w=v.mcc.w$tree,
            tree.s=v.mcc.s$tree,
            tree.m=v.mcc.m$tree,
            tree.l=v.mcc.l$tree,
```

```
            tree.bagg.l=v.mcc.l$tree.bagg,
            nb=v.mcc.b$nb,
            nb.s=v.mcc.s$nb,
            nb.m=v.mcc.m$nb,
            nb.l=v.mcc.l$nb)

  # misclassification error
v.err <- c(tree=v.err.b$tree,
            tree.w=v.err.w$tree,
            tree.s=v.err.s$tree,
            tree.m=v.err.m$tree,
            tree.l=v.err.l$tree,
            tree.bagg.l=v.err.l$tree.bagg,
            nb=v.err.b$nb,
            nb.s=v.err.s$nb,
            nb.m=v.err.m$nb,
            nb.l=v.err.l$nb)

v01.err <- c(tree=v01.err.b$tree,
            tree.w=v01.err.w$tree,
            tree.s=v01.err.s$tree,
            tree.m=v01.err.m$tree,
            tree.l=v01.err.l$tree,
            tree.bagg.l=v01.err.l$tree.bagg,
            nb=v01.err.b$nb,
            nb.s=v01.err.s$nb,
            nb.m=v01.err.m$nb,
            nb.l=v01.err.l$nb)

barplot(v.mc, ylab="Mean cost (matrix)", las=2)
lines(c(0, 12), rep(v.mc[1], 2), lty=2)
lines(c(0, 12), rep(v.mc[7], 2), lty=3)

barplot(v.mcc, ylab="Mean cost (per-class)", las=2)
lines(c(0, 12), rep(v.mcc[1], 2), lty=2)
lines(c(0, 12), rep(v.mcc[7], 2), lty=3)

barplot(v.err, ylab="Error", las=2)
lines(c(0, 12), rep(v.err[1], 2), lty=2)
lines(c(0, 12), rep(v.err[7], 2), lty=3)

barplot(v01.mc, ylab="Mean cost (two-class)", las=2)
lines(c(0, 12), rep(v01.mc[1], 2), lty=2)
lines(c(0, 12), rep(v01.mc[7], 2), lty=3)

barplot(v01.err, ylab="Error (two-class)", las=2)
lines(c(0, 12), rep(v01.err[1], 2), lty=2)
lines(c(0, 12), rep(v01.err[7], 2), lty=3)
```

The obtained barplots are presented in Figures 6.1 and 6.2. It can be immediately seen that all misclassification cost incorporation techniques increase the misclassification error, which is to be expected. The simplest weighting and resampling wrappers turn out to be the most successful in misclassification cost reduction for decision trees, followed by the relabeling wrapper with bagging-estimated class probabilities. The minimum-cost rule and plain relabeling are much less effective, but they still improve over the baseline cost-insensitive tree. The cost reduction observed for naïve Bayes models is less significant and the relabeling techniques appear not to work at all.

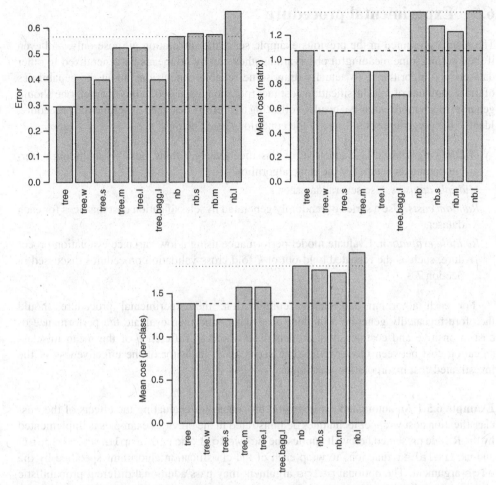

Figure 6.1   Misclassification cost incorporation for the original *Vehicle Silhouettes* data.

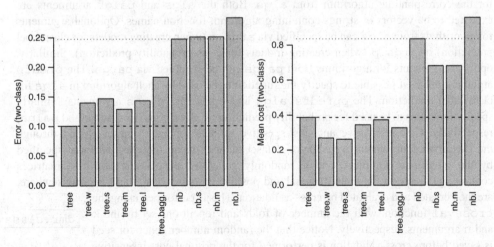

Figure 6.2   Misclassification cost incorporation for the two-class *Vehicle Silhouettes* data.

6.5   Experimental procedure

The results presented in the previous example serve the illustration purpose only, and even if they permit some meaningful observations, they can by no means be generalized to other datasets and algorithms. To actually draw some reliable conclusions about the capabilities of the techniques of misclassification cost incorporation discussed in this chapter, one should generate much more extensive results, following a carefully designed experimental procedure. Ideally, the procedure should satisfy the conditions listed below.

Multiple algorithms. Use a number of classification algorithms, possibly including different parameter setups for the same algorithm.

Multiple datasets. Use several datasets.

Random costs. Use a series of randomly generated misclassification cost matrices for each dataset.

Reliable evaluation. Evaluate model performance using a low variance evaluation procedure, such as the repeated hold-out or k-fold cross-validation procedures discussed in Section 7.3.

For each algorithm and dataset combination, the experimental procedure should therefore repeatedly generate a random cost matrix and then evaluate the performance of cost-insensitive and cost-sensitive models. The observed difference of the mean misclassification cost between those would then serve as an indicator of the effectiveness of the investigated cost incorporation techniques.

Example 6.5.1 An automated experimental procedure for evaluating the effects of the misclassification cost wrapper techniques demonstrated in the previous examples is implemented by the R code presented below. It applies the mc.weight, mc.resample, mc.mincost, and mc.relabel functions to wrap each of the classification algorithms specified by the algs argument. The optional palgs argument may pass additional different probabilistic classification algorithms to be used for probability estimation by the relabeling technique for the corresponding algorithm from algs. Both the algs and palgs arguments are expected to be vectors of strings containing algorithm function names. Optional arguments for algorithms from algs can be specified via args.c (when creating models for class label prediction) or args.p (when creating models for class probability prediction). Similarly, optional arguments for algorithms from palgs can be specified via pargs. The predfs argument makes it possible to specify the function to be used by each algorithm in algs for class label prediction. The ppredfs.algs and ppredfs.palgs arguments similarly specify functions to be used for probability prediction by algorithms from algs and palgs, respectively. Cost-insensitive and cost-sensitive models for each dataset passed through the datasets argument (a vector of dataset names), using the class attribute specified by the classes argument, and m randomly generated misclassification cost matrices containing 0 costs on the main diagonal and positive costs between 1 and rmax elsewhere, are then created and evaluated by cross-validation. This is performed using the crossval function, with the number of folds and repetitions set by the k and n arguments, respectively. Notice that the random number generator seed is saved before cross-validation is performed for the original cost-insensitive

Ex. 7.3.2
dmr.claseval

algorithm and restored when cross-validating the cost-sensitive wrappers. This ensures that the same data splitting is used for the evaluation of each of the compared algorithms.

```
## run misclassification costs experiments
mc.experiment <- function(algs, datasets, classes, rmax=10, m=25, k=10, n=1,
                          palgs=NULL, args.c=NULL, args.p=NULL, pargs=NULL,
                          predfs=predict, ppredfs.algs=predict,
                          ppredfs.palgs=predict)
{
  crossval.rs <- function(rs, ...) { .Random.seed <<- rs; crossval(...) }

  results <- NULL
  for (dn in 1:length(datasets))
  {
    res <- NULL

    data <- get(datasets[dn])
    class <- classes[dn]
    formula <- make.formula(class, ".")

    for (i in 1:m)
    {
      rho <- matrix(round(runif(nlevels(data[[class]])^2, min=1, max=rmax)-0.5),
                    nrow=nlevels(data[[class]]),
                    ncol=nlevels(data[[class]]),
                    dimnames=list(predicted=levels(data[[class]]),
                                  true=levels(data[[class]])))
      diag(rho) <- 0
      rhoc <- rhom2c(rho)

      for (an in 1:length(algs))
      {
        alg <- get(algs[[an]])
        palg <- if (!is.null(palgs[[an]])) get(palgs[[an]])
        arg.c <- args.c[[an]]
        arg.p <- args.p[[an]]
        parg <- pargs[[an]]
        predf <- if (is.vector(predfs)) predfs[[an]] else predfs
        ppredf.alg <- if (is.vector(ppredfs.algs)) ppredfs.algs[[an]]
                      else ppredfs.algs
        ppredf.palg <- if (is.vector(ppredfs.palgs)) ppredfs.palgs[[an]]
                       else ppredfs.palgs
        aname <- paste(algs[an], palgs[an], sep=".")

        if (is.null(palg))
        {
          alg.w <- mc.weight(alg, predf)
          alg.s <- mc.resample(alg, predf)
          alg.m <- mc.mincost(alg, ppredf.alg)
          alg.l <- mc.relabel(alg, pargs=arg.p, predf=predf, ppredf=ppredf.alg)
        }
        else
          alg.lp <- mc.relabel(alg, palg, pargs=parg,
                               predf=predf, ppredf=ppredf.palg)

        rs <- .Random.seed
        cv.b <- crossval(alg, formula, data, args=arg.c, predf=predf, k=k, n=n)
        if (is.null(palg))
        {
```

```
        cv.w <- crossval.rs(rs, alg.w$alg, formula, data,
                            args=c(list(rhoc), arg.c),
                            predf=alg.w$predict, k=k, n=n)
        cv.s <- crossval.rs(rs, alg.s$alg, formula, data,
                            args=c(list(rhoc), arg.c),
                            predf=alg.s$predict, k=k, n=n)
        cv.m <- crossval.rs(rs, alg.m$alg, formula, data,
                            args=c(list(rho), arg.p),
                            predf=alg.m$predict, k=k, n=n)
        cv.l <- crossval.rs(rs, alg.l$alg, formula, data,
                            args=c(list(rho), arg.c),
                            predf=alg.l$predict, k=k, n=n)
        mc <- data.frame(b=mean.cost(cv.b$pred, cv.b$true, rho),
                         w=mean.cost(cv.w$pred, cv.w$true, rho),
                         s=mean.cost(cv.s$pred, cv.s$true, rho),
                         m=mean.cost(cv.m$pred, cv.m$true, rho),
                         l=mean.cost(cv.l$pred, cv.l$true, rho))
        mc$d.w <- (mc$b-mc$w)/mc$b
        mc$d.s <- (mc$b-mc$s)/mc$b
        mc$d.m <- (mc$b-mc$m)/mc$b
        mc$d.l <- (mc$b-mc$l)/mc$b
        e <- data.frame(b=err(cv.b$pred, cv.b$true),
                        w=err(cv.w$pred, cv.w$true),
                        s=err(cv.s$pred, cv.s$true),
                        m=err(cv.m$pred, cv.m$true),
                        l=err(cv.l$pred, cv.l$true))
      }
      else
      {
        cv.lp <- crossval.rs(rs, alg.lp$alg, formula, data,
                             args=c(list(rho), arg.c),
                             predf=alg.lp$predict, k=k, n=n)
        mc <- data.frame(b=mean.cost(cv.b$pred, cv.b$true, rho),
                         lp=mean.cost(cv.lp$pred, cv.lp$true, rho))
        e <- data.frame(b=err(cv.b$pred, cv.b$true),
                        lp=err(cv.lp$pred, cv.lp$true))
        mc$d.lp <- (mc$b-mc$lp)/mc$b
      }

      res[[aname]]$mc <- rbind(res[[aname]]$mc, mc)
      res[[aname]]$e <- rbind(res[[aname]]$e, e)
    }
  }
  results <- c(results, list(res))
 }
 `names<-`(results, datasets)
}


 # experiments with decision trees and naive Bayes
 # for the Vehicle and Vehicle01 datasets
mc.res <- mc.experiment(c("rpart", "rpart", "naiveBayes"),
                        c("Vehicle", "Vehicle01"), c("Class", "Class"),
                        palgs=list(NULL, "bagging", NULL),
                        args.p=list(list(cp=0.025), list(cp=0.025), NULL),
                        pargs=list(NULL, list(control=rpart.control(cp=0.025)),
                                   NULL),
                        predfs=c(function(...) predict(..., type="c"),
                                 function(...) predict(..., type="c"), predict),
```

```
                    ppredfs.algs=c(predict, predict,
                                  function(...) predict(..., type="r")),
                    ppredfs.palgs=list(NULL,
                                       function(...) predict(..., type="p",
                                                           aggregation="a"),
                                  NULL))

barplot(colMeans(cbind(mc.res$Vehicle$rpart.NULL$mc[,6:9],
                   mc.res$Vehicle$rpart.bagging$mc[,3])),
       main="Four-class, rpart", ylab="Cost reduction",
       las=2, ylim=c(-0.01, 0.11),
       names.arg=c("weight", "resample", "mincost", "relabel", "relabel.b"))
barplot(colMeans(mc.res$Vehicle$naiveBayes.NULL$mc[,7:9]),
       main="Four-class, naiveBayes", ylab="Cost reduction",
       las=2, ylim=c(-0.01, 0.11),
       names.arg=c("resample", "mincost", "relabel"))

barplot(colMeans(cbind(mc.res$Vehicle01$rpart.NULL$mc[,6:9],
                   mc.res$Vehicle01$rpart.bagging$mc[,3])),
       main="Two-class, rpart", ylab="Cost reduction",
       las=2, ylim=c(-0.26, 0.15),
       names.arg=c("weight", "resample", "mincost", "relabel", "relabel.b"))
barplot(colMeans(mc.res$Vehicle01$naiveBayes.NULL$mc[,7:9]),
       main="Two-class, naiveBayes", ylab="Cost reduction",
       las=2, ylim=c(-0.26, 0.15),
       names.arg=c("resample", "mincost", "relabel"))
```

The experimental procedure implemented by the `mc.experiment` function may be used to conduct systematic extensive experiments with all the techniques for misclassification cost incorporation discussed in this chapter. The example call included above is just a simple demonstration using the same decision tree and naïve Bayes algorithms as in the previous examples and the same two datasets, *Vehicle Silhouettes* and its two-class version. For the relabeling decision tree wrapper bagging is additionally used for class probability prediction. The generated results are stored in a list containing named components for each dataset. These are themselves lists with named components for each algorithm, and then there are dataframes containing the mean misclassification cost and misclassification error, as well as the relative reduction of the former, for each of randomly generated cost matrices. As shown in the calls to `barplot`, they can be averaged to produce compact, readable summarized results. Barplots displaying the averaged relative mean misclassification cost decrease obtained by each algorithm for the two datasets are presented in Figure 6.3.

On the original four-class data instance, relabeling definitely yields the greatest misclassification cost reduction for decision trees, as long as sufficiently reliable bagging class probability predictions are used to determine the new class labels of training instances. The minimum-cost rule turns out clearly the best technique for the naïve Bayes classifier. This suggests that, with sufficiently good class probability predictions, their direct application to cost-sensitive class label prediction may outperform a separate cost-sensitive model trained using relabeled data. Not surprisingly, none of the more refined cost incorporation techniques beats simple instance weighting or resampling on the two-class data version. For the naïve Bayes classifier, the minimum-cost rule with two classes is equivalent to instance resampling, but instance relabeling also fails entirely.

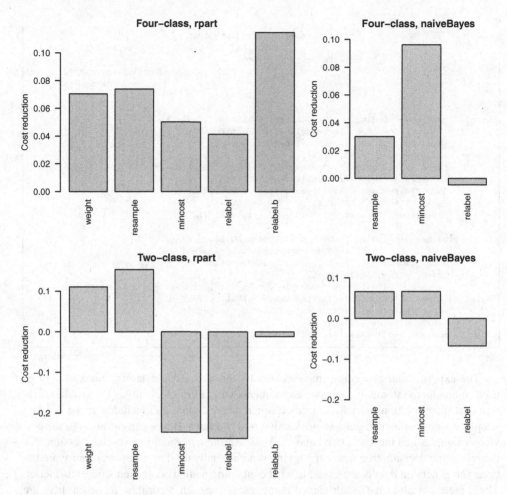

Figure 6.3 Experimental results with misclassification cost incorporation techniques using the original and two-class versions of the *Vehicle Silhouettes* data.

6.6   Conclusion

This chapter has shown that misclassification costs not only provide classifier evaluation criteria, but can also be incorporated into the model creation or prediction process. This capability is extremely valuable in applications where the usual error minimization objective does not lead to useful models due to substantially different costs of different mistakes. The instance weighting and instance replication techniques affect the created model and can be used whenever the costs to be incorporated can be specified (or at least approximated) by a vector of per-class misclassification costs. Of those, the latter can be used to make arbitrary classification algorithms cost-sensitive, but the former is the preferred (more elegant, more memory and time-efficient) approach for weight-sensitive algorithms. It is noteworthy that these techniques are fully sufficient for two-class concepts, for which misclassification costs are particularly often specified and need to be incorporated. For such concepts, the matrix and vector cost representations are equivalent.

The minimum-cost rule, which can be used to incorporate an arbitrary misclassification cost matrix at the time of prediction when using probabilistic classification models, might appear theoretically superior. It is not necessarily always so in practice, because the minimum-cost rule does fulfill its promise of misclassification cost minimization only if class probabilities predicted by the model are sufficiently reliable. This can be hardly assumed, particularly for models that are not "inherently" and "deeply" probabilistic (i.e., derived by probabilistic data analysis). It appears therefore a reasonable recommendation to use the minimum-cost rule only if the model is believed to provide reliable class probability estimates or the other techniques cannot be applied. The latter is the case when the misclassification costs to be incorporated cannot be reasonably well represented by a per-class cost vector and require a full cost matrix representation, or when a previously created model has to be used, without the possibility of re-building it in a cost-sensitive way.

The instance relabeling technique – the most refined of all the simple algorithm-independent approaches to handling misclassification costs discussed here – creates a separate model for class probability prediction and a separate final cost-sensitive model based on the relabeled training set. This makes it possible to have the former focused on reliable probability prediction (e.g., logit classification on probabilistic bagging) and the latter on accurate class label prediction, and possibly overcome the limitation of the plain minimum-cost rule approach.

An additional advantage is that the resulting cost-sensitive model is explicitly available and can be directly used for either prediction or inspection. This is arguably the most important reason to prefer instance relabeling over the minimum-cost rule. While the latter wraps prediction, the former wraps model creation. Whenever an explicitly represented cost-sensitive model is not required and just cost-sensitive predictions are needed, the minimum-cost rule may be a simpler and less expensive alternative. It becomes particularly attractive if adaptation to different or dynamically changing misclassification costs is required, which may be the case for some specific applications.

6.7   Further readings

The issue of misclassification costs is not extensively discussed in an algorithm-independent perspective by most data mining books. Those that do pay more attention to misclassification costs are rather concerned with incorporating them to model evaluation than to model creation (Han *et al.*, 2011; Witten *et al.*, 2011), or – when discussing the latter – address the related problem of imbalanced classes (Tan *et al.*, 2013). This unfortunately also causes the importance of nonuniform misclassification costs to be underappreciated in practical data mining projects. They have received significant research interest, though, and the awareness of the resulting techniques will likely increase. Unlike the corresponding sections in most other chapters, this refers mainly to original research articles.

Some work on misclassification costs incorporation has been done in the context of particular classification algorithms. Instance weighting as an internal mechanism of achieving cost-sensitivity for decision trees is used by the CART algorithm (Breiman *et al.*, 1984) and its R re-implementation (Therneau and Atkinson, 1997). Knoll *et al.* (1994) presented a cost-sensitive pruning technique for decision trees. Draper *et al*. (1994) addressed misclassification cost minimization by both pruning and split selection for multivariate decision trees (for which it actually makes more sense to speak of "split creation" than "split selection"). Similar approaches have been considered for rule-based classification models (Pazzani *et al*., 1994; Provost 1994). As later investigated by Drummond *et al*. (2000), decision tree split conditions provide rather little space for achieving cost-sensitivity, which makes stop criteria, class label assignment to leaves, and pruning much more important. Masnadi-Shirazi and Vasconcelos (2010) demonstrated how misclassification costs can be minimized by appropriately modified versions of the boosting family of ensemble modeling algorithms.

A general approach to misclassification cost incorporation proposed by Webb (1996) consists in modifying internal algorithm mechanisms rather than cost-sensitive wrapping. This is achieved by preferring more general class membership conditions for classes that are safe to predict (inexpensive when incorrectly assigned to instances of other true classes) and more specific class membership conditions for classes that are risky to predict (costly when incorrectly assigned to instances of other true classes). The instance relabeling technique was introduced by Domingos (1999) as the *MetaCost* algorithm. He also suggested that reliable class probability predictions can be obtained using bagging (Breiman 1996). The latter was partially questioned by further work (Zadrozny and Elkan 2001a, b), suggesting that single decision tree or naïve Bayes models can deliver better class probabilities if appropriately adjusted or calibrated. In particular, Laplace or m-estimation applied at decision tree leaves instead of plain frequency probability estimation appeared to provide a substantial improvement. As demonstrated by Provost and Domingos (2003), this may be still improved by combination with bagging. Zadrozny and Elkan (2002) proposed combining calibrated binary probability predictions from multiple binary models for multiclass tasks. According to Margineantu (2002), accurate probability predictions are not actually necessary to for the minimum-cost rule to successfully minimize misclassification costs. It can be modified to use confidence levels for probability predictions, empirically determined by bagging. Niculescu-Mizil and Caruana (2005) examined universal techniques for transforming scoring classifier predictions to reliable probability predictions, extending the previous work of Platt (2000) for the SVM algorithm. Loss functions for probabilistic predictions, which can be considered counterparts of misclassification costs for discrete predictions, are discussed by Reid and Williamson (2010).

The theoretical properties of instance weighting and resampling techniques were examined by Elkan (2001), who also formulated "reasonableness" conditions for cost matrices. Japkowicz and Stephen (2002) discussed different resampling schemes in the context of imbalanced classes. The relative merits of oversampling and undersampling were investigated by Drummond and Holte (2003). The two resampling approaches were combined by Chawla *et al*. (2002), with oversampling performed indirectly by synthetic instance generation. More refined forms of resampling and weighting have also been considered (Abe *et al*. 2004; Chan and Stolfo 1999; Zadrozny *et al*. 2003; Zhou and Liu 2006). Several issues related to creating and evaluating cost-sensitive classification models were investigated by Margineantu (2001). Jan *et al*. (2012) presented the idea of soft cost-sensitivity, achieved by multicriteria model performance optimization with respect to both misclassification costs and the misclassification error.

On few occasions, cost-sensitive classification is referred to in a different meaning, concerning attribute costs rather than misclassification costs (e.g., Tan 1993; Turney 1995). This assumes that determining the values of different attributes may be associated with different costs and a classification model should achieve a good balance between its prediction performance and the cost of attributes used.

References

Abe N, Zadrozny B and Langford J 2004 An iterative method for multi-class cost-sensitive learning *Proceedings of the Tenth ACM SIGKDD International Conference on Knowledge Discovery and Data Mining*. ACM Press.

Breiman L 1996 Bagging predictors. *Machine Learning* **24**, 123–140.

Breiman L, Friedman JH, Olshen RA and Stone CJ 1984 *Classification and Regression Trees*. Chapman and Hall.

Chan P and Stolfo SJ 1999 Toward scalable learning with non-uniform distributions: Effects and a multi-classifier approach *Proceedings of the Fourth International Conference on Knowledge Discovery and Data Mining*. AAAI Press.

Chawla NV, Bowyer, KW Hall LO and Kegelmeyer WP 2002 SMOTE: Synthetic minority over-sampling technique. *Journal of Artificial Intelligence Research* **16**, 321–357.

Domingos P 1999 MetaCost: A general method for making classifiers cost-sensitive *Proceedings of the Fifth ACM SIGKDD International Conference on Knowledge Discovery and Data Mining*. ACM Press.

Draper BA, Brodley CE and Utgoff PE 1994 Goal-directed classification using linear machine decision trees. *IEEE Transactions on Pattern Recognition and Artificial Intelligence* **16**, 888–893.

Drummond C and Holte RC 2000 Exploiting the cost (in)sensitivity of decision tree splitting criteria *Proceedings of the Seventeenth International Conference on Machine Learning (ICML-2000)*. Morgan Kaufmann.

Drummond C and Holte RC 2003 C4.5, class imbalance and cost sensitivity: Why under-sampling beats over-sampling *Proceedings of the ICML-2003 Workshop on Learning from Imbalanced Datasets*.

Elkan C 2001 Foundations of cost-sensitive learning *Proceedings of the Seventeenth International Joint Conference on Artificial Intelligence (IJCAI-01)*. Morgan Kaufmann.

Han J, Kamber M and Pei J 2011 *Data Mining: Concepts and Techniques* 3rd edn. Morgan Kaufmann.

Jan TK, Lin CH, Wang DW and Lin HS 2012 A simple methodology for cost-sensitive classification *Proceedings of the Eighteenth ACM SIGKDD International Conference on Knowledge Discovery and Data Mining*. ACM Press.

Japkowicz N and Stephen S 2002 The class imbalance problem: A systematic study. *Intelligent Data Analysis* **6**, 429–450.

Knoll U, Nakhaeizadeh G and Tausend B 1994 Cost-sensitive pruning of decision trees *Proceedings of the Seventh European Conference on Machine Learning (ECML-94)*. Springer.

Margineantu DD 2001 *Methods for Cost-Sensitive Learning*. PhD thesis, Oregon State University.

Margineantu DD 2002 Class probability estimation and cost-sensitive classification decisions *Proceedings of the Thirteenth European Conference on Machine Learning (ECML-02)*. Springer.

Masnadi-Shirazi H and Vasconcelos N 2010 Cost-sensitive boosting. *IEEE Transactions on Pattern Analysis and Machine Intelligence* **33**, 294–309.

Niculescu-Mizil A and Caruana R 2005 Predicting good probabilities with supervised learning *Proceedings of the Twenty-Second International Conference on Machine learning (ICML-05)*. ACM Press.

Pazzani MJ, Merz C, Murphy P, Ali K, Hume T and Brunk C 1994 Reducing misclassification costs *Proceedings of the Eleventh International Conference on Machine Learning (ICML-94)*. Morgan Kaufmann.

Platt JC 2000 Probabilistic outputs for support vector machines and comparison to regularized likelihood methods In *Advances in Large Margin Classifiers* (eds Smola AJ, Barlett P, Schölkopf B and Schuurmans D) MIT Press.

Provost FJ 1994 Goal-directed inductive learning: Trading off accuracy for reduced error cost *Proceedings of the AAAI Spring Symposium on Goal Directed Learning*. AAAI Press.

Provost FJ and Domingos P 2003 Tree induction for probability-based ranking. *Machine Learning* **52**, 199–215.

Reid MD and Williamson RC 2010 Composite binary losses. *Journal of Machine Learning Research* **11**, 2387–2422.

Tan M 1993 Cost-sensitive learning of classification knowledge and its application in robotics. *Machine Learning* **13**, 7–33.

Tan PN, Steinbach M and Kumar V 2013 *Introduction to Data Mining* 2nd edn. Addison-Wesley.

Therneau TM and Atkinson EJ 1997 An introduction to recursive partitioning using the RPART routines. Technical Report 61, Mayo Clinic.

Turney P 1995 Cost-sensitive classification: Empirical evaluation of a hybrid genetic decision tree induction algorithm. *Journal of Artificial Intelligence Research* **2**, 369–409.

Webb GI 1996 Cost-sensitive specialization *Proceedings of the 1996 Pacific Rims International Conference on Artificial Intelligence*. Springer.

Witten IH, Frank E and Hall MA 2011 *Data Mining: Practical Machine Learning Tools and Techniques* 3rd edn. Morgan Kaufmann.

Zadrozny B and Elkan C 2001a Learning and making decisions when the costs and probabilites are both unknown *Proceedings of the Seventh ACM SIGKDD International Conference on Knowledge Discovery and Data Mining*. ACM Press.

Zadrozny B and Elkan C 2001b Obtaining calibrated probability estimates from decision trees and naive Bayesian classifiers *Proceedings of the Eighteenth International Conference on Machine Learning (ICML-01)*. Morgan Kaufmann.

Zadrozny B and Elkan C 2002 Transforming classifier scores into accurate multiclass probability estimates *Proceedings of the Eighth ACM SIGKDD International Conference on Knowledge Discovery and Data Mining*. ACM Press.

Zadrozny B, Langford J and Abe N 2003 Cost-sensitive learning by cost-proportionate example weighting *Proceedings of the Third IEEE International Conference on Data Mining*. IEEE Press.

Zhou ZH and Liu XY 2006 On multi-class cost-sensitive learning *Proceedings of the Twenty-First National Conference on Artificial Intelligence (AAAI-06)*. AAAI Press.

7

Classification model evaluation

7.1 Introduction

The purpose of the evaluation of a classification model is to get a reliable assessment of the quality of the target concept's approximation represented by the model, which will be called the model's *predictive performance*. Different performance measures can be used, depending on the intended application of the model. Given the fact that the model is created based on a training set, which is a usually small subset of the domain, it is its generalization properties that are essential for the approximation quality. For any performance measure, it is important to distinguish between its value for a particular dataset (*dataset performance*), especially the training set (*training performance*), and its expected performance on the whole domain (*true performance*).

7.1.1 Dataset performance

The dataset performance of a model is assessed by calculating the value of one or more selected performance measures on a particular dataset with true class labels available. It describes the degree of match between the model and the target concept on this dataset.

7.1.2 Training performance

Evaluating a model on the training set that was used to create the model determines the model's training performance. Whereas it is sometimes useful to better understand the model and diagnose the operation of the employed classification algorithm, it is usually not of significant interest, since the purpose of classification models is not to classify the training data.

7.1.3 True performance

The true performance of a model is its expected performance (with respect to one or more selected performance measures) on the whole domain. This reflects the model's predictive

Data Mining Algorithms: Explained Using R, First Edition. Paweł Cichosz.
© 2015 John Wiley & Sons, Ltd. Published 2015 by John Wiley & Sons, Ltd.

utility, i.e., its capability to correctly classify arbitrary new instances from the given domain. Since the true class labels are generally unavailable for the domain, the true performance always remains unknown and can only be estimated by dataset performance.

Appropriate evaluation procedures are needed to reliably estimate the unknown values of the adopted performance measures on the whole domain, containing mostly previously unseen instances, i.e., to assess the true performance. Performance measures and evaluation procedures are the two major topics related to classification model evaluation, covered in the corresponding sections below.

Example 7.1.1 The classifier performance measures presented below will be illustrated in R by applying them to the evaluation of decision tree models created using the `rpart` package for the *Soybean* dataset, available in the `mlbench` package. The same dataset and classification algorithm will also be used to illustrate evaluation procedures. Some utility functions from the `dmr.util` package will be called. The following R code prepares the demonstration by loading the packages and the dataset, splitting the dataset randomly into training and test subsets, and creating a decision tree classifier based on the training set. The random generator seed is explicitly initialized to make the results easily reproducible.

```
library(dmr.util)
library(rpart)

data(Soybean, package="mlbench")

set.seed(12)
rs <- runif(nrow(Soybean))
s.train <- Soybean[rs>=0.33,]
s.test <- Soybean[rs<0.33,]

s.tree <- rpart(Class~., s.train)
```

7.2   Performance measures

Classification performance measures are calculated by comparing the predictions generated by the classifier on a dataset S with the true class labels of the instances from this dataset. The latter may be an arbitrary subset of the available labeled dataset, including the training set. As it will be more extensively discussed later, though, it is usually separate from the training set and referred to as the *validation set* or *test set*. The distinction between these terms is mostly based on the purpose of the model evaluation process. When the evaluation is performed to make some decisions that may affect the final model (e.g., select a classification algorithm, adjust its parameters, select attributes, etc.), which may be called *intermediate evaluation*, it is a common convention to speak of a validation set. Whenever the performance of the ultimately created model is to be evaluated (*final evaluation*), one would rather speak of a test set. This terminological distinction is purely conventional and has no impact on performance measures. They can be applied to any dataset, including the training set used to create the model (which determines the model's training performance), although they would not reliably estimate the model's generalization properties in that case. We will follow the convention to designate an arbitrary dataset by S and a validation/test set separate from the training set by Q.

Intermediate evaluation is closely related to model selection, which is based on evaluation results. Regardless of the exact decision to be made (algorithm selection, parameter adjustment, attribute selection, etc.), what is needed is the selection of one model from a set of candidate models. This apparently trivial task (given model evaluations, just select the best one) may become quite difficult when multiple performance measures have to be used.

Some performance measures are not only used to evaluate classification models, but also serve as – explicit or implicit – optimization criteria for the search for models performed by classification algorithms. In this role, they are sometimes referred to as *loss functions* to be minimized on the training set.

7.2.1   Misclassification error

The most common way of characterizing the performance of a classification model is by its error or accuracy. One of these directly related basic performance measures is probably always calculated when evaluating a model, although it is not always sufficient.

The *misclassification error* of model $h : X \rightarrow C$ with respect to the concept $c : X \rightarrow C$ on dataset $S \subset X$ is calculated as the relative frequency of the model's mistakes for instances from S:

$$e_{c,S}(h) = \frac{|S_{h \neq c}|}{|S|} \tag{7.1}$$

Alternatively, one can refer to the *accuracy* of model h with respect to the target concept c on dataset S defined as $1 - e_{c,S}(h)$. The misclassification error and accuracy are extremely easy to interpret: they tell us how often the model is wrong or right when applied to a dataset. Under an appropriate evaluation procedure (i.e., using one or more appropriately selected validation/test sets), they would also tell us how often the model is likely to be wrong or right when applied to new, previously unseen data, estimating the true misclassification error (the probability of an arbitrary instance from the domain being misclassified) or the true accuracy (the probability of an arbitrary instance from the domain being classified correctly).

The misclassification error is the most common loss function for classification algorithms, also referred to as the 0–1 loss.

Example 7.2.1 The following R code defines a function for calculating the misclassification error for given vectors of predicted and true class labels and demonstrates its application to the decision tree model for the *Soybean* dataset. The evaluation is performed on the test subset, for which model predictions are generated by calling the `predict` function, with the `type="c"` argument used to request class label prediction.

```
err <- function(pred.y, true.y) { mean(pred.y!=true.y) }

err(predict(s.tree, s.test, type="c"), s.test$Class)
```

7.2.2   Weighted misclassification error

Similarly as weight-sensitive algorithms use weighted training instances, one can use a set of weighted instances for model evaluation. Assuming a weight w_x is assigned to each $x \in S$, the weighted misclassification error of model h with respect to the concept c on dataset S can be calculated as follows:

$$e_{c,S,w}(h) = \frac{\sum_{x \in S_{h \neq c}} w_x}{\sum_{x \in S} w_x} \tag{7.2}$$

Example 7.2.2 The following R code defines a modified version of the function from the previous example that optionally accepts a weight vector as an additional argument and calculates the weighted misclassification error. The function is then called with a weight vector that doubles the importance of instances of the least frequent class `herbicide-injury` and then another weight vector that contains random integer weights between 1 and 5 assigned to particular classes. Of course, the random weights are totally meaningless and serve the demonstration purpose only.

```
werr <- function(pred.y, true.y, w=rep(1, length(true.y)))
{ weighted.mean(pred.y!=true.y, w) }

  # double weight for the least frequent class
s.w2test <- ifelse(s.test$Class=="herbicide-injury", 2, 1)
werr(predict(s.tree, s.test, type="c"), s.test$Class, s.w2test)

  # random per-class weights 1..5
s.wctest <- round(runif(nlevels(s.test$Class), min=1, max=5))
s.w3test <- s.wctest[s.test$Class]
werr(predict(s.tree, s.test, type="c"), s.test$Class, s.w3test)
```

7.2.3   Mean misclassification cost

The error (or accuracy) implicitly assumes that each wrong (or correct) prediction counts the same. This is not necessarily the case for several applications, where some model mistakes may be more severe than the others, i.e., different misclassification costs can be assigned to different possible mistakes. They can be represented by a pairwise cost matrix or a per-class cost vector, as extensively discussed in Section 6.2.

To take into account misclassification costs during model evaluation, one can replace the error and accuracy with a cost-sensitive performance measure. One obvious candidate is the *mean misclassification cost*, which can be calculated on the dataset S using the misclassification cost matrix ρ by summing up the misclassification costs for all instances from S:

$$r_{c,S,\rho}(h) = \frac{\sum_{x \in S} \rho[h(x), c(x)]}{|S|} \tag{7.3}$$

where $\rho[d_1, d_2]$ is the cost of the predicting class d_1 for an instance of true class d_2. This is clearly the same as the misclassification error if $\rho[d_1, d_2] = 1$ for all $d_1, d_2 \in C$, $d_1 \neq d_2$ and $\rho[d, d] = 0$ for all $d \in C$.

Assuming a simplified cost matrix assigning the same misclassification cost to each possible mistake for the same true class – i.e., a cost vector ρ, where $\rho[d]$ is the cost of failing to correctly predict class d – we can rewrite the definition of the mean misclassification cost as

$$r_{c,S,\rho}(h) = \frac{\sum_{x \in S_{h \neq c}} \rho[c(x)]}{|S|} \tag{7.4}$$

By comparing the latter with the definition of the weighted misclassification error it can be immediately noticed that

$$r_{c,S,\rho}(h) = \frac{\sum_{x \in S} w_x}{|S|} e_{c,S,w}(h) \tag{7.5}$$

as long as $w_x = \rho[c(x)]$ for all $x \in S$. This shows that the mean misclassification cost in the simplified cost vector case is equivalent (subject to a model-independent coefficient) to the weighted error, when instance weights are set to the misclassification cost associated with the corresponding classes. The same observation is used in Section 6.3.1 to justify instance weighting as a technique for misclassification cost incorporation during model creation.

Example 7.2.3 The following R code defines a function for the calculation of the mean misclassification cost for a given cost matrix. The function is then applied to evaluate the model for the *Soybean* data using four different cost matrices: one that assigns the same cost of 1 to all mistakes, one that doubles the cost of misclassifying an instance of the least frequent class `herbicide-injury`, a simplified cost matrix (i.e., a cost vector) containing the random class weights from the previous example, and a full cost matrix generated randomly. The random misclassification costs are clearly used for the demonstration purpose only and are totally meaningless.

```
mean.cost <- function(pred.y, true.y, rho) { mean(diag(rho[pred.y,true.y])) }

  # uniform cost matrix
s.r1test <- matrix(1, nrow=nlevels(s.test$Class), ncol=nlevels(s.test$Class))
diag(s.r1test) <- 0
mean.cost(predict(s.tree, s.test, type="c"), s.test$Class, s.r1test)

  # double cost for misclassifying the least frequent class
s.r2test <- matrix(1, nrow=nlevels(s.test$Class), ncol=nlevels(s.test$Class))
s.r2test[,levels(s.test$Class)=="herbicide-injury"] <- 2
diag(s.r2test) <- 0
mean.cost(predict(s.tree, s.test, type="c"), s.test$Class, s.r2test)
  # this should give the same result
sum(s.w2test)/nrow(s.test)*werr(predict(s.tree, s.test, type="c"),
                    s.test$Class, s.w2test)

  # random per-class costs 1..5
s.r3test <- matrix(s.wctest, nrow=nlevels(s.test$Class), ncol=nlevels(s.test$Class),
                byrow=TRUE)
diag(s.r3test) <- 0
mean.cost(predict(s.tree, s.test, type="c"), s.test$Class, s.r3test)
  # this should give the same result
sum(s.w3test)/nrow(s.test)*werr(predict(s.tree, s.test, type="c"),
                    s.test$Class, s.w3test)

  # random costs 1..5
s.r4test <- matrix(round(runif(nlevels(s.test$Class)*nlevels(s.test$Class),
                    min=1, max=5)),
                nrow=nlevels(s.test$Class), ncol=nlevels(s.test$Class))
diag(s.r4test) <- 0
mean.cost(predict(s.tree, s.test, type="c"), s.test$Class, s.r4test)
```

Notice that for the uniform cost matrix, the mean misclassification cost is exactly equal to the misclassification error calculated in Example 7.2.1. For the next two cost matrices, we can easily verify that the mean misclassification cost is equal to the corresponding weighted error from Example 7.2.2, scaled by an appropriate coefficient.

7.2.4   Confusion matrix

In many applications, it may not be sufficient to know how often the evaluated model is wrong or even how costly its mistakes are on average. It may be similarly or even more important to know how often it fails to predict correctly particular classes. This is especially true whenever classes of the target concept have different predictability (i.e., some are harder to predict than the others) or have different occurrence rates (i.e., some occur more frequently than the others), which is very common in practical classification tasks. This may also (but does not have to) coincide with nonuniform misclassification costs, discussed above (i.e., failing to correctly predict some of them is more costly than for the others). In such cases, model performance can be more deeply evaluated based on the *confusion matrix*.

The confusion matrix for the model $h : X \to C$ with respect to the concept $c : X \to C$ on a dataset $S \subset X$ is a $|C| \times |C|$ matrix $\Xi_{c,S}(h)$ such that

$$\Xi_{c,S}(h)[d_1, d_2] = |S_{h=d_1, c=d_2}| \tag{7.6}$$

According to this definition, rows of the confusion matrix correspond to class labels predicted by the model h and columns correspond to true class labels assigned by the concept c (which is a purely arbitrary convention, and an alternative convention of rows corresponding to true classes and columns corresponding to the predicted ones may be adopted as well). The entry of the confusion matrix corresponding to the predicted class label d_1 and true class label d_2 contains the number of instances from S for which the model predicts d_1 whereas their true class label is d_2. Sometimes the relative confusion matrix is used, with all entries divided by the size of the dataset used for the evaluation, but we will continue to refer to the absolute version as defined above.

Using the confusion matrix, we can rewrite the definition of the misclassification error as

$$e_{c,S}(h) = \frac{\sum_{d_1 \in C} \sum_{d_2 \in C-\{d_1\}} \Xi_{c,S}(h)[d_1, d_2]}{\sum_{d_1 \in C} \sum_{d_2 \in C} \Xi_{c,S}(h)[d_1, d_2]} \tag{7.7}$$

and the definition of the mean misclassification cost as

$$r_{c,S,\rho}(h) = \frac{\sum_{d_1 \in C} \sum_{d_2 \in C} \Xi_{c,S}(h)[d_1, d_2] \rho[d_1, d_2]}{\sum_{d_1 \in C} \sum_{d_2 \in C} \Xi_{c,S}(h)[d_1, d_2]} \tag{7.8}$$

Example 7.2.4 The following R code defines a function for creating the confusion matrix based on given vectors of the predicted and true class labels. The function is then applied to the decision tree model for the *Soybean* data. Unfortunately, due to a large number of classes and long class names for this dataset the printed confusion matrix is hardly readable. It is subsequently demonstrated how the confusion matrix can be used to calculate the misclassification error and the mean misclassification cost (with respect to the last cost matrix created in the previous example). The resulting values can be verified to agree with those produced by the `err` and `mean.cost` functions.

```
confmat <- function(pred.y, true.y)
{ table(pred.y, true.y, dnn=c("predicted", "true")) }

s.cm <- confmat(predict(s.tree, s.test, type="c"), s.test$Class)
  # error
(sum(s.cm)-sum(diag(s.cm)))/(sum(s.cm))
  # mean misclassification cost
sum(s.cm*s.r4test)/sum(s.cm)
```

7.2.4.1 Confusion matrix-derived performance measures

The confusion matrix gives an extremely useful insight into the model's capability to predict particular classes, and – under a proper evaluation procedure – into its generalization properties. It does not directly provide, however, the often desirable capability of comparing and ranking different models with respect to their performance, so that the best of several candidate models can be selected. A number of different performance measures can be derived from the confusion matrix that make it possible, but they can be reasonably applied only when dealing with two-class (single) concepts, when we can assume $C = \{0, 1\}$. Then we can refer to 1 as the *positive* class and to 0 as the *negative* class. Similarly, one can refer to an instance x as a *positive* instance when $c(x) = 1$ and as a *negative* instance when $c(x) = 0$. Since the confusion matrix analysis is needed only when there is some asymmetry between classes (e.g., one is more interesting, more important, and more difficult to predict or more costly to incorrectly predict than the other), some of the confusion matrix-based performance favor the positive class, which is conventionally assumed to be the more interesting one. It takes some additional effort to apply similar performance measures to multiclass models (with more than two classes), which will be briefly discussed separately later.

Example 7.2.5 The following R code creates modified two-class copies of the *Soybean* dataset, as well as of its training and test subsets. The modification is performed by selecting one class (brown-spot) and aggregating all remaining classes into a single other class. A decision tree classifier is then created based on the modified training set. It will be used in subsequent examples to illustrate performance measures specifically designed for two-class classification tasks.

```
s01.labels <- c("other", "brown-spot")
Soybean01 <- Soybean
Soybean01$Class <- factor(ifelse(Soybean$Class=="brown-spot", "brown-spot", "other"),
                          levels=s01.labels)

s01.train <- Soybean01[rs>=0.33,]
s01.test <- Soybean01[rs<0.33,]

s01.tree <- rpart(Class~., s01.train)
```

Notice that by explicitly providing the levels argument to factor when creating the modified class column, a specific ordering of classes was ensured (other first, brown-spot second). The brown-spot class will be considered positive, and the other class will be considered negative in subsequent examples.

Table 7.1   Confusion matrix notation for two-class tasks.

	c	
h	0	1
0	TN	FN
1	FP	TP

In the two-class setting with $C = \{0, 1\}$, the confusion matrix $\Xi_{c,S}(h)$ is a 2×2 matrix that can be presented as in Table 7.1, with the following shortcuts used to denote its entries:

TP is the number of *true positives* $\Xi_{c,S}(h)[1, 1]$,

TN is the number of *true negatives* $\Xi_{c,S}(h)[0, 0]$,

FP is the number of *false positives* $\Xi_{c,S}(h)[1, 0]$ (also known as *type I errors*),

FN is the number of *false negatives* $\Xi_{c,S}(h)[0, 1]$ (also known as *type II errors*).

The convention behind these terms is that "positive" or "negative" refers to the class labels predicted by the model, and "true" or "false" refers to the correctness of this prediction. Some of the most popular performance measures calculated for 2×2 confusion matrices are defined below, using the shortcuts to refer to confusion matrix entries.

Misclassification error. The ratio of incorrectly classified instances to all instances:

$$\frac{FP + FN}{TP + TN + FP + FN} \tag{7.9}$$

Accuracy. The ratio of correctly classified instances to all instances:

$$\frac{TP + TN}{TP + TN + FP + FN} \tag{7.10}$$

True positive rate. The ratio of instances correctly classified as positive to all positive instances:

$$\frac{TP}{TP + FN} \tag{7.11}$$

False positive rate. The ratio of instances incorrectly classified as positive to all negative instances:

$$\frac{FP}{TN + FP} \tag{7.12}$$

Precision. The ratio of instances correctly classified as positive to all instances classified as positive:

$$\frac{TP}{TP + FP} \tag{7.13}$$

Recall. The same as the true positive rate.

Sensitivity. The same as the true positive rate.

Specificity. The ratio of instances correctly classified as negative to all negative instances (the same as 1 − false positive rate):

$$\frac{TN}{TN + FP} \tag{7.14}$$

The misclassification error and accuracy are the same class-insensitive performance measures that were defined above, and their confusion matrix-based definitions are given here for the sake of completeness only. The remaining performance measures are much more interesting class-sensitive indicators that describe the level at which the evaluated classifier succeeds or fails to correctly detect the positive class. It does not make much sense to use them all simultaneously, which would result in considerable informational redundancy. On the other hand, no single indicator from this set can be considered sufficient, as it would be usually trivial to optimize. This is why these performance measures are usually considered in the following complementary pairs:

- true positive rate and false positive rate,
- precision and recall,
- sensitivity and specificity.

The true positive rate (aka recall, aka sensitivity) is a member of all these pairs, under different names. It represents the share of positive instances the classifier correctly detects. It obviously should be maximized, but it could be made equal to 1 by a trivial and totally useless classifier that always predicts the positive class. This is why it has to be accompanied by a complementary indicator. It could be the false positive rate, representing the share of negative instances that are incorrectly reported as positive (i.e., false alarms). This should be obviously minimized, and a trivial classifier achieving the perfect 0 false positive rate would be the one issuing no alarms at all (i.e., always predicting the negative class). Alternatively, we could look at the precision – the share of all instances predicted as positive that are truly positive – which should be maximized (and can be maximized by the same totally useless classifier that never predicts the positive class to avoid false alarms). Yet another possibility is to consider the specificity, which is the same as 1's complement of the false positive rate.

The indicators in the above pairs are complementary, since one of them represents the capability to detect positive instances, and the other the capability to avoid mis-detecting negative instances. As we have seen, each indicator in a pair can be separately optimized by a trivial and useless classifier. Moreover, there is a clear tradeoff between the indicators from the same pair – improving one is likely to worsen the other, and may at best leave it unchanged.

Example 7.2.6 The following R code defines functions for calculating the confusion matrix-based performance indicators presented above and applies them to the confusion matrix obtained for the two-class version of the *Soybean* dataset. The functions assume that class labels (i.e., the levels of the factor representing the class column) are ordered such that the negative class comes first and the positive class comes second. The previous example that prepared the modified datasets took care of forcing this ordering. Not-a-number results from the 0/0 division are replaced by 1.

```
tpr <- function(cm) { if (is.nan(p <- cm[2,2]/(cm[2,2]+cm[1,2]))) 1 else p }
fpr <- function(cm) { if (is.nan(p <- cm[2,1]/(cm[2,1]+cm[1,1]))) 1 else p }
precision <- function(cm) { if (is.nan(p <- cm[2,2]/(cm[2,2]+cm[2,1]))) 1 else p }
recall <- tpr
sensitivity <- tpr
specificity <- function(cm) { if (is.nan(p <- cm[1,1]/(cm[2,1]+cm[1,1]))) 1 else p }

s01.cm <- confmat(predict(s01.tree, s01.test, type="c"), s01.test$Class)
```

```
list(tpr=tpr(s01.cm),
     fpr=fpr(s01.cm),
     precision=precision(s01.cm),
     recall=recall(s01.cm),
     sensitivity=sensitivity(s01.cm),
     specificity=specificity(s01.cm))
```

It takes a complete pair of complementary indicators to adequately measure the performance of a classifier based on its confusion matrix in a class-sensitive way. This makes model selection much harder, though, since there is no single criterion to rank candidate models one could be required to choose from. Some measures have been proposed that try to fold two complementary indicators into a single one, to facilitate this task. One well-known example is the *F-measure* defined as the harmonic mean of the precision and recall

$$F\text{-measure} = \frac{1}{\frac{\frac{1}{\text{precision}} + \frac{1}{\text{recall}}}{2}} = \frac{2 \cdot \text{precision} \cdot \text{recall}}{\text{precision} + \text{recall}} \tag{7.15}$$

The rationale behind this definition is to seek for a value that is between these two, closer to the less of them (particularly if it is low). This implies a compromise between maximizing precision and maximizing recall that may be quite reasonable in some cases, but is no less arbitrary than any other we could think about. Luckily, such folding of indicator pairs into single indicators, that clearly cannot be lossless, is more often needed for academic research and publications (where ranking classification models and algorithms may be necessary) than in practical applications, where the domain knowledge and task requirements usually imply some additional preference criteria and guide model selection (e.g., maximize recall with subject to a constraint specifying the minimum acceptable precision, minimize false positive rate subject to a constraint specifying the minimum acceptable true positive rate). Ideally, misclassification costs are explicitly available and can be used for model evaluation and selection.

Example 7.2.7 The following R code defines a function for calculating the *F*-measure and applies it to evaluate the decision tree classifier for the two-class version of the *Soybean* dataset.

```
f.measure <- function(cm) { 1/mean(c(1/precision(cm), 1/recall(cm))) }

f.measure(s01.cm)
```

7.2.4.2 Handling more than two classes

The performance measures based on the confusion matrix can be used for the evaluation of multiclass classifiers in one of the following two ways:

1-vs-1. By measuring the capability to discriminate between instances of one class, considered positive, from instances of another classes, considered negative,

1-vs-rest. By measuring the capability to discriminate between instances of one class, considered positive, from instances of all remaining classes, considered negative.

The 1-vs-1 approach yields a separate 2×2 confusion matrix for each pair of classes, with a corresponding set of performance measure values. The 1-vs-rest approach yields a separate 2×2 confusion matrix for each class, again with a corresponding set of performance measure values. The choice between these two approaches mostly depends on application-specific requirements, in particular (implicitly assumed or explicitly expressed) misclassification costs. Whereas, in principle, a different misclassification cost can be assigned to each possible pair of (predicted and true) class labels, which would favor the 1-vs-1 approach to classifier evaluation, it is in fact quite common that the same single cost value is assigned to all possible misclassifications for a given true class. This is where the 1-vs-rest approach would be more appropriate.

In any case, we end up with several confusion matrices, each described by its own set (usually a pair, as discussed above) of indicators. This makes model selection considerably harder, making it in fact a truly multiobjective optimization problem. Unlike with a pair of indicators for a single confusion matrix, where there is a clear tradeoff and a good balance is needed, there is not necessarily any tradeoff among indicators corresponding to different confusion matrices. Simple averaging may be heplful, but not necessarily sufficient, and whereas the domain knowledge or application requirements may provide constraints to guide model selection, in general it may involve multicriteria optimization.

Example 7.2.8 The `confmat01` function defined by the R code presented below applies the 1-vs-rest approach to multiclass confusion matrix analysis. Based on the supplied predicted and true class label vectors, it returns a list of two-class confusion matrices, with each class considered positive and the remaining classes considered negative. Performance indicators can then be calculated for each of these matrices. This is demonstrated for the *Soybean* dataset, for which the average per-class true positive rate, false positive rate, and F-measure are calculated.

```
## per-class 1 vs. rest confusion matrices
confmat01 <- function(pred.y, true.y)
{
  `names<-`(lapply(levels(true.y),
                 function(d)
                 {
                     cm <- confmat(factor(as:integer(pred.y==d), levels=0:1),
                                   factor(as.integer(true.y==d), levels=0:1))
                 }), levels(true.y))
}

s.cm01 <- confmat01(predict(s.tree, s.test, type="c"), s.test$Class)
  # average TP rate, FP rate, and f-measure
rowMeans(sapply(s.cm01, function(cm) c(tpr=tpr(cm), fpr=fpr(cm), fm=f.measure(cm))))
```

7.2.4.3 Weighted confusion matrix

Similarly as for the misclassification error, a weighted version of the confusion matrix may sometimes be needed to incorporate instance weights when evaluating classifier performance. Assuming again that each instance x is assigned a weight w_x, the definition of the confusion matrix entry for the predicted class d_1 and true class d_2 can be rewritten as follows:

$$\Xi_{c,S,w}(h)[d_1, d_2] = \sum_{x \in S_{h=d_1, c=d_2}} w_x \qquad (7.16)$$

This is clearly equivalent to the ordinary confusion matrix when all weights are equal to 1.

Example 7.2.9 The following R code defines a function for calculating the weighted confusion matrix, using the `weighted.table` utility function, which is a weight-sensitive equivalent of the standard `table` function. Weights are assumed to be equal to `dmr.util` 1 if unspecified, which makes it equivalent to the function presented before in Example 7.2.4. The function is applied to evaluate the decision tree for the two-class version of the *Soybean* data, with two different weighting schemes. One gives a weight of 2 to instances of the `brown-spot` class and a weight of 1 to instances of the `other` class, and the other gives a weight of 0.1 to instances with the `plant.stand` attribute taking value 1, with the weights of the remaining instances kept at 1. There is no hidden meaning of the latter, which serves just the illustration purpose.

```
wconfmat <- function(pred.y, true.y, w=rep(1, length(true.y)))
{ weighted.table(pred.y, true.y, w=w, dnn=c("predicted", "true")) }

  # double weight for the brown-spot class
s01.w1test <- ifelse(s01.test$Class=="brown-spot", 2, 1)
s01.w1cm <- wconfmat(predict(s01.tree, s01.test, type="c"),
                     s01.test$Class, s01.w1test)
tpr(s01.w1cm)
fpr(s01.w1cm)

  # 10 times less weight for instances with plant.stand=1
s01.w2test <- ifelse(!is.na(s01.test$plant.stand) & s01.test$plant.stand=="1",
                     0.1, 1)
s01.w2cm <- wconfmat(predict(s01.tree, s01.test, type="c"),
                     s01.test$Class, s01.w2test)
tpr(s01.w2cm)
fpr(s01.w2cm)
```

Clearly, with instance weights assigned on a per-class basis, the true positive rate and the false positive rate remain the same as for the unweighted confusion matrix. They may differ only if instances of the same class receive varying weights, as in the second demonstrated weighting scheme.

7.2.5   ROC analysis

The performance measures presented above are all based on comparing a model's predicted class labels with true class labels. For scoring classifiers introduced in Section 1.3.3, which can predict different class labels in different operating points, depending on the cutoff value, this yields a separate set of performance indicators for each possible operating point. One convenient tool that facilitates classifier performance evaluation in multiple operating points, operating point comparison, and operating point selection, is the ROC analysis. The term ROC stands for *receiver operating characteristic* and refers to the methodology developed during the World War II for radar signal detection that turned out to be similarly suitable for classifier evaluation.

7.2.5.1   ROC plane

The ROC analysis considers a Cartesian coordinate system where the y-axis represents the true positive rate and the x-axis represents the false positive rate. This is called the ROC plane.

The performance of a discrete classifier is represented by a single point on the ROC plane, which visualizes the underlying tradeoff between true positives and false positives. The same is the case for a single operating point of a scoring classifier, which is also represented as a point on the ROC plane.

The $(0, 1)$ point, with a true positive rate of 1 and a false positive rate of 0, is the perfect operating point, with all instances classified correctly. The $(1, 0)$ point, with a true positive rate of 0 and a false positive rate of 1, is the worst operating point, with all instances classified incorrectly. The $(0, 0)$ point corresponds to a classifier that always predicts class 0, yielding no (true or false) positives, and the $(1, 1)$ point corresponds to a classifier that always predicts class 1.

7.2.5.2   ROC curve

By joining all possible operating points of a scoring classifier on the ROC plane with line segments, we receive a visual representation of its performance independent of the cutoff value. This is called the *ROC curve*. It shows the whole range of different operating points, with the corresponding different levels of the true positives vs. false positives tradeoff, in a single plot. It can be considered a graphical performance indicator for a scoring classifier that depends only on its scoring function component. It is the scoring function that captures the knowledge about the relationship between classes and attribute values.

To produce the ROC curve for a scoring classifier based on its scores generated for a dataset, it is necessary to identify its all possible operating points. Since a different operating point arises whenever the predicted class label changes for at least one instance, one can identify all possible operating points by considering all such and only such cutoff values that yield different class predictions for at least one instance. In particular, after sorting instances with respect to their scores (nondecreasingly or nonincreasingly), exactly one cutoff value separating two consecutive scores for instances of different classes yields a distinct operating point.

Example 7.2.10 The basic idea of the ROC curve is illustrated by the R code presented below, which identifies all operating points based on a table of predicted scores and true class labels. Its small size makes it easy to verify the calculations manually. For all possible cutoff values, the confusion matrix is determined and used to calculate the true positive rate and the false positive. The obtained eight operating points make it possible to generate the ROC curve plot presented in Figure 7.1.

```
  # predicted scores and true class labels
sctab <- data.frame(score=c(1, 1, 3, 4, 5, 5, 6, 7, 9, 9),
                 class=factor(c(0, 0, 1, 0, 0, 1, 1 ,0, 1, 1)))

  # operating point identification
scroc <- t(sapply(c(sort(unique(sctab$score)), Inf),
                function(sc)
                {
                  pred <- factor(as.numeric(sctab$score<sc),
                                 levels=levels(sctab$class))
                  list(tpr=tpr(cm <- confmat(pred, sctab$class)),
                        fpr=fpr(cm))
                }))
```

```
# ROC curve
plot(scroc, type="l", xlab="FP rate", ylab="TP rate")
```

Consider a classification model with a random scoring function that has no predictive utility at all. Whenever the cutoff value is increased (or decreased) for such a classifier, the expected resulting decrease (or increase) of the true positive rate and the false positive rate is the same, i.e., the additional instances classified positively will come both from the positive and the negative class in the same proportion. The ROC curve for such a classifier would therefore be a diagonal straight line through the ROC plane from the $(0, 0)$ point to the $(1, 1)$ point. More precisely, this would be the *expected* ROC curve. Of course, for a particular dataset and a random model, a perfectly straight diagonal will not be obtained, since the actual increase of the true positive rate and false positive rate may not always be equal, as expected. Unlike the "ideal" random ROC curve, a "real" one is in particular likely to be composed mostly of horizontal or vertical line segments, corresponding to exactly one instance changing its predicted class label when increasing or decreasing the cutoff value.

The diagonal line divides the ROC plane into the upper and lower halves. All reasonable classifiers should operate in the upper half. All reasonable scoring classifiers should have

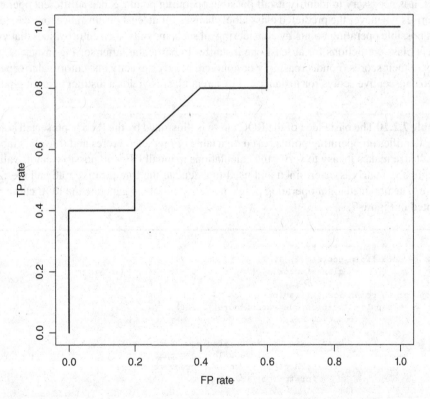

Figure 7.1   The ROC curve for a simple scoring model with eight operating points.

their ROC curve entirely above the diagonal, which means that the true positive rate should be always above the false positive rate. Only classifiers with this property indeed capture some predictively useful relationship between the target concept and attribute values. Notice also that the ROC curve for any scoring classifier is always nondecreasing, which means that the true positive rate and the false positive rate always change in the same direction when shifting the cutoff value (i.e., decrease when the cutoff value is increased and increase when the cutoff value is decreased).

Example 7.2.11 The following R code defines a function for performing all calculations necessary to plot the ROC curve for a scoring classifier. The function takes vectors of predicted scores and true class labels on input and produces a data frame with all identified operating points, each described by the corresponding true positive rate, false positive rate, and cutoff value. It is a more refined implementation than presented in the previous example that avoids calculating the true positive rate and the false positive rate for all cutoff values from scratch, but rather modifies them incrementally while scanning the data in the order of scores. The function is applied to produce the ROC curves for the decision tree model created before and a random model for the two-class version of the *Soybean* data. Calling the `rpart` prediction method function without the `type="c"` argument makes it return a matrix of class probabilities for each classified instance, with the order of columns corresponding to the ordering of class labels, so that the second column has to be selected to get the "positive" class probability.

```
roc <- function(pred.s, true.y)
{
  cutoff <- Inf  # start with all instances classified as negative
  tp <- fp <- 0
  tn <- sum(2-as.integer(true.y))  # all negative instances
  fn <- sum(as.integer(true.y)-1)  # all positive instances
  rt <- data.frame()

  sord <- order(pred.s, decreasing=TRUE)  # score ordering
  for (i in 1:length(sord))
  {
    if (pred.s[sord[i]] < cutoff)
    {
      rt <- rbind(rt, data.frame(tpr=tp/(tp+fn), fpr=fp/(fp+tn), cutoff=cutoff))
      cutoff <- pred.s[sord[i]]
    }

    p <- as.integer(true.y[sord[i]])-1  # next positive classified as positive
    n <- 2-as.integer(true.y[sord[i]])  # next negative classified as positive
    tp <- tp+p
    fp <- fp+n
    tn <- tn-n
    fn <- fn-p
  }
  rt <- rbind(rt, data.frame(tpr=tp/(tp+fn), fpr=fp/(fp+tn), cutoff=cutoff))
}

  # ROC curve for the decision tree model
s01.roc <- roc(predict(s01.tree, s01.test)[,2], s01.test$Class)
plot(s01.roc$fpr, s01.roc$tpr, type="l", xlab="FP rate", ylab="TP rate")
```

```
# ROC curve for a random model
s01rand <- runif(nrow(s01.test))
s01rand.roc <- roc(s01rand, s01.test$Class)
lines(s01rand.roc$fpr, s01rand.roc$tpr, lty=2)
```

Figure 7.2 presents the obtained plots. The ROC curve for the random model is not a perfectly straight diagonal line, which is to be expected, particularly for a relatively small dataset.

7.2.5.3 Shifting operating points

If a scoring classifier is equipped with a default labeling function, i.e., a default cutoff value, the corresponding operating point is called its *default operating point*. For probabilistic classifiers, this can be the 0.5 cutoff value corresponding to the maximum-probability rule or a value determined using the minimum-cost rule described in Section 6.3.3, if misclassification costs are specified. While this default operating point can be reasonably expected to optimize the performance measure assumed by the classification algorithm that created the model (e.g., the misclassification error or the mean misclassification cost), it is not necessarily always the case. And even if the operating point appears to be optimal with respect

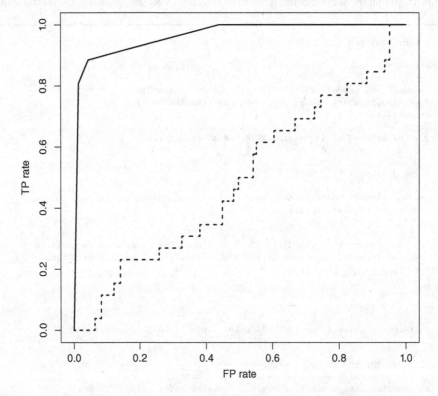

Figure 7.2 The ROC curves for the decision tree and random models for the two-class *Soybean* dataset.

to the algorithm's assumed performance measure, it does not necessarily correspond to the actual application-specific requirements, which may include, e.g., specific minimum or maximum acceptable levels of selected performance indicators. In such cases, a different operating point might happen to be more preferable. It can be easily identified on the ROC curve, and the corresponding cutoff value can be used to replace the model's default operating function. In particular, it is straightforward to identify the operating point corresponding to the maximum possible true positive rate with the false positive rate not exceeding a specified maximum level, or to the minimum possible false positive rate with the true positive rate not falling below a specified minimum level. These are the simplest, but highly useful scenarios for operating point shifting.

Example 7.2.12 The following R code defines a class labeling function and an ROC point shifting function. The former generates class labels based on a score vector and a cutoff value, using the `ustep` and `label` utility functions. The latter identifies the best cutoff value subject to the specified constraint: either yielding the maximum true positive rate with the false positive rate no greater than the maximum specified via the `max.fpr` argument, or yielding the minimum false positives rate with the true positive rate no less than the minimum specified via the `min.tpr` argument. The two functions are applied to shift the operating point of the decision tree for the *Soybean* data so as to achieve the minimum possible false positive rate with the true positive rate above or equal to 0.85, and then to achieve the maximum possible true positive rate with the false positive rate below or equal to 0.5. The ROC curve for the decision tree is plotted again, with the default operating point marked by an asterisk and the shifted operating points marked by a circle and a triangle. The obtained plot is presented in Figure 7.3.

dmr.util

```
## assign class labels according to the given cutoff value
cutclass <- function(s, cutoff, labels) { factor(ustep(s, cutoff), labels=labels) }

## identify the best cutoff value
## satisfying the minimum tpr or maximum fpr constraint
roc.shift <- function(r, min.tpr=NULL, max.fpr=NULL)
{
  if (!is.null(min.tpr))
    max(r$cutoff[r$tpr>=min.tpr])
  else if (!is.null(max.fpr))
    min(r$cutoff[r$fpr<=max.fpr])
  else
    0.5
}

  # shift to achieve tpr>0.85 at minimum fpr
s01.t085 <- roc.shift(s01.roc, min.tpr=0.85)
s01.cm.t085 <- confmat(cutclass(predict(s01.tree, s01.test)[,2],
                       s01.t085, s01.labels),
                   s01.test$Class)
  # shift to achieve maximum tpr at fpr<0.5
s01.f05 <- roc.shift(s01.roc, max.fpr=0.5)
s01.cm.f05 <- confmat(cutclass(predict(s01.tree, s01.test)[,2], s01.f05, s01.labels),
                   s01.test$Class)
  # the ROC curve
plot(s01.roc$fpr, s01.roc$tpr, type="l", xlab="FP rate", ylab="TP rate")
```

```
   # the default operating point
points(fpr(s01.cm), tpr(s01.cm), pch=8)
   # the shifted operating points
points(fpr(s01.cm.t085), tpr(s01.cm.t085), pch=1)
points(fpr(s01.cm.f05), tpr(s01.cm.f05), pch=2)
```

7.2.5.4 Interpolating between operating points

When shifting operating points, we can select from all different operating points on the same ROC curve. Sometimes the preferred operating point might lie between two neighboring points on the curve, or – even – two points from two different ROC curves, corresponding to two different models (with different scoring functions). To achieve the desired, but not directly available operating point, one can interpolate between the two selected points using the technique of *model mixing*.

The idea is to wrap the models corresponding to the selected operating points with a random selection rule that draws a model with an appropriate probability distribution. Consider

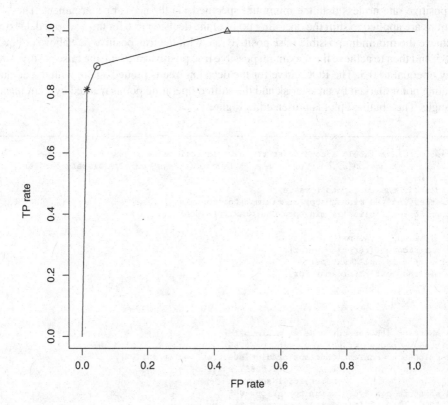

Figure 7.3 Shifting the operating point for the decision tree model for the two-class *Soybean* dataset.

models h_1 and h_2 with their corresponding operating points (FP_1, TP_1) and (FP_2, TP_2) and let (FP_*, TP_*) denote the desired operating point, on the straight line between these two. The latter can be obviously presented as

$$FP_* = \alpha FP_1 + (1 - \alpha)FP_2 \qquad (7.17)$$

$$TP_* = \alpha TP_1 + (1 - \alpha)TP_2 \qquad (7.18)$$

for some $\alpha \in (0, 1)$. Then the desired operating point can be achieved by the following rule, that probabilistically mixes the predictions of the two models:

$$h_*(x) = \begin{cases} h_1(x) & \text{with probability } \alpha \\ h_2(x) & \text{with probability } 1 - \alpha \end{cases} \qquad (7.19)$$

It is straightforward, although rarely needed in practice, to extend the idea of operating point interpolation and model mixing to more than two operating points.

Example 7.2.13 The following R code defines a function for mixing the class label predictions of two models, with a given probability of the first model. This function is subsequently used to mix the predictions of the decision tree model for the *Soybean* dataset corresponding to the two shifted operating points from the previous example. The interpolated operating point is obtained by mixing the two operating points in the 3 : 1 proportion, which corresponds to the first operating point (the true positive rate above 0.85, the minimum false positive rate) being selected with probability 0.75. The ROC curve is plotted again, with the default operating point marked by an asterisk, the shifted operating points marked by a circle and a triangle, and the interpolated operating point marked by a diamond. This is presented in Figure 7.4. Because of the randomness of the model mixing process the interpolated point may not lie exactly where expected if the dataset is not sufficiently large.

```
mixclass <- function(c1, c2, p)
{ factor(ifelse(p<runif(length(c1)), c1, c2), labels=levels(c1)) }

  # interpolate between the two shifted operating points
s01.mix <- mixclass(cutclass(predict(s01.tree, s01.test)[,2], s01.t085, s01.labels),
                cutclass(predict(s01.tree, s01.test) [,2], s01.f05, s01.labels),
                0.75)
s01.cmi <- confmat(s01.mix, s01.test$Class)

  # the ROC curve
plot(s01.roc$fpr, s01.roc$tpr, type="l", xlab="FP rate", ylab="TP rate")
  # the default operating point
points(fpr(s01.cm), tpr(s01.cm), pch=8)
  # the 1st shifted operating point
points(fpr(s01.cm.t085), tpr(s01.cm.t085), pch=1)
  # the 2nd shifted operating point
points(fpr(s01.cm.f05), tpr(s01.cm.f05), pch=2)
  # the interpolated operating point
points(fpr(s01.cmi), tpr(s01.cmi), pch=5)
```

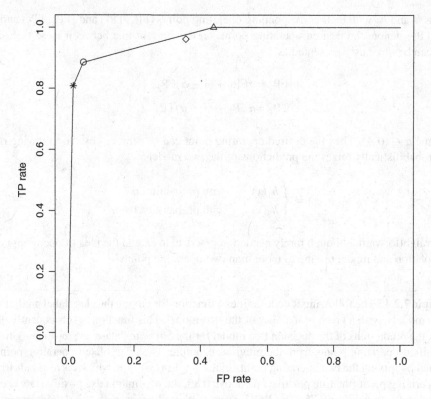

Figure 7.4 Interpolating between operating points for the decision tree model for the two-class *Soybean* dataset.

7.2.5.5   Area under the curve

The ROC analysis can not only be used to compare different operating points and to identify the best operating point but also can be used to compare scoring classifiers irrespective of their labeling functions, i.e., the scoring functions alone. This requires comparing different ROC curves. It is quite trivial when one curve is entirely above another, which clearly means that the former is better. This does not have to always imply, as it might appear, that for each operating point of the worse curve there is a superior operating point on the second curve (e.g., one with a greater true positive rate and with a less or equal false positive rate), because such an operating point does not have to be achievable on the better curve. In such cases, however, a superior operating point can be obtained by interpolation, as discussed above. With this disclaimer, the situation of one curve entirely above another allows one to unambiguously judge the model corresponding to the former as superior to the model corresponding to the latter.

With intersecting ROC curves the situation is no longer so clear. The intersection means that some parts of one curve are above the other and some parts are below the other, i.e., in some ranges of the false positive rate, one model achieves a higher true positive rate, and in some other ranges the other model has more true positives. One might be preferable to the other depending on which range we are most concerned with, i.e., where the ultimate desired operating point is most likely to lie. Sometimes a simple comparison criterion is needed even in such more complex cases, though. Whenever we have a variety of models, produced using different algorithms or parameter settings, we may need a quick and easy way of ranking them with respect to their predictive utility without considering any particular operating points. One such commonly used criterion is the *area under the ROC curve*, sometimes referred to as AUC (*area under curve*). During model comparison, the model with a greater AUC value can be roughly considered superior with respect to its overall predictive performance potential, even if another model with a less AUC value could actually achieve a more preferable operating point than any point achievable by the former. This performance measure is typically used to easily assess the effect of different parameter settings for classification algorithms and select a subset of most promising models before proceeding with operating point selection or optimization. This is also handy whenever the best scoring model needs to be selected to be subsequently used at several different operating points (e.g., minimizing misclassification costs with respect to several different or dynamically changing cost matrices).

Example 7.2.14 The following R code defines a function for calculating the area under the ROC curve, represented as a data frame with columns `tpr` and `fpr` – like that created by the `roc` function from Example 7.2.11. The function is applied to calculate the area under the two ROC curves plotted before – for the decision tree model and for a random model. For the latter, the area under the ROC curve may substantially differ from the expected value of 0.5 due to the small size of the test subset.

```
auc <- function(roc)
{ n <- nrow(roc); sum((roc$tpr[1:n-1]+roc$tpr[2:n])*diff(roc$fpr)/2) }

  # area under the ROC curve for the decision tree model
auc(s01.roc)
  # area under the ROC curve for a random model
auc(s01rand.roc)
```

7.2.5.6 Weighted ROC

Nothing prevents us from adopting the idea of weight-sensitive evaluation for ROC analysis. Similarly as for calculating the error or the confusion matrix, instance weights can be incorporated for generating points and curves on the ROC plane. Again, the only required change is to replace instance counting by summing up instance weights. Since the true positive rate and the false positive rate – the two indicators spanning the ROC place – are defined in terms of confusion matrix entries, using a weighted confusion matrix to calculate them yields appropriately weighted ROC operating points and curves.

Example 7.2.15 The following R code defines a function for calculating the indicators needed for the ROC curve with instance weights taken into account. It differs from the unweighted version presented before only marginally, just by summing up instance weights instead of instance counts. The function is applied to produce the ROC curves for the decision tree for the two-class version of the *Soybean* data, with two different weighting schemes, the same as previously used in Example 7.2.9. One gives twice more weight to instances of the `brown-spot` class than to the instances of the `other` class, and the other gives 10 times less weight to instances with the `plant.stand` attribute taking value 1. The plots are presented in Figure 7.5.

```
wroc <- function(pred.s, true.y, w=rep(1, length(true.y)))
{
  cutoff <- Inf  # start with all instances classified as negative
  tp <- fp <- 0
  tn <- sum((2-as.integer(true.y))*w)  # all negative instances
  fn <- sum((as.integer(true.y)-1)*w)  # all positive instances
  rt <- data.frame()

  sord <- order(pred.s, decreasing=TRUE)  # score ordering
  for (i in 1:length(sord))
  {
    if (pred.s[sord[i]] < cutoff)
    {
      rt <- rbind(rt, data.frame(tpr=tp/(tp+fn), fpr=fp/(fp+tn), cutoff=cutoff))
      cutoff <- pred.s[sord[i]]
    }

    p <- (as.integer(true.y[sord[i]])-1)*w[sord[i]]  # next positive
    n <- (2-as.integer(true.y[sord[i]]))*w[sord[i]]  # next negative
    tp <- tp+p
    fp <- fp+n
    tn <- tn-n
    fn <- fn-p
  }
  rt <- rbind(rt, data.frame(tpr=tp/(tp+fn), fpr=fp/(fp+tn), cutoff=cutoff))
}

  # ROC curve with double weight for the brown-spot class
s01.w1roc <- wroc(predict(s01.tree, s01.test)[,2], s01.test$Class, s01.w1test)
plot(s01.w1roc$fpr, s01.w1roc$tpr, type="l", xlab="FP rate", ylab="TP rate")
auc(s01.w1roc)

  # ROC curve with 10 times less weight for instances with plant.stand=1
s01.w2roc <- wroc(predict(s01.tree, s01.test)[,2], s01.test$Class, s01.w2test)
lines(s01.w2roc$fpr, s01.w2roc$tpr, lty=2)
legend("bottomright", c("brown-spot x2", "plant.stand=1 x10"), lty=1:2)
auc(s01.w2roc)
```

Since per-class instance weights have no impact on the true positive rate and the false positive rate, the ROC curve for the first weighting scheme is the same as presented before for the unweighted case. Under the second weighting scheme, where the weights of instances of the same class may vary, the ROC curve is different.

7.2.6   Probabilistic performance measures

For probabilistic classification models, it may be desirable to measure the quality of their class probability estimates rather than or in addition to the resulting class label assignments. For two-class models, this may be indirectly accomplished to some extent by ROC analysis (in particular, using the area under the ROC curve), but this approach focuses on the discriminative power of class probabilities only and not on their general reliability. Evaluating the latter requires appropriate probabilistic performance measures, assessing the degree of match between class labels in the dataset used for evaluation and class probabilities generated by the model.

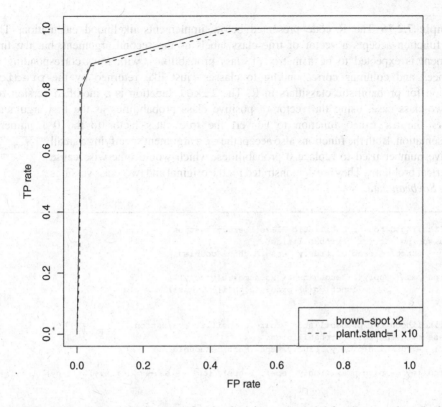

Figure 7.5 The weighted ROC curves for the decision tree model for the two-class *Soybean* dataset.

7.2.6.1 Likelihood

Regardless of a particular dataset, clearly there are no "true class probabilities" available to which the predicted class probabilities be simply compared. As usual, the dataset can only be assumed to come with true class labels attached. What one can do, however, is determining the probability of the latter using the model to estimate class probabilities for all instances. This leads to the *likelihood* of dataset S with respect to the target concept c given model probability estimates π defined as follows:

$$P(S, c|\pi) = \prod_{x \in S} P(x, c|\pi) = \prod_{x \in S} \pi(x, c(x)) \tag{7.20}$$

where

$$\pi(x, d) = P(d|x) \tag{7.21}$$

is the probability of class d for instance x predicted by the model. The higher the likelihood of the data, the better the probability predictions are considered, which makes the likelihood a performance measure to be maximized. For two-class classification tasks, we can simplify the notation using $\pi(x) = P(1|x)$ and rewrite the above definition as follows:

$$P(S, c|\pi) = \prod_{x \in S} \pi(x)^{c(x)}(1 - \pi(x))^{1-c(x)} \tag{7.22}$$

assuming class labels from the $\{0, 1\}$ set can be used as numbers.

Example 7.2.16 The R code presented below implements likelihood calculation. The `lik` function accepts a vector of true class labels as the second argument, but the first argument is expected to be a matrix of class probabilities, with rows corresponding to instances and columns corresponding to classes – just like returned by the `predict` function for probabilistic classifiers in R. The `lik01` function is a modified version for the two-class case, using the vector of positive class probabilities as the first argument. It uses the `as.num0` function to convert the true class factor to the 0–1 numeric representation. Both the functions also accept the `eps` argument specifying a small ┌──────────┐ positive number used to replace 0 probabilities, which would otherwise cause `dmr.util` numerical problems. They are demonstrated for the original and two-class versions └──────────┘ of the *Soybean* data.

```
## likelihood for probabilistic classifier evaluation
## assuming eps for 0 probabilities
lik <- function(prob.y, true.y, eps=.Machine$double.eps)
{
  prod(pmax(sapply(1:length(tn <- as.numeric(true.y)),
                   function(i) prob.y[i,tn[i]]), eps))
}

## likelihood for probabilistic binary classifier evaluation
## assuming eps for 0 probabilities
lik01 <- function(prob.y, true.y, eps=.Machine$double.eps)
{
  prod((py <- pmin(pmax(prob.y, eps), 1-eps))^(t01 <- as.num0(true.y))*(1-py)^(1-t01))
}

# likelihood for the Soybean data
lik(predict(s.tree, s.test), s.test$Class)
lik(predict(s01.tree, s01.test), s01.test$Class)
lik01(predict(s01.tree, s01.test)[,2], s01.test$Class)
```

7.2.6.2 Loglikelihood

On several occasions, it may be convenient to use the logarithm of the likelihood due to its analytic or computational properties. The resulting quantity is called the *loglikelihood* of dataset S with respect to the target concept c given model probability estimates π and defined as

$$L(S, c|\pi) = \log \prod_{x \in S} P(x, c|\pi) = \sum_{x \in S} \log \pi(x, c(x)) \qquad (7.23)$$

The base of the logarithm is irrelevant, with the natural logarithm being a particularly popular choice. An equivalent alternative definition for two-class tasks can be written as

$$L(S, c|\pi) = \sum_{x \in S} (c(x) \log \pi(x) + (1 - c(x)) \log(1 - \pi(x))) \qquad (7.24)$$

Loglikelihood values are negative, with greater values (closer to 0) preferred. The negated loglikelihood can serve as a loss function for probabilistic classification algorithms, referred to as the *logarithmic loss* or *log-loss*. It takes positive values, with less values preferred, which is consistent with the misclassification error or the mean misclassification cost.

Example 7.2.17 The loglikelihood classifier performance measure is implemented and demonstrated by the following R code, following the pattern of the previous example.

```
## loglikelihood for probabilistic classifier evaluation
## assuming eps for 0 probabilities
loglik <- function(prob.y, true.y, eps=.Machine$double.eps)
{
  sum(log(pmax(sapply(1:length(tn <- as.numeric(true.y)),
                function(i) prob.y[i,tn[i]]), eps)))
}

## loglikelihood for probabilistic binary classifier evaluation
## assuming eps for 0 probabilities
loglik01 <- function(prob.y, true.y, eps=.Machine$double.eps)
{
  sum((t01 <- as.num0(true.y))*log(py <- pmin(pmax(prob.y, eps), 1-eps))+
    (1-t01)*log(1-py))
}

  # loglikelihood for the Soybean data
loglik(predict(s.tree, s.test), s.test$Class)
loglik(predict(s01.tree, s01.test), s01.test$Class)
loglik01(predict(s01.tree, s01.test)[,2], s01.test$Class)
```

7.3   Evaluation procedures

The objective of model evaluation procedures is to determine how to apply selected performance measures in order to obtain the reliable assessment of the model's expected performance on new data, i.e., to determine its generalization properties. This is not possible as long as entirely or partially the same data is used for both model creation and model evaluation. For many modeling algorithms that would lead to overoptimistic performance estimates, which is sometimes referred to as *evaluation overfitting*. The main effort in designing evaluation procedures is therefore to ensure the separation of the validation or the test set Q from the training set T without degrading the model quality due to insufficient training data.

7.3.1   Model evaluation vs. modeling procedure evaluation

Contrary to a common misconception, there is nothing wrong in using the whole available labeled dataset as the training set for building a model. What is deeply wrong is using some part of this dataset to evaluate the same model. So one can train and use all-data models as long as one does not evaluate them. This is, of course, totally unacceptable for research, which is all about evaluating, comparing, and benchmarking, but for practical applications it is also completely unimaginable to accept any model without reliable performance estimates, The solution to this dilemma may be to evaluate one or more models built on a smaller training subset using a separate validation or test set, and use the obtained indicators as performance estimates for another model, built on the whole dataset, using exactly the same modeling procedure (i.e., the classification algorithm, its parameter settings, and all other details that impact the produced model being unchanged). The right way to look at the evaluation procedures is therefore often as methods of evaluating a *modeling procedure* rather than the actual model delivered for the application, where the term "modeling procedure" encompasses the classification algorithm and everything else other than the dataset that affects the generated model.

When evaluating a modeling procedure, an appropriate evaluation procedure is needed to keep training and validation or test sets separate. This evaluation procedure can be applied to calculate one or more performance indicators that will serve as performance estimators for the model built on the complete dataset using exactly the same modeling procedure. The final model not only can, but also should be built using as much data as is available (and can be handled within the existing computational constraints, if any), to maximize performance on new data.

7.3.2   Evaluation caveats

While the basic idea to separate the test or validation set from the training set is more than straightforward, there are some issues related to its application that deserve most careful attention:

> *Bias vs. variance.* A too small validation or test set does not provide reliable performance estimates due to high variance, but sparing a considerable part of data for the purpose of evaluation is likely to deteriorate the performance of the model being evaluated, since it would have to be created based on a smaller than possible training set, thereby introducing a pessimistic bias that also makes the performance estimates unreliable,
>
> *Intermediate vs. final evaluation.* Whenever any decisions affecting the final model are made based on the intermediate evaluation of several candidate models, the validation set used for this intermediate evaluation does not provide reliable estimates of the final model's performance, since it has been implicitly and indirectly included in the model creation process.

Both issues are more thoroughly discussed below.

7.3.2.1   Bias vs. variance

Bias and variance are the two possible reasons of the unreliability of performance estimates produced by model evaluation procedures. To understand them correctly, consider performance indicators as random variables, with realizations depending on the particular dataset from the domain that was available and used by the evaluation procedure. These random variables are supposed to be estimators of the unknown values of the corresponding performance indicators on the whole domain (i.e., on mostly unseen data). The *bias* of such an estimator is the difference between its expected value and the unknown indicator being estimated, i.e., the expected difference between the estimator's realization and the true value of the estimated indicator. An estimator's *variance* is the variance of its realizations for different possible datasets from the domain. The bias and variance of an evaluation procedure are the corresponding properties of estimators calculated using the procedure.

A reliable performance estimator should have low bias and variance. A highly biased estimator considerably systematically differs from the true performance indicator. A high-variance estimator is likely to yield substantially different results depending on the particular dataset. It may be a challenge for evaluation procedures to produce performance estimators with both low bias and low variance, since there is a natural tradeoff between these two. To reduce the variance, one should use a sufficiently large validation or test set, since this can be expected to smooth out the random effects of the particular dataset. Unfortunately, this prevents a considerable number of labeled instances from being used for creating the model. A model created based on a smaller subset of data is likely to have less predictive power than another model created in the same way using more data. As a result, one will obtain quite reliable performance estimators for a model that is probably worse than possible to obtain using a larger subset of the available data. As performance estimators for a model that could be built on the whole dataset they are therefore pessimistically biased and unreliable.

Such a pessimistic bias is something one should be definitely aware of, although not always and not necessarily worry about. The biased performance measures for the model built on the training subset calculated on the validation or test subset can serve as nonoverestimating performance estimates for another model built with the same modeling procedure on the whole available dataset, and it is the latter that would be actually deployed for the application. As long as the performance estimates are satisfactory with respect to the application's requirements, this is perfectly acceptable, since the deployed all-data model is likely to outperform the one that was actually evaluated. If the performance estimates do not meet the expectations, though, we are left in uncertainty: this may not be the effect of the pessimistic evaluation bias, but also indicate the classification algorithm's inability to produce a sufficiently good model.

High evaluation variance may be more problematic, particularly in intermediate evaluation which is used for model selection or making other decisions that affect the final model. Using high-variance performance estimators for such decisions is likely to yield suboptimal results.

7.3.2.2   Intermediate vs. final evaluation

The difference between intermediate and final evaluation is somewhat subtle and, although more widely acknowledged recently, it still tends to be overlooked or underappreciated both in research and in practical data mining projects. Whereas the latter is only supposed to provide information about the evaluated model's expected performance on unseen data, which allows one to judge about its suitability for a given application or estimate the expected gain or loss resulting from using the model to make real decisions, the former is also supposed to direct the search for a better model.

Typically several candidate models, obtained by different modeling procedures (e.g., by running possibly different algorithms, with possibly different parameter settings, using possibly different attribute subsets, etc.), go through intermediate evaluation, which ranks them with respect to their expected generalization performance and makes it possible to choose the most promising ones. These may be subject to some other refinement yielding again a set of new candidate models which should be evaluated again, until a satisfactory final model is found or no further improvement seems possible. In such cases, the validation set used to evaluate the performance of candidate models is effectively used in the overall modeling process, since it affects the ultimate choice of several details of the modeling procedure (like the choice of algorithm, parameter settings, attribute selection, etc.) that usually have considerable impact on the final model. This is a very reasonable way to proceed when seeking for the best possible model quality, but unfortunately the same validation set cannot be used to achieve reliable performance estimate for the final model, since it has already been used for training in a wider sense, which includes all possible decision-making processes (automated or human) based on the data.

Table 7.2   Training and test set illustration.

D		
T	Q	
T'	Q'	
T''		Q''

This is not to say that the final model should be expected to be poor or there is something wrong with the model selection process. This is only to say that we should not expect the final model to perform as well on new data as it performs on the validation set, on which it was found to perform best out of several examined candidates. It is likely to perform worse on previously unseen data, and one needs a separate test set to find out how. Otherwise a severe risk of evaluation overfitting arises.

The phenomenon of validation sets becoming effectively parts of *generalized training sets* and the need for separate test sets can be illustrated as presented in Table 7.2. If we build a model on the training set T and evaluate it on the validation set Q, but then use the evaluation results to make some decisions that impact a subsequently created model (e.g., to select a classification algorithm, set parameters, select attributes, etc.), then we need a separate test set Q' to reliably evaluate the new model, since $T \cup Q$ effectively becomes a generalized training set T'. But if the search for a satisfactory model continues and another decision-making process takes place to improve the model based on the evaluation performed on Q' used as the validation set, then again another separate test set Q'' should be used for the evaluation of the next resulting model, obtained based on $T' \cup Q'$ comprising a next stage generalized training set T''. In principle, several such iterations of training, evaluation, and improvement decision making are possible, although it is very uncommon to see more than two or three in practice.

Note how the necessity to separate intermediate evaluation(s) and the final evaluation apparently reinforces the bias vs. variance issue. If model creation is a multistage process that needs two or more separate data subsets for model evaluation, then the risk of too little being left for training becomes more severe. It is not necessarily that bad, luckily, since on each stage the model can and should be built using all data that cannot reliably serve for the purpose of evaluation. In the illustration above, after the first-stage models are built on T and evaluated on Q to make some decisions that affect the second-stage models, the latter can be actually built on $T' = T \cup Q$ rather than on T, since it depends on Q anyway and cannot be reliably evaluated on Q. Similarly, the final model can and should be built using all data except the final test set. In the above illustration, the final model could actually be built using T'' rather than T as the training set and then evaluated on Q''. And as noted above, in any case the ultimately deployed model (if the modeling process is performed for a real-world application rather than research or algorithm benchmarking) can be built using *all available labeled data*, as long as we are fine with pessimistically biased performance estimates obtained by evaluating its counterpart built on a smaller training set. As we will see, some sufficiently refined evaluation procedure can actually reduce the bias to a negligible level.

Table 7.3 Predicted class vector generation for the repeated hold-out procedure.

	Predicted	True
For all $x \in Q_1$	$h_1(x)$	$c(x)$
For all $x \in Q_2$	$h_2(x)$	$c(x)$
...	...	...
For all $x \in Q_n$	$h_n(x)$	$c(x)$

7.3.3   Hold-out

The hold-out evaluation procedure is the most straightforward way of separating training and validation or test data: a subset of the available labeled dataset is selected randomly as the training set and the remaining instances are *held out* for the purpose of model evaluation. This approach is clearly prone to the bias vs. variance tradeoff as mentioned above: with sufficiently many instances left for low-variance evaluation there may be too little training instances to ensure adequate model quality and one may end up with a considerable pessimistic bias resulting from quite reliable performance estimates of a quite poor model. It is common to partition the data in a $2:1$ proportion (i.e., 2/3 for training and 1/3 for evaluation), which may be hoped to keep both the bias and variance within reasonable bounds, but usually the risk of high bias and high variance remains substantial. The risk is limited when a plentiful of data is available. In particular, for very large datasets one may be forced to use a small sample as the training set anyway because of computational constraints. This is where the hold-out procedure can be safely applied.

Regardless of the dataset size, it is always a good idea to repeat the hold-out procedure a number of times, with different training sets $T_1, T_2, \dots, T_n$ and validation or test sets $Q_1, Q_2, \dots, Q_n$ drawn at random. This is likely to reduce the variance substantially and permit using smaller validation or test subsets to reduce the bias as well. Instead of averaging the results of multiple hold-out runs, which is typically recommended, it may be more convenient to create extended vectors of predicted and true class labels for performance measure calculation, containing the predictions for all randomly drawn test or validation subsets generated by models created on the corresponding training subsets, along with their true class labels, as illustrated in Table 7.3. Such a pair of vectors with predicted and true class labels can be used to calculate whatever performance measures are of interest in a given application.

As noted above, it makes sense to think about evaluation procedures as providing performance estimates not for individual models, but rather for modeling procedures. This applies in particular to the hold-out procedure, for which the performance estimates obtained for the model built on the subset of training instances can be considered as assessing the quality of the modeling procedure that created the model. The same procedure can be reapplied to the whole available dataset to hopefully create a better model.

Example 7.3.1 Note that all the examples of performance measures presented above used the hold-out procedure, by randomly dividing the *Soybean* dataset into training and test subsets. By repeating the random partitioning used for these examples several times and recalculating the performance measures we would likely observe considerably different results, due to the high variance of this evaluation procedure. The following R code implements an automated hold-out procedure that repeats the random dataset partitioning, model building, and prediction several times, and collects the observed predicted and true class labels. When provided with a probabilistic prediction function, it can return predicted class probabilities instead of labels (this requires setting `prob=TRUE` to prevent converting predictions to a factor). Note, by the way, that it would also handle numeric predictions correctly, if applied to evaluating a regression model. This is also true for other evaluation procedure implementations presented in subsequent examples, which makes them reusable for regression model evaluation examples in Chapter 10.

The `holdout` function is applied to perform 10-times repeated hold-out evaluation of decision tree models for the two-class version of the *Soybean* dataset (2/3 of which is used for training with the remaining 1/3 used for testing), using both discrete and probabilistic predictions. In the former case the error and confusion-matrix are calculated, and in the latter case the ROC curve is plotted and the area under the curve calculated. The plot is presented in Figure 7.6.

```
holdout <- function(alg, formula, data, args=NULL, predf=predict, prob=FALSE,
                    p=0.33, n=1)
{
  yn <- as.character(formula)[2]   # class column name
  ylabs <- levels(data[[yn]])      # class labels
  pred.y <- NULL  # predictions
  true.y <- NULL  # true class labels

  for (t in 1:n)
  {
    r <- runif(nrow(data))
    train <- data[r>=p,]
    test <- data[r<p,]
    model <- do.call(alg, c(list(formula, train), args))
    pred.y <- c(pred.y, predf(model, test))
    true.y <- c(true.y, test[[yn]])
  }

  if (!is.null(ylabs))
  {
    if (!prob)
      pred.y <- factor(pred.y, levels=1:length(ylabs), labels=ylabs)
    true.y <- factor(true.y, levels=1:length(ylabs), labels=ylabs)
  }

  return(data.frame(pred=pred.y, true=true.y))
}

  # hold-out evaluation of discrete predictions
s01ho <- holdout(rpart, Class~., Soybean01,
                 predf=function(...) predict(..., type="c"), n=10)
err(s01ho$pred, s01ho$true)
confmat(s01ho$pred, s01ho$true)

  # hold-out evaluation of probabilistic predictions
s01hop <- holdout(rpart, Class~., Soybean01,
                  predf=function(...) predict(..., type="p")[,2], prob=TRUE, n=10)
s01hop.roc <- roc(s01hop$pred, s01hop$true)
plot(s01hop.roc$fpr, s01hop.roc$tpr, type="l", xlab="FP rate", ylab="TP rate")
auc(s01hop.roc)
```

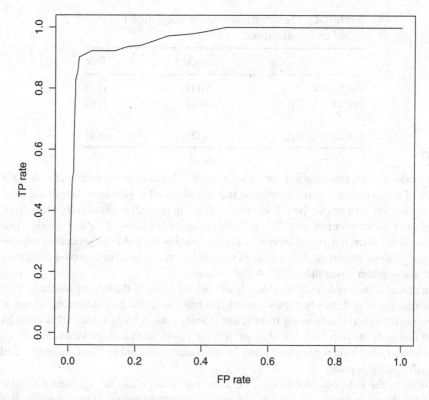

Figure 7.6 The ROC curve for the two-class *Soybean* decision tree model obtained using hold-out evaluation.

The ROC curve, which now contains several times more points, is considerably smoother than the one presented in Example 7.2.11, based on a single (manual) execution of the hold-out procedure.

7.3.4 Cross-validation

A more refined evaluation procedure that handles the bias vs. variance tradeoff better is k-fold cross-validation. It splits the available dataset at random into k disjoint subsets of (roughly) the same size $D_1, D_2, \ldots, D_k$ and then iterates over these subsets. On the ith iteration a model is built using $T_i = \bigcup_{j \neq i} D_j$ as the training set, and applied to generate predictions on $Q_i = D_i$. Once all k iterations are completed, a predicted class label (or score, for a scoring model) has been generated for each instances in the dataset, using the model built without this instance in the training set. This is schematically illustrated in Table 7.4. The resulting vector of predictions can be compared to the true class labels using one or more selected performance measures.

A single iteration of k-fold cross-validation is equivalent to the hold-out procedure, with $\frac{k-1}{k}$ of data selected for training and $\frac{1}{k}$ of data selected for evaluation. For sufficiently large k, this does not reduce the training set size to an extent that would be likely to severely

Table 7.4   Predicted class vector generation for k-fold cross-validation.

	Predicted	True
For all $x \in D_1$	$h_1(x)$	$c(x)$
For all $x \in D_2$	$h_2(x)$	$c(x)$
...	...	...
For all $x \in D_k$	$h_k(x)$	$c(x)$

impact model quality, since the validation set is small. This does not substantially increase the variance of performance estimates because in k iterations all available instances are used for model evaluation. In a sense, the k-fold cross-validation procedure effectively virtualizes the training and validation or test sets. All available instances can be used for both model creation and evaluation, albeit not simultaneously. This resembles the hold-out procedure repeated k times, but with one important difference: all the validation sets from consecutive iterations are disjoint and together cover the whole available dataset.

The proper choice of k involves a couple of tradeoffs. One is the bias vs. variance tradeoff discussed above. A sufficiently large k is needed to make sure that the evaluation process is not pessimistically biased due to using insufficient training sets. A large value of k is often likely to yield higher evaluation variance, though. It is also associated with considerably increased computational cost. Values between 3 and 20 are usually selected in practice, with 5 and 10 being particularly popular.

Similar to the hold-out procedure, k-fold cross-validation can be repeated a number of times for reduced variance, as long as the available computational power permits. With n repetitions, the resulting procedure is called the $n \times k$-fold cross-validation. Assuming a fixed number of model training cycles available for the evaluation procedure, it may be often a better idea to allocate them into $n > 1$ repetitions of the k-fold cross-validation with a smaller k than using a single run with a correspondingly larger k. In particular, 4×5-fold cross-validation may be preferred to 20-fold cross-validation, and 5×10-fold cross-validation will likely yield more reliable performance estimates than 50-fold cross-validation because they may have substantially less variance without a substantially higher bias.

Example 7.3.2 The following R code defines a function for performing k-fold cross-validation. The function is applied to evaluate discrete predictions of decision tree models for the two-class version of the *Soybean* dataset with a few different k values. Then, with $k = 10$, it is applied again in the probabilistic prediction case to produce the ROC curve and calculate the area under the curve. The plot is presented in Figure 7.7.

```
crossval <- function(alg, formula, data, args=NULL, predf=predict, prob=FALSE,
                     k=10, n=1)
{
  yn <- as.character(formula)[2]   # class column name
  ylabs <- levels(data[[yn]])       # class labels
  pred.y <- NULL   # predictions
  true.y <- NULL   # true class labels

  for (t in 1:n)
```

```
{
  ind <- sample(k, size=nrow(data), replace=TRUE)  # index of k random subsets
  for (i in 1:k)
  {
    train <- data[ind!=i,]
    test <- data[ind==i,]
    model <- do.call(alg, c(list(formula, train), args))
    pred.y <- c(pred.y, predf(model, test))
    true.y <- c(true.y, test[[yn]])
  }
}

if (!is.null(ylabs))
{
  if (!prob)
    pred.y <- factor(pred.y, levels=1:length(ylabs), labels=ylabs)
  true.y <- factor(true.y, levels=1:length(ylabs), labels=ylabs)
}

return(data.frame(pred=pred.y, true=true.y))
}


# 3-fold cross-validation for discrete predictions
s01cv3 <- crossval(rpart, Class~., Soybean01,
                   predf=function(...) predict(..., type="c"), k=3)
err(s01cv3$pred, s01cv3$true)
confmat(s01cv3$pred, s01cv3$true)


# 10-fold cross-validation for discrete predictions
s01cv10 <- crossval(rpart, Class~., Soybean01,
                    predf=function(...) predict(..., type="c"), k=10)
err(s01cv10$pred, s01cv10$true)
confmat(s01cv10$pred, s01cv10$true)


# 20-fold cross-validation for discrete predictions
s01cv20 <- crossval(rpart, Class~., Soybean01,
                    predf=function(...) predict(..., type="c"), k=20)
err(s01cv20$pred, s01cv20$true)
confmat(s01cv20$pred, s01cv20$true)


# 4x5-fold cross-validation for discrete predictions
s01cv4x5 <- crossval(rpart, Class~., Soybean01,
                     predf=function(...) predict(..., type="c"), k=5, n=4)
err(s01cv20$pred, s01cv4x5$true)
confmat(s01cv4x5$pred, s01cv4x5$true)


# 10-fold cross-validation for probabilistic predictions
s01cv10p <- crossval(rpart, Class~., Soybean01,
                     predf=function(...) predict(..., type="p")[,2], prob=TRUE, k=10)
s01cv10p.roc <- roc(s01cv10p$pred, s01cv10p$true)
plot(s01cv10p.roc$fpr, s01cv10p.roc$tpr, type="l", xlab="FP rate", ylab="TP rate")
auc(s01cv10p.roc)
```

7.3.5 Leave-one-out

The leave-one-out validation procedure takes the idea of k-fold cross-validation to the extreme, by using the number of instances in the dataset as the value of k. The procedure iterates over all instances, using the model built on the dataset with one instance removed to make prediction

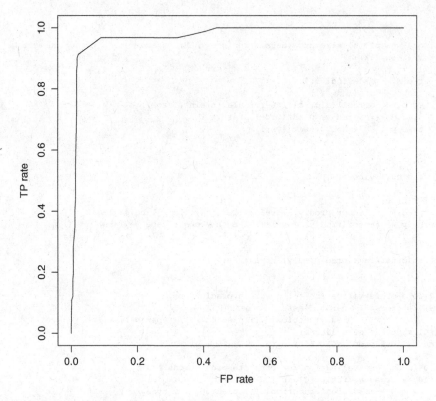

Figure 7.7   The ROC curve for the two-class *Soybean* decision tree model obtained using *k*-fold cross-validation.

for this instance. This might appear an ultimate noncompromise form of cross-validation, given the major advantage of large *k* discussed above; no pessimistic bias. This is not necessarily the case in reality, however, since, besides the variance increase due to large *k* (which cannot be reduced by multiple runs), the leave-one-out evaluation procedure has been found to sometimes yield overoptimistic estimates. This is because, particularly for larger datasets, all the individual models created on subsequent iterations are quite unlikely to differ considerably from one another, as well as from the model that would be built using the complete dataset – since their training sets differ just in a single instance. Nevertheless, it is a reasonable evaluation procedure when building classification models based on small datasets, where a single instance still matters a lot. It is hardly applicable to large datasets anyway due to the computational expense of building as many models as instances available.

Unlike hold-out and cross-validation (for *k* less than the dataset size), leave-one-out leaves no space for randomness in the evaluation process and is perfectly deterministic and reproducible.

***

Example 7.3.3 The following R code defines a function for performing the leave-one-out evaluation procedure that follows the same design pattern that was used to implement the hold-out and cross-validation procedures in the preceding examples. There is no argument for specifying the number of repetitions this time, as it would make no sense given the full determinism of leave-one-out model evaluation. As before, the function is applied to the two-class version of the *Soybean* data, both with discrete and probabilistic predictions. The ROC curve is plotted for the latter and presented in Figure 7.8.

```
leavelout <- function(alg, formula, data, args=NULL, predf=predict, prob=FALSE)
{
  yn <- as.character(formula)[2]  # class column name
  ylabs <- levels(data[[yn]])     # class labels
  pred.y <- NULL  # predictions
  true.y <- NULL  # true class labels

  for (i in 1:nrow(data))
  {
    train <- data[-i,]
    test <- data[i,]
    model <- do.call(alg, c(list(formula, train), args))
    pred.y <- c(pred.y, predf(model, test))
    true.y <- c(true.y, test[[yn]])
  }

  if (!is.null(ylabs))
  {
    if (!prob)
      pred.y <- factor(pred.y, levels=1:length(ylabs), labels=ylabs)
    true.y <- factor(true.y, levels=1:length(ylabs), labels=ylabs)
  }

  return(data.frame(pred=pred.y, true=true.y))
}

  # leave-one-out for discrete predictions
s01l1o <- leavelout(rpart, Class~., Soybean01,
                    predf=function(...) predict(..., type="c"))
err(s01l1o$pred, s01l1o$true)
confmat(s01l1o$pred, s01l1o$true)

  # leave-one-out for probabilistic predictions
s01l1op <- leavelout(rpart, Class~., Soybean01,
                    predf=function(...) predict(..., type="p")[,2], prob=TRUE)
s01l1op.roc <- roc(s01l1op$pred, s01l1op$true)
plot(s01l1op.roc$fpr, s01l1op.roc$tpr, type="l", xlab="FP rate", ylab="TP rate")
auc(s01l1op.roc)
```

Notice that, unlike those produced in the examples for the hold-out and cross-validation procedures, the ROC curve obtained with the leave-one-out procedure is perfectly stepwise. It can be easily verified that using the crossval function with the k=nrow(Soybean01) argument produces nearly identical results (although not necessarily exactly the same due to its nondeterminism).

7.3.6 Bootstrapping

Bootstrapping is a family of general purpose estimation techniques that are based on drawing multiple *bootstrap samples* at random *with replacement*, typically of the same size as the

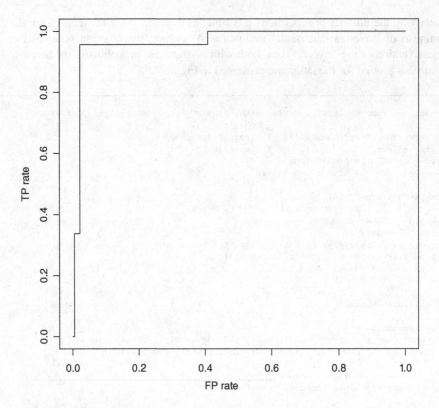

Figure 7.8 The ROC curve for the two-class *Soybean* decision tree model obtained using leave-one-out evaluation.

original dataset. Each bootstrap sample therefore represents its perturbed version with some instances replicated and some instances removed. For a dataset of size N, the probability of a single instance not being selected to a bootstrap sample of the same size is

$$\left(1 - \frac{1}{N}\right)^{N} \approx 0.368 \tag{7.25}$$

and accordingly the probability of a single instance being selected at least once is

$$1 - \left(1 - \frac{1}{N}\right)^{N} \approx 0.632 \tag{7.26}$$

The approximation holds for sufficiently large N. A bootstrap sample can therefore be expected to contain about 63.2% of instances from the original dataset, some replicated. The missing instances, standing for about 36.8% of the dataset, are referred to as *out-of-bag* (OOB) instances. The idea of bootstrapping as en evaluation procedure is to use a bootstrap sample as the training set and to evaluate the resulting model on the out-of-bag instances.

The bootstrapping procedure, similar to the other procedures discussed above, faces the challenge of reducing the bias and variance, but – unlike for the other procedures – there is no real tradeoff between these two goals, since the bias and variance of bootstrapping are controlled independently. To reduce the variance, a sufficient number of independent bootstrap samples and corresponding OOB sets are needed. Numbers ranging from several dozens to several hundreds are most typical. A model is trained for each bootstrap sample and evaluated on the corresponding OOB set. This may provide low-variance performance estimators, but one can expect a high negative bias resulting from the fact that only about 63.2% of available instances are used for training. One way to compensate for this bias is to produce the final estimator as the weighted average of the (overly pessimistic) estimator obtained on OOB instances and the corresponding (overly optimistic) estimator that can be obtained by training and evaluating the model on the full dataset.

The most common instantiation of the above idea is known as the *.632 bootstrap* procedure. Assuming the estimated performance indicator is the misclassification error, it can be presented as

$$e_D^{.632} = 0.632 \frac{1}{M} \sum_{i=1}^{M} e_{c,D_i'}(h_i) + 0.368 e_{c,D}(h) \qquad (7.27)$$

where D is the available dataset from which M bootstrap samples $D_1, D_2, \ldots, D_M$ are drawn, h_i is the model built using D_i as the training set, D_i' is the corresponding set of out-of-bag instances on which the model is evaluated, and h is the model built on the full dataset D. The resulting .632 bootstrap error estimator is the weighted average of the mean out-of-bag error and of the training error on the whole dataset, with the weights equal 0.632 and 0.368, respectively. This is hoped to remove the pessimistic bias resulting from bootstrap samples containing about 63.2% of all available training instances, with 36.8% missing. This is also likely to yield less variance that the plain bootstrap procedure with the same number of bootstrap samples, by incorporating the predictions of the all-data model.

The .632 bootstrap estimator has been found to work quite well when evaluating classification models created by algorithms that do not heavily overfit. For models that fit the training set to a great extent the estimator tends to be optimistically biased. It should therefore be avoided with models that effectively "memorize" much of the training set, such as the nearest-neighbor model or unpruned full-depth decision trees.

Whereas typical presentations of bootstrapping model evaluation in general and of the .632 bootstrap in particular assume that they are used for error estimation, there are no substantial obstacles preventing using the same techniques to estimate other performance measures, such as confusion matrix-based indicators and the ROC curve. Similarly as for other evaluation procedures, we can use bootstrapping to generate vectors of predicted class labels (or scores for scoring classifier evaluation) and true class labels which can be used to calculate arbitrary performance measure. This is straightforward with just one caveat: to apply the .632 bootstrap or another similar weighting scheme, the vectors of predictions and true class labels must be accompanied by a vector of weights, and the calculation of selected performance indicators must incorporate these weights.

Specifically, for the .632 bootstrap we would expect the evaluation procedure to produce a vector of predictions for all out-of-bag instances corresponding to all generated bootstrap samples as well as for the complete dataset, a vector of corresponding true class labels, and a vector of weights, containing a weight of $0.632/M$ for each out-of-bag instance, and a weight of 0.368 for each instance from the complete dataset. This is schematically presented, using the same notation as in the definition of the .632 bootstrap error estimator, in Table 7.5. Such output produced by the .632 bootstrap evaluation procedure can be used to calculate the weighted error, the weighted confusion matrix, and related indicators, as well as to perform weighted ROC analysis.

Table 7.5 The weighting scheme for the .632 bootstrap estimator.

	Predicted	True	Weight
For all $x \in D_1'$	$h_1(x)$	$c(x)$	$0.632/M$
For all $x \in D_2'$	$h_2(x)$	$c(x)$	$0.632/M$
...	...	...	...
For all $x \in D_M'$	$h_M(x)$	$c(x)$	$0.632/M$
For all $x \in D$	$h(x)$	$c(x)$	0.368

Example 7.3.4 To illustrate the bootstrapping approach to model evaluation, the following R code defines a function that implements the bootstrap evaluation procedure. It draws a specified number of bootstrap samples from the provided dataset and uses them to build models which it subsequently applies to generate predictions for out-of-bag instances. If the w parameter is set to a value less than 1, it also creates a full-data model and applies it to generate predictions on the full dataset. These receive a weight of $1 - w$, whereas the former receives a weight of w, divided by the number of bootstrap samples. For $w = 0.632$ this is equivalent to the .632 bootstrap procedure. The procedure returns a data frame with predictions, true class labels, and the corresponding weights. The function is applied to evaluate the decision tree model for the two-class version of the *Soybean* data, both with discrete and probabilistic predictions, with w set to 1 (plain bootstrap, likely to be pessimistically biased) and to 0.632. A relatively small number of 20 bootstrap samples is used. For discrete predictions, the produced output is used to calculate the error and confusion matrix (weighted, in the .632 case), and for probabilistic predictions – to plot the ROC curves. The plots are presented in Figure 7.9.

```
bootstrap <- function(alg, formula, data, args=NULL, predf=predict, prob=FALSE,
                 w=0.632, m=100)
{
  yn <- as.character(formula)[2]  # class column name
  ylabs <- levels(data[[yn]])     # class labels
  pred.y.w <- NULL  # predictions
  true.y.w <- NULL  # true class labels

  for (t in 1:m)
  {
    bag <- sample(nrow(data), size=nrow(data), replace=TRUE)
    train <- data[bag,]
    test <- data[-bag,]
    model <- do.call(alg, c(list(formula, train), args))
    pred.y.w <- c(pred.y.w, predf(model, test))
    true.y.w <- c(true.y.w, test[[yn]])
  }

  if (w<1)
  {
    model <- do.call(alg, c(list(formula, data), args))
    pred.y.1w <- predf(model, data)
    true.y.1w <- data[[yn]]
    w <- c(rep(w/m, length(pred.y.w)), rep(1-w, nrow(data)))
  }
  else
  {
```

```
    pred.y.1w <- true.y.1w <- NULL
    w <- rep(w/m, length(pred.y.w))
  }

  pred.y <- c(pred.y.w, pred.y.1w)
  true.y <- c(true.y.w, true.y.1w)

  if (!is.null(ylabs))
  {
    if (!prob)
      pred.y <- factor(pred.y, levels=1:length(ylabs), labels=ylabs)
    true.y <- factor(true.y, levels=1:length(ylabs), labels=ylabs)
  }

  return(data.frame(pred=pred.y, true=true.y, w=w))
}

# 20x bootstrap for discrete predictions
s01bs20 <- bootstrap(rpart, Class~., Soybean01,
                  predf=function(...) predict(..., type="c"), w=1, m=20)
err(s01bs20$pred, s01bs20$true)
confmat(s01bs20$pred, s01bs20$true)

# 20x .632 bootstrap for discrete predictions
s01.632bs20 <- bootstrap(rpart, Class~., Soybean01,
                      predf=function(...) predict(..., type="c"), m=20)
# weighted error
werr(s01.632bs20$pred, s01.632bs20$true, s01.632bs20$w)
# weighted confusion matrix
wconfmat(s01.632bs20$pred, s01.632bs20$true, s01.632bs20$w)

# 20x bootstrap for probabilistic predictions
s01bs20p <- bootstrap(rpart, Class~., Soybean01,
                  predf=function(...) predict(..., type="p")[,2], prob=TRUE,
                  w=1, m=20)
s01bs20p.roc <- roc(s01bs20p$pred, s01bs20p$true)
plot(s01bs20p.roc$fpr, s01bs20p.roc$tpr, type="l", xlab="FP rate", ylab="TP rate")
auc(s01bs20p.roc)

# 20x .632 bootstrap for probabilistic predictions
s01.632bs20p <- bootstrap(rpart, Class~., Soybean01,
                      predf=function(...) predict(..., type="p")[,2], prob=TRUE,
                      m=20)
# weighted ROC
s01.632bs20p.roc <- wroc(s01.632bs20p$pred, s01.632bs20p$true, s01.632bs20p$w)
lines(s01.632bs20p.roc$fpr, s01.632bs20p.roc$tpr, lty=2)
legend("bottomright", c("plain", ".632"), lty=1:2)
auc(s01.632bs20p.roc)
```

The .632 bootstrap performance estimators suggest a slightly better performance level than the plain bootstrap ones (with $w = 1$), which agrees with the expectation of the pessimistic bias of the latter.

7.3.7   Choosing the right procedure

The choice of the right evaluation procedure for a given application usually depends on the accepted level of the bias vs. variance tradeoff, the size of the dataset, the classification

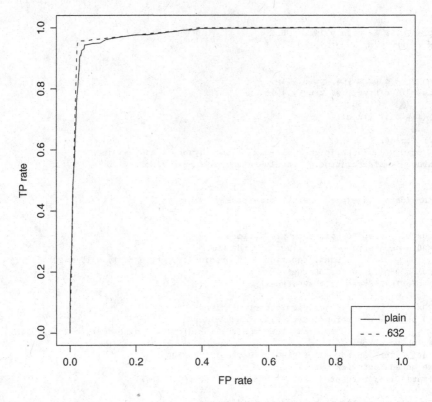

Figure 7.9 The ROC curve for the two-class *Soybean* decision tree model obtained using bootstrap evaluation.

algorithm, and the available computational resources. Most evidence suggests that the bias vs. variance tradeoff is best handled by the k-fold cross-validation procedure with k set to 10 or 20, particularly repeated several times. The .632 bootstrap procedure can also yield nearly unbiased and low-variance performance estimators for classification algorithms that are not prone to overfitting, with the possibility of making the variance arbitrarily low by using a sufficiently large number of bootstrap samples. This ultimate variance reduction makes it an attractive evaluation procedure for model selection. The leave-one-out procedure should be avoided for intermediate evaluation due to its high variance possibly leading to suboptimal decisions, except for small dataset where all other evaluation procedures would be considerably biased. The hold-out procedure is best suited to very large datasets, for which other evaluation procedures would be too expensive and for which considerably smaller data samples would have to be used for model training anyway due to computational constraints.

Example 7.3.5 To illustrate the properties of the different evaluation procedures discussed above, the following R code defines a function that implements a simple experiment to observe their bias and variance. To simulate different possible datasets from the same domain, a random 2/3 subset is drawn from the provided dataset and considered a simulated "available" dataset, with the remaining 1/3 subset considered a simulated "new" dataset. The "new" dataset is used to calculate an estimate of the true error of the model built on the "available" dataset. A number of evaluation procedures (hold-out, cross-validation, leave-one-out, and bootstrap with different parameter settings) are then run on the "available" dataset to produce their error estimates. This experiment is repeated number of times with all the obtained results collected, to finally calculate the estimated bias and variance of each evaluation procedure. The former is obtained as the mean difference between the error estimated by particular evaluation procedures and the true error estimate obtained on the "new" dataset. The latter is obtained as the variance of the error estimated by particular evaluation procedures on different runs.

The function is applied to observe the bias and variance of different evaluation procedures when applied to evaluate decision tree models for the two-class version of the *Soybean* dataset. The results are used to produce a boxplot of the error estimates produced by particular evaluation procedures, with a red horizontal line designating the mean true error estimated on the "new" dataset. Barplots of the bias and variance of all the evaluation procedures are also produced. The plots presented in Figure 7.10 are based on 200 evaluation repetitions, which takes some considerable time. The line that runs this full experiment is commented out and another one that runs a 10-times repeated evaluation experiment is recommended instead for a quick illustration.

```
eval.bias.var <- function(alg, formula, data, args=NULL, predf=predict,
                          perf=err, wperf=werr, p=0.66, n=100)
{
  yn <- as.character(formula)[2]  # class column name
  performance <- data.frame()
  for (i in 1:n)
  {
    r <- runif(nrow(data))
    data.avail <- data[r<p,]   # pretend this is the available dataset
    data.new <- data[r>=0.7,]  # pretend this a new dataset
    model <- do.call(alg, c(list(formula, data.avail), args))

    cv3 <- crossval(alg, formula, data.avail, args=args, predf=predf, k=3)
    cv5 <- crossval(alg, formula, data.avail, args=args, predf=predf, k=5)
    cv10 <- crossval(alg, formula, data.avail, args=args, predf=predf, k=10)
    cv20 <- crossval(alg, formula, data.avail, args=args, predf=predf, k=20)
    cv5x4 <- crossval(alg, formula, data.avail, args=args, predf=predf, k=5, n=4)
    ho <- holdout(alg, formula, data.avail, args=args, predf=predf)
    hox10 <- holdout(alg, formula, data.avail, args=args, predf=predf, n=10)
    llo <- leaveout(alg, formula, data.avail, args=args, predf=predf)
    bs10 <- bootstrap(alg, formula, data.avail, args=args, predf=predf, w=1, m=10)
    bs50 <- bootstrap(alg, formula, data.avail, args=args, predf=predf, w=1, m=50)
    bs10.632 <- bootstrap(alg, formula, data.avail, args=args, predf=predf, m=10)
    bs50.632 <- bootstrap(alg, formula, data.avail, args=args, predf=predf, m=50)


    performance <- rbind(performance,
                    data.frame(perf(predf(model, data.new), data.new[[yn]]),
                               perf(cv3$pred, cv3$true),
                               perf(cv5$pred, cv5$true),
                               perf(cv10$pred, cv10$true),
                               perf(cv20$pred, cv20$true),
                               perf(cv5x4$pred, cv5x4$true),
                               perf(ho$pred, ho$true),
                               perf(hox10$pred, hox10$true),
                               perf(llo$pred, llo$true),
                               perf(bs10$pred, bs10$true),
                               perf(bs50$pred, bs50$true),
                               wperf(bs10.632$pred, bs10.632$true, bs10.632$w),
                               wperf(bs50.632$pred, bs50.632$true, bs50.632$w)))
  }
}
```

```
  names(performance) <- c("true", "3-CV", "5-CV", "10-CV", "20-CV", "4x5-CV",
                          "HO", "10xHO", "L1O", "10-BS", "50-BS",
                          "10-BS.632", "50-BS.632")
  bias <- apply(performance[,-1]-performance[,1], 2, mean)
  variance <- apply(performance[,-1], 2, var)

  list(performance=performance, bias=bias, variance=variance)
}

  # the commented lines run a 200-repetition experiment, which takes a long time
#s01.ebv <- eval.bias.var(rpart, Class~., Soybean01,
#                         predf=function(...) predict(..., type="c"), n=200)
  # this can be used for a quick illustration
s01.ebv <- eval.bias.var(rpart, Class~., Soybean01,
                         predf=function(...) predict(..., type="c"), n=10)

boxplot(s01.ebv$performance[,-1], main="Error", las=2)
lines(c(0, 13), rep(mean(s01.ebv$performance[,1]), 2))
barplot(s01.ebv$bias, main="Bias", las=2)
barplot(s01.ebv$variance, main="Variance", las=2)
```

Of course, an experiment with a single dataset and classification algorithm is by no means conclusive, but at least some of these observations confirm the findings from more widespread and thorough studies as well as theoretical investigations described in the literature. Notice the high bias of hold-out (both single and repeated) and 3-fold cross-validation, and nearly nonexistent bias of 20-fold cross-validation and leave-one-out. The plain bootstrap procedure has a high pessimistic bias, as expected, but the .632 bootstrap appears to be slightly optimistically biased. With respect to variance, 4×5-fold cross-validation is the clear winner within the cross-validation procedures, but easily outperformed by the .632 bootstrap, even with just 10 bootstrap samples, and single hold-out is by far the worst. The 3-fold cross-validation and leave-one-out procedures also demonstrate high variance. The repetition reduces the variance of hold-out considerably, as expected. The 10-fold cross-validation procedure appears to achieve a reasonable compromise between bias and variance. It is particularly noteworthy that both a greater and less number of folds yields higher variance. The former is not surprising, but the latter may be somewhat unexpected and could be attributed to the instability of decision tree classifiers.

7.3.8   Evaluation procedures for temporal data

While all the evaluation procedures discussed in this section, with their specific advantages and disadvantages, remain general-purpose techniques applicable to many different instantiations of the classification task, it is worthwhile to mention one specific situation where they are more than likely to yield misleading, overoptimistic results. This is the case of temporal data, where different instances come from different points in time, and there may be some hidden impact of time on the target concept. Then with all procedures based on random data subset selection some instances on which a model is evaluated will be older than some of its training instances, possibly leading to a better observed performance. This is as if a model supposed to predict the future has been trained on some observations from the future rather than only on those from the past. This is totally unrealistic and does not match the actual model exploitation conditions, where it will only be applied to instances newer than those used for training.

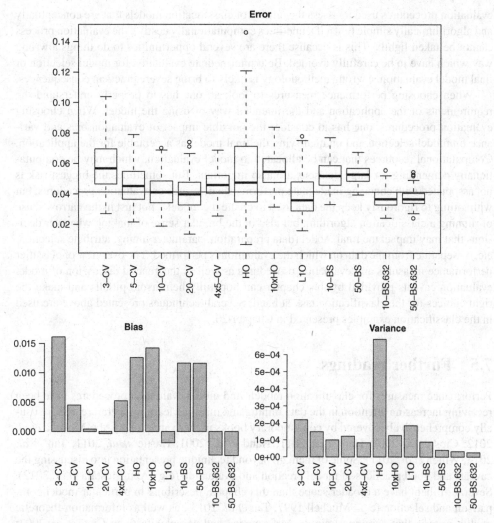

Figure 7.10 The misclassification error distribution, bias, and variance for different evaluation procedures.

For the hold-out procedure, a straightforward modification is sufficient to make it applicable to temporal data. Basically, the partitioning into the training and test subsets must preserve the temporal order of instances, with all selected for model evaluation being more recent than those used for model creation. Technically, the data would be partitioned by cutting time rather than at random. This is not possible for the remaining evaluation procedures.

7.4   Conclusion

The importance of model evaluation in the practice of data mining cannot be overestimated, and the classification task is no exception. Although both the performance measures and evaluation procedures used to assess the quality of classification models that are conceptually and algorithmically simple (even if sometimes computationally costly), the evaluation process cannot be taken lightly. This is because there are several opportunities to do things a wrong way which have to be carefully avoided. Be it intermediate evaluation for model selection or final model evaluation, a wrong methodology is likely to bring severe practical consequences.

When choosing performance measures to look at, one has to properly understand the requirements of the application and the intended way of using the model. When choosing evaluation procedures, one has to consider the possible impact of evaluation bias and variance on model selection and on qualifying the final model as acceptable for the application. Computational resources that can be allocated to model evaluation, which may be computationally demanding for large datasets, are also important. But sometimes the biggest risk is not associated with choosing inadequate performance measures or evaluation procedures, but with failing to rigorously keep the data used to create the model – not just in the narrow sense of running a classification algorithm, but also in the broader sense of making whatever decisions that may impact the final model (data preparation, parameter tuning, attribute selection, etc.) – separate from the data on which the evaluation is performed. The overview of classifier performance measures and evaluation procedures as well as the general discussion of model evaluation caveats provided by this chapter can hopefully help avoid pitfalls and make the right choices for the classification task at hand. Several techniques presented above are used in the classification examples presented in Chapter 20.

7.5   Further readings

Performance measures for classification models and model evaluation procedures have been receiving increasing attention in the data mining and machine learning literature and are usually comprehensively covered by contemporary books in these areas (e.g., Abu-Mostafa et al. 2012; Cios et al. 2007; Han et al. 2011; Hand et al. 2001; Hastie et al. 2011; Tan et al. 2013; Witten et al. 2011). Some of them go beyond technique presentation into discussing the caveats and good practices of model creation and evaluation (e.g., Abu-Mostafa et al. 2012). Several of them have a broader scope than this chapter, describing in particular model comparison and selection (e.g., Mitchell 1997; Tan et al. 2013), as well as information-theoretic quality criteria that combine training performance and complexity (e.g., Cios et al. 2007; Hand et al. 2001; Hastie et al. 2011). The latter include the minimum description length principle (MDL, Grünwald 2007; Rissanen 1978), the Akaike information criterion (AIC, Akaike 1974), and the Bayesian information criterion (BIC, Schwarz 1978). A thorough overview of these and other information-theoretic model quality and selection criteria is given by Burnham and Anderson (2002) or Claeskens and Hjort (2008).

A comprehensive survey of several variations of the cross-validation procedure (including hold-out as a simple special case) with discussion of their properties is given by Arlot and Celisse (2010). The procedures actually started to be investigated and applied several decades before (Devroye and Wagner 1979; Geisser 1975; Stone 1974), and the roots of the hold-out procedure can be traced back to Larson (1931).

The more modern .632 bootstrap procedure was proposed by Efron (1983), based on his earlier work on bootstrap estimation methods (Efron 1979). Its utility was investigated and compared to that of cross-validation by several experimental studies (e.g., Bailey and Elkan 1993; Chernick et al. 1985; Kohavi 1995). Efron and Tibshirani (1997) subsequently presented the more refined .632+ bootstrap procedure, which avoids the optimistic bias exhibited by the original .632 bootstrap procedure when used with overfitting modeling algorithms. Margineantu and Dietterich (2000) developed bootstrap evaluation procedures that can be used to evaluate and compare classification models with respect to the mean misclassification cost.

The false positive rate and the false negative rate are also known as the type I and type II errors in statistics, where they were introduced by Neyman and Pearson (1933) to characterize the properties of statistical tests. Other confusion matrix-based classifier performance measure originated from the field of information retrieval (Van Rijsbergen 1979). Dietterich (1998) presented an insightful overview of statistical questions related to learning and modeling tasks and recommended statistical tests appropriate for comparing the predictive performance of classification algorithms.

The ROC analysis originates from radar signal analysis during World War II and was then adopted by signal detection theory (Egan 1975). It was first brought to the context of classifier evaluation by Spackman (1989). A comprehensive overview of the ROC methodology and its applications is given by Fawcett (2006). Its extensions to multiclass classification tasks were discussed by Everson and Fieldsend (2005).

References

Abu-Mostafa YS, Magdon-Ismail M and Lin HT 2012 *Learning from Data*. AMLBook.

Akaike H 1974 A new look at the statistical model identification. *IEEE Transactions on Automatic Control* **19**, 716–723.

Arlot S and Celisse A 2010 A survey of cross-validation procedures for model selection. *Statistics Surveys* **4**, 40–79.

Bailey TL and Elkan C 1993 Estimating the accuracy of learned concepts *Proceedings of the Thirteenth International Joint Conference on Artificial Intelligence (IJCAI-93)*. Morgan Kaufmann.

Burnham KP and Anderson DR 2002 *Model Selection and Multimodel Inference: A Practical Information-Theoretic Approach* 2nd edn. Springer.

Chernick MR, Murthy VK and Nealy CD 1985 Application of bootstrap and other resampling techniques: Evaluation of classifier performance. *Pattern Recognition Letters* **3**, 167–178.

Cios KJ, Pedrycz W, Swiniarski RW and Kurgan L 2007 *Data Mining: A Knowledge Discovery Approach*. Springer.

Claeskens G and Hjort NL 2008 *Model Selection and Model Averaging*. Cambridge University Press.

Devroye L and Wagner TJ 1979 Distribution-free performance bounds for potential function rules. *IEEE Transaction on Information Theory* **25**, 601–604.

Dietterich TG 1998 Approximate statistical tests for comparing supervised classification learning algorithms. *Neural Computation* **10**, 1895–1924.

Efron B 1979 Bootstrap methods: Another look at the jackknife. *The Annals of Statistics* **7**, 1–26.

Efron B 1983 Estimating the error rate of a prediction rule: Improvement on cross-validation. *Journal of the American Statistical Association* **78**, 316–331.

Efron B and Tibshirani R 1997 Improvements on cross-validation: The .632+ bootstrap method. *Journal of the American Statistical Association* **92**, 548–560.

Egan JP 1975 *Signal Detection Theory and ROC Analysis*. Academic Press.

Everson RM and Fieldsend JE 2005 Multi-class ROC analysis from a multi-objective optimization perspective. *Pattern Recognition Letters* **27**, 918–927.

Fawcett T 2006 An introduction to ROC analysis. *Pattern Recognition Letters* **27**, 861–874.

Geisser S 1975 The predictive sample reuse method with applications. *Journal of the American Statistical Association* **70**, 320–328.

Grünwald P 2007 *The Minimum Description Length Principle*. MIT Press.

Han J, Kamber M and Pei J 2011 *Data Mining: Concepts and Techniques* 3rd edn. Morgan Kaufmann.

Hand DJ, Mannila H and Smyth P 2001 *Principles of Data Mining*. MIT Press.

Hastie T, Tibshirani R and Friedman J 2011 *The Elements of Statistical Learning: Data Mining, Inference, and Prediction* 2nd edn. Springer.

Kohavi R 1995 A study of cross-validation and bootstrap for accuracy estimation and model selection *Proceedings of the Fourteenth International Joint Conference on Artificial Intelligence (IJCAI-95)*. Morgan Kaufmann.

Larson SC 1931 The shrinkage of the coefficient of multiple correlation. *Journal of Educational Psychology* **22**, 45–55.

Margineantu DD and Dietterich TG 2000 Bootstrap methods for the cost-sensitive evaluation of classifiers *Proceedings of the Seventeenth International Conference on Machine Learning (ICML-2000)*. Morgan Kaufmann.

Mitchell T 1997 *Machine Learning*. McGraw-Hill.

Neyman J and Pearson ES 1933 On the problem of the most efficient tests of statistical hypotheses. *Philosophical Transactions of the Royal Society of London A* **231**, 289–337.

Rissanen J 1978 Modeling by shortest data description. *Automatica* **14**, 465–658.

Schwarz GE 1978 Estimating the dimension of a model. *Annals of Statistics* **6**, 461–464.

Spackman KA 1989 Signal detection theory: Valuable tools for evaluating inductive learning *Proceedings of the Sixth International Workshop on Machine Learning (ICML-89)*. Morgan Kaufmann.

Stone M 1974 Cross-validatory choice and assessment of statistical predictions. *Journal of the Royal Statistical Society B* **36**, 111–147.

Tan PN, Steinbach M and Kumar V 2013 *Introduction to Data Mining* 2nd edn. Addison-Wesley.

van Rijsbergen CJ 1979 *Information Retrieval* 2nd edn. Butterworths.

Witten IH, Frank E and Hall MA 2011 *Data Mining: Practical Machine Learning Tools and Techniques* 3rd edn. Morgan Kaufmann.

Part III
REGRESSION

8

Linear regression

8.1 Introduction

Parametric regression is the most direct instantiation of the idea of a parametric model representation, in which the model is represented by a finite number of parameters with a fixed functional form assumed. This is also the most frequently used approach to the regression task, to which such a representation is particularly well suited. Parametric regression algorithms can deliver successful regression models by themselves or in combination with other techniques, including those borrowed from algorithms used for the classification task.

Linear regression is the simplest approach to the regression task based on a parametric model representation. Despite its obvious and unquestionable linearity limitation (being capable of directly approximating linear target functions only), it deserves particular attention due to its algorithmic and computational advantages. Interestingly, it is possible, at least to some extent, to overcome the limitation while retaining the advantages. This chapter covers both plain linear regression and augmented versions thereof, breaking the linearity limitation. The presented discussion of model representation and creation techniques maintains a higher level of generality whenever possible, presenting the particular linear representation as an instantiation of the more general parametric regression approach. Linear model representation and gradient-based parameter estimation have already appeared in Chapter 5 in the context of the classification task, but it is here where they are thoroughly discussed.

Example 8.1.1 Demonstrating linear regression algorithms with R code examples will require the use of DMR packages providing functions for model evaluation, attribute transformation, and simple utilities. Parameter estimation for linear regression models will be illustrated using the *Boston Housing* data, available in the `mlbench` package. The *weatherr* data will be used to illustrate discrete attribute processing. The packages and the datasets are ``` dmr.data ``` loaded by the R code presented below. The larger of those is partioned into training and tests subsets, with the fourth column – containing a single discrete attribute – skipped.

Data Mining Algorithms: Explained Using R, First Edition. Paweł Cichosz.
© 2015 John Wiley & Sons, Ltd. Published 2015 by John Wiley & Sons, Ltd.

```
library(dmr.regeval)
library(dmr.trans)
library(dmr.util)

data(weatherr, package="dmr.data")
data(BostonHousing, package="mlbench")

set.seed(12)
rbh <- runif(nrow(BostonHousing))
bh.train <- BostonHousing[rbh>=0.33,-4]
bh.test  <- BostonHousing[rbh<0.33,-4]
```

Additionally a small artificial dataset will be used in several examples. It is generated by the following R code:

```
set.seed(12)

  # generate artificial data
lrdat <- data.frame(a1=floor(runif(400, min=1, max=5)),
                    a2=floor(runif(400, min=1, max=5)),
                    a3=floor(runif(400, min=1, max=5)),
                    a4=floor(runif(400, min=1, max=5)))
lrdat$f1 <- 3*lrdat$a1+4*lrdat$a2-2*lrdat$a3+2*lrdat$a4-3
lrdat$f2 <- tanh(lrdat$f1/10)
lrdat$f3 <- lrdat$a1^2+2*lrdat$a2^2-lrdat$a3^2-2*lrdat$a4^2+
            2*lrdat$a1-3*lrdat$a2+2*lrdat$a3-3*lrdat$a4+1
lrdat$f4 <- 2*tanh(lrdat$a1-2*lrdat$a2+3*lrdat$a3-lrdat$a4+1)-
            3*tanh(-2*lrdat$a1+3*lrdat$a2-2*lrdat$a3+lrdat$a4-1)+2

  # training and test subsets
lrdat.train <- lrdat[1:200,]
lrdat.test <- lrdat[201:400,]
```

It generates a dataset of 300 instances described by four continuous attributes, with integer values drawn uniformly at random from the $[1, 5]$ interval, and four different target functions all of which have known functional relationships to these attributes (with only one being linear). This is, of course, completely unrealistic and serves the illustration purpose only. The dataset is then partitioned into training and test subsets, to evaluate the performance of subsequently created models using the simple hold-out procedure. The partitioning is based on instance numbers, which would be normally unacceptable, but is perfectly fine with a randomly generated dataset. A fixed seed of the random number generator is used to ensure the reproducibility of presented results. The mean square error (MSE), defined in Section 10.2.3, will be used as the performance measure for these demonstrations, calculated using the mse function.

Ex. 10.2.3
dmr.regeval

8.2   Linear representation

Linear model representation is a particularly simple and popular instantiation of the more general parametric representation family. While focusing on the former, this will also discuss the more general case, which helps one better understand the advantages and limitations of the linear special case.

8.2.1   Parametric representation

A parametric regression model $h : X \to \mathcal{R}$ is described by the following formula, which specifies how to calculate its prediction for instance $x \in X$:

$$h(x) = F(a_1(x), a_2(x), \ldots, a_n(x), w_1, w_2, \ldots, w_N) \tag{8.1}$$

where $a_1, a_2, \ldots, a_n$ are attributes defined on the domain X, $w_1, w_2, \ldots, w_N$ are *model parameters* (also called weights), and F is a predetermined *representation function* that maps attribute value and parameter vectors to real-valued model predictions. This can also be written in a vector form as

$$h(x) = F(\mathbf{a}(x), \mathbf{w}) \tag{8.2}$$

where $\mathbf{a}(x)$ denotes the vector of attribute values for instance x and $\mathbf{w}$ denote the parameter vector. A parametric model is therefore represented by a hypersurface in an $(n + 1)$-dimensional space.

The essential feature of parametric representation is using a predetermined representation function, which reduces model creation to estimating model parameters from data. Model representations for which this property does not hold are considered *nonparametric*. This is not supposed to say that they do not have parameters – since they do – but that their representation function needs to be derived from data as well. This is the case, in particular, for regression trees that will be presented in the next chapter.

Example 8.2.1 To illustrate the parametric representation of regression models, the R code presented below defines a function that applies a parametric regression model to generate predictions for a given dataset. The model is assumed to be represented by a list containing two components, named `repf` and `w`. The former is the model's representation function and the latter is its parameter vector. Setting the `class` attribute of such an object to `par` enables appropriate prediction method dispatching.

The representation function takes an attribute value vector and a parameter vector on input and returns the resulting prediction on output. The particular representation function `repf.perf4` defined in this example can be immediately seen to match the `f4` target function in the artificial dataset generated in the previous example:

$$F_4(\mathbf{a}(x), \mathbf{w}) = w_{2n+3} \tanh\left(\sum_{i=1}^{n} w_i a_i(x) + w_{n+1} \right)$$
$$+ w_{2n+4} \tanh\left(\sum_{i=1}^{n} w_{i+n+1} a_i(x) + w_{2n+2} \right) + w_{2n+5} \tag{8.3}$$

where $n = 4$ is the number of attributes. It can therefore be referred to as the *perfect representation function* for `f4`, which – with appropriate parameters – matches the target function exactly. Indeed, when combined with the same parameters as actually used for target function generation it yields the perfect model, achieving a 0 test set error. Note that the `repf.perf4` function is implemented so that it can be applied not only to single instances, but also to complete datasets as well. This is why it uses the `rowSums` function instead of the sum function. The cmm utility function is used to multiply all columns of the dataset by the corresponding elements of the parameter vector. Subsequent examples will similarly define functions capable of handling both single and multiple instances whenever possible.

```
## parametric regression prediction for a given model and dataset
predict.par <-  function(model, data) { model$repf(data, model$w) }

  # perfect representation function for f4
repf.perf4 <- function(data, w)
{
  w[2*(n <- ncol(data))+3]*tanh(rowSums(cmm(data, w[1:n]))+w[n+1]) +
    w[2*n+4]*tanh(rowSums(cmm(data, w[(n+2):(2*n+1)]))+w[2*n+2]) + w[2*n+5]
}

  # perfect parameters for f4
w.perf4 <- c(1, -2, 3, -1, 1, -2, 3, -2, 1, -1, 2, -3, 2)
  # perfect model for f4
mod.perf4 <- `class<-`(list(w=w.perf4, repf=repf.perf4), "par")
  # test set error
mse(predict(mod.perf4, lrdat.test[,1:4]), lrdat.test$f4)
```

Of course, it is completely unrealistic to assume that either the true representation function or its true parameters are known, as in this example. This assumption is only adopted to illustrate the parametric representation of regression models.

8.2.2 Linear representation function

Linear regression is based on the following special form of a parametric model representation:

$$h(x) = \sum_{i=1}^{n} w_i a_i(x) + w_{n+1} \tag{8.4}$$

The formula specifies how the prediction of a linear model h is calculated for instance $x \in X$. This implicitly assumes that attributes are continuous (so that they can be used for arithmetics). With such a representation, h is linear with respect to both attribute values and parameters and corresponds to a hyperplane in an $(n + 1)$-dimensional space. Unless combined with some additional enhancements, linear regression models can accurately represent only target functions that are linear with respect to attribute values.

The representation function used for linear regression is defined as a linear combination of attribute values and model parameters, with an additional *intercept* parameter. It is a common and convenient practice to avoid explicitly referring to the latter in equations related to linear regression by assuming that there is an additional a_{n+1} attribute defined, always taking a value of 1, which makes it possible to rewrite the model representation formula as

$$h(x) = \sum_{i=1}^{n+1} w_i a_i(x) \tag{8.5}$$

On several occasions, it is also convenient to adopt a vector notation. With $\mathbf{a}(x)$ denoting the vector of attribute values for instance x and $\mathbf{w}$ denoting the parameter vector, we can present the linear model representation in each of the following two equivalent vector forms:

$$h(x) = \mathbf{w} \circ \mathbf{a}(x) \tag{8.6}$$

$$h(x) = \mathbf{w}^T \mathbf{a}(x) \tag{8.7}$$

where $\circ$ is the dot product operator and T is the matrix transpose operator. The latter assumes that vectors are treated as one-column matrices.

Example 8.2.2 The following R code defines a linear representation function that takes an attribute value vector and a parameter vector on input and returns a linear combination thereof on output. To make it applicable not only to single instances, but also complete datasets, the `rowSums` function is used instead of `sum` and the `cmm` function is used to multiply all columns of the dataset by the corresponding elements of the parameter vector. This `dmr.util` representation function is then combined with an appropriate parameter vector to create the perfect model for the `f1` target function from the artificial dataset created in Example 8.1.1, which was indeed generated as a linear combination of attribute values. The perfect model can be verified to achieve a 0 mean square error.

```
## linear representation function
repf.linear <- function(data, w)
{ rowSums(cmm(data, w[1:(n <- ncol(data))])) + w[n+1] }

  # perfect parameter vector for f1
w.perf1 <- c(3, 4, -2, 2, -3)
  # perfect model for f1
mod.perf1 <- `class<-`(list(w=w.perf1, repf=repf.linear), "par")
  # test set error
mse(predict(mod.perf1, lrdat.test[,1:4]), lrdat.test$f1)
```

Representation functions that are nonlinear with respect to attribute values, but linear with respect to model parameters, yield regression models that are still linear in a modified attribute space. They can be presented as

$$h(x) = \sum_{i=1}^{N} w_i a'_i(x) + w_{N+1} \tag{8.8}$$

where $a'_1, a'_2, \ldots, a'_N$ are new modified attributes that are obtained via some nonlinear transformations of the original attributes. This approach is referred to as an enhanced representation and will be further discussed in Section 8.6.2 as one of the possible ways of using linear regression to approximate nonlinear target functions.

A generalized linear representation assumes that the output of a linear model is transformed nonlinearly to generate predictions. A model of this form, although not intrinsically nonlinear, is therefore capable of approximating some nonlinear target functions, as will be further discussed in Section 8.6.1.

8.2.3 Nonlinear representation functions

Only when the representation function is nonlinear with respect to model parameters and cannot be linearized in a straightforward way, we are faced with intrinsically *nonlinear regression*. Typical nonlinear representation functions include:

- polynomial functions,

- exponential functions,

- logarithmic functions,

- trigonometric functions,

as well as various combinations or superpositions thereof. These are beyond our interest in this chapter.

8.3   Parameter estimation

Parameter estimation is the process of identifying model parameters based on a given training set that are likely to yield good prediction performance. This can be viewed as an optimization process in which the space of possible parameter vectors is searched for one that optimizes an adopted performance measure. In general, several different performance measures and optimization methods could be used for this purpose. The mean square error, defined in Section 10.2.3 as the mean squared difference between true target function values and model predictions, is a particularly convenient performance measure to adopt. In this role, it is also referred to as the quadratic loss. Gradient descent methods belong to the simplest approaches to optimization that can be employed for both linear and (some) nonlinear representations. Most of this section is focused on the combination of these two, which is simple enough to be explained and understood with just elementary maths and illustrate with a plain-vanilla implementation. An alternative least-squares method – similarly simple, easier to use, and usually much more efficient (except for excessively large data), but inapplicable to nonlinear representations – will also be discussed.

8.3.1   Mean square error minimization

The definition of the mean square error of the regression model h with respect to the target function f on the training set T can be rewritten in a slightly modified form as follows:

$$E_{T,f}(h) = \frac{1}{2} \sum_{x \in T} (f(x) - h(x))^2 \tag{8.9}$$

The modification of the original definition from Section 10.2.3 consists in replacing the $\frac{1}{|T|}$ coefficient by $\frac{1}{2}$, which will turn out to serve the purpose of aesthetics only. Clearly, any model that minimizes $E_{T,f}(h)$ does also minimize the training set mean square error, so this modification has no impact on the model that could be identified.

The MSE-like function defined above will serve as the objective function for minimization. It may appear at first unreasonable to optimize the training performance, which could easily lead to overfitting, but we actually have no choice here, since the true performance can only be estimated when a model is already built and not during the training process. An appropriately selected model representation function (e.g., without too many parameters) and an optimization technique will have to take the responsibility for overfitting prevention rather than the objective function. An alternative approach to linear model parameter estimation that relaxes the objective of training set error minimization to increase the resistance to overfitting is presented in Section 16.3.

The idea of gradient descent function minimization is to gradually modify the parameter vector by shifting it in the direction indicated by the negated gradient of the function being minimized with respect to the parameters. In our case, it can be written as follows:

$$\mathbf{w} := \mathbf{w} + \beta\left(-\nabla_{\mathbf{w}}E_{T,f}(h)\right) \tag{8.10}$$

where $\beta > 0$ is a *step-size* parameter that controls the amount of the update performed. The gradient $\nabla_{\mathbf{w}}E_{T,f}(h)$ is the vector of partial derivatives of $E_{T,f}(h)$ with respect to model parameters. The above parameter update rule can therefore be rewritten for a single parameter w_i in the following form:

$$w_i := w_i + \beta\left(-\frac{\partial E_{T,f}(h)}{\partial w_i}\right) \tag{8.11}$$

Both the sign and the size of the update performed for each parameter depend on the corresponding partial derivative. A positive derivative value indicates that increasing the parameter would increase the error, and therefore the parameter should be decreased. A negative derivative value similarly leads to increasing parameter. The smaller the absolute derivative value, which may indicate approaching a local minimum, the smaller the update.

8.3.2 Delta rule

The partial derivative of $E_{T,f}(h)$ with respect to w_i can be calculated as follows:

$$\frac{\partial E_{T,f}(h)}{\partial w_i} = \frac{1}{2}\sum_{x\in T}2(f(x)-h(x))\left(-\frac{\partial h(x)}{\partial w_i}\right)$$

$$= \sum_{x\in T}(f(x)-h(x))\left(-\frac{\partial h(x)}{\partial w_i}\right) \tag{8.12}$$

since the dependence of $E_{T,f}(h)$ on parameters is through the model representation. This makes it clear why the $\frac{1}{2}$ coefficient rather than $\frac{1}{|T|}$ is used in the definition of $E_{T,f}(h)$ for the sake of aesthetics, since it simplified with the 2 from the derivative. After substituting to the update rule this yields

$$w_i := w_i + \beta\sum_{x\in T}(f(x)-h(x))\frac{\partial h(x)}{\partial w_i} \tag{8.13}$$

or, in the equivalent vector form

$$\mathbf{w} := \mathbf{w} + \beta\sum_{x\in T}(f(x)-h(x))\nabla_{\mathbf{w}}h(x) \tag{8.14}$$

The obtained update rule shows how to modify the parameters of a parametric regression model based on the training set in order to decrease the mean square error. It is often convenient to decompose these updates into contributions of single training instances, which is achieved by simply removing the summation with respect to $x \in T$:

$$\mathbf{w} := \mathbf{w} + \beta(f(x)-h(x))\nabla_{\mathbf{w}}h(x) \tag{8.15}$$

This final update rule shows how to modify the parameters of a parametric regression model based on a single training instance. It is referred to as the *incremental delta rule*, whereas the previous rule given by Equation 8.14, aggregating the contributions of all training instances, will be called the *batch delta rule*. They can be instantiated for particular representation functions by calculating $\nabla_w h(x)$. Whenever the distinction between the batch and incremental formulation is immaterial, we will simply speak of the delta rule.

The $\nabla_w h(x)$ gradient, which is the vector of per-parameter partial derivatives $\frac{\partial h(x)}{\partial w_i}$, is more than straightforward to calculate for the linear case. Clearly we have

$$\frac{\partial h(x)}{\partial w_i} = a_i(x) \tag{8.16}$$

and

$$\nabla_w h(x) = \mathbf{a}(x) \tag{8.17}$$

accordingly. This makes it possible to write down a specialized linear version of the delta rule (assuming the incremental version) as follows:

$$\mathbf{w} := \mathbf{w} + \beta(f(x) - h(x))\mathbf{a}(x) \tag{8.18}$$

The linear delta rule is also referred to as the LMS rule or the Widrow–Hoff rule. Similarly to its general counterpart, it can be applied in either an incremental or batch mode. The former performs the updates based on single training instances immediately after they are calculated, whereas the latter accumulates the updates resulting from all training instances and applies them after the complete training set has been processed. In any case, multiple iterations are needed with an appropriately selected step-size parameter β to approach a minimum of the mean square error. Such iterative incremental or batch updates are performed by the gradient descent algorithm.

Example 8.3.1 Determining the gradient of the linear representation function is implemented and demonstrated by the R code presented below. The w argument is not actually used, since the linear representation gradient does not depend on model parameters. It appears on the argument list only to make the call interface of the gradient function ready for nonlinear representations as well.

```
## gradient of the linear representation function
grad.linear <- function(data, w) { cbind(data, 1) }

 # gradient for the first 10 instances
grad.linear(lrdat.train[1:10,1:4], rep(0, 5))
```

Example 8.3.2 The following R code implements the delta rule for mean square error minimization according to Equation 8.18. It takes vectors of true and predicted target function values, the gradient, and the step size value as arguments, and returns the resulting model parameter update vector. Its application to the f1 target function of the example dataset is demonstrated, using the linear representation gradient implemented in the previous example. The first demonstration call uses the target value predictions produced by the perfect model from Example 8.2.2 and the corresponding parameter vector, which yields parameter updates of 0, as expected. In the second call, the true target function values are modified to force the delta rule to determine nonzero updates.

```
## calculate parameter update based on given true and predicted values,
## gradient, and step-size using the delta rule for MSE minimization
delta.mse <- function(true.y, pred.y, gr, beta)
{ colSums(beta*rmm(gr, (true.y-pred.y))) }

  # parameter updates for the perfect model for f1
delta.mse(lrdat.train$f1, predict(mod.perf1, lrdat.train[,1:4]),
          grad.linear(lrdat.train[,1:4], w.perf1), 0.1)
  # parameter updates for the perfect model for f1
  # with modified target function values
delta.mse(lrdat.train$f1+0.1, predict(mod.perf1, lrdat.train[,1:4]),
          grad.linear(lrdat.train[,1:4], w.perf1), 0.1)
```

8.3.3   Gradient descent

The gradient descent algorithm for training a parametric regression model performs a number of delta rule iterations to reach a parameter vector that yields a sufficiently small training set error. Similarly as for the delta rule, one can consider a batch or incremental version of the algorithm.

The batch gradient descent algorithm can be formulated as presented below.

```
1: repeat
2:     for i = 1, 2, ..., N do
3:         Δw_i := 0;
4:     end for
5:     for all training instances x ∈ T do
6:         for i = 1, 2, ..., N do
7:             Δw_i := Δw_i + β(f(x) − h(x)) ∂h(x)/∂w_i;
8:         end for
9:     end for
10:    for i = 1, 2, ..., N do
11:        w_i := w_i + Δw_i;
12:    end for
13: until stop criteria are satisfied;
```

In each iteration of the algorithm the entire training set is used to calculate the updates for each parameter, which is then applied. The incremental version does not have to accumulate the updates resulting from each instance before applying them to actually modify parameters, so it is even simpler. The incremental gradient descent algorithm is also referred to as *online* or *stochastic gradient descent*.

```
1: repeat
2:     select an instance x ∈ T;
3:     for i = 1, 2, ..., N do
4:         w_i := w_i + β(f(x) − h(x)) ∂h(x)/∂w_i;
5:     end for
6: until stop criteria are satisfied;
```

Both versions of the algorithm are capable of reaching a (local) minimum of the mean square error if using a sufficiently small (or appropriately adapted) step-size value. The batch version, processing multiple instances at a time, may be more efficient if implemented using optimized vector and matrix operations. The incremental version may often be more convenient, however, particularly if the training set is not fixed and completely available, but consists of continuously arriving instances, which requires continuous parameter updates. It may also perform better computationally for very large data. It usually behaves similarly well or better than the batch version even when working with fixed and completely available training sets, as long as all training instances keep being selected with equal average frequency, but in varying order. An easy way to achieve this when using a fixed training set is to perform an internal iteration in which all training instances are processed in a randomized order, as in the following alternative formulation of the algorithm.

1: **repeat**
2:    **for all** training instances $x \in T$ in randomized order **do**
3:        **for** $i = 1, 2, ..., N$ **do**
4:            $w_i := w_i + \beta(f(x) - h(x))\frac{\partial h(x)}{\partial w_i}$;
5:        **end for**
6:    **end for**
7: **until** stop criteria are satisfied;

Unless some background knowledge is available that could recommend a good initial guess of model parameters, the safest approach is to initialize them to small random numbers. The more "obvious" initialization to 0 may be inappropriate for some nonlinear representation functions which may then have a 0 gradient, which would clearly prevent the algorithm from making any parameter update, but is perfectly fine for linear models.

Typical stop criteria used to terminate the gradient descent process include:

Error. Reaching a sufficiently low error level.

Duration. Completing a specified number of iterations.

No improvement. No error improvement observed during a specified number of iterations.

Example 8.3.3 The following R code contains a simple implementation of the gradient descent algorithm for parametric regression models. The `gradient.descent` function receives a formula specifying the attributes and the target function, the training dataset, the initial parameter vector to start with, as well as the representation function and its gradient. Both of these take an attribute value vector or a dataset and a parameter vector on input. One can also specify the step-size value, the training mode, and the stop criterion (via an acceptable mean square error level or a maximum number of iterations). The training mode defaults to incremental (`batch=FALSE`), which processes training instances in randomized order. The latter, almost always desirable behavior, can be turned off by specifying the `randomize=FALSE` argument. The `delta` and `perf` arguments, defaulting to `delta.mse` and `mse`, respectively, specify the functions for model parameter update and performance measure calculation. Two auxiliary functions are used to extract attribute and target function names from the input formula, `x.vars` and `y.var`. The `as.num0` function is used to convert the target attribute values to a numerical representation, which makes it possible to apply the `gradient.descent` function to datasets with discrete target attributes as

`dmr.util`

long as the conversion is successful. The gradient descent algorithm is applied to estimating the parameters of a linear model for the `f1` target function of the example artificial dataset, as well as for the real *Boston Housing* data. The implementations of the linear representation function and its gradient from Examples 8.2.2 and 8.3.1, respectively, are used for this purpose.

```
## perform gradient descent iterative parameter estimation
## for parametric regression models
gradient.descent <- function(formula, data, w, repf, grad, delta=delta.mse, perf=mse,
                    beta=0.001, batch=FALSE, randomize=!batch,
                    eps=0.001, niter=1000)
{
  f <- y.var(formula)
  attributes <- x.vars(formula, data)
  aind <- names(data) %in% attributes
  true.y <- as.num0(data[[f]])
  model <- `class<-`(list(repf=repf, w=w), "par")
  iter <- 0

  repeat
  {
    if (batch)
    {
      pred.y <- predict.par(model, data[,aind])
      model$w <- model$w + delta(true.y, pred.y, grad(data[,aind], model$w), beta)
    }
    else
    {
      pred.y <- numeric(nrow(data))
      xind <- if (randomize) sample.int(nrow(data)) else 1:nrow(data)
      for (i in 1:length(xind))
      {
        av <- data[xind[i], aind]
        pred.y[xind[i]] <- predict.par(model, av)
        model$w <- model$w +
                   delta(true.y[xind[i]], pred.y[xind[i]], grad(av, model$w), beta)
      }
    }
    iter <- iter+1

    cat("iteration ", iter, ":\t", p <- perf(pred.y, true.y), "\n")
    if (p < eps || iter >= niter)
      return(list(model=model, perf=p))
  }
}


  # linear model for f1
gd1 <- gradient.descent(f1~a1+a2+a3+a4, lrdat.train, w=rep(0, 5),
                        repf=repf.linear, grad=grad.linear, beta=0.01, eps=0.0001)


  # linear model for the Boston Housing data
bh.gd <- gradient.descent(medv~., bh.train, w=rep(0, ncol(bh.train)),
                        repf=repf.linear, grad=grad.linear, beta=1e-6, eps=25,
                        niter=5000)


  # test set error
mse(predict(gd1$model, lrdat.test[,1:4]), lrdat.test$f1)
mse(predict(bh.gd$model, bh.test[,-13]), bh.test$medv)
```

When applied to the small artificial data, the algorithm converges quite fast to a low error level. In particular, the 0.0001 mean square error boundary is crossed within less than 50 iterations in most runs. The test set error is marginally above the training error, which is to be expected, and the estimated parameters can be seen to differ from the true ones (used for data generation) only slightly. Even better results would probably be possible with a longer training process or more carefully fine-tuned step-size value. Setting it too large, by the way, results in numerical explosion with model parameters falling out of bounds permitted by the arithmetic precision. The real *Boston Housing* data requires a very small step-size value, which results in slow convergence, but a reasonably good mean square error level is ultimately reached.

8.3.4   Least squares

An alternative approach to training linear regression models is also possible, using the well-known linear algebra *least-squares* method. It treats the parameter estimation task as a linear system solving task rather than an optimization task. Indeed, consider the following equation:

$$a_1(x)w_1 + a_2(x)w_2 + \cdots + a_n(x)w_n + a_{n+1}(x)w_{n+1} = f(x) \tag{8.19}$$

which simply demands that the model's prediction agrees with the true target function value for instance x. By writing such equations for all instances from the training set we receive a linear system, with model parameters playing the role of unknown variables. If $n + 1 < |T|$, which is natural to assume (it is in fact common to request $n \ll |T|$), the system is overdetermined and it may not have an exact solution, but a least-squares solution (i.e., such that it minimizes the mean square error of the predictions made using the obtained parameter vector with respect to the true target function values) can be found using a simple algebraic procedure.

Switching to a matrix notation for compactness and convenience, the linear system under consideration can be presented as

$$\mathbf{a}(T)\mathbf{w} = \mathbf{f}(T) \tag{8.20}$$

where $\mathbf{a}(T)$ is the $|T| \times (n + 1)$ attribute value matrix for the training set (with one row per instance and one column per attribute, including the artificial a_{n+1} attribute) and $\mathbf{f}(T)$ is the target function value vector for the training set, with one value per instance. As before, $\mathbf{w}$ is the model parameter vector and all vectors are treated as single-column matrices. The above matrix equation cannot be directly solved for $\mathbf{w}$ because $\mathbf{a}(T)$ is (usually) not square. It can be made square, however, by multiplying both sides of the equation by its transpose $\mathbf{a}^T(T)$:

$$\mathbf{a}^T(T)\mathbf{a}(T)\mathbf{w} = \mathbf{a}^T(T)\mathbf{f}(T) \tag{8.21}$$

Now, unless the $\mathbf{a}^T(T)\mathbf{a}(T)$ matrix turns out to be singular, it can be inverted to achieve the desired solution:

$$\mathbf{w} = \left(\mathbf{a}^T(T)\mathbf{a}(T)\right)^{-1}\mathbf{a}^T(T)\mathbf{f}(T) \tag{8.22}$$

Otherwise a pseudo-inversion operation can be applied to achieve an approximate solution. Thus, the "multiply-by-transpose" trick turns an undetermined linear system into an ordinary linear system with the same number of unknowns and equations which can be solved in the usual way.

This is rather a conceptual description of the least-squares method than a directly applicable algorithm, since it relies on the matrix inverse operation, which is not necessarily trivial and involves some advanced numerical techniques to be performed accurately and efficiently. Discussing these is beyond the scope of this chapter, though. Assuming a numerically correct and efficient implementation, this approach would be usually preferred to the gradient descent algorithm as more efficient (at least as long as the number of attributes, which determines the dimensions of the matrix that has to be inverted, is not exceedingly large), more reliable (since the optimal parameter vector is obtained directly rather than gradually approached), and easier to use (since there is no need to adjust the step-size parameter). This is actually the most basic form of the least-squares method, often referred to as *ordinary least squares* (OLS), with other versions having been developed for some extensions of the basic linear regression model representation.

Where the gradient descent algorithm wins is the incremental learning capability, which may be important for some applications, where there is a stream of training instances arriving one at a time (or a portion at a time) and the model needs to be continuously updated to incorporate new data.

Example 8.3.4 The R code presented below implements the ordinary least-squares method for linear model parameter estimation. It uses the `x.vars` and `y.var` functions to extract the attribute and target function names from the supplied formula. Then it uses the built-in `solve` function to solve the linear system obtained after applying the [dmr.util] "multiply-by-transpose" trick, which takes care of the required matrix inversion. The least squares method is subsequently used to estimate linear model parameters for the `f1` target function.

```
## estimate linear model parameters using the OLS method
ols <- function(formula, data)
{
  f <- y.var(formula)
  attributes <- x.vars(formula, data)
  aind <- names(data) %in% attributes

  amat <- cbind(as.matrix(data[,aind]), intercept=rep(1, nrow(data)))
  fvec <- data[[f]]
  `class<-`(list(repf=repf.linear, w=solve(t(amat)%*%amat, t(amat)%*%fvec)), "par")
}

  # linear model for f1
ols1 <- ols(f1~a1+a2+a3+a4, lrdat.train)
  # linear model for the Boston Housing data
bh.ols <- ols(medv~., bh.train)

  # test set error
mse(predict(ols1, lrdat.test[,1:4]), lrdat.test$f1)
mse(predict(bh.ols, bh.test[,-13]), bh.test$medv)
```

The least-squares method is lightning fast compared to the gradient descent algorithm demonstrated in the previous example. This is, in part, due to implementational reasons (the `gradient.descent` function was entirely implemented in R, whereas the most computationally expensive matrix inverse operation of the OLS method is actually performed by an efficiently implemented R's built-in function), but the latter is inherently more efficient anyway as long as the dimension of the matrix to inverse – i.e., the number of attributes – remains within reasonable limits. For the artificial data the estimated parameters match the true ones exactly, which makes the test set mean square error practically 0. For the real *Boston Housing* data the mean square error level is somewhat less than the one previously achieved with gradient descent.

8.4   Discrete attributes

Unlike for the classification task and many classification algorithms, where it is often convenient to assume that attributes are mostly discrete and it is nearly always necessary to assume that some attributes may be discrete, it is not untypical for regression algorithms to implicitly assume that attributes are continuous. In particular, parametric regression is usually presented with this assumption. Whereas the representation function mapping attribute value and parameter vectors into model predictions can, in general, handle arbitrary attributes, the linear representation function can be directly applied to continuous attributes only. This is also the case for nearly all nonlinear parametric representations that are in common use, which employ representation functions defined via arithmetic operations on attribute values.

To make the linear representation function and other arithmetic representation functions applicable to discrete attributes, the latter can be transformed by the simple binary encoding transformation described in Section 17.3.5. For the linear representation, after a k-valued discrete attribute $a_i : X \to \{v_{i,1}, v_{i,2}, \dots, v_{i,k}\}$ has been replaced by $k - 1$ binary attributes, its contribution to the representation function is

$$a_{i,1}(x)w_{i,1} + a_{i,2}(x)w_{i,2} + \cdots + a_{i,k-1}(x)w_{i,k-1} = \sum_{j=1}^{k-1} a_{i,j}(x)w_{i,j} \qquad (8.23)$$

where

$$a_{i,j}(x) = \begin{cases} 1 & \text{if } a_i(x) = v_{i,j} \\ 0 & \text{otherwise} \end{cases} \qquad (8.24)$$

for $j = 1, 2, \dots, k - 1$. In practice, any set of two substantially different values can be used for this encoding instead of $\{0, 1\}$, with $\{-1, 1\}$ being the second typical choice.

The gradient descent algorithm can be directly used with discrete attributes as long as this binary encoding is applied when calculating the representation function and its gradient. The least-squares method requires that the encoding be applied when creating the attribute value matrix for the training set. Other than that, both the algorithms can operate without modifications.

Example 8.4.1 The R code presented below defines a wrapper that takes a representation function on input and returns its version equipped with the discrete attribute handling capability. This is achieved by applying the `discode` encoding function that leaves continuous attribute values unchanged, but replaces discrete attribute values with appropriate binary sequences, as discussed above. An analogous wrapper is also defined for the representation function gradient. This makes it possible to directly apply

Ex. 17.3.5
dmr.trans

the gradient descent algorithm to estimate linear model parameters for data with discrete attributes. Discrete attribute-enabled implementation of the least-squares regression algorithm is also presented that apply the same encoding to the training set when creating the attribute value matrix. Then the two algorithms are applied to estimate linear model parameters for the *weatherr* data. For the gradient descent algorithm, the last (intercept) parameter is initialized to 1 rather than to 0, which appears to be reasonable given the fact that all target function values are positive numbers below 1. Note that there are six linear model parameters: two for the originally continuous attributes `temperature` and `humidity`, two for the transformed `outlook` attribute (which has originally three discrete values), and one for the transformed `windy` attribute (which has originally two discrete values).

```
## representation function wrapper to handle discrete attributes
repf.disc <- function(repf)
{ function(data, w) { repf(discode(~., data, b=c(-1,1)), w) } }

## representation function gradient wrapper to handle discrete attributes
grad.disc <- function(grad)
{ function(data, w) { grad(discode(~., data, b=c(-1,1)), w) } }

## estimate linear model parameters using the OLS method
## with discrete attributes
ols.disc <- function(formula, data)
{
  f <- y.var(formula)
  attributes <- x.vars(formula, data)
  aind <- names(data) %in% attributes

  amat <- cbind(as.matrix(discode(~., data[,aind], b=c(-1,1))),
                intercept=rep(1, nrow(data)))
  fvec <- data[[f]]
  `class<-`(list(repf=repf.disc(repf.linear),
                 w=solve(t(amat)%*%amat, t(amat)%*%fvec)), "par")
}

  # gradient descent for the weatherr data
w.gdl <- gradient.descent(playability~., weatherr, w=c(rep(0, 5), 1),
                          repf=repf.disc(repf.linear), grad=grad.disc(grad.linear),
                          beta=0.0001, eps=0.005)
mse(weatherr$playability, predict(w.gdl$model, weatherr[,1:4]))

  # OLS for the weatherr data
w.ols <- ols.disc(playability~., weatherr)
mse(predict(w.ols, weatherr[,1:4]), weatherr$playability)
```

Both the algorithms work with a discrete attribute as expected. Not surprisingly, the least-squares algorithm beats gradient descent both with respect to speed and model accuracy.

8.5   Advantages of linear models

There is only one disadvantage of linear models, which is both very obvious and very severe: they cannot directly represent nonlinear relationships and hence approximate nonlinear target functions. Before discussing whether and how this major limitation could be overcome, it is worthwhile to consider the advantages of linear models that justify that effort.

One advantage that may sometimes be important is the computational efficiency of creating and using linear models. The linear representation function is inexpensive to calculate, and even more so is its derivative, which makes the cost of both prediction and gradient descent training relatively small. Moreover, the least-squares method makes it possible to calculate the optimal parameter vector directly, which may often take less time than required by the gradient descent algorithm to converge. However, with the increasing computational power available, this advantage tends to become less important.

What remains more unquestionable is the benefits resulting from the shape of the mean square error function, which – under a linear representation – is quadratic with respect to model parameters. This makes the optimization process easy and free of misleading local optima at which the gradient descent algorithm could get stuck with a nonlinear representation. Linear models are therefore not only more efficient to create and use, but also much more easy to fit accurately to the data, as long as the relationship between the target function and attributes is indeed linear.

Last but not least, the representation simplicity makes linear models much easier to understand, explain, and verify than most nonlinear parametric models. The parameter corresponding to each attribute directly reflects its contribution to the model's predictions.

All this justifies the interest that linear regression receives and the efforts made toward enhancing its capabilities toward modeling nonlinear relationships.

8.6   Beyond linearity

To retain the advantages of linear models while overcoming the linearity limitation one should seek for regression models that use the same representation and parameter estimation algorithms as linear models internally, but are wrapped with some add-on techniques that make it possible to represent nonlinear relationships. One natural way to achieve this is to transform the output of a plain linear model nonlinearly, i.e., adopting a generalized linear representation. This form of introducing nonlinearity may often be insufficient, though. In such cases two other strategies can be employed, enhanced representation and piecewise-linear regression.

8.6.1   Generalized linear representation

A *generalized linear representation* assumes that there is a nonlinear *link function* that transforms the linear combination of attribute values and model parameters into the final model prediction. This is written as

$$h(x) = L^{-1}\left(\sum_{i=1}^{n} w_i a_i(x) + w_{n+1}\right) = L^{-1}(g(x)) \tag{8.25}$$

The resulting representation function is a composite of the linear representation function (as the inner function):

$$g(x) = \sum_{i=1}^{n} w_i a_i(x) + w_{n+1} = \mathbf{w} \circ \mathbf{a}(x) \tag{8.26}$$

and the nonlinear inverse link function L^{-1} (as the outer function). The link function applied to model predictions makes them linear:

$$L(h(x)) = \mathbf{w} \circ \mathbf{a}(x) \tag{8.27}$$

It is noteworthy that the linear threshold and logit representations for linear classification presented in Sections 5.2.3 and 5.2.4 are instantiations of the generalized linear representation.

Using a generalized linear representation is one of the features of *generalized linear models*, which also include other important enhancements over plain linear models. In particular, they can incorporate a specified target function probability distribution and a performance measure to be optimized. For some specific instantiations, dedicated parameter estimation algorithms have been developed. All these issues are beyond the scope of this chapter.

8.6.1.1   Delta rule for generalized linear representation

The delta rule for the generalized linear representation with a differentiable link function can be directly derived from Equation 8.15 by calculating the gradient

$$\nabla_{\mathbf{w}} h(x) = (L^{-1})'(\mathbf{w} \circ \mathbf{a}(x))\mathbf{a}(x) \tag{8.28}$$

where $(L^{-1})'$ is the derivative of the inverse link function. This yields the following model parameter update:

$$\mathbf{w} := \mathbf{w} + \beta(f(x) - h(x))(L^{-1})'(\mathbf{w} \circ \mathbf{a}(x))\mathbf{a}(x) \tag{8.29}$$

which makes it possible to apply the gradient descent descent algorithm.

Example 8.6.1 The following R code defines the `repf.gen` and `grad.gen` functions that create the generalized linear representation function and its gradient for a given link function. These are wrappers around `repf.linear` and `grad.linear` (although another inner representation function and gradient can be specified). The inverse link function has to be supplied as an argument to `repf.gen` and its derivative as an argument to `grad.gen`. Such a pair of functions are then defined to exactly match the `f2` target function, which cannot be expected in practice for unknown target functions, but serves well the illustration purpose. This makes it possible to estimate generalized linear model parameters for `f2` using the gradient descent algorithm.

```
## generalized representation function
repf.gen <- function(link.inv, repf=repf.linear)
{ function(data, w) { link.inv(repf(data, w)) } }

## generalized representation function gradient
grad.gen <- function(link.inv.deriv, repf=repf.linear, grad=grad.linear)
{ function(data, w) { rmm(grad(data, w), link.inv.deriv(repf(data, w))) } }

  # perfect inverse link function for f2
link2.inv <- function(v) { tanh(v/10) }
  # and its derivative
link2.inv.deriv <- function(v) { (1-tanh(v/10)^2)/10 }

  # perfect generalized linear representation function for f2
repf.gen2 <- repf.gen(link2.inv)
  # and its gradient
grad.gen2 <- grad.gen(link2.inv.deriv)

  # gradient descent estimation of generalized linear model parameters for f2
gd2g <- gradient.descent(f2~a1+a2+a3+a4, lrdat.train, w=rep(0, 5),
                    repf=repf.gen2, grad=grad.gen2,
                    beta=0.5, eps=0.0001)
  # test set error
mse(predict(gd2g$model, lrdat.test[,1:4]), lrdat.test$f2)
```

The gradient descent algorithms works as expected, arriving at a low error. Interestingly, unlike in the plain linear case, a large step-size value can be used without running at numerical explosion problems. This is a consequence of the particular link function, with values bound to the $(-1, 1)$ interval.

8.6.1.2 Least squares for generalized linear representation

Unfortunately there is no simple modification of the OLS algorithm that would be it applicable to estimating generalized linear representation parameters, because there is no closed-form solution of the mean square error minimization problem for arbitrary nonlinear link functions. A naïve approach that may give reasonable results could be applying the link function to the target values in the dataset and minimizing the mean square error of the linear inner representation function with respect to such transformed target values:

$$\frac{1}{|T|} \sum_{x \in T} (L(f(x)) - g(x))^2 \tag{8.30}$$

Clearly the resulting parameter vector is not guaranteed to minimize the mean square error of h with respect to f in general. It may still be quite good, however, particularly if the link function is monotonic.

Example 8.6.2 The naïve application of the ordinary least-squares method to parameter estimation for a generalized linear representation is implemented and demonstrated by the following R code. The specified link function – matching the `f2` target function – is internally used to transform true target function values. The corresponding inverse link function from the previous example is used for prediction.

```
## a naive application of OLS to a generalized linear representation
ols.gen <- function(formula, data, link=function(v) v, link.inv=function(v) v)
{
  f <- y.var(formula)
  attributes <- x.vars(formula, data)
  aind <- names(data) %in% attributes

  amat <- cbind(as.matrix(data[,aind]), intercept=rep(1, nrow(data)))
  fvec <- link(data[[f]])
  'class<-'(list(repf=repf.gen(link.inv), w=solve(t(amat)%*%amat, t(amat)%*%fvec)),
          "par")
}

  # perfect link function for f2
link2 <- function(v) { 10*atanh(v) }

  # estimate of generalized linear model parameters for f2
ols2g <- ols.gen(f2~a1+a2+a3+a4, lrdat.train, link=link2, link.inv=link2.inv)
  # test set error
mse(predict(ols2g, lrdat.test[,1:4]), lrdat.test$f2)
```

Since the link function applied in this illustration matches the target function perfectly (which is completely unrealistic), a parameter vector is found that achieves a near-zero mean square error of the inner linear representation function with respect to the transformed target function. It should therefore be not surprising that the mean square error of the inversely transformed model predictions with respect to the target function is near-zero as well. Unfortunately such successful performance cannot be expected in practice.

8.6.2 Enhanced representation

The idea of enhanced representation is to replace the original attributes $a_1, a_2, \ldots, a_n$ defined on the domain by new attributes $a'_1, a'_2, \ldots, a'_N$ related to them nonlinearly. More exactly, these new enhanced attributes are defined by nonlinear functions of single or multiple original attributes, and typically (but not necessarily) $N \gg n$. The nonlinear relationship between the target function and the original attributes is expected to become linear in the enhanced representation.

Once the transformation that generates new attribute values has been determined, no changes to parameter estimation algorithms are required. They can operate in the usual way, just using the enhanced set of attributes. The definitions of enhanced attributes have to be retained with the model, since new data has to be transformed to the same representation before the model is applied for prediction.

There are many possible approaches to defining an enhanced representation, some of which yield sophisticated and powerful regression algorithms. Sometimes sufficient background knowledge may be available about the domain and the target function to directly suggest appropriate enhanced attribute definitions. Otherwise one of the general-purpose enhanced representations can be employed, such as

- randomized representation, in which enhanced attributes are defined using nonlinear random transformations,

- tile coding, which is based on a series of grids partitioning the domain into overlapping boxes,

- kernel methods, using an implicit nonlinear transformation based on attribute vector dot products.

Of those, the last approach is the most interesting and widely used, and will be presented in Chapter 16.

Example 8.6.3 To illustrate the basic idea of enhanced representation, the following R code defines a general enhanced representation function and its gradient, which can be used to apply an enhancement transformation performed by a function specified as the `enhance` argument to an arbitrary base representation function specified by the `repf` argument and its gradient specified by the `grad` argument. An enhanced representation function and its gradient are sufficient to use the gradient descent algorithm, which requires no changes by itself. The least squares method needs a slightly modified implementation, provided by the `ols.enh` function, that applies the enhancement internally when creating the attribute value matrix. Then a representation enhancement function is defined that implicitly creates a new set of attributes, consisting of both the original four attributes and their squares. This quite trivial enhanced representation can be immediately seen to match the `f3` target function and can therefore be considered the perfect enhanced representation for this target function. This is what could be hardly possible in practice with an unknown target function. Linear model parameters for the `f3` target function with this enhanced representation are then estimated using both the gradient descent algorithm and the least-squares method.

```
## enhanced representation function
repf.enh <- function(enhance, repf=repf.linear)
{ function(data, w) { repf(enhance(data), w) } }

## enhanced representation function gradient
grad.enh <- function(enhance, grad=grad.linear)
{ function(data, w) { grad(enhance(data), w) } }

## estimate linear model parameters using the OLS method
## with enhanced representation
ols.enh <- function(formula, data, enhance=function(data) data)
{
  f <- y.var(formula)
  attributes <- x.vars(formula, data)
  aind <- names(data) %in% attributes

  amat <- cbind(as.matrix(enhance(data[,aind])), intercept=rep(1, nrow(data)))
  fvec <- data[[f]]
  'class<-'(list(repf=repf.enh(enhance), w=solve(t(amat)%*%amat, t(amat)%*%fvec)),
           "par")
}

  # perfect representation enhancement for f3
enhance3 <- function(data) { cbind(data, sq=data^2) }

  # gradient descent estimation for f3
gd3e <- gradient.descent(f3~a1+a2+a3+a4, lrdat.train, w=rep(0, 9),
                         repf=repf.enh(enhance3), grad=grad.enh(enhance3),
                         beta=0.001, eps=0.005)
  # test set error
mse(predict(gd3e$model, lrdat.test[,1:4]), lrdat.test$f3)

  # ols estimation for f3
ols3e <- ols.enh(f3~a1+a2+a3+a4, lrdat.train, enhance3)
  # test set error
mse(predict(ols3e, lrdat.test[,1:4]), lrdat.test$f3)
```

The gradient descent algorithm turns out to perform noticeably worse in the enhanced representation than observed before in the original four-attribute representation for the f1 target function, which was indeed linear. Additional attributes not only increase the computational time, but also appear to make the optimization task more complex, a smaller step-size value has to be used, and several hundred iterations are required to reach a rather unimpressive error level of 0.01. Still, the algorithm does work and a longer training process would likely lead to a better estimated parameter vector. The OLS method remains lightning fast and accurate.

8.6.3 Polynomial regression

A popular simple special case of enhanced representation is obtained if new attributes are polynomial functions of the original attributes. The resulting representation function can be presented as

$$h(x) = F(\mathbf{a}(x), w) = \sum_{j=1}^{p} \sum_{i=1}^{n} w_{i+(j-1)n} a_i^j(x) + w_{pn+1} \tag{8.31}$$

where p is the maximum degree of polynomials used. Models using such a representation are called *polynomial regression* models.

Example 8.6.4 The `repf.enh` and `grad.enh` functions from the previous example can be used to generate polynomial regression representation functions and gradients. This is demonstrated by the following R code, which defines the `enhance.poly` function for polynomial representation enhancement, as well as the `repf.poly` and `grad.poly` functions calculating the polynomial representation function and its gradient. They are illustrated by reproducing the gradient descent and OLS parameter estimation process from the previous example.

```
## polynomial representation enhancement
enhance.poly <- function(data, p=2)
{ do.call(cbind, lapply(1:p, function(j) data^j)) }

## polynomial regression representation function
repf.poly <- function(p=2)
{ repf.enh(function(data) enhance.poly(data, p), repf.linear) }

## polynomial regression representation function gradient
grad.poly <- function(p=2)
{ grad.enh(function(data) enhance.poly(data, p), grad.linear) }

  # gradient descent polynomial regression estimation for f3
gd3p <- gradient.descent(f3~a1+a2+a3+a4, lrdat.train, w=rep(0, 9),
                         repf=repf.poly(p=2), grad=grad.poly(p=2),
                         beta=0.001, eps=0.005)
  # test set error
mse(predict(gd3p$model, lrdat.test[,1:4]), lrdat.test$f3)

  # OLS polynomial regression estimation for f3
ols3p <- ols.enh(f3~a1+a2+a3+a4, lrdat.train, enhance.poly)
  # test set error
mse(predict(ols3p, lrdat.test[,1:4]), lrdat.test$f3)
```

8.6.4 Piecewise-linear regression

Piecewise-linear regression is based on partitioning the domain into disjoint regions such that the target function can be sufficiently well approximated by a linear model in each of them. The overall regression model therefore consists of multiple linear models, each applicable in an appropriate region, and the description of the underlying partitioning that can be applied to appropriately select a linear model for any instance for which prediction would be requested. These individual linear models can be created in the usual way, using subsets of the training set consisting of instances from the corresponding regions.

There are several possible ways of decomposing the domain into regions and describing the obtained partitioning. Not surprisingly, some of them borrow their essential ideas from classification and clustering algorithms, which explicitly or implicitly partition the domain into regions corresponding to different classes or clusters. Arguably the simplest approach is to use a clustering algorithm, such as k-means or another member of the k-centers family presented in Chapter 12, to create a clustering model and then create separate linear regression models for each cluster. This would be based on the hope that the relationship between the target function and attributes, nonlinear in general, would be linear within similarity-based clusters. More refined approaches include:

- model trees, which apply the hierarchical partitioning idea borrowed from decision trees to the regression task,

- local regression, which can be considered an extension of nearest-neighbor regression.

The former is described separately in Section 9.8 and the latter is beyond the scope of this book.

8.7   Conclusion

Parametric regression is the most commonly applied approach to the regression task and, at the same time, the most common instantiation of the parametric model representation. It is not surprising given the fact that regression models predict continuous values based on usually also continuous attributes. It is much more likely to encounter discrete attributes in classification tasks, and whenever all or most attributes are discrete, the parametric model representation becomes considerably less natural and convenient.

A variety of diverse regression algorithms belong to the parametric regression family. The purpose of this chapter was to provide a common background for them and to present more details on those that employ a linear model representation. This helps us to understand various practical parametric regression algorithms and makes it possible to train domain-specific models based on custom representation functions, suggested by the available background knowledge.

Linear regression, either in its plain form or equipped with enhancements that help it to overcome the linearity limitation (generalized linear representation, enhanced representation, piecewise-linear representation), is the most commonly used instantiation of parametric regression for data mining applications. Whereas nonlinear parametric models have also gained considerable popularity and proved successful whenever model accuracy is of extreme importance (with neural networks being the most typical example), linear models and their enhancements remain unbeatable whenever the computation time needed to create the model matters or large volumes of data have to be used. This is because of the efficiency of their parameter estimation algorithms and their robustness against local error minima. They are also relatively easy to interpret, which may be important in some application areas where regression models have to be compared to or combined with existing background expert knowledge.

8.8   Further readings

Similar to linear classification, linear regression has its roots in both machine learning and statistics, with the former traditionally being associated with the gradient descent approach to parameter estimation and the latter with least squares and other more advanced methods not covered in this chapter. This tradition distinction tends to disappear, by the way, with linear regression algorithms usually presented in a similar way by contemporary textbooks in both these areas. From the large set of available machine learning and data mining books, Hand *et al.* (2001) and Bishop (2007) may be particularly worthwhile to consult for a much broader review of linear regression parameter estimation techniques and their theoretical foundations. For an even greater scope and depth, dedicated books on regression and statistical modeling can be referred to (e.g., Draper and Smith 1998; Freedman 2009; Glantz and Slinker 2000). Faraway (2004, 2005) complements an extensive discussion of linear regression by R language demonstrations.

The gradient-based delta or LMS parameter update rule for linear regression models was presented by Widrow and Hoff (1960). Related nonlinear regression algorithms studied in the field of neural networks (e.g., Hertz *et al.* 1991), such as error backpropagation for multilayer nonlinear perceptrons, are also covered by some data mining and machine learning books (Bishop 2007; Hand *et al.* 2001; Mitchell 1997; Tan *et al.* 2013). More refined optimization methods that can be applied to parameter estimation, particularly in the nonlinear case, include the Newton–Raphson, Gauss–Newton, and conjugate gradient algorithms (e.g., Björck 1996; Snyman 2005). Different variations of the online (stochastic) gradient descent algorithm can be seen as instantiations of stochastic approximation, initialized by Robbins and Monro (1951). Bottou (1998) discussed links between these related families of algorithms and showed how the theory of stochastic approximation can be used to prove the convergence of online gradient descent.

The ordinary least-squares method belongs to the oldest modeling algorithms still in widespread use, with its first description published nearly 200 years ago (Legendre 1805). There are of course several contemporary reviews, covering different variations of least-squares parameter estimation (e.g. Hansen *et al.* 2012; Lawson and Hanson 1987).

Generalized linear models (GLM) introduced by Nelder and Wedderburn (1972) are extensively described by McCullagh and Nelder (1989). This discussion covers, in particular, a general approach to parameter estimation and a variety of different link functions and target distributions that yield specific GLM instantiations. Two specific types of enhanced linear representations mentioned in this chapter, but not discussed in the book, are tile coding and random representation. The former was proposed by Albus (1975a,b) as the CMAC function approximator (*cerebellar model articulation controller*) and the latter by Sutton and Whitehead (1993), inspired by the earlier work of Kanerva (1988).

References

Albus JS 1975a Data storage in the cerebellar model articulation controller (CMAC). *Transactions of the ASME Journal of Dynamic Systems, Measurement, and Control* **97**, 228–233.

Albus JS 1975b New approach to manipulator control: The cerebellar model articulation controller (CMAC). *Transactions of the ASME Journal of Dynamic Systems, Measurement, and Control* **97**, 220–227.

Bishop CM 2007 *Pattern Recognition and Machine Learning*. Springer.

Björck A 1996 *Numerical Methods for Least Squares Problems*. Society for Industrial and Applied Mathematics.

Bottou L 1998 Online algorithms and stochastic approximations In *Online Learning and Neural Networks* (ed. Saad D) Cambridge University Press.

Draper NR and Smith H 1998 *Applied Regression Analysis* 3rd edn. Wiley.

Faraway JJ 2004 *Linear Models in R*. Chapman and Hall.

Faraway JJ 2005 *Extending the Linear Model with R: Generalized Linear, Mixed Effects, and Nonparametric Regression Models*. Chapman and Hall.

Freedman DA 2009 *Statistical Models: Theory and Practice*. Cambridge University Press.

Glantz SA and Slinker BS 2000 *Primer of Applied Regression and Analysis of Variance* 2nd edn. McGraw-Hill.

Hand DJ, Mannila H and Smyth P 2001 *Principles of Data Mining*. MIT Press.

Hansen PC, Pereyra V and G. S 2012 *Least Squares Data Fitting with Applications Hardcover*. John Hopkins University Press.

Hertz J, Krogh A and Palmer RG 1991 *Introduction to the Theory of Neural Computation*. Addison-Wesley.

Kanerva P 1988 *Sparse Distributed Memory*. MIT Press.

Lawson CL and Hanson RJ 1987 *Solving Least Squares Problems*. Society for Industrial and Applied Mathematics.

Legendre AM 1805 *Nouvelles méthodes pour la détermination des orbites des comètes*. Didot.

McCullagh P and Nelder JA 1989 *Generalized Linear Models* 2nd edn. Chapman and Hall.

Mitchell T 1997 *Machine Learning*. McGraw-Hill.

Nelder J and Wedderburn R 1972 Generalized linear models. *Journal of the Royal Statistical Society A* **135**, 370–384.

Robbins H and Monro S 1951 A stochastic approximation method. *The Annals of Mathematical Statistics* **22**, 400–407.

Snyman JA 2005 *Practical Mathematical Optimization: An Introduction to Basic Optimization Theory and Classical and New Gradient-Based Algorithms*. Springer.

Sutton RS and Whitehead SD 1993 Online learning with random representations *Proceedings of the Tenth International Conference on Machine Learning (ICML-93)*. Morgan Kaufmann.

Tan PN, Steinbach M and Kumar V 2013 *Introduction to Data Mining* 2nd edn. Addison-Wesley.

Widrow B and Hoff ME 1960 Adaptive switching circuits *Western Electronic Show and Convention (WesCon-60)*, vol. 4. Institute of Radio Engineers.

9

Regression trees

9.1 Introduction

Similarly as regression is sometimes referred to as "classification with continuous classes," regression trees can be considered decision trees with continuous classes. Actually, the latter is much more justified than the former, since the differences between regression trees and decision trees are pretty minor compared to the differences between regression algorithms and classification algorithms in general. They can be thought of as slightly different instantiations of the same family of modeling algorithms based on hierarchical domain decomposition. They are discussed separately in this chapter rather than in Chapter 3 to preserve the task-oriented organization of this book and to better expose their specificity. To highlight the differences between regression trees and decision trees, the latter will be referred to for whatever they have in common, and this chapter will mostly focus on model representation and algorithm modifications that are necessary for the regression task.

Example 9.1.1 A number of examples will be presented throughout this chapter illustrating the major operations performed during regression tree with simple R code snippets. They closely follow the pattern of decision tree examples from Chapter 3. The *weatherr* data will be used for these examples. Besides auxiliary functions from DMR packages, the rpart package providing the standard R regression tree implementation will be used. The following code loads the packages and the data.

Ex. 1.4.1
dmr.data

```
library(dmr.regeval)
library(dmr.stats)
library(dmr.util)

library(rpart)

data(weatherr, package="dmr.data")
```

Data Mining Algorithms: Explained Using R, First Edition. Paweł Cichosz.
© 2015 John Wiley & Sons, Ltd. Published 2015 by John Wiley & Sons, Ltd.

9.2   Regression tree model

Similarly as decision trees, regression tree models are hierarchical structures consisting of nodes, branches, and leaves that represent the decomposition of the domain into a number of regions in which target function values can be trivially approximated with sufficient accuracy. Such a tree can be used to identify the corresponding leaf for an arbitrary instance from the domain and generate a target function prediction. It therefore represents a regression model.

9.2.1   Nodes and branches

The purpose and representation of regression tree nodes and branches is exactly the same as for decision tree nodes and branches. Nodes correspond to domain regions that need to be decomposed into smaller regions by splits. Branches link to descendant nodes or leaves corresponding to particular split outcomes. The notation used to refer to splits, their outcomes, domain regions, and data subsets corresponding to particular nodes will be exactly the same as introduced in Section 3.2 for decision trees. Just like for the latter, the branches linking the parent node and its descendant nodes do not always have to be explicitly represented in the regression tree data structure. In particular, when binary splits are used, the branches can be implicitly represented by an appropriate node numbering scheme, e.g., the descendants of node numbered k can be numbered $2k$ and $2k + 1$.

9.2.2   Leaves

Leaves represent domain regions where no further splits are applied, like for decision tree leaves. Instead of discrete class labels or probability distributions, though, they contain some information useful to generate numerical predictions. In the simplest but most common case there are fixed target function values assigned to leaves. For a leaf $\mathbf{l}$, the associated target function value will be denoted by $v_{\mathbf{l}}$. In practice, it may also be convenient to assign target function values to internal nodes. For a node $\mathbf{n}$ the associated target function value will be denoted by $v_{\mathbf{n}}$.

9.2.3   Split types

The types of splits that can be used for regression splits are exactly the same as possible decision tree splits, presented in Section 3.2.3. Since for many practical regression tasks there continuous attributes only, inequality-based (or interval-based) splits are particularly common.

9.2.4   Piecewise-constant regression

As described above, regression trees are piecewise-constant regression models – they decompose the domain into a number of regions and the model's prediction is constant in each region. If the partitioning into regions is sufficiently dense and their boundaries are properly selected, this representation may yield accurate predictions. A more refined piecewise-linear form of regression trees will be discussed later.

Example 9.2.1 To illustrate the piecewise-constant regression tree model representation, the following R code generates a simple two-attribute artificial dataset with a known

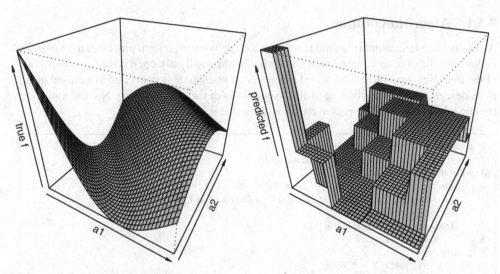

Figure 9.1   An illustration of the piecewise-constant regression tree model representation.

pre-established target function and then creates a regression tree model based on this dataset using the `rpart` package available in R. The latter is used here for the sake of illustration only. Both the true target function and its approximation represented by the regression tree are plotted side by side for easy comparison, as presented in Figure 9.1.

```
  # example target function
rtf <- function(a1, a2) { sin(a1+a2)/(a1+a2) }

  # artificial dataset
rtdat <- data.frame(a1=floor(runif(300, min=1, max=6)),
                    a2=floor(runif(300, min=1, max=6)))
rtdat$f <- rtf(rtdat$a1, rtdat$a2)

  # regression tree
rtf.rp <- rpart(f~., rtdat)
  # target function predictions
rtf.p <- function(a1, a2) { predict(rtf.rp, data.frame(a1, a2)) }

  # 3D plots
par(mfrow=1:2, mar=rep(0.1, 4))
a1 <- a2 <- seq(1, 5, 0.1)
  # true f
persp(a1, a2, outer(a1, a2, rtf), zlab="true f", theta=30, phi=30, col="grey")
  # predicted f
persp(a1, a2, outer(a1, a2, rtf.p), zlab="predicted f", theta=30, phi=30, col="grey")
```

9.3   Growing

The first and mandatory phase of creating regression trees is growing. The same top-down approach known from decision tree growing is adopted, which starts from a single root node and iteratively adds nodes as long as further splits are required.

9.3.1   Algorithm outline

As its decision tree counterpart, the regression tree growing algorithm processes a set of *open* nodes until they become *closed* nodes or leafs. The first and only open node is the root node of the tree. New open nodes are added whenever an open node is converted to a closed node, as its descendants (corresponding to all the outcomes of the selected split). No new nodes are created when an open node is converted to a closed leaf.

```
 1: create the root node and mark it as open;
 2: assign all training instances from T to the root node;
 3: while there are open nodes do
 4:     select an open node n;
 5:     calculate target function summary statistics Stats(f|n) based on T_n;
 6:     assign target function value v_n;
 7:     if stop criteria are satisfied for n then
 8:         mark n as a closed leaf;
 9:     else
10:         select a split t : X → R_t for n;
11:         for all split outcomes r ∈ R_t do
12:             create a descendant node n_r corresponding to r and mark it as open;
13:             assign all instances from T_{n,t=r} to n_r;
14:         end for
15:         mark n as a closed node;
16:     end if
17: end while
```

The algorithm only marginally differs from the decision tree growing algorithm, as the most important differences are actually hidden within particular steps. These will be reviewed below.

Example 9.3.1 This example starts a sequence of examples illustrating the major steps of regression tree growing, which directly corresponds to Examples 3.3.1–3.3.7 for decision trees. The following R code initializes variables used to refer to the dataset, attribute names, and the target function name that facilitate the adaptation of subsequent examples to other datasets.

```
data <- weatherr
attributes <- names(weatherr)[1:4]
target <- names(weatherr)[5]
```

The following R code defines the `init` function for regression tree initialization. It creates a single root node and assigns to it all training instances. When called with the `data` variable set as shown above, it initializes tree growing for the *weatherr* data.

```
init <- function()
{
  tree <<- data.frame(node=1, attribute=NA, value=NA, target=NA,
                      count=NA, mean=NA, variance=NA)
  nodemap <<- rep(1, nrow(data))
```

```
  n <<- 1
}

init()
```

This and subsequent examples assume that a regression tree is represented by a data frame with rows corresponding to nodes and the following columns:

node: node number (starting from 1 for the root node),

attribute: the attribute used for the split,

value: the value used for the split,

target: the assigned target function value,

count: the instance count for the node,

mean, variance: the mean and variance of target function values for the subset of training instances associated with the node (target function summary statistics).

The assignment of instances to regression tree nodes is represented by a vector containing the numbers of nodes to which the corresponding instances are assigned. Subsequent examples will illustrate the major operations of the algorithm for the node indicated by the n variable, assuming the data, attributes, and class variable assignments presented above.

9.3.2   Target function summary statistics

The purpose of this operation is to gather distribution statistics of the target function in the subset of training instances T_n corresponding to the currently processed node n. These statistics will be subsequently used for target value assignment, stop criteria verification, and split selection. The exact set of target function statistics that need to be calculated strictly depends on the particular stop- and split-selection criteria for which they are supposed to be used.

The most basic statistics used to describe the distribution of continuous attributes include location and dispersion measures. These are sufficient for most practical regression tree growing algorithms. Several different location and dispersion measures could be adopted and prove useful. It is reasonable to assume that a single location measure and a single dispersion measure of target function values for training instances associated with a particular node would be sufficient. Unless discussing particular specific examples of such measures, these will be referred to, for node n, as $\mathrm{loc}_{T_n}(f)$ and $\mathrm{disp}_{T_n}(f)$ or shortly $\mathrm{loc}_n(f)$ and $\mathrm{disp}_n(f)$, respectively.

It is most common to assume that the distribution of the target function in the current node's subset of training instances is described in the arguably simplest possible way, by its mean (as a location measure) and variance (as a dispersion measure):

$$m_n(f) = m_{T_n}(f) \tag{9.1}$$

$$s_n^2(f) = s_{T_n}^2(f) \tag{9.2}$$

The standard deviation can also be used instead of the variance.

It maybe sometimes reasonable, particularly if the training set is likely to contain some outlying values of the target function, to consider more robust location and dispersion measures presented in Section 2.4.1, such as the median and the median absolute deviation or the quartile dispersion coefficient.

Example 9.3.2 The following R code defines a function that calculates the target function mean and variance for a regression tree node. The function, which also takes care of setting the instance count, is then applied to the root node of the previously initialized regression tree for the *weatherr* data. A modified version of the standard `var` function is used for variance calculation, called `var1`, that returns 0 for single-element vectors and NaN for empty vectors.

> Ex. 2.4.9
> dmr.stats

```
target.summary <- function(n)
{
  tree$count[tree$node==n] <<- sum(nodemap==n)
  tree$mean[tree$node==n] <<- mean(data[nodemap==n,target])
  tree$variance[tree$node==n] <<- var1(data[nodemap==n,target])
}

target.summary(n)
```

9.3.3 Target value assignment

As mentioned above and assumed in the presented algorithm outline, it makes sense to perform the operation of target value assignment, although strictly necessary only for leaves, for all nodes processed during the growing process. This is what a location measure of the target function distribution in the subset of training instances associated with each node is needed for. The value of the adopted target function location measure for node **n** is used as the target value assigned to the node:

$$v_n = \mathrm{loc}_n(f) \tag{9.3}$$

Clearly, this is the best possible choice for the basic piecewise-constant version of regression trees currently presented, as long as the location measure is appropriately selected to match the performance measure that the tree is supposed to optimize in each leaf. When the resulting tree is used for prediction, all instances that end up in the same leaf will be assigned the same predicted target value, equal to the location measure value obtained for the training instances associated with the leaf. It is easy to see that in the most common case, when it is the mean square error that should be minimized for each leaf, the mean has to be used as the location measure.

There is one obvious exceptional situation of the empty instance set that needs to be considered. When there are no training instances that satisfy the conjunction of conditions represented by the sequence from the root node to the current node **n**, we will have $T_n = \emptyset$. This by no means guarantees that there will be no such instances when the tree is applied for prediction. To make the regression tree a proper regression model, we have to make sure that it will be capable of generating some reasonable predictions for such possible future instances. A simple solution is to inherit the target value assigned to the parent node. This does not have to be taken into account only if all available splits are binary, since a binary split yielding the same outcome for all training instances to which it is applied would be useless and should not be selected in the parent node anyway.

Example 9.3.3 The following R code continues the demonstration of regression tree growing for the *weatherr* data by assigning the mean target function value calculated in the previous example to the current node as its target function value. This makes the `target` column of the regression tree data frame a duplicate of the `mean` column. This obvious redundancy serves the purpose of conceptual clarity only: separating target function summary statistics from target function values assigned to particular nodes.

```
target.value <- function(n)
{
  tree$target[tree$node==n] <<- tree$mean[tree$node==n]
}
target.value(n)
```

9.3.4 Stop criteria

Stop criteria are used to decide whether a given open node requires no further split and should become a closed leaf. Some of the stop criteria presented in Section 3.3.4 for decision trees directly apply to regression tree growing as well. This is the case for the following:

No instances left. The set of training instances assigned to the node is empty, i.e., $T_n = \emptyset$.

No splits left. There is no split that can be applied to further partition the current subset of training instances T_n, either because all available splits have been already used up on the path from the root to **n**, or every split not yet used gives a single outcome for all instances from T_n, which would put them to the same branch.

Clearly, these are the strict (*most definite*) stop criteria which, when satisfied, make further splitting totally pointless. When either of these two is satisfied, any possible descendants of **n** would receive the very same target value v_n and any possible subtree starting in **n** would predict for any instances in exactly the same way as **n** alone. None of them, however, corresponds to a desirable situation. Particularly unwelcome is satisfying the "no splits left" criterion, which results in creating a leaf that will reduce the training performance of the tree, and probably its true performance as well.

The "no instances left" criterion creates a leaf that does not contribute to reducing the training performance, as it will not be used for making predictions for any training instances, and also it cannot be expected to be particularly useful in delivering good true performance. This criterion may be needed only if nonbinary splits are employed, though (otherwise we would have already stopped at the parent node due to the "no splits left" criterion). However, it is commonly used in a relaxed form, satisfied when the number of remaining training instances is sufficiently small.

The primary stop criterion that under normal conditions should be responsible for creating the majority of leaves corresponds to the desirable situation of isolating a subset of training instances T_n for which the target function can be reasonably well approximated by a single constant number v_n – the target value assigned to **n**. Such a criterion can be formulated as:

Sufficiently low dispersion. The target function value dispersion in the set of instances assigned to the node is sufficiently low, i.e., $\text{disp}_n(f) < \text{disp}_{\min}$, where $\text{disp}_{\min}$ is the specified threshold below which no further splits are applied.

This can be thought of as the counterpart of the uniform class decision tree stop criterion, albeit not so definite, since requesting a uniform target function value (i.e., setting $disp_{min} = 0$) would usually lead to heavily overgrown and probably overfitted trees.

A more technical, but sometimes useful stop criterion for regression tree growing is based on the maximum tree depth. No further splits are performed after a sufficiently long path has been created. In a sense, this can be thought of as a relaxed version of the "no splits left" criterion, as it sets the limit on the number of splits used along each tree path.

Example 9.3.4 The following R code demonstrates how to check the "sufficiently low dispersion" criterion (by comparing the target function variance against the `minvar` threshold) and the (relaxed) "no instances left" criterion (by comparing the number of instances corresponding to the node against the `minsplit` threshold). Additionally, the more technical maximum tree depth criterion is checked, by comparing the node's number to $2^{maxdepth}$. The three stop criteria parameters:

- `minvar` – the minimum target function variance required for a split,
- `minsplit` – the minimum number of instances required for a split,
- `maxdepth` – the maximum tree depth,

are specified via variables for the sake of this demonstration. All these criteria are obviously found to be unsatisfied for the root node of the regression tree for the *weatherr* data.

```
minvar <- 0.005
minsplit <- 2
maxdepth <- 8

stop.criteria <- function(n)
{
  n>=2^maxdepth || tree$count[tree$node==n]<minsplit ||
    tree$variance[tree$node==n]<minvar
}

stop.criteria(n)
```

9.3.5   Split selection

The stop criteria presented above are sufficient alone to assure good training performance of regression tree models. The role of split selection is to increase the chance that their generalization capabilities will be sufficient to deliver satisfactory true performance. Similarly as with decision tree split selection, this is achieved by following Ockham's razor principle, i.e., selecting splits that are likely to yield small trees.

9.3.5.1   Preference for simplicity

Tree simplicity is directly related to the average path length, i.e., the average number of splits needed before a leaf is reached from the root node. Under the stop criteria presented above, leafs are preferably created when a sufficiently low target function dispersion is reached.

This makes the dispersion a natural candidate for a split evaluation measure as well – splits that yield low dispersion descendant nodes should be preferred, since they are more likely to permit leaf creation soon.

9.3.5.2   Split evaluation

To evaluate a candidate split $t : X \rightarrow R_t$ for node $\mathbf{n}$ one has, similarly as for decision tree split evaluation, consider the subsets $T_{\mathbf{n},t=r}$ for all $r \in R_t$ to which the candidate split partitions the current set of training instances $T_{\mathbf{n}}$, and calculate the target function dispersion for each of them: $\mathrm{disp}_{T_{\mathbf{n},t=r}}(f)$ for $r \in R_t$. Note that, although target function summary statistics are normally calculated for each node at the beginning of its processing, the dispersion values used here need to be calculated separately, since the corresponding descendant nodes do not actually exist (and will not be created unless the evaluated split is ultimately selected).

The split can be evaluated by the weighted average of the target function dispersion values corresponding to the outcomes of split t:

$$\mathrm{disp}_{\mathbf{n}}(f|t) = \sum_{r \in R_t} \frac{|T_{\mathbf{n},t=r}|}{|T_{\mathbf{n}}|} \mathrm{disp}_{T_{\mathbf{n},t=r}}(f) \tag{9.4}$$

The weights in this average are based on the proportions of the partitioning of the current set of training instances into the subsets corresponding to the possible outcomes of t, i.e., subset dispersions are weighted proportionally to subset sizes. Split selection can then be performed by *minimizing* such weighted dispersion impurity measure over all available splits.

It may be convenient to use a slightly modified split evaluation function, measuring not just the average target function dispersion obtained after the split, but rather the improvement (decrease) of the dispersion. It is defined as the difference between the target function dispersion for the subset of instances corresponding to the node in which the split is to be applied and the weighted average dispersion of the subsets corresponding to split outcomes. This can be written as follows:

$$\Delta\mathrm{disp}_{\mathbf{n}}(f|t) = \mathrm{disp}_{\mathbf{n}}(f) - \sum_{r \in R_t} \frac{|T_{\mathbf{n},t=r}|}{|T_{\mathbf{n}}|} \mathrm{disp}_{T_{\mathbf{n},t=r}}(f) \tag{9.5}$$

Maximizing the difference will clearly select exactly the same splits as minimizing the average dispersion. Its advantage is that it not only indicates which split is the best, but also how much improvement it gives. This may be used to define an additional stop criterion that converts a node to a leaf when the improvement due to the best available split is too small.

Example 9.3.5 The R code presented below defines the `weighted.dispersion` function that calculates the weighted dispersion of two numeric vectors, representing target values for the true and false split outcomes, using the supplied dispersion measure. The latter defaults to `var1`, a modified version of variance already used above. It also optionally receives the corresponding weight vectors that can be used to weight individual instances, if permitted by the dispersion measure (these are not used here, but will come handy later). The application of the `weighted.dispersion` function to evaluating the split based on the `outlook==overcast` condition is demonstrated.

> Ex. 2.4.9
> dmr.stats

```
weighted.dispersion <- function(v1, v0, w1, w0, disp=var1)


{
  if (missing(w1) || missing(w0))
    weighted.mean(c(disp(v1), disp(v0)), c(length(v1), length(v0)))
  else
    weighted.mean(c(disp(v1, w1), disp(v0, w0)), c(length(v1), length(v0)))
}

  # weighted dispersion of playability for outlook=overcast and outlook!=overcast
weighted.dispersion(weatherr$playability[weatherr$outlook=="overcast"],
                    weatherr$playability[weatherr$outlook!="overcast"])
```

Example 9.3.6 The following R code defines functions that perform split evaluation and split selection for regression trees, using the `weighted.dispersion` function from the previous example. Equality-based splits for discrete attributes and inequality-based splits for continuous attributes are attempted as candidates. The `midbrk` auxiliary function for calculating middle breaks between consecutive values is used to get thresholds for `dmr.util` inequality-based splits. This implementation of split selection is then demonstrated for the root node of the regression tree for the *weatherr* data. It is worthwhile to notice that the `split.eval` function detects useless splits that give the same outcome for all instances and returns `Inf` for them to make sure they will not be selected. Its return value is checked by the `split.select` function which actually makes no split selection if the best available split is useless. This can be considered an implicit implementation of the "no splits left" stop criterion.

```
split.eval <- function(av, sv, tv)
{
  cond <- !is.na(av) & (if (is.numeric(av)) av<=as.numeric(sv) else av==sv)
  v1 <- tv[cond]
  n1 <- sum(cond)
  v0 <- tv[!cond]
  n0 <- sum(!cond)
  if (n1>0 && n0>0)
    weighted.dispersion(v1, v0)
  else
    Inf
}

split.select <- function(n)
{
  splits <- data.frame()
  for (attribute in attributes)
  {
    uav <- sort(unique(data[nodemap==n,attribute]))
    if (length(uav)>1)
      splits <- rbind(splits,
                      data.frame(attribute=attribute,
                                 value=if (is.numeric(uav))
                                         midbrk(uav)
                                       else as.character(uav),
```

```
                            stringsAsFactors=FALSE))
  }

  if (nrow(splits)>0)
    splits$eval <- sapply(1:nrow(splits),
                          function(s)
                          split.eval(data[nodemap==n,splits$attribute[s]],
                                     splits$value[s],
                                     data[nodemap==n,target]))
  if ((best.eval <- min(splits$eval))<Inf)
    tree[tree$node==n,2:3] <<- splits[which.min(splits$eval),1:2]
  best.eval
}

  # variance-based split selection
split.select(n)
```

The only significant difference between the `split.select` function presented above and the corresponding function for decision trees defined in Example 3.3.6 is the different split evaluation function. The `outlook=overcast` split is found to yield the lowest weighted target function variance and is selected for the root node.

9.3.6   Split application

The operation of split application for regression trees does not differ from its decision tree counterpart in any way. When a split $t : X \to R_t$ for a node **n** has been selected, new descendant nodes $\mathbf{n}_r$ are created for all $r \in R_t$, corresponding to each of its possible outcomes. These new nodes are marked as open, to get processed in subsequent iterations of the algorithm.

The set of training instances $T_\mathbf{n}$ corresponding to the parent node has to be partitioned into subsets corresponding to the newly created descendant nodes by applying the split: $T_{\mathbf{n}_r} = T_{\mathbf{n},t=r}$ for $r \in R_t$. This completes the current iteration of the algorithm and node **n** can be marked as closed.

Example 9.3.7 The following R code demonstrates how a selected split can be applied to partition the set instances assigned to the currently processed node into subsets assigned to two newly created descendant nodes. For the regression tree being grown for the *weatherr* data this results in creating two descendants of the root node, numbered 2 and 3.

```
split.apply <- function(n)
{
  tree <<- rbind(tree,
                 data.frame(node=(2*n):(2*n+1), attribute=NA, value=NA, target=NA,
                            count=NA, mean=NA, variance=NA))

  av <- data[[tree$attribute[tree$node==n]]]
  cond <- !is.na(av) & (if (is.numeric(av))
                          av<=as.numeric(tree$value[tree$node==n])
                        else av==tree$value[tree$node==n])
  nodemap[nodemap==n & cond] <<- 2*n
```

```
    nodemap[nodemap==n & !cond] <<- 2*n+1
}

split.apply(n)
```

Except for using a slightly different tree representation, this function is the same as the corresponding function presented in the example of decision tree split application.

9.3.7   Complete process

All operations performed during a single iteration of top-down decision tree growing have been presented above. After a number of iterations, when there are no more open nodes left, the growing process terminates, yielding a completely grown tree.

Example 9.3.8 The code from the previous examples which provides simple illustrative implementations of target function summary statistics calculation, target function value assignment, split selection, and split application can be applied again to process the newly created nodes, by passing the appropriate node number via the n argument of the corresponding functions. After seven iterations the growing process will complete, yielding the regression tree represented by the following data frame:

```
    node  attribute     value   target count      mean    variance
1      1    outlook  overcast 0.5671429    14 0.5671429 0.015714286
2      2       <NA>      <NA> 0.7225000     4 0.7225000 0.002825000
3      3 temperature       25 0.5050000    10 0.5050000 0.006738889
4      6       <NA>      <NA> 0.5400000     7 0.5400000 0.003133333
5      7    outlook     rainy 0.4233333     3 0.4233333 0.006633333
6     14       <NA>      <NA> 0.3300000     1 0.3300000 0.000000000
7     15       <NA>      <NA> 0.4700000     2 0.4700000 0.000200000
```

The implementations of the major regression tree growing steps presented so far can be easily supplemented with the main loop to create a very simple and inefficient, but easy to understand and working implementation of regression tree growing. Such an implementation is presented below. The grow.regtree function defined by the R code presented below organizes the iterative processing of regression tree nodes, using the init, target.summary, target.value, stop.criteria, split.eval, split.select, and split.apply functions from the previous examples as its internal functions. Two additional utility functions, x.vars and y.var, are used to extract the attribute and target function names from the input formula, to make the implemen- dmr.util
tation applicable to other datasets. Apart from the formula and dataset, the grow.regtree function receives the stop criteria parameters as its optional arguments. The class attribute of the created regression tree object is set to regtree to enable appropriate prediction method dispatching and a method for conversion to a data frame is provided.

```
## a simple regression tree growing implementation
grow.regtree <- function(formula, data, minvar=0.005, minsplit=2, maxdepth=8)
{
  init <- function()
```

```
{
  tree <<- data.frame(node=1, attribute=NA, value=NA, target=NA,
                      count=NA, mean=NA, variance=NA)
  nodemap <<- rep(1, nrow(data))
  n <<- 1
}

next.node <- function(n)
{
  if (any(opn <- tree$node>n))
    min(tree$node[opn])
  else Inf
}

target.summary <- function(n)
{
  tree$count[tree$node==n] <<- sum(nodemap==n)
  tree$mean[tree$node==n] <<- mean(data[nodemap==n,target])
  tree$variance[tree$node==n] <<- var1(data[nodemap==n,target])
}

target.value <- function(n)
{
  tree$target[tree$node==n] <<- tree$mean[tree$node==n]
}

stop.criteria <- function(n)
{
  n>=2^maxdepth || tree$count[tree$node==n]<minsplit ||
    tree$variance[tree$node==n]<minvar
}

split.eval <- function(av, sv, tv)
{
  cond <- !is.na(av) & (if (is.numeric(av)) av<=as.numeric(sv) else av==sv)
  v1 <- tv[cond]
  n1 <- sum(cond)
  v0 <- tv[!cond]
  n0 <- sum(!cond)
  if (n1>0 && n0>0)
    weighted.dispersion(v1, v0)
  else
    Inf
}

split.select <- function(n)
{
  splits <- data.frame()
  for (attribute in attributes)
  {
    uav <- sort(unique(data[nodemap==n,attribute]))
    if (length(uav)>1)
      splits <- rbind(splits,
                      data.frame(attribute=attribute,
                                 value=if (is.numeric(uav))
                                         midbrk(uav)
                                       else as.character(uav),
                                 stringsAsFactors=FALSE))
  }
  if (nrow(splits)>0)
```

```
      splits$eval <- sapply(1:nrow(splits),
                            function(s)
                            split.eval(data[nodemap==n,splits$attribute[s]],
                                       splits$value[s],
                                       data[nodemap==n,target]))
    if ((best.eval <- min(splits$eval))<Inf)
      tree[tree$node==n,2:3] <<- splits[which.min(splits$eval),1:2]
    best.eval
  }

  split.apply <- function(n)
  {
    tree <<- rbind(tree,
                   data.frame(node=(2*n):(2*n+1), attribute=NA, value=NA, target=NA,
                              count=NA, mean=NA, variance=NA))

    av <- data[[tree$attribute[tree$node==n]]]
    cond <- !is.na(av) & (if (is.numeric(av))
                            av<=as.numeric(tree$value[tree$node==n])
                          else av==tree$value[tree$node==n])
    nodemap[nodemap==n & cond] <<- 2*n
    nodemap[nodemap==n & !cond] <<- 2*n+1
  }

  tree <- nodemap <- n <- NULL
  target <- y.var(formula)
  attributes <- x.vars(formula, data)

  init()
  while (is.finite(n))
  {
    target.summary(n)
    target.value(n)
    if (!stop.criteria(n))
      if (split.select(n)<Inf)
        split.apply(n)
    n <- next.node(n)
  }
  `class<-`(tree, "regtree")
}

## convert a regtree object to a data frame
as.data.frame.regtree <- function(x, row.names=NULL, optional=FALSE, ...)
{ as.data.frame(unclass(x), row.names=row.names, optional=optional) }

  # grow a regression tree for the weatherr data
tree <- grow.regtree(playability~., weatherr)

  # data frame conversion
as.data.frame(tree)
```

9.4   Pruning

Just like with decision tree pruning discussed in Section 3.4, regression tree pruning may help prevent overfitting. It essentially retracts some growing steps by cutting off selected sub-trees believed to be overgrown and replacing them by leaves. Whereas it might appear more reasonable, at least from the computational economics perspective, to rather prevent growing such subtrees – which is often referred to as *pre-pruning* – their predictive utility cannot be reliably estimated before they are actually grown. This makes the two-phase growing and pruning process the preferred approach to assuring good generalization capabilities of decision trees, unless computational time constraints make it impractical.

Of the following major components of regression tree pruning algorithms, it is only the middle one that has to be regression-specific, with the possible choices for the remaining ones the same as for decision trees.:

Pruning operators. Which determine how the operation of cutting off nodes from the tree is exactly performed.

Pruning criterion. Which determines how to judge whether a pruning operator should be applied to a given node.

Pruning control strategy. Which determines the order in which candidate nodes for pruning are considered.

9.4.1   Pruning operators

There is no reason why operators used for regression tree pruning should differ from decision tree pruning operators in any way. The same *subtree cutoff* and *node removal* operators can therefore be applied, with the former being much more common and the latter rather "theoretically possible" than really practically used.

9.4.2   Pruning criterion

The pruning criterion is used to judge whether a given pruning operator should be applied to a given node or not. This is based on the comparison of the original subtree rooted at the node with the new subtree (in particular, the single new leaf for the most common subtree cutoff operator) with respect to some quality measures. Regression tree pruning has not been investigated as extensively as decision tree pruning, and the number of regression tree pruning criteria described in the literature is not that large as for decision tree pruning criteria. This section presents some criteria that are in fact rather straightforward adaptations of their decision tree counterparts.

9.4.2.1   Reduced error pruning

The reduced error pruning criterion for regression trees is based on the same idea as its decision tree prototype, which is in fact the simplest possible computational interpretation of the very goal of pruning – to improve the generalization capability of the regression tree. To achieve this, the true performance of the original subtree and the leaf that would be replacing it is estimated and compared using a separate *pruning set R*. Assuming the most common mean square error performance measure, the reduced error pruning criterion would recommend to replace node $\mathbf{n}$ by leaf $\mathbf{n}$ if

$$\mathrm{mse}_R(\mathbf{l}) \leq \mathrm{mse}_R(\mathbf{n}) \tag{9.6}$$

where $\mathrm{mse}_R(\mathbf{n})$ and $\mathrm{mse}_R(\mathbf{l})$ denote the pruning set mean square errors of the original node $\mathbf{n}$ and the replacing leaf $\mathbf{l}$, respectively. The calculation of error values has to be based only on $R_\mathbf{n}$ – the subset of R corresponding to $\mathbf{n}$ – to ensure that both the original subtree and the replacing leaf are evaluated in the proper context, where they actually appear in the tree.

As discussed for decision trees, reduced error pruning could be easily considered the perfect pruning criterion unless its application were so often prevented by "data economy" issues. To provide sufficiently reliable error estimates even for low-level nodes (where pruning is most likely to be needed), a considerable number of pruning instances may be required, definitely comparable to the number of training instances used for growing. Unless we have a plethora of labeled data which have to be sampled anyway, it may be unreasonable to leave out such a substantial portion of the available data from the growing process, since this may considerably impact the quality of the grown tree.

9.4.2.2   Minimum error pruning

The essential idea of decision tree minimum error pruning is to calculate the training set misclassification error at leaves based on the m-estimated dominating class probability. By incorporating fictitious instances, representing a prior probability estimate – as discussed in Section 2.4.4 – class probability m-estimation helps avoid the optimistic bias associated with training set evaluation. Such corrected error estimates of leaves are propagated upward to nodes.

A similar approach can be applied to regression tree pruning by using m-estimation for mean square error calculation. This is easily possible using the technique of variance m-estimation from Section 2.4.4, which was actually developed for this very purpose. The training set mean square error at regression tree leaves is indeed the (biased estimator of) target function variance. While the definition of the m-estimated variance given by Equation 2.61 corresponds to the unbiased variance estimator, the basic biased estimator is used for mean square error calculation:

$$\hat{\mathrm{mse}}_T(\mathbf{l}) = \frac{\sum_{x \in S}(f(x) - m_{\mathbf{l},m,m_0}(f))^2 + ms_0^2}{|T_\mathbf{l}| + m} \tag{9.7}$$

where

$$m_{\mathbf{l},m,m_0}(f) = \frac{\sum_{x \in T_\mathbf{l}} f(x) + mm_0}{|T_\mathbf{l}| + m} \tag{9.8}$$

is the m-estimated mean target function value at leaf $\mathbf{l}$ and m is the m-estimation parameter. The prior mean value m_0 and the prior variance value s_0^2 are normally calculated based on the complete training set, i.e., $m_0 = m_T(f)$ and $s_0^2 = s_T^2(f)$.

The leaf error estimates are propagated upward to receive node error estimates exactly as in decision tree minimum error pruning:

$$\hat{\mathrm{mse}}_T(\mathbf{n}) = \sum_{\mathbf{n}' \in N(\mathbf{n})} \frac{|T_{\mathbf{n}'}|}{|T_\mathbf{n}|} \hat{\mathrm{mse}}_T(\mathbf{n}') \tag{9.9}$$

where $N(\mathbf{n})$ is the set of all the descendants (nodes and leaves) of node $\mathbf{n}$. This makes it possible to use the following inequality as the pruning criterion:

$$\hat{\mathrm{mse}}_T(\mathbf{l}) \leq \hat{\mathrm{mse}}_T(\mathbf{n}) \tag{9.10}$$

9.4.2.3   Cost-complexity pruning

Instead of trying to reliably estimate the expected error on new data using the training set, which is hard to achieve, cost-complexity pruning applies Ockham's razor more directly by punishing the original subtree for its complexity when comparing its training set performance with that of the replacement leaf. The relative square error, defined in Section 10.2.6, may be more convenient to use than the mean square error when trading off performance and complexity, as it relates the differences between predicted and true target function values to the dispersion of the latter and is therefore easier to interpret directly. This yields the following criterion:

$$\text{rse}_T(\mathbf{l}) \leq \text{rse}_T(\mathbf{n}) + \alpha \mathbf{C}(\mathbf{n}) \tag{9.11}$$

The *complexity parameter* α represents the amount of performance improvement per single node required to justify retaining the original subtree.

As with decision tree cost-complexity pruning, this pruning criterion becomes truly useful when coupled with an appropriate technique for adjusting the complexity parameter value. It consists in identifying, for a given value of α, the smallest pruned tree that minimizes the sum

$$\text{rse}_T(\mathbf{n}_1) + \alpha \mathbf{C}(\mathbf{n}_1) \tag{9.12}$$

where $\mathbf{n}_1$ is the root node. Such a tree is called the optimally pruned tree with respect to α. As explained in Section 3.4.2, the sequence of all optimally pruned trees and the corresponding complexity parameter intervals can be identified after completing the growing process. Selecting one of them is possible based on true performance estimates obtained by the internally employed k-fold cross-validation technique, discussed in Section 7.3.4.

Of course, the cross-validation performed during regression tree growing increases the amount of computation considerably, but whenever the computational cost is not an issue, cost-complexity pruning belongs to the most effective pruning methods using the training set.

9.4.3   Pruning control strategy

Just like with decision tree pruning control, the control strategy for regression tree pruning in principle has to be selected accordingly for particular pruning criteria, since different control strategies can work best with different criteria. The choice is limited to the same basic *bottom-up*, *top-down*, and *best-first* strategies (or some hybrid approaches combining two or all of these three), with the first strategy being the most wide-spread. In particular, it suits well the most common pruning criteria described above.

9.5   Prediction

To apply a regression tree to generate target function predictions for a dataset, one just needs to identify leaves where particular instances from the dataset land when propagated through the tree. This requires the same instance down-propagation process as presented in Section 3.5 for decision trees, consisting in sequentially applying splits and descending along branches corresponding to their outcomes, starting from the root node, until a leaf is reached. The split application operation previously discussed for regression tree growing has to be performed multiple times, until all instances arrive at the corresponding leaves.

Example 9.5.1 The R code implements the regression tree prediction operation, assuming that the tree representation as created in the growing examples. It is then applied to the regression tree grown for the *weatherr* data, assuming it is stored in the `tree` variable.

```
## regression tree prediction
predict.regtree <- function(tree, data)
{
  descend <- function(n)
  {
    if (!is.na(tree$attribute[tree$node==n]))   # unless reached a leaf
    {
      av <- data[[tree$attribute[tree$node==n]]]
      cond <- !is.na(av) & (if (is.numeric(av))
                             av<=as.numeric(tree$value[tree$node==n])
                           else av==tree$value[tree$node==n])
      nodemap[nodemap==n & cond] <<- 2*n
      nodemap[nodemap==n & !cond] <<- 2*n+1
      descend(2*n)
      descend(2*n+1)
    }
  }

  nodemap <- rep(1, nrow(data))
  descend(1)
  tree$target[match(nodemap, tree$node)]
}

# regression tree prediction for the weatherr data
predict(tree, weatherr)
```

9.6   Weighted instances

Regression tree models can be created using weighted training instances in essentially the same way as discussed for decision trees in Section 3.6. It is possible because instance weights can be easily incorporated to all the calculations performed during regression tree growing. These are limited to location and dispersion measure calculation and instance counting.

Specifically, when growing a regression tree from a training set T, to apply stop and split selection criteria at node $\mathbf{n}$, one just needs $loc_\mathbf{n}(f)$, $disp_\mathbf{n}(f)$, $disp_{T_{\mathbf{n},t=r}}(f)$, and $|T_{\mathbf{n},t=r}|$ for all splits t and their outcomes r. Similarly, when pruning a tree based on the pruning set R, to apply pruning criteria to node $\mathbf{n}$ it is sufficient to determine the value of the selected performance measure for the subtree rooted at $\mathbf{n}$ on the corresponding subset of R.

Achieving weight sensitivity for all these calculations is straightforward, boiling down to appropriately redefining counts as weight sums:

$$|T_{\mathbf{n},t=r}| = \sum_{x \in T_{\mathbf{n},t=r}} w_x \tag{9.13}$$

and using weighted location and dispersion measures, as well as a weighted performance measure for pruning. In particular, if the mean is used as the location measure, the variance as the dispersion measure, and the mean square error as the performance measure for pruning, they should be replaced by the weighted mean, the weighted variance, and the weighted mean square error, respectively. It makes it possible for regression tree induction algorithms to use *weighted* instances, i.e., to act as weight-sensitive algorithms.

9.7   Missing value handling

Similarly as decision trees, regression trees belong to those modeling algorithms that can use special techniques to handle missing values during model building and application, to achieve the least possible degradation of model and prediction quality. The same two major approaches to missing value handling, presented in Section 3.7, are also applicable here:

Fractional instances. Whenever a split on an attribute with a missing value is considered or applied, each incomplete instance is virtually replaced by several instances corresponding to all possible split outcomes, with a fractional "copy count,"

Surrogate splits. A number additional splits are stored for each node, apart from the ordinary main split, and used to dispatch instances for which the outcome of the main split cannot be determined due to missing attribute values.

Of those, only the former needs some minor regression-specific modifications.

9.7.1   Fractional instances

The technique of fractional instances considers all possible split outcomes for an instance with a missing value of the split attribute and assigns them appropriate weights or probabilities, based on the observed distribution of outcomes for the training set. For split t at node $\mathbf{n}$, the probability of outcome $r \in R_t$ is estimated as

$$P(t = r|\mathbf{n}) = \frac{|T_{\mathbf{n}, t=r}|}{|T_{\mathbf{n}}| - |T_{\mathbf{n}, t=?}|} \tag{9.14}$$

9.7.1.1   Growing with fractional instances

Regression tree growing with fractional instances requires the split evaluation and split application operations to appropriately use and modify instance "copy counts," with $w_{x,\mathbf{n}}$ denoting the "copy count" of instance x at node $\mathbf{n}$. For the root node the "copy counts" of all instances are set to 1 or user-specified instance weights. The location and dispersion measures used for target value assignment and split evaluation also have to handle instance "copy counts" appropriately. Subset size needs to be redefined as the sum of "copy" counts" for subset members.

Split evaluation with fractional instances   Recall that split evaluation is based on the dispersion of target function values for each possible split outcome. Two modifications are necessary when using fractional instances. First, the dispersion measure has to be weight-sensitive to incorporate instance "copy counts." This could be, in particular, the weighted variance instead of the ordinary variance. Each instance with a known split outcome would be used with weight equal to its copy count. Second, if the split outcome is unknown for some attribute values, the instances with missing values of the tested attribute have to be assigned to each possible split outcome, with weights equal to their "copy counts" multiplied with the corresponding outcome probability. The dispersion corresponding to outcome r of split t for node $\mathbf{n}$ is therefore calculated using instances $x \in T_{\mathbf{n},t=r}$ with weights $w_{x,\mathbf{n}}$ and instances $x \in T_{\mathbf{n},t=?}$ with weights $P(t = r|\mathbf{n})w_{x,\mathbf{n}}$.

Split application with fractional instances Since the split application operation for regression trees does not differ from the corresponding operation for decision trees, there is no regression-specific processing required when performing this operation with fractional instances. An instance with a missing value of the split attribute is dispatched along all branches, corresponding to all split outcomes, with "copy counts" modified by calculating the corresponding split outcome probabilities.

Example 9.7.1 The previously presented regression tree growing implementation took a very primitive approach to missing value handling, considering the split condition unsatisfied for instances with missing values (i.e., always assigning such instances to the branch corresponding to the false split outcome). The following R code contains a modified implementation that includes missing value support by the technique of fractional instances. Note the following major changes:

- the mapping of instances to nodes is no longer represented by a simple vector, but by a matrix containing the `instance`, `node`, and `weight` columns, to make it possible to have (fractions of) a single instance assigned to multiple nodes,

- dispersion calculation takes instance weights into account and is performed using the `weighted.var1` function,

Ex. 2.4.11
dmr.stats

- the `split.eval` and `split.apply` functions perform instance fractionization.

The `grow.regtree.frac` is applied to modified versions of the *weatherr* dataset, with some attribute values removed.

```
## a simple regression tree growing implementation
grow.regtree.frac <- function(formula, data, minvar=0.005, minsplit=2, maxdepth=8)
{
  nmn <- function(n) { nodemap[,"node"]==n }            # nodemap entries for node n
  inn <- function(n)
  { nodemap[nodemap[,"node"]==n,"instance"] }                  # instances at node n
  wgn <- function(n) { nodemap[nodemap[,"node"]==n,"weight"] }   # weights at node n

  init <- function()
  {
    tree <<- data.frame(node=1, attribute=NA, value=NA, target=NA,
                        count=NA, mean=NA, variance=NA)
    nodemap <<- cbind(instance=1:nrow(data), node=rep(1, nrow(data)),
                      weight=rep(1, nrow(data)))
    n <<- 1
  }

  next.node <- function(n)
```

```
{
  if (any(opn <- tree$node>n))
    min(tree$node[opn])
  else Inf
}

target.summary <- function(n)
{
  tree$count[tree$node==n] <<- sum(wgn(n))
  tree$mean[tree$node==n] <<- weighted.mean(data[inn(n),target], wgn(n))
  tree$variance[tree$node==n] <<- weighted.var1(data[inn(n),target], wgn(n))
}

target.value <- function(n)
{
  tree$target[tree$node==n] <<- tree$mean[tree$node==n]
}

stop.criteria <- function(n)
{
  n>=2^maxdepth || tree$count[tree$node==n]<minsplit ||
    tree$variance[tree$node==n]<minvar
}

split.eval <- function(av, sv, tv, w)
{
  cond <- if (is.numeric(av)) av<=as.numeric(sv) else av==sv
  cond1 <- !is.na(av) & cond   # true split outcome
  cond0 <- !is.na(av) & !cond  # false split outcome

  v1 <- tv[cond1]
  n1 <- sum(w[cond1])
  w1 <- w[cond1]
  v0 <- tv[cond0]
  n0 <- sum(w[cond0])
  w0 <- w[cond0]
  vm <- tv[is.na(av)]
  nm <- sum(w[is.na(av)])
  wm <- w[is.na(av)]

  if (nm>0)
  {
    p1 <- if (n1+n0>0) n1/(n1+n0) else 0.5
    p0 <- 1-p1
    v1 <- c(v1, vm)
    w1 <- c(w1, p1*wm)
    v0 <- c(v0, vm)
    w0 <- c(w0, p0*wm)
  }

  if (n1>0 && n0>0)
    weighted.dispersion(v1, v0, w1, w0, disp=weighted.var1)
  else
    Inf
}

split.select <- function(n)
{
  splits <- data.frame()
  for (attribute in attributes)
  {
```

```
      uav <- sort(unique(data[inn(n),attribute]))
      if (length(uav)>1)
        splits <- rbind(splits,
                      data.frame(attribute=attribute,
                                 value=if (is.numeric(uav))
                                          midbrk(uav)
                                       else as.character(uav),
                                 stringsAsFactors=FALSE))
    }

    if (nrow(splits)>0)
      splits$eval <- sapply(1:nrow(splits),
                            function(s)
                            split.eval(data[inn(n),splits$attribute[s]],
                                       splits$value[s],
                                       data[inn(n),target], wgn(n)))
    if ((best.eval <- min(splits$eval))<Inf)
      tree[tree$node==n,2:3] <<- splits[which.min(splits$eval),1:2]
    best.eval
}

split.apply <- function(n)
{
  tree <<- rbind(tree,
                 data.frame(node=(2*n):(2*n+1), attribute=NA, value=NA, target=NA,
                            count=NA, mean=NA, variance=NA))

  av <- data[nodemap[,"instance"],tree$attribute[tree$node==n]]
  cond <- if (is.numeric(av)) av<=as.numeric(tree$value[tree$node==n])
          else av==tree$value[tree$node==n]
  cond1 <- !is.na(av) & cond   # true split outcome
  cond0 <- !is.na(av) & !cond  # false split outcome

  n1 <- sum(nodemap[nmn(n) & cond1,"weight"])
  n0 <- sum(nodemap[nmn(n) & cond0,"weight"])
  nm <- sum(nodemap[nmn(n) & is.na(av),"weight"])

  nodemap[nmn(n) & cond1,"node"] <<- 2*n
  nodemap[nmn(n) & cond0,"node"] <<- 2*n+1

  if (nm>0)
  {
    p1 <- if (n1+n0>0) n1/(n1+n0) else 0.5
    p0 <- 1-p1
    newnn <- nodemap[nmn(n) & is.na(av),,drop=FALSE]
    nodemap[nmn(n) & is.na(av),"weight"] <<-
      p1*nodemap[nmn(n) & is.na(av),"weight"]
    nodemap[nmn(n) & is.na(av),"node"] <<- 2*n
    newnn[,"weight"] <- p0*newnn[,"weight"]
    newnn[,"node"] <- 2*n+1
    nodemap <<- rbind(nodemap, newnn)
  }
}

tree <- nodemap <- n <- NULL
target <- y.var(formula)
attributes <- x.vars(formula, data)

init()
while (is.finite(n))
{
```

```
    target.summary(n)
    target.value(n)
    if (!stop.criteria(n))
      if (split.select(n)<Inf)
        split.apply(n)
    n <- next.node(n)
  }
  `class<-`(tree, "regtree.frac")
}

## convert a regtree.frac object to a data frame
as.data.frame.regtree.frac <- function(x, row.names=NULL, optional=FALSE, ...)
{ as.data.frame(unclass(x), row.names=row.names, optional=optional) }

  # grow a regression tree for the weatherr data with missing attribute values
weatherrm <- weatherr
weatherrm$outlook[1] <- NA
weatherrm$humidity[1:2] <- NA
treem <- grow.regtree.frac(playability~., weatherrm)

  # data frame conversion
as.data.frame(treem)
```

9.7.1.2   Prediction with fractional instances

When a regression tree is applied to an instance with missing attribute values, split application is performed as during growing until all instances reach leaves. It is important to underline that split outcome probabilities required for instance fractionization are the same as those estimated during decision tree growing, based on the training set.

Each of fractional instances replacing an original instance with missing values will clearly end up in a different leaf. To make the final prediction, a weighted averaging mechanism has to be used, with weights assigned based on instance fractions. More exactly, we can use the fraction of x arriving to l, denoted by $w_{x,l}$ to weight the leaf's predicted target function value and average over all leaves.

$$h(x) = \frac{\sum_l v_l w_{x,l}}{\sum_l w_{x,l}} \tag{9.15}$$

where the summation runs over all regression tree leaves.

Example 9.7.2 The R code presented below implements regression tree prediction with fractional instances. The internal `descend` function is a minor modification of the `split.apply` function used for growing.

```
## regression tree prediction
## with missing value support using fractional instances
predict.regtree.frac <- function(tree, data)
{
  nmn <- function(n) { nodemap[,"node"]==n }  # nodemap entries for node n

  descend <- function(n)
```

```
{
    if (!is.na(tree$attribute[tree$node==n]))   # unless reached a leaf
    {
        av <- data[nodemap[,"instance"],tree$attribute[tree$node==n]]
        cond <- if (is.numeric(av)) av<=as.numeric(tree$value[tree$node==n])
                else av==tree$value[tree$node==n]
        cond1 <- !is.na(av) & cond    # true split outcome
        cond0 <- !is.na(av) & !cond   # false split outcome

        nodemap[nmn(n) & cond1, "node"] <<- 2*n
        nodemap[nmn(n) & cond0, "node"] <<- 2*n+1

        if (sum(nodemap[nmn(n) & is.na(av), "weight"])>0)
        {
            n1 <- tree$count[tree$node==2*n]
            n0 <- tree$count[tree$node==2*n+1]
            p1 <- if (n1+n0>0) n1/(n1+n0) else 0.5
            p0 <- 1-p1

            newnn <- nodemap[nmn(n) & is.na(av),,drop=FALSE]
            nodemap[nmn(n) & is.na(av),"weight"] <<-
                p1*nodemap[nmn(n) & is.na(av),"weight"]
            nodemap[nmn(n) & is.na(av), "node"] <<- 2*n
            newnn[,"weight"] <- p0*newnn[,"weight"]
            newnn[,"node"] <- 2*n+1
            nodemap <<- rbind(nodemap, newnn)
        }

        descend(2*n)
        descend(2*n+1)
    }
}

nodemap <- cbind(instance=1:nrow(data), node=rep(1, nrow(data)),
                 weight=rep(1, nrow(data)))
descend(1)

votes <- merge(nodemap, as.data.frame(tree)[,c("node", "target")])
as.numeric(by(votes, votes$instance,
            function(v) weighted.mean(v$target, v$weight)))
}

# regression tree prediction for the weatherr data with missing attribute values
predict(treem, weatherrm)
```

9.7.2 Surrogate splits

The technique of surrogate splits assumes that a split on a missing attribute value can be replaced by a surrogate using another attribute. It can be applied to regression trees in exactly the same way as presented in Section 3.7.2 for decision trees, since it does not depend on the type of the target attribute in any way.

9.8 Piecewise linear regression

As presented so far, regression trees represent piecewise-constant regression models that decompose the domain into a number of regions and assign a fixed target function value to each region. Whereas such models can be made arbitrarily accurate on the training set by adding as many nodes as required, the resulting true performance will usually not improve above a certain level and is actually likely to severely deteriorate due to extreme overfitting. This will happen if the regions to which domain is decomposed to ensure sufficiently good training set performance become too small to yield good predictions on new data.

The inherent limitations of piecewise-constant regression trees can be overcome by an enhanced form of regression trees that replace fixed target function values with simple linear regression models. This makes the resulting piecewise-linear models capable of delivering much better accuracy, as well as providing a more smooth approximation of the target function. This form of regression trees, to avoid confusion with plain piecewise-constant regression trees, is sometimes called *model trees*. In the remainder of this section the term "regression trees" will therefore be reserved for the basic piecewise-constant version of regression tree models, unless explicitly indicated otherwise, and the enhanced piecewise-linear version will be referred to as "model trees." The following subsections review the key changes in the model creation and prediction processes that differentiate model trees from regression trees. Refer to Sections 8.2 and 8.3 for details on linear model representation and parameter estimation.

9.8.1   Growing

The general top-down growing algorithm presented for regression trees can be applied to model trees with just a single change: instead of fixed target function values linear models are used. This accounts for the primary difference between regression trees and model trees.

9.8.1.1   Linear models at nodes

Just like with target function values for regression trees, for model trees it makes sense to assign linear models not only to leaves (where they are strictly necessary), but to internal nodes as well. This makes the model tree ready for pruning – any node can be immediately replaced by a leaf without any additional calculations. An additional and actually more important reason to create linear models for internal nodes is to enable a smoothing process that may partially reduce the discontinuity of predictions, which – whereas usually not as severe as for regression trees – is anyway unavoidable for a model representation that combines multiple models corresponding to domain regions.

The linear model assigned to a leaf or node $\mathbf{n}$, designated as $h_\mathbf{n}$, is created based on the corresponding subset of training instances $T_\mathbf{n}$ using standard linear regression parameter estimation methods. The least-squares algorithm presented in Section 8.3.3 is definitely preferred for its efficiency, unless incremental model building is required.

9.8.1.2   Attribute preselection

Since linear models are created using subsets of the training set, which become smaller and smaller as descending farther from the root node, it may also be necessary to restrict the set of attributes used for these models to minimize the risk of overfitting. Natural candidates for attributes to skip are those already used for splitting on the path from the root node to the current node, since their impact on model tree predictions has been already captured. This attribute preselection scheme appears to work well in many cases. A more restrictive form thereof that may be adopted – to skip any attributes not used in the subtree rooted at the current node – may be as good or better with just the following two caveats:

- linear models for nodes cannot be created until the tree is fully grown, which is necessary to determine the sets of attributes to use at particular nodes,

- the adopted stop criteria should not be too liberal (i.e., causing premature leaf creation), since linear models for leaves actually degenerate to simple mean target function values, just like for plain regression trees (when no attributes can be used, a linear model is represented by a single intercept parameter).

The last remark clearly applies to the leaves of an unpruned model tree only, since leaves obtained after pruning inherit linear models from internal nodes that have been pruned off.

9.8.1.3   Stop criteria

The "no instances left" and "no splits left" criteria presented above for regression trees are obviously no less applicable to model trees, but they can be thought of as secondary split criteria, applied when no further splitting is possible. A primary stop criterion is still needed to indicate of when no further splits are required or desirable (although possible). The "sufficiently low dispersion" criterion that satisfies this requirement for plain regression trees is usually employed for model trees as well, particularly in combination with the more restrictive attribute preselection scheme discussed above (which permits linear models to include only attributes that are used for splitting in the subtree below the current node).

When adopting no or a less restrictive attribute preselection scheme (like the other proposed above that only skips attributes already used above the current node) and building a linear model for each newly created node immediately, an alternative stop criterion can be considered based on its observed performance:

Sufficiently good performance. The linear model created for the current node reaches a sufficiently good training performance.

This results in no further splitting attempted for a node in which the target function can be sufficiently well approximated linearly.

Different instantiations of this criterion are obtained depending on the adopted performance measure, with the mean square error or the relative square error (or, equivalently, the coefficient of determination) being the most typical choices. Clearly, to check this stop criterion for node $\mathbf{n}$ the training performance of the corresponding linear model $h_\mathbf{n}$ should be measured on the set of training instances associated with this node.

9.8.1.4   Split selection

Using the weighted target function dispersion as a split evaluation measure is an obvious choice for piecewise-constant regression trees, where a low dispersion is required to reasonably approximate the target function with constant values assigned to particular leaves. It is not so natural for model trees, though, where we are concerned with the quality of a linear rather than constant approximation. Actually, it is quite easy to see that perfectly linearly predictable values do not necessarily have to exhibit low dispersion. It would not therefore be unreasonable to replace the dispersion-based split quality measure with another measure based on the performance of linear models that could be created for the subsets of instances obtained after applying the evaluated split.

This idea, whereas well justified and consistent with the principle of model trees, is hardly applicable in practice due to the computational expense of creating linear models for each outcome of each candidate split in each node. Practical model tree growing algorithms fall back to simple and imperfect, but computationally efficient heuristics. Of those, the weighted target function dispersion, as used for plain regression trees, remains the most popular choice.

Example 9.8.1 The following R code provides a simplified implementation of the model tree growing process by defining the `grow.modtree` function, which is a modified version of the `grow.regtree` function presented before for regression trees. The target function value is dropped from the tree structure and a separate model list is maintained, containing linear models created for all tree nodes. Accordingly, the `model` internal function comes in place of the `target.value` function used previously, responsible for creating a linear model for the currently processed node using the R standard `lm` function. It uses the `make.formula` function to create a formula based on a given target function name and attribute names. A simplified attribute preselection scheme is used that drops all attributes with just a single unique value in the current subset of training instances. This is accomplished using the `drop1val` function. Other than that, the function matches its regression tree prototype precisely, including the very same stop and split selection criteria. The function returns a list with two named components, `structure` – the tree structure data frame – and `models` – the linear models list, with the `class` attribute set to `modtree`.

`dmr.util`

`dmr.util`

```
## a simple model tree growing implementation
grow.modtree <- function(formula, data, minvar=0.005, minsplit=2, maxdepth=8)
{
  init <- function()
  {
    tree <<- data.frame(node=1, attribute=NA, value=NA,
                        count=NA, mean=NA, variance=NA)
    models <<- list()
    nodemap <<- rep(1, nrow(data))
    n <<- 1
  }

  next.node <- function(n)
  {
    if (any(opn <- tree$node>n))
      min(tree$node[opn])
    else Inf
  }

  target.summary <- function(n)
  {
    tree$count[tree$node==n] <<- sum(nodemap==n)
    tree$mean[tree$node==n] <<- mean(data[[target]][nodemap==n])
    tree$variance[tree$node==n] <<- var1(data[[target]][nodemap==n])
  }

  model <- function(n)
  {
    attrs <- drop1val(attributes, data[nodemap==n,])
    models <<- c(models,
                 list(lm(make.formula(target, if (length(attrs)==0) 1 else attrs),
```

```
                           data[nodemap==n,])))
}

stop.criteria <- function(n)
{
  n>=2^maxdepth || tree$count[tree$node==n]<minsplit ||
    tree$variance[tree$node==n]<minvar
}

split.eval <- function(av, sv, tv)
{
  cond <- !is.na(av) & (if (is.numeric(av)) av<=as.numeric(sv) else av==sv)
  v1 <- tv[cond]
  n1 <- sum(cond)
  v0 <- tv[!cond]
  n0 <- sum(!cond)
  if (n1>0 && n0>0)
    weighted.dispersion(v1, v0)
  else
    Inf
}

split.select <- function(n)
{
  splits <- data.frame()
  for (attribute in attributes)
  {
    uav <- sort(unique(data[nodemap==n,attribute]))
    if (length(uav)>1)
      splits <- rbind(splits,
                      data.frame(attribute=attribute,
                                 value=if (is.numeric(uav))
                                         midbrk(uav)
                                       else as.character(uav),
                                 stringsAsFactors=FALSE))
  }

  if (nrow(splits)>0)
    splits$eval <- sapply(1:nrow(splits),
                          function(s)
                          split.eval(data[nodemap==n,splits$attribute[s]],
                                     splits$value[s],
                                     data[nodemap==n,target]))
  if ((best.eval <- min(splits$eval))<Inf)
    tree[tree$node==n,2:3] <<- splits[which.min(splits$eval),1:2]
  best.eval
}

split.apply <- function(n)
{
  tree <<- rbind(tree,
                 data.frame(node=(2*n):(2*n+1), attribute=NA, value=NA,
                            count=NA, mean=NA, variance=NA))

  av <- data[[tree$attribute[tree$node==n]]]
  cond <- !is.na(av) & (if (is.numeric(av))
                          av<=as.numeric(tree$value[tree$node==n])
                        else av==tree$value[tree$node==n])
  nodemap[nodemap==n & cond] <<- 2*n
  nodemap[nodemap==n & !cond] <<- 2*n+1
}
```

```
tree <- models <- nodemap <- n <- NULL
target <- y.var(formula)
attributes <- x.vars(formula, data)

init()
while (is.finite(n))
{
  target.summary(n)
  model(n)
  if (! stop.criteria(n))
    if (split.select(n)<Inf)
      split.apply(n)
  n <- next.node(n)
}
'class<-'(list(structure=tree, models=models), "modtree")
}

# grow a model tree for the weatherr data
mtree <- grow.modtree(playability~., weatherr)
# tree structure
mtree$structure
```

9.8.2 Pruning

As for piecewise-constant regression trees, pruning is the process of cutting off overgrown subtrees that are likely to overfit the training set to hopefully improve the model's true performance. The previously discussed regression tree pruning techniques are therefore fully applicable to model trees as well. A node that is found useless is then replaced by a leaf with the same linear model. This is the model tree interpretation of the subtree cutoff pruning operator, the only pruning operator practically used. It is worthwhile to mention, though, one model tree-specific approach to pruning that can not only replace apparently useless nodes with leaves, but also simplify linear models in those nodes and leaves that remain.

The technique combines the ideas of pessimistic and cost-complexity pruning by using a pruning criterion that compares the performance of a node and its replacement leaf on the training set. It compensates for the inherent optimistic bias of such performance estimates by using a pessimistic correction coefficient that also penalizes the complexity of the corresponding linear models. It also resembles the minimum error pruning for decision trees by defining the modified performance measure directly only for leaves and using a bottom-up propagation process to obtain performance estimates for internal nodes.

For each leaf l the following coefficient is calculated:

$$\frac{|T_l| + |h_l|}{|T_l| - |h_l|} \tag{9.16}$$

where $|h_l|$ denotes the number of attributes used by the linear model h_l assigned to leaf l and can be considered a measure of its complexity. The coefficient represents a multiplicative complexity penalty that increases with the increased number of attributes used by the linear model and decreases with the increased number of associated training instances. Its effect is particularly severe if the leaf corresponds to a small subset of training instances, yet uses a complex linear model.

The above penalty coefficient is applied as a multiplier to the leaf's training performance, calculated with whatever performance measure is chosen (with the mean absolute error being the most typical choice for this pruning technique):

$$\hat{e}_T(\mathbf{l}) = \frac{|T_\mathbf{l}| + |h_\mathbf{l}|}{|T_\mathbf{l}| - |h_\mathbf{l}|} e_T(\mathbf{l}) \tag{9.17}$$

where the leaf's training performance $e_T(\mathbf{l})$ is obviously the performance of its linear model $h_\mathbf{l}$ on the associated subset of training instances $T_\mathbf{l}$.

For each node $\mathbf{n}$ the performance estimate $\hat{e}_T(\mathbf{n})$ is obtained as the average of the performance estimates calculated for its descendants, weighted by the numbers of training instances corresponding to them. In effect, the error estimates for leaves calculated as shown above are propagated upward to nodes. The pruning criterion is based on comparing such node performance estimates with those for the corresponding replacement leaves

$$\hat{e}_T(\mathbf{l}) \le \hat{e}_T(\mathbf{n}) \tag{9.18}$$

To see why this may be satisfied, notice that a node's performance estimate obtained by propagating upward the performance estimates of the leaves from its subtree will be affected by the pessimistic correction applied in these leaves to a much greater extent than the performance estimate of its replacement leaf. This is because the number of training instances associated with the replacement leaf, equal to the number of all training instances corresponding to the original subtree, will be usually substantially larger than the number of instances corresponding to each of its individual leaves. The effect of the penalty coefficient for the replacement leaf is therefore likely to be much less than for the original leaves (and the original node, accordingly).

9.8.3   Prediction

In the most basic case, model tree prediction can be performed by down-propagating each instance until it arrives at a leaf, and applying the leaf's linear model. While there is nothing wrong with this approach, which works as expected, there is a more refined and possibly better (although more costly) alternative.

Linear models created for internal model tree nodes are useful not only for pruning. They can be useful even for unpruned tree nodes to reduce the discontinuity of predictions generated by model trees, which occurs whenever a minor change of an attribute's value makes an instance land in another leaf, yielding a substantial change of the model's prediction. This often undesirable phenomenon can be alleviated by a smoothing process that combines the predictions of all linear models encountered on the path on which an instance descends from the root node to a leaf.

Consider using a model tree to generate a prediction for instance x that arrives to leaf $\mathbf{l}$ by traversing the sequence of nodes $\mathbf{n}_1, \mathbf{n}_2, \ldots, \mathbf{n}_k$, where $\mathbf{n}_1$ is the root node and $\mathbf{n}_k = \mathbf{l}$. The linear models assigned to all these nodes, $h_{\mathbf{n}_1}, h_{\mathbf{n}_2}, \ldots, h_{\mathbf{n}_k}$, contribute to the final prediction as defined by the following recursive formula:

$$h_k(x) = h_{\mathbf{n}_k}(x) \tag{9.19}$$

$$h_i(x) = \frac{|T_{\mathbf{n}_{i+1}}| h_{i+1}(x) + m h_{\mathbf{n}_i}(x)}{|T_{\mathbf{n}_{i+1}}| + m} \tag{9.20}$$

for $i = k - 1, k - 2, \ldots, 1$, where m is a parameter that adjusts the degree of smoothing. This propagates the prediction from the leaf up to the root node, for each consecutive node averaging the value calculated below this node (weighted by the number of training instances associated with the corresponding descendant node) with this node's prediction (weighted by m). The model tree's final smoothed prediction $h(x) = h_1(x)$ is the value obtained from this propagation process in the root node. Of course, the smoothing process requires that training instance counts be stored for each node, which can be easily taken care of at the time of growing the tree.

Example 9.8.2 The R code presented below implements the prediction operation for model trees, including the smoothing process described above. It is based on the same repeated split application scheme as regression tree prediction, therefore sharing some code portions, but with additionally updating predictions when visiting each node. The resulting `predict.modtree` function is then applied to generate training set predictions using the previously grown model tree for the *weatherr* data.

```
## model tree prediction
predict.modtree <- function(tree, data, m=10)
{
  descend <- function(n)
  {
    predn <- predict(models[[which(tree$node==n)]], data[nodemap==n,])
    if (is.na(tree$attribute[tree$node==n]))  # reached a leaf
      pred[nodemap==n] <<- predn
    else
    {
      av <- data[[tree$attribute[tree$node==n]]]
      cond <- !is.na(av) & (if (is.numeric(av))
                              av<=as.numeric(tree$value[tree$node==n])
                            else av==tree$value[tree$node==n])
      nodemap[left <- nodemap==n & cond] <<- 2*n
      nodemap[right <- nodemap==n & !cond] <<- 2*n+1
      descend(2*n)
      descend(2*n+1)

      leftn <- match(which(left), which(left|right))
      rightn <- match(which(right), which(left|right))
      pred[left] <<- (tree$count[tree$node==2*n]*pred[left] + m*predn[leftn])/
                       (tree$count[tree$node==2*n]+m)
      pred[right] <<- (tree$count[tree$node==2*n+1]*pred[right] + m*predn[rightn])/
                       (tree$count[tree$node==2*n+1]+m)
    }
  }

  models <- tree$models
  tree <- tree$structure
  nodemap <- rep(1, nrow(data))
  pred <- rep(NA, nrow(data))
  descend(1)
  pred
}

# model tree prediction for the weatherr data
predict(mtree, weatherr, m=2)
```

9.9   Conclusion

The biggest power of regression trees lies in the interpretability of this model representation, which – just like decision trees – can be inspected to see what attributes are used to make predictions and what conditions they are tested for. This may be important in some applications where a model – to be considered reliable – has not only to be proved successful in evaluation, but also understood. It makes it also possible to create models in a semi-automatic, human-assisted mode, where all or selected growing and pruning steps are subject to human verification and possibly modification (e.g., an apparently slightly worse split could be selected if it has a better interpretation based on the available domain knowledge or the analyst is more confident about its true predictive utility).

Accuracy-wise, regression trees do not always belong to the best regression models that can be created. Particularly, the basic piecewise-constant trees provide a simplified representation of complex target functions and are prone to overfitting when attempting to reach good training performance levels, which limits their true performance potential. Their piecewise-linear version – model trees – tend to be better in this respect and can reduce the often undesirable prediction discontinuity (stepped representation effect), but to achieve this, they sacrifice human readability to some extent, since their predictions are no longer so easy to explain.

Regression trees are not nearly as commonly used as decision trees. Whereas the latter belong to the most successful and widespread approaches to the classification task, the former often give ground to parametric regression. This is because the advantages of human-readable model representation are not always as important for regression applications as they tend to be for classification applications and can be outweighed by their disadvantages, but also largely due to historical reasons. Regression trees definitely deserve more interest and wider popularity than they have gained so far, particularly in the more refined piecewise linear version, which retains at least some human readability with a much better accuracy potential.

9.10   Further readings

Regression trees are by far less frequently used and, correspondingly, much less frequently described in the literature than decision trees. However, they have so much in common to make many decision tree references presented in Chapter 3 useful regression tree readings as well. Regression trees or model trees as such are covered by some of them as well (e.g., Witten *et al.* 2011).

The capability to create regression models is a standard feature of the *CART* (Breiman *et al.* 1984). It represents the basic piecewise-constant form of regression trees. The idea of model trees – with linear models rather than constant values in leaves – was introduced by Quinlan (1992) in his *M5* algorithm, adding the numerical prediction capability to the *ID3*, *C4*, and *C4.5* series of algorithms (Quinlan 1986, 1993). It was further refined by Wang and Witten (1996). Their *M5'* algorithm includes, in particular, missing value handling by surrogate splits.

Several other algorithms for regression tree and model tree induction have been considered. One dimension across which they differ is the representation of leaf predictions (Torgo 1997). A more common differentiating factor is the split selection method. Karalič (1992) argued the dispersion-based split selection criterion not to be well suited to leaves with linear models and proposed an alternative criterion based on the mean square error of linear models that can be created after the split. It is unfortunately more expensive to calculate. Dobra and Gehrke (2002) demonstrated how split selection can be made more efficient by local clustering at the node and then selecting a split that best separates the clusters. Split evaluation with respect to the relationship between the split and the sign of constant or linear model residuals at the node, measured using the χ^2 statistic, was proposed by Loh (2002). The algorithm presented by Lubinsky (1994) evaluates the predictive utility of splits using internal cross-validation and, in a sense, combines the ideas of model and regression trees. This is achieved by adding a new type of nodes to the regular regression tree structure that have single-attribute linear regression models associated with them. A tree path can then be interpreted as representing stepwise linear models, with terms corresponding to regression nodes occurring on the path. A more refined version of this approach was later developed by Malerba et al. (2004). Landwehr et al. (2005) backported the idea of model trees to the context of classification, by proposing a model tree with logistic regression models in leaves.

Regression and model tree pruning has not received as much attention as the corresponding operation for decision trees, with most approaches used for the former being adapted from the latter. In particular, Karalič and Cestnik (1991) proposed the technique of mean square error m-estimation for regression tree pruning. Torgo (1998) discussed and experimentally evaluated regression tree pruning criteria based on this and other error estimators. Robnik-Šikonja and Kononenko (1998) presented a minimum description length-based regression tree pruning algorithm. Pruning and grafting for model trees has been addressed by Ceci et al. (2003) in the context of the stepwise model tree algorithm of Malerba et al. (2004).

Regression and model trees can be considered instantiations of a more general idea of using a classification model to decompose the domain into "pieces" for piecewise-constant or piecewise-linear regression (Cichosz 2007). Weiss and Indurkhya (1995) used this approach in combination with decision rules. Torgo and Gama (1997) proposed a general framework for solving the regression task using classification algorithms.

References

Breiman L, Friedman JH, Olshen RA and Stone CJ 1984 *Classification and Regression Trees*. Chapman and Hall.

Ceci M, Appice A and Malerba D 2003 Comparing simplification methods for model trees with regression and splitting nodes *Proceedings of the Fourteenth International Symposium on Methodologies for Intelligent Systems (ISMIS-03)*. Springer.

Cichosz P 2007 *Learning Systems (in Polish)* 2nd edn. WNT.

Dobra A and Gehrke JE 2002 SECRET: A scalable linear regression tree algorithm *Proceedings of the Eighth ACM SIGKDD International Conference on Knowledge Discovery and Data Mining*. ACM Press.

Karalič A 1992 Employing linear regression in regression tree leaves *Proceedings of the Tenth European Conference on Artificial Intelligence (ECAI-92)*. Wiley.

Karalič A and Cestnik B 1991 The Bayesian approach to tree-structured regression *Proceedings of the Thirteenth International Conference on Information Technology Interfaces (ITI-91)*. University Computing Centre, University of Zagreb, Croatia.

Landwehr N, Hall M and Frank E 2005 Logistic model trees. *Machine Learning* **59**, 161–205.

Loh WY 2002 Regression trees with unbiased variable selection and interaction detection. *Statistica Sinica* **12**, 361–386.

Lubinsky D 1994 Tree structured interpretable regression In *Learning from Data: Artificial Intelligence and Statistics V* (ed. Fisher D and Lenz HJ) Springer.

Malerba D, Esposito F and Ceci M 2004 Top-down induction of model trees with regression and splitting nodes. *IEEE Transactions on Pattern Analysis and Machine Intelligence* **26**, 612–625.

Quinlan JR 1986 Induction of decision trees. *Machine Learning* **1**, 81–106.

Quinlan JR 1992 Learning with continuous classes *Proceedings of the Fifth Australian Joint Conference on Artificial Intelligence*. World Scientific.

Quinlan JR 1993 *C4.5: Programs for Machine Learning*. Morgan Kaufmann.

Robnik-Šikonja M and Kononenko I 1998 Pruning regression trees with MDL *Proceedings of the Thirteenth European Conference on Artificial Intelligence (ECAI-98)*. Wiley.

Torgo L 1997 Functional models for regression tree leaves *Proceedings of the Fourteenth International Conference on Machine Learning (ICML-97)*. Morgan Kaufmann.

Torgo L 1998 Error estimators for pruning regression trees *Proceeding of the Tenth European Conference on Machine Learning (ECML-98)*. Springer.

Torgo L and Gama J 1997 Regression using classification algorithms. *Intelligent Data Analysis* **1**, 275–292.

Wang Y and Witten IH 1996 Induction of model trees for predicting continuous classes. Technical Report 96/23, University of Waikato, Department of Computer Science.

Weiss SM and Indurkhya N 1995 Rule-based machine learning methods for functional prediction. *Journal of Artificial Intelligence Research* **3**, 383–403.

Witten IH, Frank E and Hall MA 2011 *Data Mining: Practical Machine Learning Tools and Techniques* 3rd edn. Morgan Kaufmann.

10

Regression model evaluation

10.1 Introduction

Just like for classification model evaluation addressed in Chapter 7, the evaluation of a regression model is intended to provide a reliable assessment of its *predictive performance*, i.e., the quality of the target function's approximation it represents. There are several regression performance measures calculated by comparing the model's predictions and true target function values on a particular dataset. These are not only the direct indicators of *dataset performance*, but – under some conditions – can also serve as estimators of *true performance*, i.e., their expected values on the whole domain.

10.1.1 Dataset performance

Dataset performance, obtained by calculating one or more selected performance measures on a particular dataset, represents the degree of match between model predictions and target function values on this dataset.

10.1.2 Training performance

Performance measures calculated for a model on the training set used to create the model represent its training performance. It may be useful for diagnostic purposes, but does not provide information on the actual predictive utility of the model.

10.1.3 True performance

The actual predictive power of a model is reflected by its expected performance (with respect to one or more selected performance measures) on the whole domain, which is the model's true performance. Since target function values are generally unavailable, true performance always remains unknown and has to be estimated by dataset performance. The challenge of reliably estimating the unknown values of the adopted performance measures on the whole domain is addressed by the same *evaluation procedures* as presented in Section 7.3 for classification models and will therefore be discussed only very briefly. Regression model performance measures, however, have to take into account the specificity of numeric predictions and will be presented more extensively.

Example 10.1.1 The regression model performance measures and evaluation procedures presented in this chapter will be illustrated in R by applying them to the evaluation of a regression tree model created using the `rpart` package for the *Boston Housing* dataset, available in the `mlbench` package. The following R code prepares the demonstration by loading the packages and the dataset, splitting the dataset randomly into training and test subsets, and creating a model based on the training set. The random generator seed is explicitly initialized to make the results which are presented for some of the forthcoming examples easily reproducible.

```
library(dmr.claseval)
library(rpart)

data(BostonHousing, package="mlbench")

set.seed(12)
rbh <- runif(nrow(BostonHousing))
bh.train <- BostonHousing[rbh>=0.33,]
bh.test <- BostonHousing[rbh<0.33,]

bh.tree <- rpart(medv~., bh.train)
```

10.2   Performance measures

Regression performance measures share the same basic principle with classifcation performance measures: they compare the predictions generated by the model on a dataset S with the true target function values for the instances from this dataset. What has to be different, due to the specificity of the regression task, is the exact way of making the comparison.

Similarly as for classification, some of regression performance measures serve as implicit or explicit optimization criteria for regression algorithms and, in this role, they are sometimes referred to as *loss functions*.

10.2.1   Residuals

The most common regression performance measures are based on the differences between true and predicted function values. Such differences are called model *residuals*. For any instance x, the difference $f(x) - h(x)$ is the model's residual for instance x.

Apart from being used for calculating performance measures, model residuals are often analyzed statistically using distribution description and visualization techniques such as presented in Section 2.4. In particular, one simple and commonly used visual tool for residual analysis is a *residual plot*, i.e., a plot of model residuals vs. true target function values.

Example 10.2.1 The following R code produces the distribution summary of the test set residuals (as well as their absolute values) of the regression tree created for the *Boston Housing* data, as well as produces a boxplot, a histogram, and a residual plot thereof, using a simple function that calculates model residuals by subtracting model predictions from the true target function values.

```
res <- function(pred.y, true.y) { true.y-pred.y }

bh.res <- res(predict(bh.tree, bh.test), bh.test$medv)
summary(bh.res)
summary(abs(bh.res))

boxplot(bh.res, main="Residual boxplot")
hist(bh.res, main="Residual histogram")
plot(bh.test$medv, bh.res, main="Residual plot")
```

The plots are presented in Figure 10.1. Although more than a half of residuals are in the $(-3, 3)$ interval, and the absolute values of more than 75% of them do not exceed 4.5, there are some much larger residuals as well. The piecewise-constant regression tree model representation makes the residual plot contain multiple linear segments.

10.2.2   Mean absolute error

Of several types of residual-based performance measures, the *mean absolute error* (MAE) is the most straightforward. It is calculated for model $h : X \rightarrow \mathcal{R}$ with respect to target function $f : X \rightarrow \mathcal{R}$ on dataset $S \subset X$ as the mean absolute residual for instances from S:

$$\mathrm{mae}_{f,S}(h) = \frac{1}{|S|} \sum_{x \in S} |f(x) - h(x)| \tag{10.1}$$

This makes the contribution of each residual proportional to its absolute value.

The mean absolute error is also referred to as the *absolute loss*.

Example 10.2.2 The following R code defines a function for calculating the mean absolute error for given vectors of predicted and true target function values and demonstrates its application for the *Boston Housing* dataset.

```
mae <- function(pred.y, true.y) { mean(abs(true.y-pred.y)) }

mae(predict(bh.tree, bh.test), bh.test$medv)
```

10.2.3   Mean square error

The most widely employed performance measure for regression models is the *mean square error* (MSE). For model $h : X \rightarrow \mathcal{R}$ and target function $f : X \rightarrow \mathcal{R}$ it is calculated on dataset $S \subset X$ as the model's mean squared residual on this dataset:

$$\mathrm{mse}_{f,S}(h) = \frac{1}{|S|} \sum_{x \in S} (f(x) - h(x))^2 \tag{10.2}$$

Compared to the mean absolute error, the mean square error more severely "punishes" large residuals. A model with a small number of large residuals and a large number of small residuals might appear good according to the mean absolute error, but poor according to the

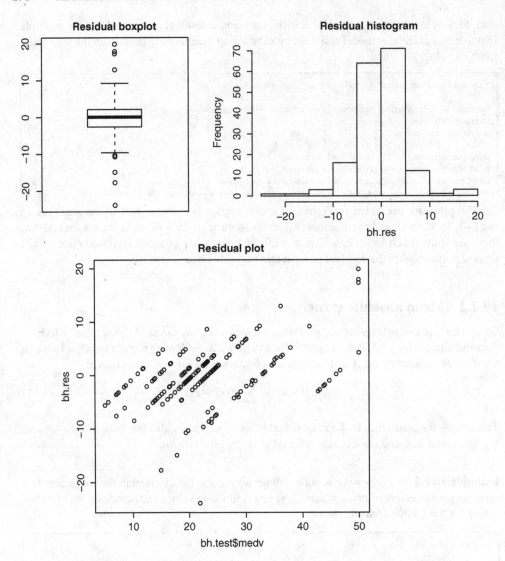

Figure 10.1   Visualization of model residuals for the *Boston Housing* data.

mean square error. Since large residuals are generally undesired, this can be considered one reason of the popularity of the mean square error. Another benefit is related to the analytical advantage of the square function over the absolute function, with the former being differentiable. This is important, in particular, for parametric regression, where gradient-based parameter estimation methods are available to minimize the mean square error on the training set, as discussed in Section 8.3.

The mean square error is the most commonly used type of loss function for regression, referred to as the *quadratic loss*.

Example 10.2.3 The following R code implements mean square error calculation and demonstrates it on the *Boston Housing* dataset.

```
mse <- function(pred.y, true.y) { mean((true.y-pred.y)^2) }

mse(predict(bh.tree, bh.test), bh.test$medv)
```

10.2.4   Root mean square error

A minor practical disadvantage of the mean square error is the effect of changed scale due to squaring, which makes the interpretation of error values harder, particularly if the target function represents quantities in some meaningful units of measurement (such as currency or physical units). This easily solved without loosing the benefits of this performance measure by applying the square root, which yields the *root mean square error* (RMSE):

$$\mathrm{rmse}_{f,S}(h) = \sqrt{\mathrm{mse}_{f,S}(h)} \tag{10.3}$$

Due to the strict monotonicity of the root square function the preference for models implied by the root mean square error is exactly the same as that of the mean square error (i.e., any model minimizing one of these measures also minimizes the other). They only differ in interpretation convenience, which makes the root mean square error popular for presenting the quality of obtained regression models in reports, particularly addressed to business-oriented recipients.

Example 10.2.4 The function defined by the following R code calculates the root mean square error. Like in the previous examples, it is demonstrated in application to the *Boston Housing* dataset.

```
rmse <- function(pred.y, true.y) { sqrt(mse(pred.y, true.y)) }

rmse(predict(bh.tree, bh.test), bh.test$medv)
```

Notice that the root mean square error, although expressed in the same scale and units as the mean absolute error calculated in Example 10.2.2, is considerably larger than the latter, which results from a small number of large residuals being more severely punished.

10.2.5   Relative absolute error

The performance measures presented above, based on absolute or squared residuals, can be directly used for comparing different models, but in order to assess a model's practical utility for a given application they have to be accompanied by some description of the target function distribution (observed on the dataset used for the evaluation), making it possible to judge whether model residuals are sufficiently small. On some occasions it may be interesting and useful to relate model residuals to the values or variability of the target function itself directly within a performance measure. This is accomplished, in particular, by the *relative absolute error* (RAE), defined as follows:

$$\mathrm{rae}_{f,S}(h) = \frac{\sum_{x \in S} |f(x) - h(x)|}{\sum_{x \in S} |f(x) - m_S(f)|} = \frac{\mathrm{mae}_{f,S}(h)}{\frac{1}{|S|} \sum_{x \in S} |f(x) - m_S(f)|} \tag{10.4}$$

where $m_S(f)$ is the mean target function value on S.

The relative absolute error indicates how the mean residual relates to the mean deviation of the target function from its mean. The latter can be thought of as the mean absolute error of a trivial mean value prediction model. The relative absolute error should be clearly less than 1 for any reasonable models, and preferably close to 0.

Example 10.2.5 The following R code defines a function for calculating the relative absolute error and demonstrates its application to the *Boston Housing* dataset.

```
rae <- function(pred.y, true.y)
{ mae(pred.y, true.y)/mean(abs(true.y-mean(true.y))) }

rae(predict(bh.tree, bh.test), bh.test$medv)
```

The obtained value of about 0.5 suggests that the quality of the evaluated model leaves somewhat to be desired, but is substantially better than simple mean value prediction.

10.2.6   Coefficient of determination

A more commonly used performance measure that is similarly motivated as the relative absolute error is the *coefficient of determination*, also referred to as R^2 or R-squared. Consider a quantity that relates to the mean square error in the same way as the relative absolute error relates to the mean absolute error. This could be called the *relative square error* (RSE) and defined as

$$\text{rse}_{f,S}(h) = \frac{\sum_{x \in S} (f(x) - h(x))^2}{\sum_{x \in S} (f(x) - m_S(f))^2} = \frac{|S| \text{mse}_{f,S}(h)}{(|S| - 1)s_S^2(f)} \tag{10.5}$$

where $s_S^2(f)$ is the variance of the target function on S. The coefficient of determination is then obtained as 1's complement of the relative square error:

$$R_{f,S}^2(h) = 1 - \frac{\sum_{x \in S} (f(x) - h(x))^2}{\sum_{x \in S} (f(x) - m_f(S))^2} \tag{10.6}$$

This performance measure is typically interpreted as the fraction of the target function's variance explained by the evaluated model. If approaching 1, it indicates a nearly perfect model. Negative values indicate a totally useless model.

Example 10.2.6 The R code presented below defines a function that implements the calculation of the coefficient of determination and demonstrates its application to the *Boston Housing* dataset.

```
r2 <- function(pred.y, true.y)
{ 1 - length(true.y)*mse(pred.y, true.y)/((length(true.y)-1)*var(true.y)) }

r2(predict(bh.tree, bh.test), bh.test$medv)
```

The obtained value shows that the model managed to explain about 70% of the target function's variance, which may be not enough to consider a model fully satisfactory, but it is definitely useful.

10.2.7   Correlation

Residual-based performance measures are not necessarily well suited to some applications of regression models, where even large differences between predicted and true target function values may be acceptable as long as they exhibit roughly the same monotonic behavior with respect to attribute values. In such applications one requires the model's predictions to react to changes in attribute values in a way that most closely mimics the reaction of the target function: whenever the latter changes slightly the former should also change slightly and whenever the latter changes vastly the former should also change vastly, with the direction of the change preserved. This is required, in particular, whenever a regression model is used to support a decision making or optimization process, where the effects of several alternative decisions or solutions need to be predicted so that the most promising choice can be made. The error of predictions may be then of less importance than their utility for distinguishing between good and poor decisions or ordering candidate decisions in the order of preference. In such cases the linear or rank correlation of predicted and true target function values becomes a natural performance measure.

Example 10.2.7 The following R code demonstrates how the model created for the *Boston Housing* dataset can be evaluated using the linear and rank correlation.

```
cor(predict(bh.tree, bh.test), bh.test$medv, method="pearson")
cor(predict(bh.tree, bh.test), bh.test$medv, method="spearman")
```

The model's predictions turn out to correlate with the true target function values on the test set quite well, with both correlation coefficients approaching 0.85.

10.2.8   Weighted performance measures

Similarly as weight-sensitive algorithms use weighted training instances, one can use a set of weighted instances for model evaluation. Assuming a weight w_x is assigned to each $x \in S$, the definitions of the residual-based performance measures presented above can be rewritten as follows:

$$\text{mae}_{f,S,w}(h) = \frac{\sum_{x \in S} w_x |f(x) - h(x)|}{\sum_{x \in S} w_x} \tag{10.7}$$

$$\text{mse}_{f,S,w}(h) = \frac{\sum_{x \in S} w_x (f(x) - h(x))^2}{\sum_{x \in S} w_x} \tag{10.8}$$

$$\text{rmse}_{f,S,w}(h) = \sqrt{\text{mse}_{f,S,w}(h)} \tag{10.9}$$

$$\text{rae}_{f,S,w}(h) = \frac{\sum_{x \in S} w_x |f(x) - h(x)|}{\sum_{x \in S} w_x |f(x) - m_S(f)|} \tag{10.10}$$

$$R^2_{f,S,w}(h) = 1 - \frac{\sum_{x \in S} w_x (f(x) - h(x))^2}{\sum_{x \in S} w_x (f(x) - m_S(f))^2} \tag{10.11}$$

These are weight-sensitive versions of the mean absolute error, the mean square error, the root mean square error, the relative absolute error, and the coefficient of determination, respectively.

Example 10.2.8 The following R code defines modified versions of the functions from the previous examples that optionally accept a weight vector as an additional argument and calculate weighted performance measures. When called without specifying weights, they behave as their weight-insensitive counterparts. The functions are applied to evaluate the performance of the model created for the *Boston Housing* dataset on the test subset, using a weight vector that doubles the importance of instances with the values of the target function above 25.

```
wmae <- function(pred.y, true.y, w=rep(1, length(true.y)))
{ weighted.mean(abs(true.y-pred.y), w) }

wmse <- function(pred.y, true.y, w=rep(1, length(true.y)))
{ weighted.mean((true.y-pred.y)^2, w) }

wrmse <- function(pred.y, true.y, w=rep(1, length(true.y)))
{ sqrt(wmse(pred.y, true.y, w)) }

wrae <- function(pred.y, true.y, w=rep(1, length(true.y)))
{ wmae(pred.y, true.y, w)/weighted.mean(abs(true.y-weighted.mean(true.y, w)), w) }

wr2 <- function(pred.y, true.y, w=rep(1, length(true.y)))
{
  1-weighted.mean((true.y-pred.y)^2, w)/
      weighted.mean((true.y-weighted.mean(true.y, w))^2, w)
}

  # double weight for medv>25
bh.wtest <- ifelse(bh.test$medv>25, 2, 1)

wmae(predict(bh.tree, bh.test), bh.test$medv, bh.wtest)
wmse(predict(bh.tree, bh.test), bh.test$medv, bh.wtest)
wrmse(predict(bh.tree, bh.test), bh.test$medv, bh.wtest)
wrae(predict(bh.tree, bh.test), bh.test$medv, bh.wtest)
wr2(predict(bh.tree, bh.test), bh.test$medv, bh.wtest)
```

10.2.9 Loss functions

The mean absolute error and the mean square error differ only in the function applied to model residuals before averaging them, which is the absolute value function for the former and the quadratic function of the latter. These two are the most common examples of *loss* functions used to measure the performance of regression models.

In general, a loss function is any function $\mathcal{L} : \mathcal{R}^2 \to \mathcal{R}$ that maps a pair consisting of a predicted and true target function value into a real number, representing the associated cost or regret, that should be considered for performance evaluation. The absolute and quadratic loss functions are simply defined as follows:

$$\mathcal{L}_{\|}(f(x), h(x)) = |f(x) - h(x)| \tag{10.12}$$

$$\mathcal{L}_2(f(x), f(x)) = (f(x) - h(x))^2 \tag{10.13}$$

For the model h, target function f, and loss function $\mathcal{L}$ the mean loss on dataset S is then calculated as

$$\text{mls}_{f,S,\mathcal{L}}(h) = \frac{1}{|S|} \sum_{x \in S} \mathcal{L}(f(x), h(x)) \tag{10.14}$$

In general, arbitrarily selected loss functions can be employed for model evaluation, incorporating task-specific requirements or preferences. In particular, some loss functions can be *asymmetric*, i.e., assigning different costs to residuals with the same absolute value, but different signs, or *ε-insensitive*, i.e., assigning zero costs to residuals below some tolerance threshold.

Example 10.2.9 The following R code defines a function to calculate the mean loss as well as three loss functions: the absolute loss, the quadratic loss, and a simple loss function wrapper that can make an arbitrary loss function asymmetric by permitting specifying different multipliers applied for the positive and negative residuals. They are all demonstrated in application to the regression tree models for the *Boston Housing* dataset.

```
mls <- function(pred.y, true.y, loss) { mean(loss(pred.y, true.y)) }

loss.abs <- function(pred.y, true.y) { abs(true.y-pred.y) }

loss.square <- function(pred.y, true.y) { (true.y-pred.y)^2 }

loss.asymmetric <- function(loss, p=1, n=1)
{
  function(pred.y, true.y)
  {
    ifelse(res(pred.y, true.y)>0, p*loss(pred.y, true.y), n*loss(pred.y, true.y))
  }
}

mls(predict(bh.tree, bh.test), bh.test$medv, loss.abs)
mls(predict(bh.tree, bh.test), bh.test$medv, loss.square)
mls(predict(bh.tree, bh.test), bh.test$medv, loss.asymmetric(loss.abs, 2, 1))
```

10.3   Evaluation procedures

Evaluation procedures for regression models address the same challenge of reliably assessing a model's expected performance on new data as discussed in Section 7.3 for classification models, which requires the separation of the validation or test set from the training set without degrading the model quality due to insufficient training data. It also involves the same difficulties and essentially the same techniques for overcoming them. The extensive discussion of those presented in Section 7.3 will not be repeated here, but it fully applies to regression model evaluation as well. In particular, it is important to keep in mind the following:

- The purpose of evaluation procedures is to evaluate *modeling procedures* (consisting of a regression algorithm, its parameters, applied data transformations and anything else other than the data itself that affects the created model) rather than individual models.

- One or more models have to be created using a given modeling procedure and their performance measured to evaluate the modeling procedure.

- The evaluation should reliably estimate the true performance of the model created using the same modeling procedure on the whole available dataset.

- The evaluation reliability may suffer from *bias* (resulting from evaluating models created on subsets of the available data, which may reduce their quality compared to all-data models) and *variance* (resulting from using limited randomly selected test/validation subsets).

The remaining contents of this section are limited to a very concise review of the same set of evaluation procedures, with examples of their application to regression models. These procedures can all be used to generate vectors of predicted and true target function values, making it possible to calculate arbitrary performance measures based on these vectors.

10.3.1   Hold-out

The hold-out evaluation procedure separates training and validation or test data in the simplest possible way: a subset of the available labeled dataset is selected randomly as the training set and the remaining instances are *held out* for the purpose of model evaluation. It does not handle the bias vs. variance tradeoff very well: with sufficiently many instances left for low-variance evaluation there may be too little training instances to ensure adequate model quality and one may end up with a considerable pessimistic bias resulting from measuring the performance of a poor model.

Example 10.3.1 Notice that all the examples of performance measures presented above used the hold-out procedure, randomly dividing the *Boston Housing* dataset into training and test subsets. By repeating the random partitioning used for these examples several times and recalculating the performance measures, we would likely observe considerably different results, due to the high variance of this evaluation procedure. The following R code uses the holdout function to perform an automated hold-out procedure that repeats the random dataset partitioning, model building, and prediction several times, and collects the observed predicted and true target function values. This procedure is applied to perform 10-times
repeated hold-out evaluation of regression tree models for the *Boston Housing* dataset (2/3 of which is used for training with the remaining 1/3 used for testing). All previously discusses performance measures are applied to the generated predictions.

> Ex. 7.3.1
> dmr.claseval

```
# hold-out regression tree evaluation for the Boston Housing data
bh.ho <- holdout(rpart, medv~., BostonHousing, n=10)
mae(bh.ho$pred, bh.ho$true)
mse(bh.ho$pred, bh.ho$true)
rmse(bh.ho$pred, bh.ho$true)
rae(bh.ho$pred, bh.ho$true)
r2(bh.ho$pred, bh.ho$true)
cor(bh.ho$pred, bh.ho$true, method="pearson")
cor(bh.ho$pred, bh.ho$true, method="spearman")
```

The holdout function, originally developed for classification, handles regression model evaluation as well. The same applies to other implementations of model evaluation procedures demonstrated in subsequent examples.

10.3.2   Cross-validation

A more refined evaluation procedure that better handles the bias vs. variance tradeoff is k-fold cross-validation that consists in splitting the available dataset at random into k

disjoint equal-size subsets $D_1, D_2, \ldots, D_k$ and then iterating over these subsets. On the ith iteration, a model is built using $T_i = \bigcup_{j \neq i} D_j$ as the training set, and applied to generate predictions on $Q_i = D_i$. For typical values of k (between 5 and 20) this requires substantially more computation than the simple hold-out procedure, but helps to reduce the bias while keeping the variance under control. It can be repeated several times for even further variance reduction. By choosing k one can tradeoff the bias, variance, and computational expense of the cross-validation procedure, as discussed in Section 7.3.4.

Example 10.3.2 The following R code applies the `crossval` function to evaluate predictions of regression tree models for the *Boston Housing* dataset using k-fold cross-validation with a few different k values. Unlike in the previous example, only a single performance measure – the mean square error – is calculated, but any other performance indicator can be used instead.

> Ex. 7.3.2
> dmr.claseval

```
  # regression tree cross-validation for the BostonHousing data
bh.cv3 <- crossval(rpart, medv~., BostonHousing, k=3)
mse(bh.cv3$pred, bh.cv3$true)
bh.cv5 <- crossval(rpart, medv~., BostonHousing, k=5)
mse(bh.cv5$pred, bh.cv5$true)
bh.cv10 <- crossval(rpart, medv~., BostonHousing, k=10)
mse(bh.cv10$pred, bh.cv10$true)
bh.cv20 <- crossval(rpart, medv~., BostonHousing, k=20)
mse(bh.cv20$pred, bh.cv20$true)
```

10.3.3 Leave-one-out

The leave-one-out validation procedure is an extreme form of k-fold cross-validation in which k is set to the number of instances in the dataset. The procedure iterates over all instances, using the model built on the dataset with one instance removed to make prediction for this instance. This is hardly applicable to large datasets due to the computational expense of building as many models as instances available, but it may be a reasonable evaluation procedure for very small datasets. It has no pessimistic bias, but its variance is high due to using single instances for evaluation.

Example 10.3.3 The following R code uses the `leaveoneout` function to perform the leave-one-out evaluation procedure for the *Boston Housing* data.

> Ex. 7.3.3
> dmr.claseval

```
  # leave-one-out regression tree evaluation for the BostonHousing data
bh.llo <- leaveoneout(rpart, medv~., BostonHousing)
mse(bh.llo$pred, bh.llo$true)
```

10.3.4 Bootstrapping

Bootstrapping estimation techniques are based on drawing multiple *bootstrap samples* from the data. Each bootstrap sample, typically of the same size as the original dataset, is drawn uniformly at random *with replacement*. A single bootstrap sample can be expected to contain about 63.2% of instances from the original dataset, some replicated. The missing instances, also called *out-of-bag* (OOB) instances, stand for about 36.8% of the dataset. The idea of bootstrapping is to use a bootstrap sample as the training set and to evaluate the resulting model on the OOB instances.

Plain bootstrap performance estimators, obtained on OOB instances only, have low variance, but are usually pessimistically biased, since only about 63.2% of available instances are used for training. This is compensated by the .632 bootstrap procedure which produces the final performance estimator as a weighted average of the (overly pessimistic) estimator obtained on OOB instances and the (overly optimistic) estimator that can be obtained by training and evaluating a model on the full dataset. Using the mean square error as the performance measure, this can be presented as follows:

$$\text{mse}_D^{.632} = 0.632 \frac{1}{M} \sum_{i=1}^{M} \text{mse}_{f, D_i'}(h_i) + 0.368 \text{mse}_{f, D}(h) \qquad (10.15)$$

where D is the available dataset from which M bootstrap samples $D_1, D_2, \ldots, D_M$ are drawn, h_i is the model built using D_i as the training set, D_i' is the corresponding set of OOB instances on which the model is evaluated, and h is the model built on the full dataset D.

The .632 bootstrap estimator has been most often employed for classification models, but it can be prove similarly useful for regression models. It can be expected to work best with regression algorithms that do not heavily overfit. For models that fit the training set to a great extent the estimator tends to be optimistically biased.

Similarly as for other evaluation procedures, we can use bootstrapping to generate vectors of predicted and true target function values and then calculate arbitrary performance measures based on these vectors. This is straightforward with just one caveat: to apply the .632 bootstrap or another similar weighting scheme, the vectors of predictions and true target function values must be accompanied by a vector of weights, and the calculation of selected performance indicators must incorporate these weights. This was more extensively discussed in Section 7.3.6.

Example 10.3.4 To illustrate the bootstrapping approach to model evaluation, the following R code uses the `bootstrap` function. It draws a specified number of bootstrap samples from the provided dataset and uses them to build models which it subsequently applies to generate predictions for OOB instances. If the w parameter is set to a value less than 1, it also creates a full-data model and applies it to generate predictions on the full dataset. These receive a weight of $1 - w$, whereas the former receive a weight of w, divided by the number of bootstrap samples. For $w = 0.632$ this is equivalent to the .632 bootstrap procedure. The procedure returns a data frame with predictions, target function values, and the corresponding weights. The function is applied to evaluate the regression tree model for the *Boston Housing* data, with w set to 1 (plain bootstrap, likely to be pessimistically biased) and to 0.632. A relatively small number of 20 bootstrap samples is used. The produced output is used to calculate the mean square error.

> Ex. 7.3.4
> dmr.claseval

```
# 20x bootstrap regression tree evaluation for the BostonHousing data
bh.bs20 <- bootstrap(rpart, medv~., BostonHousing, w=1, m=20)
mse(bh.bs20$pred, bh.bs20$true)

# 20x .632 bootstrap regression tree evaluation for the BostonHousing data
bh.632bs20 <- bootstrap(rpart, medv~., BostonHousing, m=20)
wmse(bh.632bs20$pred, bh.632bs20$true, bh.632bs20$w)
```

The .632 bootstrap performance estimators suggest a better performance level than the plain bootstrap ones (with $w = 1$), which agrees with the expectation of the pessimistic bias of the latter. The former may appear overoptimistic, though, given the error estimates previously obtained with the cross-validation and leave-one-out procedures.

10.3.5   Choosing the right procedure

As discussed in Section 7.3.7, the choice of the right evaluation procedure for a given application depends on the accepted level of the bias vs. variance tradeoff, the size of the dataset, the regression algorithm, and the available computational resources. Most evidence suggests that the bias vs. variance tradeoff is best handled by the k-fold cross-validation procedure with k set to 10 or 20, particularly repeated several times. Whereas the .632 bootstrap procedure should also yield nearly unbiased and low-variance performance estimators for algorithms that are not prone to overfitting, in practice it is often hardly possible to completely eliminate the risk of overfitting *a priori*, which result in .632 bootstrap performance estimates being overly optimistic. The leave-one-out procedure should be rather avoided, except for small datasets where all other evaluation procedures would be considerably biased. The hold-out procedure is, conversely, best suited to very large datasets, for which other evaluation procedures would be too expensive and for which considerably smaller training samples would have to be used anyway due to computational constraints.

Example 10.3.5 To illustrate the properties of the different evaluation procedures discussed above, the following R code applies the `eval.bias.var` function to perform a simple experiment to observe their bias and variance. This basically reproduces Example 7.3.5 in the regression context. A random 2/3 subset is drawn from the provided dataset and considered a simulated "available" dataset, with the remaining 1/3 subset considered a simulated "new" dataset. The "new" dataset is used to calculate an estimate of the true performance of the model built on the "available" dataset (the mean square error in this case). The hold-out, cross-validation, leave-one-out, and bootstrap evaluation procedures are then run on the "available" dataset to produce their performance estimates. This experiment is repeated a number of times with all the obtained results collected, to finally calculate the estimated bias and variance of each evaluation procedure.

> Ex. 7.3.5
> `dmr.claseval`

The function is applied to observe the bias and variance of different evaluation procedures when applied to evaluate regression tree models for the *Boston Housing* dataset. The results are used to produce a boxplot of the error estimates produced by particular evaluation procedures, with a horizontal line designating the mean true error estimated on the "new" dataset. Barplots of the bias and variance of all the evaluation procedures are also produced. The plots presented in Figure 10.2 are based on 200 evaluation repetitions, which takes some considerable time. The line that runs this full experiment is commented out and another one, that runs a 10-times repeated evaluation experiment, is recommended instead for a quick illustration.

```
    # the commented line runs a 200-repetition experiment, which takes a long time
#bh.ebv <- eval.bias.var(rpart, medv~., BostonHousing, perf=mse, wperf=wmse, n=200)
    # this can be used for a quick illustration
bh.ebv <- eval.bias.var(rpart, medv~., BostonHousing, perf=mse, wperf=wmse, n=10)

boxplot(bh.ebv$performance[,-1], main="Error", las=2)
lines(c(0, 13), rep(mean(bh.ebv$performance[,1]), 2), lty=2)
barplot(bh.ebv$bias, main="Bias", las=2)
barplot(bh.ebv$variance, main="Variance", las=2)
```

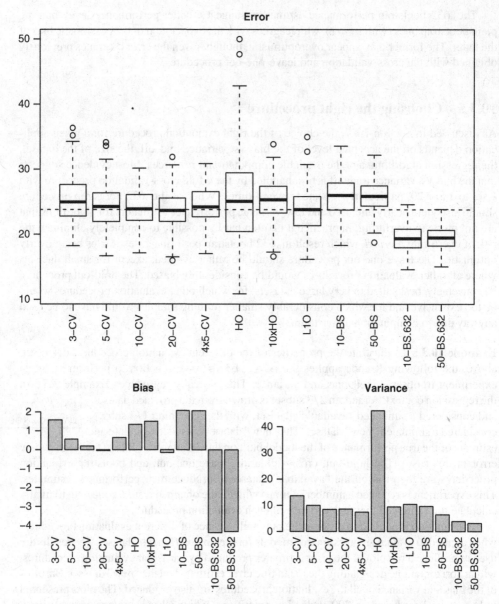

Figure 10.2 The mean square distribution, bias, and variance for different evaluation procedures.

The results agree with expectations and are consistent with those observed in Example 7.3.5, confirming the high bias of hold-out (both single and repeated) and 3-fold cross-validation, the very small bias of 20-fold cross-validation and leave-one-out, the pessimistic bias of the plain bootstrap procedure, and the optimistic bias of the .632 bootstrap procedure. With respect to variance, 4×5-fold cross-validation is the best of the cross-validation procedures, the .632 bootstrap is clearly better, even with just 10 bootstrap samples, and the single hold-out is by far the worst. The repetition reduces the variance of hold-out considerably, as expected. Overall, the 4×5-fold cross-validation procedure appears to achieve a good compromise between bias and variance.

10.4   Conclusion

Performance measures used to assess the quality of regression models differ from those used for classification models for obvious reasons. With a numeric target function the quality of model predictions is no longer assessed by counting the number of mistakes, but rather calculating the differences or correlations between them. The former leads to the most common residual-based performance measures: the mean absolute error and the mean square error, as well as their more easily interpretable counterparts (the relative absolute error, the root mean square error, and the coefficient of determination). The latter may be preferred in some special model application scenarios where capturing the monotonic behavior of the target function with respect to attribute values is more important than achieving small differences.

Despite different performance measures, the purpose and overall methodology of regression model evaluation remains the same as for classification models. The same evaluation procedures are used, facing the same challenge of controlling evaluation bias and variance. And as for classification, the biggest risk of failing to reliably evaluate regression models is not associated with choosing inadequate performance measures or evaluation procedures, but with insufficiently careful separation of the data used to create the model and the data on which the evaluation is performed. The former should be understood broadly, including all data subsets on which any decisions that may impact the final model are based. These are, in particular, data transformation, parameter tuning, and attribute selection decisions.

10.5   Further readings

With evaluation procedures for regression models being the same as for classification models, the same references as provided in Chapter 7 also apply here, including both books (e.g., Abu-Mostafa *et al.* 2012; Cios *et al.* 2007; Han *et al.* 2011; Hand *et al.* 2001; Tan *et al.* 2013; Witten *et al.* 2011) and research articles (e.g., Arlot and Celisse 2010; Efron 1983; Efron and Tibshirani 1997). Some of the former also present basic regression-specific predictive performance measures (e.g., Hand *et al.* 2001; Witten *et al.* 2011), but the statistical literature on regression modeling gives a much wider and more in-depth coverage of those (e.g., Draper and Smith 1998; Freedman 2009; Glantz and Slinker 2000). They discuss the issue of assessing the performance of regression models – referred to as goodness of fit in standard statistical terminology – much more extensively than this chapter. Even if they are often presented – as this term suggests – in application to evaluating the training performance only, they can be clearly applied to nontraining data and used, in combination with evaluation procedures, to assess the true performance. For historical reasons, this may be not such a well-established practice as for classification model evaluation, though (Picard and Cook 1984; Snee 1977).

It is worthwhile to mention the close relationship between the evaluation of regression models and that of measurements in experimental physics or engineering. The latter was extensively discussed by Taylor (1996). In particular, popular types of residual-based performance measures are common for these two domains. Evaluating the predictive performance (or the goodness of fit) of regression models is, however, only one of the several aspects of statistical regression model diagnostics, which also include statistics and tests for verifying the model's underlying assumptions or quantifying the impact of particular attributes and training instances on model parameters and predictions (Cook and Weisberg 1982; Davison and Tsai 1992).

References

Abu-Mostafa YS, Magdon-Ismail M and Lin HT 2012 *Learning from Data*. AMLBook.

Arlot S and Celisse A 2010 A survey of cross-validation procedures for model selection. *Statistics Surveys* **4**, 40–79.

Cios KJ, Pedrycz W, Swiniarski RW and Kurgan L 2007 *Data Mining: A Knowledge Discovery Approach*. Springer.

Cook RD and Weisberg S 1982 *Residuals and Influence in Regression*. Chapman and Hall.

Davison AC and Tsai CL 1992 Regression model diagnostics. *International Statistical Review* **60**, 337–353.

Draper NR and Smith H 1998 *Applied Regression Analysis* 3rd edn. Wiley.

Efron B 1983 Estimating the error rate of a prediction rule: Improvement on cross-validation. *Journal of the American Statistical Association* **78**, 316–331.

Efron B and Tibshirani R 1997 Improvements on cross-validation: The .632+ bootstrap method. *Journal of the American Statistical Association* **92**, 548–560.

Freedman DA 2009 *Statistical Models: Theory and Practice*. Cambridge University Press.

Glantz SA and Slinker BS 2000 *Primer of Applied Regression and Analysis of Variance* 2nd edn. McGraw-Hill.

Han J, Kamber M and Pei J 2011 *Data Mining: Concepts and Techniques* 3rd edn. Morgan Kaufmann.

Hand DJ, Mannila H and Smyth P 2001 *Principles of Data Mining*. MIT Press.

Picard RR and Cook RD 1984 Cross-validation of regression models. *Journal of the American Statistical Association* **79**, 575–583.

Snee RD 1977 Validation of regression models: Methods and examples. *Technometrics* **19**, 415–428.

Tan PN, Steinbach M and Kumar V 2013 *Introduction to Data Mining* 2nd edn. Addison-Wesley.

Taylor JR 1996 *An Introduction to Error Analysis: The Study of Uncertainties in Physical Measurements* 2nd edn. University Science Books.

Witten IH, Frank E and Hall MA 2011 *Data Mining: Practical Machine Learning Tools and Techniques* 3rd edn. Morgan Kaufmann.

Part IV

CLUSTERING

11

(Dis)similarity measures

11.1 Introduction

While exploring and exploiting similarity patterns in data is at the heart of the clustering task and therefore inherent for all clustering algorithms, not all of them adopt an explicit similarity measure to drive their operation. Such similarity, or actually more often dissimilarity measures (since they typically take minimum values for maximum similarity), are functions that assign real values to instance pairs from the domain and can be used by clustering algorithms both in the cluster formation and cluster modeling processes. Such algorithms can be referred to as similarity-based or – perhaps more appropriately – dissimilarity-based clustering algorithms.

This chapter presents a selection of the most commonly used general-purpose similarity and dissimilarity measures for clustering, providing a necessary common background for presenting the most widely used dissimilarity-based clustering algorithms. The latter are described in detail in Chapters 12 and 13.

Example 11.1.1 Each dissimilarity measure presented in this chapter will be illustrated with a simple R implementation, applied to the *weathercl* data. Utility functions from the `dmr.util` package as well as the standardization implementation available in the `dmr.trans` package will be also used. The R code presented below loads the packages and the dataset.

EX. 1.5.1
dmr.data

```
library(dmr.util)
library(dmr.trans)

data(weathercl, package="dmr.data")
```

11.2 Measuring dissimilarity and similarity

A number of dissimilarity and similarity measures have been proposed in the literature and are applied in practice. Their common objective is to measure and express numerically the

---

Data Mining Algorithms: Explained Using R, First Edition. Paweł Cichosz.
© 2015 John Wiley & Sons, Ltd. Published 2015 by John Wiley & Sons, Ltd.

degree to which two instances from the same domain, described by the same set of attributes, are dissimilar from or similar to each other.

The most popular general purpose dissimilarity and similarity measures presented in this chapter fall into two categories:

Difference based. Which transform and aggregate in some way attribute value differences for the two compared instances.

Correlation based. Which detect the common pattern of low and high attribute values for the two compared instances.

The distinction between them as well as types of applications to which they are best suited is presented in the subsequent sections below.

Example 11.2.1 The following R code defines a function that automates dissimilarity matrix generation with respect to a given dissimilarity measure that will be used for these subsequent examples. For the sake of simple illustration, it is applied below to generate the dissimilarity matrix for the *weathercl* data using a totally meaningless dissimilarity measure, defined as the absolute difference of instance numbers.

```
dissmat <- function(data, diss)
{
  as.dist(outer(1:nrow(data), 1:nrow(data),
           Vectorize(function(i, j)
                      if (j<=i) diss(data[i,], data[j,]) else NA)),
         diag=TRUE, upper=TRUE)
}

  # dummy dissimilarity matrix for the weathercl data
dummy.diss <- function(x1, x2)
{ abs(as.integer(row.names(x1))-as.integer(row.names(x2))) }

dissmat(weathercl, dummy.diss)
```

The function returns a `dist` object, used in R to represent dissimilarity (or distance) matrices. Only the lower triangle of the matrix is actually calculated and stored to save time and space.

11.3   Difference-based dissimilarity

Difference-based dissimilarity measures are particularly common and often adopted as default for dissimilarity-based clustering algorithms unless existing domain knowledge suggests that they may be inappropriate.

11.3.1   Euclidean distance

The *Euclidean distance* – by far the most widely and frequently applied dissimilarity measure – is well known from geometry. For two instances x_1, x_2 it is calculated as

$$\delta^{\text{euc}}(x_1, x_2) = \sqrt{\sum_{i=1}^{n} (a_i(x_1) - a_i(x_2))^2} \tag{11.1}$$

which is actually the L^2 norm of the difference of their attribute value vectors, $\|\mathbf{a}(x_1) - \mathbf{a}(x_2)\|$. This formula is directly applicable to domains with continuous attributes only, but it is not uncommon to use it when some attributes are discrete as well, by appropriately redefining the difference $a_i(x_1) - a_i(x_2)$ as $\delta_{01}(a_i(x_1), a_i(x_2))$, where

$$\delta_{01}(v_1, v_2) = \begin{cases} 0 & \text{if } v_1 = v_2 \\ 1 & \text{otherwise} \end{cases} \tag{11.2}$$

Such a binary discrete value difference makes it possible to use the Euclidean dissimilarity (as well as several other related dissimilarity measures that will be presented later) when there are both discrete and continuous attributes, but may not always be appropriate when all attributes are discrete. This is when it reduces to measuring dissimilarity by the number of different attribute values, yielding as many (or rather as little) distinct dissimilarity levels as attributes. For realistically sized datasets this may yield large numbers of equally dissimilar instances. It is not a major problem since dissimilarity-based clustering algorithms are typically applied to datasets with all or most attributes being continuous anyway, but if necessary – it could be alleviated by defining custom nonbinary per-attribute discrete value difference measures based on domain knowledge, which would not consider all different values equally different.

Example 11.3.1 The following R code defines a function to calculate the Euclidean distance. It handles both continuous and discrete attributes using an auxiliary `avdiff` function that returns attribute value differences appropriately depending on attribute types. The Euclidean distance-based dissimilarity matrix for the *weathercl* data is then generated.

```
avdiff <- function(x1, x2)
{
  mapply(function(v1, v2) ifelse(is.numeric(v1), v1-v2, v1!=v2), x1, x2)
}

euc.dist <- function(x1, x2) { sqrt(sum(avdiff(x1,x2)^2, na.rm=TRUE)) }

  # Euclidean distance dissimilarity matrix for the weathercl data
dissmat(weathercl, euc.dist)
```

11.3.2 Minkowski distance

The Euclidean distance can be considered the best known and most practically important special case of a more general parameterized distance family, known as the *Minkowski distance*. When applied to measure the dissimilarity of the two instances x_1 and x_2, it is defined as follows:

$$\delta^{\text{mink}}(x_1, x_2) = \left(\sum_{i=1}^{n} |a_i(x_1) - a_i(x_2)|^p \right)^{\frac{1}{p}} \tag{11.3}$$

where $p \geq 1$ is a parameter. The Euclidean distance is obtained for $p = 2$. It can be easily seen that with increasing p large attribute value differences receive relatively more and more weight than smaller ones. Whereas for $p = 1$ the contribution of each attribute value difference to the calculated distance is proportional to the difference, for larger p the impact of small differences diminishes and the distance mostly depends on large differences. This may be an important change when there are many attributes that differentiate the two compared instances, but only a few of them have large value differences. For small p, these instances could still be considered quite similar (as similar as two other instances with all attribute value differences roughly equal and just slightly larger than the small differences mentioned before), but for larger p they would be considered substantially dissimilar.

Example 11.3.2 The following R code demonstrates Minkowski distance calculation, in the very same manner as presented before for the Euclidean distance. Dissimilarity matrices for the *weathercl* data are generated using $p = 1$ and $p = 3$.

```
mink.dist <- function(x1, x2, p) { (sum(abs(avdiff(x1,x2))^p, na.rm=TRUE))^(1/p) }

  # Minkowski distance dissimilarity matrices for the weathercl data
dissmat(weathercl, function (x1, x2) mink.dist(x1, x2, 1))
dissmat(weathercl, function (x1, x2) mink.dist(x1, x2, 3))
```

11.3.3   Manhattan distance

The special case of the Minkowski distance obtained for $p = 1$ is also known as the *Manhattan distance*. This corresponds to a metaphor of going from one street crossing to another in a grid-arranged city by moving along streets (rather than following the straight line joining the two points, as for the Euclidean distance). As discussed above, it makes the contribution of each attribute value difference proportional to the difference. The formula for the Manhattan distance can be simplified to

$$\delta^{\mathrm{man}}(x_1, x_2) = \sum_{i=1}^{n} |a_i(x_1) - a_i(x_2)| \tag{11.4}$$

Example 11.3.3 The R code presented below defines a function for Manhattan distance calculation by simply wrapping a call to the `mink.dist` function from the previous example. As before, it is applied to obtain the dissimilarity matrix for the *weathercl* data.

```
man.dist <- function(x1, x2) { mink.dist(x1, x2, 1) }

  # Manhattan distance dissimilarity matrix for the weathercl data
dissmat(weathercl, function (x1, x2) man.dist(x1, x2))
```

11.3.4   Canberra distance

A modified form of the Manhattan distance that relates the attribute value differences to the values themselves is known as the *Canberra distance* and calculated as

$$\delta^{\mathrm{can}}(x_1, x_2) = \sum_{i=1}^{n} \frac{|a_i(x_1) - a_i(x_2)|}{|a_i(x_1)| + |a_i(x_2)|} \tag{11.5}$$

This adjusts the contribution of each attribute value difference so that the same absolute difference receives more or less weight depending on whether it is observed for small or large values. It also makes the dissimilarity measure less sensitive to differences in attribute ranges than for other difference-based dissimilarity measures. Each term in the summation obviously falls to the [0, 1] interval, regardless of the ranges of particular attributes.

A deficiency of the Canberra distance is its somewhat degenerate behavior when an attribute's value for one or both compared instances approaches 0. If one of $a_i(x_1), a_i(x_2)$ is close to 0, the contribution of a_i to the calculated distance is roughly equal to 1 regardless of the difference $a_i(x_1) - a_i(x_2)$. If both $a_i(x_1), a_i(x_2)$ are close to 0 (and, accordingly, also close to each other), the contribution of a_i may also approach 1 even if the attribute value difference is actually very small.

Since the formula refers to attribute values themselves apart from their differences, it might appear that, unlike the difference-based dissimilarity measures presented above, the Canberra distance cannot be directly applied to domains with discrete attributes simply by redefining the attribute value difference appropriately. This is easily solved though by replacing the whole ratio $\frac{|a_i(x_1) - a_i(x_2)|}{|a_i(x_1) + a_i(x_2)|}$ for discrete attributes with 1 if the attribute values differ and 0 if they are the same.

Example 11.3.4 The following R code implements Canberra distance calculation and applies it to the *weathercl* data. It uses an auxiliary function `ravdiff` to handle both continuous and discrete attributes.

```
ravdiff <- function(x1, x2)
{
  mapply(function(v1, v2) ifelse(is.numeric(v1), (v1-v2)/(abs(v1)+abs(v2)), v1!=v2),
      x1, x2)
}

can.dist <- function(x1, x2) { sum(abs(ravdiff(x1,x2)), na.rm=TRUE) }

  # Canberra distance dissimilarity matrix for the weathercl data
dissmat(weathercl, function (x1, x2) can.dist(x1, x2))
```

11.3.5   Chebyshev distance

The Manhattan distance corresponds to one extreme special case of the Minkowski distance, with small and large attribute value differences having proportionally the same impact. The other extreme special case, obtained for $p = \infty$, makes the distance equal to the single largest attribute value difference. The resulting dissimilarity measure is called the *Chebyshev distance*, also known as the *chessboard distance* (as it provides the minimum number of moves necessary for a chess king to go from one square to another) or the *maximum metric*. The definition is simply written as follows:

$$\delta^{\text{cheb}}(x_1, x_2) = \max_i |a_i(x_1) - a'_i(x_2)| \tag{11.6}$$

While the Minkowski distance is not directly computable in machine arithmetic for too large p, it approaches the Chebyshev distance even for moderate p values.

It is not particularly common to see the Chebyshev distance being used as a clustering dissimilarity measure since in many practical clustering tasks it might be inappropriate to entirely focus on a single most differentiating attribute for each pair of compared instances. It may be reasonable, though, in some specific domains.

Example 11.3.5 The R code presented below defines a function for calculating the Cheby-shev distance and demonstrates its application to the *weathercl* data. For comparison, the Minkowski distance dissimilarity matrix for $p = 10$ is also calculated. It can be verified to approximate the Chebyshev dissimilarity matrix quite closely.

```
cheb.dist <- function(x1, x2) { max(abs(avdiff(x1,x2)), na.rm=TRUE) }

  # Chebyshev distance dissimilarity matrix for the weathercl
dissmat(weathercl, cheb.dist)
  # roughly the same as
dissmat(weathercl, function (x1, x2) mink.dist(x1, x2, 10))
```

11.3.6   Hamming distance

Whereas all the difference-based dissimilarity measures presented above are primarily intended for continuous attributes and can be "hacked" to be applicable to discrete attributes when necessary, the *Hamming distance* is targeted at discrete attributes only and – whilst can be calculated – makes usually little sense for continuous attributes. It is defined as the number of attributes that take different values for the two compared instances:

$$\delta^{\text{ham}}(x_1, x_2) = \sum_{i=1}^{n} \mathbb{I}_{a_i(x_1) \neq a_i(x_2)} \tag{11.7}$$

where the $\mathbb{I}_{\text{condition}}$ notation is used to denote an indicator function that yields 1 when the *condition* is satisfied and 0 otherwise.

The biggest problem with the Hamming distance is that it takes the very limited set of values $\{0, 1, \dots, n\}$, where n is the number of attributes. This may not provide sufficient dis-similarity diversification to reasonably drive the cluster formation process, unless the number of attributes is much greater than the number of clusters to be created. This is why it is usually pointless to apply the Hamming distance unless there are no continuous attributes defined on the domain. If all attributes are discrete, the previously presented dissimilarity measures suf-fer from the very same problem, but with at least some continuous attributes they yield much more diversified dissimilarity matrices.

Example 11.3.6 The following R code implements the Hamming distance and applies it to the *weathercl* data. For the sake of illustration, the fact that the presence of two continuous attributes makes it a questionable choice is ignored.

```
ham.dist <- function(x1, x2) { sum(x1!=x2, na.rm=TRUE) }

  # Hamming distance dissimilarity matrix for the weathercl
dissmat(weathercl, ham.dist)
```

11.3.7   Gower's coefficient

The most commonly used difference-based dissimilarity measures presented so far favor either continuous (as for the whole Minkowski distance-based family) or discrete (as for the Hamming distance) attributes. *Gower's coefficient* is a dissimilarity measure specifically designed with the intention of handling mixed attribute types. The coefficient is calculated as the weighted average of individual attribute contributions, with weights usually used only to indicate which attribute values could actually be compared meaningfully:

$$\delta^{\text{gow}}(x_1, x_2) = \frac{\gamma_i \delta^{\text{gow},i}(a_i(x_1), a_i(x_2))}{\sum_{i=1}^{n} \gamma_i} \tag{11.8}$$

where γ_i is the weight assigned to attribute a_i and $\delta^{\text{gow},i}$ is the attribute value similarity measure for attribute i, defined as

$$\delta^{\text{gow},i}(v_1, v_2) = \begin{cases} \frac{|v_1-v_2|}{\max_{x \in X} a_i(x) - \min_{x \in X} a_i(x)} & \text{if } a_i \text{ is continuous} \\ 0 & \text{if } a_i \text{ is discrete and } v_1 = v_2 \\ 1 & \text{if } a_i \text{ is discrete and } v_1 \neq v_2 \end{cases} \tag{11.9}$$

Attribute weights are normally equal to 1 unless an attribute's value for one or both instances is missing, when the corresponding weight is set to 0. It is also possible to incorporate some domain-specific attribute weights. It is probably more common to encounter a slightly different version of Gower's coefficient for measuring similarity rather than dissimilarity, in which the contributions of all attributes are 1s complements of those defined above.

Although often recommended as the best approach to handling mixed attribute types in dissimilarity measuring, Gower's coefficient is not actually very different from the Manhattan or Canberra distances, modified to accommodate discrete attributes by redefining attribute value differences. It is particularly closely related to the latter, differing only in the divisor applied to continuous attribute value differences. In Gower's coefficient, they are divided by the attribute's range, formally defined above as the difference between the maximum and minimum value of the attribute in the domain, but in practice determined based on the training set used for clustering. This is another and arguably the better approach to ensuring that the contribution of each continuous attribute is between 0 and 1, since there is no risk of anomalies for near-zero values the Canberra distance suffers from. It does not necessarily make Gower's coefficient superior to the members of the Minkowski distance family, though, since – as will be discussed below – it is a common practice to have continuous attributes standardized or normalized prior to their application.

Example 11.3.7 The following R code implements a simplified version of Gower's dissimilarity coefficient in which – instead of explicit attribute weighting – terms corresponding to missing attribute values are removed when averaging. The implementation is then applied to the *weathercl* data, with the `ranges` function used to determine continuous attribute ranges.

`dmr.util`

```
gower.coef <- function(x1, x2, rngs)
{
  mean(mapply(function(v1, v2, r) ifelse(is.numeric(v1), abs(v1-v2)/r, v1!=v2),
              x1, x2, rngs), na.rm=TRUE)
}


  # Gower's coefficient dissimilarity matrix for the weathercl data
dissmat(weathercl, function (x1, x2) gower.coef(x1, x2, ranges(weathercl)))
```

11.3.8   Attribute weighting

Difference-based dissimilarity measures make it possible and straightforward to incorporate any available domain knowledge about the relative importance of particular attributes for instance dissimilarity assessment. It can be achieved by assigning numerical weights to attributes and weighting the corresponding attribute values accordingly. Assuming that γ_i denotes the weight assigned to attribute a_i, the dissimilarity measure definitions presented above can be rewritten in the following weighted form:

Weighted Euclidean distance.

$$\delta^{euc}(x_1, x_2) = \sqrt{\sum_{i=1}^{n} \gamma_i^2 (a_i(x_1) - a_i(x_2))^2} \tag{11.10}$$

Weighted Minkowski distance.

$$\delta^{mink}(x_1, x_2) = \left(\sum_{i=1}^{n} \gamma_i^p |a_i(x_1) - a_i(x_2)|^p \right)^{\frac{1}{p}} \tag{11.11}$$

Weighted Manhattan distance.

$$\delta^{man}(x_1, x_2) = \sum_{i=1}^{n} \gamma_i |a_i(x_1) - a_i(x_2)| \tag{11.12}$$

Weighted Canberra distance.

$$\delta^{can}(x_1, x_2) = \sum_{i=1}^{n} \gamma_i \frac{|a_i(x_1) - a_i(x_2)|}{|a_i(x_1) + a_i(x_2)|} \tag{11.13}$$

Weighted Chebyshev distance.

$$\delta^{cheb}(x_1, x_2) = \max_i \gamma_i |a_i(x_1) - a_i(x_2)| \tag{11.14}$$

Weighted Hamming distance.

$$\delta^{ham}(x_1, x_2) = \sum_{i=1}^{n} \gamma_i \mathbb{I}_{a_i(x_1) \neq a_i(x_2)} \tag{11.15}$$

Gower's dissimilarity is missing in this list only because it already incorporates weights in its original form. Although typically only used to eliminate the impact of attributes with missing values, they can also be used to express domain-specific knowledge on attribute importance.

11.3.9   Attribute transformation

One common problem with the most difference-based dissimilarity or similarity measures is their obvious sensitivity to differences in attribute ranges and distributions. With the exception of the Canberra distance and Gower's coefficient, which compensate for such differences in some ways, the other measures presented above may be easily fooled when applied to domains with continuous attributes that represent quantities expressed in substantially different scales. The same absolute value difference always yields the same contribution to the calculated measure, whereas it may actually indicate similarity for attributes with large or relatively diversified values and dissimilarity for attributes with small or relatively uniform values.

A typical solution to this problem for dissimilarity or similarity measures that do not address it somehow internally (as in the case of the Canberra distance or Grower's coefficient) is to preprocess the data before performing any dissimilarity calculations by applying appropriate attribute transformations. Two transformations that may be useful in this context are standardization and normalization, presented in Sections 17.3.1 and 17.3.2, with the former being particularly popular. It is not uncommon for implementations of dissimilarity-based clustering algorithms to include such a transformation within their functional scope and perform it either by default or at the user's request. If this is not the case, it can be easily applied before running a clustering algorithm. It is usually a good idea to do so, unless some domain-specific background knowledge suggests otherwise. In any case, the decision needs to be taken consciously and the default behavior of the particular clustering algorithm implementation should be carefully checked and understood.

As will be discussed in Section 17.2.5, whenever an attribute transformation is performed based on some parameters derived from a dataset, which is subsequently used for creating a predictive model, the very same transformation parameters should be used to transform new data prior to applying the model for prediction. This is the case, in particular, for a clustering model that can be used to predict cluster membership for new instances. If the training set used for model creation is standardized or normalized, the underlying transformation parameters (the mean and standard deviation for the former, the range for the latter) should be retained and re-applied when transforming new data before generating predictions.

Example 11.3.8 The R code presented below performs the standardization of continuous attributes in the *weathercl* data using the `std.all` and `predict.std` functions and then generates the Euclidean distance dissimilarity matrix based on the standardized dataset.

> EX. 17.3.1
> dmr.trans

```
weathercl.std <- predict.std(std.all(.~., weathercl), weathercl)
dissmat(weathercl.std, euc.dist)
```

11.4   Correlation-based similarity

Measuring instance similarity based on attribute value differences appears perfectly reasonable and is indeed the right way to follow in most situations, but for some domains it may yield misleading results. This is the case whenever instances that differ substantially with respect to attribute values should still be considered similar based on domain knowledge as sharing roughly the same relative value pattern. Consider attributes that represent frequencies of some events, word occurrence counts in text documents, performance or quality evaluations, ratings or preferences expressed by some individuals, etc. It is not necessarily value differences that really matter for them when comparing two instances but rather patterns of "highs" and "lows": Are mostly the same events particularly frequent; do mostly the same words dominate; do mostly the same people, organizations, devices, or whatever other entities achieve top performance or quality; are mostly the same items highly rated or preferred? This considerably different kind of similarity can be captured by correlation-based measures, which – unlike difference-based measures – typically indeed measure similarity rather than dissimilarity, assigning high values to similar instances and low values to dissimilar ones.

11.4.1   Discrete attributes

Correlation-based similarity measures are even more oriented toward continuous attributes than most difference-based measures. Actually, the whole justification behind them is strongly based on the assumption that attributes are continuous (or at least ordered, which can be treated as continuous when calculating these measures simply by assigning consecutive integer numbers to their ordered values). This does not limit their utility in any serious way since – whenever it makes sense to consider their application – the assumption is almost always satisfied. The only situation that may need some workaround is that of a small number of discrete attributes accompanying a larger number of continuous attributes that – based on the available domain knowledge – ask for correlation-based similarity measures.

The workaround is fortunately easily available and the same as used for modeling algorithms based on a parametric representation, such as linear classification and linear regression. This is the binary encoding technique described in Section 17.3.5 that replaces a discrete k-valued attribute $a : X \rightarrow \{v_1, v_2, \ldots , v_k\}$ with k (or – actually – $k - 1$, to avoid redundancy) binary attributes. These new attributes are then treated as continuous for similarity calculation.

Example 11.4.1 The following R code demonstrates how the `discode` function for binary discrete attribute encoding can be applied to transform selected instances from the *weathercl* data to an all-continuous representation. This transformation will be used when calculating correlation-based similarity measures in subsequent examples.

> EX. 17.3.5
> dmr.trans

```
discode(~., weathercl[1,])
discode(~., weathercl[5,])
```

11.4.2   Pearson's correlation similarity

The most straightforward way to measure similarity based on attribute value correlation is to use Pearson's linear correlation, defined in Section 2.5.2. Maximally similar instances are assigned values approaching 1 and maximally dissimilar instances are assigned values approaching -1.

Example 11.4.2 The following R code demonstrates the application of Pearson's correlation, calculated using a simple wrapper around the standard R `corr` function, with the discrete attribute handling capability added, to the *weathercl* data. Despite using the `dissmat` function, the obtained output is actually a similarity rather than dissimilarity matrix.

```
pearson.sim <- function(x1, x2)
{
  cor(unlist(discode(~., x1)), unlist(discode(~., x2)), method="pearson",
      use="pairwise.complete.obs")
}

  # Pearson similarity matrix for the weathercl data
dissmat(weathercl, pearson.sim )
```

11.4.3   Spearman's correlation similarity

According to Pearson's correlation, two instances, to be considered highly similar, not only need to share the same low-high pattern of attribute values, but also to be linearly related. The latter requirement is relaxed with Spearman's rank correlation, also defined in Section 2.5.2. It may therefore be a better choice unless one deliberately seeks for linear relationships between instances as indications of similarity.

Example 11.4.3 Following the pattern of the previous example, the following R code implements and demonstrates the similarity measure based on Spearman's correlation.

```
spearman.sim <- function(x1, x2)
{
  cor(unlist(discode(~., x1)), unlist(discode(~., x2)), method="spearman",
      use="pairwise.complete.obs")
}

  # Spearman similarity matrix for the weathercl data
dissmat(weathercl, spearman.sim )
```

11.4.4   Cosine similarity

Another related similarity measure that has gained high popularity, particularly in text clustering applications, is the *cosine similarity*, which considers two compared instances as attribute value vectors and calculates the cosine of the angle between them:

$$\cos(x_1, x_2) = \frac{\sum_{i=1}^{n} a_i(x_1)a_i(x_2)}{\sqrt{\sum_{i=1}^{n} a_i^2(x_1)}\sqrt{\sum_{i=1}^{n} a_i^2(x_2)}} \tag{11.16}$$

or more readably,

$$\cos(x_1, x_2) = \frac{\mathbf{a}(x_1) \circ \mathbf{a}(x_2)}{\|\mathbf{a}(x_1)\| \cdot \|\mathbf{a}(x_2)\|} \tag{11.17}$$

where $\mathbf{a}(x_1) \circ \mathbf{a}(x_2)$ denotes the dot (inner) product of attribute value vectors representing instances x_1 and x_2, and $\|\mathbf{a}(x_1)\|$ and $\|\mathbf{a}(x_2)\|$ are their Euclidean norms. High cosine values correspond to attribute value vectors pointing roughly in the same direction in the n-dimensional attribute space, indicating high similarity.

Example 11.4.4 The R code presented below implements cosine similarity calculation and demonstrates its application to the *weathercl* data. The `cosine` function performs the actual cosine calculation for two continuous vectors, with the discrete attribute encoding applied.

`dmr.util`

```
cos.sim <- function(x1, x2) { cosine(discode(~., x1), discode(~., x2)) }

  # cosine similarity matrix for the weathercl data
dissmat(weathercl, cos.sim)
```

11.5   Missing attribute values

Missing attribute values, a common problem for real-world datasets, have an obvious impact on instance similarity assessment. Both difference-based and correlation-based measures presented above cannot meaningfully deal with attributes that have missing values for one or both instances being compared and thus are unable to estimate their contribution to the overall instance similarity or dissimilarity. Possible workarounds for this problem include:

Omit. Skip attributes with missing values for one or both instances in dissimilarity calculation (and possibly scale up the obtained dissimilarity accordingly, if using a difference-based measure),

Impute. Fill-in missing attribute values in a data preprocessing phase using some imputation techniques,

Process internally. Use some internal techniques to estimate the contribution of attributes with missing values to the calculated dissimilarity measure.

The omitting approach is probably the most commonly applied in practice for its simplicity, at least unless the number of missing attribute values in the data is prevailing. The functions implementing dissimilarity measure calculation presented in the above examples adopt a simplified version of this approach by including the `na.rm=TRUE` argument in calls to aggregating functions such as `sum` or `max`, but without subsequent scaling. However, to keep the obtained dissimilarity consistent with those calculated without missing values, the aggregated contributions of attributes with nonmissing values should be scaled up proportionally to the number of attributes actually used, i.e., multiplied by the ratio of the number of all attributes to the number of attributes with nonmissing values for both instances. Gower's dissimilarity coefficient assumes that attribute weighting is used to implement this scheme.

The last approach may be sometimes worthwhile to consider when there are many missing attribute values. Clearly it only makes sense for attributes that have missing values for just one of the two instances being compared. The available value for the other instances can then contribute to the calculated dissimilarity based on how typical or untypical it is for the attribute in question. More precisely, whenever the exact contribution of an attribute cannot be determined due to one missing value, its *expected* contribution can be used instead. For continuous attributes, this reduces to substituting the mean attribute value for the missing one, which is equivalent to the most common type of imputation. For discrete attributes, which represent a somewhat more interesting case, the expected contribution can be calculated by considering all possible values in place of the missing one and averaging the corresponding contributions with weights set to their probabilities estimated from the data. With the contribution of a discrete attribute to a difference-based dissimilarity measure being 0 for equal values and 1 otherwise, this reduces to the probability of the attribute's value being different than its actual value for the instance for which it is not missing.

11.6   Conclusion

Using explicit dissimilarity measures to guide the cluster formation and modeling processes is a very popular approach, adopted by many widely used clustering algorithms. This includes the following major families of clustering algorithms:

- k-centers clustering,

- agglomerative hierarchical clustering,

- divisive clustering (since it is usually based on the repeated application of k-centers algorithms).

These are discussed in the next two chapters.

The different dissimilarity and similarity measures presented in this chapter can be considered for use with all (dis)similarity-based algorithms that are flexible enough not to be tied to a particular single measure. It makes a thoughtful choice possible, based on the available domain knowledge and task requirements, as well as experimental verification of the effects of several candidate measures. It is also possible to incorporate domain knowledge by appropriate attribute transformation, attribute selection, or attribute weighting.

While it is hard to choose the most appropriate (dis)similarity measure for a given clustering task without at least some preliminary experiments, the general choice between difference-based and correlation-based measures is usually much easier if at least some domain knowledge (including, in particular, attribute interpretation) is available. It should be sufficient to judge what actually makes instances similar: small attribute value differences or high attribute value correlations. With the former considered the obvious default, the latter may be more appropriate for attributes that represent frequencies of some events, performance or quality of some evaluated entities, ratings, or preferences expressed by some individuals, etc.

11.7   Further readings

Clustering falls within the scope of most data mining and many machine learning books and algorithms based on instance (dis)similarity belong to the most popular ones. This makes brief descriptions of basic dissimilarity and similarity measures, accompanying the presentation of dissimilarity-based clustering algorithms, readily available (e.g., Cios *et al.* 2007; Witten *et al.* 2011). Theodoridis and Koutroumbas (2008), devoting their book almost entirely to the classification and clustering tasks, thoroughly discuss the issue of measuring dissimilarity or similarity for both single instances and sets of instances. A similar discussion is provided by Han *et al.* (2011) and Tan *et al.* (2013), as well as in an appendix of the book by Webb (2002). Survey articles and books dedicated to clustering cover a greater variety of measures (Everitt *et al.* 2011; Jain and Dubes 1988; Jain *et al.* 1999; Kaufman and Rousseeuw 1990), including both general-purpose ones for instances represented by attribute value vectors and those designed for special nonvector data representations.

With the Euclidean dissimilarity measure, as well as other measures from the Minkowski family, being distances for real vector spaces, they are presented and their properties are extensively discussed in mathematical topology literature (e.g., Engelking 1989). The Hamming distance, on the other hand, originates from the study of error-detecting and error-correcting codes (Hamming 1950). These and many other types of distances with different roots and scopes of applications – reaching far beyond clustering and data mining – are systematically presented by Deza and Deza (2013).

Gower's similarity, explicitly permitting mixed discrete-continuous attribute sets, was introduced by Gower (1971). The issue of measuring instance similarity of dissimilarity in the presence of discrete attributes has been addressed by several studies since then, in the context of both clustering (Diday and Simon 1976; Ichino and Yaguchi 1994; Ng *et al.* 2007; San *et al.* 2004) and memory-based learning (Cheng *et al.* 2004; Stanfill and Waltz 1986; Wilson and Martinez 1997). They represent various approaches to overcoming the limitation of simple value equality tests as in Gower's coefficient or the Hamming distance by incorporating attribute value distribution.

Simple difference-based dissimilarity measures for continuous attributes have also some limitations. Apart from the sensitivity to differences in ranges and distributions, these include ignoring possible relationships among attributes. The Mahalanobis distance is a generalization of the Euclidean distance that addresses these issues (Mahalanobis 1936). Other possible improvements include incorporating the context of surrounding instances into dissimilarity calculation (Gowda and Krishna 1978; Jarvis and Patrick 1973).

This chapter – as the whole book – assumes the attribute-value representation of instances, with a fixed number of discrete or continuous attributes. Dissimilarity measures for several other data representations that may be more appropriate for some domains have been proposed, including text strings (Baeza-Yates 1992), text documents (Wajeed and Adilakshmi 2011), tree structures (Zhang 1995), composite symbolic objects (Gowda and Diday 1991), and images (Dubuisson and Jain 1994; Huttenlocher *et al.* 1993).

References

Baeza-Yates RA 1992 Introduction to data structures and algorithms related to information retrieval In *Information Retrieval: Data Structures and Algorithms* (ed. Frakes WB and Baeza-Yates RA) Prentice-Hall.

Cheng V, Li CH, Kwok JT and Li CK 2004 Dissimilarity learning for nominal data. *Pattern Recognition* **37**, 1471–1477.

Cios KJ, Pedrycz W, Swiniarski RW and Kurgan L 2007 *Data Mining: A Knowledge Discovery Approach*. Springer.

Deza MM and Deza E 2013 *Encyclopedia of Distances* 2nd edn. Springer.

Diday E and Simon JC 1976 Clustering analysis In *Digital Pattern Recognition* (ed. Fu KS) Springer.

Dubuisson MP and Jain AK 1994 A modified Hausdorff distance for object matching *Proceedings of the Twelfth International Conference on Pattern Recognition (ICPR-94)*. IEEE Computer Society Press.

Engelking R 1989 *General Topology*. Heldermann.

Everitt BS, Landau S, Leese M and Stahl D 2011 *Cluster Analysis* 5th edn. Wiley.

Gowda KC and Diday E 1991 Symbolic clustering using a new dissimilarity measure. *Pattern Recognition* **24**, 567–578.

Gowda KC and Krishna G 1978 Agglomerative clustering using the concept of mutual nearest neighborhood. *Pattern Recognition* **10**, 105–112.

Gower JC 1971 A general coefficient of similarity and some of its properties. *Biometrics* **27**, 857–874.

Hamming RW 1950 Error detecting and error correcting codes. *Bell System Technical Journal* **29**, 147–160.

Han J, Kamber M and Pei J 2011 *Data Mining: Concepts and Techniques* 3rd edn. Morgan Kaufmann.

Huttenlocher DP, Klanderman GA and Rucklidge WJ 1993 Comparing images using the Hausdorff distance. *IEEE Transactions on Pattern Analysis and Machine Intelligence* **15**, 850–863.

Ichino M and Yaguchi H 1994 Generalized Minkowski metrics for mixed feature-type data analysis. *IEEE Transactions on System, Man and Cybernetics* **24**, 698–708.

Jain AK and Dubes RC 1988 *Algorithms for Clustering Data*. Prentice-Hall.

Jain AK, Murty MN and Flynn PJ 1999 Data clustering: A review. *ACM Computing Surveys* **31**, 264–323.

Jarvis RA and Patrick EA 1973 Clustering using a similarity method based on shared near neighbors. *IEEE Transactions on Computing* **22**, 1025–1034.

Kaufman L and Rousseeuw PJ 1990 *Finding Groups in Data: An Introduction to Cluster Analysis*. Wiley.

Mahalanobis PC 1936 On the generalised distance in statistics. *Proceedings of the National Institute of Sciences of India* **2**, 49–55.

Ng MK, Junjie M, Joshua L, Huang Z and He Z 2007 On the impact of dissimilarity measure in k-modes clustering algorithm. *IEEE Transactions on Pattern Analysis and Machine Intelligence* **29**, 503–507.

San OM, Huynh VM and Naamori Y 2004 An alternative extension of the *k*-means algorithm for clustering categorical data. *International Journal of Applied Mathematics and Computer Science* **14** (2), 241–247.

Stanfill C and Waltz D 1986 Toward memory-based reasoning. *Communications of the ACM* **29**, 1213–1228.

Tan PN, Steinbach M and Kumar V 2013 *Introduction to Data Mining* 2nd edn. Addison-Wesley.

Theodoridis S and Koutroumbas K 2008 *Pattern Recognition* 4th edn. Academic Press.

Wajeed MA and Adilakshmi T 2011 Different similarity measures for text classification using kNN *Proceedings of the Second International Conference on Computer and Communication Technology (ICCCT-2011*. IEEE Press.

Webb AR 2002 *Statistical Pattern Recognition* 2nd edn. Wiley.

Wilson DR and Martinez TR 1997 Improved heterogeneous distance functions. *Journal of Artificial Intelligence Research* **6**, 1–34.

Witten IH, Frank E and Hall MA 2011 *Data Mining: Practical Machine Learning Tools and Techniques* 3rd edn. Morgan Kaufmann.

Zhang K 1995 Algorithms for the constrained editing distance between ordered labeled trees and related problems. *Pattern Recognition* **28**, 463–474.

12

k-Centers clustering

12.1 Introduction

The clustering task was presented in Section 1.5 as the combination of cluster formation, which identifies similarity-based groups in the training set, and cluster modeling, which creates a model for cluster membership prediction. Dissimilarity-based clustering algorithms address both of these subtasks using measures of instance dissimilarity or similarity. The family of *k*-centers clustering algorithms represents not only the conceptually simplest but also the most popular approach to dissimilarity-based clustering. Of all algorithms using explicit similarity or dissimilarity measures, *k*-centers algorithms employ these measures in the most direct and straightforward way to determine cluster membership.

12.1.1 Basic principle

Algorithms from the *k*-centers family share the same basic operation principle that can be outlined as follows:

1. the number of clusters is predetermined and referred to as *k* (hence the "*k*-" in algorithm names),

2. clusters are represented by single attribute value vectors, generically called cluster *centers*, but also referred to with more specific terms in the context of particular algorithms,

3. the combined cluster formation and modeling process is performed by iteratively assigning training instances to clusters with the closest (i.e., least dissimilar) centers and then shifting the centers to reflect the actual content's of particular clusters.

The most widely known, studied, and applied algorithm of the *k*-centers family is the *k*-means algorithm, using vectors of attribute value means as cluster centers. Other *k*-centers algorithms adopt not only more robust but also more computationally demanding cluster center representations.

Data Mining Algorithms: Explained Using R, First Edition. Paweł Cichosz.
© 2015 John Wiley & Sons, Ltd. Published 2015 by John Wiley & Sons, Ltd.

12.1.2   (Dis)similarity measures

Several different instance dissimilarity or similarity measures can be used for k-centers clustering, including all of those presented in Chapter 11. The k-means algorithm originally assumed and has best interpretation with the Euclidean distance and difference-based dissimilarity measures from the Minkowski family remain the most popular choice for k-centers clustering applied to datasets with continuous attributes only. As discussed in Section 11.3.9, they may require performing attribute standardization or normalization to prevent different attribute distributions or ranges from distorting dissimilarities.

While it is not uncommon for practical implementations of k-centers algorithms to apply such transformations internally at the user's request, they do not necessarily support the predictive modeling view of clustering. If a clustering model is supposed not only to describe similarity patterns discovered in the training set, but also to predict cluster membership for new instances, those new instances have to be transformed using the very same transformation parameters that were determined on the training set. This is the essence of modeling attribute transformations, more extensively discussed in Section 17.2.5. Standardization and normalization as modeling transformations – applicable when preprocessing data for predictive modeling – are presented in Sections 17.3.1 and 17.3.2.

Example 12.1.1 A series of examples will be presented in this chapter illustrating different instantiations of k-centers clustering with simplified R implementations. They will be applied to the tiny *weathercl* data, as well as two larger datasets: *Iris* from the standard `datasets` package and *Glass* from the `mlbench` package. The

> EX. 1.5.1
> dmr.data

R code presented below loads these datasets and partitions the two larger ones randomly into training and test subsets (using a fixed random generator seed for result reproducibility), to make it possible to demonstrate the application of clustering models to predicting cluster membership for new instances. The standardization transformation is applied to the data using the `std.all` and `predict.std` functions to ensure

> EX. 17.3.1
> dmr.trans

more meaningful dissimilarity calculations. Notice that standardization parameters determined on the training sets are then applied to both the training and test sets. The `dmr.dissim` package with dissimilarity measure implementations from Chapter 11 and other DMR packages providing auxiliary functions are also loaded. So are the `rpart` and `rpart.plot` packages for decision tree model creation and visualization, as they will be employed in an example illustrating explicit cluster membership modeling.

```
library(dmr.claseval)
library(dmr.dissim)
library(dmr.stats)
library(dmr.trans)

library(rpart)
library(rpart.plot)

data(weathercl, package="dmr.data")
data(iris)
data(Glass, package="mlbench")

set.seed(12)
```

```
ri <- runif(nrow(iris))
i.train <- iris[ri>=0.33,]
i.test <- iris[ri<0.33,]

rg <- runif(nrow(Glass))
g.train <- Glass[rg>=0.33,]
g.test <- Glass[rg<0.33,]

wcl.std <- predict.std(std.all(.~., weathercl), weathercl)

i.stdm <- std.all(Species~., i.train)
i.std.train <- predict.std(i.stdm, i.train)
i.std.test <- predict.std(i.stdm, i.test)

g.stdm <- std.all(Type~., g.train)
g.std.train <- predict.std(g.stdm, g.train)
g.std.test <- predict.std(g.stdm, g.test)
```

12.2   Algorithm scheme

The basic operation principle of *k*-centers clustering informally introduced above is more precisely described by the following scheme, which is instantiated by particular *k*-centers algorithms.

```
1: select initial cluster centers ζ₁, ζ₂, ..., ζₖ;
2. repeat
3.     for all training instances x ∈ T do
4.         assign instance x to cluster dₓ = arg minₐ δ(x, ζₐ);
5.     end for
6.     for d = 1, 2, ..., k do
7.         modify cluster center ζₐ based on cluster member set Tₐ;
8.     end for
9: until stop criteria are satisfied;
```

The clustering process starts from k initially selected cluster centers $\zeta_1, \zeta_2, \ldots, \zeta_k$, each of which can be assumed to be represented by a vector of attribute values, just like instances from the domain from which the dataset comes. It makes it possible to compare them to instances from the training set T using the adopted dissimilarity measure δ (it can be a similarity measure as well, provided the minimization operation is changed to maximization). This is used to identify the closest cluster for each instance, defined to be the cluster with the least dissimilar/most similar center. Although originally defined for instances, it is assumed to be applicable to arbitrary attribute value vectors, since cluster centers, in general, do not necessarily correspond to any existing instances.

For instance x, the cluster to which it is assigned will be designated by d_x and T^d denotes the subset of the training set assigned to cluster d:

$$T^d = \{x \in T \mid d_x = d\} \tag{12.1}$$

After all instances have been assigned to their respective closest clusters, clusters centers are modified to make sure that they are "true centers" for cluster members assigned to them. This is repeated until stop criteria are satisfied.

It is noteworthy that the instance assignment operation, which may be computationally demanding for large datasets, can be easily parallelized by partitioning the training set into subsets and processing each subset on a separate processor. This is definitely a desirable property of the k-centers algorithms, in the era of increasingly large data on one hand and increasingly available parallel or distributed processing environments on the other hand.

12.2.1   Initialization

A universal approach to cluster center initialization that is commonly used with different k-centers algorithms is to randomly select k instances $x_1, x_2, \ldots, x_k$ from the training set and use their attribute value vectors as initial cluster centers by setting $\zeta_d = \mathbf{a}(x_d)$.

What may be considered a deficiency of this simple approach is that it makes the clustering process nondeterministic: different final cluster centers and hence different clustering models may be obtained on repeated algorithm invocations, of possibly different quality. This is not necessarily as serious disadvantage as it might appear, since the simplicity and efficiency of most k-centers clustering algorithms make it usually computationally affordable to run them several times with different random generator seeds and choose the best of multiple models obtained. Quality measures that can be used for this selection are described in Chapter 14.

Still, some more refined deterministic initialization techniques may be worth using, particularly when working with large datasets or applying more computationally demanding algorithms. Sufficiently good initial cluster centers may substantially reduce the necessary number of k-centers iterations, so the effort invested to initialization may return in savings on instance assignment and center adjustment. Techniques used for this purpose typically consist either in identifying a set of the most mutually dissimilar instances in a sample drawn from the training set or selecting points from a multidimensional grid spanning the attribute value space. One simple approach that has been found to yield good results is to select initial centers sequentially, with the first one selected uniformly at random, and each subsequent selected according to a probability distribution that assigns higher selection probabilities to instances more dissimilar to the centers selected so far.

12.2.2   Stop criteria

A natural and ultimate stop criterion for any k-centers clustering algorithm is reaching *convergence*, defined as the situation of no instances changing cluster membership (or, equivalently, cluster centers remaining unchanged) from one iteration to another. This can be guaranteed after a finite number of iterations, as long as both instance assignment and cluster adjustment reduce the total dissimilarity between training instances and their respective cluster centers, but is not necessarily always required. In practice, it is often sufficient to have cluster centers nearly converged, i.e., with a small number of instances changing cluster membership. This is usually achieved within a relatively small number of iterations (between a few and a dozen or not much more), and it is not uncommon to see just the number of iterations adopted not only as a supplementary, but also as the only stop criterion for k-centers algorithms.

12.2.3   Cluster formation

According to the scheme presented above, k-centers algorithms identify a partitioning of the training set into a fixed number of disjoint clusters, with each instance being assigned to the closest cluster. This approach to cluster formation can be easily seen to minimize the total dissimilarity of training instances to their respective cluster centers:

$$\Delta_T(h) = \sum_{x \in T} \delta(x, \zeta_{h(x)}) \qquad (12.2)$$

where $h(x)$ denotes the cluster to which instance x is assigned. The minimization is only local and therefore dependent on cluster center initialization (the best set of cluster centers that can be reached from the initial set of cluster centers is identified).

12.2.4 Implicit cluster modeling

The algorithm outline presented above explicitly performs cluster formation only, but it is implicitly combined with cluster modeling as well. Basically, the center vector approach to cluster representation makes it possible to assign any new instances to the identified clusters based on their dissimilarity to cluster centers, just the same as with training instances during the cluster formation process (except that no adjustment of cluster centers would be performed). Since cluster centers can be used for prediction, they constitute a valid clustering model representation that can be applied to any instance $x \in X$ as follows:

$$h(x) = \arg \min_d \delta(x, \zeta_d) \qquad (12.3)$$

with minimization replaced by maximization if δ is a similarity rather than dissimilarity measure.

12.2.5 Instantiations

The presented algorithm scheme contains the following major operations:

Initialization. Selecting initial cluster centers.

Instance assignment. Assigning training instances to the closest clusters (with respect to dissimilarity to cluster centers).

Center adjustment. Shifting cluster centers based on assigned cluster members.

Stop criteria. Verifying whether more iterations are required.

Of those, the second is fully specified by the adopted dissimilarity (or similarity) measure and, other than that, needs no further concretization. The first and the last ones are mostly independent of particular algorithms and have already been discussed above. Instantiating the scheme therefore requires that only the other two operations be appropriately specified. This makes it possible to focus the discussion of particular *k*-centers clustering algorithms around center adjustment, which heavily depends on particular cluster representation, i.e., on what cluster centers actually are.

Example 12.2.1 The following R code defines the `k.centers` function which implements the generic *k*-centers clustering algorithm as described in this section. It accepts the training dataset, the number of clusters as required input arguments. The dissimilarity measure defaults to the Euclidean dissimilarity, calculated by the `euc.dist` function. The clustering model is represented by a list containing cluster centers and the assignment of training instances to them, with the `class` attribute set to `k.centers` to enable appropriate prediction method dispatching.

EX. 11.3.1
dmr.dissim

```
k.centers <- function(data, k, diss=euc.dist, max.iter=10,
                      init=k.centers.init.rand,
                      assign=k.centers.assign,
                      adjust=k.centers.adjust.dummy)
{
  dm <- dissmat(data, diss)
  centers <- init(data, k)
  clustering <- NULL
  iter <- 0
  repeat
  {
    iter <- iter+1
    clustering.old <- clustering
    clustering <- assign(centers, data, diss, dm)
    centers <- adjust(clustering, data, k, diss, dm)
    if (iter >= max.iter || all(clustering==clustering.old))
      break
  }

  `class<-`(list(centers=centers, clustering=clustering), "k.centers")
}

k.centers.init.rand <- function(data, k) { data[sample(1:nrow(data), k),] }

k.centers.assign <- function(centers, data, diss, dm)
{
  center.diss <- function(i)
  { sapply(1:nrow(centers), function(d) diss(data[i,], centers[d,])) }
  assign1 <- function(i) { which.min(center.diss(i)) }
  sapply(1:nrow(data), assign1)
}

predict.k.centers <- function(model, data, diss=euc.dist)
{
  k.centers.assign(model$centers, data, diss)
}

k.centers.adjust.dummy <- function(clustering, data, k, diss, dm)
{
  do.call(rbind, lapply(1:k, function(d) data[which.max(clustering==d),]))
}

  # dummy k-centers clustering for the weathercl data
k.centers(wcl.std, 3)
```

The `k.centers` function can be instantiated to perform a specific algorithm from the k-centers family by specifying the `init`, `assign`, and `adjust` arguments as functions that perform the initialization, instance assignment, and center adjustment operations, respectively. The stop criteria are hard coded and satisfied on convergence or reaching a maximum number of iterations, specified via the `max.iter` argument. In practice, only the `adjust` argument needs to be specified for most algorithms, as it strongly and directly depends on the particular representation of cluster centers. As we will see below, the other operations have basic but usually satisfactory defaults.

The algorithm starts by calculating the dissimilarity matrix for the provided data, using the specified dissimilarity measure, which is then passed, along with that measure, to the `assign` and `adjust` functions. Passing both the dissimilarity matrix and the underlying dissimilarity measure is an obvious redundancy, justified only by the generic nature of the `k.centers` function. For some instantiations of the k-centers algorithm, it may be more convenient or efficient to perform the instance assignment or center adjustment operations using one or the other. Needless to say, precalculating the dissimilarity matrix and not using it afterward would be a striking computational efficiency loss, but it is forgivable in a simple illustration-only implementation.

The default and usually sufficient initialization operation, based on random training instance selection, is defined by the R code presented below:

```
k.centers.init.rand <- function(data, k) { data[sample(1:nrow(data), k),] }
```

The default instance assignment operation, that should not require any changes for commonly used *k*-centers algorithms, simply chooses the closest cluster for each instance and is defined by the following R code:

```
k.centers.assign <- function(centers, data, diss, dm)
{
    center.diss <- function(i)
    { sapply(1:nrow(centers), function(d) diss(data[i,], centers[d,])) }
    assign1 <- function(i) { which.min(center.diss(i)) }
    sapply(1:nrow(data), assign1)
}
```

A simple wrapper around the `k.centers.assign` function defined by the following R code can be used as the prediction function for *k*-centers models:

```
predict.k.centers <- function(model, data, diss=euc.dist)
{
    k.centers.assign(model$centers, data, diss)
}
```

Finally, the following R code implements a dummy center adjustment function that simply selects the first instance from each cluster as its center. This clearly makes little sense and serves only the purpose of illustration, since it makes it possible to fully instantiate the generic *k*-centers algorithm implemented in this example and demonstrate its application to the *weathercl* data.

```
k.centers.adjust.dummy <- function(clustering, data, k, diss, dm)
{
    do.call(rbind, lapply(1:k, function(d) data[which.max(clustering==d),]))
}

# dummy k-centers clustering for the weathercl data
k.centers(wcl.std, 3)
```

12.3 *k*-Means

The name of the most basic and common member of the *k*-centers algorithm family says (nearly) all about its operation principle. It uses vectors of mean attribute values (often referred to as *centroids*) as cluster centers, around which clustering formation proceeds.

12.3.1   Center adjustment

As stated above, the vector of mean attribute values of cluster members serves as each cluster's center vector. The center adjustment operation boils down therefore to recalculating the means, which can be written as

$$\zeta_{d,i} = m_{T^d}(a_i) = \frac{1}{|T^d|} \sum_{x \in T^d} a_i(x) \tag{12.4}$$

Needless to say, this is only applicable to continuous attributes. Discrete attributes may be numerically encoded (which also makes dissimilarity or similarity measures for continuous attributes directly applicable) using the technique presented in Section 17.3.5. A simple alternative that may be reasonable when there are just a few discrete attributes and when the employed dissimilarity measure can handle them directly is to use modes instead of means for them.

The simplicity of the *k*-means algorithm makes it easy to implement and efficient as well as straightforward to apply. This justifies its role of a typical default first attempt clustering algorithm for datasets with continuous attributes. It is not as common in application to datasets with mixed attribute sets, and definitely not a good choice for datasets with discrete attributes only. It usually requires a small number of iterations to reach or closely approach convergence and can be definitely considered a fast clustering algorithm, although in some rare worst cases the number of iterations required to fully converge may grow exponentially with the training set size.

The *k*-means algorithm is usually presented and implemented as coupled with the Euclidean dissimilarity. Although it is indeed particularly natural to combine mean attribute values for center representation with sums of squared attribute differences for measuring dissimilarity, technically nothing prevents us from using the algorithm with any other dissimilarity measures appropriate for a given domain. This requires care, however, since convergence guarantees are lost if the center adjustment operation does not reduce the total dissimilarity between training instances and their cluster centers under the adopted dissimilarity measure.

Example 12.3.1 The following R code implements the *k*-means version of the center adjustment operation and demonstrates the application of the correspondingly instantiated *k*-centers algorithm to the *weathercl*, *Iris*, and *Glass* data (with target concepts skipped for the latter two, but *k* set to the corresponding numbers of classes). Discrete attributes are handled by using modes, determined by the `modal` function, instead of means. The auxiliary `attr.mm` function not only generates a one-row dataframe containing the means of continuous attributes and the modes of discrete attributes for the supplied dataset, but can also be used to calculate other statistics, specified using the `mc` argument for the former and `md` argument for the latter.

> EX. 2.4.19
> `dmr.stats`

The previously defined cluster membership prediction method is also demonstrated. For the two larger datasets, training set cluster assignments and test set cluster membership predictions are compared to class labels to see to what extent the discovered instance similarity patters match the target concepts.

```
## attribute value means, medians, or modes
attr.mm <- function(data, mc=mean, md=modal)
{
   data.frame('names<-'(lapply(data, function(v)
```

```
                              if (is.numeric(v)) mc(v) else md(v)),
                  names(data)))
}

## k-means center adjustment
k.centers.adjust.mean <- function(clustering, data, k, diss, dm)
{
  do.call(rbind, lapply(1:k, function(d) attr.mm(data[clustering==d,])))
}

  # k-means clustering
w.kmeans <- k.centers(wcl.std, 3, adjust=k.centers.adjust.mean)
w.kmeans$centers

i.kmeans <- k.centers(i.std.train[,-5], 3, adjust=k.centers.adjust.mean)
g.kmeans <- k.centers(g.std.train[,-10], 7, adjust=k.centers.adjust.mean)

  # k-means prediction
w.kmeans$clustering
predict(w.kmeans, wcl.std)

i.pred.kmeans <- predict(i.kmeans, i.std.test[,-5])
g.pred.kmeans <- predict(g.kmeans, g.std.test[,-10])

  # clusters vs. classes on the training set
table(i.kmeans$clustering, i.std.train$Species)
table(g.kmeans$clustering, g.std.train$Type)

  # clusters vs. classes on the test set
table(predict(i.kmeans, i.std.test[,-5]), i.std.test$Species)
table(predict(g.kmeans, g.std.test[,-10]), g.std.test$Type)

  # attribute distribution within clusters for the Iris data
par(mfrow=c(2, 2))
for (attr in names(i.std.train)[1:4])
  boxplot(i.std.train[[attr]]~i.kmeans$clustering, main=attr)
```

While a clear relationship between dissimilarity-based clusters and target concept classes can be observed for the *Iris* and *Glass* dataset, it is not nearly as strong as one could expect from a classification model, with instances of different classes occurring in the same clusters with comparable frequency. It is interesting to notice, by the way, that for both the datasets one of the identified clusters is particularly small compared to the others, which might indicate that the values of *k* used in the example calls exceed the numbers of clusters that can be actually discovered.

For the *Iris* dataset, which has four continuous attributes, boxplots illustrating their distribution within clusters are generated and presented in Figure 12.1. High attribute value dispersion can be observed in cluster 3, which also differs from the other two clusters with respect to the attribute value location. The latter appear more homogeneous and similar, but can be easily distinguished by the Sepal.Width attribute.

12.3.2 Minimizing dissimilarity to centers

One reason to particularly prefer the Euclidean dissimilarity to other dissimilarity measures for *k*-means is that it makes the operation and results of the algorithm particularly clearly interpretable. It is easy to verify that adjusting cluster centers to become cluster member

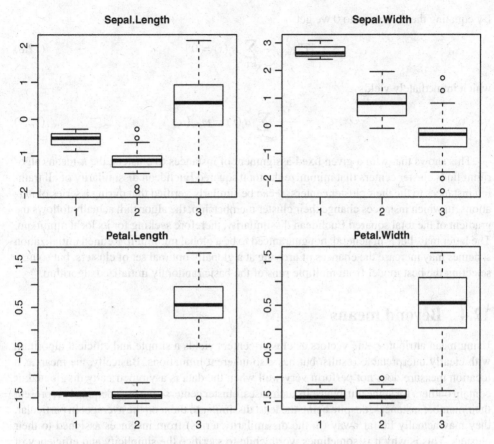

Figure 12.1   Attribute distribution within *k*-means clusters for the *Iris* data.

mean attribute value vectors guarantees minimizing the total squared Euclidean dissimilarity of all training instances to the corresponding cluster centers over all possible cluster center adjustment schemes. Consider the following criterion to be minimized:

$$\Delta_T^{(2)}(h) = \sum_{x \in T} \delta_{\text{euc}}^2(x, \zeta_{h(x)}) = \sum_{x \in T} \sum_{i=1}^{n} (a_i(x) - \zeta_{d,i})^2 \tag{12.5}$$

where *h* denotes the clustering model represented by a set of cluster centers $\zeta_1, \zeta_2, \ldots, \zeta_k$. The criterion can be reorganized as follows:

$$\Delta_T^{(2)}(h) = \sum_{d=1}^{k} \sum_{x \in T^d} \sum_{i=1}^{n} (a_i(x) - \zeta_{d,i})^2 \tag{12.6}$$

It is then easy to see that its derivative with respect to the *i*th element of the center vector of cluster *d* can be calculated as

$$\frac{\partial \Delta_T^{(2)}(h)}{\partial \zeta_{d,i}} = 2 \sum_{x \in T^d} (\zeta_{d,i} - a_i(x)) \tag{12.7}$$

By equating the derivative to 0 we get

$$|T^d|\zeta_{d,i} - \sum_{x \in T^d} a_i(x) = 0 \qquad (12.8)$$

which immediately yields

$$\zeta_{d,i} = \frac{1}{|T^d|} \sum_{x \in T^d} a_i(x) = m_{T^d}(a_i) \qquad (12.9)$$

This shows that with a given fixed assignment of instances to clusters, the *k*-means algorithm finds cluster centers that minimize the total squared Euclidean dissimilarity of all training instances to the their cluster centers. It can be similarly verified that during a series of iterations, in which instances change their cluster membership, the algorithm actually follows the gradient of the total squared Euclidean dissimilarity, therefore seeking for its local minimum. The latter may, but is in general, not guaranteed to be a global minimum. Refined initialization schemes may increase the chances of arriving at a globally optimal set of clusters, but so may selecting the best model from multiple runs of the basic randomly initialized algorithm.

12.4   Beyond means

Using mean attribute value vectors as cluster centers yields a simple and efficient algorithm with clearly interpretable results, but has also inherent limitations. Basically, the mean as a location measure does not perform very well when the data is asymmetrically distributed or contain outliers for some attributes. In such cases, cluster centers may not adequately represent their member instances, despite being the least dissimilar to them on the average. In particular, they may actually lie far away (in the dissimilarity terms) from instances assigned to their clusters. This is why it is sometimes worthwhile to sacrifice the simplicity and efficiency of means for better representation properties.

12.4.1   *k*-Medians

The immediately self-suggesting idea is to replace attribute value means with attribute value medians. The median, while more costly to obtain, is often the preferred location measure. The *k*-centers algorithm instantiated to use attribute value median vectors as cluster centers yields the *k-medians* algorithm.

Like *k*-means, the *k*-medians algorithm can be used with arbitrary dissimilarity measures that are believed to be meaningful for a particular domain, but with the risk of losing convergence guarantees if the center adjustment operation does not reduce the total dissimilarity between training instances and cluster centers. It can be verified to be best suited to the Manhattan dissimilarity, for which it minimizes the total the total dissimilarity between training instances and cluster centers. With other dissimilarity measures this interpretation of *k*-medians clustering loses validity.

Example 12.4.1 The following R code defines the *k*-medians center adjustment operation, as a marginally modified version of the corresponding *k*-means adjustment function from the previous example (reusing the auxiliary `attr.mm` function defined there) and demonstrates the k-medians algorithm obtained by passing it to the `k.centers` function strictly following the pattern of the previous k-means demonstrations. The center vectors of the created clustering models differ from those observed for the latter noticeably, but the relationships between clusters and classes appear mostly similar. The smallest clusters contain more instances than before, particularly for the *Iris* dataset, making the k-medians clustering more balanced.

```
## k-medians center adjustment
k.centers.adjust.median <- function(clustering, data, k, diss, dm)
{
  do.call(rbind, lapply(1:k, function(d) attr.mm(data[clustering==d,], mc=median)))
}

  # k-medians clustering
w.kmedians <- k.centers(wcl.std, 3, adjust=k.centers.adjust.median)
w.kmedians$centers

i.kmedians <- k.centers(i.std.train[,-5], 3, adjust=k.centers.adjust.median)
g.kmedians <- k.centers(g.std.train[,-10], 7, adjust=k.centers.adjust.median)

  # k-medians prediction
w.kmedians$clustering
predict(w.kmedians, wcl.std)

i.pred.kmedians <- predict(i.kmedians, i.std.test[,-5])
g.pred.kmedians <- predict(g.kmedians, g.std.test[,-10])

  # clusters vs. classes on the training set
table(i.kmedians$clustering, i.std.train$Species)
table(g.kmedians$clustering, g.std.train$Type)

  # clusters vs. classes on the test set
table(predict(i.kmedians, i.std.test[,-5]), i.std.test$Species)
table(predict(g.kmedians, g.std.test[,-10]), g.std.test$Type)

  # attribute distribution within clusters for the Iris data
par(mfrow=c(2, 2))
for (attr in names(i.std.train)[1:4])
  boxplot(i.std.train[[attr]]~i.kmedoids$clustering, main=attr)
```

The boxplots illustrating attribute distribution within clusters for the *Iris* data, presented in Figure 12.2, substantially differ from those obtained in the previous example for k-means. While one cluster exhibits higher diversity than the two other clusters, similarly as before, this time the difference is smaller and it is the `Sepal.Length` attribute that best discriminates between them all.

12.4.2 k-Medoids

While the k-medians algorithm is usually more robust to asymmetric distributions and outliers than the k-means algorithm, it still does not guarantee that cluster centers – although minimizing the total dissimilarity to their cluster members – are actually similar to any instances. This is because, unless attributes are independent, median attribute value vectors may not resemble any existing instances from the training set or the whole domain at all.

This is where another, more refined approach to cluster center representation steps in. It consists in using *medoids* – selected cluster members that are the least dissimilar to other

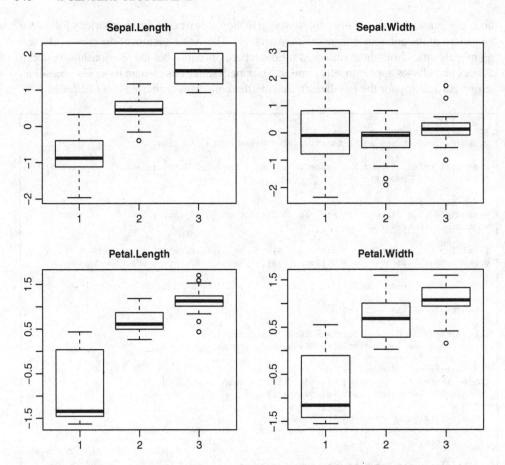

Figure 12.2   Attribute distribution within *k*-medians clusters for the *Iris* data.

cluster members, on the average – as cluster centers. The corresponding instantiation of the *k*-centers algorithm is referred to as the *k-medoids* algorithm. The identification of cluster medoids is a computationally intensive operation, as it requires that all pairwise dissimilarities within each cluster are calculated and aggregated, to obtain per-instance averages, which are then searched for minima. It usually makes sense to have the full dissimilarity matrix for the training set precalculated and use appropriately selected submatrices thereof for center adjustment. The algorithm is still considerably more expensive computationally than the simpler *k*-means and *k*-medians algorithms, but in some applications or for some domains the advantage of having existing instances used as cluster center vectors is worth the extra expense. It makes the algorithm particularly robust with respect to noise and outliers. It is also completely safe to use in combination with arbitrary dissimilarity measures, without losing convergence guarantees and the meaningfulness of the results, since it explicitly minimizes the total dissimilarity between training instances and their respective cluster centers regardless of the adopted dissimilarity measure, with the constraint that cluster centers are instances from the training set themselves.

Example 12.4.2 The R code presented below implements the *k*-medoids cluster center adjustment operation and demonstrates the application of the resulting algorithm in the very same way as before for *k*-means and *k*-medians. The observed distribution of instances of different classes across clusters resembles that observed for the latter rather than for the former. As we can see from the boxplots of attribute values within clusters displayed in Figure 12.3, there is one cluster with particularly low attribute value dispersion. No single attribute appears to clearly separate all the clusters.

```
## medoid for data with respect to dissimilarity matrix dm
medoid <- function(data, dm)
{
  data[which.min(colMeans(dm)),]
}

## k-medoids center adjustment
k.centers.adjust.medoid <- function(clustering, data, k, diss, dm)
{
  do.call(rbind, lapply(1:k, function(d)
                   medoid(data[clustering==d,],
                       as.matrix(dm)[clustering==d, clustering==d])))
}

  # k-medoids clustering
w.kmedoids <- k.centers(wcl.std, 3, adjust=k.centers.adjust.medoid)
w.kmedoids$centers

i.kmedoids <- k.centers(i.std.train[,-5], 3, adjust=k.centers.adjust.medoid)
g.kmedoids <- k.centers(g.std.train[,-10], 7, adjust=k.centers.adjust.medoid)

  # k-medoids prediction
w.kmedoids$clustering
predict(w.kmedoids, wcl.std)

i.pred.kmedoids <- predict(i.kmedoids, i.std.test[,-5])
g.pred.kmedoids <- predict(g.kmedoids, g.std.test[,-10])

  # clusters vs. classes on the training set
table(i.kmedoids$clustering, i.std.train$Species)
table(g.kmedoids$clustering, g.std.train$Type)

  # clusters vs. classes on the test set
table(predict(i.kmedoids, i.std.test[,-5]), i.std.test$Species)
table(predict(g.kmedoids, g.std.test[,-10]), g.std.test$Type)

  # attribute distribution within clusters for the Iris data
par(mfrow=c(2, 2))
for (attr in names(i.std.train)[1:4])
  boxplot(i.std.train[[attr]]~i.kmedoids$clustering, main=attr)
```

12.4.2.1 Partitioning around medoids

One particularly popular variation of the *k*-medoids algorithm is known as the PAM algorithm (*Partitioning Around Medoids*). It departs from the common *k*-centers clustering pattern by

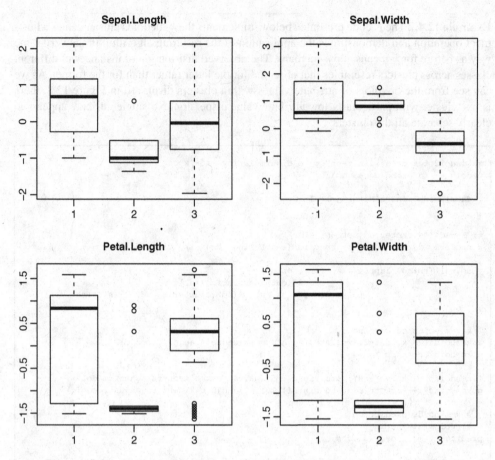

Figure 12.3   Attribute distribution within *k*-medoids clusters for the *Iris* data.

replacing the instance assignment and center adjustment operations with a medoid swap operation which, when iterated over all clusters and all nonmedoid training instances, turns out to achieve roughly the same effect more efficiently. For a given cluster d and instance x, the swap operation consists in making x a new center ζ_d of cluster d if it reduces the total dissimilarity between training instances and their respective closest cluster centers.

12.5   Beyond (fixed) *k*

Assuming a fixed predetermined number of clusters helps the *k*-centers algorithm to maintain its simplicity and efficiency, but can also be considered a substantial deficiency. For many applications the right number of clusters does not follow from the domain knowledge and cannot be reliably guessed. While this situation is indeed uncomfortable, it is not as serious disadvantage as it first appears. This is because in the vast majority of clustering applications the potential range of *k* values to consider is quite small and one can try them all. A more refined – but not necessarily more useful – approach is to have the value of *k* adapted during the cluster formation process.

12.5.1   Multiple runs

For a clustering model to be a meaningful and useful description of similarity patterns present in the data, the number of clusters should usually not exceed a dozen or two. In practice, it is often possible to limit the range of k to just a few candidate values, and have multiple runs of the selected k-centers algorithm that produce different models, the best of which can be ultimately selected based on evaluation results. Chapter 14 presents a variety of clustering quality measures that can be used for this purpose.

12.5.2   Adaptive k-centers

Rather than trying out several candidate k values, one can start from a predetermined "best guess" k, but permit changing it on the go whenever it appears too large or too small for a given dataset. These situations are relatively straightforward to identify, although there is room for several specific approaches that differ in some details:

Too many clusters. The two closest cluster centers are too close and can be joined (decrease k).

Too little clusters. The most diverse cluster is too diverse and can be split (increase k).

The exact criteria used are usually based on the dissimilarity between the cluster centers or the average pairwise intercluster instance dissimilarity for the former, and on the average dissimilarity from the cluster center or on the average pairwise intracluster similarity for the latter.

12.6   Explicit cluster modeling

The cluster modeling process performed by k-centers clustering algorithms was described above as *implicit*, since it can be viewed as a side effect of cluster formation. The same center vectors that are used to identify clusters subsequently serve the purpose of cluster membership prediction. Arbitrary instances from the domain can be assigned their closest clusters based on the same dissimilarity or similarity measure that is employed in the instance assignment phase of the algorithm. While this perfectly matches the requirements of most applications, it may be sometimes desirable to have an *explicit* model for cluster membership prediction that does not depend either on the center vectors of the original k-centers model or on the underlying dissimilarity measure.

The main motivation for creating an explicit cluster model is the incomprehensibility of k-centers implicit cluster membership prediction. Thus makes it impossible to explain why a particular instance is assigned to a given cluster other than by referring to the dissimilarity between the former and the center vector of the latter. Providing alternative, human-readable criteria for cluster membership makes it possible to much better understand what particular clusters have in common, how they differ, and what makes an instance assigned to one cluster or another. This may be important especially if the primary purpose of the clustering model is to describe the similarity patterns detected in the training set and, hopefully, occurring in the whole domain.

An explicit cluster membership model can be created as a classification model, by applying a classification algorithm with cluster membership assignment used as the target concept. An arbitrary classification algorithm can be applied to the original training set with an additional attribute, representing such a target concept, added. Clearly, to meet the comprehensibility expectations presented above as the primary motivation for explicit cluster modeling, a classification algorithm employing a human-readable model representation should be employed. As in most other situations where this is required, decision trees extensively discussed in Chapter 3 tend to be the most common approach.

It is noteworthy that an explicit cluster membership model can only be expected to approximate, but not perfectly mimic the original cluster membership assignments based on the dissimilarity to the center vectors of the underlying *k*-centers model. The simplicity and interpretability of the explicit model representation may be often more important than the degree of match between its predictions and the original *k*-centers assignments, since – unlike in a "normal" classification task – the latter do not represent any objective property of instances anyway.

Example 12.6.1 The idea of explicit cluster modeling is illustrated by the R code presented below, which uses the `rpart` package to create decision tree cluster membership prediction models approximating cluster assignments of the *k*-means clustering model for the *Iris* and *Glass* datasets from Example 12.3.1. Notice that the `minsplit` argument, specifying the minimum number of instances required for a split to be considered, is set to the smallest cluster size. The cost-complexity parameter `cp` is adjusted to achieve relatively simple trees.

```
# explicit decision tree representations of k-means models
i.kmeans.tree <- rpart(cluster~.,
                  cbind(i.train, cluster=as.factor(i.kmeans$clustering)),
                  minsplit=min(table(i.kmeans$clustering)), cp=0.05)
g.kmeans.tree <- rpart(cluster~.,
                  cbind(g.train, cluster=as.factor(g.kmeans$clustering)),
                  minsplit=min(table(g.kmeans$clustering)), cp=0.05)

# cluster membership prediction tree plots
prp(i.kmeans.tree, varlen=0, faclen=0, main="Iris")
prp(g.kmeans.tree, varlen=0, faclen=0, main="Glass")

# predicted vs. true clusters
confmat(predict(i.kmeans.tree, i.train, type="c"), i.kmeans$clustering)
confmat(predict(i.kmeans.tree, i.test, type="c"),
      predict(i.kmeans, i.std.test[,-5], euc.dist))

confmat(predict(g.kmeans.tree, g.train, type="c"), g.kmeans$clustering)
confmat(predict(g.kmeans.tree, g.test, type="c"),
      predict(g.kmeans, g.std.test[,-10], euc.dist))
```

The resulting models, visualized as presented in Figure 12.4 using the `prp` function from the `rpart.plot` package, confirm that cluster membership can indeed be represented in a human-readable way. The confusion matrices generated using the `confmat` function show a nearly perfect match between the original and predicted cluster assignments for the *Iris* data. For the *Glass* data the accuracy of cluster membership predictions may leave somewhat to be desired, but is not bad given the much higher number of clusters.

EX. 7.2.4
dmr.claseval

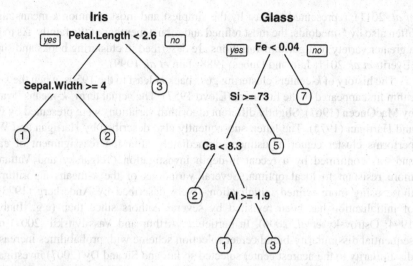

Figure 12.4 Decision tree cluster membership models.

12.7 Conclusion

The family of k-centers clustering algorithms provides a good balance of simplicity, efficiency, and customizability that makes them quick and easy to apply to many real-world clustering tasks. With the quality of results being often on par with that achieved by more refined algorithms, such as probabilistic clustering, they have become extremely popular and widely used. The quality comes at a usually unprohibitive computational cost, which can be further reduced by parallel or distributed implementations. This makes it possible to cluster large datasets, the capability that is becoming increasingly essential for applications in several areas, such as clustering customers in the retail or service industries, text documents, or biological data.

Despite their simplicity, algorithms discussed in this chapter belong to the most universal clustering algorithms that can be applied in a variety of domains, incorporating custom domain-specific dissimilarity measures where necessary or using the most appropriate of standard dissimilarity measures otherwise. While standard k-centers algorithms are crisp and flat clustering algorithms, they can be modified or wrapped appropriately to achieve fuzzy or hierarchical clustering. Their two apparent disadvantages – fixed k and the dependence of results on cluster initialization – are largely ameliorated by their efficiency, which makes it possible to run them multiple times and select the most satisfactory result. The latter is possible given appropriate clustering quality measures.

12.8 Further readings

k-Centers clustering algorithms belong not only to the most often used clustering algorithms, but also to the most often described ones. They are covered by most data mining and machine learning books that have clustering within their scope (e.g., Cios *et al.* 2007; Han *et al.* 2011; Hastie *et al.* 2011; Tan *et al.* 2013; Theodoridis and Koutroumbas 2008; Webb 2002; Witten *et al.* 2011), represented at least by the simplest and most common *k*-means algorithm, but often also by *k*-medoids, the most refined and robust member of the family. As to be expected, a greater variety of algorithm variations are described in clustering books and survey articles (Everitt *et al.* 2011; Jain and Dubes, 1988; Jain *et al.* 1999).

The history of *k*-centers clustering goes back at least to the 1950s, when the *k*-means algorithm first appeared in the literature (Lloyd 1957). The actual term "*k*-means" was introduced by MacQueen (1967). Slightly different algorithm variations were presented by Forgy (1965) and Hartigan (1975). The latter, subsequently also described by Hartigan and Wong (1979), performs cluster center adjustment immediately after the reassignment of each instance and – as confirmed by a recent in-depth investigation (Telgarsky and Vattani 2010) – is more resistant to local optima. Several variations of the *k*-means algorithm – including those using more refined initialization – were described by Anderberg (1973). The issue of initialization has been revisited by several authors since then (e.g., Inaba and Imai, 1994; Ostrovsky *et al.* 2006). In particular, Arthur and Vassilvitskii (2007) proposed the sequential dissimilarity-based center selection scheme with probabilities increasing with the dissimilarity to the nearest center selected so far, and Su and Dy (2007) investigated possible deterministic initialization methods, including PCA- and variance-based partitioning. The *ISODATA* algorithm proposed by Ball and Hall (1965) adjusts the number of clusters by merging and splitting. Some guidelines for the choice of *k* follow from the study performed by Dubes (1987).

k-Centers algorithms have been originally designed and remain to be the most frequently used for numerical-only data. Extending them to discrete attributes, using modes instead of means or medians, is conceptually straightforward, but several enhancements may be needed for the resulting *k*-modes algorithm to be sufficiently efficient and deliver good results (Chaturvedi *et al.* 2001; Huang, 1998). In particular, the quality of clusters created by *k*-modes can be improved by a more refined dissimilarity measure (Ng *et al.* 2007). An alternative approach to *k*-centers clustering with discrete attributes, incorporating attribute value distribution to cluster representation and dissimilarity calculation, was proposed by San *et al.* (2004).

The PAM algorithm was introduced by Kaufman and Rousseeuw (1987) and subsequently described in their book (Kaufman and Rousseeuw 1990), along with several other clustering algorithms that have acronyms coinciding with female names. An alternative efficient but simpler form of *k*-medoids clustering that actually better matches the common operation scheme of *k*-centers algorithms was presented by Park and Jun (2009). The idea of fuzzy clustering was first proposed by Ruspini (1969) and then extensively discussed by Bezdek (1981). Several studies addressed the issue of achieving comprehensible symbolic representations of clustering models (e.g., Diday and Simon, 1976; Michalski *et al.* 1981).

As this books covers only clustering algorithms based on explicit dissimilarity or similarity measures, it makes sense to mention here at least some of the most noteworthy approaches to clustering that do not fit in this category. These include, in particular, distribution-based clustering, in which clusters are identified as groups of instances most likely to come from the same probability distribution. A clustering model can then be represented by a mixture of probability distributions, the parameters of which have to be identified from the data. This can be achieved using the expectation-maximization (EM) algorithm (Dempster *et al.* 1977; McLachlan and Peel, 2000). Another well-known clustering algorithm is DBSCAN (Ester *et al.* 1996), a prominent representative of density-based approaches which identify clusters as data regions of increased density.

References

Anderberg MR 1973 *Cluster Analysis for Applications*. Academic Press.

Arthur D and Vassilvitskii S 2007 k-means++: The advantages of careful seeding *Proceedings of the Eighteenth Annual ACM-SIAM Symposium on Discrete Algorithms (SODA-2007)*. Society for Industrial and Applied Mathematics.

Ball G and Hall DJ 1965 ISODATA, a novel method of data analysis and classification. Technical report, Stanford University.

Bezdek JC 1981 *Pattern Recognition with Fuzzy Objective Function Algorithms*. Plenum Press.

Chaturvedi A, Green PE and Carroll JD 2001 K-modes clustering. *Journal of Classification* **18**, 35–55.

Cios KJ, Pedrycz W, Swiniarski RW and Kurgan L 2007 *Data Mining: A Knowledge Discovery Approach*. Springer.

Dempster AP, Laird NM and Rubin DB 1977 Maximum likelihood from incomplete data via the EM algorithm. *Journal of the Royal Statistical Society B* **39**, 1–38.

Diday E and Simon JC 1976 Clustering analysis In *Digital Pattern Recognition* (ed. Fu KS) Springer.

Dubes RC 1987 How many clusters are best? – an experiment. *Pattern Recognition* **20**, 645–663.

Ester M, Kriegel HP, Sander J and Xu X 1996 A density-based algorithm for discovering clusters in large spatial databases with noise *Proceedings of the Second International Conference on Knowledge Discovery and Data Mining (KDD-96)*. AAAI Press.

Everitt BS, Landau S, Leese M and Stahl D 2011 *Cluster Analysis* 5th edn. Wiley.

Forgy EW 1965 Cluster analysis of multivariate data: Efficiency versus interpretability of classifications. *Biometrics* **21**, 768–769.

Han J, Kamber M and Pei J 2011 *Data Mining: Concepts and Techniques* 3rd edn. Morgan Kaufmann.

Hartigan JA 1975 *Clustering Algorithms*. Wiley.

Hartigan JA and Wong MA 1979 Algorithm AS 136: A k-means clustering algorithm. *Applied Statistics* **28**, 100–108.

Hastie T, Tibshirani R and Friedman J 2011 *The Elements of Statistical Learning: Data Mining, Inference, and Prediction* 2nd edn. Springer.

Huang Z 1998 Extensions to the k-means algorithm for clustering large data sets with categorical values. *Data Mining and Knowledge Discovery* **2**, 283–304.

Inaba, M. Katoh N and Imai H 1994 Applications of weighted Voronoi diagrams and randomization to variance-based k-clustering *Proceedings of the Tenth Annual Symposium on Computational Geometry (SCG-94)*. ACM Press.

Jain AK and Dubes RC 1988 *Algorithms for Clustering Data*. Prentice-Hall.

Jain AK, Murty MN and Flynn PJ 1999 Data clustering: A review. *ACM Computing Surveys* **31**, 264–323.

Kaufman L and Rousseeuw PJ 1987 Clustering by means of medoids In *Statistical Data Analysis Based on the L_1 Norm and Related Methods* (ed. Dodge Y) North-Holland.

Kaufman L and Rousseeuw PJ 1990 *Finding Groups in Data: An Introduction to Cluster Analysis*. Wiley.

Lloyd SP 1957 Least squares quantization in PCM. Technical report, Bell Laboratories. Reprinted in 1982 in *IEEE Transactions on Information Theory*, **28**, 128–137.

MacQueen J 1967 Some methods for classification and analysis of multivariate observations *Proceedings of the Fifth Berkeley Symposium on Mathematical Statistics and Probability*. University of California Press.

McLachlan GJ and Peel DA 2000 *Finite Mixture Models*. Wiley.

Michalski RS, Stepp RE and Diday E 1981 A recent advance in data analysis: Clustering objects into classes characterized by conjunctive concepts In *Progress in Pattern Recognition* (ed. Kanal L and Rosenfeld A) vol. 1, North-Holland.

Ng MK, Junjie M, Joshua L, Huang Z and He Z 2007 On the impact of dissimilarity measure in *k*-modes clustering algorithm. *IEEE Transactions on Pattern Analysis and Machine Intelligence* **29**, 503–507.

Ostrovsky R, Rabani Y, Schulman LJ and Swamy C 2006 The effectiveness of Lloyd-type methods for the *k*-means problem *Proceedings of the Fourty-Seventh Annual IEEE Symposium on Foundations of Computer Science (FOCS-2006)*. IEEE Press.

Park HS and Jun CH 2009 A simple and fast algorithm for K-medoids clustering. *Expert Systems with Applications* **36**, 3336–3341.

Ruspini EH 1969 A new approach to clustering. *Information and Control* **15**, 22–32.

San OM, Huynh VM and Naamori Y 2004 An alternative extension of the *k*-means algorithm for clustering categorical data. *International Journal of Applied Mathematics and Computer Science* **14**, pp. 241–247.

Su T and Dy JG 2007 In search of deterministic methods for initializing *k*-means and Gaussian mixture clustering. *Intelligent Data Analysis* **11**, 319–338.

Tan PN, Steinbach M and Kumar V 2013 *Introduction to Data Mining* 2nd edn. Addison-Wesley.

Telgarsky M and Vattani A 2010 Hartigan's method: *k*-means clustering without Voronoi *Proceedings of the Thirteenth International Conference on Artificial Intelligence and Statistics (AISTATS-10)*. JMLR Workshop and Conference Proceedings.

Theodoridis S and Koutroumbas K 2008 *Pattern Recognition* 4th edn. Academic Press.

Webb AR 2002 *Statistical Pattern Recognition* 2nd edn. Wiley.

Witten IH, Frank E and Hall MA 2011 *Data Mining: Practical Machine Learning Tools and Techniques* 3rd edn. Morgan Kaufmann.

13

Hierarchical clustering

13.1   Introduction

Hierarchical clustering extends the basic clustering task by requesting that the created clustering model is hierarchical. With nodes of a cluster hierarchy representing clusters, and their descendants representing their subclusters, such a model can be viewed as a combination of multiple clustering models, applicable to different domain regions. It is therefore not surprising that hierarchical clustering is much more computationally demanding than flat clustering. However, this increased computational complexity does not coincide with increased conceptual or algorithmic complexity, since the process of cluster hierarchy formation can be organized as a sequence of basic cluster merging or partitioning operations.

13.1.1   Basic approaches

This chapter reviews two approaches to creating hierarchical clustering models, both of which have very simple formulations:

Agglomerative clustering. A bottom-up approach which starts with many small clusters and iteratively merges selected clusters until a single root cluster is reached.

Divisive clustering. A top-down approach which starts with a single root cluster and iteratively partitions existing clusters into subclusters.

For both these approaches, we will assume that cluster merging or partitioning decisions are made based on an instance dissimilarity (or similarity) measure. While they explicitly perform cluster hierarchy formation only, we will see that clustering trees can also be viewed as model representations and applied to predict cluster membership for new data.

13.1.2   (Dis)similarity measures

Most hierarchical clustering algorithms can be combined with arbitrary dissimilarity or similarity measures, such as those presented in Chapter 11. Similarly as for k-centers clustering,

---

Data Mining Algorithms: Explained Using R, First Edition. Paweł Cichosz.
© 2015 John Wiley & Sons, Ltd. Published 2015 by John Wiley & Sons, Ltd.

it is the Euclidean distance for continuous attributes, the Hamming distance for discrete attributes, and Gower's coefficient for mixed attribute types, which are the most common. While the latter compensates for different ranges of continuous attributes internally, the former usually require data preprocessing by standardization or normalization.

The precautions discussed in Sections 11.3.9 and 12.1.2 also apply here. Whenever a hierarchical clustering model is supposed to be used to predict cluster membership for new instances (i.e., clustering is performed as a form of predictive modeling), the transformation parameters determined on the training set have to be retained and re-applied to any new instances prior to prediction. The general idea of modeling transformations is more extensively discussed in Section 17.2.5. Standardization and normalization are presented as modeling transformations in Sections 17.3.1 and 17.3.2.

Example 13.1.1 R code examples illustrating hierarchical clustering algorithms presented in this chapter use the toy *weathercl* dataset, the very small size of which makes it possible to manually verify the algorithms' operation, and two more real- | Ex. 1.5.1
istic datasets: *Iris* from the standard `datasets` package and *Glass* from the | dmr.data
`mlbench` package. Dissimilarity measures will be calculated using functions from Chapter 11 available in the `dmr.dissim` package, standardization will be performed using functions from the `dmr.trans` package, and the `cluster` package will provide a flat clustering algorithm for divisive hierarchical clustering. The environment for the examples is prepared below by loading required R packages and the datasets and partitioning the two larger ones into training and test subsets. The datasets are standardized using the | Ex. 17.3.1
`std.all` and `predict.std` functions. Notice that standardization parameters | dmr.trans
determined on the training sets are applied to both the training and test sets.

```
library(dmr.dissim)
library(dmr.trans)

library(cluster)

data(weathercl, package="dmr.data")
data(iris)
data(Glass, package="mlbench")

set.seed(12)

ri <- runif(nrow(iris))
i.train <- iris[ri>=0.33,]
i.test <- iris[ri<0.33,]

rg <- runif(nrow(Glass))
g.train <- Glass[rg>=0.33,]
g.test <- Glass[rg<0.33,]

wcl.std <- predict.std(std.all(.~., weathercl), weathercl)

i.stdm <- std.all(Species~., i.train)
i.std.train <- predict.std(i.stdm, i.train)
i.std.test <- predict.std(i.stdm, i.test)
```

```
g.stdm <- std.all(Type~., g.train)
g.std.train <- predict.std(g.stdm, g.train)
g.std.test <- predict.std(g.stdm, g.test)
```

13.2   Cluster hierarchies

A hierarchical clustering model is a multilevel hierarchy of clusters. Each internal node of the hierarchy represents a cluster that is partitioned into subclusters represented by its descendants. Leaves correspond to clusters that are not further divided, and the root node – to the whole domain. Before proceeding with the presentation of hierarchical clustering algorithms, it makes sense to spend a while on discussing why such models could be useful and how they could be represented.

13.2.1   Motivation

Since creating cluster hierarchies takes some considerable additional effort compared to flat clustering, it is more than reasonable to consider the additional motivation behind it that would make this effort justified. While the motivation for clustering in general, discussed in Section 1.5, remains valid, for each of previously discussed application domains, there may be indeed situations where flat clustering is insufficient.

In applications where the primary purpose of clustering is to discover and describe similarity patterns in the data the need for hierarchical clustering models may arise if those patterns are believed to be too complex to be adequately captured by a flat clustering model. This is the case when the diversity in the data is so high that a small number of clusters (typical for flat clustering) does not provide sufficient intracluster similarity and a large number of independent, unorganized clusters does not enable meaningful interpretation. Hierarchical clustering makes it possible to organize these many clusters into a hierarchy, and use a varying level of resolution to analyze them. Instead of trying to find a compromise number of clusters, not too little to hide any useful patterns and not too large to make one lost in overwhelming details, a cluster tree is produced with the possibility of adjusting the resolution on an *ad hoc* basis. One domain for which this is particularly desirable is that of text documents (e.g., press articles, web pages, discussion group messages, blog posts, etc.), where the variety of topics or styles may be huge and a hierarchical clustering model may be extremely helpful in understanding this variety and putting some order into it. Similar needs arise for several biological or medical domains, where similarity structures for organisms or molecular sequences needs to be identified. This application domain is where hierarchical clustering is originated from.

The variable resolution capability provided by hierarchical clustering that makes it possible to adjust the number of clusters to the current needs may also be useful in other domains where the proper level of tradeoff between the number of clusters, their size, and cohesivity cannot be determined *a priori* or may dynamically change. A set of flat clustering models with different numbers of clusters may not fit the bill due to the possible inconsistency between them.

If the purpose of clustering is not only to provide a useful insights into the domain, but also to predict hidden attribute values based on cluster membership, then choosing the right number of clusters may be critical. With too little clusters, the high intracluster diversity will enable only rough predictions. With too many clusters, they may not be sufficiently representative to enable reliable predictions. Different clustering resolution levels may be appropriate for predicting different hidden attributes. With a hierarchical clustering model it becomes possible not to decide about the number of clusters once forever, but to retain the capability of using different numbers of clusters depending on what and how precisely and reliably has to be predicted. Again, the variable clustering resolution is the key benefit.

For anomaly detection clustering applications, where instances not matching any cluster are considered suspicious, a hierarchical clustering model makes it possible to much better control the trade-off between missed anomalies and false alarms, and use different alarm thresholds in different domain regions. The resulting anomaly detection system may be then better adjusted to the actual needs of the particular application.

When clustering is used as a form of preparation for other data mining tasks, to decompose the domain, the capability of dynamically adjusting the resolution of this decomposition may also be a benefit. It makes it possible to examine the effects of decomposing into regions of varying levels of homogeneity. The typical scenario for such clustering applications is to create separate classification or regression models for all clusters. Adjusting the level of clustering resolution may have strong impact on the quality of these models.

All these possible benefits by no means imply that hierarchical clustering should always be preferred to flat clustering. The latter is not only more computationally efficient (for the most popular algorithms, at least), which makes it applicable to large datasets, but may also produce better models than a flat clustering model extracted from a hierarchical model.

13.2.2 Model representation

As discussed in Section 1.5.5, a hierarchical clustering model can be thought of as a set of ordinary (flat) clustering models organized in a tree structure. Each nonleaf node of such a tree is associated with a flat clustering, with its descendant nodes corresponding to subclusters, and at the same time represents a cluster of its parent's model. Formally, a hierarchical clustering model $h : X \times \mathcal{N} \rightarrow C_h$ assigns to each instance $x \in X$ and the hierarchy level $l \in \mathcal{N}$ (where $\mathcal{N}$ is the set of nonnegative integer numbers) the l-level cluster of x, $h(x, l)$ determined as follows:

$$h(x, l) = \begin{cases} d_0 & \text{if } l = 0 \\ h_{h(x,l-1)}(x) & \text{otherwise} \end{cases} \tag{13.1}$$

where d_0 denotes the root node cluster, corresponding to the whole domain, and h_d denotes the flat clustering model associated with cluster d for any $d \in C_h$. Simply speaking, instances descend the tree, on each level using the current node's clustering model to identify the next level node.

As we will see in subsequent sections, common approaches to creating hierarchical clustering models assume a *strictly binary tree* representation, with every nonleaf node having exactly two descendants, corresponding to two subclusters into which the cluster represented by the node is partitioned. With this assumption, it is easy to see than a hierarchical clustering tree with K leaf clusters contains $K - 1$ nonleaf clusters.

Example 13.2.1 To illustrate the idea of hierarchical clustering, the following R code generates a simple dendrogram plot, which is a visual representation of a complete binary clustering tree with four levels and eight leaves. An R dendrogram object is created as a nested list: a tree is represented by a list containing lists representing its subtrees, with R object attributes used to store properties required for dendrogram plotting. The obtained plot is presented in Figure 13.1.

```
dg.14 <- lapply(1:8, function(i)
                  {
                    d <- list(i)
                    attr(d, "members") <- 1
                    attr(d, "height") <- 0
                    attr(d, "leaf") <- TRUE
                    attr(d, "label") <- i
                    'class<-'(d, "dendrogram")
                  })

dgmerge <- function(dg, i1, i2)
{
  d <- list(dg[[i1]], dg[[i2]])
  attr(d, "members") <- attr(d[[1]], "members")+attr(d[[2]], "members")
  attr(d, "height") <- 1+max(attr(d[[1]], "height"), attr(d[[2]], "height"))
  attr(d, "leaf") <- FALSE
  lab <- if (is.null(attr(d[[1]], "edgetext"))) "label" else "edgetext"
  attr(d, "edgetext") <- paste(attr(d[[1]], lab), attr(d[[2]], lab), sep="+")
  'class<-'(d, "dendrogram")
}

dg.13 <- lapply(seq(1, length(dg.14)-1, 2), function(i) dgmerge(dg.14, i, i+1))
dg.12 <- lapply(seq(1, length(dg.13)-1, 2), function(i) dgmerge(dg.13, i, i+1))
dg.11 <- lapply(seq(1, length(dg.12)-1, 2), function(i) dgmerge(dg.12, i, i+1))

plot(dg.11[[1]], center=TRUE)
```

13.3   Agglomerative clustering

The family of agglomerative hierarchical clustering (AHC) algorithms adopts the most natural and direct method of discovering multilevel instance similarity patterns. Starting from an initial bottom-level clustering with many tiny (usually singleton) clusters, it builds a clustering tree by performing multiple merge operations. Each of them creates a new parent cluster for the two most similar existing clusters. It does not make much sense to consider merging more than two clusters in one step, since any other highly similar cluster can be merged-in on a subsequent iteration anyway. This is why clustering trees created by all practically used agglomerative hierarchical clustering algorithms are binary.

Operating by bottom-up cluster merging gives cluster hierarchies produced by agglomerative hierarchical clustering a straightforward interpretation. Each tree node represents a union of the most similar lower level clusters. If starting from singleton bottom-level clusters, it therefore fully describes the similarity structure identified in the training set.

13.3.1   Algorithm scheme

The operation of agglomerative hierarchical clustering is more precisely described by the algorithm scheme presented below. It starts from identifying a set of bottom-level clusters and assigning training instances to them. This is usually a trivial operation, since either singleton

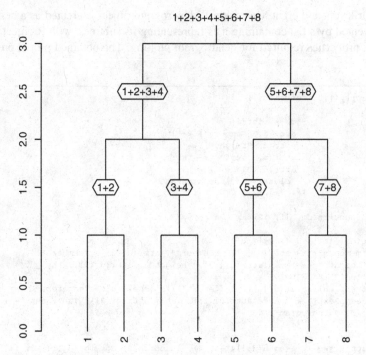

Figure 13.1 An illustrative dendrogram plot.

bottom-level clustering is assumed – with a separate cluster for each training instance – or an initial clustering is provided on input. The set of bottom-level clusters is also used to initialize the set of top-level clusters, i.e., the remaining candidates for merging. The algorithm then proceeds by iteratively selecting and merging top-level cluster pairs. The merge operation consists of assigning all the training instances from the selected two clusters to their newly created parent and modifying the set of top-level clusters accordingly. This is continued until only one top-level cluster remains, which is then the root of the created hierarchical clustering tree.

1: identify the set of bottom-level clusters C_{bottom};
2: **for all** $d \in C_{\text{bottom}}$ **do**
3:     identify the corresponding subset of training instances T^d;
4: **end for**
5: $C_{\text{top}} := C_{\text{bottom}}$;
6: **while** $|C_{\text{top}}| > 1$ **do**
7:     $d_{1*}, d_{2*} := \arg\min_{d_1, d_2 \in C_{\text{top}}} \Lambda_{\delta, T}(d_1, d_2)$;
8:     create new cluster d_* as parent of d_{1*} and d_{2*};
9:     $T^{d_*} := T^{d_{1*}} \cup T^{d_{2*}}$;
10:     $C_{\text{top}} := C_{\text{top}} - \{d_{1*}, d_{2*}\} \cup \{d_*\}$;
11: **end while**

The criterion used to select two clusters for merging in agglomerative hierarchical clustering is called the *linkage*. It is assumed above to be represented by a real-valued evaluation function $\Lambda_{\delta, T}$ the minimum value of which indicates the best candidates for merging. It is calculated using an instance dissimilarity measure δ and can be thought of as an extension of the former, applicable not only to single instances, but also to multi-instance clusters.

No clustering tree representation is explicitly created in the above pseudocode to keep it simple. It should be implicitly understood, however, that creating a new cluster as a parent of two existing clusters does actually perform an update of some structure which – upon the algorithm's completion – represents the created clustering tree.

Example 13.3.1 The R code presented below defines the `ahc` function, which is a simple implementation of agglomerative hierarchical clustering. Its operation is parameterized by a linkage function and a dissimilarity measure. The latter defaults to the Euclidean distance calculated using the `euc.dist` function, which also handles discrete attributes reasonably. There is no default for the former and the example call for

> Ex. 11.3.1
> `dmr.dissim`

the *weathercl* data uses a purely illustrative and totally useless cluster-size-based linkage that always merges the clusters with the least total size. More reasonable linkage functions will be defined by subsequent examples. Notice that the linkage matrix created at the beginning is then updated rather than recalculated on each iteration.

The representation of a hierarchical clustering tree created by the `ahc` function and passed on output via the `merge` component of the returned model object is the same as used by the popular R implementation of agglomerative hierarchical clustering: the `hclust` function from the standard `stats` package. It is a matrix with rows corresponding to all merged (parent) clusters and two columns, specifying the identifiers of their descendants. Merged clusters are identified by consecutive positive integers starting at 1 (which serve as indices of the corresponding rows of the `merge` matrix), and the initial bottom-level clusters (which have no corresponding rows in the `merge` matrix) – by consecutive negative integers starting from -1. The bottom-level clustering is supplied via the `bottom` argument and returned as the `clustering` component. The linkage function is also stored as the `link` component of the model object. Its remaining components – `height` and `order` – are of less interest, since they only serve the purpose of producing dendrograms, i.e., graphical representations of clustering trees. The former is a vector which assigns to each merged cluster a height value at which it should be drawn. It should reflect the order in which clusters are merged (with those merged later receiving higher height) and is typically set based on the linkage function value for the two merged clusters. The latter is a vector of bottom-level cluster numbers that specifies their ordering for dendrogram drawing. The `ahc` function sets it arbitrarily to the order of merging.

The class attribute of the returned hierarchical clustering model object is set to `hcl`, to permit dispatching some methods that will be defined later (including the prediction method). Due to the compatibility of its representation with that used by the `hclust` function a conversion method is also defined which simply alters the class attribute.

```
## agglomerative hierarchical clustering
ahc <- function(data, linkf=ahc.size, diss=euc.dist, bottom=1:nrow(data))
{
    # hclust-compatible cluster id scheme
    clid <- function(d)
    { if (d>length(bottom.clusters)) d-length(bottom.clusters) else -d }

    dm <- as.matrix(dissmat(data, diss))  # instance dissimilarity matrix for linkage

    bottom.clusters <- unique(bottom)    # bottom-level clusters
    clustering <- bottom                 # current cluster assignment
    clusters <- bottom.clusters          # current set of clusters

    merge <- NULL    # merge matrix
```

```
height <- NULL  # height vector
order <- NULL   # order vector

links <- outer(1:length(clusters), 1:length(clusters),
               Vectorize(function(i1, i2)
                         if (i1<i2)
                             linkf(clustering, clusters[i1], clusters[i2], data,
                                   diss, dm)
                         else NA))
while(length(clusters)>1)
{
  mli <- arrayInd(which.min(links), dim(links))  # minimum link index
  d1 <- clusters[i1 <- mli[1]]
  d2 <- clusters[i2 <- mli[2]]
  d12 <- max(clusters)+1
      # merge d1 and d2 into d12
  merge <- rbind(merge, c(clid(d1), clid(d2)))
  height <- c(height, if (is.null(height) || links[i1,i2]>height[length(height)])
                        links[i1,i2]
                      else height[length(height)]+height[1])  # height correction
  clustering[clustering==d1 | clustering==d2] <- d12
  clusters <- clusters[-c(i1, i2)]
  links <- links[-c(i1, i2),,drop=FALSE]  # remove links for d1
  links <- links[,-c(i1, i2),drop=FALSE]  # remove links for d2.
  if (length(clusters)>0)
  {
    links <- cbind(links, sapply(clusters,
                                 function(d) linkf(clustering, d, d12, data,
                                             diss, dm)))
    links <- rbind(links, NA)  # keep the matrix square
  }
  clusters <- c(clusters, d12)
}

'class<-'(list(clustering=bottom, link=linkf, data=data,
               merge=merge, height=height, order=-t(merge)[t(merge)<0]),
          "hcl")
}

## convert to hclust
as.hclust.hcl <- function(model) { 'class<-'(unclass(model), c("hclust")) }

## size linkage (dummy)
ahc.size <- function(clustering, d1, d2, data, diss, dm)
{ sum(clustering==d1) + sum(clustering==d2) }

  # agglomerative clustering for the weathercl data
wcl.ahc.d <- ahc(wcl.std, linkf=ahc.size)
as.hclust(wcl.ahc.d)
```

Notice that the ahc function preserves the bottom-level cluster identifiers of the supplied initial clustering. They are simply instance numbers with the default singleton initialization.

13.3.2   Cluster linkage

Several types of linkage are used for agglomerative hierarchical clustering. Their common purpose is to extend an instance dissimilarity measure to multi-instance clusters. The specific way of achieving this extension has of course a substantial impact on the properties of the resulting model, just like the underlying dissimilarity measure. Some of the most popular linkage types are presented below.

13.3.2.1 Single linkage

Single linkage uses the minimum dissimilarity value between instances from two clusters as the dissimilarity between the clusters:

$$\Lambda_{\delta,T}^{\text{single}}(d_1, d_2) = \arg \min_{\substack{x_1 \in T^{d_1} \\ x_2 \in T^{d_2}}} \delta(x_1, x_2) \tag{13.2}$$

where T^d designates the set of training instances assigned to cluster d. This makes the two closest instances determine how close the clusters will be considered. Even clusters with many largely dissimilar instances may be selected for merging with single linkage. This tends to yield internally diverse clusters.

Example 13.3.2 Single linkage for agglomerative hierarchical clustering is implemented and demonstrated by the following R code. Notice that the `ahc.single` function, although receives both the dissimilarity function and the dissimilarity matrix for the training set, uses only the latter. It makes sense for linkage functions which are based on dissimilarities between training instances only. As we will see later, this is not the case for all linkage types, which justifies passing the dissimilarity function as well. Single-linkage clustering models are created for the *weathercl*, *Iris*, and `Glass` data.

```
## single linkage
ahc.single <- function(clustering, d1, d2, data, diss, dm)
{ min(dm[clustering==d1, clustering==d2]) }

  # agglomerative hierarchical single-linkage clustering for the weathercl data
wcl.ahc.sl <- ahc(wcl.std, linkf=ahc.single)

  # agglomerative hierarchical single-linkage clustering for the iris data
i.ahc.sl <- ahc(i.std.train[,-5], linkf=ahc.single)

  # agglomerative hierarchical single-linkage clustering for the Glass data
g.ahc.sl <- ahc(g.std.train[,-10], linkf=ahc.single)
```

13.3.2.2 Complete linkage

Despite its name suggesting otherwise, *complete linkage* is based on the dissimilarity between single instances, just like single linkage. It is, however, the two most distant rather than the two closest ones that determine the dissimilarity between clusters:

$$\Lambda_{\delta,T}^{\text{complete}}(d_1, d_2) = \arg \max_{\substack{x_1 \in T^{d_1} \\ x_2 \in T^{d_2}}} \delta(x_1, x_2) \tag{13.3}$$

For two clusters to be merged under complete linkage even the two most dissimilar instances from these clusters must be sufficiently similar. If this is the case for the two most dissimilar instances, this is also the case for all the remaining ones, which is the true justification of the term "complete." This may be expected to yield much more compact clusters than single linkage.

Example 13.3.3 The following code implements complete linkage and demonstrates its application to agglomerative hierarchical clustering using the same datasets as in the previous example.

```
## complete linkage
ahc.complete <- function(clustering, d1, d2, data, diss, dm)
{ max(dm[clustering==d1, clustering==d2]) }

  # agglomerative hierarchical complete-linkage clustering for the weathercl data
wcl.ahc.cl <- ahc(wcl.std, linkf=ahc.complete)

  # agglomerative hierarchical complete-linkage clustering for the iris data
i.ahc.cl <- ahc(i.std.train[,-5], linkf=ahc.complete)

  # agglomerative hierarchical complete-linkage clustering for the Glass data
g.ahc.cl <- ahc(g.std.train[,-10], linkf=ahc.complete)
```

13.3.2.3   Average linkage

With both single and complete linkage being entirely dependent on the dissimilarity between just two particular instances – the most similar or the most dissimilar ones – both of them can be misled by outlying or otherwise unreliable attribute values, leading the former to unjustified merge or the latter to unjustified not-merge decisions. This makes their use with potentially noisy data problematic. One self-suggesting more robust linkage type is *average linkage*, in which the average dissimilarity between instances from two clusters serves as the dissimilarity between the clusters:

$$\Lambda_{\delta,T}^{\text{average}}(d_1, d_2) = \frac{1}{|T^{d_1}| \cdot |T^{d_2}|} \sum_{\substack{x_1 \in T^{d_1} \\ x_2 \in T^{d_2}}} \delta(x_1, x_2) \tag{13.4}$$

What makes this linkage popular, besides its increased noise resistance, is that it achieves a compromise between single and complete linkage, without being ready to merge clusters based on just two highly similar instances and without refusing to merge clusters based on just two highly dissimilar instances.

Example 13.3.4 Following the pattern of the previous two examples, the R code presented below implements and demonstrates average linkage.

```
## average linkage
ahc.average <- function(clustering, d1, d2, data, diss, dm)
{ mean(dm[clustering==d1, clustering==d2]) }

  # agglomerative hierarchical average-linkage clustering for the weathercl data
wcl.ahc.al <- ahc(wcl.std, linkf=ahc.average)

  # agglomerative hierarchical average-linkage clustering for the iris data
i.ahc.al <- ahc(i.std.train[,-5], linkf=ahc.average)

  # agglomerative hierarchical average-linkage clustering for the Glass data
g.ahc.al <- ahc(g.std.train[,-10], linkf=ahc.average)
```

13.3.2.4   Center linkage

Another and somewhat less popular approach to finding compromise between the extremes of single and complete linkage is *center linkage*. It borrows the idea of center vectors for cluster representation from k-centers clustering algorithms discussed in Chapter 12 and measures the dissimilarity between clusters by the dissimilarity between their centers:

$$\Lambda_{\delta,T}^{\text{center}}(d_1, d_2) = \delta(\zeta_{d_1}, \zeta_{d_2}) \tag{13.5}$$

where ζ_d denotes the center of cluster d. The most common instantiation of center linkage uses vectors of attribute value means as centers:

$$\zeta_{d,i} = m_{T^d}(a_i) = \frac{1}{|T^d|} \sum_{x \in T^d} a_i(x) \tag{13.6}$$

In this version, it is also known as *centroid linkage*. For discrete attributes, means can be replaced by modes. Other types of centers presented in Section 12.4 (vectors of attribute value medians or cluster medoids) are much less frequently adopted for center linkage.

Unlike the previously presented linkage types, center linkage does not have the monotonic property. The dissimilarity between clusters may sometimes drop when ascending the tree, i.e., a merged cluster may turn out closer to other clusters than its descendants. This is usually undesirable.

Example 13.3.5 The `ahc.center` function defined by the R code presented below implements average linkage, with vectors of attribute value means (for continuous attributes) or modes (for discrete attributes) used as cluster centers. These are calculated using the `attr.mm` function, first appearing the implementation of k-means clustering. Unlike the previous linkage implementations, it does not use the supplied instance dissimilarity matrix, but rather the instance dissimilarity function. This is because cluster centers are not (in general) training instances. Of course, it is a severe performance loss to precalculate the instance dissimilarity matrix which is not used later, but we can accept this striking inefficiency in illustrative code.

> Ex. 12.3.1
> dmr.kcenters

```
## center (mean/mode) linkage
ahc.center <- function(clustering, d1, d2, data, diss, dm)
{ diss(attr.mm(data[clustering==d1,]), attr.mm(data[clustering==d2,])) }

  # agglomerative hierarchical center-linkage clustering for the weathercl data
wcl.ahc.ml <- ahc(wcl.std, linkf=ahc.center)

  # agglomerative hierarchical center-linkage clustering for the iris data
i.ahc.ml <- ahc(i.std.train[,-5], linkf=ahc.center)

  # agglomerative hierarchical center-linkage clustering for the Glass data
g.ahc.ml <- ahc(g.std.train[,-10], linkf=ahc.center)
```

13.3.2.5   Ward linkage

Ward linkage, also referred to as Ward's method, is an example of more refined linkage types that are not directly based on dissimilarities between instances from the two clusters considered for merging, but rather on an explicit clustering quality criterion or objective function. Clusters to be merged are selected as those that yield the best top-level clustering after merging with respect to this criterion. The specific objective function adopted by Ward linkage is the total sum of squared dissimilarities between cluster members and cluster centers for all clusters. If used with the Euclidean dissimilarity and mean vectors as cluster centers, as originally intended, it represents the total intracluster variance and is also called minimum-variance linkage.

Ward's criterion depends on all current top-level clusters, not just on the two candidates for merging being evaluated, but – since a single merge operation does not affect any clusters other that the two being merged – it is actually equivalent to the following link function:

$$\Lambda_{\delta,T}^{\text{ward}}(d_1, d_2) = \sum_{x \in T^{d_1} \cup T^{d_2}} \delta^2(x, \zeta_{d_{12}})$$

$$- \left(\sum_{x \in T^{d_1}} \delta^2(x, \zeta_{d_1}) + \sum_{x \in T^{d_2}} \delta^2(x, \zeta_{d_2}) \right) \tag{13.7}$$

where d_{12} denotes the cluster that would be obtained after merging d_1 and d_2. This compares the squared sum of member-center dissimilarities before and after merging clusters d_1 and d_2.

Example 13.3.6 An implementation of Ward linkage for agglomerative hierarchical clustering is presented and demonstrated below. Like the center linkage implementation from the previous example, it uses the instance dissimilarity function rather than the dissimilarity matrix, which only provides dissimilarities between training instances.

```
## Ward linkage
ahc.ward <- function(clustering, d1, d2, data, diss, dm)
{
  c1 <- attr.mm(data[clustering==d1,])
  c2 <- attr.mm(data[clustering==d2,])
  c12 <- attr.mm(data[clustering==d1 | clustering==d2,])

  sum(sapply(which(clustering==d1 | clustering==d2),
             function(i) diss(data[i,], c12)^2)) -
    sum(sapply(which(clustering==d1), function(i) diss(data[i,], c1)^2)) -
    sum(sapply(which(clustering==d2), function(i) diss(data[i,], c2)^2))
}

  # agglomerative hierarchical Ward-linkage clustering for the weathercl data
wcl.ahc.wl <- ahc(wcl.std, linkf=ahc.ward)

  # agglomerative hierarchical Ward-linkage clustering for the iris data
i.ahc.wl <- ahc(i.std.train[,-5], linkf=ahc.ward)

  # agglomerative hierarchical Ward-linkage clustering for the Glass data
g.ahc.wl <- ahc(g.std.train[,-10], linkf=ahc.ward)
```

13.3.2.6   Choosing linkage type

Of the two basic "extreme" linkage types, complete linkage is definitely preferred to single linkage in most applications, as it promotes the compactness and homogeneity of clusters. Single linkage implicitly assumes that similarity should always be considered transitive, which may be justified only for some domains. When searching for a simple compromise between the single and complete linkage, one would usually prefer average linkage to center linkage, as the latter loses the desirable linkage monotonicity.

While more complex than complete linkage and less intuitive than average or center linkage, Ward linkage is believed to usually yield superior cluster hierarchies. This superiority, confirmed by published experimental studies, may be due to its explicit objective function approach.

13.4   Divisive clustering

While the agglomerative approach to hierarchical clustering may be considered the most natural and therefore preferred, it is computationally expensive and therefore hardly applicable to large datasets. The divisive approach may not have equally clean and straightforward interpretation, but can be usually performed more efficiently, particularly given the fact there is usually no reason to go down to singleton bottom-level clusters. Rather than repeatedly merging the most similar clusters, it repeatedly partitions selected clusters into similarity-based subclusters.

13.4.1   Algorithm scheme

A more detailed description of divisive hierarchical clustering is given by the algorithm scheme presented below. It starts from a single root cluster, to which all training instances are assigned, and on each iteration considers dividing one existing cluster into subclusters. Clusters that remain to be considered are marked as open and those that already have been divided or decided to remain leaves are marked as closed. The latter is determined by some stop criteria. Dividing a cluster is performed by partitioning the corresponding set of training instances into similarity-based subsets and creating descendant clusters corresponding to these subsets.

1: create the root cluster and mark it as *open*;
2: assign all training instances to the root cluster;
3: **while** there are open clusters **do**
4:     select an open cluster d;
5:     **if** stop criteria for d are not satisfied **then**
6:         partition T^d into similarity-based subsets $T_1^d, T_2^d, ...$;
7:         **for all** $i = 1, 2, ...$ **do**
8:             create new cluster d_i as an descendant of d;
9:             $T^{d_i} := T_i^d$;
10:        **end for**
11:        mark d as a *closed node*;
12:    **else**
13:        mark d as a *closed leaf*;
14:    **end if**
15: **end while**

13.4.2   Wrapping a flat clustering algorithm

Partitioning clusters into similarity-based subclusters is most naturally achieved by employing a flat clustering algorithm. Divisive hierarchical clustering can therefore be considered a wrapper approach to creating cluster hierarchies which turns a flat clustering algorithm into a hierarchical clustering algorithm by its repeated application. While this is possible, in principle, with arbitrary flat clustering algorithms, most instantiations of this idea use *k*-centers clustering algorithms, such as presented in Chapter 12. What sometimes appears a disadvantage of theirs – the requirement to predetermine the number of clusters – is a very convenient feature in this application. It makes it possible to enforce *bi-partitioning*, i.e., always creating two subclusters, resulting in a binary clustering tree, such as those created by the agglomerative approach. While the algorithm scheme presented above does not explicitly assume either using a flat clustering algorithm for cluster division or bi-partitioning, most of its practical instantiations do.

13.4.3   Stop criteria

The most ultimate stop criterion for divisive clustering could be reaching singleton bottom-level clusters, such as those that agglomerative clustering starts from. It is often unnecessary, and, for larger datasets, hardly possible to go down so deeply. This is why additional stop criteria are usually employed, such as

- the maximum clustering tree depth,

- the minimum number of instances in a cluster (sufficiently small clusters are not further divided),

- the minimum intracluster dissimilarity (sufficiently homogeneous clusters are not further divided).

Of those, the first is by far the most popular and usually sufficient, since the main purpose of stop criteria is to avoid the computational expense of building overly deep clustering trees.

Example 13.4.1 The dhc function defined by the R code presented below is a simple implementation of divisive hierarchical clustering. It uses a flat clustering algorithm specified via the alg argument to divide clusters. The algorithm is assumed to belong to the *k*-centers clustering family and create a flat clustering model that has at least two named components: cluster membership assignment for training instances and cluster centers. The names of these components are specified via the cls and cnt arguments, with defaults matching the output of the pam function from the cluster package, which is the default for alg. Clusters are bi-partitioned until singleton bottom-level clusters are obtained or the maximum depth specified via the maxdepth argument is reached. The demonstration calls for the *weathercl*, *Iris*, and Glass data are presented, using both default and – for the two larger datasets – alternative parameter settings (using the kmeans function from the standard stats package for flat clustering, limiting tree depth to 3).

```
## divisive clustering using alg, which is assumed to be called:
##   alg(data, 2, ...)
dhc <- function(data, alg=pam, cls="clustering", cnt="medoids", centf=as.numeric,
                maxdepth=16, ...)
{
  clustering <- rep(1, nrow(data))  # cluster membership assignment
  centers <- NULL  # cluster centers
  merge <- NULL
  height <- NULL
```

```
while (any(clustering>0))
{
  d <- min(clustering[clustering>0])  # cluster to process
  if ((m <- sum(clustering==d))>1 && d<2^maxdepth)
  {
    cls.d <- if (m>2) (mod.d <- alg(data[clustering==d,], 2, ...))[[cls]] else 1:2
    centers <- c(list(if (m>2) mod.d[[cnt]]
                      else sapply(data[clustering==d,], centf)), centers)
    clustering[clustering==d] <- 2*d + (cls.d-1)
    merge <- rbind(c(2*d, 2*d+1), merge)
    height <- c(height, length(height)+1)
  }
  else
  {
    clustering[clustering==d] <- -d    # mark as leaf
    merge[merge==d] <- -d
  }
}

bottom <- unique(clustering)
clustering <- (1:length(bottom))[match(clustering, bottom)]  # re-assign ids
merge[merge<0] <- - (1:length(bottom))[match(merge[merge<0], bottom)]
merge[merge>0][order(merge[merge>0])] <- sum(merge>0):1
`class<-`(list(clustering=clustering, centers=centers, merge=merge, height=height,
              order=-t(merge)[t(merge)<0]),
          "hcl")
}

# divisive clustering for the weathercl data
wcl.dhc <- dhc(wcl.std)

# divisive hierarchical clustering for the iris data
i.dhc <- dhc(i.std.train[,-5])
i.dhc.km <- dhc(i.std.train[,-5], alg=kmeans, cls="cluster", cnt="centers")
i.dhc.d3 <- dhc(i.std.train[,-5], maxdepth=3)
i.dhc.km.d3 <- dhc(i.std.train[,-5], alg=kmeans, cls="cluster", cnt="centers",
                   maxdepth=3)

# divisive hierarchical clustering for the Glass data
g.dhc <- dhc(g.std.train[,-10]   )
g.dhc.km <- dhc(g.std.train[,-10], alg=kmeans, cls="cluster", cnt="centers")
g.dhc.d3 <- dhc(g.std.train[,-10], maxdepth=3)
g.dhc.km.d3 <- dhc(g.std.train[,-10], alg=kmeans, cls="cluster", cnt="centers",
                   maxdepth=3)
```

As with the previously presented implementation of agglomerative hierarchical clustering, the model representation created by the dhc function is compatible with that adopted by the standard R implementation of hierarchical clustering. The merge matrix represents the cluster–subcluster relationship and the height vector assigns height values in the reverse order of cluster dividing (using consecutive integers rather than cluster dissimilarities). The mysterious-looking cluster number reassignments performed after the completion of the main loop make sure that the cluster numbers occurring in the merge matrix are consecutive integers, negative for bottom-level clusters and positive for higher level clusters. Bottom level clusters are numbered in the order of their first appearances in the cluster membership assignments for the training set. In the case of reaching singleton leaves this makes their numbering consistent with training instance numbers. Unlike for the ahc function, the returned model object does not store the linkage function (since there is none), but it contains the `centers` component instead, which is a list of cluster centers for all nonleaf clusters.

13.5 Hierarchical clustering visualization

Hierarchical clustering models – much more often and to a much greater extent than flat clustering models – represent knowledge *per se*, describing the similarity patterns discovered in the data. To make them truly useful in this role it is therefore essential to have a comprehensible visualization technique cluster hierarchies. This is achieved using *dendrograms* – graphical representations of hierarchical clustering trees.

A dendrogram is a schematically drawn tree, with the vertical position of nodes corresponding to the order of cluster merge or divide operations and their horizontal position adjusted to avoid edge intersections. For agglomerative clustering – for which dendrograms were originally developed – node heights are proportional to the dissimilarities between their descendants. For most linkage types, this is guaranteed to put clusters created later on higher heights, since they are monotone (linkage function values do not decrease during clustering). For divisive clustering, which normally does not require calculating intercluster dissimilarities, heights may simply correspond to the reverse order of cluster dividing operations. To remain readable, dendrograms may not have too many leaves and therefore it hardly makes sense to draw them for cluster hierarchies with more than a few levels.

Example 13.5.1 The following R code defines the plot method for hierarchical clustering objects. The standard dendrogram plot method is used after converting the clustering model to a dendrogram object. This is possible due to the compatibility of the model representation adopted by the agglomerative and divisive clustering implementations presented before with that of the `hclust` class (conversion to dendrogram is achieved via the `as.dendrogram` method for `hclust` objects). The dendrograms of the hierarchical clustering models for the *weathercl* data created in the previous examples are then plotted. Since both the `ahc` and `dhc` functions use training instance numbers as singleton bottom-level cluster identifiers, the dendrograms are easily interpretable.

```
## convert to dendrogram
as.dendrogram.hcl <- function(model) { as.dendrogram(as.hclust(model)) }

## plot a hierarchical clustering dendrogram
plot.hcl <- function(model, ...)
{
  plot(as.dendrogram(model), center=TRUE, ...)
}

  # dendrogram plots for the weathercl data
par(mfrow=c(3, 2))
plot(wcl.ahc.sl, main="Single linkage")
plot(wcl.ahc.cl, main="Complete linkage")
plot(wcl.ahc.al, main="Average linkage")
plot(wcl.ahc.ml, main="Center linkage")
plot(wcl.ahc.wl, main="Ward linkage")
plot(wcl.dhc, main="Divisive clustering")
```

The produced dendrogram plots are presented in Figure 13.2. Despite minor differences, all the hierarchical clustering models roughly represent the same similarity patterns. The single-linkage dendrogram is less balanced than the others, with a single instance often being merged to a larger cluster. Average and center linkage produced exactly the same clustering tree structures. The results of Ward linkage and divisive clustering are only slightly different from those. For the former, node heights set based on linkage function values show which merge operations increased the cluster variance the most. The resulting dendrogram suggests that there may be three natural clusters in the *weathercl* data: $\{1, 3, 4, 8\}$, $\{5, 6, 7, 9, 10\}$, and $\{2, 11, 12, 13, 14\}$.

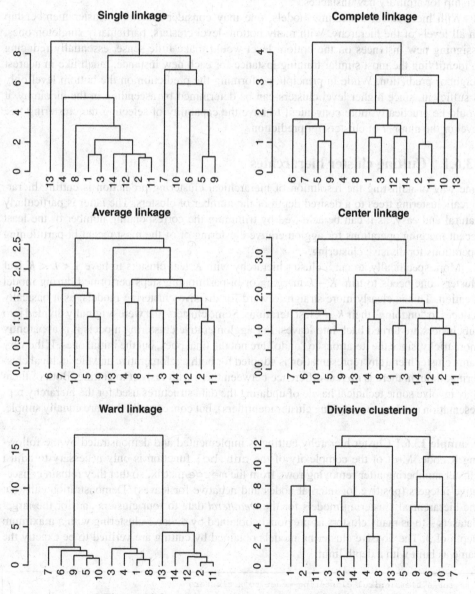

Figure 13.2   Dendrograms for the *weathercl* data.

13.6   Hierarchical clustering prediction

Hierarchical clustering is most commonly presented as serving the purpose of discovering and presenting the similarity structure of the training set and the domain from which it comes. This book's central theme of predictive modeling makes us always more interested in applying models to new data than using them to understand the training data. Just like for *k*-centers clustering models presented in Chapter 12, essentially the same similarity-based mechanism that determines cluster membership for training instances may be used to predict cluster membership for arbitrary new instances.

With hierarchical clustering models, one may consider assigning cluster membership on all levels of the hierarchy. With many bottom-level clusters, particularly singleton ones, assigning new instances on the bottom level would make little sense, essentially reducing to identifying the most similar training instance for each new instance, much like in nearest neighbor prediction. While in principle performing the prediction on the bottom level only is sufficient, since higher level clusters can be determined by ascending in the hierarchy, it would be practically more convenient to have the capability of selecting the clustering tree level or the number of clusters for prediction.

13.6.1   Cutting cluster hierarchies

One way of adjusting the resolution of hierarchical clustering prediction is cutting hierarchical clustering trees to a desired depth or the number of clusters. The latter is particularly natural and easy, as it can be achieved by trimming the corresponding number of the least recent merging operations for agglomerative clustering or of the most recent bi-partitioning operations for divisive clustering.

More specifically, to cut a cluster hierarchy with K leaf clusters to have $1 < k < K$ leaf clusters, one needs to trim $K - k$ merging or bi-partitioning steps performed during model creation. This is clearly more straightforward for divisive clustering models, as it basically reduces to "undoing" their $K - k$ last iterations. Some clusters that were originally divided into subclusters are turned back into leaves. For agglomerative clustering models it is apparently more tricky, since the iterations to "undo" are not the final ones, but the initial ones. If the very same cluster hierarchy representation is adopted for both agglomerative and hierarchical clustering, though, there is no real difference between them with respect to cutting. Both of them may involve some technical hassle of updating the data structures used for the hierarchy representation (such as re-assigning cluster identifiers), but conceptually they are equally simple.

Example 13.6.1 Cluster hierarchy cutting is implemented and demonstrated by the following R code. Most of the complexity of the `cut.hcl` function is only necessary to adjust cluster numbering after removing rows from the `merge` matrix, so that they remain consecutive integers (positive for internal nodes and negative for leaves). Demonstration calls cut the hierarchical clustering models for the *weathercl* data to four clusters and for the larger datasets – to as many clusters as previously obtained by divisive clustering with a maximum depth of 3. The divisive clustering models obtained by cutting are verified to be exactly the same as built with a depth limit.

```
## cut a hierarchical clustering model to k clusters
cut.hcl <- function(model, k)
{
```

```
  nc <- maxc <- nrow(model$merge)+1  # number of clusters
  k <- clip.val(k, 2, nc)            # make sure k is in the valid range

  clustering <- model$clustering
  merge <- model$merge
  height <- model$height
  while (nc>k)
  {
    mr <- merge[1,]     # merge to remove
    merge <- merge[-1,]
    clustering[clustering %in% -mr] <- (maxc <- maxc + 1)  # id for new leaf
    merge[merge>0] <- merge[merge>0]-1  # shift node numbers
    merge[merge==0] <- -maxc
    height <- height[-1]-(height[2]-height[1])
    nc <- nc-1
  }

  bottom <- unique(clustering)
  clustering <- (1:length(bottom))[match(clustering, bottom)]  # re-assign ids
  merge[merge<0] <- -(1:length(bottom))[match(merge[merge<0], -bottom)]

  model$clustering <- clustering
  model$centers <- model$centers[-(1:(length(model$centers)-k+1))]
  model$merge <- merge
  model$height <- height
  model$order <- -t(merge)[t(merge)<0]
  model
}

  # cutting hierarchical clustering trees for the weathercl data
wcl.ahc.sl.c4 <- cut(wcl.ahc.sl, 4)
wcl.ahc.cl.c4 <- cut(wcl.ahc.cl, 4)
wcl.dhc.c4 <- cut(wcl.dhc, 4)

  # cutting hierarchical clustering trees for the iris data
i.ahc.sl.cd3 <- cut(i.ahc.sl, max(i.dhc.d3$clustering))
i.ahc.cl.cd3 <- cut(i.ahc.cl, max(i.dhc.d3$clustering))
i.ahc.al.cd3 <- cut(i.ahc.al, max(i.dhc.d3$clustering))
i.ahc.ml.cd3 <- cut(i.ahc.ml, max(i.dhc.d3$clustering))
i.ahc.wl.cd3 <- cut(i.ahc.wl, max(i.dhc.d3$clustering))
i.dhc.cd3 <- cut(i.dhc, max(i.dhc.d3$clustering))
  # verify i.dhc.cd3 and i.dhc.d3 are the same
all(i.dhc.cd3$clustering==i.dhc.d3$clustering)
all(i.dhc.cd3$merge==i.dhc.d3$merge)
all(sapply(1:length(i.dhc.cd3$centers),
           function(d) all(i.dhc.cd3$centers[[d]]==i.dhc.d3$centers[[d]])))

  # cutting hierarchical clustering trees for the Glass data
g.ahc.sl.cd3 <- cut(g.ahc.sl, max(g.dhc.d3$clustering))
g.ahc.cl.cd3 <- cut(g.ahc.cl, max(g.dhc.d3$clustering))
g.ahc.al.cd3 <- cut(g.ahc.al, max(g.dhc.d3$clustering))
g.ahc.ml.cd3 <- cut(g.ahc.ml, max(g.dhc.d3$clustering))
g.ahc.wl.cd3 <- cut(g.ahc.wl, max(g.dhc.d3$clustering))
g.dhc.cd3 <- cut(g.dhc, max(g.dhc.d3$clustering))
  # verify g.dhc.cd3 and g.dhc.d3 are the same
all(g.dhc.cd3$clustering==g.dhc.d3$clustering)
all(g.dhc.cd3$merge==g.dhc.d3$merge)
all(sapply(1:length(g.dhc.cd3$centers),
           function(d) all(g.dhc.cd3$centers[[d]]==g.dhc.d3$centers[[d]])))
```

13.6.2   Cluster membership assignment

With the capability of cutting cluster hierarchies available, we only need to address bottom-level cluster membership assignment to provide a complete hierarchical clustering prediction capability. This can be directly achieved by determining the level of dissimilarity between each instance and the bottom-level clusters of the hierarchical clustering model and choosing the least dissimilar instance for each cluster.

It makes sense to measure the instance/cluster dissimilarity differently for agglomerative and divisive hierarchical models, to match the different ways in which they are created. For the former, this is clearly a special case of the general intercluster dissimilarity measurement problem that is solved using linkage functions. Assuming the linkage function used for model creation is retained in model representation, it can be applied to measure instance-cluster dissimilarity by temporarily treating new instances as singleton clusters. For divisive clustering models, created with a k-centers clustering algorithm used for dividing clusters, it is definitely more reasonable to apply the cluster membership assignment scheme of the latter. In this case, bottom-level cluster centers determined during model creation have to be retained and used during prediction.

Example 13.6.2 Hierarchical clustering prediction is implemented and demonstrated by the below R code. The prediction method for `hcl` objects determines whether the model was created by the `ahc` or `dhc` function and calls either the `predict.ahc` or `predict.dhc` function to do the real job of assigning instances from the dataset to bottom-level clusters. Demonstration calls generate the predictions of selected hierarchical clustering models created before. For the *Iris* and *Glass* data, the models cut down in the previous example are applied to the test subsets.

```
## hierarchical clustering prediction
predict.hcl <- function(model, data, ...)
{
  if (!is.null(model$data) && !is.null(model$link))
    predict.ahc(model, data, ...)
  else if (!is.null(model$centers))
    predict.dhc(model, data, ...)
}

## agglomerative hierarchical clustering prediction
predict.ahc <- function(model, data, diss=euc.dist)
{
  ext.data <- rbind(model$data, data)
  dm <- as.matrix(dissmat(ext.data, diss))  # dissimilarity matrix for linkage

  clusters <- sort(unique(model$clustering))
  x.clusters <- length(clusters) + 1:nrow(data)
  ext.clustering <- c(model$clustering, x.clusters)

  links <- outer(clusters, x.clusters,
              Vectorize(function(d1, d2)
                          model$link(ext.clustering, d1, d2, ext.data, diss, dm)))
  apply(links, 2, which.min)
}

## divisive hierarchical clustering prediction
predict.dhc <- function(model, data, diss=euc.dist)
{
```

```
    centers <- do.call(rbind,
                        lapply(1:nrow(model$merge),
                               function(i)
                               model$centers[[i]][model$merge[i,]<0,]))
    clusters <- -t(model$merge)[t(model$merge)<0]
    centers <- centers[match(1:length(clusters), clusters),]  # reorder centers
    k.centers.assign(centers, data, diss)
}

  # hierarchical clustering prediction for the weathercl data
predict(wcl.ahc.cl, wcl.std)
predict(wcl.dhc, wcl.std)

  # hierarchical clustering prediction for the iris data
i.ahc.cl.cd3.pred <- predict(i.ahc.cl.cd3, i.std.test[,-5])
i.ahc.sl.cd3.pred <- predict(i.ahc.sl.cd3, i.std.test[,-5])
i.ahc.al.cd3.pred <- predict(i.ahc.al.cd3, i.std.test[,-5])
i.ahc.ml.cd3.pred <- predict(i.ahc.ml.cd3, i.std.test[,-5])
i.ahc.wl.cd3.pred <- predict(i.ahc.wl.cd3, i.std.test[,-5])
i.dhc.cd3.pred <- predict(i.dhc.cd3, i.std.test[,-5])

  # hierarchical clustering prediction for the Glass data
g.ahc.cl.cd3.pred <- predict(g.ahc.cl.cd3, g.std.test[,-10])
g.ahc.sl.cd3.pred <- predict(g.ahc.sl.cd3, g.std.test[,-10])
g.ahc.al.cd3.pred <- predict(g.ahc.al.cd3, g.std.test[,-10])
g.ahc.ml.cd3.pred <- predict(g.ahc.ml.cd3, g.std.test[,-10])
g.ahc.wl.cd3.pred <- predict(g.ahc.wl.cd3, g.std.test[,-10])
g.dhc.cd3.pred <- predict(g.dhc.cd3, g.std.test[,-10])
```

13.7 Conclusion

The need for hierarchical clustering naturally emerges in domains where it is not only required to discover similarity-based groups, but also organize them. This is a valuable capability wherever the complexity of similarity patterns exceeds the limited representation power of flat clustering models. Cluster hierarchies, organized in the natural and intuitive general to specific order, help understand the domain. They also make it possible to postpone the decision about the proper granularity of clustering till the time of its actual application and effortlessly revise this decision as often as desirable. These capabilities come at an increased computational expense, particularly when using the dominating agglomerative approach, but this has been at least partially ameliorated by improved computing hardware performance. The computational expense for larger datasets can be reduced by starting from a moderate number of nonsingleton bottom-level clusters or using the divisive approach with a limited maximum depth.

Hierarchical clustering, more than other types of clustering, tends to be presented as a form of descriptive rather than predictive modeling. Even if this view is indeed fully adequate for the majority of its applications, cluster hierarchies can serve as predictive models as well. As explained and demonstrated above, one can assign new data instances to clustering tree leaves or nodes. This makes hierarchical clustering useful for applications where it is not sufficient to determine the cluster membership of training instances, but the prediction capability is essential. This is the case when clustering is used for inferring about hidden attribute values, detecting isolated anomalous instances, or decomposing the domain for other types of predictive modeling. While it is by far more common to employ flat clustering models in such applications, the enhanced variable-resolution model representation power may be a worthwhile to consider advantage.

13.8   Further readings

Hierarchical clustering algorithms may not be similarly widespread as k-centers algorithms in practical applications, but they belong to the set of core data mining techniques that are discussed in most data mining books, sometimes rather briefly (e.g., Cios *et al.* 2007; Hastie *et al.* 2011; Witten *et al.* 2011), but sometimes also quite extensively (e.g., Han *et al.* 2011; Tan *et al.* 2013; Theodoridis and Koutroumbas 2008; Webb 2002). Clustering survey books and articles usually cover the major hierarchical approaches as well (Everitt *et al.* 2011; Gordon 1999; Jain and Dubes 1988; Jain *et al.* 1999). Kaufman and Rousseeuw (1990) included representatives of the agglomerative and divisive hierarchical clustering approaches in their collection of female-named clustering algorithms. These are the AGNES (agglomerative nesting) and DIANA (divisive analysis) algorithms, the R implementations of which are available in the `cluster` package. A brief survey of hierarchical clustering visualization techniques was presented by Freeman (1994). Since hierarchical clustering algorithms are often applied to text documents, they are also covered by the text clustering literature (e.g., Aggarwal and Zhai 2012).

The original primary motivation behind hierarchical clustering that drove its initial development was automated biological taxonomy creation (Sneath and Sokal 1973). This led Sneath (1957) to propose the first single-linkage clustering algorithm, although the underlying idea can be traced back to the earlier work of Florek *et al.* (1951). Other linkage types, still popular today, were proposed within a decade since then: average and center linkage, (Sokal and Michener 1958), complete linkage (McQuitty 1960), and Ward linkage (Ward 1963). Another early linkage type, not described in this chapter, is known as McQuitty linkage (McQuitty 1966). It determines the dissimilarity between a newly merged cluster, obtained by merging clusters d_1 and d_1, and any other cluster d, as the average of the dissimilarities between d_1 and d, and d_2 and d. Experimental comparisons suggest that out of those classic linkage types it is Ward linkage that usually performs best (Blashfield 1976; Ferreira and Hitchcock 2009; Kuiper and Fisher 1975). It was more recently generalized by Székely and Rizzo (2005) by permitting the adjustment of the power of the Euclidean distance used in the objective function. Gowda and Krishna (1978) presented an algorithm that uses multiple nearest neighbors for agglomerative clustering, which can be viewed as an extension of single linkage. Recently, Zhang *et al.* (2013) proposed a graph-structural linkage type.

Agglomerative hierarchical clustering with specific linkage types can often be performed more efficiently than suggested by the generic algorithm scheme and illustrative implementation presented in this chapter. Such an efficient algorithm was first proposed for single linkage by Sibson (1973) and then, following a similar pattern, for complete linkage by Defays (1977). Even if sticking with the generic agglomerative algorithm, there are some noteworthy improvement possibilities. Lance and Williams (1967) derived recursive dissimilarity update formulae that make it possible to avoid re-calculating linkage function values after cluster merging. Wishart (1969) combined this approach with Ward linkage.

There are at least two noteworthy hierarchical clustering algorithms that do not follow either the agglomerative or divisive approaches presented in this chapter. One is the Cobweb algorithm (Fisher 1987) which incrementally constructs a clustering tree by processing one training instance at a time and using a probabilistic objective function to choose one of a few available tree modification operators. The other is the BIRCH algorithm (Zhang *et al.* 1996), designed to efficiently handle large datasets and resist noise. It creates an initial clustering tree in a single data scan, adjusting its resolution to memory limits, then further reduces its size if possible and needed, and finally performs clustering of its leaves.

References

Aggarwal CC and Zhai C 2012 A survey of text clustering algorithms In *Mining Text Data* (ed. Aggarwal CC and Zhai CX) Springer.

Blashfield RK 1976 Mixture model tests of cluster analysis: Accuracy of four agglomerative hierarchical methods. *The Psychological Bulletin* **83**, 377–388.

Cios KJ, Pedrycz W, Swiniarski RW and Kurgan L 2007 *Data Mining: A Knowledge Discovery Approach*. Springer.

Defays D 1977 An efficient algorithm for a complete link method. *The Computer Journal* **20**, 364–366.

Everitt BS, Landau S, Leese M and Stahl D 2011 *Cluster Analysis* 5th edn. Wiley.

Ferreira L and Hitchcock DB 2009 A comparison of hierarchical methods for clustering functional data. *Communications in Statistics: Simulation and Computation* **38**, 1925–1949.

Fisher DH 1987 Knowledge acquisition via incremental conceptual clustering. *Machine Learning* **2**, 139–172.

Florek K, Łukaszewicz J, Perkal J, Steinhaus H and Zubrzycki S 1951 Sur la liaison et la division des points d'un ensemble fini. *Colloquium Mathematicae* **2**, 282–285.

Freeman LC 1994 Displaying hierarchical clusters. *Connections* **17**, 48–52.

Gordon AD 1999 *Classification* 2nd edn. Chapman and Hall.

Gowda KC and Krishna G 1978 Agglomerative clustering using the concept of mutual nearest neighborhood. *Pattern Recognition* **10**, 105–112.

Han J, Kamber M and Pei J 2011 *Data Mining: Concepts and Techniques* 3rd edn. Morgan Kaufmann.

Hastie T, Tibshirani R and Friedman J 2011 *The Elements of Statistical Learning: Data Mining, Inference, and Prediction* 2nd edn. Springer.

Jain AK and Dubes RC 1988 *Algorithms for Clustering Data*. Prentice-Hall.

Jain AK, Murty MN and Flynn PJ 1999 Data clustering: A review. *ACM Computing Surveys* **31**, 264–323.

Kaufman L and Rousseeuw PJ 1990 *Finding Groups in Data: An Introduction to Cluster Analysis*. Wiley.

Kuiper FK and Fisher L 1975 A Monte Carlo comparison of six clustering procedures. *Biometrics* **31**, 777–783.

Lance GN and Williams WT 1967 A general theory of classificatory sorting strategies: 1. hierarchical systems. *The Computer Journal* **9**, 373–380.

McQuitty LL 1960 Hierarchical linkage analysis for the isolation of types. *Educational and Psychological Measurement* **20**, 55–67.

McQuitty LL 1966 Similarity analysis by reciprocal pairs for discrete and continuous data. *Educational and Psychological Measurement* **26**, 825–831.

Sibson R 1973 SLINK: An optimally efficient algorithm for the single-link cluster method. *The Computer Journal* **16**, 30–34.

Sneath PHA 1957 The application of computers to taxonomy. *Journal of General Microbiology* **17**, 201–226.

Sneath PHA and Sokal RR 1973 *Numerical Taxonomy: The Principles and Practice of Numerical Classification*. Freeman.

Sokal RR and Michener CD 1958 A statistical method for evaluating systematic relationships. *University of Kansas Science Bulletin* **38**, 1409–1438.

Székely GJ and Rizzo ML 2005 Hierarchical clustering via joint between-within distances: Extending Ward's minimum variance method. *Journal of Classification* **22**, 151–183.

Tan PN, Steinbach M and Kumar V 2013 *Introduction to Data Mining* 2nd edn. Addison-Wesley.

Theodoridis S and Koutroumbas K 2008 *Pattern Recognition* 4th edn. Academic Press.

Ward JH 1963 Hierarchical grouping to optimize an objective function. *Journal of the American Statistical Association* **58**, 236–244.

Webb AR 2002 *Statistical Pattern Recognition* 2nd edn. Wiley.

Wishart D 1969 An algorithm for hierachical classifications. *Biometrics* **25**, 165–170.

Witten IH, Frank E and Hall MA 2011 *Data Mining: Practical Machine Learning Tools and Techniques* 3rd edn. Morgan Kaufmann.

Zhang T, Ramakrishnan R and Livny M 1996 BIRCH: An efficient data clustering method for very large databases *Proceedings of the 1996 ACM SIGMOD International Conference on Management of Data (SIGMOD-96)*. ACM Press.

Zhang W, Zhao D and Wang X 2013 Agglomerative clustering via maximum incremental path integral. *Pattern Recognition* **46**, 3056–3065.

14

Clustering model evaluation

14.1 Introduction

The challenge of reliable model evaluation, discussed for classification and regression models in Chapters 7 and 10, respectively, is similarly important for clustering models. Unlike for the former, though, where there are certain natural quality criteria, for the latter it is not so clear how to assess their quality in an objective way. This results in a much greater number of different performance measures being proposed and used on one hand, and in some considerable reserve with which their outcomes tend to be taken on the other hand.

Even if it is not so widely realized as for more common classification and regression model evaluation, when evaluating clustering models one may also be concerned with their generalization properties. For any performance measure its value on a particular dataset (dataset performance) is therefore a possibly imperfect estimator of the corresponding value on the whole domain (true performance).

Clustering quality measures may, but do not have to, explicitly use instance dissimilarity or similarity measures presented in Chapter 11. Those that do are often applied to evaluate models created by dissimilarity-based clustering algorithms and then it usually makes most sense to adopt the same dissimilarity measure for model creation and evaluation.

14.1.1 Dataset performance

The dataset performance of a clustering model is assessed directly by calculating one or more selected performance measures on a particular dataset. Since there is no predefined target attribute to be approximated, they are primarily supposed to measure how well the model captures similarity patterns exhibited by the data. This is quite different from measuring the predictive performance of classification and regression models by comparing predicted and true target attribute values. To underline this difference, clustering performance measures will be referred to as *quality measures* in this chapter.

14.1.2   Training performance

Evaluating a clustering model on the training set that was used to create the model determines the model's training performance. This is where clustering model evaluation usually begins. Unlike for classification and regression models, it may also end here sometimes, particularly when there is no intention to ever apply the evaluated model to new data, but its sole purpose is to provide insights about the training set.

14.1.3   True performance

The true performance of a clustering model is represented by the expected values of one or more selected clustering quality measures on the whole domain. This reflects the quality of cluster membership predictions for arbitrary new instances from the given domain and can be estimated by dataset performance for appropriately selected datasets. This is the responsibility of model evaluation procedures, which handle data splitting into training and evaluation or test subsets and use the latter to produce reliable true performance estimates. The same procedures as presented in Section 7.3 for classification models are applicable for clustering models as well and therefore do not need to be discussed here.

However, it is not uncommon for clustering models to be evaluated on the training set only. This is because, unlike performance measures for classification or regression models, clustering quality measures can be hardly interpreted or used other than by comparing their values for a number of models. Whereas the misclassification error or the mean square error, for example, have straightforward algorithm-independent interpretation that makes it possible to use them as final model acceptance criteria for applications, most if not all measures presented in this chapter yield numbers that are only meaningful when compared across a set of candidate models and do not serve any other purpose than model selection. The latter (including the choice of clustering algorithms and their parameters) is indeed quite often possible based on training performance only. It may be optimistically biased and unreliable due to overfitting, but – if not used mechanically, but combined with some reasonable overfitting-prevention constraints (in particular, for the maximum number of clusters) – may successfully identify the best models.

Sparing some nontraining data for model evaluation always remains a good idea and should be preferred to measuring training performance only whenever possible. The simplest hold-out evaluation procedure should be sufficient in most cases, since its major problem – the pessimistic evaluation bias due to limited training set size – is not a severe disadvantage for clustering model evaluation. As discussed in Section 7.3.1, what is actually evaluated by any evaluation procedure is a modeling procedure rather than a particular model, and when the best modeling procedure has been identified, it can be re-applied to the whole available dataset to create the final model.

Example 14.1.1 Clustering quality measures described in this chapter will be illustrated in a series of R language examples by applying them to a set of models created using the PAM algorithm, which is a variation of k-medoids clustering mentioned in Section 12.4.2. The environment for these demonstrations is set up by the R code presented below. It loads several DMR packages that will be used, the `cluster` package which provides an implementation of the PAM algorithm, and the *Iris* dataset available in the standard `datasets` package. Then it splits the dataset into training and test subsets, applies the standardization transformation using the `std.all` and `predict.std` functions, and creates several PAM models (with different numbers of clusters and dissimilarity measures) on

Ex. 17.3.1
dmr.trans

the training subset. To ensure the exact reproducibility of results, the random generator seed is explicitly set.

```
library(dmr.claseval)
library(dmr.stats)
library(dmr.trans)
library(dmr.util)

library(cluster)

data(iris)

set.seed(12)
ri <- runif(nrow(iris))
i.train <- iris[ri>=0.33,]
i.test <- iris[ri<0.33,]

i.stdm <- std.all(Species~., i.train)
i.std.train <- predict.std(i.stdm, i.train)
i.std.test <- predict.std(i.stdm, i.test)

i.pam2.euc <- pam(i.std.train[,-5], 2, metric="euclidean")
i.pam3.euc <- pam(i.std.train[,-5], 3, metric="euclidean")
i.pam5.euc <- pam(i.std.train[,-5], 5, metric="euclidean")
i.pam7.euc <- pam(i.std.train[,-5], 7, metric="euclidean")

i.pam2.man <- pam(i.std.train[,-5], 2, metric="manhattan")
i.pam3.man <- pam(i.std.train[,-5], 3, metric="manhattan")
i.pam5.man <- pam(i.std.train[,-5], 5, metric="manhattan")
i.pam7.man <- pam(i.std.train[,-5], 7, metric="manhattan")
```

Example 14.1.2 There is no prediction method for pam objects available in the cluster package and therefore cluster membership can be directly determined for training instances only. To make clustering quality measures implemented by subsequent examples applicable to arbitrary datasets, the R code presented below defines the prediction method for pam clustering models. The prediction process boils down to dissimilarity calculation for new instances and cluster centers, and choosing the closest cluster for each instance, exactly as in the instance assignment phase of k-centers algorithms. This is performed using the k.centers.assign function. The daisy function from the cluster package is used for dissimilarity calculation by default, for maximum compatibility with pam. The same prediction method is also applicable to clara objects – models created by the clara function from the same package, which an approximate k-medoids clustering algorithm using internal subsampling for large data. It is demonstrated by generating test set predictions for the models created in the previous example.

> Ex. 12.2.1
> dmr.kcenters

```
## prediction for pam clustering models (only if created with stand=FALSE)
## using daisy or dist (selected via the dmf argument) for dissimilarity calculation
predict.pam <- function(model, data, dmf=daisy, ...)
{
  k.centers.assign(model$medoids, data,
                   function(x1, x2) dmf(rbind(x1, x2), ...))
}
```

```
## the same is applicable to clara clustering models
predict.clara <- predict.pam

 # test set predictions
i.pam2.euc.pred <- predict(i.pam2.euc, i.std.test[,-5])
i.pam3.euc.pred <- predict(i.pam3.euc, i.std.test[,-5])
i.pam5.euc.pred <- predict(i.pam5.euc, i.std.test[,-5])
i.pam7.euc.pred <- predict(i.pam7.euc, i.std.test[,-5])


i.pam2.man.pred <- predict(i.pam2.man, i.std.test[,-5])
i.pam3.man.pred <- predict(i.pam3.man, i.std.test[,-5])
i.pam5.man.pred <- predict(i.pam5.man, i.std.test[,-5])
i.pam7.man.pred <- predict(i.pam7.man, i.std.test[,-5])
```

14.2   Per-cluster quality measures

There are a number of quality measures designed to describe the level of a particular cluster's cohesion and/or its separation from other clusters. They make it possible to get insight into the properties of individual clusters and differences between them, which may be interesting even if it does not directly translate into the evaluation of the whole clustering model. Some of these per-cluster measures can be appropriately aggregated into overall clustering quality measures, though.

14.2.1   Diameter

The *diameter* of a cluster is the maximum dissimilarity between its members. For cluster d on dataset S, with respect to dissimilarity measure δ, it can be written as

$$\text{diam}_{\delta,S}(d) = \max_{x_1, x_2 \in S^d} \delta(x_1, x_2) \tag{14.1}$$

where S^d denotes the subset of S assigned to cluster d.

Compact clusters, which are clearly preferred, achieve small diameter values. It is usually also desirable for clusters of the same clustering model not to have considerably different diameters. When this is not the case, i.e., the diameters of some clusters are substantially below average, one could suspect that the number of clusters is too high. On the other hand, a single high-diameter cluster or a small number of clusters with considerably above-average diameter values may indicate too little clusters. For k-centers clustering algorithms these can be used as rough heuristics for the selection of k. In particular, one possible approach to choosing k could be to identify a range in which the maximum diameter remains stable, and then select k from that range that yields the most uniform diameter values.

Unfortunately, conclusions drawn from analyzing cluster diameters may not be quite reliable if the dataset suffers from noise, and in particular contains outliers, since in such circumstances the maximum intracluster dissimilarity is likely to be disrupted. It is therefore recommended to carefully examine and possibly fix data quality issues before using the diameter to make clustering model selection decisions.

Example 14.2.1 The following R code implements diameter calculation, using the `daisy` function from the `cluster` package to create the dissimilarity matrix, used to find the maximum dissimilarity in each cluster. It is then applied to determine the diameters of all clusters in all the PAM clustering models created in the previous example, both on the training and test sets. The optional `stand` parameter may be used to instruct the `daisy` function to perform internal standardization prior to dissimilarity calculation, but this capability will not be used in this and subsequent examples, since the parameters of internal standardization are not retained and therefore cannot be re-applied to new data. This has been discussed in Sections 11.3.9 and 12.1.2.

```
diameter <- function(clustering, data, metric="euclidean", stand=FALSE)
{
  clusters <- sort(unique(clustering))
  dm <- as.matrix(daisy(data, metric, stand))
  `names<-`(sapply(clusters, function(d) max(dm[clustering==d,clustering==d])),
        clusters)
}


  # training set diameter
diameter(i.pam2.euc$clustering, i.std.train[,-5])
diameter(i.pam3.euc$clustering, i.std.train[,-5])
diameter(i.pam5.euc$clustering, i.std.train[,-5])
diameter(i.pam7.euc$clustering, i.std.train[,-5])

diameter(i.pam2.man$clustering, i.std.train[,-5], metric="manhattan")
diameter(i.pam3.man$clustering, i.std.train[,-5], metric="manhattan")
diameter(i.pam5.man$clustering, i.std.train[,-5], metric="manhattan")
diameter(i.pam7.man$clustering, i.std.train[,-5], metric="manhattan")

  # test set diameter
diameter(i.pam2.euc.pred, i.std.test[,-5])
diameter(i.pam3.euc.pred, i.std.test[,-5])
diameter(i.pam5.euc.pred, i.std.test[,-5])
diameter(i.pam7.euc.pred, i.std.test[,-5])

diameter(i.pam2.man.pred, i.std.test[,-5], metric="manhattan")
diameter(i.pam3.man.pred, i.std.test[,-5], metric="manhattan")
diameter(i.pam5.man.pred, i.std.test[,-5], metric="manhattan")
diameter(i.pam7.man.pred, i.std.test[,-5], metric="manhattan")
```

Notice how the maximum cluster diameter is reduced at first when adding more clusters and then it tends to remain stable, but further increasing the number of clusters introduces more variability into cluster diameters (some low-diameter clusters are introduced). This is observed when evaluating on both the training set and the test set.

14.2.2 Separation

The *separation* of a cluster measures the degree to which it is distinct from other clusters. It is typically defined as the minimum dissimilarity between an instance from this cluster and an instance from another cluster:

$$\mathrm{sep}_{\delta,S}(d) = \min_{\substack{x_1 \in S^d \\ x_2 \in S - S^d}} \delta(x_1, x_2) \tag{14.2}$$

Ideally, one would prefer relatively uniform and high cluster separation values, but this is not always possible for realistic datasets. Substantial differences in cluster separation do not necessarily indicate a poor clustering model (e.g., an incorrectly chosen number of clusters for *k*-centers algorithms), since they may result from the nature of the data. Like the diameter, the separation is prone to disruptions due to outlying attribute values, being based on a single minimum dissimilarity.

Example 14.2.2 Separation calculation is illustrated by the R code presented below, which defines an appropriate function and applies it to evaluate each cluster of all the PAM models created before. As in the previous example, the `daisy` function is used to create the dissimilarity matrix, this time for the complete dataset.

```
separation <- function(clustering, data, metric="euclidean", stand=FALSE)
{
  clusters <- sort(unique(clustering))
  dm <- as.matrix(daisy(data, metric, stand))
  `names<-`(sapply(clusters, function(d) min(dm[clustering==d,clustering!=d])),
           clusters)
}

  # training set separation
separation(i.pam2.euc$clustering, i.std.train[,-5])
separation(i.pam3.euc$clustering, i.std.train[,-5])
separation(i.pam5.euc$clustering, i.std.train[,-5])
separation(i.pam7.euc$clustering, i.std.train[,-5])

separation(i.pam2.man$clustering, i.std.train[,-5], metric="manhattan")
separation(i.pam3.man$clustering, i.std.train[,-5], metric="manhattan")
separation(i.pam5.man$clustering, i.std.train[,-5], metric="manhattan")
separation(i.pam7.man$clustering, i.std.train[,-5], metric="manhattan")

  # test set separation
separation(i.pam2.euc.pred, i.std.test[,-5])
separation(i.pam3.euc.pred, i.std.test[,-5])
separation(i.pam5.euc.pred, i.std.test[,-5])
separation(i.pam7.euc.pred, i.std.test[,-5])

separation(i.pam2.man.pred, i.std.test[,-5], metric="manhattan")
separation(i.pam3.man.pred, i.std.test[,-5], metric="manhattan")
separation(i.pam5.man.pred, i.std.test[,-5], metric="manhattan")
separation(i.pam7.man.pred, i.std.test[,-5], metric="manhattan")
```

As we can see, the separation tends to decrease with increasing number of clusters, which is to be expected, although in most cases some highly separated clusters remain. The test set separation is usually below the training set one.

14.2.3   Isolation

Isolation is not a quality measure strictly speaking, but rather a property that can be expected for particularly compact and distinctive clusters. It is usually defined based on the diameter and separation. Basically, it is desirable (but not necessarily very common) that the diameter of a good cluster is less than its separation. Clusters with this property are considered *isolated*. A stronger form of isolation occurs if for any cluster member its maximum dissimilarity to any other member of the same cluster is less than its minimum dissimilarity to any instance from other clusters.

Example 14.2.3 The R code presented below implements and demonstrates the application of a function that determines cluster isolation, returning L for isolated clusters (with the diameter below the separation), `L*` for more strongly isolated clusters, and `no` for clusters that are not isolated.

```
isolation <- function(clustering, data, metric="euclidean", stand=FALSE)
{
  clusters <- sort(unique(clustering))
  dm <- as.matrix(daisy(data, metric, stand))
  diam <- diameter(clustering, data, metric, stand)
  sep <- separation(clustering, data, metric, stand)

  is <- sapply(clusters,
              function(d)
              if (all(apply(dm[clustering==d,clustering==d,drop=FALSE], 1, max)<
                      apply(dm[clustering==d,clustering!=d,drop=FALSE], 1, min)))
                "L*"
              else if (diam[d]<sep[d])
                "L"
              else
                "no")
  'names<-'(factor(is, levels=c("no", "L", "L*")), clusters)
}

  # training set isolation
isolation(i.pam2.euc$clustering, i.std.train[,-5])
isolation(i.pam3.euc$clustering, i.std.train[,-5])
isolation(i.pam5.euc$clustering, i.std.train[,-5])
isolation(i.pam7.euc$clustering, i.std.train[,-5])

isolation(i.pam2.man$clustering, i.std.train[,-5], metric="manhattan")
isolation(i.pam3.man$clustering, i.std.train[,-5], metric="manhattan")
isolation(i.pam5.man$clustering, i.std.train[,-5], metric="manhattan")
isolation(i.pam7.man$clustering, i.std.train[,-5], metric="manhattan")

  # test set isolation
isolation(i.pam2.euc.pred, i.std.test[,-5])
isolation(i.pam3.euc.pred, i.std.test[,-5])
isolation(i.pam5.euc.pred, i.std.test[,-5])
isolation(i.pam7.euc.pred, i.std.test[,-5])

isolation(i.pam2.man.pred, i.std.test[,-5], metric="manhattan")
isolation(i.pam3.man.pred, i.std.test[,-5], metric="manhattan")
isolation(i.pam5.man.pred, i.std.test[,-5], metric="manhattan")
isolation(i.pam7.man.pred, i.std.test[,-5], metric="manhattan")
```

No isolated clusters are found based on the training set, but some of the evaluated models contain isolated clusters on the test set.

14.2.4 Silhouette width

The *silhouette width* is related to the isolation and can be thought of as a more subtle, numerically expressed measure thereof. It is actually calculated on a per-instance basis, and then aggregated over all cluster members. The silhouette width of instance x from cluster d with respect to dissimilarity measure δ on dataset S is defined as follows:

$$\text{sw}_{\delta,S}(x,d) = \frac{\min_{d'} \Delta_{\delta,S}(x,d') - \Delta_{\delta,S}(x)}{\max\{\min_{d'} \Delta_{\delta,S}(x,d'), \Delta_{\delta,S}(x)\}} \tag{14.3}$$

where $\Delta_{\delta,S}(x, d')$ is the average dissimilarity between x and all instances from another cluster d':

$$\Delta_{\delta,S}(x, d') = \frac{1}{|S^{d'}|} \sum_{x' \in S^{d'}} \delta(x, x') \tag{14.4}$$

and $\Delta_{\delta,S}(x)$ is the average dissimilarity between x and other instances from its cluster d:

$$\Delta_{\delta,S}(x) = \frac{1}{|S^d| - 1} \sum_{x' \in S^d - \{x\}} \delta(x, x') \tag{14.5}$$

The silhouette width of an instance is a number from the $[-1, 1]$ interval that represents the difference between its average dissimilarity to instances from the closest (according to average dissimilarity) other cluster and other instances from the same cluster, normalized by dividing by the greater of these two.

Significantly positive values, particularly approaching 1, indicate that an instance is very well placed in its cluster, being much closer to its "fellow" members than to any "strangers" from other clusters. Values close to 0 and negative values indicate a possibly misplaced instance that is not much (or not at all) closer to its "fellows" than to some "strangers." A cluster in which all instances have positive silhouette width values is likely but not guaranteed to be isolated. This is because – unlike the diameter and separation used to judge isolation – the silhouette width is based on average dissimilarities. The silhouette width averaged over all members of a cluster can be used as a measure of its quality, taking into account both its cohesion and separation:

$$sw_{\delta,S}(d) = \frac{1}{|S^d|} \sum_{x \in S^d} sw_{\delta,S}(x, d) \tag{14.6}$$

Contrary to the diameter and separation, it does not depend on single maximally dissimilar or similar instances, but averages the dissimilarity between all instances within a cluster and between two clusters. High values indicate good clusters. When using k-centers clustering algorithms, there is probably no good reason to increase k if this yields additional clusters with significantly reduced silhouette width values.

Nonincreasingly ordered per-instance silhouette width values for each cluster, presented graphically as horizontal lines or bars, yield so-called *silhouette plots*. They make it immediately visible how many high and low silhouette width values occur in each cluster and whether clusters differ substantially with respect to the silhouette width distribution. This is a very convenient and popular visual approach to assessing the quality of clusters.

Example 14.2.4 The following R code implements and demonstrates silhouette width calculation. The `silwidth` function applied to a cluster returns the vector of its members' silhouette width values. Then the `silwidth.cluster` function gets per-cluster averages thereof. These two are applied to determine the per-cluster average silhouette width for the clustering models created in Example 14.1.1, as well as to generate silhouette plots for selected models ($k = 2$ and $k = 3$, Euclidean dissimilarity). The plots are presented in Figures 14.1 and 14.2.

```
silwidth <- function(clustering, d, data, metric="euclidean", stand=FALSE)
{
  if (sum(clustering==d)==1)
    1  # singleton cluster
  else
  {
    clusters <- unique(clustering)
    other <- clusters[! clusters %in% d]
    dm <- as.matrix(daisy(data, metric, stand))
    avg.intra <- apply(dm[clustering==d,clustering==d,drop=FALSE], 1, sum)/
                   (sum(clustering==d)-1)
    avg.inter <- apply(sapply(other,
                           function(d1)
                           apply(dm[clustering==d,clustering==d1,drop=FALSE],
                               1, mean)),
                      1, min)
    (avg.inter-avg.intra)/pmax(avg.inter, avg.intra)
  }
}

silwidth.cluster <- function(clustering, data, metric="euclidean", stand=FALSE)
{
  clusters <- sort(unique(clustering))
  `names<-`(sapply(clusters, function(d)
                       mean(silwidth(clustering, d, data, metric, stand))),
         clusters)
}

  # training set per-cluster silhouette width
silwidth.cluster(i.pam2.euc$clustering, i.std.train[,-5])
silwidth.cluster(i.pam3.euc$clustering, i.std.train[,-5])
silwidth.cluster(i.pam5.euc$clustering, i.std.train[,-5])
silwidth.cluster(i.pam7.euc$clustering, i.std.train[,-5])

silwidth.cluster(i.pam2.man$clustering, i.std.train[,-5], metric="manhattan")
silwidth.cluster(i.pam3.man$clustering, i.std.train[,-5], metric="manhattan")
silwidth.cluster(i.pam5.man$clustering, i.std.train[,-5], metric="manhattan")
silwidth.cluster(i.pam7.man$clustering, i.std.train[,-5], metric="manhattan")

  # test set per-cluster silhouette width
silwidth.cluster(i.pam2.euc.pred, i.std.test[,-5])
silwidth.cluster(i.pam3.euc.pred, i.std.test[,-5])
silwidth.cluster(i.pam5.euc.pred, i.std.test[,-5])
silwidth.cluster(i.pam7.euc.pred, i.std.test[,-5])

silwidth.cluster(i.pam2.man.pred, i.std.test[,-5], metric="manhattan")
silwidth.cluster(i.pam3.man.pred, i.std.test[,-5], metric="manhattan")
silwidth.cluster(i.pam5.man.pred, i.std.test[,-5], metric="manhattan")
silwidth.cluster(i.pam7.man.pred, i.std.test[,-5], metric="manhattan")

  # training set silhouette plots
par(mfrow=c(2, 1), mar=c(2, 6, 0, 1), oma=c(0, 0, 2, 0))
barplot(sort(silwidth(i.pam2.euc$clustering, 1, i.std.train[,-5])),
      xlim=c(-0.2, 1), yaxt="n", horiz=TRUE)
barplot(sort(silwidth(i.pam2.euc$clustering, 2, i.std.train[,-5])),
      xlim=c(-0.2, 1), yaxt="n", horiz=TRUE)
title("Training set, k=2", outer=TRUE)
```

```
par(mfrow=c(3, 1), mar=c(2, 6, 0, 1), oma=c(0, 0, 2, 0))
barplot(sort(silwidth(i.pam3.euc$clustering, 1, i.std.train[,-5])),
        xlim=c(-0.2, 1), yaxt="n", horiz=TRUE)
barplot(sort(silwidth(i.pam3.euc$clustering, 2, i.std.train[,-5])),
        xlim=c(-0.2, 1), yaxt="n", horiz=TRUE)
barplot(sort(silwidth(i.pam3.euc$clustering, 3, i.std.train[,-5])),
        xlim=c(-0.2, 1), yaxt="n", horiz=TRUE)
title("Training set, k=3", outer=TRUE)

 # test set silhouette plot
par(mfrow=c(2, 1), mar=c(2, 6, 0, 1), oma=c(0, 0, 2, 0))
barplot(sort(silwidth(i.pam2.euc.pred, 1, i.std.test[,-5])),
        xlim=c(-0.2, 1), yaxt="n", horiz=TRUE)
barplot(sort(silwidth(i.pam2.euc.pred, 2, i.std.test[,-5])),
        xlim=c(-0.2, 1), yaxt="n", horiz=TRUE)
title("Test set, k=2", outer=TRUE)

par(mfrow=c(3, 1), mar=c(2, 6, 0, 1), oma=c(0, 0, 2, 0))
barplot(sort(silwidth(i.pam3.euc.pred, 1, i.std.test[,-5])),
        xlim=c(-0.2, 1), yaxt="n", horiz=TRUE)
barplot(sort(silwidth(i.pam3.euc.pred, 2, i.std.test[,-5])),
        xlim=c(-0.2, 1), yaxt="n", horiz=TRUE)
barplot(sort(silwidth(i.pam3.euc.pred, 3, i.std.test[,-5])),
        xlim=c(-0.2, 1), yaxt="n", horiz=TRUE)
title("Test set, k=3", outer=TRUE)
```

Notice that the average per-cluster silhouette width values tend to decrease with increasing number of clusters, or more precisely, increasing the number of clusters appears to introduce more low silhouette width clusters. This might indicate that going above $k = 2$ is not well justified. The same can be concluded from the comparison of the silhouette plots obtained for $k = 2$ and $k = 3$, with the latter showing more not-so-well placed instances. Of course, well-justified k selection requires more thorough investigation and comparison of a wider range of values.

14.2.5   Davies–Bouldin index

One deficiency of the silhouette width is the computational cost associated with its calculation for large datasets. The *Davies–Bouldin index* adopts a simpler and less computationally demanding approach of averaging the dissimilarity to cluster centers instead of averaging the dissimilarity for all pairs of instances. This does not necessarily limit its applicability to evaluating models created using k-centers clustering algorithms, since cluster centers can also be identified for models created with other algorithms, even if they are not used by these algorithms themselves.

The Davies–Bouldin index for two clusters d_1 and d_2 can be considered a measure of their mutual separation. Its definition with respect to dissimilarity measure δ and dataset S can be written as

$$\mathrm{db}_{\delta,S}(d_1, d_2) = \frac{\Delta_{\delta,S}(d_1) + \Delta_{\delta,S}(d_2)}{\delta(\zeta_{d_1}, \zeta_{d_2})} \tag{14.7}$$

where

$$\Delta_{\delta,S}(d) = \frac{1}{|S^d|} \sum_{x \in S^d} \delta(x, \zeta_d) \tag{14.8}$$

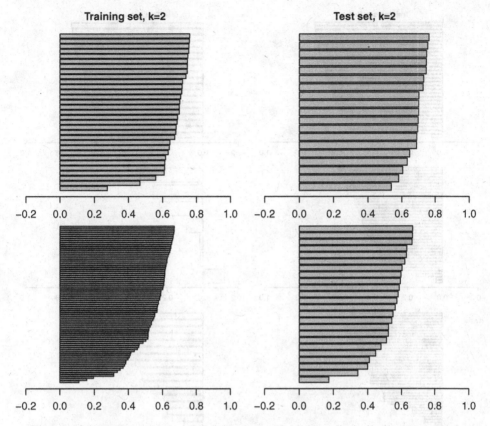

Figure 14.1   Silhouette plots for the 2-cluster Euclidean-dissimilarity pam model.

is the mean dissimilarity between the members of cluster d and its center. This yields small values for well separated clusters, for which the dissimilarity between their centers is large compared to the average dissimilarity between their members and the corresponding centers. To evaluate a single cluster d, its Davies–Bouldin index is calculated as the maximum index value obtained for all cluster pairs containing d:

$$\mathrm{db}_{\delta,S}(d) = \max_{d' \neq d} \mathrm{db}_{\delta,S}(d, d') \tag{14.9}$$

This yields small values for clusters that are well separated from all other clusters. Similarly as the silhouette width, it can be considered a more subtle cluster isolation indicator. Unlike the diameter and separation, it does not depend on single maximally dissimilar or similar instances, but uses the dissimilarity between cluster centers and the average dissimilarity between cluster members and cluster centers instead. It makes it more robust with respect to noisy data, in particular, containing outliers.

Example 14.2.5 The following R code defines a function for calculating the Davies–Bouldin index and demonstrates its application. The `dbindex` function returns the matrix of Davies–Bouldin index values for all cluster pairs. The row (or column) maxima of this matrix represent

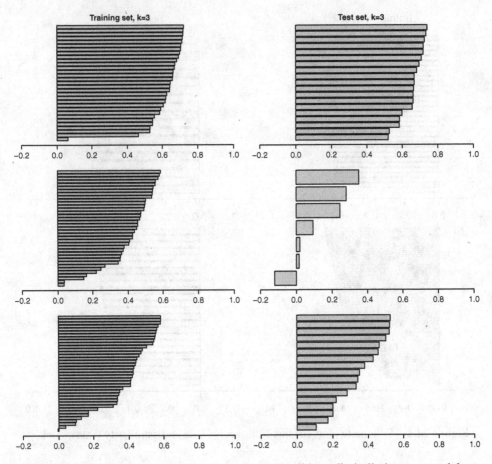

Figure 14.2   Silhouette plots for the 3-cluster Euclidean-dissimilarity pam model.

per-cluster Davies–Bouldin index values, which are returned by the dbindex.cluster function.

```
dbindex <- function(clustering, centers, data, metric="euclidean", stand=FALSE)
{
  clusters <- sort(unique(clustering))
  ds <- as.matrix(daisy(rbind(data, centers), metric, stand))

  to.center <- sapply(clusters, function(d) mean(ds[clustering==d,nrow(data)+d]))
  between.centers <- ds[(nrow(data)+1):(nrow(data)+length(clusters)),
                        (nrow(data)+1):(nrow(data)+length(clusters))]
  diag(between.centers) <- NA

  `dimnames<-`(outer(clusters, clusters, Vectorize(function(d1, d2)
                                   (to.center[d1]+to.center[d2])/
                                   between.centers[d1,d2])),
          list(clusters, clusters))
}
```

```
dbindex.cluster <- function(clustering, centers, data,
                           metric="euclidean", stand=FALSE)
{ apply(dbindex(clustering, centers, data, metric, stand), 1, max, na.rm=TRUE) }

  # training set Davies-Bouldin index for cluster pairs
dbindex(i.pam3.euc$clustering, i.pam3.euc$medoids, i.std.train[,-5])

  # training set Davies-Bouldin index
dbindex.cluster(i.pam2.euc$clustering, i.pam2.euc$medoids, i.std.train[,-5])
dbindex.cluster(i.pam3.euc$clustering, i.pam3.euc$medoids, i.std.train[,-5])
dbindex.cluster(i.pam5.euc$clustering, i.pam5.euc$medoids, i.std.train[,-5])
dbindex.cluster(i.pam7.euc$clustering, i.pam7.euc$medoids, i.std.train[,-5])

dbindex.cluster(i.pam2.man$clustering, i.pam2.euc$medoids, i.std.train[,-5],
               metric="manhattan")
dbindex.cluster(i.pam3.man$clustering, i.pam3.euc$medoids, i.std.train[,-5],
               metric="manhattan")
dbindex.cluster(i.pam5.man$clustering, i.pam5.euc$medoids, i.std.train[,-5],
               metric="manhattan")
dbindex.cluster(i.pam7.man$clustering, i.pam7.euc$medoids, i.std.train[,-5],
               metric="manhattan")

  # test set Davies-Bouldin index
dbindex.cluster(i.pam2.euc.pred, i.pam2.euc$medoids, i.std.test[,-5])
dbindex.cluster(i.pam3.euc.pred, i.pam3.euc$medoids, i.std.test[,-5])
dbindex.cluster(i.pam5.euc.pred, i.pam5.euc$medoids, i.std.test[,-5])
dbindex.cluster(i.pam7.euc.pred, i.pam7.euc$medoids, i.std.test[,-5])

dbindex.cluster(i.pam2.man.pred, i.pam2.euc$medoids, i.std.test[,-5],
               metric="manhattan")
dbindex.cluster(i.pam3.man.pred, i.pam3.euc$medoids, i.std.test[,-5],
               metric="manhattan")
dbindex.cluster(i.pam5.man.pred, i.pam5.euc$medoids, i.std.test[,-5],
               metric="manhattan")
dbindex.cluster(i.pam7.man.pred, i.pam7.euc$medoids, i.std.test[,-5],
               metric="manhattan")
```

Considerably larger Davies–Bouldin index values are obtained for models with more clusters. Whereas the minimum value (for the best cluster) does not always change much, additional clusters that are introduced exhibit much greater values of the index. This again suggests that $k = 2$ may be the right choice for the *Iris* dataset.

14.3   Overall quality measures

Although per-cluster quality measures provide useful insights into the quality of the evaluated clustering model, what is usually desired is a single quality indicator, that can be used to directly compare a number of alternative models (obtained using different algorithms or the same algorithm with different parameter settings). This is the purpose of overall quality measures that take into account all clusters, and combine some indicators of their cohesion and separation in a meaningful way. For some of them, it is as simple as averaging per-cluster measures (particularly those that take into account other clusters when evaluating a cluster).

14.3.1   Dunn index

The *Dunn index* measures the quality of a clustering model by relating the minimum separation to the maximum diameter exhibited by its clusters:

$$\text{dunn}_{\delta,S}(h) = \frac{\min_{d \in C_h} \text{sep}_{\delta,S}(d)}{\max_{d \in C_h} \text{diam}_{\delta,S}(d)} \qquad (14.10)$$

Clearly, high values are preferred, as they correspond to intercluster dissimilarities that are large in comparison to intracluster dissimilarities. Being based on the diameter and separation, the Dunn index inherits its sometimes problematic feature of being overly sensitive to single outlying instances. It is not uncommon to see some modifications of the Dunn index being proposed and used, based on different – more robust – definitions of the diameter and separation.

Example 14.3.1 The following R code implements the Dunn index calculation, using the `separation` and `diameter` procedures defined in the preceding examples, and applies it to evaluate all clusters of the same PAM clustering models.

```
dunn <- function(clustering, data, metric="euclidean", stand=FALSE)
{
  min(separation(clustering, data, metric, stand))/
    max(diameter(clustering, data, metric, stand))
}

  # training set Dunn index
dunn(i.pam2.euc$clustering, i.std.train[,-5])
dunn(i.pam3.euc$clustering, i.std.train[,-5])
dunn(i.pam5.euc$clustering, i.std.train[,-5])
dunn(i.pam7.euc$clustering, i.std.train[,-5])

dunn(i.pam2.man$clustering, i.std.train[,-5], metric="manhattan")
dunn(i.pam3.man$clustering, i.std.train[,-5], metric="manhattan")
dunn(i.pam5.man$clustering, i.std.train[,-5], metric="manhattan")
dunn(i.pam7.man$clustering, i.std.train[,-5], metric="manhattan")

  # test set Dunn index
dunn(i.pam2.euc.pred, i.std.test[,-5])
dunn(i.pam3.euc.pred, i.std.test[,-5])
dunn(i.pam5.euc.pred, i.std.test[,-5])
dunn(i.pam7.euc.pred, i.std.test[,-5])

dunn(i.pam2.man.pred, i.std.test[,-5], metric="manhattan")
dunn(i.pam3.man.pred, i.std.test[,-5], metric="manhattan")
dunn(i.pam5.man.pred, i.std.test[,-5], metric="manhattan")
dunn(i.pam7.man.pred, i.std.test[,-5], metric="manhattan")
```

The Dunn index turns out to clearly favor $k = 2$ for the evaluated PAM models, with the values obtained for more clusters being considerably lower. One exception to this pattern can be observed for $k = 7$ with the Manhattan dissimilarity, which appears to be better than $k = 2$ on the training set.

14.3.2   Average Davies–Bouldin index

The Davies–Bouldin clustering quality index is calculated by averaging the values of per-cluster Davies–Bouldin indices as follows:

$$\text{db}_{\delta,S}(h) = \frac{1}{|C_h|} \sum_{d \in C_h} \text{db}_{\delta,S}(d) \tag{14.11}$$

As discussed in Section 14.2.5, small values are preferred, as they indicate clustering models with compact and separated clusters. The average index value can be used to compare different clustering models, including those obtained using the same k-centers clustering algorithms for different k values. The measure is not easily misled by outliers. It may be inappropriate or inconvenient, though, when there are not sufficiently representative cluster centers or they are expensive to identify.

Example 14.3.2 The R code presented below defines a simple wrapper around the `dbindex.cluster`, presented in the example of the per-cluster Davies–Bouldin index that performs the averaging needed to obtain the overall clustering index and demonstrates its application to the same PAM clustering models.

```
dbindex.avg <- function(clustering, centers, data, metric="euclidean", stand=FALSE)
{ mean(dbindex.cluster(clustering, centers, data, metric, stand)) }

# training set average Davies-Bouldin index
dbindex.avg(i.pam2.euc$clustering, i.pam2.euc$medoids, i.std.train[,-5])
dbindex.avg(i.pam3.euc$clustering, i.pam3.euc$medoids, i.std.train[,-5])
dbindex.avg(i.pam5.euc$clustering, i.pam5.euc$medoids, i.std.train[,-5])
dbindex.avg(i.pam7.euc$clustering, i.pam7.euc$medoids, i.std.train[,-5])

dbindex.avg(i.pam2.man$clustering, i.pam2.euc$medoids, i.std.train[,-5],
            metric="manhattan")
dbindex.avg(i.pam3.man$clustering, i.pam3.euc$medoids, i.std.train[,-5],
            metric="manhattan")
dbindex.avg(i.pam5.man$clustering, i.pam5.euc$medoids, i.std.train[,-5],
            metric="manhattan")
dbindex.avg(i.pam7.man$clustering, i.pam7.euc$medoids, i.std.train[,-5],
            metric="manhattan")

# test set average Davies-Bouldin index
dbindex.avg(i.pam2.euc.pred, i.pam2.euc$medoids, i.std.test[,-5])
dbindex.avg(i.pam3.euc.pred, i.pam3.euc$medoids, i.std.test[,-5])
dbindex.avg(i.pam5.euc.pred, i.pam5.euc$medoids, i.std.test[,-5])
dbindex.avg(i.pam7.euc.pred, i.pam7.euc$medoids, i.std.test[,-5])

dbindex.avg(i.pam2.man.pred, i.pam2.euc$medoids, i.std.test[,-5], metric="manhattan")
dbindex.avg(i.pam3.man.pred, i.pam3.euc$medoids, i.std.test[,-5], metric="manhattan")
dbindex.avg(i.pam5.man.pred, i.pam5.euc$medoids, i.std.test[,-5], metric="manhattan")
dbindex.avg(i.pam7.man.pred, i.pam7.euc$medoids, i.std.test[,-5], metric="manhattan")
```

The average Davies–Bouldin index is clearly minimized for the $k = 2$ models, both on the training and test set.

14.3.3   C index

The *C index* evaluates clustering models by comparing intracluster pairwise dissimilarities with those directly observed in the dataset without taking cluster membership into account. For clustering model h and with respect to dissimilarity measure δ it is calculated on dataset S as follows:

$$\text{cind}_{\delta,S}(h) = \frac{\sum_{d \in C_h} \sum_{\substack{x_1,x_2 \in S^d \\ x_1 \neq x_2}} \delta(x_1,x_2) - \sum_{x_1,x_2 \in \Gamma_{\delta,S}^{\min}(h)} \delta(x_1,x_2)}{\sum_{\langle x_1,x_2 \rangle \in \Gamma_{\delta,S}^{\max}(h)} \delta(x_1,x_2) - \sum_{\langle x_1,x_2 \rangle \in \Gamma_{\delta,S}^{\min}(h)} \delta(x_1,x_2)} \tag{14.12}$$

where $\Gamma_{\delta,S}^{\min}(h)$ and $\Gamma_{\delta,S}^{\max}$ are the sets of $N_S(h)$ minimally dissimilar and maximally dissimilar pairs of different instances from S, respectively, and $N_S(h)$ is the number of all different instance pairs from the same cluster:

$$N_S(h) = \sum_{d \in C_h} |S^d|(|S^d| - 1) \tag{14.13}$$

The complex-looking formula basically calculates the sum of all pairwise intracluster dissimilarities and compares it to the sum of exactly the same number of the smallest dissimilarities in the dataset, irrespective of cluster membership. The difference is given an appropriate scale by dividing it by the analogous difference of the summed largest and smallest dissimilarities in the dataset, irrespective of cluster membership, again taking into account exactly the same number of instance pairs. The resulting index is guaranteed to fall in the [0, 1] interval and its smaller values are preferred, as they indicate clustering models with more internally cohesive clusters.

While it makes sense to compare the C index of different clustering models (e.g., created using different clustering algorithms and their parameter settings), it is not necessarily a good tool for selecting the number of clusters for k-centers clustering algorithms. This is because it is likely to decrease with the increase of k simply because intracluster dissimilarities tend to become lower with more clusters.

Example 14.3.3 C index calculation is implemented and demonstrated by the R code presented below.

```
cindex <- function(clustering, data, metric="euclidean", stand=FALSE)
{
  clusters <- unique(clustering)
  dm <- as.matrix(daisy(data, metric, stand))
  dm[lower.tri(dm)] <- diag(dm) <- NA
  sdm <- sort(dm)
  cc <- table(clustering)
  m <- sum(cc*(cc-1)/2)

  s <- sum(sapply(clusters,
              function(d) sum(dm[clustering==d,clustering==d], na.rm=TRUE)))
  smin <- sum(sdm[1:m])
  smax <- sum(sdm[(length(sdm)-m+1):length(sdm)])
  (s-smin)/(smax-smin)
}
```

```
  # training set C index
cindex(i.pam2.euc$clustering, i.std.train[,-5])
cindex(i.pam3.euc$clustering, i.std.train[,-5])
cindex(i.pam5.euc$clustering, i.std.train[,-5])
cindex(i.pam7.euc$clustering, i.std.train[,-5])

cindex(i.pam2.man$clustering, i.std.train[,-5], metric="manhattan")
cindex(i.pam3.man$clustering, i.std.train[,-5], metric="manhattan")
cindex(i.pam5.man$clustering, i.std.train[,-5], metric="manhattan")
cindex(i.pam7.man$clustering, i.std.train[,-5], metric="manhattan")

  # test set C index
cindex(i.pam2.euc.pred, i.std.test[,-5])
cindex(i.pam3.euc.pred, i.std.test[,-5])
cindex(i.pam5.euc.pred, i.std.test[,-5])
cindex(i.pam7.euc.pred, i.std.test[,-5])

cindex(i.pam2.man.pred, i.std.test[,-5], metric="manhattan")
cindex(i.pam3.man.pred, i.std.test[,-5], metric="manhattan")
cindex(i.pam5.man.pred, i.std.test[,-5], metric="manhattan")
cindex(i.pam7.man.pred, i.std.test[,-5], metric="manhattan")
```

Unlike the Dunn index illustrated in the previous example, or per-cluster quality measures presented before, the C index does not follow a strict monotonic pattern when applied to evaluate the PAM models created for the *Iris* dataset. It does not suggest $k = 2$ as the best number of clusters, either. The maximum C index value is obtained for $k = 7$ or, when using the Manhattan distance and evaluating the training set, for $k = 5$.

14.3.4 Average silhouette width

The silhouette width has been presented in Section 14.2.4 as a measure of goodness of particular instances' cluster membership that, averaged on a per-cluster basis, provides a cluster quality measure. It can also serve as an overall clustering model quality measure if averaged over all instances in the dataset:

$$\mathrm{sw}_{\delta,S}(h) = \frac{1}{|S|} \sum_{x \in S} \mathrm{sw}_{\delta,S}(x, h(x)) \qquad (14.14)$$

High values are obtained if many instances appear to be well placed in their clusters, which should be the case for good clustering models.

Example 14.3.4 The following R code implements and demonstrates average silhouette width calculation.

```
silwidth.avg <- function(clustering, data, metric="euclidean", stand=FALSE)
{
  clusters <- unique(clustering)
  mean(unlist(sapply(clusters,
                function(d) silwidth(clustering, d, data, metric, stand))))
}
```

```
    # training set average silhouette width
silwidth.avg(i.pam2.euc$clustering, i.std.train[,-5])
silwidth.avg(i.pam3.euc$clustering, i.std.train[,-5])
silwidth.avg(i.pam5.euc$clustering, i.std.train[,-5])
silwidth.avg(i.pam7.euc$clustering, i.std.train[,-5])

silwidth.avg(i.pam2.man$clustering, i.std.train[,-5], metric="manhattan")
silwidth.avg(i.pam3.man$clustering, i.std.train[,-5], metric="manhattan")
silwidth.avg(i.pam5.man$clustering, i.std.train[,-5], metric="manhattan")
silwidth.avg(i.pam7.man$clustering, i.std.train[,-5], metric="manhattan")

    # test set average silhouette width
silwidth.avg(i.pam2.euc.pred, i.std.test[,-5])
silwidth.avg(i.pam3.euc.pred, i.std.test[,-5])
silwidth.avg(i.pam5.euc.pred, i.std.test[,-5])
silwidth.avg(i.pam7.euc.pred, i.std.test[,-5])

silwidth.avg(i.pam2.man.pred, i.std.test[,-5], metric="manhattan")
silwidth.avg(i.pam3.man.pred, i.std.test[,-5], metric="manhattan")
silwidth.avg(i.pam5.man.pred, i.std.test[,-5], metric="manhattan")
silwidth.avg(i.pam7.man.pred, i.std.test[,-5], metric="manhattan")
```

Observe that the average silhouette width falls with increasing k, indicating the smallest $k = 2$ as the best choice.

14.3.5   Loglikelihood

The *loglikelihood* measure of clustering model quality is quite different from the other quality measures presented before. Unlike all of them, it does not – either explicitly or implicitly – incorporate any dissimilarity measure defined on the domain. Instead, it probabilistically assesses the degree of match between the evaluated model and the dataset used for the evaluation, treating the former as a representation of a probability distribution mixture. While this approach is particularly natural for probability distribution-based clustering, it can actually be used for arbitrary clustering models, including those obtained using dissimilarity-based algorithms.

The loglikelihood is based on the assumption that a clustering model h can be used to determine the probability of any instance x, $P(x|h)$ in the following way:

$$P(x|h) = \sum_{d \in C_h} P(d)P(x|d) \tag{14.15}$$

where $P(d)$ is the probability of cluster d, i.e., the probability that an arbitrarily selected instance from the domain is assigned to this cluster:

$$P(d) = P(h = d) \tag{14.16}$$

and $P(x|d)$ is the probability that cluster d contains an instance exactly like x (i.e., with the same attribute values), which can be calculated as

$$P(x|d) = \prod_{i=1}^{n} P(a_i = a_i(x) \mid h = d) \tag{14.17}$$

assuming the conditional independence of attribute values given clusters membership, which is the same independence assumption, discussed in Section 4.3.3, that is adopted by the naïve Bayes classifier. The probabilities $P(h = d)$ and $P(a_i = v_i \mid h = d)$ referred to above can be estimated based on the training set as follows:

$$P(h = d) = \frac{|T^d|}{|T|} \tag{14.18}$$

$$P(a_i = v_i \mid h(x) = d) = \frac{|T^d_{a_i=v_i}|}{|T^d|} \tag{14.19}$$

where T^d is the subset of the training set assigned to cluster d and $T^d_{a_i=v_i}$ is the subset with additionally the value of attribute a_i equal to v_i. Following the practice common for the naïve Bayes classifier, the technique of m-estimation presented in Section 2.4.4 can be employed to make attribute value probability estimates more reliable and safer to use:

$$P(a_i = v_i \mid h(x) = d) = \frac{|T^d_{a_i=v_i}| + mp}{|T^d| + m} \tag{14.20}$$

Unless some specific domain knowledge of attribute value probabilities is available, $p = \frac{1}{|A_i|}$ is assumed and often combined with $m = |A_i|$.

The latter obviously makes sense for discrete attributes only. For continuous attributes, much more common for practical clustering tasks, the corresponding density function value $g_i^d(v_i)$ should be used, where g_i^d denotes the probability density function of attribute a_i within cluster d. Unless explicitly available, it is usually assumed to be normal, and its parameters are estimated from the training set:

$$m_i^d = m_{T^d}(a_i) \tag{14.21}$$

and

$$s_i^d = s_{T^d}(a_i) \tag{14.22}$$

Similarly as for attribute value probabilities, it may be a good idea to use the m-estimated mean and variance, as presented in Section 2.4.4, to better handle small clusters.

Having explained how cluster and attribute value within cluster probabilities are calculated, the definition of the loglikelihood for clustering model h on dataset S can be written as follows:

$$L_{\delta,S}(h) = \log P(S|h)$$
$$= \sum_{x \in S} \log \left(\sum_{d \in C_h} P(h = d) \prod_{i=1}^{n} P(a_i = a_i(x) \mid h = d) \right) \tag{14.23}$$

The loglikelihood represents the degree to which the clustering model explains the dataset. High values are obtained if the model, treated as a "data generator," would be likely to generate the dataset. This indicates that the model suits the data well.

Despite the close relationship between the loglikelihood as a clustering quality measure and the loglikelihood as a classification model performance measure which was discussed in Section 7.2.6, these two should not be confused. While the former refers to the probabilities of attribute value vectors representing instances, the latter refers to class probabilities. When evaluating a classification model, we measure the degree to which class labels in a dataset are likely given the classifier. When evaluating a clustering model, we measure the degree to which attribute value vectors representing instances in a dataset are likely given the clustering.

Example 14.3.5 The following R code implements and demonstrates loglikelihood calculation for arbitrary clustering models. The `mest`, `mmean`, and `mvar` functions from Examples 2.4.31, 2.4.33, and 2.4.34 are used for probability, mean, and variance *m*-estimation, respectively. The `clusloglik` function needs the cluster assignments on the training set and the training set itself for probability estimation, even if the evaluation is performed on another dataset.

> Ex. 2.4.31, 2.4.34
> dmr.stats

```
clusloglik <- function(train.clustering, train.data,
                       eval.clustering=train.clustering, eval.data=train.data)
{
  clusters <- unique(train.clustering)
  prob.d <- function(d) { sum(train.clustering==d)/nrow(train.data) }
  prob.avd <- function(a, v, d)
  {
    ifelse(is.numeric(v),
           dnorm(v, mmean(train.data[train.clustering==d,a],
                          m0=mean(train.data[,a])),
                 sqrt(mvar(train.data[train.clustering==d,a],
                           m0=mean(train.data[,a]), s02=var(train.data[,a])))),
           mest(sum(train.data[,a]==v & train.clustering==d),
                sum(train.clustering==d),
                nlevels(train.data[,a]), 1/nlevels(train.data[,a])))
  }

  sum(sapply(1:nrow(eval.data),
             function(i)
             log(sum(sapply(clusters,
                            function(d)
                            prob.d(d)*prod(mapply(function(a, v) prob.avd(a, v, d),
                                                  1:ncol(eval.data),
                                                  eval.data[i,]))))))))
}

  # training set loglikelihood
clusloglik(i.pam2.euc$clustering, i.std.train[,-5])
clusloglik(i.pam3.euc$clustering, i.std.train[,-5])
clusloglik(i.pam5.euc$clustering, i.std.train[,-5])
clusloglik(i.pam7.euc$clustering, i.std.train[,-5])

clusloglik(i.pam2.man$clustering, i.std.train[,-5])
clusloglik(i.pam3.man$clustering, i.std.train[,-5])
clusloglik(i.pam5.man$clustering, i.std.train[,-5])
clusloglik(i.pam7.man$clustering, i.std.train[,-5])

  # test set loglikelihood
clusloglik(i.pam2.euc$clustering, i.std.train[,-5], i.pam2.euc.pred,
           i.std.test[,-5])
clusloglik(i.pam3.euc$clustering, i.std.train[,-5], i.pam3.euc.pred,
```

```
                     i.std.test[,-5])
clusloglik(i.pam5.euc$clustering, i.std.train[,-5], i.pam5.euc.pred,
                     i.std.test[,-5])
clusloglik(i.pam7.euc$clustering, i.std.train[,-5], i.pam7.euc.pred,
                     i.std.test[,-5])

clusloglik(i.pam2.man$clustering, i.std.train[,-5], i.pam2.man.pred,
                     i.std.test[,-5])
clusloglik(i.pam3.man$clustering, i.std.train[,-5], i.pam3.man.pred,
                     i.std.test[,-5])
clusloglik(i.pam5.man$clustering, i.std.train[,-5], i.pam5.man.pred,
                     i.std.test[,-5])
clusloglik(i.pam7.man$clustering, i.std.train[,-5], i.pam7.man.pred,
                     i.std.test[,-5])
```

As expected, loglikelihood values increase with increasing number of clusters. The models using the Manhattan dissimilarity appear inferior to those using the Euclidean dissimilarity. The test set loglikelihood values are greater than the training set ones simply due to the smaller size of the former.

14.4   External quality measures

Clustering quality measures presented above assume no available external information about a "correct" or "desired" way of clustering the data. Instead, they adopt various approaches to assess the cohesion and separation of clusters or their degree of match to the data. In contrast, external measures assume the existence of externally provided class labels, like in the classification task, which – while not used for model creation – can be used for model evaluation. These class labels represent some partitioning of the data into subsets against which the cluster assignments generated by the evaluated model are compared.

The primary application of external quality measures is for clustering algorithm benchmarking, which is a common research task. Any datasets suitable for the classification task can be used for this purpose, although it makes particular sense for those where class labels are indeed related to instance similarity. It is much less common for such measures to be useful for practical applications, since clustering tends to be needed for domains where no appropriately labeled datasets are available. Sometimes they may become handy, though, if at least a small data subsets can be provided with human-assigned labels, to complement other hardly interpretable relative performance measures with some directly meaningful indicators.

14.4.1   Misclassification error

The most straightforward approach to clustering model evaluation with external quality measures is to adopt classifier performance measures presented in Section 7.2, such as the misclassification error, for this purpose. This can be done by treating the evaluated clustering model as a classification model that associates the majority class with each cluster based on the training set and then for each instance predicts the class associated with its cluster.

Example 14.4.1 The R code presented below demonstrates how classification performance measures can be applied to assess the quality of clustering models. The clustclas function generates class label predictions on a dataset with given cluster membership assignments on any datasets, based on the training set cluster membership and class labels. The obtained predicted class labels are then compared to the corresponding true class labels using the `err` function for misclassification error calculation.

> Ex. 7.2.1
> `dmr.claseval`

```
clustclas <- function(train.clustering, train.classes,
                       eval.clustering=train.clustering)
{
  clusters <- unique(train.clustering)
  labels <-
    sapply(clusters,
           function(d)
           levels(train.classes)[which.max(pdisc(train.classes[train.clustering==d]))])
  factor(labels[eval.clustering], levels=levels(train.classes))
}

  # training set error
err(clustclas(i.pam2.euc$clustering, i.std.train[,5]), i.std.train[,5])
err(clustclas(i.pam3.euc$clustering, i.std.train[,5]), i.std.train[,5])
err(clustclas(i.pam5.euc$clustering, i.std.train[,5]), i.std.train[,5])
err(clustclas(i.pam7.euc$clustering, i.std.train[,5]), i.std.train[,5])

err(clustclas(i.pam2.man$clustering, i.std.train[,5]), i.std.train[,5])
err(clustclas(i.pam3.man$clustering, i.std.train[,5]), i.std.train[,5])
err(clustclas(i.pam5.man$clustering, i.std.train[,5]), i.std.train[,5])
err(clustclas(i.pam7.man$clustering, i.std.train[,5]), i.std.train[,5])

  # test set error
err(clustclas(i.pam2.euc$clustering, i.std.train[,5], i.pam2.euc.pred),
    i.std.test[,5])
err(clustclas(i.pam3.euc$clustering, i.std.train[,5], i.pam3.euc.pred),
    i.std.test[,5])
err(clustclas(i.pam5.euc$clustering, i.std.train[,5], i.pam5.euc.pred),
    i.std.test[,5])
err(clustclas(i.pam7.euc$clustering, i.std.train[,5], i.pam7.euc.pred),
    i.std.test[,5])

err(clustclas(i.pam2.man$clustering, i.std.train[,5], i.pam2.man.pred),
    i.std.test[,5])
err(clustclas(i.pam3.man$clustering, i.std.train[,5], i.pam3.man.pred),
    i.std.test[,5])
err(clustclas(i.pam5.man$clustering, i.std.train[,5], i.pam5.man.pred),
    i.std.test[,5])
err(clustclas(i.pam7.man$clustering, i.std.train[,5], i.pam7.man.pred),
    i.std.test[,5])
```

The misclassification error is reduced or unchanged with increasing number of clusters.

14.4.2   Rand index

Rather than forcing a clustering model to work as a classifier, one could examine the degree of concordance between the provided class labels and model-assigned clusters. The *Rand index* is a quality measure that accomplishes this by considering all pairs of instances from the dataset used for the evaluation. Each such pair $\langle x_1, x_2 \rangle$ is assigned to one of the following four categories, closely related to confusion matrix entries used for classification model evaluation presented in Section 7.2.4:

True positives. $c(x_1) = c(x_2)$ and $h(x_1) = h(x_2)$,

True negatives. $c(x_1) \neq c(x_2)$ and $h(x_1) \neq h(x_2)$,

False positives. $c(x_1) \neq c(x_2)$ and $h(x_1) = h(x_2)$,

False negatives. $c(x_1) = c(x_2)$ and $h(x_1) \neq h(x_2)$.

The Rand index is then calculated based on dataset S as the ratio of the total number of true positives and true negatives to the total number of instance pairs from S. This can be written as

$$\text{rand}_{c,S}(h) = \frac{\text{TP}_{c,S}(h) + \text{TN}_{c,S}(h)}{\text{TP}_{c,S}(h) + \text{TN}_{c,S}(h) + \text{FP}_{c,S}(h) + \text{FN}_{c,S}(h)} \quad (14.24)$$

where

$$\text{TP}_{c,S}(h) = \left| \{ \langle x_1, x_2 \rangle \in S^2 | x_1 \neq x_2 \wedge c(x_1) = c(x_2) \wedge h(x_1) = h(x_2) \} \right| \quad (14.25)$$

$$\text{TN}_{c,S}(h) = \left| \{ \langle x_1, x_2 \rangle \in S^2 | x_1 \neq x_2 \wedge c(x_1) \neq c(x_2) \wedge h(x_1) \neq h(x_2) \} \right| \quad (14.26)$$

$$\text{FP}_{c,S}(h) = \left| \{ \langle x_1, x_2 \rangle \in S^2 | x_1 \neq x_2 \wedge c(x_1) \neq c(x_2) \wedge h(x_1) = h(x_2) \} \right| \quad (14.27)$$

$$\text{FN}_{c,S}(h) = \left| \{ \langle x_1, x_2 \rangle \in S^2 | x_1 \neq x_2 \wedge c(x_1) = c(x_2) \wedge h(x_1) \neq h(x_2) \} \right| \quad (14.28)$$

High values indicate that cluster membership assignments match external class labels well.

Example 14.4.2 The following R code implements and demonstrates Rand index calculation.

```
randindex <- function(clustering, classes)
{
  mean(outer(1:length(clustering), 1:length(classes),
             function(i, j)
             ifelse(i!=j,
                    clustering[i]==clustering[j] & classes[i]==classes[j] |
                    clustering[i]!=clustering[j] & classes[i]!=classes[j], NA)),
        na.rm=TRUE)
}

  # training set Rand index
randindex(i.pam2.euc$clustering, i.std.train[,5])
randindex(i.pam3.euc$clustering, i.std.train[,5])
randindex(i.pam5.euc$clustering, i.std.train[,5])
randindex(i.pam7.euc$clustering, i.std.train[,5])

randindex(i.pam2.man$clustering, i.std.train[,5])
randindex(i.pam3.man$clustering, i.std.train[,5])
randindex(i.pam5.man$clustering, i.std.train[,5])
randindex(i.pam7.man$clustering, i.std.train[,5])

  # test set Rand index
randindex(i.pam2.man$clustering, i.std.test[,5])
randindex(i.pam3.man$clustering, i.std.test[,5])
randindex(i.pam5.man$clustering, i.std.test[,5])
randindex(i.pam7.man$clustering, i.std.test[,5])
```

```
randindex(i.pam2.man.pred, i.std.test[,5])
randindex(i.pam3.man.pred, i.std.test[,5])
randindex(i.pam5.man.pred, i.std.test[,5])
randindex(i.pam7.man.pred, i.std.test[,5])
```

According to the degree of match between cluster assignments and class labels measured by the Rand index on both the training and test subsets of the *Iris* data, the 3-cluster PAM models outperform those with less and greater k values.

14.4.3   General relationship detection measures

The Rand index can be considered a special case of a more general approach to clustering model evaluation, based on examining the relationship between the provided class labels and model-assigned clusters. While the Rand index is a particularly popular way of doing this in the context of clustering, using any general-purpose relationship measures or tests, such as the mutual information or the χ^2 test, is also perfectly reasonable. A strong and significant relationship would then be interpreted as indicating a good quality model.

Example 14.4.3 The R code presented below applies the χ^2 test, using the standard `chisq.test` function, to test the relationship between cluster assignments and class labels for the previously created PAM models, both on the training and test set.

```
  # training set chi-square
chisq.test(i.pam2.euc$clustering, i.std.train[,5])
chisq.test(i.pam3.euc$clustering, i.std.train[,5])
chisq.test(i.pam5.euc$clustering, i.std.train[,5])
chisq.test(i.pam7.euc$clustering, i.std.train[,5])

chisq.test(i.pam2.man$clustering, i.std.train[,5])
chisq.test(i.pam3.man$clustering, i.std.train[,5])
chisq.test(i.pam5.man$clustering, i.std.train[,5])
chisq.test(i.pam7.man$clustering, i.std.train[,5])

  # test set chi-square
chisq.test(i.pam2.man.pred, i.std.test[,5])
chisq.test(i.pam3.man.pred, i.std.test[,5])
chisq.test(i.pam5.man.pred, i.std.test[,5])
chisq.test(i.pam7.man.pred, i.std.test[,5])

chisq.test(i.pam2.man.pred, i.std.test[,5])
chisq.test(i.pam3.man.pred, i.std.test[,5])
chisq.test(i.pam5.man.pred, i.std.test[,5])
chisq.test(i.pam7.man.pred, i.std.test[,5])
```

The relationships turn out to be definitely significant for all the evaluated models. This makes it possible to to consider all of them reasonable, but does not help in choosing the best one. Looking at the χ^2 statistic values one can observe a mostly monotonic increase when adding more clusters.

The mutual information can also be used for the same purpose, as demonstrated by the following code, which uses the `mutinfo` function:

Ex. 2.5.6
dmr.stats

```
# training set mutual information
mutinfo(i.pam2.euc$clustering, i.std.train[,5])
mutinfo(i.pam3.euc$clustering, i.std.train[,5])
mutinfo(i.pam5.euc$clustering, i.std.train[,5])
mutinfo(i.pam7.euc$clustering, i.std.train[,5])

mutinfo(i.pam2.man$clustering, i.std.train[,5])
mutinfo(i.pam3.man$clustering, i.std.train[,5])
mutinfo(i.pam5.man$clustering, i.std.train[,5])
mutinfo(i.pam7.man$clustering, i.std.train[,5])

# test set mutual information
mutinfo(i.pam2.man.pred, i.std.test[,5])
mutinfo(i.pam3.man.pred, i.std.test[,5])
mutinfo(i.pam5.man.pred, i.std.test[,5])
mutinfo(i.pam7.man.pred, i.std.test[,5])

mutinfo(i.pam2.man.pred, i.std.test[,5])
mutinfo(i.pam3.man.pred, i.std.test[,5])
mutinfo(i.pam5.man.pred, i.std.test[,5])
mutinfo(i.pam7.man.pred, i.std.test[,5])
```

Again, in most cases the mutual information increases with the increased number of clusters. This might suggest choosing the smallest k value above which no significant further improvement can be observed. This appears to be $k = 3$ or $k = 5$.

14.5   Using quality measures

For most classification and regression performance measures models with appropriately extreme values (minimum or maximum, depending on the particular measure) are always preferred, as long as they are calculated with sufficient care needed to reliably estimate true performance based on dataset performance. It is not necessarily the case for clustering quality measures, for the following reasons:

1. clustering models tend to be evaluated on the training set only, particularly if they are not supposed to be ever applied to new data, but just represent similarity patterns identified by the available data,

2. several clustering quality measures exhibit monotonic or nearly monotonic behavior when increasing the number of clusters, which makes the straightforward optimization of their values pointless.

When performing clustering model selection (e.g., selecting the value of k for k-centers clustering algorithms) it is therefore not always the best idea to strictly maximize (or minimize, as appropriate) the adopted quality measure. It often makes more sense to use one or several quality measures as guidance rather than definite citeria, combined with some additional general or task-specific preferences. To compensate for the monotonicity with respect to the number of clusters, one may prefer, in particular, the smallest number of clusters with a sufficiently good quality or the smallest number of clusters above which the quality does not improve substantially.

14.6   Conclusion

The major difference between the clustering task and the classification and regression tasks consists in the lack of a predefined target attribute. This has a direct and decisive impact on clustering model evaluation, which differs from classification and regression model evaluation in important ways. In particular, the lack of a target attribute to be approximated by the model removes the tension from assessing the expected quality of such approximation on new data. While this does not necessarily make model evaluation procedures, which handle data splitting into training and validation or test subsets, useless for clustering model evaluation, they are not similarly essential for classification or regression model evaluation. This is because the primary purpose of evaluation is no longer to answer the burning question of how well the model will predict the target attribute when applied to previously unseen instances, but rather to find which of several candidate models appear most successful at identifying the similarity patterns in the training set. While clustering is not free of the risk of overfitting, which makes training performance possibly optimistically biased, it may still serve for model selection when used with care and not followed blindly. Using separate data for model evaluation always remains a good idea, but the resulting pessimistic evaluation bias is usually not harmful. Refined evaluation procedures designed for bias reduction, such as k-fold cross-validation or bootstrapping, are often unnecessary, with the simple hold-out procedure, possibly repeated, sufficient in most cases. This is why this chapter has not addressed the issue of evaluation procedures at all.

Limiting our attention to quality measures, there are two main closely related problems with those adopted for clustering model evaluation:

- the lack of unquestionable objective quality criteria that can be used to for candidate model ranking or as final model acceptance criteria,

- the existence of a variety of quality indicators that are not necessarily consistent with one another may be difficult to match the requirements of particular.

It is rather an inherent feature of the clustering task than a fault of existing clustering model performance measures, and therefore one should rather find a practically acceptable solution than complain. What should usually make the most sense is to apply multiple performance measures, but not to trust their values blindly. Clustering model performance measures should be considered as providing hints or recommendations rather than definitive answers to model selection or acceptance dilemmas. Combining several such recommendations, based on different quality indicators, with some domain knowledge and common sense, may lead to much better results than mechanically applying even the most refined performance measures. This is the way to go in most cases.

14.7   Further readings

Data mining books that present clustering algorithms – particularly from the k-centers family where a quality measure is necessary for the choice of k – usually allocate some space to reviewing selected approaches to clustering model evaluation (e.g., Cios *et al.* 2007; Han *et al.* 2011; Webb 2002). The books by Theodoridis and Koutroumbas (2008) and Tan *et al.* (2013) stand out with respect to the scope and depth of their discussion of clustering quality measures. These measures – more commonly referred to as clustering validity indices – are even more extensively discussed by Jain and Dubes (1988) in their book on clustering algorithms. There are also dedicated reviews and comparative experimental studies that include a much wider selection of quality measures than this chapter (e.g., Arbelaitz *et al.* 2013; Dubes 1987; Guerra *et al.* 2012; Halkidi *et al.* 2001; Milligan and Cooper 1985), with particular emphasis on their application to determining the right number of clusters.

The importance of clustering model evaluation has been realized early and most quality measures presented in this chapter can be already considered classics, dating back to the 1970s. In particular, the Dunn index was introduced by Dunn (1974a,b), the C index by Hubert and Levin (1976), and the Davies–Bouldin index by Davies and Bouldin (1979). The slightly newer silhouette width measure was proposed by Rousseeuw (1987). The idea of measuring clustering quality by loglikelihood comes from mixture models, representing clusters by probability distributions, usually identified using some version of the expectation minimization algorithm, introduced by Dempster *et al.* (1977) and then extensively described by several authors (e.g., McLachlan and Peel 2000). Bezdek and Pal (1998) came up with modifications to the Dunn index, reducing its sensitivity to outliers. The Rand index for external clustering evaluation was proposed by Rand (1971).

References

Arbelaitz O, Gurrutxaga I, Muguerza J, Pérez JM and Perona I 2013 An extensive comparative study of cluster validy indices. *Pattern Recognition* **46**, 243–256.

Bezdek JC and Pal NR 1998 Some new indexes of cluster validity. *IEEE Transactions on Systems, Man and Cybernetics* **28**, 301–315.

Cios KJ, Pedrycz W, Swiniarski RW and Kurgan L 2007 *Data Mining: A Knowledge Discovery Approach*. Springer.

Davies DL and Bouldin DW 1979 A cluster separation measure. *IEEE Transactions on Pattern Analysis and Machine Intelligence* **1**, 224–227.

Dempster AP, Laird NM and Rubin DB 1977 Maximum likelihood from incomplete data via the EM algorithm. *Journal of the Royal Statistical Society B* **39**, 1–38.

Dubes RC 1987 How many clusters are best? – -an experiment. *Pattern Recognition* **20**, 645–663.

Dunn JC 1974a A fuzzy relative of the ISODATA process and its use in detecting compact well-separated clusters. *Journal of Cybernetics* **3**, 32–57.

Dunn JC 1974b Well separated clusters and optimal fuzzy partitions. *Journal of Cybernetics* **4**, 95–104.

Guerra L, Robles V, Bielza C and Larrañaga P 2012 A comparison of clustering quality indices using outliers and noise. *Intelligent Data Analysis* **16**, 703–715.

Halkidi M, Batistakis Y and Vazirgiannis M 2001 On clustering validation techniques. *Journal of Intelligent Information Systems* **17**, 107–145.

Han J, Kamber M and Pei J 2011 *Data Mining: Concepts and Techniques* 3rd edn. Morgan Kaufmann.

Hubert LJ and Levin JR 1976 A general statistical framework for assesing categorical clustering in free recall. *Psychological Bulletin* **83**, 1072–1080.

Jain AK and Dubes RC 1988 *Algorithms for Clustering Data*. Prentice-Hall.

McLachlan GJ and Peel DA 2000 *Finite Mixture Models*. Wiley.

Milligan GW and Cooper MC 1985 An examination of procedures for determining the number of clusters in a data set. *Psychometrika* **50**, 157–179.

Rand WM 1971 Objective criteria for the evaluation of clustering methods. *Journal of the American Statistical Association* **66**, 846–850.

Rousseeuw PJ 1987 Silhouettes: A graphical aid to the interpretation and validation of cluster analysis. *Computational and Applied Mathematics* **20**, 53–65.

Tan PN, Steinbach M and Kumar V 2013 *Introduction to Data Mining* 2nd edn. Addison-Wesley.

Theodoridis S and Koutroumbas K 2008 *Pattern Recognition* 4th edn. Academic Press.

Webb AR 2002 *Statistical Pattern Recognition* 2nd edn. Wiley.

Part V

GETTING BETTER MODELS

15

Model ensembles

15.1 Introduction

The idea of ensemble modeling is to create and combine multiple inductive models for the same domain, possibly obtaining better prediction quality than most or all of them. For this improvement to be possible, the strengths of individual models should be retained or reinforced and their weaknesses should be canceled out or reduced. It turns out that dozens or hundreds of models, even of rather mediocre quality, may produce top-notch predictions as a team.

Ensemble modeling is applicable to the two major predictive modeling tasks, classification and regression. In each case, it may yield substantial improvement over single models at the cost of investing considerably more computation time for multiple model creation and loosing overall human readability, even if each individual model is perfectly human readable. To exploit this potential for better predictive power, appropriate techniques for base model generation and aggregation are required, the most common of which will be discussed in this chapter. The former are mostly task independent and the task-specific aspects of the latter are sufficiently simple and isolated to make most of this discussion applicable both to the classification and regression tasks. It is the former, though, where model ensembles are most often and most successfully used, and some ensemble modeling techniques developed specifically for classification will also be discussed.

Example 15.1.1 Ensemble modeling techniques presented in this chapter will be illustrated with simple R code examples. They will use the *HouseVotes84* and *Boston Housing* datasets from the `mlbench` package, for the classification task and regression task, respectively. The decision tree and naïve Bayes algorithms will be used for classification base model creation, with their R implementations provided by the `rpart` and `e1071` packages. The regression tree and linear regression algorithms will be used for regression base model creation, with their R implementations provided again by the `rpart` package and the `lm` function from the standard `stats` package. Several DMR packages, containing functions defined in other chapters and simple utilities, will also be used. The following R code sets up the environment for these demonstrations by loading the packages and the datasets. The latter are then

---

Data Mining Algorithms: Explained Using R, First Edition. Paweł Cichosz.
© 2015 John Wiley & Sons, Ltd. Published 2015 by John Wiley & Sons, Ltd.

randomly partitioned into training and test subsets, to apply a manually performed hold-out evaluation procedure. A fixed initial seed of the random number generator is used to make the results strictly reproducible. Single models are created based on the training subsets, using the decision tree, naïve Bayes, regression tree, and linear regression algorithms, the same that will be subsequently used to create base models for ensembles. These will serve for the comparison of misclassification error and mean square error levels possible to achieve. The two quality indicators are calculated using the `err` function, defined in Example 7.2.1, and the `mse` function, defined in Example 10.2.3.

> `dmr.claseval`

> `dmr.regeval`

```
library(dmr.claseval)
library(dmr.dectree)
library(dmr.regeval)
library(dmr.regtree)
library(dmr.stats)
library(dmr.util)

library(rpart)
library(e1071)

data(HouseVotes84, package="mlbench")
data(BostonHousing, package="mlbench")

set.seed(12)

rhv <- runif(nrow(HouseVotes84))
hv.train <- HouseVotes84[rhv>=0.33,]
hv.test <- HouseVotes84[rhv<0.33,]

rbh <- runif(nrow(BostonHousing))
bh.train <- BostonHousing[rbh>=0.33,]
bh.test <- BostonHousing[rbh<0.33,]

hv.tree <- rpart(Class~., hv.train)
hv.nb <- naiveBayes(Class~., hv.train)

hv.err.tree <- err(predict(hv.tree, hv.test, type="c"), hv.test$Class)
hv.err.nb <- err(predict(hv.nb, hv.test), hv.test$Class)

bh.tree <- rpart(medv~., bh.train)
bh.lm <- lm(medv~., bh.train)

bh.mse.tree <- mse(predict(bh.tree, bh.test), bh.test$medv)
bh.mse.lm <- mse(predict(bh.lm, bh.test), bh.test$medv)
```

15.2   Model committees

The expectation of model ensembles to improve over individual models is sometimes explained by the common-sense idea of a committee consisting of multiple "experts," making better decisions collectively than individually. For this justification to remain valid, all of these "experts" must possess at least some reasonable level of competences and – at the same time – exhibit sufficiently diverse opinions to make their collective behavior different from that each of them would exhibit alone. Applying this metaphor to predictive modeling, we would expect combined models to predict better as a "model committee" than individually if they are all sufficiently good and sufficiently different from one another. These conditions create space for improvement by reinforcing strengths and compensating weaknesses (rather than compensating both, reinforcing both, or reinforcing weaknesses and compensating strengths).

As an elementary illustration, consider three classification models h_1, h_2, h_3, with the corresponding true misclassification error values $e_c(h_1)$, $e_c(h_2)$, $e_c(h_3)$ with respect to a target concept c. As explained in Section 7.2.1, these are the probabilities that the corresponding models would produce incorrect predictions for a randomly chosen instance from the domain. Now consider a simple combined model h_* that aggregates the predictions of base models h_1, h_2, h_3 by voting. Assuming a two-class classification task, the true error of this model is then the probability that majority of base models (i.e., two or three in our case) make mistakes, i.e.,

$$e_c(h_*) = e_c(h_1)e_c(h_2)e_c(h_3)$$
$$+ (1 - e_c(h_1))e_c(h_2)e_c(h_3) + e_c(h_1)(1 - e_c(h_2))e_c(h_3) \qquad (15.1)$$
$$+ e_c(h_1)e_c(h_2)(1 - ee_c(h_3))$$

Let us assume that all base models are sufficiently good and different from one another. The former may be represented by setting an upper bound ϵ for their error values and the latter –in the unrealistically idealized case – by considering their mistakes independent. Under these assumptions the above error may be bound as follows:

$$e_c(h_*) \leq \epsilon^3 + 3\epsilon^2(1 - \epsilon) \qquad (15.2)$$

What one might be interested to see is how this compares to the base model error bound ϵ. The corresponding inequality

$$\epsilon^3 + 3\epsilon^2(1 - \epsilon) < \epsilon \qquad (15.3)$$

may be easily solved, yielding $0 < \epsilon < 0.5$. This is a pretty weak requirement, which means (with the two-class assumption) that, for the ensemble to give an improvement, base models have to perform better than random (although not perfectly, as the latter clearly would leave no space for improvement). Thus, with just three base models, if they are just minimally reasonable, but fully independent, a simple voting-based combined model will perform better.

While the discussion above is an illustrative special case rather than a general argument, it at least demonstrates that the expectation of improved prediction performance by ensemble modeling is justified and which are the conditions necessary to actually make it happen. Its main limitation is not the small number of base models, since adding more models – which makes the error more complex to calculate – may only improve the prediction quality. It is the idealized assumption of model mistake independence that makes the derivation of error bounds practically inapplicable. It is still useful as a source of insights, though. In practice, creating multiple totally independent models for the same domain may be next to impossible, but it remains possible and worthwhile to approximate this ideal situation with models that are as diverse as possible, without sacrificing too much of their quality. The more such reasonable quality base models are available to combine and more diverse they are, more prediction quality improvement may be expected from the resulting model ensemble. While the actual results may also differ significantly depending on the particular model aggregation method, at least the available improvement potential depends on the base model portfolio.

15.3   Base models

As discussed above, the main challenge for successful ensemble modeling is creating sufficiently many sufficiently diverse and sufficiently good base models. Since any deterministic
algorithm will yield the same model when applied to the same training set with the same
parameter setup, the following approaches to ensuring diversity may be considered:

Different training sets. Use a different training set from the same domain to create each
base model.

Different algorithms. Use a different algorithm to create each base model.

Different parameter setups. Use a different algorithm parameter setup to create each base
model.

Algorithm randomization. Use independent runs of a nondeterministic algorithm to create
each base model.

For reinforced effect, two or more of these approaches can also be applied in combination.
Each of them must be used with care, though, as pressing too much on the diversity of base
models may ruin their quality, which must remain at some reasonable level.

15.3.1   Different training sets

The most popular approach to creating multiple base models relies on the assumption that
applying the same algorithm to different training sets for the same task from the same domain
will yield models that are sufficiently diverse and sufficiently good at the same time. Ideally,
we should be able to draw these training sets from the domain independently at random. In
practice, no direct domain sampling is possible, though, and a number of different training
sets may only be obtained by sampling or transforming the original training set supplied for
the task at hand. This may include:

- sampling instances,
- replicating instances,
- varying instance weights (for weight-sensitive algorithms),
- sampling attributes,
- applying attribute transformations.

15.3.1.1   Instance sampling

Instance sampling is typically performed by drawing multiple bootstrap samples $T_1, T_2,$
$\ldots, T_m$ of the original training set T, i.e., uniform random samples with replacement,
usually of the same size as the former. As demonstrated in Section 7.3.6, when discussing
bootstrapping as a model evaluation procedure, such a bootstrap sample may be expected to
contain about 63.2% of instances from T. Each sample T_i is used to create a base model h_i
using the same modeling algorithm.

For instance sampling to be successful in delivering diverse models, the latter should be
created by an *unstable* algorithm, i.e., highly sensitive even to minor data variations. Decision
and regression trees are the most obvious and natural candidates, since their split selection,

stop, and pruning criteria may yield different outcomes for slightly different datasets, resulting in different models being obtained. On the other hand, modeling algorithms that use the training data to estimate numerical parameters representing their models rather than to make discrete decisions, such as linear and other parametric models or the naïve Bayes classifier, do not react excessively to data perturbations and are considered stable.

Example 15.3.1 The following R code demonstrates the instance sampling approach to base model generation. The `base.ensemble.sample.x` function applies the specified algorithm to samples drawn from the provided dataset, using the standard `sample` function. Its default settings produce bootstrap samples of the same size as the original dataset. The function is applied to create 50 base decision tree and naïve Bayes models for the *HouseVotes84* data, as well as 50 base regression tree and linear regression models for the *Boston Housing* data. As an indirect and rough means of assessing the diversity of the base models obtained by instance sampling, their training and test set misclassification error or mean square error values are determined. Test set errors are of particular interest, since – while even substantially different models can be similarly good on the training set, they are more likely to differ with respect to their performance on previously unseen data.

```
## generate base models by instance sampling
base.ensemble.sample.x <- function(formula, data, m, alg, args=NULL,
                                    size=nrow(data), replace=TRUE)
{
  lapply(1:m, function(i)
         {
             bag <- sample(nrow(data), size=nrow(data), replace=replace)
             do.call(alg, c(list(formula, data[bag,]), args))
         })
}

  # base models for the HouseVotes84 data
hv.bm.tree.sx <- base.ensemble.sample.x(Class~., hv.train, 50, rpart)
hv.bm.nb.sx <- base.ensemble.sample.x(Class~., hv.train, 50, naiveBayes)

  # base models for the BostonHousing data
bh.bm.tree.sx <- base.ensemble.sample.x(medv~., bh.train, 50, rpart)
bh.bm.lm.sx <- base.ensemble.sample.x(medv~., bh.train, 50, lm)

  # base model training set errors for the HouseVotes84 data
hv.train.err.tree.sx <- sapply(hv.bm.tree.sx,
                               function(h) err(predict(h, hv.train, type="c"),
                                               hv.train$Class))
hv.train.err.nb.sx <- sapply(hv.bm.nb.sx,
                             function(h) err(predict(h, hv.train), hv.train$Class))

  # base model training set MSE values for the BostonHousing data
bh.train.mse.tree.sx <- sapply(bh.bm.tree.sx,
                               function(h) mse(predict(h, bh.train), bh.train$medv))
bh.train.mse.lm.sx <- sapply(bh.bm.lm.sx,
                             function(h) mse(predict(h, bh.train), bh.train$medv))

  # base model test set errors for the HouseVotes84 data
hv.test.err.tree.sx <- sapply(hv.bm.tree.sx,
                              function(h) err(predict(h, hv.test, type="c"),
                                              hv.test$Class))
hv.test.err.nb.sx <- sapply(hv.bm.nb.sx,
                            function(h) err(predict(h, hv.test), hv.test$Class))
```

```
# base model test set MSE values for the BostonHousing data
bh.test.mse.tree.sx <- sapply(bh.bm.tree.sx,
                            function(h) mse(predict(h, bh.test), bh.test$medv))
bh.test.mse.lm.sx <- sapply(bh.bm.lm.sx,
                            function(h) mse(predict(h, bh.test), bh.test$medv))
```

As it could be expected, decision trees and regression trees obtained for different bootstrap samples appear to exhibit more variability, at least with respect to their performance, than naïve Bayes and linear models.

15.3.1.2   Instance replication

Closely related to instance sampling, this approach modifies the training set by replicating some instances (usually selected uniformly at random). With an unstable algorithm, this may lead to different models being obtained. It is listed here as a possibility for the sake of completeness, but it does not appear to offer any important advantages over sampling and is not commonly used.

15.3.1.3   Instance weighting

If the modeling algorithm used for base model creation is not only unstable, but also weight-sensitive, as discussed in Section 1.3.7, varying nonuniform instance weights applied to the fixed training set may be an attractive alternative to sampling or replication. Rather than generating modified copies of the data, just the vector of weights is modified. It may be therefore an elegant and efficient technique for base model creation. Base models $h_1, h_2,$ $\ldots, h_m$ would then be generated using the same training set, but different vectors of per-instance weights $w^{(1)}, w^{(2)}, \ldots, w^{(m)}$, with each vector $w^{(i)}$ containing weight $w_x^{(i)}$ for each instance $x \in T$.

In the simplest case, weights could be generated at random, but it is more common to see this technique combined with some more refined weight adjustment schemes. Some of them will be presented below when discussing boosting, one of the most popular and successful instantiations of ensemble modeling.

Example 15.3.2 The following R code implements and demonstrates the instance weighting approach to base model generation. The `base.ensemble.weight.x` function assumes that the specified modeling algorithm accepts the `weights` argument. The weight vector is by default initialized uniformly at random in the $[0.3, 3]$ interval and then randomly re-generated for each base model, which only serves the illustration purpose. Other initial vectors and more refined reweighting schemes may be specified. In particular, the reweighting function specified via the `reweight` argument obtains the last model's predictions as its second argument, to make it possible to alter instance weights based on how the model predicted for each of them. The reweighting function may return `NULL` to instruct the `base.ensemble.weight.x` function not to include the last created model in the ensemble and to stop creating base models (before reaching the maximum number thereof specified via the m argument). This capability is not actually used in the subsequent demonstrations, but will come handy later. The `skip.cond` utility function is used to skip `NULL` models from the obtained list. | dmr.util |

Base model generation by instance weighting is demonstrated similarly as in the previous example, but with the naïve Bayes classifier skipped, since the R implementation used does not support instance weights.

```
## generate base models by instance weighting
base.ensemble.weight.x <- function(formula, data, m, alg, args=NULL,
                             weights=runif(nrow(data), min=0.3, max=3),
                             reweight=function(w, p=NULL)
                                     runif(nrow(data), min=0.3, max=3),
                             predf=predict)
{
  skip.cond(lapply(1:m,
               function(i)
               {
                 if (!is.null(weights))
                 {
                   h <- do.call(alg, c(list(formula, data, weights=weights),
                                       args))
                   pred <- predf(h, data)
                   if (!is.null(weights <<- reweight(weights, pred)))
                     h
                 }
               }),
             is.null)
}

  # base models for the HouseVotes84 data
hv.bm.tree.wx <- base.ensemble.weight.x(Class~., hv.train, 50, rpart)

  # base models for the BostonHousing data
bh.bm.tree.wx <- base.ensemble.weight.x(medv~., bh.train, 50, rpart)
bh.bm.lm.wx <- base.ensemble.weight.x(medv~., bh.train, 50, lm)

  # base model training set errors for the HouseVotes84 data
hv.train.err.tree.wx <- sapply(hv.bm.tree.wx,
                          function(h) err(predict(h, hv.train, type="c"),
                                          hv.train$Class))

  # base model training set MSE values for the BostonHousing data
bh.train.mse.tree.wx <- sapply(bh.bm.tree.wx,
                          function(h) mse(predict(h, bh.train), bh.train$medv))
bh.train.mse.lm.wx <- sapply(bh.bm.lm.wx,
                        function(h) mse(predict(h, bh.train), bh.train$medv))

  # base model test set errors for the HouseVotes84 data
hv.test.err.tree.wx <- sapply(hv.bm.tree.wx,
                         function(h) err(predict(h, hv.test, type="c"),
                                         hv.test$Class))

  # base model test set MSE values for the BostonHousing data
bh.test.mse.tree.wx <- sapply(bh.bm.tree.wx,
                         function(h) mse(predict(h, bh.test), bh.test$medv))
bh.test.mse.lm.wx <- sapply(bh.bm.lm.wx,
                       function(h) mse(predict(h, bh.test), bh.test$medv))
```

The distribution of the training and test performance of the generated base models provide some insights into their diversity. While the simple and arbitrary instance reweighting mechanism used for this example does produce some base model variability, but it is less effective than the instance sampling approach demonstrated in the previous example.

15.3.1.4   Attribute sampling

Unlike the previous approaches, attribute sampling does not require that the modeling algorithm be unstable. It requires instead that the set of attributes is sufficiently large to permit drawing sufficiently many smaller samples. Any reasonable algorithm is likely to yield considerably different models when supplied with different attribute subsets. This is why the attribute sampling approach might appear attractive, but it may require careful sample size tuning to properly tradeoff between base model diversity and quality.

With this approach, multiple random samples $A_1, A_2, \ldots, A_m$ are drawn (without replacement, as this would make no sense for attributes) from the original set of attributes A and base models $h_1, h_2, \ldots, h_m$ are then created using the same set of training instances, each time casted to the selected attribute subset. If using this approach with decision trees or regression trees, it makes sense to use strict stop criteria that result in growing large trees, better fitted (or possibly overfitted) to the training set, since with more splits selected there are more opportunities for trees using different attribute samples to differ.

Example 15.3.3 The following R code implements the attribute sampling approach to base model generation and presents a demonstration thereof, following the pattern of the previous examples. The `frac` parameter may be used to specify the number of attributes to use as the fraction of the number of all available attributes. The `clip.val` function is used to make sure that it is in the [0, 1] interval. The default value of 0 triggers ┌───────────┐ `dmr.util` └───────────┘ a heuristic setting the attribute sample size to the square root of the number of attributes. The `x.vars` and `y.var` functions are used to extract the input and target attribute names from the supplied R formula, and the `make.formula` function is used ┌───────────┐ `dmr.util` └───────────┘ to construct a modified formula with a sample of attributes only. Notice the `minsplit=2` and `cp=0` parameters passed to `rpart`, that result in growing maximally fitted decision or regression trees.

```
## generate base models by attribute sampling
base.ensemble.sample.a <- function(formula, data, m, alg, args=NULL,
                                    frac=0, replace=TRUE)
{
  attributes <- x.vars(formula, data)
  target <- y.var(formula)
  ns <- ifelse(clip.val(frac, 0, 1)>0, ceiling(frac*length(attributes)),
                                       ceiling(sqrt(length(attributes))))
  lapply(1:m, function(i)
             {
               sa <- sample(length(attributes), ns)
               do.call(alg, c(list(make.formula(target, attributes[sa]), data),
                           args))
             })
}

  # base models for the HouseVotes84 data
hv.bm.tree.sa <- base.ensemble.sample.a(Class~., hv.train, 50, rpart,
                                    args=list(minsplit=2, cp=0))
hv.bm.nb.sa <- base.ensemble.sample.a(Class~., hv.train, 50, naiveBayes)
```

```
 # base models for the BostonHousing data
bh.bm.tree.sa <- base.ensemble.sample.a(medv~., bh.train, 50, rpart,
                                  args=list(minsplit=2, cp=0))
bh.bm.lm.sa <- base.ensemble.sample.a(medv~., bh.train, 50, lm)

 # base model training set errors for the HouseVotes84 data
hv.train.err.tree.sa <- sapply(hv.bm.tree.sa,
                          function(h) err(predict(h, hv.train, type="c"),
                                          hv.train$Class))
hv.train.err.nb.sa <- sapply(hv.bm.nb.sa,
                        function(h) err(predict(h, hv.train), hv.train$Class))

 # base model training set MSE values for the BostonHousing data
bh.train.mse.tree.sa <- sapply(bh.bm.tree.sa,
                          function(h) mse(predict(h, bh.train), bh.train$medv))
bh.train.mse.lm.sa <- sapply(bh.bm.lm.sa,
                        function(h) mse(predict(h, bh.train), bh.train$medv))

 # base model test set errors for the HouseVotes84 data
hv.test.err.tree.sa <- sapply(hv.bm.tree.sa,
                         function(h) err(predict(h, hv.test, type="c"),
                                         hv.test$Class))
hv.test.err.nb.sa <- sapply(hv.bm.nb.sa,
                       function(h) err(predict(h, hv.test), hv.test$Class))

 # base model test set MSE values for the BostonHousing data
bh.test.mse.tree.sa <- sapply(bh.bm.tree.sa,
                         function(h) mse(predict(h, bh.test), bh.test$medv))
bh.test.mse.lm.sa <- sapply(bh.bm.lm.sa,
                       function(h) mse(predict(h, bh.test), bh.test$medv))
```

It is worthwhile to note how attribute sampling, in combination with strict stop criteria, not only promotes the variability of tree models, observed indirectly via their test set performance, but also reduces their quality. In particular, for the *HouseVotes84* data decision trees loose most of their accuracy advantage over the naïve Bayes classifier.

15.3.1.5   Attribute transformation

Another attribute-oriented approach to base model generation is to transform attributes in such ways that would affect models created by the adopted modeling algorithm. Not all common attribute transformations discussed in Chapter 17 may be appropriate for the purpose of creating multiple base models, though. In particular, standardization and normalization are of rather limited usefulness here, as they can only transform each attribute in a single way. Discretization for continuous attributes may be more useful, as different discretization algorithms and parameter setups may yield different discretized attributes. What is actually the most interesting possibility is to use transformations specifically designed to stimulate model diversity. They may be custom (possibly randomized) versions of the discretization or discrete attribute encoding transformations that – unlike their standard counterparts – are not just trying to preserve the predictive utility of the original attribute, but also to introduce diversity. Such techniques are not employed by the most commonly used ensemble modeling algorithms and therefore are not discussed here.

One specific example of attribute transformation that is particularly relevant to ensemble modeling is multiclass encoding, normally used to handle more than two classes with algorithms capable of delivering two-class classification models only. As extensively discussed and demonstrated in Section 17.4, multiclass encoding techniques create and combine multiple two-class models, and therefore can also be viewed as ensemble modeling techniques.

15.3.2   Different algorithms

The approach of using different algorithms to create base models assumes that the very same training set is passed to a number of modeling algorithms that hopefully produce sufficiently good and sufficiently diverse base models. This rarely makes it possible to create more than a few or a dozen base models, as this is how many algorithms for the same modeling task are typically available in analytic toolboxes. This limits the utility of this approach, at least in its pure form. It may become more attractive, though, in combination with the two related techniques discussed below.

15.3.3   Different parameter setups

The same algorithm may sometimes deliver substantially different models based on the same data if used with different parameter setups. The corresponding approach to base model creation makes sense for modeling algorithms that have parameters altering sufficiently important aspects of their operation to yield diverse models. These could be, e.g., different split selection or pruning criteria for decision or regression trees or different kernel functions for support vector machines and support vector regression modeling algorithms that will be presented in Chapter 16. This technique alone does not usually make it possible to generate a large number of different base models and is of limited usefulness.

15.3.4   Algorithm randomization

The same algorithm with the same parameter setup may yield different models for the same dataset if some of its processing steps are nondeterministic. While some algorithms may be nondeterministic by nature, it is much more common and useful to deliberately randomize deterministic (or nearly deterministic) algorithms.

Algorithm randomization consists in incorporating a nondeterministic modification to the standard algorithm operation that does not degrade model quality too severely, but makes different algorithm invocations likely to produce noticeably different models. The choice of algorithm steps to modify and the exact modifications is obviously algorithm specific. For algorithms that make internal decisions using certain criteria, based on evaluations of multiple possible candidate decisions, it usually makes sense to randomize such decision-making steps, e.g., by adding random noise to decision evaluations or randomly sampling the space of candidate decisions. In particular, for decision or regression tree growing the split selection operation is the natural candidate for randomization. It can be achieved either by randomly disturbing the split evaluation function, or limiting the set of candidate splits to consider at a given node to a randomly selected subset of all available splits or attributes. The latter resembles attribute sampling, but here an independent sample of attributes is drawn in each node instead of having a single fixed attribute sample for all nodes, as with the latter.

Notice that algorithm randomization can be easily applied in combination with any of the training set modification techniques for base model creation, such as instance sampling or instance weighting. Basically, multiple models for modified training sets can be created using a randomized instead of a deterministic algorithm.

Example 15.3.4 To illustrate algorithm randomization as an approach to base model creation, the following R code implements a simple randomized decision tree growing algorithm. The `grow.randdectree` function is actually a slightly modified version of the `grow.dectree` function from Example 3.3.8. The modification consists

`dmr.dectree`

in restricting the split selection process (performed by the internally defined `split.select` function) to splits based on a random subset of available attributes. The number of attributes to use is passed via the `ns` argument. If unspecified, the square root of the number of all attributes is assumed. The `clip.val` function is applied to ensure the value of `ns` is in the proper range. Notice that the `class` attribute of

`dmr.util`

the created tree object is set to `dectree`, to enable prediction method dispatching (using the `predict.dectree` function defined in Example 3.5.1).

```
## randomized decision tree growing
## with split selection using ns randomly chosen attributes at each node
## (if unspecified or 0, it defaults to the square root of the number of attributes)
grow.randdectree <- function(formula, data, ns=0,
                             imp=entropy.p, maxprob=0.999, minsplit=2, maxdepth=8)
{
  init <- function()
  {
    clabs <<- factor(levels(data[[class]]),
                     levels=levels(data[[class]]))  # class labels
    tree <<- data.frame(node=1, attribute=NA, value=NA, class=NA, count=NA,
                        `names<-`(rep(list(NA), length(clabs)),
                                  paste("p", clabs, sep=".")))
    cprobs <<- (ncol(tree)-length(clabs)+1):ncol(tree)  #.class probability columns
    nodemap <<- rep(1, nrow(data))
    n <<- 1
  }

  next.node <- function(n)
  {
    if (any(opn <- tree$node>n))
      min(tree$node[opn])
    else Inf
  }

  class.distribution <- function(n)
  {
    tree[tree$node==n,"count"] <<- sum(nodemap==n)
    tree[tree$node==n,cprobs] <<- pdisc(data[nodemap==n,class])
  }

  class.label <- function(n)
  {
    tree$class[tree$node==n] <<- which.max(tree[tree$node==n,cprobs])
  {

  stop.criteria <- function(n)
  {
    n>=2^maxdepth || tree[tree$node==n,"count"]<minsplit ||
      max(tree[tree$node==n,cprobs])>maxprob
```

```
}

split.eval <- function(av, sv, cl)
{
  cond <- !is.na(av) & (if (is.numeric(av)) av<=as.numeric(sv) else av==sv)

  pd1 <- pdisc(cl[cond])
  n1 <- sum(cond)
  pd0 <- pdisc(cl[!cond])
  n0 <- sum(!cond)

  if (n1>0 && n0>0)
    weighted.impurity(pd1, n1, pd0, n0, imp)
  else
    Inf
}

split.select <- function(n)
{
  splits <- data.frame()
  for (attribute in sample(attributes, ns))
  {
    uav <- sort(unique(data[nodemap==n,attribute]))
    if (length(uav)>1)
      splits <- rbind(splits,
                     data.frame(attribute=attribute,
                                value=if (is.numeric(uav))
                                        midbrk(uav)
                                      else as.character(uav),
                                stringsAsFactors=F))
  }

  if (nrow(splits)>0)
    splits$eval <- sapply(1:nrow(splits),
                          function(s)
                          split.eval(data[nodemap==n,splits$attribute[s]],
                                     splits$value[s],
                                     data[nodemap==n,class]))
  if ((best.eval <- min(splits$eval))<Inf)
    tree[tree$node==n,2:3] <<- splits[which.min(splits$eval),1:2]
  return(best.eval)
}

split.apply <- function(n)
{
  tree <<- rbind(tree,
                 data.frame(node=(2*n):(2*n+1),
                            attribute=NA, value=NA, class=NA, count=NA,
                            `names<-`(rep(list(NA), length(clabs)),
                                      paste("p", clabs, sep="."))))

  av <- data[[tree$attribute[tree$node==n]]]
  cond <- !is.na(av) & (if (is.numeric(av))
                          av<=as.numeric(tree$value[tree$node==n])
                        else av==tree$value[tree$node==n])
  nodemap[nodemap==n & cond] <<- 2*n
  nodemap[nodemap==n & !cond] <<- 2*n+1
}

tree <- nodemap <- n <- NULL
clabs <- cprobs <- NULL
```

```
class <- y.var(formula)
attributes <- x.vars(formula, data)
ns <- ifelse(ns==0, round(sqrt(length(attributes))),
                    clip.val(ns, 1, length(attributes)))

init()
while (is.finite(n))
{
  class.distribution(n)
  class.label(n)
  if (!stop.criteria(n))
    if (split.select(n)<Inf)
      split.apply(n)
  n <- next.node(n)
}
tree$class <- clabs[tree$class]
'class<-'(tree, "dectree")
}
```

To similarly randomize regression tree growing, the following R code defines a modified version of the `grow.regtree` function, defined in Example 9.3.8. The `ns` argument, if unspecified, is set to one-third of the number of attributes.

```
## randomized regression tree growing
## with split selection using ns randomly chosen attributes at each node
## (if unspecified or 0, it defaults to the square root of the number of attributes)
grow.randregtree <- function(formula, data, ns=0,
                             minvar=0.005, minsplit=2, maxdepth=8)
{
  init <- function()
  {
    tree <<- data.frame(node=1, attribute=NA, value=NA, target=NA,
                        count=NA, mean=NA, variance=NA)
    nodemap <<- rep(1, nrow(data))
    n <<- 1
  }

  next.node <- function(n)
  {
    if (any(opn <- tree$node>n))
      min(tree$node[opn])
    else Inf
  }

  target.summary <- function(n)
  {
    tree$count[tree$node==n] <<- sum(nodemap==n)
    tree$mean[tree$node==n] <<- mean(data[nodemap==n,target])
    tree$variance[tree$node==n] <<- var1(data[nodemap==n,target])
  }

  target.value <- function(n)
  {
    tree$target[tree$node==n] <<- tree$mean[tree$node==n]
  }

  stop.criteria <- function(n)
  {
```

```
    n>=2^maxdepth || tree$count[tree$node==n]<minsplit ||
      tree$variance[tree$node==n]<minvar
}

split.eval <- function(av, sv, tv)
{
  cond <- !is.na(av) & (if (is.numeric(av)) av<=as.numeric(sv) else av==sv)
  v1 <- tv[cond]
  n1 <- sum(cond)
  v0 <- tv[!cond]
  n0 <- sum(!cond)
  if (n1>0 && n0>0)
    weighted.dispersion(v1, v0)
  else
    Inf
}

split.select <- function(n)
{
  splits <- data.frame()
  for (attribute in sample(attributes, ns))
  {
    uav <- sort(unique(data[nodemap==n,attribute]))
    if (length(uav)>1)
      splits <- rbind(splits,
                      data.frame(attribute=attribute,
                                 value=if (is.numeric(uav))
                                         midbrk(uav)
                                       else as.character(uav),
                                 stringsAsFactors=F))
  }

  if (nrow(splits)>0)
    splits$eval <- sapply(1:nrow(splits),
                          function(s)
                          split.eval(data[nodemap==n,splits$attribute[s]],
                                     splits$value[s],
                                     data[nodemap==n,target]))
  if ((best.eval <- min(splits$eval))<Inf)
    tree[tree$node==n,2:3] <<- splits[which.min(splits$eval),1:2]
  best.eval
}

split.apply <- function(n)
{
  tree <<- rbind(tree,
                 data.frame(node=(2*n):(2*n+1), attribute=NA, value=NA, target=NA,
                            count=NA, mean=NA, variance=NA))

  av <- data[[tree$attribute[tree$node==n]]]
  cond <- !is.na(av) & (if (is.numeric(av))
                          av<=as.numeric(tree$value[tree$node==n])
                        else av==tree$value[tree$node==n])
  nodemap[nodemap==n & cond] <<- 2*n
  nodemap[nodemap==n & !cond] <<- 2*n+1
}

tree <- nodemap <- n <- NULL
target <- y.var(formula)
attributes <- x.vars(formula, data)
ns <- ifelse(ns==0, round(length(attributes)/3),
```

```
            clip.val(ns, 1, length(attributes)))

  init()
  while (is.finite(n))
  {
    target.summary(n)
    target.value(n)
    if (!stop.criteria(n))
      if (split.select(n)<Inf)
        split.apply(n)
    n <- next.node(n)
  }
  `class<-`(tree, "regtree")
}
```

The application of these randomized decision tree and regression tree growing implementations is demonstrated by the following R code, following the pattern of the previous examples. Notice that base model creation in this case simply reduces to multiple invocations of the same randomized algorithm for the same training set. This is performed by the `base.ensemble.simple` function. The `predict.dectree` and `predict.regtree` functions, defined in Examples 3.5.1 and 9.5.1, are used to generate base model predictions. The resulting training and test set misclassification error and mean square error values are calculated to make it possible to roughly assess the diversity of the randomized tree models.

```
## generate base models by simple multiple algorithm application
base.ensemble.simple <- function(formula, data, m, alg, args=NULL)
{
  lapply(1:m, function(i) do.call(alg, c(list(formula, data), args)))
}

  # base models for the HouseVotes84 data
hv.bm.tree.rnd <- base.ensemble.simple(Class~., hv.train, 50, grow.randdectree)

  # base models for the BostonHousing data
bh.bm.tree.rnd <- base.ensemble.simple(medv~., bh.train, 50, grow.randregtree,
                              args=list(minvar=5))

  # base model training set errors for the HouseVotes84 data
hv.train.err.tree.rnd <- sapply(hv.bm.tree.rnd,
                          function(h) err(predict(h, hv.train),
                                          hv.train$Class))

  # base model training set MSE values for the BostonHousing data
bh.train.mse.tree.rnd <- sapply(bh.bm.tree.rnd,
                          function(h) mse(predict(h, bh.train),bh.train$medv))

  # base model test set errors for the HouseVotes84 data
hv.test.err.tree.rnd <- sapply(hv.bm.tree.rnd,
                          function(h) err(predict(h, hv.test), hv.test$Class))

  # base model test set MSE values for the BostonHousing data
bh.test.mse.tree.rnd <- sapply(bh.bm.tree.rnd,
                          function(h) mse(predict(h, bh.test), bh.test$medv))
```

Decision tree growing randomization applied to the *HouseVotes84* data appears to give little effect when looking at training set error distribution, but this is to be expected given their default strict stop criteria leading to fitting the training set exactly. On the test set they exhibit randomized trees substantially more variability than those obtained using instance sampling, with somewhat reduced accuracy. Similar observations can be made for randomized regression trees on the *Boston Housing* data.

15.3.5   Base model diversity

While all the base model generation techniques presented in this section serve the same purpose of generating multiple diverse, but at the same time reasonably good, models for the same domain, they are not equally effective in achieving this goal. Instance sampling has the widest applicability and should be capable of delivering substantial diversity if used with an unstable modeling algorithm. Instance weighting makes most sense when it is desirable to somehow adjust subsequent base models to the performance exhibited by those created previously. Otherwise it offers no advantages over instance sampling. Attribute sampling is only applicable when there are sufficiently many attributes. Otherwise eliminating some of them may excessively degrade base model performance. Varying algorithm parameters may be effective only for some algorithms that are sufficiently sensitive to their parameter values. It has to be used with care to avoid destroying base model quality. Usually only a small number of different but sufficiently good base models can be created using this technique alone and it can be truly useful only in combination with one of the others, usually instance sampling. Algorithm randomization, whenever applicable, is easier to control and has much greater base model diversity potential. It can be applied as a standalone base model generation technique or as a diversity-stimulating companion to instance sampling.

Example 15.3.5 To illustrate the base model diversity potential of the techniques presented in this section, the following R code produces boxplots visualizing the training and test set performance of the base models created in the previous examples.

```
# base model training set errors for the HouseVotes84 data
boxplot(list(tree.sx=hv.train.err.tree.sx,
        tree.wx=hv.train.err.tree.wx,
        tree.sa=hv.train.err.tree.sa,
        tree.rnd=hv.train.err.tree.rnd,
        nb.sx=hv.train.err.nb.sx,
        nb.sa=hv.train.err.nb.sa),
        main="HouseVotes84 (train)", las=2, col="grey", ylim=c(0, 0.26))

# base model test set errors for the HouseVotes84 data
boxplot(list(tree.sx=hv.test.err.tree.sx,
        tree.wx=hv.test.err.tree.wx,
        tree.sa=hv.test.err.tree.sa,
        tree.rnd=hv.test.err.tree.rnd,
        nb.sx=hv.test.err.nb.sx,
        nb.sa=hv.test.err.nb.sa),
        main="HouseVotes84 (test)", las=2, col="grey", ylim=c(0, 0.26))

# base model training set MSE values for the BostonHousing data
```

```
boxplot(list(tree.sx=bh.train.mse.tree.sx,
         tree.wx=bh.train.mse.tree.wx,
         tree.sa=bh.train.mse.tree.sa,
         tree.rnd=bh.train.mse.tree.rnd,
         lm.sx=bh.train.mse.lm.sx,
         lm.wx=bh.train.mse.lm.wx,
            lm.sa=bh.train.mse.lm.sa),
       main="BostonHousing (train)", las=2, col="grey", ylim=c(0, 130))

  # base model test set MSE values for the BostonHousing data
boxplot(list(tree.sx=bh.test.mse.tree.sx,
         tree.wx=bh.test.mse.tree.wx,
         tree.sa=bh.test.mse.tree.sa,
         tree.rnd=bh.test.mse.tree.rnd,
         lm.sx=bh.test.mse.lm.sx,
         lm.wx=bh.train.mse.lm.wx,
         lm.sa=bh.test.mse.lm.sa),
       main="BostonHousing (test)", las=2, col="grey", ylim=c(0, 130))
```

The obtained boxplots are presented in Figures 15.1 and 15.2. Clearly attribute sampling yields the most diverse base models (judging based on their test set performance), but this is at the cost of considerably worse error levels. Of the remaining techniques, instance sampling and algorithm randomization appear to offer acceptable levels of tradeoff between diversity and quality.

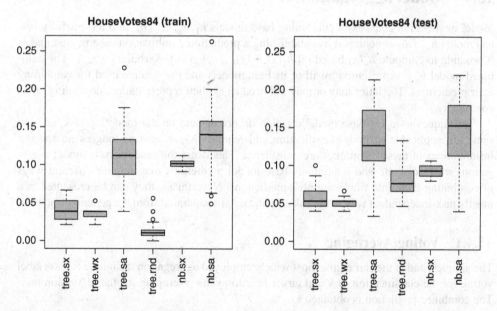

Figure 15.1   The boxplots of the training and test set misclassification errors for base models created for the *HouseVotes84* data.

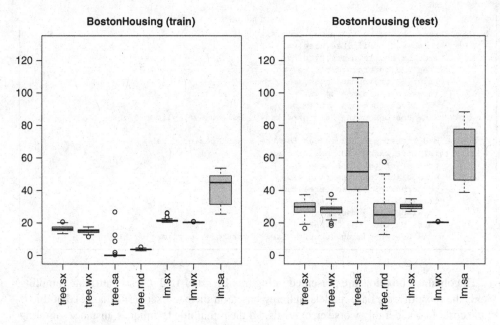

Figure 15.2 The boxplots of the training and test set mean square errors for base models created for the Boston Housing data.

15.4 Model aggregation

Model aggregation consists in combining base models $h_1, h_2, \ldots, h_m$ into a (hopefully better) model h_*. This is achieved by establishing a prediction combination scheme that makes it possible to compute $h_*(x)$ based on $h_1(x), h_2(x), \ldots, h_m(x)$ for arbitrary $x \in X$. The combined model h_* is represented by all of its base models and the scheme used for combining their predictions. The latter may or may not need an explicit representation, depending on its complexity.

Techniques used for base model creation do not depend on the modeling task, and the same are applicable to both classification and regression – at least as long as no specific instantiations of these techniques are considered. This is not the case for base model aggregation, where discrete and continuous base model predictions may require different ways of combining them into final ensemble predictions. Nevertheless, they can be presented in a mostly task-independent way, with task-specific details separated from the general principles.

15.4.1 Voting/Averaging

The simplest and at the same time most widely employed aggregation technique is class label voting for the classification task and target function value averaging for the regression task. The combined prediction is obtained as

$$h_*(x) = \arg \max_{d \in C} \sum_{i=1}^{n} \mathbb{I}_{h_i(x)=d} \tag{15.4}$$

where the $\mathbb{I}_{condition}$ notation is used to denote an indicator function that yields 1 when the *condition* is satisfied and 0 otherwise, or

$$h_*(x) = \frac{1}{m} \sum_{i=1}^{m} h_i(x) \qquad (15.5)$$

respectively.

In any case, while we formally speak of the combined model, what happens in reality is just combining base model predictions, without creating any other model representation. Base models are therefore just wrapped by the voting/averaging scheme for prediction combination. There is no need therefore to access the training set in the model combination phase.

Example 15.4.1 Basic voting or averaging model combination is implemented by the `predict.ensemble.basic` function defined by the following R code. It applies each base model to the obtained dataset and, depending on the type of their predictions, uses one of the two combination variants. Voting is performed using the `modal` function. The `predict.ensemble.basic` function is demonstrated by applying it to combine all the base models generated in the previous examples. The misclassification error or mean square error values on the test sets are calculated for the combined models and visually presented using barplots, with the corresponding indicators for single models also included for comparison.

> Ex. 2.4.19
> dmr.stats

```
## combine base models by voting/averaging
predict.ensemble.basic <- function(models, data, predf=predict)
{
  bp <- data.frame(lapply(models, function(h) predf(h, data)))
  combf <- if (is.numeric(bp[,1])) mean else modal  # combination scheme
  cp <- sapply(1:nrow(bp), function(i) combf(as.vector(as.matrix(bp[i,]))))
  if (is.factor(bp[,1]))
    factor(cp, levels=levels(bp[,1]))
  else
    cp
}

  # combine base models for the HouseVotes84 data
hv.pred.tree.sx.b <- predict.ensemble.basic(hv.bm.tree.sx, hv.test,
                                    predf=function(...)
                                      predict(..., type="c"))
hv.pred.nb.sx.b <- predict.ensemble.basic(hv.bm.nb.sx, hv.test)
hv.pred.tree.wx.b <- predict.ensemble.basic(hv.bm.tree.wx, hv.test,
                                    predf=function(...)
                                      predict(..., type="c"))
hv.pred.tree.sa.b <- predict.ensemble.basic(hv.bm.tree.sa, hv.test,
                                    predf=function(...)
                                      predict(..., type="c"))
hv.pred.nb.sa.b <- predict.ensemble.basic(hv.bm.nb.sa, hv.test)
hv.pred.tree.rnd.b <- predict.ensemble.basic(hv.bm.tree.rnd, hv.test)

  # combine base models for the BostonHousing data
bh.pred.tree.sx.b <- predict.ensemble.basic(bh.bm.tree.sx, bh.test)
bh.pred.lm.sx.b <- predict.ensemble.basic(bh.bm.lm.sx, bh.test)
bh.pred.tree.wx.b <- predict.ensemble.basic(bh.bm.tree.wx, bh.test)
bh.pred.lm.wx.b <- predict.ensemble.basic(bh.bm.lm.wx, bh.test)
```

```
bh.pred.tree.sa.b <- predict.ensemble.basic(bh.bm.tree.sa, bh.test)
bh.pred.lm.sa.b <- predict.ensemble.basic(bh.bm.lm.sa, bh.test)
bh.pred.tree.rnd.b <- predict.ensemble.basic(bh.bm.tree.rnd, bh.test)

 # ensemble model test set errors for the HouseVotes84 data
hv.err.b <- c(tree = hv.err.tree,
              tree.sx = err(hv.pred.tree.sx.b, hv.test$Class),
              tree.wx = err(hv.pred.tree.wx.b, hv.test$Class),
              tree.sa = err(hv.pred.tree.sa.b, hv.test$Class),
              tree.rnd = err(hv.pred.tree.rnd.b, hv.test$Class),
              nb = hv.err.nb,
              nb.sx = err(hv.pred.nb.sx.b, hv.test$Class),
              nb.sa = err(hv.pred.nb.sa.b, hv.test$Class))

 # ensemble model test set MSE values for the BostonHousing data
bh.mse.b <- c(tree = bh.mse.tree,
              tree.sx = mse(bh.pred.tree.sx.b, bh.test$medv),
              tree.wx = mse(bh.pred.tree.wx.b, bh.test$medv),
              tree.sa = mse(bh.pred.tree.sa.b, bh.test$medv),
              tree.rnd = mse(bh.pred.tree.rnd.b, bh.test$medv),
              lm = bh.mse.lm,
              lm.sx = mse(bh.pred.lm.sx.b, bh.test$medv),
              lm.wx = mse(bh.pred.lm.wx.b, bh.test$medv),
              lm.sa = mse(bh.pred.lm.sa.b, bh.test$medv))

barplot(hv.err.b, main="HouseVotes84", ylab="Error", las=2)
lines(c(0, 10), rep(hv.err.b[1], 2), lty=2)
lines(c(0, 10), rep(hv.err.b[6], 2), lty=3)

barplot(bh.mse.b, main="Boston Housing", ylab="MSE", las=2)
lines(c(0, 11), rep(bh.mse.b[1], 2), lty=2)
lines(c(0, 11), rep(bh.mse.b[6], 2), lty=3)
```

The barplots illustrating the performance of the created model ensembles are presented in Figure 15.3. Only one of the decision tree ensembles, the one using base models obtained by instance sampling, outperforms the single decision tree for the *HouseVotes84* data. The ensembles with base models obtained by attribute sampling are particularly poor, suggesting that this base model generation methods may be not very useful if used alone. No improvement can be observed for the naïve Bayes models. Somewhat better results are obtained for the *Boston Housing* dataset, with most regression tree ensembles (except that using base models obtained by attribute sampling) outperforming the single tree. The ensemble consisting of randomized regression trees turns out particularly successful. None of linear model ensembles brings any improvement, though. This is because the averaged predictions of multiple linear models remain linear, i.e., they could have been generated by a single linear model.

15.4.2   Probability averaging

For probabilistic classification models, generating class probabilities rather than or apart from class labels, an alternative probabilistic prediction combination scheme is possible. Let $P_{h_i}(d|x)$ denote the probability of class d for instance x delivered by base model h_i. Then the combined probability of class d for instance x is calculated by probability averaging as

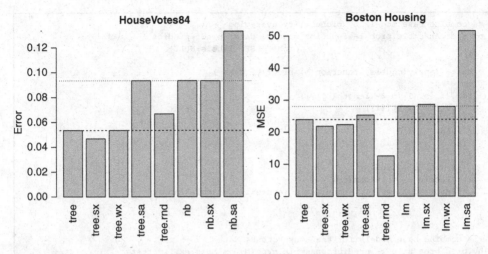

Figure 15.3   The barplots of error values for base models combined by voting/averaging.

follows:

$$P_{h_*}(d|x) = \sum_{i=1}^{m} P(h_i)P_{h_i}(d|x) \tag{15.6}$$

where $P(h_i)$ is the probability of model h_i in the ensemble, assumed to be $\frac{1}{m}$ for all $i = 1, 2, \ldots, m$. This preserves the probability prediction capability of base models in the ensemble, with all the related advantages, including the possibility of misclassification cost minimization, as discussed in Section 6.3.3, or operating point tuning by ROC analysis, as discussed in Section 7.2.5. If these are not needed, class label predictions can be generated by simple probability maximization.

As a matter of fact, probabilistic predictions are possible even with model ensembles comprising nonprobabilistic base models, by taking

$$P_{h_i}(d|x) = \begin{cases} 1 & \text{if } h_i(x) = d \\ 0 & \text{otherwise} \end{cases} \tag{15.7}$$

The predicted probability of each class becomes then the number of votes for this class divided by the number of base models. Even though the quality of such probability predictions is likely to be inferior to that possible with proper probabilistic models, it may still be useful.

Example 15.4.2 The R code presented below defines the `predict.ensemble.prob` function that performs base classification model aggregation by probability averaging. The prediction function for base models passed via the `prob` argument is expected to produce class probability predictions. Such probabilities obtained for all base models are averaged and either returned directly, if the `prob` argument is set to `TRUE`, or used to assign maximum-probability class labels. The latter is the default behavior demonstrated by example calls which combine all the previously created decision tree and naïve Bayes base models for the *HouseVotes84* data, except those obtained using decision tree randomization, since its simple implementation lacks the probabilistic prediction functionality.

```
## combine base models by probability averaging
predict.ensemble.prob <- function(models, data, predf=predict,
                                  prob=FALSE, labels=NULL)
{
  bp <- lapply(models, function(h) predf(h, data))
  cp <- 0
  for (i in (1:(m <- length(bp))))
    cp <- cp + bp[[i]]
  if (prob)
    cp/m
  else
  {
    if (is.null(labels))
      labels <- colnames(cp)
    factor(apply(cp, 1, which.max), levels=1:2, labels=labels)
  }
}


  # combine base models for the HouseVotes84 data
hv.pred.tree.sx.p <- predict.ensemble.prob(hv.bm.tree.sx, hv.test)
hv.pred.nb.sx.p <- predict.ensemble.prob(hv.bm.nb.sx, hv.test,
                            predf=function(...) predict(..., type="r"))
hv.pred.tree.wx.p <- predict.ensemble.prob(hv.bm.tree.wx, hv.test)
hv.pred.tree.sa.p <- predict.ensemble.prob(hv.bm.tree.sa, hv.test)
hv.pred.nb.sa.p <- predict.ensemble.prob(hv.bm.nb.sa, hv.test,
                            predf=function(...) predict(..., type="r"))

  # ensemble model test set errors for the HouseVotes84 data
hv.err.p <- c(tree = hv.err.tree,
              tree.sx = err(hv.pred.tree.sx.p, hv.test$Class),
              tree.wx = err(hv.pred.tree.wx.p, hv.test$Class),
              tree.sa = err(hv.pred.tree.sa.p, hv.test$Class),
            nb = hv.err.nb,
            nb.sx = err(hv.pred.nb.sx.p, hv.test$Class),
            nb.sa = err(hv.pred.nb.sa.p, hv.test$Class))

barplot(hv.err.p, main="HouseVotes84", ylab="Error", las=2)
lines(c(0, 9), rep(hv.err.p[1], 2), lty=2)
lines(c(0, 9), rep(hv.err.p[5], 2), lty=3)
```

Figure 15.4 displays the barplots of the obtained misclassification error values. They exactly match those observed in Example 15.4.1 using basic voting.

15.4.3   Weighted voting/averaging

It may be sometimes a good idea to weight base models depending on their training set performance or estimated true performance, with a weighting scheme that allows better models to have more prediction impact. Incorporating model weights W_i for each base model h_i leads to the following prediction combination schemes:

$$h_*(x) = \arg\max_{d \in C} \sum_{i=1}^{n} W_i \mathbb{I}_{h_i(x)=d} \tag{15.8}$$

for classification, and

$$h_*(x) = \frac{\sum_{i=1}^{m} W_i h_i(x)}{\sum_{i=1}^{m} W_i} \tag{15.9}$$

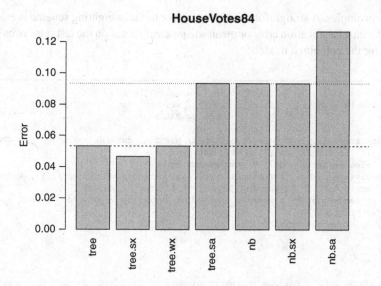

Figure 15.4   The barplot of error values for base models combined by probability averaging.

for regression, where the uppercase W_i is used to avoid confusion of model weights with instance weights w_x, also referred to in this chapter. Sometimes it may be more convenient and natural to use the weighted sum rather than the weighted average:

$$h_*(x) = \sum_{i=1}^{m} W_i h_i(x) \qquad (15.10)$$

which is clearly the same if model weights are normalized to sum up to 1. Finally, the weighted version of class probability averaging is defined as follows:

$$P_{h_*}(d|x) = \frac{\sum_{i=1}^{m} W_i P(h_i) P_{h_i}(d|x)}{\sum_{i=1}^{m} W_i P(h_i)} \qquad (15.11)$$

which can further be simplified, if all base models are assumed to have the same probability of $\frac{1}{m}$, to the following form:

$$P_{h_*}(d|x) = \frac{\sum_{i=1}^{m} W_i P_{h_i}(d|x)}{\sum_{i=1}^{m} W_i} \qquad (15.12)$$

Model weighting schemes are usually specific to particular ensemble modeling techniques.

Example 15.4.3 Weighted voting/averaging model combination is implemented by the `pre-dict.ensemble.weighted` function defined by the following R code. Weighted voting is performed using the `weighted.modal` function, and weighted averaging using the standard `weighted.mean` function. Optionally, summing may be requested instead of averaging. The `predict.ensemble.weighted` function is demonstrated by applying it to combine all the base models generated in

> Ex. 2.4.20
> dmr.stats

the previous examples. A straightforward inverse error model weighting scheme is employed. As before, the misclassification error or mean square error values on the test sets are calculated and plotted for the combined models.

```
## combine base models by weighted voting/averaging/summing
predict.ensemble.weighted <- function(models, weights, data, predf=predict,
                                       summing=FALSE)
{
  bp <- data.frame(lapply(models, function(h) predf(h, data)))
  combf <- if (is.numeric(bp[,1])) weighted.mean
           else weighted.modal  # combination scheme
  cp <- sapply(1:nrow(bp), function(i) combf(as.vector(as.matrix(bp[i,])), weights))
  if (is.numeric(bp[,1]) && summing)
    cp <- cp*sum(weights)  # summing instead of averaging requested
  if (is.factor(bp[,1]))
    factor(cp, levels=levels(bp[,1]))
  else
    cp
}


# combine base models for the HouseVotes84 data
hv.pred.tree.sx.w <- predict.ensemble.weighted(hv.bm.tree.sx,
                                        1/(hv.train.err.tree.sx+0.01),
                                        hv.test,
                                        predf=function(...)
                                              predict(..., type="c"))
hv.pred.nb.sx.w <- predict.ensemble.weighted(hv.bm.nb.sx,
                                      1/(hv.train.err.nb.sx+0.01),
                                      hv.test)
hv.pred.tree.wx.w <- predict.ensemble.weighted(hv.bm.tree.wx,
                                        1/(hv.train.err.tree.wx+0.01),
                                        hv.test,
                                        predf=function(...)
                                              predict(..., type="c"))
hv.pred.tree.sa.w <- predict.ensemble.weighted(hv.bm.tree.sa,
                                        1/(hv.train.err.tree.sa+0.01),
                                        hv.test,
                                        predf=function(...)
                                              predict(..., type="c"))
hv.pred.nb.sa.w <- predict.ensemble.weighted(hv.bm.nb.sa,
                                      1/(hv.train.err.nb.sa+0.01), hv.test)
hv.pred.tree.rnd.w <- predict.ensemble.weighted(hv.bm.tree.rnd,
                                         1/(hv.train.err.tree.rnd+0.01),
                                         hv.test)


# combine base models for the BostonHousing data
bh.pred.tree.sx.w <- predict.ensemble.weighted(bh.bm.tree.sx,
                                        1/(bh.train.mse.tree.sx+1), bh.test)
bh.pred.lm.sx.w <- predict.ensemble.weighted(bh.bm.lm.sx,
                                      1/(bh.train.mse.lm.sx+1), bh.test)
bh.pred.tree.wx.w <- predict.ensemble.weighted(bh.bm.tree.wx,
                                        1/(bh.train.mse.tree.wx+1), bh.test)
bh.pred.lm.wx.w <- predict.ensemble.weighted(bh.bm.lm.wx,
                                      1/(bh.train.mse.lm.wx+1), bh.test)
bh.pred.tree.sa.w <- predict.ensemble.weighted(bh.bm.tree.sa,
                                        1/(bh.train.mse.tree.sa+1), bh.test)
bh.pred.lm.sa.w <- predict.ensemble.weighted(bh.bm.lm.sa,
                                      1/(bh.train.mse.lm.sa+1), bh.test)
bh.pred.tree.rnd.w <- predict.ensemble.weighted(bh.bm.tree.rnd,
                                         1/(bh.train.mse.tree.rnd+1), bh.test)
```

```
# ensemble model test set errors for the HouseVotes84 data
hv.err.w <- c(tree = hv.err.tree,
              tree.sx = err(hv.pred.tree.sx.w, hv.test$Class),
              tree.wx = err(hv.pred.tree.wx.w, hv.test$Class),
              tree.sa = err(hv.pred.tree.sa.w, hv.test$Class),
              tree.rnd = err(hv.pred.tree.rnd.w, hv.test$Class),
              nb = hv.err.nb,
              nb.sx = err(hv.pred.nb.sx.w, hv.test$Class),
              nb.sa = err(hv.pred.nb.sa.w, hv.test$Class))

# ensemble model test set MSE values for the BostonHousing data
bh.mse.w <- c(tree = bh.mse.tree,
              tree.sx = mse(bh.pred.tree.sx.w, bh.test$medv),
              tree.wx = mse(bh.pred.tree.wx.w, bh.test$medv),
              tree.sa = mse(bh.pred.tree.sa.w, bh.test$medv),
              tree.rnd = mse(bh.pred.tree.rnd.w, bh.test$medv),
              lm = bh.mse.lm,
              lm.sx = mse(bh.pred.lm.sx.w, bh.test$medv),
              lm.wx = mse(bh.pred.lm.wx.w, bh.test$medv),
              lm.sa = mse(bh.pred.lm.sa.w, bh.test$medv))

barplot(hv.err.w, main="HouseVotes84", ylab="Error", las=2)
lines(c(0, 10), rep(hv.err.w[1], 2), lty=2)
lines(c(0, 10), rep(hv.err.w[6], 2), lty=3)

barplot(bh.mse.w, main="Boston Housing", ylab="MSE", las=2)
lines(c(0, 11), rep(bh.mse.w[1], 2), lty=2)
lines(c(0, 11), rep(bh.mse.w[6], 2), lty=3)
```

The barplots illustrating the performance of the created model ensembles are presented in Figure 15.5. There is no significant impact of base model weighting on the prediction quality of most of the model ensembles, which remains the same as with basic voting/averaging. Only for the worst attribute sampling ensembles some improvement due to weighting may be observed. The applied weighting scheme may not sufficiently vary the contributions of better and worse base models, or the quality of base models may not sufficiently differ.

15.4.4 Using as attributes

A more refined approach than plain or weighted voting or averaging consists in using a modeling algorithm to create the aggregated model h_*, with base models $h_1, h_2, \ldots, h_m$ playing the role of (the only or additional) attributes. Technically, this means that their predictions for the training set are generated and used instead of or apart form the original attribute values. Such data is passed to the modeling algorithm used to create the aggregated model, which may, but does not have to, be the same as (possibly one of those) used for base model creation. It is more common to use rather simple algorithms for model combination, but more refined ones for base model generation.

It is also possible to consider multiple levels of such model aggregation, leading to a hierarchical model ensemble. In this approach, base models created using the original set of attributes, that may be referred to as level 0 models, are used as attributes to create multiple level 1 models, which then in turn are used to create level 2 models, etc. The same techniques for base model creation as discussed above may be used on each level to obtain multiple

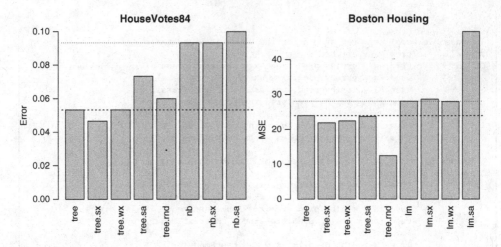

Figure 15.5 The barplots of error values for base models combined by weighted voting/ averaging.

aggregated models – what changes is only the set of attributes used, which constitutes of or is supplemented by the lower level models.

Despite its refinement and conceptual elegance, the approach of using base models as attributes for model aggregation is not necessarily superior to simple voting or averaging. Relationships between particular base models and the target attribute may not be predictively useful enough to outperform the latter. This is not to say that this aggregation technique is universally poor and useless, but rather warn that it is not necessarily superior to the much simpler alternatives discussed previously.

Unlike basic or weighted voting/averaging, using base models as attributes means that an actual representation of the combined model is created. The training set therefore needs to remain available in the model combination phase.

Example 15.4.4 Base model combination by using them as attributes is implemented and demonstrated by the R code. The implementation comprises two functions, `combine.ensemble.attributes` and `predict.ensemble.attributes`. The former creates the combined model using the training set, with base models used instead of or apart from the original attributes (depending on the value of the `append` argument). The latter applies such a combined model for prediction. The presented implementation assumes that the target attribute name is available in the `terms` component of the model object structure used to represent base models. This is true for some, but not all modeling algorithms available in R – in particular, for the `rpart` and `lm` models, but neither for the `naiveBayes` model nor for the randomized decision and regression trees created by the `grow.randdectree` and `grow.randregtree` functions. The demonstrations presented below are actually limited to combining `rpart` decision tree and regression tree models only, using the naïve Bayes and linear regression algorithms on the second level.

```
## combine base models by using as attributes
## create a model using the specified algorithm and training data
combine.ensemble.attributes <- function(models, data, alg, args=NULL, predf=predict,
                                        append=FALSE)
{
  target <- as.character(models[[1]]$terms[[2]])
  tind <- match(target, names(data))
  data.base <- `names<-`(cbind(data.frame(lapply(models,
                                                 function(h) predf(h, data))),
                              data[[target]]),
                         c(paste("h", 1:length(models), sep=""), target))
  if (append)
    data.base <- cbind(data[,-tind], data.base)
  do.call(alg, c(list(make.formula(target, "."), data.base), args))
}


## combine base models by using as attributes
## predict using the specified base models and combined model
predict.ensemble.attributes <- function(combined.model, base.models, data,
                                        combined.predf=predict, base.predf=predict)
{
  data.pred <- `names<-`(data.frame(lapply(base.models,
                                           function(h) base.predf(h, data))),
                         paste("h", 1:length(base.models), sep=""))
  data.pred <- cbind(data, data.pred)  # make the original attributes available
  combined.predf(combined.model, data.pred)
}

  # combine base models for the HouseVotes84 data
hv.tree.sx.nb <- combine.ensemble.attributes(hv.bm.tree.sx, hv.train, naiveBayes,
                                             predf=function(...)
                                                   predict(..., type="c"))
hv.pred.tree.sx.nb <- predict.ensemble.attributes(hv.tree.sx.nb, hv.bm.tree.sx,
                                                  hv.test,
                                                  base.predf=function(...)
                                                        predict(..., type="c"))
hv.tree.wx.nb <- combine.ensemble.attributes(hv.bm.tree.wx, hv.train, naiveBayes,
                                             predf=function(...)
                                                   predict(..., type="c"))
hv.pred.tree.wx.nb <- predict.ensemble.attributes(hv.tree.wx.nb, hv.bm.tree.wx,
                                                  hv.test,
                                                  base.predf=function(...)
                                                        predict(..., type="c"))
hv.tree.sa.nb <- combine.ensemble.attributes(hv.bm.tree.sa, hv.train, naiveBayes,
                                             predf=function(...)
                                                   predict(..., type="c"))
hv.pred.tree.sa.nb <- predict.ensemble.attributes(hv.tree.sa.nb, hv.bm.tree.sa,
                                                  hv.test,
                                                  base.predf=function(...)
                                                        predict(..., type="c"))

  # combine base models for the BostonHousing data
bh.tree.sx.lm <- combine.ensemble.attributes(bh.bm.tree.sx, bh.train, lm)
bh.pred.tree.sx.lm <- predict.ensemble.attributes(bh.tree.sx.lm, bh.bm.tree.sx,
                                                  bh.test)
bh.tree.wx.lm <- combine.ensemble.attributes(bh.bm.tree.wx, bh.train, lm)
bh.pred.tree.wx.lm <- predict.ensemble.attributes(bh.tree.wx.lm, bh.bm.tree.wx,
                                                  bh.test)
```

```
bh.tree.sa.lm <- combine.ensemble.attributes(bh.bm.tree.sa, bh.train, lm)
bh.pred.tree.sa.lm <- predict.ensemble.attributes(bh.tree.sa.lm, bh.bm.tree.sa,
                                                  bh.test)

# ensemble model test set errors for the HouseVotes84 data
hv.err.a <- c(tree = hv.err.tree,
              tree.sx.nb = err(hv.pred.tree.sx.nb, hv.test$Class),
              tree.wx.nb = err(hv.pred.tree.wx.nb, hv.test$Class),
              tree.sa.nb = err(hv.pred.tree.sa.nb, hv.test$Class))

# ensemble model test set MSE values for the BostonHousing data
bh.mse.a <- c(tree = bh.mse.tree,
              tree.sx.lm = mse(bh.pred.tree.sx.lm, bh.test$medv),
              tree.wx.lm = mse(bh.pred.tree.wx.lm, bh.test$medv),
              tree.sa.lm = mse(bh.pred.tree.sa.lm, bh.test$medv))

barplot(hv.err.a, main="HouseVotes84", ylab="Error", las=2)
lines(c(0, 5), rep(hv.err.a[1], 2), lty=2)

barplot(bh.mse.a, main="Boston Housing", ylab="MSE", las=2)
lines(c(0, 5), rep(bh.mse.a[1], 2), lty=2)
```

The barplots illustrating the performance of the created model ensembles are presented in Figure 15.6. Combining base models by using them as attributes does not appear to provide any advantages over the simple voting/averaging approaches demonstrated in Example 15.4.1.

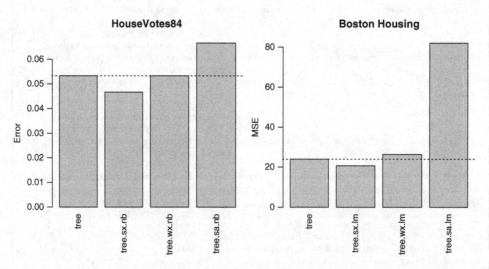

Figure 15.6 The barplots of error values for base models combined by using them as attributes.

15.5   Specific ensemble modeling algorithms

Various combinations of all the possible approaches to base model creation and aggregation discussed above may be used, yielding a variety of ensemble modeling techniques. Some of them have proved particularly useful and become extremely popular. These most noteworthy specific instantiations of model ensembles are overviewed in this section.

15.5.1   Bagging

Bagging (standing for *bootstrap aggregating*) is definitely the simplest ensemble modeling algorithm that combines the very basic approaches to base model creation and aggregation:

- base models are created using bootstrap samples of the training set,
- combined by plain (unweighted) voting for the classification task or averaging for the regression task.

If using probabilistic base classification models, class label voting can be replaced by class probability averaging, leading to a probabilistic version of bagging.

This technique may not promise extreme prediction quality, but is likely to give an improvement compared to single models created using the same algorithm as base models, as long as the algorithm is unstable. For stable algorithms, with base models not sufficiently diverse, there may be no improvement or even minor degradation of prediction quality. There are no particular requirements for the modeling algorithm other than instability. Actually, it may be simplified compared to what would be normally used for single model creation, if this makes it more unstable. This may include, in particular, giving up overfitting precautions used in some algorithms, such as pruning decision or regression trees. Models overfitted to their particular bootstrap samples are more likely to differ. The overfitting of base models will not entail the overfitting of the ensemble, as their aggregation will effectively cancel it out. Similarly, there is usually no need to bother with attribute selection, as more attributes provide more opportunities to create many diverse models. This is a striking difference compared to what is typical when single models are created.

Bagging may be thought of as a means of stabilizing unstable algorithms. Single models obtained using such algorithms may be subject to considerable variation depending on a particular training set. There is always a possibility that for a slightly different training set a better or worse model would be created. Creating multiple models based on different data samples without combining them into an ensemble, and simply selecting one of them that appears the best does not provide a valuable solution. This is because model selection would have to based on model evaluation and the latter, as discussed in Section 7.3.1, only makes sense for a repeatable modeling procedure. In particular, producing low-variance performance estimates that could serve for model selection requires repeating training and evaluation cycles multiple times. This is completely impossible for models that only differ in their training samples.

Bagging, with sufficiently many base models, allows one to be pretty confident that the final model is at least as good as a single model in the optimistic case, and possibly even improve over that. This is enough to justify the use of bagging if computational resources permit creating dozens or more models, as typically used for this technique and if model human readability is not required. The bagging ensemble performance tends to improve with increasing the number of base models up to a certain point, after which it stabilizes. This is where the limit of model diversity possible using bootstrap samples is achieved. Additional models are too much similar to the other ones to make any difference.

Example 15.5.1 The bagging ensemble modeling technique is implemented and demonstrated by the R code presented below. Since bagging is the most straightforward combination of instance sampling for base model creation and voting/averaging for ensemble prediction, the corresponding functions are simple wrappers around the functions defined in Examples 15.3.1 and 15.4.1. The demonstrations also follow the same pattern and include the application of bagging with decision trees and the naïve Bayes classifier to the *HouseVotes84* data, and with regression trees and linear regression to the *Boston Housing* data.

```
## bagging ensemble modeling using m base models created with algorithm alg
## with arguments arg
bagging <- function(formula, data, m, alg, args=NULL)
{
  `class<-`(base.ensemble.sample.x(formula, data, m, alg, args), "bagging")
}

## bagging prediction
predict.bagging <- function(models, data, predf=predict)
{
  predict.ensemble.basic(models, data, predf)
}

# bagging for the HouseVotes84 data
hv.bagg.tree <- bagging(Class~., hv.train, 50, rpart, args=list(minsplit=2, cp=0))
hv.bagg.nb <- bagging(Class~., hv.train, 50, naiveBayes)

hv.pred.bagg.tree <- predict(hv.bagg.tree, hv.test,
                             predf=function(...) predict(..., type="c"))
hv.pred.bagg.nb <- predict(hv.bagg.nb, hv.test)

# bagging for the BostonHousing data
bh.bagg.tree <- bagging(medv~., bh.train, 50, rpart, args=list(minsplit=2, cp=0))
bh.bagg.lm <- bagging(medv~., bh.train, 50, lm)

bh.pred.bagg.tree <- predict(bh.bagg.tree, bh.test)
bh.pred.bagg.lm <- predict(bh.bagg.lm, bh.test)

# bagging test set errors for the HouseVotes84 data
hv.err.bagg <- list(tree = err(hv.pred.bagg.tree, hv.test$Class),
                    nb = err(hv.pred.bagg.nb, hv.test$Class))

# bagging test set MSE values for the BostonHousing data
bh.mse.bagg <- list(tree = mse(bh.pred.bagg.tree, bh.test$medv),
                    lm = mse(bh.pred.bagg.lm, bh.test$medv))
```

The bagging ensembles for the *HouseVotes84* bring no improvement over the corresponding single models. For the *Boston Housing* data regression tree bagging ensemble outperforms the single tree considerably, though. The lack of improvement for the ensemble of linear models is not at all surprising, as bagging cannot overcome their linearity limitation in any way.

15.5.2   Stacking

The combination of using different algorithms (possibly with instance sampling to enable a greater number of diverse models) for base model creation and using base models as attributes for their aggregation yields the technique known as *stacking*. This term suggests a multilevel hierarchy of models could be created, as discussed in Section 15.4.4. The number of levels (no more than a few), the number of models, and the choice of algorithms used on particular levels are design decisions that may have a significant impact on the final ensemble quality. This makes stacking much more difficult to properly use than bagging, where just one algorithm and the number of base models need to be selected. Even in the simplest one-level setting, stacking is actually more refined than just using base model outputs as attributes for creating an aggregated model. It employs an internal data splitting technique related to the k-fold evaluation procedure presented in Section 7.3.4 which makes sure that predictions serving as attribute values for any instance x are produced by base models created with x excluded from the training set.

Using a modeling algorithm instead of simple voting or averaging to combine base models might appear a much more powerful and promising approach, capable of delivering at least as good, and likely better prediction quality. This is not necessarily the case, though, since base models may not be sufficiently good attributes for typical modeling algorithms. This is because the latter are usually designed to search for relationship patterns between the target attribute and other attributes. Such patterns may not exist, or may be not predictively be strong enough to outperform simple voting or averaging. In other words, using detailed information how particular base models predicted may not permit any improvement over simply using the very basic summary statistics: mode or mean. While there is definitely evidence of the usefulness of stacking in some cases, this ensemble modeling technique has not become nearly as popular as the other techniques reviewed in this section.

15.5.3   Boosting

Boosting can be best explained as an enhancement of bagging that attempts to include base model diversity by shifting the focus during base model creation toward instances that turn out the most "predictively difficult." This effectively makes consecutive base models specialized in different domain regions.

15.5.3.1   Base models

The shift of focus that underlies boosting is most naturally achieved by instance weighting. A single modeling algorithm is applied to the same original training set T using a sequence of varying weight vectors $w^{(1)}, w^{(2)}, \ldots, w^{(n)}$. It does not necessarily rule out the application of boosting with modeling algorithms that are not weight sensitive, though, since weighting can be approximated by random sampling with replacement, using weights – normalized to sum up to 1 – as instance selection probabilities. This sampling-based form of boosting most directly corresponds to bagging and makes the view of boosting as a bagging enhancement the most natural, but – as an approximation to the ordinary weighting-based boosting – is usually not used unless necessary.

Starting from uniform initial weights $w^{(1)}$, the weight vector is modified after each base model has been created and applied to the training set T. Instances for which the model yields poor predictions have their weights increased and/or those for which it yields good predictions have their weights decreased. For the classification task, this means simply raising the weights of misclassified training instances and/or reducing the weights of correctly classified training instances. For the regression task weight modifications would depend on model residuals.

For the regression task, it is also possible to use the previous models' residuals as target function values for subsequent base model creation instead of instance weighting. This will make the regression algorithm attempt to compensate the previous models' deficiencies rather than optimize its own training performance. The first model h_1 is created in the usual way. For $i > 1$, after models $h_1, \ldots, h_{i-1}$ have been created, their combined residuals are used instead of the target function values to create model h_i. This is another way of achieving the shift of focus effect that is at the heart of boosting.

15.5.3.2   Model aggregation

Base models are combined using weighted voting, with model weights $W_1, W_2, \ldots, W_m$ based on their training performance (and of course better models assigned higher voting weights). This is necessary, since (unlike for bagging) – due to the shift of focus during their creation – base models may exhibit considerably different training performance levels. In particular, if sufficiently many of them are created, the most recent ones may be entirely focused on the "most difficult" instances and yield poor predictions. It is important to underline that weighted model performance measures need to be adopted, with the same instance weights vector $w^{(i)}$ previously used to create model h_i also used to evaluate it and assign its voting weight W_i. This can be, in particular, the weighted misclassification error defined in Section 7.2.2 or any of weighted residual-based regression performance measures defined in Section 10.2.8.

15.5.3.3   Properties

Notice that the shift of focus during base model creation in boosting not only stimulates their diversity, but also drives the overall prediction quality, since instances that turned out to be "predictively difficult" keep receiving increasingly more attention. This is expected to boost the ensemble performance, as reflected by the term "boosting." Indeed, boosting model ensembles often belong to the most accurate models that can be achieved, at least for the classification task, on which boosting research and applications are mostly focused. Interestingly, the performance of even very simple and imperfect base models may be boosted substantially. It is particularly common to apply this technique with simple decision or regression trees limited to just a few levels, or even just a single split. No overfitting prevention, parameter tuning, or attribute selection is then necessary or desirable. It is actually sufficient that base models are just better than random guessing. Algorithms with parameters set up to yield such just-above-random models are referred to as *weak learners*. All base models are then nearly useless individually, yet they still form a powerful ensemble collectively. Each of them is much more likely to be "underfitted" than overfitted, and it is the boosting process, with its instance and model weight adjustments, that is responsible for most of the actual "fitting" to the training set. This is in contrast to bagging, where the overfitting of individual base models is normal and even desired for greater diversity, but canceled out by aggregation.

One possible disadvantage of boosting in comparison to bagging is that base models are not independent and have to be created sequentially. For bagging, all base models can be created in parallel, which enables efficient parallel implementations. This may be important for such computation-intensive algorithms.

15.5.3.4   Instantiations

Different schemes for instance weight modifications (or other focus shift techniques) and model weighting may be used to instantiate boosting, which makes it actually a family of ensemble modeling algorithms. The most noteworthy of these boosting instantiations for classification and regression are briefly reviewed below.

AdaBoost   The AdaBoost (standing for *adaptive boosting*) algorithm is the best-known instantiation of boosting, applicable to two-class classification tasks. As usual in this book, we will assume that the set of classes is $C = \{0, 1\}$, although it is more common to present the algorithm for the $\{-1, 1\}$ set of classes, which makes some steps easier to write by implicitly exploiting the numerical nature of class labels. The essential specific features that AdaBoost brings to the generic boosting techniques are its instance and model weighting schemes. The weight of the model h_i depends on its training set weighted misclassification error $e_{c,T,w^{(i)}}(h_i)$, calculated according to the definition presented in Section 7.2.2, in the following way:

$$W_i = \frac{1}{2} \ln \frac{1 - e_{c,T,w^{(i)}}(h_i)}{e_{c,T,w^{(i)}}(h_i)} \tag{15.13}$$

This weighting scheme is a decreasing function of error values, which gives more weight to more accurate models. The weight of model h_i is not only used for voting during prediction, but also to control the degree of instance weight modifications. The latter is performed as follows:

$$w_x^{(i+1)} = w_x^{(i)} e^{W_i(2\mathbb{I}_{h_i(x) \neq c(x)} - 1)} \tag{15.14}$$

where the indicator function $\mathbb{I}_{h_i(x) \neq c(x)}$ returns 0 is the model predicts correctly for instance x and 1 otherwise. The $2\mathbb{I}_{h_i(x) \neq c(x)} - 1$ expression is therefore equal to -1 if x is classified correctly and 1 if x is misclassified. This increases the weights of misclassified instances and decreases the weights of correctly classified instances to a degree that depends on the weight of model h_i. More accurate (higher weighted) models result in more extensive instance weight updates.

Example 15.5.2 The following R code produces plots that illustrate the AdaBoost weighting schemes.

```
curve(0.5*log((1-x)/x), from=0, to=0.5,
    xlab="model error", ylab="model weight")
curve(exp(0.5*log((1-x)/x)), from=0, to=0.5,
    xlab="model error", ylab="instance weight multiplier", ylim=c(0, 10), lty=2)
curve(exp(-0.5*log((1-x)/x)), from=0, to=0.5, lty=3, add=TRUE)
legend("topright", legend=c("misclassified", "correctly classified"), lty=2:3)
```

The obtained plots are presented in Figure 15.7. The first plot represents the dependence of model weight on model error and the other the dependence of the multiplier applied to modify instance weights on model error. The latter contains two curves, a gashed one for misclassified instances, and a dotted one for correctly classified instances. In all the cases, the range of model error values is limited to the $[0, 0.5]$ interval, assuming base models have above-random training performance.

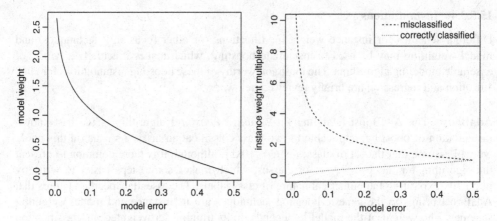

Figure 15.7   Model and instance weighting in AdaBoost.

As we can see, model weights may considerably exceed 1 for good models (with error of about 0.1 and below) and approach 0 for poor models with near-random training performance. Model weight changes are more rapid for small error values than for large ones. The instance weight multiplier applied for incorrectly classified instances grows rapidly with model error dropping below about 0.1, correspondingly. Then it near-linearly drops from about 3 to about 2 for model error increasing from 0.1 to 0.2, and also near-linearly goes down from about 2 to about 1 for model error raising from 0.2 to 0.5. The multiplier applied to the weights of correctly classified instances changes from 0 for perfectly accurate models to 1 for near-random models in a mostly linear manner, except for small-error models, when it drops toward 0 faster.

The complete AdaBoost algorithm is presented below. It assumes that the modeling algorithm used, referred to as $\mathcal{M}$, is weight sensitive and does not require instance weights to sum up to 1 (if the latter is not true, the normalization of the weight vector is required). Similarly the weighted misclassification error is assumed to be calculated correctly without requiring instance weights to sum up to 1. The algorithm performs at most m iterations, with m designating the specified maximum number of base models, but may terminate earlier after obtaining a base model that is not sufficiently better than random. To check this condition, the model's weighted misclassification error on the training set is compared against the expected random guess error 0.5, using a specified margin $\epsilon > 0$. Receiving such a poor model before reaching the maximum number of base models indicates that no further improvement is probably possible, and putting more weight on misclassified instances would result in further degradation rather than improvement. The algorithm returns the set of created models and their weights, to be used for weighted voting prediction.

```
1: for all x ∈ T do
2:      w_x^(1) := 1;
3: end for
4: for i = 1, 2, ..., m do
5:      h_i := M(T, w^(i));
6:      if e_{c,T,w^(i)}(h_i) > 0.5 − ε then
7:          return ⟨h_1, W_1⟩, ..., ⟨h_{i−1}, W_{i−1}⟩;
```

```
 8:    end if
 9:    W_i := 1/2 ln (1−e_{c,T,w^(i)}(h_i)) / (e_{c,T,w^(i)}(h_i));
10:    for all x ∈ T do
11:        w_x^(i+1) := w_x^(i) e^{W_i(2·𝟙_{h_i(x)≠c(x)}−1)};
12:    end for
13: end for
14: return ⟨h_1, W_1⟩, ..., ⟨h_m, W_m⟩;
```

It can be shown that AdaBoost solves the optimization problem consisting in minimizing the *exponential loss* of the created ensemble model h_* on the training set

$$\sum_{x \in T} e^{2\mathbb{1}_{h_*(x) \neq c(x)} - 1} \qquad (15.15)$$

which clearly leads to minimizing the training misclassification error as well. Despite perfectly fitting to the training set, it is not prone to overfitting – although not completely overfitting resistant. This resistance is reinforced if base models are indeed severely underfitted, just above random. Hence the popularity of *decision stumps*, i.e., one-split decision trees, in this role. On the other hand, the risk of overfitting is increased for noisy data. These intuitively "obvious" statements are not necessarily fully supported by empirical evidence, which sometimes provide surprising counter arguments, but in general – boosting does manage to avoid overfitting in most practical classification tasks much better than most nonensemble classification algorithms.

Example 15.5.3 The R code presented below implements the AdaBoost algorithm, using the `base.ensemble.weight.x` function from Example 15.3.2 for base model generation and the `predict.ensemble.weighted` function from Example 15.4.3 for prediction combination. The former requires the instance reweighting function to be provided, which does most of the work. Notice that the function takes care, in particular, of calculating and retaining base model weights. The model weighting function applied includes an additional term that depends on the number of classes and is equal to 0 for the two-class setting assumed by AdaBoost. This actually implements one of its possible multiclass extensions, as discussed in the next subsection. The algorithm is demonstrated in the same way as before, though, using the two-class *HouseVotes84* dataset. Decision trees of fixed maximum depth equal to 1, 3, and 5 are used as base models.

```
## AdaBoost ensemble modeling using up to m base models created using algorithm alg
## with arguments arg and maximum allowed base model error 0.5-eps
adaboost <- function(formula, data, m, alg, eps=0.01, args=NULL, predf=predict)
{
  class <- y.var(formula)
  nc <- nlevels(data[[class]])
  model.weights <- NULL

  abst.reweight <- function(weights, pred)
  {
    e <- werr(pred, data[[class]], weights)
    if (e<=0.5-eps && is.finite(mw <- 0.5*(log((1-e)/e)+log(nc-1))))
```

```
  {
    model.weights <<- c(model.weights, mw)
    weights*exp(mw*(2*(pred!=data[[class]])-1))
  }
  else
    NULL
}

'class<-'(list(models=base.ensemble.weight.x(formula, data, m, alg, args,
                                             weights=rep(1, nrow(data)),
                                             abst.reweight,
                                             predf=predf),
             model.weights=model.weights), "adaboost")
}


## AdaBoost prediction
predict.adaboost {- function(boost, data, predf=predict)
{
  predict.ensemble.weighted(boost$models, boost$model.weights, data, predf)
}


  # AdaBoost for the HouseVotes84 data
hv.abst.tree1 <- adaboost(Class~., hv.train, 50, rpart,
                          args=list(minsplit=2, cp=0, maxdepth=1),
                          predf=function(...) predict(..., type="c"))
hv.abst.tree3 <- adaboost(Class~., hv.train, 50, rpart,
                          args=list(minsplit=2, cp=0, maxdepth=3),
                          predf=function(...) predict(..., type="c"))
hv.abst.tree5 <- adaboost(Class~., hv.train, 50, rpart,
                          args=list(minsplit=2, cp=0, maxdepth=5),
                          predf=function(...) predict(..., type="c"))

hv.pred.abst.tree1 <- predict(hv.abst.tree1, hv.test,
                              predf=function(...) predict(..., type="c"))
hv.pred.abst.tree3 <- predict(hv.abst.tree3, hv.test,
                              predf=function(...) predict(..., type="c"))
hv.pred.abst.tree5 <- predict(hv.abst.tree5, hv.test,
                              predf=function(...) predict(..., type="c"))

  # AdaBoost test set errors for the HouseVotes84 data
hv.err.abst <- list(tree1 = err(hv.pred.abst.tree1, hv.test$Class),
                    tree3 = err(hv.pred.abst.tree3, hv.test$Class),
                    tree5 = err(hv.pred.abst.tree5, hv.test$Class))
```

Notice that depth-1 decision trees (i.e., decision stumps) achieve the least misclassification error, improving over that obtained for bagging in Example 15.5.1. Larger trees give worse results.

Multiclass AdaBoost The AdaBoost algorithm strongly relies on the assumption that the error of all base models does not exceed 0.5. This is perfectly reasonable if there are two classes, for which this is the random guess performance level, but cannot be expected otherwise. With errors above 0.5 the AdaBoost model weighting scheme is no longer useful, as it may deliver negative weights.

The restriction to two-class classification tasks limits the practical utility of the AdaBoost algorithm in an important way. There have been several attempts to overcome this restriction. They have different levels of complexity, theoretical justifications, and practical advantages.

One self-suggesting approach is to apply one of the binary multiclass encoding techniques described in Section 17.4. The simplest of them would be to decompose a multiclass classification task into multiple two-class tasks using the "1 vs. rest" approach. Conceptually, it consists in replacing the original target concept $c : X \rightarrow C$ with $|C|$ concepts c_d for each $d \in C$, where

$$c_d(x) = \begin{cases} 1 & \text{if } c(x) = d \\ 0 & \text{otherwise} \end{cases} \tag{15.16}$$

For each of these a separate AdaBoost binary model ensemble can be created in the usual way. This applies the 1-of-k encoding presented in Section 17.4.2.

A more refined incarnation of this "1 vs. rest" idea is also possible. Basically, instead of multiple applications of the AdaBoost algorithm, one may apply the algorithm once, but with each training instance x replaced by its $|C|$ copies $\langle x, d \rangle$, each with one of the original class labels $d \in C$ appended. The weights vector used for base model creation is correspondingly extended, to assign a numerical weight $w_{x,d}$ to instance $\langle x, d \rangle$. Base models created for the resulting extended training set and weight vector are assumed, correspondingly, to make binary predictions for instance-class pairs: $h_i : X \times C \rightarrow \{0, 1\}$. Weights for extended instances (i.e., instance-class pairs) are modified using the same formula as in the original algorithm. The ensemble's prediction for instance x would be then obtained by weighted voting:

$$h_*(x, d) = \arg \max_{b \in \{0,1\}} \sum_{i=1}^{m} W^i \mathbb{I}_{h_i(x,d)=b} \tag{15.17}$$

$$h_*(x) = \arg \max_{d \in C} h_*(x, d) \tag{15.18}$$

This technique is known as the AdaBoost.MH algorithm.

Another approach, known as the SAMME algorithm (*stagewise additive modeling using an exponential loss function*) proceeds in a completely different way, by directly creating an ensemble multiclass base models. It uses a modified model weighting scheme, incorporating an apparently minor, but important change:

$$W_i = \frac{1}{2} \ln \frac{1 - e_{c,T,w_i}(h)}{e_{c,T,w_i}(h)} + \frac{1}{2} \ln (|C| - 1) \tag{15.19}$$

It is equivalent to AdaBoost for $|C| = 2$ and for $|C| > 2$ it incorporates an adjustment term that preserves the exponential loss minimization property.

Gradient boosting   Gradient boosting applies the idea of boosting to the regression task. A sequence of regression models is created, with each model trying to contribute an improvement to the training set performance achieved by its predecessors. Unlike for AdaBoost or other instantiations of classification boosting, this is achieved not by instance weighting, but rather by using the residuals of the ensemble of previously created base models instead of the original target function values when creating a subsequent base model. Base models are then combined using weighted averaging (or, actually, summation). This technique is presented below in its most generic form, although a randomized version thereof that additionally applies instance sampling for greater base model diversity, referred to as *stochastic gradient boosting*, may often perform better.

The first model h_1 is created in the usual way. For $i = 2, \ldots, m$ model h_i is created to predict the so-called pseudoresiduals:

$$r_x^{(i)} = -\frac{\partial \mathcal{L}(f(x), h_{1:i-1}(x))}{\partial h_{1:i-1}(x)} \tag{15.20}$$

of the partial ensemble $h_{1:i-1}$, consisting of the previously created models $h_1, \ldots, h_{i-1}$, for each $x \in T$. In this equation, $\mathcal{L}$ is the adopted loss function to be minimized, as discussed in Section 10.2.9. Predicting its negated derivative with respect to the previous iteration's predictions is expected to decrease the total loss. For the most popular quadratic loss (i.e., mean square error minimization), we would take $r_x^{(i)} = f(x) - h_{1:i-1}(x)$. These pseudoresiduals are passed to the regression algorithm instead of the original target function values. Base models are combined by weighted summation to achieve both the partial and final ensemble, i.e.,

$$h_{1:i-1}(x) = \sum_{j=1}^{i-1} W_j h_j(x) \tag{15.21}$$

$$h_*(x) = h_{1:m}(x) = \sum_{j=1}^{m} W_j h_j(x) \tag{15.22}$$

The weight W_i of model h_i is selected to minimize the total loss, under the adopted loss function, for the ensemble extended to include the model, which may be written as follows:

$$W_i = \beta \arg\min_{W} \sum_{x \in T} \mathcal{L}(f(x), h_{1:i-1}(x) + W h_i(x)) \tag{15.23}$$

where $0 < \beta \leq 1$ is a step-size parameter. While the basic version of gradient boosting assumes $\beta = 1$, using a smaller β value may help to reduce the risk of overfitting and improve the generalization capabilities of the resulting model ensemble. This form of overfitting prevention is referred to as *shrinkage*. The complete gradient boosting algorithm is presented below.

1: $h_1 := \mathcal{M}(T, f)$; $W_1 := 1$;
2: **for** $i = 2, 3, \ldots, m$ **do**
3:     **for all** $x \in T$ **do**
4:         $r_x^{(i)} := -\frac{\partial \mathcal{L}(f(x), h_{1:i-1}(x))}{\partial h_{1:i-1}(x)}$;
5:     **end for**
6:     $h_i := \mathcal{M}(T, r^{(i)})$;
7:     $W_i := \beta \arg\min_W \sum_{x \in T} \mathcal{L}(f(x), h_{1:i-1}(x) + W h_i(x))$;
8: **end for**
9: **return** $\langle h_1, W_1 \rangle, \ldots, \langle h_m, W_m \rangle$;

The regression algorithm $\mathcal{M}$ used for base model creation is assumed to receive the training set as well as the corresponding target values, with the original target function values used for h_1 and the current residuals used afterward. It is not uncommon, though, to create a particularly simple first model that predicts a constant value, chosen to minimize the adopted loss function. In particular, for the quadratic loss, this constant model would simply predict the mean target function value for the training set:

$$h_1(x) = \frac{1}{|T|} \sum_{x \in T} f(x) \tag{15.24}$$

as this clearly minimizes the mean square error over all possible constant models.

For this most common special case of the quadratic loss (mean square error minimization) we have

$$\sum_{x \in T} \mathcal{L}(f(x), h_{1:i-1}(x) + Wh_i(x)) = \sum_{x \in T} \left(f(x) - (h_{1:i-1}(x) + Wh_i(x)) \right)^2 \qquad (15.25)$$

Minimizing this quantity with respect to W yields, by equating the corresponding derivative to 0:

$$\sum_{x \in T} (f(x) - h_{1:i-1}(x) - Wh_i(x))h_i(x) = 0 \qquad (15.26)$$

from which one can obtain

$$W = \frac{\sum_{x \in T} (f(x) - h_{1:i-1}(x))h_i(x)}{\sum_{x \in T} h_i^2(x)} = \frac{\sum_{x \in T} r_x^{(i)} h_i(x)}{\sum_{x \in T} h_i^2(x)} \qquad (15.27)$$

This will clearly yield 1 if model h_i does indeed perfectly predict the previous ensemble's residuals, but can be verified to be also equal to 1 even for completely imperfect regression trees with target value means assigned to leaves. To see why, consider a leaf l assigned a target function value equal to the mean target value for the corresponding subset of training instances:

$$v_l = \frac{1}{|T_l|} \sum_{x \in T_l} f(x) \qquad (15.28)$$

Then the sum of target function value and prediction products for training instances assigned to leaf l can be written as

$$\sum_{x \in T_l} f(x)h(x) = \sum_{x_1 \in T_l} \left(f(x_1) \frac{1}{|T_l|} \sum_{x_2 \in T_l} f(x_2) \right) = \frac{1}{|T_l|} \sum_{x_1 \in T_l} \sum_{x_2 \in T_l} f(x_1)f(x_2) \qquad (15.29)$$

On the other hand, the sum of squared predictions for the same subset of training instances can be transformed in the following way:

$$\sum_{x \in T_l} h^2(x) = \sum_{x \in T_l} \left(\frac{1}{|T_l|} \sum_{x_1 \in T_l} f(x_1) \right) \left(\frac{1}{|T_l|} \sum_{x_2 \in T_l} f(x_2) \right)$$

$$= |T_l| \frac{1}{|T_l|^2} \sum_{x_1 \in T_l} \sum_{x_2 \in T_l} f(x_1)f(x_2) = \frac{1}{|T_l|} \sum_{x_1 \in T_l} \sum_{x_2 \in T_l} f(x_1)f(x_2) \qquad (15.30)$$

which yields

$$\sum_{x \in T_l} f(x)h(x) = \sum_{x \in T_l} h^2(x) \qquad (15.31)$$

This immediately implies

$$\sum_{x \in T} f(x)h(x) = \sum_{x \in T} h^2(x) \qquad (15.32)$$

since summation over all training instances can be decomposed into summation over leaves and then training instances assigned to these leaves. This property holds for an arbitrary target function, including, in particular, the previous ensemble's residuals in gradient boosting:

$$\sum_{x \in T} r_x^{(i)} h_i(x) = \sum_{x \in T} h_i^2(x) \qquad (15.33)$$

from which $W = 1$.

Example 15.5.4 The following R code implements and demonstrates gradient boosting. The `gradboost` function creates the first base model in the usual way and then for each subsequent base model it creates a modified dataset copy, with target function values replaced by the residuals of the partial ensemble created so far. The `predict.gradboost` function is a simple wrapper around `predict.ensemble.weighted`, with the summing argument used to request base model combination by weighted summation. The demonstrations use the *Boston Housing* dataset, with regression tree and linear regression base models. For the former, a fixed maximum depth of 1, 3, and 5 is used.

```
## gradient boosting ensemble modeling using up to m base models
## created with algorithm alg with arguments arg
gradboost <- function(formula, data, m, alg, beta=0.1, args=NULL, predf=predict)
{
  attributes <- x.vars(formula, data)
  aind <- match(attributes, names(data))
  f <- y.var(formula)
  find <- match(f, names(data))

  models <- list(do.call(alg, c(list(formula, data), args)))
  model.weights <- 1

  for (i in (2:m))
  {
    res <- data[,find]-predict.gradboost(list(models=models,
                                  model.weights=model.weights),
                              data, predf=predf)
    data.i <- eval(parse(text=paste("cbind(data[,aind],", f, "=res)")))
    models <- c(models, list(h <- do.call(alg, c(list(formula, data.i), args))))
    model.weights <- c(model.weights,
                    beta*sum(res*(pred <- predf(h, data)))/sum(pred^2))
  }
  `class<-`(list(models=models, model.weights=model.weights), "gradboost")
}

## gradient boosting prediction
predict.gradboost <- function(boost, data, predf=predict)
{
  predict.ensemble.weighted(boost$models, boost$model.weights, data, predf,
                          summing=TRUE)
}

  # gradient boosting for the BostonHousing data
bh.gbst.tree1 <- gradboost(medv~., bh.train, 50, rpart,
                          args=list(minsplit=2, cp=0, maxdepth=1))
bh.gbst.tree3 <- gradboost(medv~., bh.train, 50, rpart,
                          args=list(minsplit=2, cp=0, maxdepth=3))
bh.gbst.tree5 <- gradboost(medv~., bh.train, 50, rpart,
                          args=list(minsplit=2, cp=0, maxdepth=5))
bh.gbst.lm <- gradboost(medv~., bh.train, 50, lm)

bh.pred.gbst.tree1 <- predict(bh.gbst.tree1, bh.test)
bh.pred.gbst.tree3 <- predict(bh.gbst.tree3, bh.test)
bh.pred.gbst.tree5 <- predict(bh.gbst.tree5, bh.test)
bh.pred.gbst.lm <- predict(bh.gbst.lm, bh.test)

  # gradient boosting test set MSE values for the BostonHousing data
bh.mse.gbst <- list(tree1 = mse(bh.pred.gbst.tree1, bh.test$medv),
```

```
tree3 = mse(bh.pred.gbst.tree3, bh.test$medv),
tree5 = mse(bh.pred.gbst.tree5, bh.test$medv),
lm = mse(bh.pred.gbst.lm, bh.test$medv))
```

Notice that model weights for regression trees are all equal to 1, as expected. Unfortunately, the mean square error levels for gradient boosting with regression trees are worse than obtained for regression tree bagging in Example 15.5.1, although – with a maximum depth of 3 – better than for the single regression tree model.

15.5.4   Random forest

The random forest technique of ensemble modeling can be viewed as another enhancement of bagging. This view is even more justified than that of boosting, since random forests actually use bootstrap data samples as training sets for base model creation, just like bagging. The enhancement consists in stimulating greater base model diversity by randomizing the modeling algorithm applied to these samples, which is – as the name of the technique suggests – a decision tree or regression tree algorithm. Being tied to a particular modeling algorithm (or a family of algorithms) is not such a distinctive feature of random forests as it might appear, though, given the prevailing practice of using (the standard unrandomized versions of) the very same algorithms with other ensemble modeling techniques.

15.5.4.1   Base models

Random forests combine two approaches to base model creation: instance sampling (using bootstrap samples) and algorithm nondeterminism. The latter is achieved by randomizing the split selection operation used for decision tree or regression tree growing. The randomization consists in drawing a random subset of available attributes in each node and restricting the subsequent split selection process to splits using attributes from that subset. The usual split evaluation criteria for decision trees or for regression trees are then applied. Otherwise the growing process remains unchanged. Stop criteria for decision or regression tree growing are set up to yield relatively large, accurately fitted (more than likely overfitted) trees and no pruning is applied. This setup, resulting in many splits being selected (and not pruned off), permits a very high level of base model diversity, at least unless the number of available splits (directly implied by the number of attributes) is overly small. A standard heuristic is to use the square root of the number of all available attributes as the size of the randomly drawn subset of attributes. Typically at least several hundred base models are created. Their individual overfitting is canceled out by the aggregation process, which makes the random forest ensemble highly resistant to overfitting.

15.5.4.2   Model aggregation

Randomized decision trees or regression trees used as base models for random forests are aggregated via plain (unweighted) voting or averaging/summation.

Example 15.5.5 The R code presented below implements a simple version of random forest ensemble modeling, using the grow.randdectree, grow.randregtree, and base.ensemble.simple functions from Example 15.3.4 and the predict.ensemble

.basic function from Example 15.4.1. The predict.dectree and predict. regtree functions, defined in Examples 3.5.1 and 9.5.1, are used to generate base model predictions. The random forest algorithm is demonstrated for the *HouseVotes84* and *Boston Housing* datasets, using three different maximum tree depth settings: 3, 5, and 8.

```
## random forest ensemble modeling using m randomized decision or regression trees
## with ns randomly selected attributes at each node used for splitting
randforest <- function(formula, data, m, ns=0, args=NULL)
{
  target <- y.var(formula)
  alg <- if (!is.numeric(data[[target]])) grow.randdectree else grow.randregtree

  `class<-`(base.ensemble.sample.x(formula, data, m, alg, args=c(list(ns=ns), args)),
          "randforest")
}

## random forest prediction
predict.randforest <- function(rf, data)
{
  predict.ensemble.basic(rf, data)
}

  # random forest for the HouseVotes84 data
hv.rf.tree3 <- randforest(Class~., hv.train, 50, args=list(maxdepth=3))
hv.rf.tree5 <- randforest(Class~., hv.train, 50, args=list(maxdepth=5))
hv.rf.tree8 <- randforest(Class~., hv.train, 50, args=list(maxdepth=8))

hv.pred.rf.tree3 <- predict(hv.rf.tree3, hv.test)
hv.pred.rf.tree5 <- predict(hv.rf.tree5, hv.test)
hv.pred.rf.tree8 <- predict(hv.rf.tree8, hv.test)

  # random forest for the BostonHousing data
bh.rf.tree3 <- randforest(medv~., bh.train, 50, args=list(maxdepth=3))
bh.rf.tree5 <- randforest(medv~., bh.train, 50, args=list(maxdepth=5))
bh.rf.tree8 <- randforest(medv~., bh.train, 50, args=list(maxdepth=8))

bh.pred.rf.tree3 <- predict(bh.rf.tree3, bh.test)
bh.pred.rf.tree5 <- predict(bh.rf.tree8, bh.test)
bh.pred.rf.tree8 <- predict(bh.rf.tree8, bh.test)

  # random forest test set errors for the HouseVotes84 data
hv.err.rf <- list(tree3 = err(hv.pred.rf.tree3, hv.test$Class),
              tree5 = err(hv.pred.rf.tree5, hv.test$Class),
              tree8 = err(hv.pred.rf.tree8, hv.test$Class))

  # random forest test set MSE values for the BostonHousing data
bh.mse.rf <- list(tree3 = mse(bh.pred.rf.tree3, bh.test$medv),
              tree5 = mse(bh.pred.rf.tree5, bh.test$medv),
              tree8 = mse(bh.pred.rf.tree8, bh.test$medv))
```

For the *HouseVotes84* data, the evaluated random forest ensembles improve over a single decision tree, unless using the greatest maximum tree depth. This may appear surprising, since increased tree depth permits more base model diversity and should therefore offer better improvement potential. Attribute sampling may be too aggressive or the small number of trees may be insufficient to compensate for their accuracy reduction due to randomized split selection. The inefficiency of the presented illustrative implementation prevents creating larger random forests in reasonable time. For the *Boston Housing* data, the observations better match the expectations, with the random forest model using the least maximum depth performing worse and the other two models – better than single regression trees.

15.5.4.3 Side effects

Apart from delivering ensemble models, the random forest technique also has some quite useful "side effects," obtained by appropriately using individual decision or regression trees from the grown forest as well as the corresponding training sets or out-of-bag (OOB) instances. The most useful of those are briefly discussed below.

Performance estimates Since each tree in the forest is grown using a bootstrap sample drawn from the original training set, there is also the corresponding subset of instances not used for growing. These are the OOB instances that were not drawn to the training sample. For the particular tree these instances are therefore perfectly usable for the purpose of model evaluation, i.e., can be used to calculate true performance estimates as if the standard hold-out procedure were employed. It is the performance of the complete forest rather than that of individual trees that is to be estimated, though. This is possible by combining the OOB predictions of base models.

Let $T'_1, T'_2, \ldots, T'_m$ denote the sets of OOB instances for base models (trees) $h_1, h_2, \ldots, h_m$, grown using training sets $T_1, T_2, \ldots, T_m$, respectively. For any instance $x \in T$ let

$$I_x = \left\{ i \in \{1, 2, \ldots, m\} \mid x \in T'_i \right\} \tag{15.34}$$

designate the set of base model numbers for which x is an OOB instance. The OOB prediction for instance x, denoted by $h_{OOB}(x)$, is then obtained by combining the predictions of all models h_i for $i \in I_x$ via plain voting (for classification) or averaging (for regression). These predictions for all $x \in T$ may be then compared against true class labels $c(x)$ or target function values $f(x)$ to calculate any performance measure of interest. The misclassification error and the mean square error are of course the typical choices.

Notice that such OOB-based performance estimation technique is by no means the same as or a variation of the bootstrapping evaluation procedure discussed in Section 7.3.6, to which it is only superficially similar. This is because the latter estimates the performance of single models whereas the former estimates the performance of a complete ensemble. The resulting estimate is quite reliable and comparable to standard cross-validation with respect to its bias and variance. Given the computational cost of random forest growing (with hundreds or more trees), performing a standard cross-validation loop might easily become computationally prohibitive.

Instance proximity Random forests make it straightforward to measure instance proximity based on instance co-occurring statistics in individual trees. Basically, for each instance $x \in T$ and each tree h_i for $i = 1, 2, \ldots, n$, the corresponding tree leaf $\mathbf{l}_{i,x}$ may be determined, by passing down the training set through the tree. Then the proximity of instances $x_1, x_2 \in T$ is calculated as the number of trees where they both end up in the same leaf:

$$\varepsilon(x_1, x_2) = \sum_{i=1}^{m} \mathbb{I}_{\mathbf{l}_{i,x_1} = \mathbf{l}_{i,x_2}} \tag{15.35}$$

While definitely related to dissimilarity and similarity measures discussed in Chapter 11, random forest-based instance proximity is calculated in an entirely different way and may serve different purposes. While the former are based on attribute-value differences or correlations, the latter represent rather the property of (usually) falling into the same domain regions, with boundaries determined by the values of selected, predictively useful attributes. One step toward relating these two quantities would be therefore to consider proximity as similarity restricted only to predictively important attributes. Moreover, proximity is not necessarily sensitive to attribute value differences that usually have no high impact on model predictions, regardless of their scale. It may therefore not be appropriate for typical instantiations of the clustering task, but may become useful for other purposes, such as domain decomposition for modeling tasks or data preprocessing. In particular, one natural application of such a proximity measure is missing value imputation where missing attribute values for an instance may be imputed based on the known values observed for other instances with the highest proximity to that instance.

Attribute utility   Out-of-bag instances are useful not only for estimating model quality, but also (for estimating) attribute utility, which may be viewed as particular attributes' impact on the former. With a set of spared nontraining instances for each tree one can observe how crucial particular attributes are for the obtained predictive performance level. One way to do this is to simulate "corrupting" each attribute (separately) by randomly permuting its values in each tree's OOB set and measure the effect of this "corruption."

Assuming the same notation as introduced earlier in this section, one would compare regular OOB predictions, yielding $h_{\text{OOB}}(x)$ for each $x \in T$, with the corresponding predictions $h_{\text{OOB},a}(x)$ obtained with the values of attribute a randomly permuted in each of the OOB sets $T_1', T_2', \ldots, T_m'$. The permutation, performed independently for each tree, will have a considerable impact on the predictions of those trees which use attribute a for splitting, particularly on high levels, and little or no impact otherwise. With some base models yielding different predictions, the combined predictions will change to some extent. For any selected performance measure – with the misclassification error typically used for classification and mean square error typically used for regression – the degradation observed for $h_{\text{OOB},a}$ in comparison to h_{OOB} may be considered a measure of the predictive utility of attribute a.

Attribute utility estimation is arguably the most useful of random forest side effects that sometimes becomes the main or only reason of creating a random forest. The estimated attribute utilities may be then used for attribute selection, as discussed in Section 19.4.5, and the final model may be created using another modeling algorithm based on the selected subset. Such a usage scenario basically treats a random forest as an attribute selection filter, and – setting the computational cost apart – it turns out to belong to the best filtering attribute selection algorithms.

15.5.5   Random Naïve Bayes

The success of the random forest ensemble has become motivation for exploring a similar combination of base model creation techniques (i.e., instance sampling and attribute sampling, which is closely related to decision tree randomization in random forests) with other modeling algorithms. One particularly interesting candidate is the naïve Bayes classifier. The resulting ensemble modeling algorithm is referred to as the random naïve Bayes classifier.

15.5.5.1   Base models

Basically, the random naïve Bayes ensemble consists of multiple naïve Bayes models $h_1, h_2,$ $\dots, h_m$, each created from an independent bootstrap training set sample T_i and permitted to use only a randomly selected subset $\mathbf{A}_i$ of attributes. Being a stable algorithm, the naïve Bayes classifier does not deliver sufficient base model diversity when created using bootstrap samples, as discussed above. Incorporating random attribute sampling apart from bootstrap instance sampling overcomes this deficiency, though.

15.5.5.2   Model combination

Unlike in most versions of bagging or random forests, the base models of the random naïve Bayes ensemble are not combined via simple voting. The inherent probabilistic prediction capability of the naïve Bayes classifier makes it much more reasonable to apply class probability averaging to aggregate base model predictions. The resulting combined class probabilities may be used for class label prediction in the usual way.

15.5.5.3   Properties

On one hand, random naïve Bayes is just one out of many possible random forest-like bagging extensions, combining instance sampling and attribute sampling for base model generation. There are some reasons, however, to consider it particularly interesting. This is because the simplicity of the naïve Bayes classifier makes it possible to create multiple base models with a relatively low computational expense, in particular much below that of decision trees. It may be therefore more practical to apply to large datasets. This is also because using independent attribute samples not only stimulates base model diversity, but additionally makes each of them less prone to harmful effects of the unsatisfied independence assumption on which naïve Bayesian classification is based. In smaller attribute subsets attribute dependences are less likely to occur. Therefore, each base model, while possibly inaccurate due to using incomplete information, is less likely to be fooled by attribute dependences. Random naïve Bayes may be therefore successful whenever the "naïvety" of the naïve Bayes classifier becomes a problem.

Example 15.5.6 The random naïve Bayes algorithm is implemented and demonstrated by the following R code, using the naïve Bayes classifier provided by the `e1071` package. The `randnaiveBayes` function is essentially a combination of the `base.ensemble.samle.x` function from Example 15.3.1 (instance sampling) and the `base.ensemble.sample.a` function from Example 15.3.3 (attribute sampling). The `predict.randnaiveBayes` function is basically a wrapper around the implementation of probability averaging from Example 15.4.2.

```
## random naive Bayes ensemble modeling using m base models
## each with ns randomly selected attributes
## (if unspecified, it defaults to the square root of the number of attributes)
randnaiveBayes <- function(formula, data, m, ns=0)
{
  attributes <- x.vars(formula, data)
  target <- y.var(formula)
  ns <- ifelse(ns==0, round(sqrt(length(attributes))),
```

```
                         clip.val(ns, 1, length(attributes)))

   `class<-`(lapply(1:m, function(i)
                       {
                           bag <- sample(nrow(data), size=nrow(data), replace=TRUE)
                           sa <- sample(length(attributes), ns)
                           naiveBayes(make.formula(target, attributes[sa]),
                                       data[bag,])
                       }), "randnaiveBayes")
}

## random naive Bayes prediction
predict.randnaiveBayes <- function(rnb, data, prob=FALSE)
{
   predict.ensemble.prob(rnb, data, predf=function(...) predict(..., type="r"),
                       prob=prob, labels=rnb[[1]]$levels)
}

   # random naive Bayes for the HouseVotes84 data
hv.rnb <- randnaiveBayes(Class~., hv.train, 500)
hv.pred.rnb <- predict(hv.rnb, hv.test)
   # random naive Bayes test set error for the HouseVotes84 data
hv.err.rnb <- list(nb = err(hv.pred.rnb, hv.test$Class))
```

Unfortunately the random naïve Bayes algorithm performs slightly worse than the original deterministic algorithm. This may be due to a relatively small number of attributes, for which sampling deteriorates model quality too much.

15.6 Quality of ensemble predictions

Model ensembles may be often expected to outperform single models, even created using refined and carefully tuned algorithms. Sometimes, particularly for boosting and random forests, the improvement may be quite substantial. It is, however the combination of data, base model creation algorithms, and ensemble modeling techniques that is ultimately responsible for the final prediction quality.

Example 15.6.1 The misclassification error or mean square error values for the model ensembles created in the series of previous examples – using bagging and boosting with decision tree, naïve Bayes, regression tree, and linear base models, as well as the random forest and random naïve Bayes algorithms – are collected and compared to one another, as well as to those achieved by single models, by the R code presented below. For each of the two datasets used a barplot of error values is produced.

```
hv.err <- c(tree=hv.err.tree,
            bagg.tree=hv.err.bagg$tree,
            abst.tree1=hv.err.abst$tree1,
            abst.tree3=hv.err.abst$tree3,
            abst.tree5=hv.err.abst$tree5,
            rf.tree3=hv.err.rf$tree3,
            rf.tree5=hv.err.rf$tree5,
```

```
            rf.tree8=hv.err.rf$tree8,
            nb=hv.err.nb,
            bagg.nb=hv.err.bagg$nb,
            rnb=hv.err.rnb$nb)

bh.mse <- c(tree=bh.mse.tree,
            bagg.tree=bh.mse.bagg$tree,
            gbst.tree1=bh.mse.gbst$tree1,
            gbst.tree3=bh.mse.gbst$tree3,
            gbst.tree5=bh.mse.gbst$tree5,
            rf.tree3=bh.mse.rf$tree3,
            rf.tree5=bh.mse.rf$tree5,
            rf.tree8=bh.mse.rf$tree8,
            lm=bh.mse.lm,
            bagg.lm=bh.mse.bagg$lm,
            gbst.lm=bh.mse.gbst$lm)

barplot(hv.err, main="HouseVotes84", ylab="Error", las=2)
lines(c(0, 13), rep(hv.err[1], 2), lty=2)
lines(c(0, 13), rep(hv.err[9], 2), lty=3)

barplot(bh.mse, main="Boston Housing", ylab="MSE", las=2)
lines(c(0, 13), rep(bh.mse[1], 2), lty=2)
lines(c(0, 13), rep(bh.mse[9], 2), lty=3)
```

The obtained barplots are presented in Figure 15.8. On the *HouseVotes84* data the AdaBoost and random forest ensembles gives an improvement over single decision tree models, unless using excessive tree depth. Bagging produces worse predictions than single models in the case of decision trees and gave no effect for the naïve Bayes classifier. Introducing random attribute sampling to the latter turns out to be harmful rather than beneficial. On the *Boston Housing* dataset bagging applied to regression trees is the most successful, with the random forest ensemble approaching a similar performance level with sufficiently deep trees. Linear model ensembles all perform on the very same level as a single model, which is to be expected, since the averaged predictions of a multiple linear models remain linear.

15.7   Conclusion

There is a lot to be excited about in the idea of ensemble modeling. It is a conceptually appealing and extremely successful practically approach to improving the predictive power of inductive models. It not only makes it possible to get better predictive performance, but it also makes the modeling process easier for the human analyst. Ensemble modeling usually means no or little risk of overfitting, no or little parameter tuning, and no or little need for attribute selection (although it may provide useful tools for the latter, as in the case of random forests). This is because, when aggregating dozens or hundreds of base models, one may be much less concerned about their individual quality. Actually, base models that would be quite poor individually – in particular, overfitted due to lack of any overfitting prevention or underfitted due to using simplified modeling algorithms – are likely to be useful ensemble components.

These unquestionable benefits are not received without a price. What has to be paid is the vastly increased computational expense of creating many base models and using them

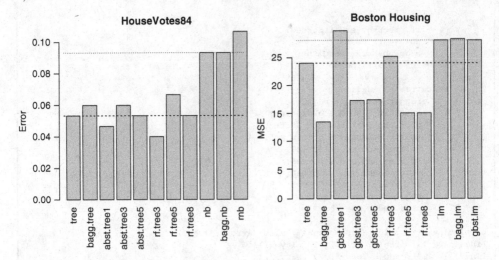

Figure 15.8 The barplots of error values for bagging, boosting, and random forest model ensembles.

for prediction (although it could be partially ameliorated for some ensemble modeling techniques by appropriate parallel implementations) and the loss of human readability, even if individual base models are perfectly human readable. Despite some efforts toward developing human-readable representations of model ensembles, the latter may remain the primary limitation of their applicability in some domains.

15.8   Further readings

Ensemble modeling has been one of the hottest topics of machine learning research over the last two decades, also becoming increasingly popular in practical applications in which the predictive performance is of top priority. It has also found its way into recent comprehensive data mining and machine learning books (e.g., Bishop 2007; Han *et al.* 2011; Hastie *et al.* 2011; Tan *et al.* 2013; Witten *et al.* 2011). There are also several survey articles on model ensembles (e.g., Dietterich 2000a; Rokach 2010).

The idea of combining multiple models for improved predictive performance can be traced back to early financial forecasting research (e.g., Bates and Granger 1969; Clemen 1989; Reid 1968), but it became a hot topic in the area of machine learning in the 1990s. Bagging was introduced by Breiman (1996a) as an approach to stabilizing unstable algorithms and improving their model quality using the technique of bootstrapping (Efron 1979; Efron and Tibshirani 1994). Schapire (1990) developed theoretical foundations of boosting and an early boosting algorithm, following earlier theoretical work on weak and strong learnability (Kearns and Valiant 1989). Freund and Schapire (1995) subsequently introduced the more refined AdaBoost algorithm that remains the most widely used boosting algorithm. In the same article a regression version of AdaBoost was also presented, which has not reached similar popularity. Quinlan (1996) combined both bagging and boosting with his C4.5 decision tree induction algorithm. Dietterich (2000b) compared bagging and boosting with an ensemble of decision trees obtained by split selection randomization. Friedman *et al.* (2000) presented a statistical view of boosting as a form of additive logistic regression, leading to the LogitBoost algorithm. Friedman (2001, 2002) also developed the gradient boosting regression algorithm. Several other boosting algorithms have been proposed, including multiclass versions of AdaBoost (Schapire and Singer 1999; Zhu *et al.* 2009) and cost-sensitive modifications to boosting algorithms (Masnadi-Shirazi and Vasconcelos 2010). Mukherjee and Schapire (2013) developed a theory of multiclass boosting. Schapire and Freund (2012) described the theoretical foundations of boosting and state-of-the-art boosting algorithms.

Breiman (2001) presented the random forest algorithm, discussing its theoretical properties and demonstrating practical capabilities. Liaw and Wiener (2002) developed an R implementation of the algorithm. The idea of OOB performance estimation for model ensembles was first proposed by Wolpert and Macready (1999), along with other related evaluation techniques. Prinzie and Van den Poel (2007), inspired by the success of the random forest algorithm, proposed re-using the combination of instance and attribute sampling with other base model creation algorithms. In particular, they presented the random naïve Bayes classifier, as well as random multiclass logistic regression.

Stacking introduced by Wolpert (1992) has never reached the popularity of bagging, boosting, and random forest, since its additional complexity of multilevel models appeared not to pay off. According to the empirical investigation performed by Džeroski and Ženko (2004), a stacking ensemble is usually no better than the best base model that could be selected using cross-validated performance estimates, although they also presented refinements to the stacking algorithm that may deliver better results. Breiman (1996b) applied the idea of stacking to the regression task, with base regression tree models combined via linear regression. Another related ensemble modeling technique that has not become very widely used is Bayesian model averaging, the idea of which was first introduced by Leamer (1978), but became practically applicable several years later when better theoretical foundations and computational resources were available (Chatfield 1995; Clyde 1999; Draper 1995; Hoeting *et al.* 1999). It combines base models by weighted averaging, with their estimated posterior probabilities used as weights, and was compared to stacking by Clarke (2003).

Opitz and Maclin (1999) reported systematic comparative experiments with bagging and boosting on many datasets (but with boosting base models created by modifying selection probabilities for instance sampling rather than by instance weighting). Kuncheva and Whitaker (2003a,b) investigated possible approaches to measuring base model diversity and discussed its impact on the predictive performance of model ensembles. There have been some attempts to create comprehensible representations of model ensembles (Park and Kargupta 2002; Triviño Rodriguez *et al.* 2008; Van Assche and Blockeel 2007).

References

Bates JM and Granger CWJ 1969 The combination of forecasts. *Operational Research Quarterly* **20**, 451–468.

Bishop CM 2007 *Pattern Recognition and Machine Learning*. Springer.

Breiman L 1996a Bagging predictors. *Machine Learning* **24**, 123–140.

Breiman L 1996b Stacked regression. *Machine Learning* **24**, 49–64.

Breiman L 2001 Random forests. *Machine Learning* **45**, 5–32.

Chatfield C 1995 Model uncertainty, data mining, and statistical inference. *Journal of the Royal Statistical Society A* **158**, 419–466.

Clarke B 2003 Comparing Bayes model averaging and stacking when model approximation error cannot be ignored. *Journal of Machine Learning Research* **4**, 683–712.

Clemen RT 1989 Combining forecasts: A review and annotated bibliography. *International Journal of Forecasting* **5**, 559–583.

Clyde M 1999 Bayesian model averaging and model search strategies *Proceedings of the Sixth Valencia International Meeting on Bayesian Statistics*. Oxford University Press.

Dietterich TG 2000a Ensemble methods in machine learning *Proceedings of the First International Workshop on Multiple Classifier Systems*. Springer.

Dietterich TG 2000b An experimental comparison of three methods for construction ensembles of decision trees: bagging, boosting and randomization. *Machine Learning* **40**, 139–157.

Draper D 1995 Assessment and propagation of model uncertainty. *Journal of the Royal Statistical Society B* **57**, 45–97.

Džeroski S and Ženko B 2004 Is combining classifiers with stacking better than selecting the best one. *Machine Learning* **54**, 255–273.

Efron B 1979 Bootstrap methods: Another look at the jackknife. *The Annals of Statistics* **7**, 1–26.

Efron B and Tibshirani R 1994 *An Introduction to the Bootstrap*. Chapman and Hall.

Freund Y and Schapire RE 1995 A decision-theoretic generalization of on-line learning and an application to boosting. *Journal of Computer and System Sciences* **55**, 119–139.

Friedman JH 2001 Greedy function approximation: A gradient boosting machine. *Annals of Statistics* **29**, 1189–1232.

Friedman JH 2002 Stochastic gradient boosting. *Computational Statistics and Data Analysis* **38**, 367–378.

Friedman JH, Hastie T and Tibshirani R 2000 Additive logistic regression: A statistical view of boosting. *Annals of Statistics* **28**, 337–407.

Han J, Kamber M and Pei J 2011 *Data Mining: Concepts and Techniques* 3rd edn. Morgan Kaufmann.

Hastie T, Tibshirani R and Friedman J 2011 *The Elements of Statistical Learning: Data Mining, Inference, and Prediction* 2nd edn. Springer.

Hoeting JA, Madigan D, Raftery AE and Volinsky CT 1999 Bayesian model averaging: A tutorial. *Statistical Science* **14**, 382–401.

Kearns M and Valiant LG 1989 Cryptographic limitations on learning Boolean formulae and finite automata *Proceedings of the Twenty-First Annual ACM Symposium on Theory of Computing*. ACM Press.

Kuncheva LI and Whitaker CJ 2003a Measures of diversity in classifier ensembles. *Machine Learning* **51**, 181–207.

Kuncheva LI and Whitaker CJ 2003b Measures of diversity in classifier ensembles and their relationship with the ensemble accuracy. *Machine Learning* **51**, 181–207.

Leamer EE 1978 *Specification Searches: Ad Hoc Inference with Nonexperimental Data*. Wiley.

Liaw A and Wiener M 2002 Classification and regression by randomForest. *R News*.

Masnadi-Shirazi H and Vasconcelos N 2010 Cost-sensitive boosting. *IEEE Transactions on Pattern Analysis and Machine Intelligence* **33**, 294–309.

Mukherjee I and Schapire RE 2013 A theory of multiclass boosting. *Journal of Machine Learning Research* **14**, 437–497.

Opitz D and Maclin R 1999 Popular ensemble methods: An empirical study. *Journal of Artificial Intelligence Research* **11**, 169–198.

Park BH and Kargupta H 2002 Constructing simpler decision trees from ensemble models using fourier analysis *Proceedings of the Seventh ACM SIGMOD Workshop on Research Issues in Data Mining and Knowledge Discovery*. ACM Press.

Prinzie A and Van den Poel D 2007 Random multiclass classification: Generalizing random forests to random MNL and random NB *Proceedings of the Eighteenth International Conference on Database and Expert Systems Applications (DEXA-2007)*. Springer.

Quinlan JR 1996 Bagging, boosting, and C4.5 *Proceedings of the Thirteenth National Conference on Artificial Intelligence (AAAI-96)*. AAAI Press.

Reid DJ 1968 Combining three estimates of gross domestic product. *Economica* **35**, 431–444.

Rokach L 2010 Ensemble-based classifiers. *Artificial Intelligence Review* **33**, 1–39.

Schapire RE 1990 The strength of weak learnability. *Machine Learning* **5**, 197–227.

Schapire RE and Freund Y 2012 *Boosting: Foundations and Algorithms*. MIT Press.

Schapire RE and Singer Y 1999 Improved boosting algorithms using confidence-rated prediction. *Machine Learning* **37**, 297–336.

Tan PN, Steinbach M and Kumar V 2013 *Introduction to Data Mining* 2nd edn. Addison-Wesley.

Triviño Rodriguez JL, Ruiz-Sepúlveda A and Morales-Bueno R 2008 How an ensemble method can compute a comprehensible model *Proceedings of the Tenth international conference Data Warehousing and Knowledge Discovery (DaWaK-08)*. Springer.

Van Assche A and Blockeel H 2007 Seeing the forest through the trees: Learning a comprehensible model from an ensemble *Proceedings of the Eighteenth European Conference on Machine Learning (ECML-07)*. Springer.

Witten IH, Frank E and Hall MA 2011 *Data Mining: Practical Machine Learning Tools and Techniques* 3rd edn. Morgan Kaufmann.

Wolpert DH 1992 Stacked generalization. *Neural Networks* **5**, 241–260.

Wolpert DH and Macready WG 1999 An efficient method to estimate bagging's generalization error. *Machine Learning* **35**, 41–45.

Zhu J, Zou H, Rosset S and Hastie T 2009 Multi-class Adaboost. *Statistics and Its Interface* **2**, 349–360.

16

Kernel methods

16.1 Introduction

Kernel methods make it possible to overcome the linearity limitation of linear classification and linear regression models, as discussed in Chapters 5 and 8, respectively, so that they can be successfully applied to linearly inseparable concepts and nonlinear target functions. This is highly desirable due to the advantages of linear models: efficient parameter estimation algorithms and no risk of false local optima. They represent a specific, but particularly convenient and highly successful approach to enhancing input representation, i.e., transforming the original set of attributes into an enhanced set of attributes, with the hope that the originally nonlinear relationship to be modeled will become linear in the enhanced representation. Unlike other enhanced representation techniques, discussed in Section 8.6.2, kernel methods make it possible to use the enhanced representation implicitly, without actually transforming any instances and calculating new attribute values. They can only be combined, however, with linear modeling algorithms that have one important property: the capability to work without accessing data other than within dot products, both during model creation and prediction. Two noteworthy closely related algorithms that belong to this category and have proved practically successful are known as support vector machines (SVM) and support vector regression (SVR).

Even without the representation enhancement provided by kernel methods, the SVM and SVR algorithms have some advantages over ordinary linear classification and regression. They adopt alternative, more refined parameter estimation methods for linear threshold classification models and linear regression models than those presented in Sections 5.3 and 8.3. By departing from the simple error minimization objective assumed by the former they may achieve better generalization capabilities and overfitting resistance.

While the plain linear versions of support vector machines and support vector regression do not use kernel methods, they are most naturally presented in this chapter rather than in Chapters 5 and 8. However, we will refer to the basic principles of linear models and reuse the notational conventions introduced in these two chapters. For any instance x we will again

Data Mining Algorithms: Explained Using R, First Edition. Paweł Cichosz.
© 2015 John Wiley & Sons, Ltd. Published 2015 by John Wiley & Sons, Ltd.

write $\mathbf{a}(x)$ to refer to the vector of its all attribute values, i.e., $a_1(x), a_2(x), \ldots, a_n(x), a_{n+1}(x)$, where a_{n+1} is the fictitious constant 1 attribute used to conveniently handle the intercept term. Unlike in Chapters 5 and 8, it will be usually necessary to refer to the vectors of real attribute values only, with a_{n+1} skipped, which will be reflected by adding the $1 : n$ subscript in the notation, i.e., writing $\mathbf{a}_{1:n}(x)$ to designate the vector of $a_1(x), a_2(x), \ldots, a_n(x)$. Similarly, the symbol $\mathbf{w}$ will denote the vector of linear model parameters $w_1, w_2, \ldots, w_n, w_{n+1}$, i.e., with the intercept term w_{n+1} included. Whenever the latter has to be excluded, $\mathbf{w}_{1:n}$ will denote the vector of $w_1, w_2, \ldots, w_n$. Under these conventions, a linear representation function can be presented as

$$g(x) = \mathbf{w} \circ \mathbf{a}(x) = \mathbf{w}_{1:n} \circ \mathbf{a}_{1:n}(x) + w_{n+1} \tag{16.1}$$

where the $\circ$ symbol designates the dot product operator. Like in Chapter 5, it will often be convenient to exploit the numerical representation of class labels for two-class tasks. As always in this book, these are assumed to come from the $\{0, 1\}$ set. To exploit the possibility of simplifying some mathematical expressions, a transformed target concept defined as $c_-(x) = 2c(x) - 1$ will be used wich maps the original class labels to the $\{-1, 1\}$ set.

Example 16.1.1 Demonstrating the SVM and SVR algorithms in examples presented in this chapter will require the use of quadratic programming solvers available in the `quadprog` and `kernlab` packages, as well as an auxiliary function from the `Matrix` package. The `wireframe` function from the `lattice` package will be employed for producing surface plots illustrating kernel functions. Functions from other chapters and R utility functions provided by several DMR packages will also be used. The following R code sets up the environment for these demonstrations by loading the packages, as well as two datasets used for model creation and evaluation. These are the *Pima Indians Diabetes* dataset for classification and the *Boston Housing* dataset for regression. They are partitioned into training and test sets and standardized using the `std.all` and `predict.std` functions.

Ex. 17.3.1
dmr.trans

```
library(dmr.claseval)
library(dmr.linclas)
library(dmr.regeval)
library(dmr.util)

library(lattice)
library(quadprog)
library(kernlab)
library(Matrix)

data(PimaIndiansDiabetes, package="mlbench")
data(BostonHousing, package="mlbench")

set.seed(12)

rpid <- runif(nrow(PimaIndiansDiabetes))
pid.train <- PimaIndiansDiabetes[rpid>=0.33,]
pid.test <- PimaIndiansDiabetes[rpid<0.33,]
```

```
rbh <- runif(nrow(BostonHousing))
bh.train <- BostonHousing[rbh>=0.33,-4]
bh.test <- BostonHousing[rbh<0.33,-4]

pid.stdm <- std.all(diabetes~., pid.train)
pid.std.train <- predict.std(pid.stdm, pid.train)
pid.std.test <- predict.std(pid.stdm, pid.test)

bh.stdm <- std.all(medv~., bh.train)
bh.std.train <- predict.std(bh.stdm, bh.train)
bh.std.test <- predict.std(bh.stdm, bh.test)
```

Besides the two real datasets loaded above, a simple artificial four-attribute dataset for both classification and regression will be used, divided into training and test subsets. Some plots will be produced using an even simpler two-attribute dataset. The artificial data are generated by the R code presented below. The `ustep` function is used for the unit step calculation. Linearly separable data subsets are identified using the `linsep.sub` function.

dmr.util

Ex. 5.2.5.
dmr.linclas

```
set.seed(12)

  # dataset for plots
kmf.plot <- function(a1, a2) { 2*a1-3*a2+4 }
kmdat.plot <- 'names<-'(expand.grid(seq(1, 5, 0.05), seq(1, 5, 0.05)), c("a1", "a2"))
kmdat.plot$f <- kmf.plot(kmdat.plot$a1, kmdat.plot$a2)
kmdat.plot$c <- as.factor(ustep(kmdat.plot$f))

  # datasets for parameter estimation examples
kmg <- function(a1, a2, a3, a4) { a1^2+2*a2^2-a3^2-2*a4^2+2*a1-3*a2+2*a3-3*a4+1 }
kmf <- function(a1, a2, a3, a4) { 3*a1+4*a2-2*a3+2*a4-3 }
kmdat <- data.frame(a1=runif(400, min=1, max=5), a2=runif(400, min=1, max=5),
                    a3=runif(400, min=1, max=5), a4=runif(400, min=1, max=5))
kmdat$g <- kmg(kmdat$a1, kmdat$a2, kmdat$a3, kmdat$a4)
kmdat$c <- as.factor(ustep(kmdat$g))
kmdat$f <- kmf(kmdat$a1, kmdat$a2, kmdat$a3, kmdat$a4)

kmdat.train <- kmdat[1:200,]
kmdat.test <- kmdat[201:400,]

  # linearly separable training and test subsets
kmdat.ls <- linsep.sub(c~a1+a2+a3+a4, kmdat)
kmdat.train.ls <- kmdat[1:200,][kmdat.ls[1:200],]
kmdat.test.ls <- kmdat[201:400,][kmdat.ls[201:400],]
```

Notice that the artificial dataset is generated essentially the very same way as shown in Example 5.1.1 for linear classification illustrations, with just the inner representation function g additionally included as a nonlinear target function. The linear target function f is also added, which can be verified to be the same as the f1 target function from the linear regression artificial dataset generated in Example 8.1.1.

16.2 Support vector machines

Two approaches to parameter estimation for linear threshold classification models were presented in Section 5.3:

- the gradient descent algorithm with the delta rule for minimizing the distance of misclassified instances to the decision boundary,

- the least-squares algorithm for mean square error minimization with respect to $\{-1, 1\}$ target function values.

These are not the only possible ways of identifying a separating hyperplane for linear threshold models, though. One alternative that is particularly worthwhile to consider and that often yields models with better generalization properties is based on *classification margin maximization*.

16.2.1 Classification margin

As discussed in Section 5.3.3, the following quantity represents the signed distance between instance x and the separating hyperplane for a linear threshold model:

$$\delta_{\mathbf{w}}(x) = \frac{\mathbf{w} \circ \mathbf{a}(x)}{\|\mathbf{w}_{1:n}\|} \qquad (16.2)$$

with the sign indicating whether the point represented by $\mathbf{a}_{1:n}(x)$ lies on the positive or negative side of the hyperplane. The following slightly modified form of this distance, negated for instances of class 0 by introducing the $c_-(x)$ multiplier:

$$\gamma_{\mathbf{w}}(x) = c_-(x)\frac{\mathbf{w} \circ \mathbf{a}(x)}{\|\mathbf{w}_{1:n}\|} \qquad (16.3)$$

is called the *geometric margin* of the parameter vector $\mathbf{w}$ with respect to instance x. This is positive if x lies on the correct side of the decision hyperplane (is classified correctly), negative if it lies on the wrong side of the decision hyperplane, and 0 if it lies exactly on the decision hyperplane. With the normalizing denominator skipped, it is referred to as the *functional margin* of the parameter vector $\mathbf{w}$ with respect to instance x:

$$\hat{\gamma}_{\mathbf{w}}(x) = c_-(x)\mathbf{w} \circ \mathbf{a}(x) = \|\mathbf{w}_{1:n}\|\gamma_{\mathbf{w}}(x) \qquad (16.4)$$

The minimum values of the geometric and functional margins of the parameter vector $\mathbf{w}$ with respect to all training instances are called, respectively, the geometric and functional margins of $\mathbf{w}$ with respect to the training set:

$$\gamma_{\mathbf{w}}(T) = \min_{x \in T} \gamma_{\mathbf{w}}(x) \qquad (16.5)$$

$$\hat{\gamma}_{\mathbf{w}}(T) = \min_{x \in T} \hat{\gamma}_{\mathbf{w}}(x) \qquad (16.6)$$

If the hyperplane represented by $\mathbf{w}$ correctly separates all training instances of different classes, we have $\gamma_{\mathbf{w}}(T) > 0$ and $\hat{\gamma}_{\mathbf{w}}(T) > 0$.

Adding and subtracting a constant value $b > 0$ to the intercept term of the parameter vector **w** representing a separating hyperplane identifies two parallel hyperplanes (on the negative and positive side of the former) such that all correctly classified instances lying on these hyperplanes have functional margin of b. Indeed, for any instance lying on the hyperplane corresponding to the $\mathbf{w} \circ \mathbf{a}(x) + b = 0$ equation we have

$$\hat{\gamma}_\mathbf{w}(x) = c_-(x)\mathbf{w} \circ \mathbf{a}(x) = -c_-(x)b \qquad (16.7)$$

For $b > 0$ the $\mathbf{w} \circ \mathbf{a}(x) + b = 0$ equation therefore represents the hyperplane corresponding to functional margin b on the negative side of the separating hyperplane and the $\mathbf{w} \circ \mathbf{a}(x) - b = 0$ equation represents the hyperplane corresponding to functional margin b on the positive side of the separating hyperplane. These hyperplanes will be referred to as the (negative and positive) *margin-b hyperplanes* for **w** and, if $b = 1$ – simply as the *margin hyperplanes* for **w**. Instances lying on the margin hyperplanes will be shortly referred to as *lying on the margin*.

Example 16.2.1 A simple graphical illustration of the linear threshold classification margin is produced by the following R code. It defines the `fmarg` and `gmarg` functions for functional and geometric margin calculation. The latter uses the `l2norm` function to determine the norm of the parameter vector (with the intercept skipped). The `dmr.util` `plot.margin` function is also defined for plotting the separating line and the margin line in the two-dimensional case (i.e., for two-attribute datasets). The latter also returns the minimum functional and geometric margins. The `kmdat.m` dataset is created as a subset of randomly selected instances from `kmdat.plot` which are sufficiently distant from the hyperplane represented by the linear function used for class label assignment during data generation in Example 16.1.1. A parameter vector `w.m` is then chosen that actually represents a hyperplane correctly separating positive and negative instances, with an intercept term adjusted so that the minimum margin with respect to instances of each class is the same ("symmetric margin"). The parameter vector is subsequently scaled so that its functional margin with respect to the dataset is 1.

```
## functional margin of w with respect to instances from data
## using the cvec vector of {-1, 1} class labels
fmarg <- function(w, data, cvec)
{ cvec*predict.par(list(repf=repf.linear, w=w), data) }

## geometric margin of w with respect to instances from data
## using the cvec vector of {-1, 1} class labels
gmarg <- function(w, data, cvec) { fmarg(w, data, cvec)/l2norm(w[-length(w)]) }

## plot separating and b-margin lines for linear threshold classification
## with 2 attributes
plot.margin <- function(w, data, cvec, b=1, add=FALSE,
                        col.sep="black", col.pos="grey70", col.neg="grey30", ...)
```

```
{
  # y value corresponding to x on the regression line represented by w
  lry <- function(x, w) {sum(-w[c(1,3)]/w[2]*c(x, 1)) }

  if (!add)
  {
    plot(data[,1][cvec==1], data[,2][cvec==1], col=col.pos,
      xlab="a1", ylab="a2", xlim=range(data[,1]), ylim=range(data[,2]), ...)
    points(data[,1][cvec!=1], data[,2][cvec!=1], col=col.neg, ...)
  }

  lines(range(data[,1]), c(lry(min(data[,1]), w),
                           lry(max(data[,1]), w)), col=col.sep, ...)
  lines(range(data[,1]), c(lry(min(data[,1]), w-c(0, 0, b)),
                           lry(max(data[,1]), w-c(0, 0, b))), col=col.pos, ...)
  lines(range(data[,1]), c(lry(min(data[,1]), w+c(0, 0, b)),
                           lry(max(data[,1]), w+c(0, 0, b))), col=col.neg, ...)
  list(fmargin=min(fmarg(w, data, cvec)), gmargin=min(gmarg(w, data, cvec)))
}

  # dataset for margin illustration (skip near-boundary instances from kmdat.plot)
kmdat.m <- kmdat.plot[abs(kmdat.plot$f)>2,c("a1", "a2", "c")]
kmdat.m <- kmdat.m[sample(nrow(kmdat.m), 100),]

  # parameter vector for margin demonstration
w.m <- c(1, -2)
  # predictions with intercept 0
p0.m <- predict.par(list(repf=repf.linear, w=c(w.m, 0)), kmdat.m[,1:2])
  # symmetric-margin intercept
w.m <- c(w.m, -(max(p0.m[kmdat.m$c==0])+min(p0.m[kmdat.m$c==1]))/2)

  # minimum functional margin
min(fmarg(w.m, kmdat.m[,1:2], 2*as.num0(kmdat.m$c)-1))
  # minimum geometric
min(gmarg(w.m, kmdat.m[,1:2], 2*as.num0(kmdat.m$c)-1))

  # scale parameters to get minimum functional margin of 1
w.m <- w.m/min(fmarg(w.m, kmdat.m[,1:2], 2*as.num0(kmdat.m$c)-1))
  # minimum functional margin after parameter scaling (1)
min(fmarg(w.m, kmdat.m[,1:2], 2*as.num0(kmdat.m$c)-1))
  # minimum geometric margin after parameter scaling (unchanged)
min(gmarg(w.m, kmdat.m[,1:2], 2*as.num0(kmdat.m$c)-1))

plot.margin(w.m, kmdat.m[,1:2], 2*as.num0(kmdat.m$c)-1)
```

The separating line (in black) and the positive and negative (in lighter and darker gray, respectively) margin 1-lines corresponding to w.m are then plotted, with the obtained plot presented in Figure 16.1. Since the parameter vector was set to ensure that the minimum functional margin with respect to positive and negative instances is the same and equal to 1, the distance between the separating line and either of the margin lines is the geometric margin of the parameter vector.

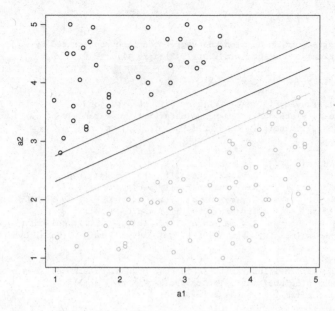

Figure 16.1   Linear classification margin.

16.2.2   Maximum-margin hyperplane

Selecting the maximum-margin hyperplane (out of all separating hyperplanes) for a target concept linearly separable on the training set may be expected to increase the resistance to overfitting and therefore improve the true classification performance. This is because a decision boundary that is as far as possible from training instances of different classes is more likely to correctly separate new instances during prediction. More confident predictions are possible if instances are distant from the decision boundary. The *support vector machines* (SVM) algorithm is based on this very idea of margin maximization for linear threshold classification.

More precisely, it is the geometric margin of the parameter vector with respect to the training set the maximization of which would be desirable. This is because the functional margin can be made arbitrarily large by simple parameter scaling. Indeed it can be easily verified that

$$\hat{\gamma}_{\tau\mathbf{w}}(x) = c_-(x)\tau\mathbf{w} \circ \mathbf{a}(x) = \tau\hat{\gamma}_{\mathbf{w}}(x) \tag{16.8}$$

for any instance x and $\tau > 0$, while the separating hyperplanes represented by $\mathbf{w}$ and $\tau\mathbf{w}$ are the same.

16.2.3   Primal form

Unfortunately, the geometric margin is hard to maximize directly. Assuming the linear separability of the target concept on the training set, which guarantees the existence of a parameter vector with a positive margin value, a much more tractable maximization problem is obtained by restricting one's interest to parameter vectors in the *canonical form* that satisfy

$$\hat{\gamma}_{\mathbf{w}}(T) = 1 \tag{16.9}$$

This is possible with no loss of generality since the geometric margin can be scaled up or down arbitrarily without affecting the maximum-margin hyperplane. With this restriction we have

$$\gamma_{\mathbf{w}}(T) = \frac{\hat{\gamma}_{\mathbf{w}}(T)}{\|\mathbf{w}_{1:n}\|} = \frac{1}{\|\mathbf{w}_{1:n}\|} \tag{16.10}$$

and thus the maximization of the geometric margin $\gamma_{\mathbf{w}}(T)$ becomes equivalent to maximizing $\frac{1}{\|\mathbf{w}_{1:n}\|}$. The latter can be replaced by the minimization of $\|\mathbf{w}_{1:n}\|^2$ (with the square introduced to get rid of the square root in the parameter vector norm), subject to the $\hat{\gamma}_{\mathbf{w}}(T) = 1$ restriction, which can be written as

minimize

$$\frac{1}{2}\|\mathbf{w}_{1:n}\|^2 \tag{16.11}$$

subject to

$$(\forall x \in T) \; c_-(x)\mathbf{w} \circ \mathbf{a}(x) \geq 1 \tag{16.12}$$

with the $\frac{1}{2}$ multiplier introduced for mathematical convenience only. This is an instantiation of the standard quadratic programming problem that is referred to as the *primal form* of margin maximization. It can be solved by readily available quadratic programming algorithms. Clearly no solution satisfying the constraints exists unless the target concept is indeed linearly separable on the training set.

Note that $\hat{\gamma}_{\mathbf{w}}(T) = 1$ restriction is represented by a set of inequality constraints for each training instance. Instances for which the functional margin is equal to 1 are called *support vectors* and are said to *lie on the margin* whereas the remaining instances, with functional margin greater than 1, are said to *lie outside the margin*. Also note that the intercept term w_{n+1} does not occur in the minimization objective and is only included in the constraints. It remains, nevertheless, an element of the solution vector that is searched for by quadratic programming.

Example 16.2.2 The following R code demonstrates parameter estimation for the linear SVM model by solving the primal form of the margin maximization problem. Two popular quadratic programming solvers available in R, `solve.QP` from the `quadprog` package and `ipop` from the `kernlab` package, may be used for this purpose. The former requires that its `Dmat` argument, used to specify the maximization objective, is positive definite, which is enforced using the `nearPD` function from the `Matrix` package. The latter assumes a box constraints specification (two-sided inequalities) the boundaries of which have to be finite. The `inf` argument to the `svm.linear.prim` function specifies a finite number sufficiently large not to constrain the solution and sufficiently small not to cause numerical problems. The `svthres` argument is the tolerance threshold used when identifying support vectors, i.e., checking which instances lie on the margin. The function is demonstrated using the linearly separable subset of the example artificial data. The `err` function is used for misclassification error calculation.

Ex. 7.2.5
dmr.claseval

```
## linear SVM parameter estimation using primal-form quadratic programming
## solvers: "solve.QP" or "ipop"
svm.linear.prim <- function(formula, data, svthres=1e-9, inf=1e3, solver="solve.QP")
{
  class <- y.var(formula)
  attributes <- x.vars(formula, data)
  aind <- names(data) %in% attributes

  cvec <- 2*as.num0(data[[class]])-1  # class vector using {-1, 1} labels
  amat <- cbind(as.matrix(data[,aind]), intercept=1)  # attribute value matrix

  if (solver=="solve.QP")
    args <- list(Dmat=nearPD(rbind(cbind(diag(sum(aind)), 0), 0))$mat,
                 dvec=rep(0, sum(aind)+1),
                 Amat=t(rmm(amat, cvec)),
                 bvec=rep(1, nrow(data)))
  else if (solver=="ipop")
    args <- list(c=rep(0, sum(aind)+1),
                 H=rbind(cbind(diag(sum(aind)), 0), 0),
                 A=rmm(amat, cvec),
                 b=rep(1, nrow(data)),
                 l=rep(-inf, sum(aind)+1),
                 u=rep(inf, sum(aind)+1),
                 r=rep(inf, nrow(data)))
  else stop("Unknown solver: ", solver)

  qp <- do.call(solver, args)
  w <- if (solver=="solve.QP") qp$solution else if (solver=="ipop") qp@primal
  sv <- unname(which(cvec*predict.par(list(repf=repf.linear, w=w),
                                      data[,aind,drop=FALSE])<=1+svthres))
  list(model='class<-'(list(repf=repf.threshold(repf.linear), w=w), "par"), sv=sv)
}

  # estimate linear SVM model parameters
svm.p.ls <- svm.linear.prim(c~a1+a2+a3+a4, kmdat.train.ls)

  # misclassification error
err(predict(svm.p.ls$model, kmdat.train.ls[,1:4]), kmdat.train.ls$c)
err(predict(svm.p.ls$model, kmdat.test.ls[,1:4]), kmdat.test.ls$c)
```

The two solver functions, `solve.QP` and `ipop`, assume different somewhat representations of the quadratic programming problem to be solved and require different parameters,
accordingly. While this can be easily understood by comparing the parameter setup in the
above code with the documentation of these two functions, the following hints may be useful.

For the `solve.QP` function:

1. `Dmat` is the matrix specifying the coefficients of the quadratic term of the optimization objective – here an $(n + 1) \times (n + 1)$ matrix with 1's on the first n elements of
   the main diagonal (corresponding to $w_1, w_2, \ldots, w_n$) and 0's elsewhere, since the
   quadratic term is the norm of $\mathbf{w}_{1:n}$ only,

2. `dvec` is the vector specifying the coefficients of the linear term of the optimization
   objective – here an $(n + 1)$-element vector of 0's, since there is no linear term,

3. `Amat` is the (transpose of) the matrix specifying the coefficients of the inequal-
   ity constraints – here a $(n + 1) \times |T|$ matrix, containing columns $c_-(x)\mathbf{a}(x)$ for each
   $x \in T$,

4. `bvec` is the vector specifying the values for the *greater than* inequality constraints represented by `Amat` – here a $|T|$-element vector of 1's.

For the `ipop` function:

1. `c` is the vector specifying the coefficients of the linear term of the optimization objective – here an $(n + 1)$-element vector of 0's (the same as `dvec` for `solve.QP`),

2. `H` is the matrix specifying the coefficients of the quadratic term of the optimization objective – here an $(n + 1) \times (n + 1)$ matrix with 1's on the first n elements of the main diagonal (corresponding to $w_1, w_2, \ldots, w_n$) and 0's elsewhere (the same as `Dmat` for `solve.QP`),

3. `A` is the matrix specifying the coefficients of the box constraints – here a $|T| \times (n + 1)$ matrix, containing rows $c_-(x)\mathbf{a}(x)$ for each $x \in T$ (the same as the transpose of `Amat` for `solve.QP`),

4. `b` is the vector specifying the values for the *greater than* inequality constraints represented by `A` – here a $|T|$-element vector of 1's (the same as `bvec` for `solve.QP`),

5. `l` is the vector specifying the lower bounds for the solution – here not needed and hence set to an $(n + 1)$-element vector of `-inf`,

6. `u` is the vector specifying the upper bounds for the solution – here not needed and hence set to an $(n + 1)$-element vector of `inf`,

7. `r` is the vector that, added to `b`, specifies the values for the *less than* inequality constraints represented by `A` – here not needed and hence set to a $|T|$-element vector of `inf`.

Note that while the `c`, `H`, `A`, and `b` parameters of the `ipop` function have their counterparts for the `solve.QP` function, the remaining `r`, `l`, and `u` parameters are unique to the former. This is because it a adopts different constraint representation, including both box constraints for linear combinations of the solution vector, obtained by multiplying it by the constraint coefficient matrix `A` (greater than `b`, less than `b+r`), and simple per-parameter box constraints (greater than `l`, less than `u`).

Example 16.2.3 The following R code uses the implementation of SVM from the previous example to graphically illustrate the maximum geometric margin. Following the pattern of the margin illustration presented in Example 16.2.1, the separating line as well as the positive and negative margin lines are plotted for the maximum-margin parameter vector found by SVM for the `kmdat.m` margin illustration dataset. The separating line and the margin lines corresponding to the previously used suboptimal parameter vector `w.m` are also plotted for comparison. The result is presented in Figure 16.2. Clearly the SVM algorithm successfully maximized the geometric margin while keeping the functional margin at 1.

```
 # hard-margin SVM
svm.mh <- svm.linear.prim(c~., kmdat.m, solver="ipop")

 # optimal separating and margin lines
plot.margin(svm.mh$model$w, kmdat.m[,1:2], 2*as.num0(kmdat.m$c)-1)
```

```
# suboptimal separating and margin lines for comparison
plot.margin(w.m, kmdat.m[,1:2], 2*as.num0(kmdat.m$c)-1, add=TRUE, lty=3)
```

16.2.4 Dual form

One technique that can be used to handle constraints in optimization tasks is that of Lagrange multipliers. Using this technique makes it possible to transform the problem presented above to the following equivalent alternative formulation, called the *dual form* of the margin maximization problem:

maximize

$$-\frac{1}{2}\sum_{x_1 \in T}\sum_{x_2 \in T} c_-(x_1)c_-(x_2)\alpha_{x_1}\alpha_{x_2}\mathbf{a}_{1:n}(x_1) \circ \mathbf{a}_{1:n}(x_2) + \sum_{x \in T}\alpha_x \qquad (16.13)$$

subject to

$$\sum_{x \in T} c_-(x)\alpha_x = 0 \qquad (16.14)$$

$$(\forall x \in T)\quad \alpha_x \geq 0 \qquad (16.15)$$

This is again a quadratic programming problem, the solution thereof delivers Lagrange multipliers α_x for all $x \in T$. The solution can be obtained using general-purpose quadratic programming solvers, but some more efficient dedicated algorithms, exploiting the specific

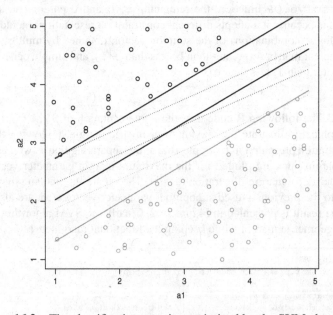

Figure 16.2   The classification margin maximized by the SVM algorithm.

properties of the problem, also exist that make it possible to use larger datasets. These are not discussed here.

The following relationship between model parameters and Lagrange multipliers:

$$\mathbf{w}_{1:n} = \sum_{x \in T} c_-(x)\alpha_x \mathbf{a}_{1:n}(x) \tag{16.16}$$

makes it possible to obtain the solution of the primal form, i.e., the model parameter vector being searched for. This reveals that model parameters can be viewed as weighted averages of training instance attribute values, and only instances with nonzero α values actually matter. From the Lagrangian used to derive the dual form problem, not presented here, it follows that nonzero α values correspond to those constraints of the primal form for which equality occurred, i.e.,

$$\alpha_x > 0 \quad \equiv \quad c_-(x)\mathbf{w} \circ \mathbf{a}(x) = 1 \tag{16.17}$$

Support vectors – as instances with a functional margin of 1 – are therefore also identified by nonzero α values and can be seen to be the only training instances with impact on model parameters. Of course, strictly comparing Lagrange multipliers with 0 is inappropriate in an actual implementation for numerical reasons.

Notice that the relationship between model parameters and Lagrange multipliers does not hold for the intercept term. It can be obtained, however, by plugging $\mathbf{w}_{1:n}$ to the constraints of the primal form. For any instance x_s with functional margin equal to 1, i.e., with a nonzero α value (a support vector), we have

$$c_-(x_s)\mathbf{w} \circ \mathbf{a}(x_s) = c_-(x_s)\mathbf{w}_{1:n} \circ \mathbf{a}_{1:n}(x_s) + c_-(x_s)w_{n+1} = 1 \tag{16.18}$$

This yields

$$w_{n+1} = \frac{1}{c_-(x_s)} - \mathbf{w}_{1:n} \circ \mathbf{a}_{1:n}(x_s) = c_-(x_s) - \mathbf{w}_{1:n} \circ \mathbf{a}_{1:n}(x_s) \tag{16.19}$$

$$= c_-(x_s) - \sum_{x \in T} c_-(x)\alpha_x \mathbf{a}_{1:n}(x) \circ \mathbf{a}_{1:n}(x_s) \tag{16.20}$$

In practice, it may be reasonable to use the instance with the maximum Lagrange multiplier or average the result over multiple support vectors to increase resistance to numerical inaccuracy. In particular, we might choose to average over the two instances of different classes closest to the decision boundary:

$$w_{n+1} = \frac{(1 - \min_{x_1 \in T^1} \mathbf{w}_{1:n} \circ \mathbf{a}_{1:n}(x_1)) + (-1 - \max_{x_0 \in T^0} \mathbf{w}_{1:n} \circ \mathbf{a}_{1:n}(x_0))}{2} \tag{16.21}$$

$$= -\frac{\min_{x_1 \in T^1} \mathbf{w}_{1:n} \circ \mathbf{a}_{1:n}(x_1) + \max_{x_0 \in T^0} \mathbf{w}_{1:n} \circ \mathbf{a}_{1:n}(x_0)}{2} \tag{16.22}$$

$$= -\frac{1}{2}\left(\min_{x_1 \in T^1} \sum_{x \in T} c_-(x)\alpha_x \mathbf{a}_{1:n}(x) \circ \mathbf{a}_{1:n}(x_1) \right.$$

$$\left. + \max_{x \in T^0} \sum_{x \in T} c_-(x)\alpha_x \mathbf{a}_{1:n}(x) \circ \mathbf{a}_{1:n}(x_0) \right) \tag{16.23}$$

Interestingly, all attribute values used in the optimization objective given by Equation 16.13 appear inside dot products. Expressing model predictions in terms of Lagrange multipliers

$$h(x_*) = \sum_{x \in T} c_-(x)\alpha_x \mathbf{a}_{1:n}(x) \circ \mathbf{a}_{1:n}(x_*) + c_-(x_s) - \sum_{x \in T} c_-(x)\alpha_x \mathbf{a}_{1:n}(x) \circ \mathbf{a}_{1:n}(x_s) \qquad (16.24)$$

reveals another noteworthy observation that all attribute value vectors used for prediction appear inside dot products as well, and there is no need to explicitly calculate the primal form parameter vector $\mathbf{w}$ at all. Using the dual form, both parameter estimation and prediction for SVM models can therefore be performed without using any attribute values other than inside dot products. This also makes it possible not to explicitly calculate model parameters and just use Lagrange multipliers.

Example 16.2.4 Linear SVM parameter estimation based on the dual form representation of the margin-maximization problem is demonstrated by the R code presented below. It uses the same quadratic programming solvers as in Example 16.2.2, but with appropriately changed input parameters. It then transforms the dual form solution (Lagrange multipliers) to model parameters. The previous demonstrations are repeated and yield approximately same results.

```
## linear SVM parameter estimation using dual-form quadratic programming
## solvers: "solve.QP" or "ipop"
svm.linear.dual <- function(formula, data, svthres=1e-3, inf=1e3, solver="solve.QP")
{
  class <- y.var(formula)
  attributes <- x.vars(formula, data)
  aind <- names(data) %in% attributes

  cvec <- 2*as.num0(data[[class]])-1  # class vector using {-1, 1} labels
  ccmat <- outer(cvec, cvec)          # class-class product matrix
  amat <- as.matrix(data[,aind])      # attribute value matrix
  dpmat <- amat%*%t(amat)             # dot product matrix

  if (solver=="solve.QP")
    args <- list(Dmat=nearPD(dpmat*ccmat)$mat,
                 dvec=rep(1, nrow(data)),
                 Amat=matrix(c(cvec, diag(1, nrow(data))), nrow=nrow(data)),
                 bvec=rep(0, nrow(data)+1),
                 meq=1)
  else if (solver=="ipop")
    args <- list(c=rep(-1, nrow(data)),
                 H=dpmat*ccmat,
                 A=cvec,
                 b=0,
                 l=rep(0, nrow(data)),
                 u=rep(inf, nrow(data)),
                 r=0)
  else
    stop("Unknown solver: ", solver)

  qp <- do.call(solver, args)
  alpha <- if (solver=="solve.QP") qp$solution else if (solver=="ipop") qp@primal
  sv <- which(alpha>svthres)
  w <- c(colSums(rmm(amat[sv,], cvec[sv]*alpha[sv])))  # no intercept yet
```

```
p0 <- predict.par(list(repf=repf.linear, w=c(w, 0)), data[,aind,drop=FALSE])
w <- c(w, intercept=-(max(p0[cvec==-1])+min(p0[cvec==1]))/2)
list(model='class<-'(list(repf=repf.threshold(repf.linear), w=w), "par"), sv=sv)
}

# estimate linear SVM model parameters
svm.d.ls <- svm.linear.dual(c~a1+a2+a3+a4, kmdat.train.ls)

# misclassification error
err(predict(svm.d.ls$model, kmdat.train.ls[,1:4]), kmdat.train.ls$c)
err(predict(svm.d.ls$model, kmdat.test.ls[,1:4]), kmdat.test.ls$c)
```

As before, it may be helpful to take a closer look at the parameters passed to the two quadratic programming solvers.

For the `solve.QP` function:

1. `Dmat` is the matrix specifying the coefficients of the quadratic term of the optimization objective – here an $|T| \times |T|$ matrix that for all $x_1, x_2 \in T$ contains $c_-(x_1)c_-(x_2)\mathbf{a}_{1:n}(x_1) \circ \mathbf{a}_{1:n}(x_2)$ in the cell corresponding to x_1 and x_2,

2. `dvec` is the vector specifying the coefficients of the linear term of the optimization objective – here an $|T|$-element vector of 1's,

3. `Amat` is the (transpose of) the matrix specifying the coefficients of the equality and inequality constraints – here a $|T| \times (|T| + 1)$ matrix obtained by prepending a vector of $c_-(x)$ for each $x \in T$ as the first column (representing the coefficients of the equality constraint) to the $|T| \times |T|$ identity matrix (representing the coefficients of the inequality constraints),

4. `bvec` is the vector specifying the values for equality and *greater than* inequality constraints represented by `Amat` – here a $(|T| + 1)$-element vector of 0's,

5. `meq` is the number of the first columns in `Amat` that represent equality constraints – here set to 1.

For the `ipop` function:

1. `c` is the vector specifying the coefficients of the linear term of the optimization objective – here a $|T|$-element vector of -1's (the same as negated `dvec` for `solve.QP`),

2. `H` is the matrix specifying the coefficients of the quadratic term of the optimization objective – here an $|T| \times |T|$ matrix that for all $x_1, x_2 \in T$ contains $c_-(x_1)c_-(x_2)\mathbf{a}_{1:n}(x_1) \circ \mathbf{a}_{1:n}(x_2)$ in the cell corresponding to x_1 and x_2 (the same as `Dmat` for `solve.QP`),

3. `A` is the matrix specifying the coefficients of the box constraints – here a $|T|$-element vector of $c_-(x)$ for each $x \in T$, representing the coefficients of the single equality constraint (the same as the first column of `Amat` for `solve.QP`),

4. `b` is the vector specifying the values for the *greater than* inequality constraints represented by `A` – here a single 0 (the same as the first element of `bvec` for `solve.QP`),

5. l is the vector specifying the lower bounds for the solution – here a $|T|$-element vector of 0's (the same as the remaining elements of bvec for solve.QP),

6. u is the vector specifying the upper bounds for the solution – here not needed and hence set to an $|T|$-element vector of inf,

7. r is the vector that, added to b, specifies the values for the *less than* inequality constraints represented by A – here set to 0, to make the constraint actually an equality constraint.

Notice that while the c and H parameters of the ipop function – used to specify the optimization objective – directly correspond to the bvec and Dmat parameters of the solve.QP function, it is not the case for the remaining parameters, used to specify the constraints. For solve.QP both the equality and inequality constraints are specified using Amat (coefficients) and bvec (values). The meq=1 argument instructs the function to treat the first constraint as an equality constraint and the remaining ones as inequality constraints. For ipop the equality constraint is represented by A and b (with r=0) and the inequality constraints – by l (with u=inf).

16.2.5   Soft margin

As presented above, SVM heavily relies on the linear separability assumption, which is necessary to make the primal form inequality constraints, representing the $\hat{\gamma}_w(T) = 1$ restriction, satisfiable. This version of the algorithm is referred to as *hard-margin* SVM. It is highly desirable in practical classification tasks, however, to produce reasonable quality linear models even if the target concept is not linearly separable on the training set. To make it possible with the maximum margin approach, the underlying margin maximization problem has to be relaxed.

The most popular modified version of SVM capable of identifying parameter vectors that fail to correctly separate all training instances of different classes is *soft-margin* SVM. It is based on replacing the original primal form maximization problem's inequality constraints, that may be impossible to satisfy, by their relaxed counterparts:

$$(\forall x \in T) \quad c_-(x)\mathbf{w} \circ \mathbf{a}(x) \geq 1 - \xi_x \tag{16.25}$$

$$(\forall x \in T) \quad \xi_x \geq 0 \tag{16.26}$$

where ξ_x is the so-called *slack variable* for instance x, specifying the permitted difference between the functional margin of $\mathbf{w}$ with respect to x and 1. It is therefore no longer guaranteed that $\hat{\gamma}_w(x) \geq 1$ – we may also have $\hat{\gamma}_w(x) < 1$ or even $\hat{\gamma}_w(x) < 0$, i.e., some instances may lie *within the margin* (rather than on or outside the margin) or even be misclassified.

It is of course desirable to keep the level of slackness relatively low. This requires incorporating slack variables to the minimization criterion. The primal form soft-margin problem can be then formulated as follows:

maximize

$$\frac{1}{2}\|\mathbf{w}_{1:n}\|^2 + \Gamma \sum_x \xi_x \tag{16.27}$$

subject to

$$(\forall x \in T) \quad c_-(x)\mathbf{w} \circ \mathbf{a}(x) \geq 1 - \xi_x \tag{16.28}$$

$$(\forall x \in T) \quad \xi_x \geq 0 \tag{16.29}$$

where $\Gamma > 0$ (more commonly designated by C, but the latter is reserved to denote the set of classes in this book's other chapters on classification) is a cost parameter that controls the tradeoff between margin maximization and slackness minimization, i.e., trying to get a possibly large geometric margin and possibly few misclassified instances (or more precisely, instances with insufficient classification margin).

For small Γ violating the functional margin constraint costs little and geometric margin maximization is nearly unconstrained. This may yield a parameter vector with a large geometric margin with respect to correctly classified instances, but with several instances misclassified. Large Γ assigns high margin violation cost and may result in a parameter vector with a small geometric margin, but no or few instances misclassified.

Soft-margin SVM has a particularly simple and elegant dual form representation. After applying the Lagrangian transformation, it turns out that the minimization criterion remains unchanged and single-sided inequality constrains for Lagrange multipliers are just replaced by two-sided (box) inequality constraints:

maximize

$$-\frac{1}{2} \sum_{x_1 \in T} \sum_{x_2 \in T} c_-(x_1)c_-(x_2)\alpha_{x_1}\alpha_{x_2}\mathbf{a}_{1:n}(x_1) \circ \mathbf{a}_{1:n}(x_2) + \sum_{x \in T} \alpha_x \qquad (16.30)$$

subject to

$$\sum_{x \in T} c_-(x)\alpha_x = 0 \qquad (16.31)$$

$$(\forall x \in T) \quad 0 \leq \alpha_x \leq \Gamma \qquad (16.32)$$

As before, support vectors are identified as instances with positive α values. These are no longer only instances lying exactly on the margin, though. The set of support vectors is larger and also includes instances lying within the margin or on the wrong side of the hyperplane. More specifically, we have

$$c_-(x)\mathbf{w} \circ \mathbf{a}(x) = 1 \quad \equiv \quad 0 < \alpha_x < \Gamma \qquad (16.33)$$

$$c_-(x)\mathbf{w} \circ \mathbf{a}(x) > 1 \quad \equiv \quad \alpha_x = 0 \qquad (16.34)$$

$$c_-(x)\mathbf{w} \circ \mathbf{a}(x) < 1 \quad \equiv \quad \alpha_x = \Gamma \qquad (16.35)$$

Of course, testing for exact equalities is inappropriate in an actual implementation for numerical reasons.

Support vectors lying on the margin (i.e., those with Lagrange multipliers positive but less than Γ) can be used to calculate the intercept term in the same way as presented for the hard-margin case, i.e., by applying Equation 16.20 for any x_s such that $0 < \alpha_{x_s} < \Gamma$. For greater reliability, this can be averaged over multiple instances satisfying the condition or one can choose an instance with an α value possibly far from both the lower and upper bound (i.e., near $\frac{\Gamma}{2}$). Notice, however, that the approach based on averaging over the instances closest to the decision boundary, presented for the hard-margin case, is not valid here, as these instances may lie within the margin rather than on the margin.

To reasonably compare the geometric margin of parameter vectors obtained using soft-margin and hard-margin SVM, we can define the *soft geometric margin* of the parameter vector $\mathbf{w}$ with respect to the training set as follows:

$$\tilde{\gamma}_{\mathbf{w}}(T) = \frac{1}{\|\mathbf{w}_{1:n}\|} \tag{16.36}$$

This can be seen as the geometric margin value corresponding to the functional margin of 1, i.e., the geometric margin of instances that "lie on the margin." It does not actually depend directly on T and therefore makes sense only if $\mathbf{w}$ satisfies the soft-margin constraints on T, i.e., if $\mathbf{w}$ is actually a valid SVM parameter vector for T. Increasing the cost parameter tends to decrease the soft geometric margin.

It is worthwhile to underline that soft-margin SVM may be a good choice even for linearly separable data. Not being forced to achieve perfect training set accuracy, it may be able to find better generalizing separating hyperplanes. This increases the algorithm's resistance to overfitting. Unfortunately the cost parameter usually requires some careful tuning and there is no reasonable data-independent default value.

Example 16.2.5 The following R code defines the `svm.linear` function, implementing the soft-margin linear SVM algorithm using dual-form quadratic programming. The `cost` parameter corresponds to Γ in the above equations. The algorithm is demonstrated using both the full artificial dataset and its linearly separable subset, as well as the real *Pima Indians Diabetes* data.

```
## linear soft-margin SVM parameter estimation using quadratic programming
## solvers: "solve.QP" or "ipop"
svm.linear <- function(formula, data, cost=1, svthres=1e-3, solver="solve.QP")
{
  class <- y.var(formula)
  attributes <- x.vars(formula, data)
  aind <- names(data) %in% attributes

  cvec <- 2*as.num0(data[[class]])-1  # class vector using {-1, 1} labels
  ccmat <- outer(cvec, cvec)          # class-class product matrix
  amat <- as.matrix(data[,aind])      # attribute value matrix
  dpmat <- amat%*%t(amat)             # dot product matrix

  if (solver=="solve.QP")
    args <- list(Dmat=nearPD(dpmat*ccmat)$mat,
                 dvec=rep(1, nrow(data)),
                 Amat=matrix(c(cvec, diag(1, nrow(data)), diag(-1, nrow(data))),
                             nrow=nrow(data)),
                 bvec=c(0, rep(0, nrow(data)), rep(-cost, nrow(data))),
                 meq=1)
  else if (solver=="ipop")
    args <- list(c=rep(-1, nrow(data)),
                 H=dpmat*ccmat,
                 A=cvec,
                 b=0,
                 l=rep(0, nrow(data)),
                 u=rep(cost, nrow(data)),
                 r=0)
  else
    stop("Unknown solver: ", solver)

  qp <- do.call(solver, args)
  alpha <- if (solver=="solve.QP") qp$solution else if (solver=="ipop") qp@primal
```

```
sv <- which(alpha>svthres)
w <- c(colSums(rmm(amat[sv,], cvec[sv]*alpha[sv])))   # no intercept yet
i <- which.min(abs(alpha-cost/2))
w <- c(w, intercept=cvec[i]-unname(predict.par(list(repf=repf.linear, w=c(w, 0)),
                                        data[i,aind,drop=FALSE])))
list(model='class<-'(list(repf=repf.threshold(repf.linear), w=w), "par"), sv=sv)
}


# linear SVM for the artificial data
svm.s <- svm.linear(c~a1+a2+a3+a4, kmdat.train)
svm.s.ls <- svm.linear(c~a1+a2+a3+a4, kmdat.train.ls)

svm.s.01 <- svm.linear(c~a1+a2+a3+a4, kmdat.train, cost=0.1)
svm.s.ls.01 <- svm.linear(c~a1+a2+a3+a4, kmdat.train.ls, cost=0.1)

svm.s.10 <- svm.linear(c~a1+a2+a3+a4, kmdat.train, cost=10)
svm.s.ls.10 <- svm.linear(c~a1+a2+a3+a4, kmdat.train.ls, cost=10)

# linear SVM for the Pima Indians Diabetes data
pid.svm.s <- svm.linear(diabetes~., pid.std.train)
pid.svm.s.01 <- svm.linear(diabetes~., pid.std.train, cost=0.1)
pid.svm.s.10 <- svm.linear(diabetes~., pid.std.train, cost=10)

# training set misclassification error
err(predict(svm.s$model, kmdat.train[,1:4]), kmdat.train$c)
err(predict(svm.s.01$model, kmdat.train[,1:4]), kmdat.train$c)
err(predict(svm.s.10$model, kmdat.train[,1:4]), kmdat.train$c)

err(predict(svm.s.ls$model, kmdat.train.ls[,1:4]), kmdat.train.ls$c)
err(predict(svm.s.ls.01$model, kmdat.train.ls[,1:4]), kmdat.train.ls$c)
err(predict(svm.s.ls.10$model, kmdat.train.ls[,1:4]), kmdat.train.ls$c)

err(factor(predict(pid.svm.s$model, pid.std.train[,-9]),
           levels=0:1, labels=levels(pid.std.train$diabetes)),
    pid.std.train$diabetes)
err(factor(predict(pid.svm.s.01$model, pid.std.train[,-9]),
           levels=0:1, labels=levels(pid.std.train$diabetes)),
    pid.std.train$diabetes)
err(factor(predict(pid.svm.s.10$model, pid.std.train[,-9]),
           levels=0:1, labels=levels(pid.std.train$diabetes)),
    pid.std.train$diabetes)


# test set misclassification error
err(predict(svm.s$model, kmdat.test[,1:4]), kmdat.test$c)
err(predict(svm.s.01$model, kmdat.test[,1:4]), kmdat.test$c)
err(predict(svm.s.10$model, kmdat.test[,1:4]), kmdat.test$c)

err(predict(svm.s.ls$model, kmdat.test.ls[,1:4]), kmdat.test.ls$c)
err(predict(svm.s.ls.01$model, kmdat.test.ls[,1:4]), kmdat.test.ls$c)
err(predict(svm.s.ls.10$model, kmdat.test.ls[,1:4]), kmdat.test.ls$c)

err(factor(predict(pid.svm.s$model, pid.std.test[,-9]),
           levels=0:1, labels=levels(pid.std.train$diabetes)),
    pid.test$diabetes)
err(factor(predict(pid.svm.s.01$model, pid.std.test[,-9]),
           levels=0:1, labels=levels(pid.std.train$diabetes)),
    pid.test$diabetes)
err(factor(predict(pid.svm.s.10$model, pid.std.test[,-9]),
           levels=0:1, labels=levels(pid.std.train$diabetes)),
    pid.test$diabetes)
```

Notice that the new *less than* inequality constraints are specified for the solve.QP function by additional $|T|$ columns in Amat and additional $|T|$ elements in bvec. Since solve.QP only supports "greater than" inequality constraints, the corresponding new constraint coefficients and values are negated. For ipop this is achieved in a much more straightforward way using the u parameter, set to a $|T|$-element vector of cost.

As expected, the soft SVM algorithm no longer achieves a 0 training set error for the linearly separable artificial data subset (although it could be forced to do so by specifying a sufficiently large cost value). It delivers reasonable models for both the full artificial dataset and the real *Pima Indians Diabetes* data, though.

Example 16.2.6 The R code presented below uses the soft-margin SVM implementation from the previous example to graphically illustrate soft margin maximization. Using the same kmdat.m dataset used before for margin illustrations in Example 16.2.1, the separating line and the margin lines obtained for two values of the cost parameter are plotted. For $\Gamma = 1$ (solid lines) the results are nearly the same as in the hard-margin case presented before. For $\Gamma = 0.1$ (dotted lines), we obtain a substantially larger soft geometric margin, with several instances lying within the margin. The same demonstration is repeated using a modified linearly inseparable version of the data, with class labels of some instances flipped. The margin plots are presented in Figure 16.3.

```
# soft-margin SVM
svm.ms.1 <- svm.linear(c~., kmdat.m, solver="ipop", cost=1)
w.ms.1 <- svm.ms.1$model$w

svm.ms.01 <- svm.linear(c~., kmdat.m, solver="ipop", cost=0.1)
w.ms.01 <- svm.ms.01$model$w

  # soft margin: geometric margin corresponding to functional margin of 1
1/l2norm(w.ms.1[-length(w.ms.1)])
1/l2norm(w.ms.01[-length(w.ms.01)])

  # separating and margin lines for cost=1
plot.margin(w.ms.1, kmdat.m[,1:2], 2*as.num0(kmdat.m$c)-1, main="Linearly separable")
  # separating and margin lines for cost=0.1
plot.margin(w.ms.01, kmdat.m[,1:2], 2*as.num0(kmdat.m$c)-1, add=TRUE, lty=3)

  # the same for linearly inseparable data

kmdat.m.nls <- kmdat.m
kmdat.m.nls$c <- as.factor(ifelse(runif(nrow(kmdat.m))<0.1,
                    1-as.numchar(kmdat.m$c), as.numchar(kmdat.m$c)))

svm.ms.nls.1 <- svm.linear(c~., kmdat.m.nls, solver="ipop", cost=1)
w.ms.nls.1 <- svm.ms.nls.1$model$w

svm.ms.nls.01 <- svm.linear(c~., kmdat.m.nls, solver="ipop", cost=0.1)
w.ms.nls.01 <- svm.ms.nls.01$model$w

  # soft margin: geometric margin corresponding to functional margin of 1
1/l2norm(w.ms.nls.1[-length(w.ms.nls.1)])
1/l2norm(w.ms.nls.01[-length(w.ms.nls.01)])
```

```
  # separating and margin lines for cost=1
plot.margin(w.ms.nls.1, kmdat.m.nls[,1:2], 2*as.num0(kmdat.m.nls$c)-1,
            main="Linearly inseparable")
  # separating and margin lines for cost=0.1
plot.margin(w.ms.nls.01, kmdat.m.nls[,1:2], 2*as.num0(kmdat.m.nls$c)-1,
            add=TRUE, lty=3)
```

16.3 Support vector regression

The support vector machines algorithm, originally proposed for linear classification, can be modified to permit application to linear regression as well. One of such modifications, known as *support vector regression* (SVR), attempts to achieve better generalization and overfitting-prevention by relaxing the error minimization objective adopted by the gradient descent and least-squares algorithms presented in Chapter 8. More specifically, it finds a parameter vector acceptable as long as it yields predictions that differ from the corresponding true target function values by no more than a fixed maximum deviation ϵ:

$$|f(x) - h(x)| \le \epsilon \tag{16.37}$$

for each $x \in T$. The magnitude of differences between the predicted and true values is immaterial as long as they are below the permitted maximum. Among all parameter vectors that satisfy this condition, support vector regression prefers those with the minimum Euclidean norm $\|\mathbf{w}_{1:n}\|$, calculated with the intercept term w_{n+1} skipped (as it is only responsible for shifting the hyperplane represented by $\mathbf{w}$ without changing its slope).

The minimization of $\|\mathbf{w}_{1:n}\|$ happens to be the same optimization objective as that is adopted by SVM for classification, as presented in Section 16.2.3. In that case, it was used

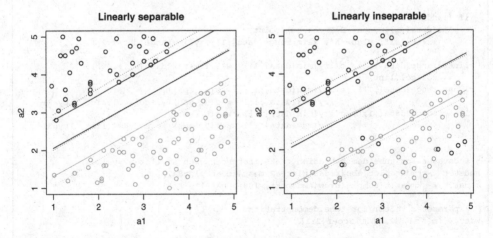

Figure 16.3 The soft classification margin maximized by the soft-margin SVM algorithm using the full example dataset and the linearly separable subset.

to maximize the classification geometric margin (with the functional margin kept fixed at 1). In the case of the regression task it can be interpreted as ensuring the "flatness" of the model prediction function represented by $\mathbf{w}$. The smaller the model parameters, the less the function represented thereby is sensitive to attributes (approaching a constant in the extreme case), which makes it also less prone to overfitting. By specifying $\epsilon > 0$ and minimizing the parameters one therefore agrees to accept a certain level of the model's deviation from the target function if it makes it possible to keep the predictions more "flat."

16.3.1   Regression tube

The ϵ tolerance threshold for a linear regression model's predictions may be thought of as determining a *tube* around the hyperplane represented by the model's parameter vector. An instance x for which $|f(x) - h(x)| = \epsilon$ is said to *lie on the ϵ-tube*. If $|f(x) - h(x)| < \epsilon$, the instance is said to *lie within the ϵ-tube* and $|f(x) - h(x)| > \epsilon$, it is said to *lie outside the ϵ-tube*.

Example 16.3.1 The idea of regression tube is illustrated by the R code presented below. It defines the `plot.tube` function to plot the regression line and the corresponding tube lines in the basic two-dimensional case (i.e., with a single attribute). A small subset of the `kmdat.plot` dataset is used for the purpose of tube illustration. The plot is presented in Figure 16.4.

```
## plot regression tube lines for linear regression
## with a single attributes
plot.tube <- function(w, data, eps, add=FALSE,
                      col.point="black", col.line="black", ...)
{
  # y value corresponding to x on the regression line represented by w
  lry <- function(x, w) {sum(w*c(x, 1)) }

  if (!add)
    plot(data[,1], data[,2], col=col.point,
         xlab="a1", ylab="h", xlim=range(data[,1]), ylim=range(data[,2]), ...)

  lines(range(data[,1]), c(lry(min(data[,1]), w), lry(max(data[,1]), w)),
        col=col.line)
  lines(range(data[,1]), c(lry(min(data[,1]), w-c(0, eps)),
                           lry(max(data[,1]), w-c(0, eps))), col=col.line, lty=3)
  lines(range(data[,1]), c(lry(min(data[,1]), w+c(0, eps)),
                           lry(max(data[,1]), w+c(0, eps))), col=col.line, lty=3)
}

  # dataset for tube demonstration (take instances with similar a2 values)
kmdat.t <- kmdat.plot[abs(kmdat.plot$a2-mean(kmdat.plot$a2))<1,]
kmdat.t <- kmdat.t[sample(nrow(kmdat.t), 100), c("a1", "f")]

  # parameter vector for tube demonstration
w.t <- lm(f~a1, kmdat.t)$coef[2:1]

plot.tube(w.t, kmdat.t, eps=1)
```

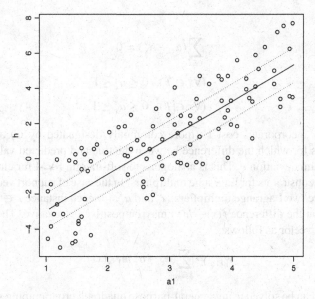

Figure 16.4   Linear regression tube.

16.3.2   Primal form

Model parameters for support vector regression can be identified by solving the following quadratic programming problem:

minimize

$$\frac{1}{2}\|\mathbf{w}_{1:n}\|^2 \tag{16.38}$$

subject to

$$(\forall x \in T) \; f(x) - \mathbf{w} \circ \mathbf{a}(x) \le \epsilon \tag{16.39}$$

$$(\forall x \in T) \; \mathbf{w} \circ \mathbf{a}(x) - f(x) \le \epsilon \tag{16.40}$$

with the $\frac{1}{2}$ multiplier introduced for mathematical convenience only. This is the *primal form* of the SVR optimization problem, which assumes that indeed there is a parameter vector that yields no deviations above ϵ.

16.3.3   Dual form

As with SVM for classification, it is more convenient to solve the problem obtained by applying a Lagrangian transformation rather than the original problem directly. It is also more realistic to assume that the maximum deviation of ϵ may be impossible to achieve and permit relaxing it. The resulting *dual form* of SVR optimization can be stated as follows:

minimize

$$\frac{1}{2}\sum_{x_1 \in T}\sum_{x_2 \in T}(\alpha_{x_1} - \alpha_{x_1}^*)(\alpha_{x_2} - \alpha_{x_2}^*)\mathbf{a}_{1:n}(x_1) \circ \mathbf{a}_{1:n}(x_2)$$
$$+ \epsilon \sum_{x \in T}(\alpha_x + \alpha_x^*) - \sum_{x \in T} f(x)(\alpha_x - \alpha_x^*) \tag{16.41}$$

subject to

$$\sum_{x \in T} (\alpha_x - \alpha_x^*) = 0 \tag{16.42}$$

$$(\forall x \in T) \quad 0 \le \alpha_x \le \Gamma \tag{16.43}$$

$$(\forall x \in T) \quad 0 \le \alpha_x^* \le \Gamma \tag{16.44}$$

This additionally incorporates cost Γ (more commonly designated by C) associated with training instances for which the difference between the true and predicted value exceeds the specified maximum deviation ϵ. This is analogous to soft-margin SVM for classification and leads to the same constraints for Lagrange multipliers. In the case of support vector regression, however, there are two Lagrange multipliers, α_x and α_x^*, for each instance $x \in T$, corresponding to the fact that the difference $f(x) - h(x)$ may be positive or negative. They relate to the model parameter vector as follows:

$$\mathbf{w}_{1:n} = \sum_{x \in T} (\alpha_x - \alpha_x^*) \mathbf{a}_{1:n}(x) \tag{16.45}$$

The problem can be solved using general-purpose quadratic programming solvers or more efficient dedicated algorithms, not discussed here. The solution may also be represented by $\beta_x = \alpha_x - \alpha_x^*$ for convenience.

It can be verified that at most one of α_x, α_x^* can be nonzero for any $x \in T$, and instances for which one of the corresponding Lagrange multipliers is positive (or β is nonzero) are *support vectors*. Clearly

$$\beta_x = \begin{cases} \alpha_x & \text{if } \alpha_x > 0 \\ -\alpha_x^* & \text{otherwise} \end{cases} \tag{16.46}$$

and $|\beta_x| = \max\{\alpha_x, \alpha_x^*\}$ for all $x \in T$. The parameter vector calculated as

$$\mathbf{w}_{1:n} = \sum_{x \in T} \beta_x \mathbf{a}_{1:n}(x) \tag{16.47}$$

depends on support vectors only. The solution can also be verified to exhibit the following properties:

$$f(x) - \mathbf{w} \circ \mathbf{a}(x) = \epsilon \quad \equiv \quad 0 < \alpha_x < \Gamma \tag{16.48}$$

$$f(x) - \mathbf{w} \circ \mathbf{a}(x) < \epsilon \quad \equiv \quad \alpha_x = 0 \tag{16.49}$$

$$f(x) - \mathbf{w} \circ \mathbf{a}(x) > \epsilon \quad \equiv \quad \alpha_x = \Gamma \tag{16.50}$$

and, analogously:

$$\mathbf{w} \circ \mathbf{a}(x) - f(x) = \epsilon \quad \equiv \quad 0 < \alpha_x^* < \Gamma \tag{16.51}$$

$$\mathbf{w} \circ \mathbf{a}(x) - f(x) < \epsilon \quad \equiv \quad \alpha_x^* = 0 \tag{16.52}$$

$$\mathbf{w} \circ \mathbf{a}(x) - f(x) > \epsilon \quad \equiv \quad \alpha_x^* = \Gamma \tag{16.53}$$

Support vectors are therefore instances for which the difference between the predicted and true target function value is equal or exceeds ϵ.

Notice that Equation 16.45 used to calculate model parameters based on Lagrange multipliers does not apply to the intercept term. It can be calculated, though, based on any support vector for which the absolute difference between the predicted and true target function value is equal to ϵ. Combining the conditions given by Equations 16.48 and 16.49 yields

$$f(x) - \mathbf{w} \circ \mathbf{a}(x) = \text{sgn}(\beta_x)\epsilon \quad \equiv \quad 0 < |\beta_x| < \Gamma \qquad (16.54)$$

and therefore

$$w_{n+1} = f(x_s) - \mathbf{w}_{1:n} \circ \mathbf{a}_{1:n}(x_s) - \text{sgn}(\beta_{x_s})\epsilon \qquad (16.55)$$

for any x_s such that $0 < |\beta_{x_s}| < \Gamma$ (i.e., either $0 < \alpha_{x_s} < \Gamma$ or $0 < \alpha_{x_s}^* < \Gamma$). For greater reliability, this can be averaged over multiple instances satisfying the condition or one can choose an instance with a $|\beta|$ value possibly far from both the lower and upper bound.

It is noteworthy that, similarly as for SVM, attribute value vectors used in the dual form optimization objective appear inside dot products only. Expressing model predictions using the dual form solution

$$\begin{aligned}
h(x_*) &= \sum_{x \in T} \beta_x \mathbf{a}_{1:n}(x) \circ \mathbf{a}_{1:n}(x_*) \\
&\quad + f(x) - \sum_{x \in T} \mathbf{a}_{1:n}(x) \circ \mathbf{a}_{1:n}(x_s) - \text{sgn}(\beta_{x_s})\epsilon
\end{aligned} \qquad (16.56)$$

shows that also attribute value vectors used for prediction appear inside dot products only. Using the dual form both parameter estimation and prediction for support vector regression models can therefore be performed without using any attribute values other than inside dot products. It also makes it possible not to calculate the primal form parameter vector $\mathbf{w}$ explicitly.

Example 16.3.2 The following R code implements and demonstrates the SVR algorithm, following the pattern of the soft-margin SVM implementation by the svm.linear function from Example 16.2.5. The demonstration uses the same artificial training and test sets as the previous examples, with the target concept ignored, but two target functions f and g used instead. Additionally, support vector regression model creation and evaluation for the real *Boston Housing* data is performed. The mse function is used for mean square error calculation.

> Ex. 10.2.3
> dmr.regeval

```
## linear SVR parameter estimation using quadratic programming
## solvers: "solve.QP" or "ipop"
svr.linear <- function(formula, data, eps=0.01, cost=1, svthres=1e-3,
                       solver="solve.QP")
{
  f <- y.var(formula)
  attributes <- x.vars(formula, data)
  aind <- names(data) %in% attributes

  fvec <- data[[f]]                   # target function vector
  amat <- as.matrix(data[,aind])      # attribute value matrix
  dpmat <- amat%*%t(amat)             # dot product matrix
```

```
    if (solver=="solve.QP")
      args <- list(Dmat=nearPD(rbind(cbind(dpmat, -dpmat), cbind(-dpmat, dpmat)))$mat,
                   dvec=c(fvec-eps, -fvec-eps),
                   Amat=matrix(c(rep(1, nrow(data)), rep(-1, nrow(data)),
                                 diag(1, 2*nrow(data)), diag(-1, 2*nrow(data))),
                              nrow=2*nrow(data)),
                   bvec=c(0, rep(0, 2*nrow(data)), rep(-cost, 2*nrow(data))),
                   meq=1)
    else if (solver=="ipop")
      args <- list(c=c(-fvec+eps, fvec+eps),
                   H=rbind(cbind(dpmat, -dpmat), cbind(-dpmat, dpmat)),
                   A=c(rep(1, nrow(data)), rep(-1, nrow(data))),
                   b=0,
                   l=rep(0, 2*nrow(data)),
                   u=rep(cost, 2*nrow(data)),
                   r=0)
    else
      stop("Unknown solver: ", solver)

    qp <- do.call(solver, args)
    alpha <- if (solver=="solve.QP") qp$solution else if (solver=="ipop") qp@primal
    beta <- alpha[1:nrow(data)]-alpha[(nrow(data)+1):(2*nrow(data))]
    sv <- which(abs(beta)>svthres)
    w <- c(colSums(rmm(amat[sv,], beta[sv])))  # no intercept yet
    i <- which.min(abs(beta-cost/2))
    w <- c(w, intercept=fvec[i]-unname(predict.par(list(repf=repf.linear, w=c(w, 0)),
                                     data[i,aind,drop=FALSE]))-
                      sign(beta[i])*eps)
    list(model=`class<-`(list(repf=repf.linear, w=w), "par"), sv=sv)
}


  # linear SVR for f
svrf <- svr.linear(f~a1+a2+a3+a4, kmdat.train)
svrf.e1 <- svr.linear(f~a1+a2+a3+a4, eps=1, kmdat.train)
svrf.c01 <- svr.linear(f~a1+a2+a3+a4, cost=0.1, kmdat.train)

  # linear SVR for g
svrg <- svr.linear(g~a1+a2+a3+a4, kmdat.train)
svrg.e1 <- svr.linear(g~a1+a2+a3+a4, eps=1, kmdat.train)
svrg.c01 <- svr.linear(g~a1+a2+a3+a4, cost=0.1, kmdat.train)

  # linear SVR for the Boston Housing data
bh.svr <- svr.linear(medv~., bh.std.train)
bh.svr.e1 <- svr.linear(medv~., eps=1, bh.std.train)
bh.svr.c01 <- svr.linear(medv~., cost=0.1, bh.std.train)

  # training set MSE
mse(predict(svrf$model, kmdat.train[,1:4]), kmdat.train$f)
mse(predict(svrf.e1$model, kmdat.train[,1:4]), kmdat.train$f)
mse(predict(svrf.c01$model, kmdat.train[,1:4]), kmdat.train$f)

mse(predict(svrg$model, kmdat.train[,1:4]), kmdat.train$g)
mse(predict(svrg.e1$model, kmdat.train[,1:4]), kmdat.train$g)
mse(predict(svrg.c01$model, kmdat.train[,1:4]), kmdat.train$g)

mse(predict(bh.svr$model, bh.std.train[,-13]), bh.std.train$medv)
mse(predict(bh.svr.e1$model, bh.std.train[,-13]), bh.std.train$medv)
mse(predict(bh.svr.c01$model, bh.std.train[,-13]), bh.std.train$medv)
```

```
# test set MSE
mse(predict(svrf$model, kmdat.test[,1:4]), kmdat.test$f)
mse(predict(svrf.e1$model, kmdat.test[,1:4]), kmdat.test$f)
mse(predict(svrf.c01$model, kmdat.test[,1:4]), kmdat.test$f)

mse(predict(svrg$model, kmdat.test[,1:4]), kmdat.test$g)
mse(predict(svrg.e1$model, kmdat.test[,1:4]), kmdat.test$g)
mse(predict(svrg.c01$model, kmdat.test[,1:4]), kmdat.test$g)

mse(predict(bh.svr$model, bh.std.test[,-13]), bh.std.test$medv)
mse(predict(bh.svr.e1$model, bh.std.test[,-13]), bh.std.test$medv)
mse(predict(bh.svr.c01$model, bh.std.test[,-13]), bh.std.test$medv)
```

The parameters passed to the two quadratic programming solvers are explained below.

For the `solve.QP` function:

1. `Dmat` is the matrix specifying the coefficients of the quadratic term of the optimization objective – here an $2|T| \times 2|T|$ matrix composed of four quarter submatrices:

   upper left $|T| \times |T|$ containing data dot products (with $\mathbf{a}_{1:n}(x_1) \circ \mathbf{a}_{1:n}(x_2)$ in the cell corresponding to x_1 and x_2),

   upper right $|T| \times |T|$ containing negated data dot products (with $-\mathbf{a}_{1:n}(x_1) \circ \mathbf{a}_{1:n}(x_2)$ in the cell corresponding to x_1 and x_2),

   lower left the same as the upper right,

   lower right the same as the upper left.

2. `dvec` is the vector specifying the coefficients of the linear term of the optimization objective – here a vector containing $f(x) - \epsilon$ for all $x \in T$ as the first $|T|$ elements and $-f(x) - \epsilon$ for all $x \in T$ as the remaining $|T|$ elements,

3. `Amat` is the (transpose of) the matrix specifying the coefficients of the equality and inequality constraints – here a $2|T| \times (4|T| + 1)$ matrix with a vector of $|T|$ 1's followed by $|T|$ −1's as the first column (representing the coefficients of the equality constraint), the $2|T| \times 2|T|$ identity matrix (representing the coefficients of the *greater than* inequality constraints) as the subsequent $2|T|$ columns, and the negated $2|T| \times 2|T|$ identity matrix (representing the coefficients of the *less than* inequality constraints) as the last $2|T|$ columns,

4. `bvec` is the vector specifying the values for equality and inequality constraints represented by `Amat` – here a vector of $(2|T| + 1)$ 0's followed by a $2|T|$-time repetition of `-cost`,

5. `meq` is the number of the first columns in `Amat` that represent equality constraints – here set to 1.

For the `ipop` function:

1. `c` is the vector specifying the coefficients of the linear term of the optimization objective – here a vector containing $\epsilon - f(x)$ for all $x \in T$ as the first $|T|$ elements and $\epsilon + f(x)$ for all $x \in T$ as the remaining $|T|$ elements (the same as negated `dvec` for `solve.QP`),

2. H is the matrix specifying the coefficients of the quadratic term of the optimization objective – here an $2|T| \times 2|T|$ matrix (the same as Dmat for solve.QP),

3. A is the matrix specifying the coefficients of the box constraints – here a vector of $|T|$ 1's followed by $|T|$ −1's, representing the coefficients of the single equality constraint (the same as the first column of Amat for solve.QP),

4. b is the vector specifying the values for the *greater than* inequality constraints represented by A – here a single 0 (the same as the first element of bvec for solve.QP),

5. l is the vector specifying the lower bounds for the solution – here a $2|T|$-element vector of 0's (the same as the $2 : |T| + 1$ elements of bvec for solve.QP),

6. u is the vector specifying the upper bounds for the solution – here a $2|T|$-element vector of cost (the same as the negated $|T| + 2 : 2|T| + 1$ elements of bvec for solve.QP),

7. r is the vector that, added to b, specifies the values for the *less than* inequality constraints represented by A – here set to 0, to make the constraint actually an equality constraint.

Only *greater than* constraints are originally supported by the solve.QP function and the *less than* constraints have to be indirectly specified by using negated constraint coefficients and values. For ipop the equality constraint is represented by A and b (with r=0) and the inequality constraints – by l and u.

With default settings, the SVR algorithm identifies a nearly perfect model for the f target function, which is not surprising given its truly linear dependence on attributes in the example artificial dataset. Decreasing Γ (cost) has little effect, but increasing ϵ (eps) results in a flatter model with a substantially larger error value. Somewhat different behavior is observed for the nonlinear g target function in the artficial data. While the default settings appear to yield a good linear approximation, this time larger ϵ gives no effect and smaller Γ degrades performance. All the three parameter setups perform similarly well for the *Boston Housing* data.

Example 16.3.3 The following R code illustrates the effect of the ϵ and Γ parameters of the support vector regression algorithm on the flatness of the hyperplane represented by model parameters. The svr.linear function from the previous example is applied to the simple single-attribute dataset previously generated for tube illustration, using different parameter setups. The resulting regression tubes are plotted using the plot.tube function and presented in Figure 16.5. The norms of the parameter vectors obtained in each case are calculated using the l2norm function to quantify the effect of larger ϵ | dmr.util | or smaller Γ on model flatness.

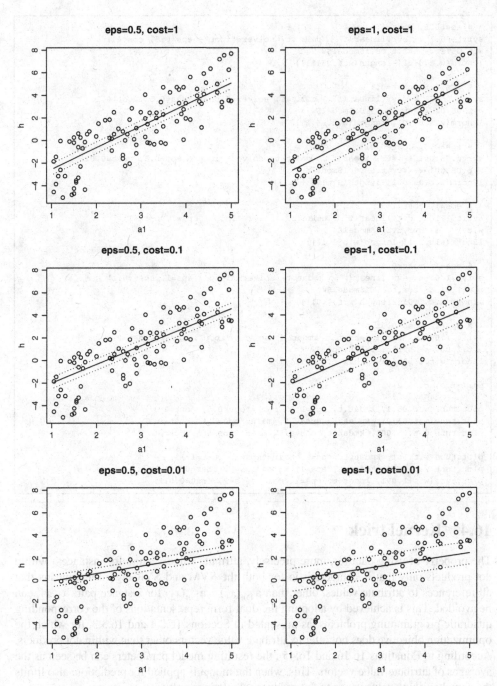

Figure 16.5   The linear regression tubes obained by the SVR algorithm.

```
  # eps=0.5, cost=1
svr.t.05.1 <- svr.linear(f~., kmdat.t, solver="ipop", eps=0.5, cost=1)
w.t.05.1 <- svr.t.05.1$model$w
l2norm(w.t.05.1[-length(w.t.05.1)])

  # eps=0.5, cost=0.1
svr.t.05.01 <- svr.linear(f~., kmdat.t, solver="ipop", eps=0.5, cost=0.1)
w.t.05.01 <- svr.t.05.01$model$w
l2norm(w.t.05.01[-length(w.t.05.01)])

  # eps=0.5, cost=0.01
svr.t.05.001 <- svr.linear(f~., kmdat.t, solver="ipop", eps=0.5, cost=0.01)
w.t.05.001 <- svr.t.05.001$model$w
l2norm(w.t.05.001[-length(w.t.05.001)])

  # eps=1, cost=1
svr.t.1.1 <- svr.linear(f~., kmdat.t, solver="ipop", eps=1, cost=1)
w.t.1.1 <- svr.t.1.1$model$w
l2norm(w.t.1.1[-length(w.t.1.1)])

  # eps=1, cost=0.1
svr.t.1.01 <- svr.linear(f~., kmdat.t, solver="ipop", eps=1, cost=0.1)
w.t.1.01 <- svr.t.1.01$model$w
l2norm(w.t.1.01[-length(w.t.1.01)])

  # eps=1, cost=0.01
svr.t.1.001 <- svr.linear(f~., kmdat.t, solver="ipop", eps=1, cost=0.01)
w.t.1.001 <- svr.t.1.001$model$w
l2norm(w.t.1.001[-length(w.t.1.001)])

par(mfcol=c(3, 2))

plot.tube(w.t.05.1, kmdat.t, eps=0.5, main="eps=0.5, cost=1")
plot.tube(w.t.05.01, kmdat.t, eps=0.5, main="eps=0.5, cost=0.1")
plot.tube(w.t.05.001, kmdat.t, eps=0.5, main="eps=0.5, cost=0.01")

plot.tube(w.t.1.1, kmdat.t, eps=1, main="eps=1, cost=1")
plot.tube(w.t.1.01, kmdat.t, eps=1, main="eps=1, cost=0.1")
plot.tube(w.t.1.001, kmdat.t, eps=1, main="eps=1, cost=0.01")
```

16.4   Kernel trick

The property of using data (or, more precisely, attribute value vectors for instances) within dot products only that is characteristic for both the SVM and SVR algorithms means that all references to attribute values, other than $\mathbf{a}_{1:n}(x_1) \circ \mathbf{a}_{1:n}(x_2)$ for instance pairs x_1, x_2, can be avoided. This is achieved by adopting the dual form representations of the corresponding quadratic programming problems, as presented in Sections 16.2.4 and 16.3.3, for which the optimization objective does not refer to attribute value vectors other than within dot products. According to Equations 16.16 and 16.45 , the resulting model parameters can be seen as the averages of attribute value vectors. This, when the model is applied for prediction, also limits the actual attribute value usage to dot product calculations only:

$$h(x_*) = H\left(\sum_{x \in T} c_-(x)\alpha_x \mathbf{a}_{1:n}(x) \circ \mathbf{a}_{1:n}(x_*) + w_{n+1}\right) \tag{16.57}$$

for SVM, where H is the unit step function, and

$$h(x_*) = \sum_{x \in T} (\alpha_x - \alpha_x^*) \mathbf{a}_{1:n}(x) \circ \mathbf{a}_{1:n}(x_*) + w_{n+1} \tag{16.58}$$

for SVR.

Using an enhanced representation with new attributes $a_1', a_2', \dots, a_N'$ and (usually) $N \gg n$ is possible by simply replacing the dot products in the original attribute space, $\mathbf{a}(x_1) \circ \mathbf{a}(x_2)$, with the corresponding dot products in the new attribute space, $\mathbf{a}'_{1:N}(x_1) \circ \mathbf{a}'_{1:N}(x_2)$. The so-called *kernel trick* makes it possible to avoid explicitly calculating the new attribute values at all and use $K(x_1, x_2)$ instead of $\mathbf{a}'_{1:N}(x_1) \circ \mathbf{a}'_{1:N}(x_2)$, where K is a *kernel function* that can be calculated based on attribute value vectors *in the original representation*. As long as there exists some representation (i.e., a set of new attributes) for which K does indeed represent a dot product, this representation would be then implicitly used, without the need to explicitly calculate new attribute values. For kernel function K and dataset S the matrix of kernel function values for all instance pairs from the dataset will be referred to as the *kernel matrix* with respect to K for S and denoted by K_S. Using instances as row and column indices we may write, for any $x_1, x_2 \in S$

$$K_S[x_1, x_2] = K(x_1, x_2) \tag{16.59}$$

Consider a particularly simple kernel function defined as

$$K(x_1, x_2) = (\mathbf{a}_{1:n}(x_1) \circ \mathbf{a}_{1:n}(x_2))^2 \tag{16.60}$$

Assuming $n = 2$ for even greater simplicity, we have

$$K(x_1, x_2) = a_1^2(x_1)a_1^2(x_2) + a_2^2(x_1)a_2^2(x_2) + 2a_1(x_1)a_2(x_1)a_1(x_2)a_2(x_2) \tag{16.61}$$

This does indeed represent a dot product for the enhanced attribute space defined as

$$a_1'(x) = a_1^2(x) \tag{16.62}$$

$$a_2'(x) = a_2^2(x) \tag{16.63}$$

$$a_3'(x) = a_1(x)a_2(x) \tag{16.64}$$

$$a_4'(x) = a_1(x)a_2(x) \tag{16.65}$$

or, to avoid two identical attributes:

$$a_1'(x) = a_1^2(x) \tag{16.66}$$

$$a_2'(x) = a_2^2(x) \tag{16.67}$$

$$a_3'(x) = \sqrt{2}a_1(x)a_2(x) \tag{16.68}$$

Any symmetric positive definite matrix is a kernel matrix with respect to some kernel function and any positive definite symmetric function is a kernel function according to Mercer's theorem. For any such K it can be then guaranteed that $K(x_1, x_2)$ is indeed equal to $\mathbf{a}'_{1:N}(x_1) \circ \mathbf{a}'_{1:N}(x_2)$ for some $a_1', a_2', \dots, a_N'$, although providing the definitions of these new attributes in terms of the original attributes may be often difficult and calculating their values is never necessary, if using algorithms designed to work with dot products only. More specifically, the kernel trick brings two closely related advantages:

- high-dimensional nonlinear enhanced representations may be used for linear models without defining new attributes explicitly,

- high-dimensional nonlinear enhanced representations may be used for linear models without calculating new attributes explicitly.

The former makes it convenient to specify general-purpose, domain-independent enhanced representations and incorporate them to dot product-based linear modeling algorithms. The latter makes it computationally efficient to use such representations for linear model creation and prediction, by avoiding the need to calculate the values of a possibly huge number of new attributes, that are instead used implicitly. In essence, we get the increased representation power – that may suffice to overcome or alleviate the linearity limitation – without the associated effort and cost.

Example 16.4.1 A simple illustration of the kernel trick is presented by the following R code. It takes a small 10-row subset of the kmdat.train dataset and applies a transformation that replaces the original 4 attributes with 16 new attributes, defined as all their pairwise products. The dot product in this enhanced representation can be verified to be equal to the squared dot product in the original representation.

```
## data transformation that generates new attributes
## defined as the products of all original attribute pairs
trans.mult2 <- function(data)
{
  t(apply(data, 1, function(d) d %o% d))
}

  # original dataset
kmdat.orig <- kmdat.train[1:10,1:4]
  # dot product matrix for the original dataset
kmdat.dp <- as.matrix(kmdat.orig) %*% t(kmdat.orig)

  # transformed dataset
kmdat.trans <- trans.mult2(kmdat.orig)
  # dot product matrix for the transformed dataset
kmdat.dpt <- kmdat.trans %*% t(kmdat.trans)

  # verify that the dot product matrix for the transformed dataset
  # is the same as the squared original dot product matrix
max(abs((kmdat.dpt-kmdat.dp^2)))
```

16.5   Kernel functions

Choosing a kernel function appropriate for a particular classification or regression task is crucial for the quality of predictions that can be obtained by models using the resulting enhanced representation. Sometimes available domain knowledge helps make a good choice, but often a trial-and-error approach may be necessary. Practical applications of kernel methods usually adopt a kernel function belonging to one of a few standard families, described below. These families are sufficiently diverse to cover most common needs.

16.5.1   Linear kernel

A linear kernel function is defined as

$$K(x_1, x_2) = \mathbf{a}_{1:n}(x_1) \circ \mathbf{a}_{1:n}(x_2) \tag{16.69}$$

This obviously leaves the original representation unchanged and does not overcome the linearity limitation of linear classification and linear regression models in any way. It makes it possible, however, to consider linear dot product-based algorithms (such as the linear support vector machines and linear support vector regression algorithms) as special cases of the corresponding kernel-based algorithms.

16.5.2   Polynomial kernel

A polynomial kernel function basically takes a power of the dot product in the original representation:

$$K(x_1, x_2) = (\gamma \mathbf{a}_{1:n}(x_1) \circ \mathbf{a}_{1:n}(x_2) + b)^p \tag{16.70}$$

where $\gamma > 0$, $b \geq 0$, and $p > 0$ are parameters. This kernel family makes it possible to easily control the enhanced representation size and degree of nonlinearity by adjusting the p parameter. Positive b can be used to adjust the relative impact of higher order and lower order terms in the resulting polynomial representation. To see how this is possible, assume $p = 2$ and $n = 2$ for simplicity and compare:

$$
\begin{aligned}
(\mathbf{a}_{1:n}(x_1) \circ \mathbf{a}_{1:n}(x_2))^2 = {} & a_1^2(x_1)a_1^2(x_2) + a_2^2(x_1)a_2^2(x_2) \\
& + 2a_1(x_1)a_2(x_1)a_1(x_2)a_2(x_2)
\end{aligned}
\tag{16.71}
$$

$$
\begin{aligned}
(\mathbf{a}_{1:n}(x_1) \circ \mathbf{a}_{1:n}(x_2) + 1)^2 = {} & a_1^2(x_1)a_1^2(x_2) + a_2^2(x_1)a_2^2(x_2) \\
& + 2a_1(x_1)a_2(x_1)a_1(x_2)a_2(x_2) + 2a_1(x_1)a_1(x_2) \\
& + 2a_2(x_1)a_2(x_2) + 1
\end{aligned}
\tag{16.72}
$$

$$
\begin{aligned}
(\mathbf{a}_{1:n}(x_1) \circ \mathbf{a}_{1:n}(x_2) + 2)^2 = {} & a_1^2(x_1)a_1^2(x_2) + a_2^2(x_1)a_2^2(x_2) \\
& + 2a_1(x_1)a_2(x_1)a_1(x_2)a_2(x_2) + 4a_1(x_1)a_1(x_2) \\
& + 4a_2(x_1)a_2(x_2) + 4
\end{aligned}
\tag{16.73}
$$

This demonstrates that increasing b increases the coefficients of lower order terms. While convenient to control and easy to understand, the polynomial kernel family may be insufficient to adequately represent more complex relationships.

16.5.3   Radial kernel

A radial kernel (also called Gaussian or RBF) depends on the Euclidean distance between the original attribute value vectors (i.e., the Euclidean norm of their difference) rather than their dot product:

$$K(x_1, x_2) = e^{-\gamma \|\mathbf{a}_{1:n}(x_1) - \mathbf{a}_{1:n}(x_2)\|^2} \tag{16.74}$$

where $\gamma > 0$ is a parameter. This type of kernel tends to be particularly popular, as it makes the contribution of instance x to the prediction calculated for instance x_*, which is proportional to $K(x, x_*)$, exponentially decay with the increasing distance between them. The resulting prediction is therefore highly localized and, as such, can be well fitted to arbitrarily complex target functions or concepts. However, it is very sensitive to the choice of the γ parameter and may be prone to overfitting.

16.5.4   Sigmoid kernel

A sigmoid kernel as defined as the hyperbolic tangent of the dot product in the original representation:

$$K(x_1, x_2) = \tanh(\gamma \mathbf{a}_{1:n}(x_1) \circ \mathbf{a}_{1:n}(x_2) + b) \tag{16.75}$$

where γ and b are the parameters (usually $\gamma > 0$). The sigmoid function is widely used as the activation function for neural networks (multilayer perceptrons) and hence has also become popular for kernel methods. Like the radial kernel family, if used with properly adjusted parameters, it can represent complex nonlinear relationships. For some parameter settings it actually becomes similar to the radial kernel. However, the sigmoid function may not be positive definite for some parameters, and therefore not actually represent a valid kernel.

Example 16.5.1 Kernel functions discussed above are implemented and demonstrated by the following R code. Each function can be applied both to single instances (assumed to be represented by single matrix or data frame rows) or to a dataset of multiple instances (assumed to be represented by a matrix or data frame, with rows corresponding to instances and columns corresponding to attributes). In the latter case, the kernel matrix is calculated, i.e., the matrix of kernel function values for all instance pairs.

```
## can be called for both single attribute value vectors and for the whole dataset
kernel.linear <- function(av1, av2=av1) { as.matrix(av1)%*%t(av2) }

## can be called for both single attribute value vectors and for the whole dataset
kernel.polynomial <- function(av1, av2=av1, gamma=1, b=0, p=3)
{ (gamma*(as.matrix(av1)%*%t(av2))+b)^p }

## can be called for both single attribute value vectors and for the whole dataset
kernel.radial <- function(av1, av2=av1, gamma=1)
{
  exp(-gamma*outer(1:nrow(av1 <- as.matrix(av1)), 1:ncol(av2 <- t(av2)),
                  Vectorize(function(i, j) l2norm(av1[i,]-av2[,j])^2)))
}

## can be called for both single attribute value vectors and for the whole dataset
kernel.sigmoid <- function(av1, av2=av1, gamma=0.1, b=0)
{ tanh(gamma*(as.matrix(av1)%*%t(av2))+b) }

  # kernel functions called for instance pairs
kernel.linear(kmdat.train[1,1:4], kmdat.train[2,1:4])
kernel.polynomial(kmdat.train[1,1:4], kmdat.train[2,1:4])
kernel.radial(kmdat.train[1,1:4], kmdat.train[2,1:4])
kernel.sigmoid(kmdat.train[1,1:4], kmdat.train[2,1:4])
```

```
# kernel functions called for the dataset (using the first 10 instances)
kernel.linear(kmdat.train[1:10,1:4])
kernel.polynomial(kmdat.train[1:10,1:4])
kernel.radial(kmdat.train[1:10,1:4])
kernel.sigmoid(kmdat.train[1:10,1:4])
```

16.6 Kernel prediction

Kernel-based prediction is basically linear prediction using kernel-based estimations of model parameters. Assuming

$$\mathbf{w}_{1:n} = \sum_{x \in T} \eta_x \mathbf{a}_{1:n}(x) \tag{16.76}$$

to match both Equations 16.16 and 16.45, we receive the following kernel-based representation function:

$$g(x_*) = \sum_{x \in T} \eta_x K(x, x_*) + w_{n+1} \tag{16.77}$$

that can be used directly as the model's prediction function for regression or as the inner representation function for classification. This covers both SVM and SVR as special cases. For the former, we have

$$\eta_x = c_-(x)\alpha_x \tag{16.78}$$

$$h(x_*) = H(g(x_*)) \tag{16.79}$$

where α_x is the Lagrange multiplier (dual form solution element) for instance x. For the latter:

$$\eta_x = \alpha_x - \alpha_x^* \tag{16.80}$$

$$h(x_*) = g(x_*) \tag{16.81}$$

In both cases, only training instances with nonzero η_x values (i.e., support vectors) contribute to prediction and the remaining instances can be skipped in the summations. A kernel-based model can therefore be represented by a vector of coefficients and a corresponding set of attribute value vectors or, equivalently, an attribute value matrix for a set of instances.

Example 16.6.1 The R code presented below implements and demonstrates kernel prediction. The `predict.kernel` function assumes the supplied model contains the following components:

- `coef` – the vector of kernel-based model coefficients,
- `intercept` – the intercept value,
- `mat` – the attribute value matrix representing the training set for kernel prediction (with rows corresponding to instances and columns corresponding to attributes),
- `kernel` – the kernel function,
- `kernel.args` – the list of kernel function arguments,
- `formula` – the formula specifying attributes used for model creation.

Then a random small subset of the `kmdat.plot` dataset is selected and used to create several kernel-based models, using different kernel functions. The models use arbitrarily selected meaningless coefficients and serve the illustration purpose only. Their predictions are then plotted as surfaces, to visualize the nonlinearity introduced by the kernel transformation. The obtained plots are presented in Figure 16.6.

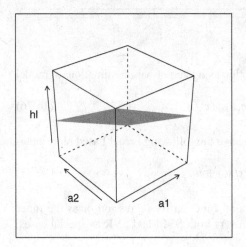
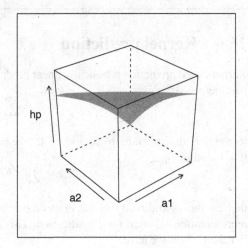
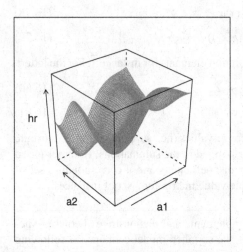
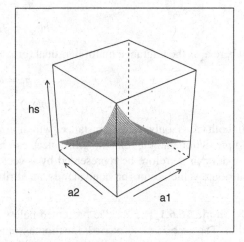

Figure 16.6   Kernel predictions for different kernel types.

```
## predict using a kernel-based model
predict.kernel <- function(model, data)
{
  attributes <- x.vars(model$formula, data)
  aind <- names(data) %in% attributes
  amat <- as.matrix(data[,aind,drop=FALSE])
  kmat <- do.call(model$kernel, c(list(amat, model$mat), model$kernel.args))
  rowSums(cmm(kmat, model$coef))+model$intercept
}
```

```
 # kernel models for producing plots
kmplot <- list(coef=c(rep(1, 50), rep(-2, 50)),
               mat=as.matrix(kmdat.plot[sample(nrow(kmdat.plot), 100),1:2]),
               intercept=1, formula=f~a1+a2)
kmplot.l <- 'class<-'(c(kmplot, kernel=kernel.linear), "kernel")
kmplot.p <- 'class<-'(c(kmplot, kernel=kernel.polynomial), "kernel")
kmplot.r <- 'class<-'(c(kmplot, kernel=kernel.radial), "kernel")
kmplot.s <- 'class<-'(c(kmplot, kernel=kernel.sigmoid), "kernel")

 # generate predictions using different kernel functions
kmdat.plot$hl <- predict(kmplot.l, kmdat.plot)
kmdat.plot$hp <- predict(kmplot.p, kmdat.plot)
kmdat.plot$hr <- predict(kmplot.r, kmdat.plot)
kmdat.plot$hs <- predict(kmplot.s, kmdat.plot)

 # plot prediction surfaces
wf.kl <- wireframe(hl~a1+a2, kmdat.plot, col="grey50", zoom=0.8)
wf.kp <- wireframe(hp~a1+a2, kmdat.plot, col="grey50", zoom=0.8)
wf.kr <- wireframe(hr~a1+a2, kmdat.plot, col="grey50", zoom=0.8)
wf.ks <- wireframe(hs~a1+a2, kmdat.plot, col="grey50", zoom=0.8)

print(wf.kl, split=c(1, 1, 2, 2), more=TRUE)
print(wf.kp, split=c(2, 1, 2, 2), more=TRUE)
print(wf.kr, split=c(1, 2, 2, 2), more=TRUE)
print(wf.ks, split=c(2, 2, 2, 2))
```

16.7   Kernel-based algorithms

Arbitrary linear modeling algorithms that use data within dot products only for both model creation and prediction can be combined with kernel methods. This section demonstrates this more specifically for the support vector machines and support vector regression algorithms.

16.7.1   Kernel-based SVM

The dual form soft-margin version of the quadratic programming problem of support vector machines parameter estimation from Section 16.2.4 can be restated as follows:

maximize

$$-\frac{1}{2} \sum_{x_1 \in T} \sum_{x_2 \in T} c_-(x_1) c_-(x_2) \alpha_{x_1} \alpha_{x_2} K(x_1, x_2) + \sum_{x \in T} \alpha_x \qquad (16.82)$$

subject to

$$\sum_{x \in T} c_-(x) \alpha_x = 0 \qquad (16.83)$$

$$(\forall x \in T) \quad 0 \leq \alpha_x \leq \Gamma \qquad (16.84)$$

This simply replaces the $\mathbf{a}_{1:n}(x_1) \circ \mathbf{a}_{1:n}(x_2)$ dot product used in the plain linear version of the algorithm with $K(x_1, x_2)$, which effectively transforms the problem to a new representation. As in the linear case, a solution can be found using general-purpose quadratic programming solvers or more efficient dedicated algorithms.

Example 16.7.1 The following R code defines the `svm.kernel` function for kernel-based soft-margin SVM model creation. It is basically a slightly modified version of the `svm.linear` function from Example 16.2.5. The `kernel` and `kernel.args` arguments specify the kernel function and its parameters. The same quadratic programming solvers, `solve.QP` or `ipop`, are employed, and their arguments are prepared in the same way as before. The only modification is that the kernel matrix replaces the dot product matrix and the resulting model is represented by Lagrange multipliers (providing the model's coefficients) and support vectors (providing the model's attribute value matrix). Unlike in the linear case, primal form model parameters are not calculated. The `predict.svm.kernel` for kernel-based SVM prediction is also defined, which simply applies the unit step function `ustep` to the output of the `predict.kernel` function from the previous example. The demonstrations use the previously defined kernel functions with very roughly tuned parameters that may not actually provide the best possible results. All of them are applied to the artificial data and two – the linear and radial kernel functions – to the real *Pima Indians Diabetes* data. For the linear kernel the results can be verified to be the same as delivered by the `svm.linear` function (but with a different, kernel-based model representation employed). The predictions are evaluated using the `err` function for misclassification error calculation.

> Ex. 7.2.1
> dmr.claseval

```
## kernel-based soft-margin SVM parameter estimation using quadratic programming
## solvers: "solve.QP" or "ipop"
svm.kernel <- function(formula, data, kernel=kernel.linear, kernel.args=NULL,
                       cost=1, svthres=1e-3, solver="solve.QP")
{
  class <- y.var(formula)
  attributes <- x.vars(formula, data)
  aind <- names(data) %in% attributes

  cvec <- 2*as.num0(data[[class]])-1  # class vector using {-1, 1} labels
  ccmat <- outer(cvec, cvec)          # class-class product matrix
  amat <- as.matrix(data[,aind])      # attribute value matrix
  kmat <- do.call(kernel, c(list(amat), kernel.args)) # kernel matrix

  if (solver=="solve.QP")
    args <- list(Dmat=nearPD(kmat*ccmat)$mat,
                 dvec=rep(1, nrow(data)),
                 Amat=matrix(c(cvec, diag(1, nrow(data)), diag(-1, nrow(data))),
                             nrow=nrow(data)),
                 bvec=c(0, rep(0, nrow(data)), rep(-cost, nrow(data))),
                 meq=1)
  else if (solver=="ipop")
    args <- list(c=rep(-1, nrow(data)),
                 H=kmat*ccmat,
                 A=cvec,
                 b=0,
                 l=rep(0, nrow(data)),
                 u=rep(cost, nrow(data)),
                 r=0)
  else
    stop("Unknown solver: ", solver)

  qp <- do.call(solver, args)
  alpha <- if (solver=="solve.QP") qp$solution else if (solver=="ipop") qp@primal
  sv <- which(alpha>svthres)
  model <- list(coef=cvec[sv]*alpha[sv], mat=amat[sv,,drop=FALSE],
                kernel=kernel, kernel.args=kernel.args, formula=formula)
```

```
   i <- which.min(abs(alpha-cost/2))
   'class<-'(c(model, intercept=cvec[i]-
                          unname(predict.kernel'(c(model, intercept=0),
                                          data[i,aind,drop=FALSE]))),
              "svm.kernel")
}

## kernel-based SVM prediction
predict.svm.kernel <- function(model, data)
{
   ustep(predict.kernel(model, data))
}

   # kernel-based SVM for the artificial data
svm.kl <- svm.kernel(c~a1+a2+a3+a4, kmdat.train)
svm.kp <- svm.kernel(c~a1+a2+a3+a4, kmdat.train,
                     kernel=kernel.polynomial, kernel.args=list(p=2, b=1))
svm.kr <- svm.kernel(c~a1+a2+a3+a4, kmdat.train,
                     kernel=kernel.radial, kernel.args=list(gamma=0.5))
svm.ks <- svm.kernel(c~a1+a2+a3+a4, kmdat.train,
                     kernel=kernel.sigmoid, kernel.args=list(gamma=0.04,b=-0.8))

   # kernel-based SVM for the Pima Indians Diabetes
pid.svm.kl <- svm.kernel(diabetes~., pid.std.train)
pid.svm.kr <- svm.kernel(diabetes~., pid.std.train,
                     kernel=kernel.radial, kernel.args=list(gamma=0.1))

   # training set misclassification error
err(predict(svm.kl, kmdat.train), kmdat.train$c)
err(predict(svm.kp, kmdat.train), kmdat.train$c)
err(predict(svm.kr, kmdat.train), kmdat.train$c)
err(predict(svm.ks, kmdat.train), kmdat.train$c)

err(factor(predict(pid.svm.kl, pid.std.train[,-9]),
           levels=0:1, labels=levels(pid.std.train$diabetes)),
    pid.std.train$diabetes)
err(factor(predict(pid.svm.kr, pid.std.train[,-9]),
           levels=0:1, labels=levels(pid.std.train$diabetes)),
    pid.std.train$diabetes)

   # test set misclassification error
err(predict(svm.kl, kmdat.test), kmdat.test$c)
err(predict(svm.kp, kmdat.test), kmdat.test$c)
err(predict(svm.kr, kmdat.test), kmdat.test$c)
err(predict(svm.ks, kmdat.test), kmdat.test$c)

err(factor(predict(pid.svm.kl, pid.std.test[,-9]),
           levels=0:1, labels=levels(pid.std.train$diabetes)),
    pid.test$diabetes)
err(factor(predict(pid.svm.kr, pid.std.test[,-9]),
           levels=0:1, labels=levels(pid.std.train$diabetes)),
    pid.test$diabetes)
```

It can be observed that nonlinear kernels may, but always have to improve predictive performance for linearly inseparable data. For the artificial data the polynomial kernel works best, which should not be surprising given the quadratic function used for data generation. The radial kernel achieves a low training set error, but appears to suffer from overfitting. The latter can also be observed for the *Pima Indians Diabetes* data.

16.7.2   Kernel-based SVR

Applying the kernel trick to the support vector regression algorithm results in the following quadratic programming problem formulation:

minimize

$$\frac{1}{2} \sum_{x_1 \in T} \sum_{x_2 \in T} (\alpha_{x_1} - \alpha_{x_1}^*)(\alpha_{x_2} - \alpha_{x_2}^*) K(x_1, x_2)$$

$$+ \epsilon \sum_{x \in T} (\alpha_x + \alpha_x^*) - \sum_{x \in T} f(x)(\alpha_x - \alpha_x^*) \tag{16.85}$$

subject to

$$\sum_{x \in T} (\alpha_x - \alpha_x^*) = 0 \tag{16.86}$$

$$(\forall x \in T) \quad 0 \le \alpha_x \le \Gamma \tag{16.87}$$

$$(\forall x \in T) \quad 0 \le \alpha_x^* \le \Gamma \tag{16.88}$$

This is directly based on the original linear SVR dual form from Section 16.3.3, with kernel function values used instead of dot products.

Example 16.7.2 Kernel-based support vector regression is implemented and demonstrated by the R code presented below. The `svr.kernel` function is a straightforward modification of the `svr.linear` function, using the kernel matrix instead of the dot product matrix and returning a kernel-based model representation. All the four kernel types from Example 16.5.1 are applied to the artificial data and two of them – the linear and radial kernel functions – to the real *Boston Housing* data. Kernel function parameters are only very roughly tuned and may not lead to the best possible results.

```
## kernel-based SVR parameter estimation using quadratic programming
## solvers: "solve.QP" or "ipop"
svr.kernel <- function(formula, data, eps=0.01,
                   kernel=kernel.linear, kernel.args=NULL,
                   cost=1, svthres=1e-3, solver="solve.QP")
{
  f <- y.var(formula)
  attributes <- x.vars(formula, data)
  aind <- names(data) %in% attributes

  fvec <- data[[f]]                   # target function vector
  amat <- as.matrix(data[,aind])      # attribute value matrix
  kmat <- do.call(kernel, c(list(amat), kernel.args)) # kernel matrix

  if (solver=="solve.QP")
    args <- list(Dmat=nearPD(rbind(cbind(kmat, -kmat), cbind(-kmat, kmat)))$mat,
              dvec=c(fvec-eps, -fvec-eps),
              Amat=matrix(c(rep(1, nrow(data)), rep(-1, nrow(data)),
                          diag(1, 2*nrow(data)), diag(-1, 2*nrow(data))),
                      nrow=2*nrow(data)),
```

```
                      bvec=c(0, rep(0, 2*nrow(data)), rep(-cost, 2*nrow(data))),
                      meq=1)
  else if (solver=="ipop")
    args <- list(c=c(-fvec+eps, fvec+eps),
                 H=rbind(cbind(kmat, -kmat), cbind(-kmat, kmat)),
                 A=c(rep(1, nrow(data)), rep(-1, nrow(data))),
                 b=0,
                 l=rep(0, 2*nrow(data)),
                 u=rep(cost, 2*nrow(data)),
                 r=0)
  else
    stop("Unknown solver: ", solver)

  qp <- do.call(solver, args)
  alpha <- if (solver=="solve.QP") qp$solution else if (solver=="ipop") qp@primal
  beta <- alpha[1:nrow(data)]-alpha[(nrow(data)+1):(2*nrow(data))]
  sv <- which(abs(beta)>svthres)
  model <- list(coef=beta[sv], mat=amat[sv,,drop=FALSE],
                kernel=kernel, kernel.args=kernel.args, formula=formula)
  i <- which.min(abs(beta-cost/2))
  'class<-'(c(model,
             intercept=fvec[i]-unname(predict.kernel(c(model, intercept=0),
                                          data[i,aind,drop=FALSE]))-
                      sign(beta[i])*eps),
           "svr.kernel")
}

## kernel-based SVR prediction
predict.svr.kernel <- predict.kernel

  # kernel-based SVR for f
svrf.kl <- svr.kernel(f~a1+a2+a3+a4, kmdat.train)
svrf.kp <- svr.kernel(f~a1+a2+a3+a4, kmdat.train,
                      kernel=kernel.polynomial, kernel.args=list(p=2, b=1))
svrf.kr <- svr.kernel(f~a1+a2+a3+a4, kmdat.train,
                      kernel=kernel.radial, kernel.args=list(gamma=0.02))
svrf.ks <- svr.kernel(f~a1+a2+a3+a4, kmdat.train,
                      kernel=kernel.sigmoid, kernel.args=list(gamma=0.2, b=0))

  # kernel-based SVR for g
svrg.kl <- svr.kernel(g~a1+a2+a3+a4, kmdat.train)
svrg.kp <- svr.kernel(g~a1+a2+a3+a4, kmdat.train,
                      kernel=kernel.polynomial, kernel.args=list(p=2, b=1))
svrg.kr <- svr.kernel(g~a1+a2+a3+a4, kmdat.train,
                      kernel=kernel.radial, kernel.args=list(gamma=0.1))
svrg.ks <- svr.kernel(g~a1+a2+a3+a4, kmdat.train,
                      kernel=kernel.sigmoid, kernel.args=list(gamma=0.02, b=-1))

  # kernel-based SVR for the Boston Housing data
bh.svr.kl <- svr.kernel(medv~., bh.std.train)
bh.svr.kr <- svr.kernel(medv~., bh.std.train,
                        kernel=kernel.radial, kernel.args=list(gamma=0.1))

  # training set MSE
mse(predict(svrf.kl, kmdat.train), kmdat.train$f)
mse(predict(svrf.kp, kmdat.train), kmdat.train$f)
mse(predict(svrf.kr, kmdat.train), kmdat.train$f)
mse(predict(svrf.ks, kmdat.train), kmdat.train$f)

mse(predict(svrg.kl, kmdat.train), kmdat.train$g)
mse(predict(svrg.kp, kmdat.train), kmdat.train$g)
```

```
mse(predict(svrg.kr, kmdat.train), kmdat.train$g)
mse(predict(svrg.ks, kmdat.train), kmdat.train$g)

mse(predict(bh.svr.kl, bh.std.train[,-13]), bh.std.train$medv)
mse(predict(bh.svr.kr, bh.std.train[,-13]), bh.std.train$medv)

  # test set MSE
mse(predict(svrf.kl, kmdat.test), kmdat.test$f)
mse(predict(svrf.kp, kmdat.test), kmdat.test$f)
mse(predict(svrf.kr, kmdat.test), kmdat.test$f)
mse(predict(svrf.ks, kmdat.test), kmdat.test$f)

mse(predict(svrg.kl, kmdat.test), kmdat.test$g)
mse(predict(svrg.kp, kmdat.test), kmdat.test$g)
mse(predict(svrg.kr, kmdat.test), kmdat.test$g)
mse(predict(svrg.ks, kmdat.test), kmdat.test$g)

mse(predict(bh.svr.kl, bh.std.test[,-13]), bh.std.test$medv)
mse(predict(bh.svr.kr, bh.std.test[,-13]), bh.std.test$medv)
```

As in the previous example, the polynomial kernel delivers by far the most accurate predictions for the nonlinear target function g, which is to be expected given the quadratic function used for data generation. Not surprisingly, it also works well for the linear target function f, with which other nonlinear kernels appear to have problems. For the *Boston Housing* data the radial kernel function does not appear to give any improvement.

16.8   Conclusion

Kernel methods constitute an elegant, convenient, and computationally efficient approach to using enhanced representation with linear regression and linear classification algorithms that restrict data usage during model creation and prediction to dot products only. The kernel trick makes it possible to use dot products in high-dimensional spaces of nonlinearly transformed attributes without increased computational effort. A few standard kernel function families are sufficient for most practical needs. Determining the right kernel family and parameters for a particular task may require extensive tuning, though, as no universally good defaults are available. Without sufficiently careful experimental adjustments the predictive potential of this approach may remain waisted and lead to poor results. Domain knowledge, if available, may help narrow down the scope of such experiments, but not replace them entirely.

The most popular and successful kernel-based algorithms – support vector machines and support vector regression – have much more to offer than the implicit representation enhancement. Separating margin maximization by the former, hypersurface flatness maximization by the latter, and relaxing the training error-minimization requirement by both, makes them less prone to overfitting. These advantages often outweigh the inconvenience and effort associated with properly tuning the cost parameter.

The kernel-based SVM and SVR algorithms belong to the most useful, but not the easiest to used modeling algorithms. The underlying optimization problem may be computationally demanding for large data, and the obtained model representation can be hardly considered comprehensible – not even comparing to symbolic decision or regression tree representations, but also to plain linear representations. Judging their utility for particular applications requires balancing their potential prediction quality capabilities and these disadvantages.

16.9   Further readings

The family of kernel-based learning algorithms made an impressive entrance on the machine learning and data mining stage in the 1990s, attracting strong research interest and rapidly gaining popularity. Their development and spreading out coincided, by the way, with that of model ensembles are presented in Chapter 15. The level of excitement that surrounded support vector machines and related algorithms at the turn of the 2000s may be even compared to the renaissance of interest in neural networks a decade earlier, after the backpropagation algorithm was re-invented and became widely known (Rumelhart *et al.* 1986). They now keep a solid position of well theoretically grounded and practically successful modeling algorithms and, as such, are presented by contemporary machine learning and data mining books (Bishop 2007; Han *et al.* 2011; Hastie *et al.* 2011; James *et al.* 2013; Tan *et al.* 2013; Theodoridis and Koutroumbas 2008; Witten *et al.* 2011). There is also a number of dedicated books on support vector algorithms and kernel methods (Cristianini and Shawe-Taylor 2000; Hamel 2009; Herbrich 2001; Schölkopf and Smola 2001). Of those, the last one offers a noteworthy combination of thoroughness and accessibility which makes it, along with its R code examples, a particularly good text for self-study. An insightful tutorial article by Burges (1998) can also be very helpful.

The idea of maximum-margin separating hyperplanes which is at the heart of support vector machines is due to Vapnik (1982). The first hard-margin version of the SVM algorithm was described by Boser *et al.* (1992), who presented both the primal and dual form of the underlying quadratic programming problem. Kernel-based soft-margin SVM, which remains the most widely used member of the support vector family of algorithms, was presented by Cortes and Vapnik (1995). Platt (1998) introduced a dedicated sequential minimal optimization (SMO) algorithm for SVM parameter estimation, substantially more efficient than general-purpose quadratic programming solvers.

The SVR algorithm, as described in this chapter, was presented by Drucker *et al.* (1996). This version of support vector regression, known as ϵ-SVR, was subsequently enhanced by Schölkopf *et al.* (2000), whose v-SVR algorithm automatically finds a possibly minimum value of the tolerance threshold ϵ. An alternative approach of adapting the idea of support vectors to the regression task, called least-squares SVM, was proposed by Suykens and Vandewalle (1999).

Support vector machine classifiers are usually applied to multiclass tasks using the general-purpose multiclass encoding approach presented in Chapter 17. The success of SVM has actually provided a substantial part of motivation for the research on such encoding techniques (Allwein *et al.* 2000). Other noteworthy approaches to multiclass classification using SVM include the directed acyclic graph technique of Platt *et al.* (1999) for combining two-class models created for every pair of classes and the multiclass quadratic optimization method of Crammer and Singer (2001). Platt (2000) showed how a logit link function can be applied to support vector machines predictions to transform them into calibrated class probability predictions. Several other enhancements and variations of support vector and kernel-based modeling are presented in research articles collections (Schölkopf *et al.* 1998; Smola *et al.* 2000). Meyer *et al.* (2003) systematically compared the predictive performance of radial kernel-based support vector models with that achieved by several other classification and regression models, created using algorihms available in R.

Support vector machines took over some part of interest earlier devoted to neural networks, and in particular multilayer nonlinear perceptrons (Hertz *et al.* 1991). While both these types of algorithms have proved useful in numerous applications and may deliver excellent results when used skillfully, the latter may suffer from local optima and do not have a similarly solid theory. Links between these two families of algorithms were highlighted by Collobert and Bengio (2004). In this context it is also worthwhile to mention the work of Freund and Schapire (1999), who presented an extension to the linear perceptron algorithm for approximate margin maximization, which is easier to implement and more efficient than parameter estimation via quadratic optimization, although does not quite match the predictive performance of SVM.

References

Allwein EL, Schapire RE and Singer Y 2000 Reducing multiclass to binary: A unifying approach for margin classifiers. *Journal of Machine Learning Research* **1**, 113–141.

Bishop CM 2007 *Pattern Recognition and Machine Learning*. Springer

Boser BE, Guyon IM and Vapnik VN 1992 A training algorithm for optimal margin classifiers *Proceedings of the Fifth Annual Workshop on Computational Learning Theory (COLT-92)*. ACM Press.

Burges CJC 1998 A tutorial on support vector machines for pattern recognition. *Data Mining and Knowledge Discovery* **2**, 121–167.

Collobert R and Bengio S 2004 Links between perceptrons, MLPs and SVMs *Proceedings of the Twenty-First International Conference on Machine Learning (ICML-2004)*. ACM Press.

Cortes C and Vapnik VN 1995 Support-vector networks. *Machine Learning* **20**, 273–297.

Crammer K and Singer Y 2001 On the algorithmic implementation of multiclass kernel-based vector machines. *Journal of Machine Learning Research* **2**, 265–292.

Cristianini N and Shawe-Taylor J 2000 *An Introduction to Support Vector Machines and Other Kernel-Based Learning Methods*. Cambridge University Press.

Drucker H, Burges CJC, Kaufman L, Smola AJ and Vapnik VN 1996 Support vector regression machines *Advances in Neural Information Processing Systems 9 (NIPS-1996)*. MIT Press.

Freund Y and Schapire RE 1999 Large margin classification using the perceptron algorithm. *Machine Learning* **37**, 277–296.

Hamel LH 2009 *Knowledge Discovery with Support Vector Machines*. Wiley.

Han J, Kamber M and Pei J 2011 *Data Mining: Concepts and Techniques* 3rd edn. Morgan Kaufmann

Hastie T, Tibshirani R and Friedman J 2011 *The Elements of Statistical Learning: Data Mining, Inference, and Prediction* 2nd edn. Springer.

Herbrich R 2001 *Learning Kernel Classifiers: Theory and Algorithms*. MIT Press.

Hertz J, Krogh A and Palmer RG 1991 *Introduction to the Theory of Neural Computation*. Addison-Wesley.

James G, Witten D, Hastie T and Tibshirani R 2013 *An Introduction to Statistical Learning with Applications in R*. Springer

Meyer D, Leisch F and Hornik K 2003 The support vector machine under test. *Neurocomputing* **55**, 169–186.

Platt JC 1998 Fast training of support vector machines using sequential minimal optimization In *Advances in Kernel Methods: Support Vector Learning* (ed. Schölkopf B, Burges CJC and Smola AJ) MIT Press.

Platt JC 2000 Probabilistic outputs for support vector machines and comparison to regularized likeli-hood methods In *Advances in Large Margin Classifiers* (ed. Smola AJ, Barlett P, Schölkopf B and Schuurmans D) MIT Press.

Platt JC, Cristianini N and Shawe-Taylor J 1999 Large margin DAGs for multiclass classification *Advances in Neural Information Processing Systems 12 (NIPS-99)*. MIT Press.

Rumelhart DE, Hinton GE and Williams RJ 1986 Learning internal representations by error propagation In *Parallel Distributed Processing: Explorations in the Microstructure of Cognition* (ed. Rumelhart DE and McClelland JL) vol. 1 MIT Press.

Schölkopf B, Burges CJC and Smola AJ (eds) 1998 *Advances in Kernel Methods: Support Vector Learning*. MIT Press.

Schölkopf B and Smola AJ 2001 *Learning with Kernels*. MIT Press.

Schölkopf B, Smola AJ, Williamson RC and Bartlett PL 2000 New support vector algorithms. *Neural Computation* **12**, 1207–1245.

Smola AJ, Barlett P, Schölkopf B and Schuurmans D (eds) 2000 *Advances in Large Margin Classifiers*. MIT Press.

Suykens JAK and Vandewalle JPL 1999 Least squares support vector machine classifiers. *Neural Processing Letters* **9**, 293–300.

Tan PN, Steinbach M and Kumar V 2013 *Introduction to Data Mining* 2nd edn. Addison-Wesley.

Theodoridis S and Koutroumbas K 2008 *Pattern Recognition* 4th edn. Academic Press.

Vapnik VN 1982 *Estimation of Dependences Based on Empirical Data*. Springer.

Witten IH, Frank E and Hall MA 2011 *Data Mining: Practical Machine Learning Tools and Techniques* 3rd edn. Morgan Kaufmann.

17

Attribute transformation

17.1 Introduction

Attribute transformation alters the data by replacing a selected attribute by one or more new attributes, functionally dependent on the original one, to facilitate further analysis. This is typically performed prior to creating classification, regression, or clustering models to bypass some limitations of the modeling algorithms used or improve model quality. The latter is possible, since – like attribute selection covered in Chapter 19 – attribute transformation modifies the space of models being searched by changing the representation of instances and may make good models more likely or easier to find. This chapter presents the description and examples illustrating the most commonly used attribute transformations.

Example 17.1.1 Attribute transformation algorithms presented in this chapter will be illustrated by R code examples, containing simple implementations and usage demonstrations thereof. The latter will be using the toy *weatherc* data, as well as the more realistic *Vehicle Silhouettes* and *Glass* datasets from the `mlbench` package. The effects of some transformations will be illustrated by creating classification models based on the original and transformed datasets. The `dmr.claseval` and `dmr.util` packages will be used to evaluate prediction quality and provide auxiliary utility functions. The popular R implementations of the decision tree and naïve Bayes algorithms, provided by the `rpart` and `e1071` packages, will be used for this purpose. The following R code sets up the environment for this chapter's demonstrations by loading the packages and datasets, partitioning the larger two of those into training and tests subsets for simple hold-out model evaluation, and creating baseline models (using the original, untransformed dataset versions).

> Ex. 1.3.2
> dmr.data

```
library(dmr.claseval)
library(dmr.util)

library(rpart)
library(e1071)
```

Data Mining Algorithms: Explained Using R, First Edition. Paweł Cichosz.
© 2015 John Wiley & Sons, Ltd. Published 2015 by John Wiley & Sons, Ltd.

```
data(weatherc, package="dmr.data")
data(Vehicle, package="mlbench")
data(Glass, package="mlbench")

set.seed(12)

rv <- runif(nrow(Vehicle))
v.train <- Vehicle[rv>=0.33,]
v.test <- Vehicle[rv<0.33,]

rg <- runif(nrow(Glass))
g.train <- Glass[rg>=0.33,]
g.test <- Glass[rg<0.33,]

  # baseline models and their prediction
v.tree <- rpart(Class~., v.train)
v.tree.pred <- predict(v.tree, v.test, type="c")

g.tree <- rpart(Type~., g.train)
g.tree.pred <- predict(g.tree, g.test, type="c")

v.nb <- naiveBayes(Class~., v.train)
v.nb.pred <- predict(v.nb, v.test, type="c")

g.nb <- naiveBayes(Type~., g.train)
g.nb.pred <- predict(g.nb, g.test, type="c")
```

17.2   Attribute transformation task

Since attribute transformation serves as a form of preprocessing for inductive learning tasks, it can be most naturally defined using the context provided by their definitions presented in Section 1.2. In particular, we will assume the same view of the data as a set of instances from a domain, described by a number of attributes.

17.2.1   Target task

The data mining task for which attribute transformation is to be performed is the *target task*. It is usually one of the three major inductive learning tasks, classification, regression, or clustering. If attribute transformation is used prior to another form of data preprocessing, such as attribute selection, they share the same target task, which is the ultimate modeling task to be addressed after they are all completed. Sometimes the same attribute transformation may be used to preprocess data subsequently used to perform multiple tasks, independently or in combination, e.g., classification for one (discrete) attribute and regression for another (continuous) attribute, or clustering and then classification or regression within clusters. In such cases we can speak of several target tasks for the same attribute transformation task.

While specific attribute transformations tend to be more useful and frequently applied for certain target tasks than for others, the actual transformation techniques described in this chapter are task independent. Discussing target tasks may only be needed to explain their typical application context or justify their utility, but not to present their computational details.

17.2.2   Target attribute

If the target task is the classification or regression task, a *target attribute* is also specified. It is completely irrelevant for most simple attribute transformations, though.

17.2.3   Transformed attribute

While it is possible for practical implementations of attribute transformation to be applied to multiple attributes in a dataset at the same time rather than sequentially one by one, for conceptual clarity we can consider transforming a single attribute as a single instance of the attribute transformation task. Transforming multiple attributes can be then viewed as multiple instances of the task. This does not necessarily match the way some more refined transformation algorithms actually work, since they may exploit inter-attribute relationships or transform one attribute within the context provided by other transformed attributes to better preserve their predictive utility. This is the case, in particular, for certain linear algebra-based transformations that map a set of original attributes into a set of new attributes in such a way that each new attribute depends on multiple (possibly all) original attributes. For the simple techniques described in this chapter, the assumption of each attribute being transformed independently holds perfectly, though.

17.2.4   Training set

The target task's training set is usually the dataset on which attribute transformation is performed, and can therefore be considered the *training set* for attribute transformation as well. A smaller subset thereof or an independent sample from a larger available dataset can be used as well.

As explained when discussing model evaluation caveats in Section 7.3.2, appropriately separating the data used in the modeling process from that used to subsequently evaluate models is crucial for avoiding the effect of evaluation overfitting, i.e., arriving at unreliably optimistic model performance estimates. It has to be underlined that whenever attribute transformation becomes a part of the modeling process, the attribute transformation training set becomes a part of the data that impacts the finally obtained model, even if it is entirely separate from the training set passed to the subsequently applied modeling algorithm. It is therefore important to strictly separate the data used for attribute transformation from that subsequently used for the evaluation of models created when performing the target task after attribute transformation. More precisely, this is the case for modeling transformations, as explained below.

17.2.5   Modeling transformations

Attribute transformation is often viewed as a simple operation that, once applied to the available dataset, can be completely ignored thereafter, with all subsequent analytical processing performed on the transformed data. This is not exactly correct if attribute transformation is performed not on a per-instance basis, but based on the whole dataset to be transformed, and is a part of a modeling process in which one or more predictive models need to be created and evaluated. The latter is actually quite often the case. In such situations one has to take care of properly transforming not only the training set, but also any other datasets to which the models may be subsequently applied (including, in particular, data subsets used for model evaluation). Two apparently natural and reasonable approaches:

All-data transformation. Apply attribute transformation to the combined dataset, including the training set and any other available datasets from the same domain that might be subsequently used for any purpose.

Separate transformation. Apply attribute transformation separately to the training set and each of the other available datasets from the same domain before using any of them for any purpose,

may be applicable in some specific circumstances, but in general they are both severely flawed. The all-data approach has two obvious downsides:

- performing attribute transformation on the whole available dataset makes the model creation process potentially benefit from indirectly having access to all that data (since attribute transformation may cause some characteristics of the whole dataset, used to determine the exact way of transforming attributes, to have impact on their transformed values in the training set), which makes it impossible to evaluate it reliably,

- the ultimate purpose of any predictive models is to be applied to new data that were not available at the time of their creation, which makes all-data transformation completely pointless.

The separate transformation approach yields other problems:

- transforming data to which a model is to be applied separately from the data used for its creation may degrade its performance due to data inconsistencies,

- datasets to which a model is to be applied may be arbitrarily small, including even single instances, for which several attribute transformations are either not applicable or unreliable.

What makes therefore most sense, even if it is surprisingly unpopular, is to consider attribute transformation as a simple degenerate form of modeling, which creates a *transformation model* from the training set and makes it applicable to any dataset from the same domain. Attribute transformations following this approach will be referred to as *modeling transformations*. The exact representation of such transformation models depends on the particular transformation and will be discussed in the corresponding subsections. According to this approach, and contrary to many popular views, replacing the original attribute with a newly created transformed attribute cannot be simply reduced to replacing the original values with new values in the training set. What attribute transformation should produce is not (or, at least, not just) a modified copy of the data, but the definition of a new attribute that can be subsequently applied to transform any dataset from the same domain from which the training set originated.

The modeling view of attribute transformation makes it clear how to apply predictive models obtained based on transformed training sets. Whenever a model is created after attribute transformation, it has to be considered tied to the transformation model that was applied to the training set beforehands. It is the very same transformation model that needs to be subsequently applied to any datasets for which the model would be used to predict.

Example 17.2.1 The R code presented below defines the `transmod.all` function to generate a wrapper around a single-attribute modeling transformation that applies it to all attributes specified by the provided formula. The transformation specified via the `transm` argument is a function that receives the vector of values of a single attribute for a dataset and returns the transformation model created therefrom. The optional `condf` argument specifies a transformation condition – a function that must return `TRUE` for an attribute value vector to be transformed. The `x.vars` function is used to extract attribute names from an R formula. The `predict.transmod` function can be similarly used to generate a multiattribute transformation model application wrapper around a single-attribute transformation application function. The wrapper calls the latter for all attributes for which there is a corresponding component in the transformation model.

`dmr.util`

The `transmod.all` function is used to generate an all-attribute wrapper around a simple centering (mean subtraction) transformation, subsequently applied to the *weatherc* data. The obtained transformation model is then applied using the `predict.transmod` function to generate the transformed version of the same dataset.

```
## wrap single-attribute modeling transformation transm
## so that it is applied to all attributes for which condf returns TRUE
transmod.all <- function(transm, condf=function(v) TRUE)
{
  function(formula, data, ...)
  {
    attributes <- x.vars(formula, data)
    sapply(attributes,
           function(a) if (condf(data[[a]])) transm(data[[a]], ...),
           simplify=FALSE)
  }
}

## apply transformation model to a dataset
predict.transmod <- function(pred.transm)
{
  function(model, data, ...)
  {
    as.data.frame(sapply(names(data),
                         function(a)
                         if (a %in% names(model) && !is.null(model[[a]]))
                           pred.transm(model[[a]], data[[a]], ...)
                         else data[[a]],
                         simplify=FALSE))
  }
}

  # simple centering (mean subtraction) transformation
center.m <- transmod.all(mean, is.numeric)
  # performed on the weatherc data
w.cm <- center.m(play~., weatherc)
  # applied to the weatherc data
predict.center.m <- predict.transmod(function(m, v) v-m)
predict.center.m(w.cm, weatherc)
```

17.2.6   Nonmodeling transformations

The above discussion applies to many, but not all commonly used attribute transformations. Those that operate on a per-instance basis may be applied to any datasets (including single instances), either combined or separately, with the very same effect. It makes no sense therefore to consider them models. We will refer to them as to *nonmodeling transformations*. For any instance, its transformed attribute values depend only on the value of the original attribute for the same instance.

Example 17.2.2 The following R code defines the `transnonmod.all` function to generate a wrapper around any specified nonmodeling transformation that applies it to all attributes in a dataset. The transformation specified via the `transnm` takes a vector of a single attribute's values and returns the transformed version thereof, represented by one or more value vectors. The optional `condf` argument specifies a transformation condition – a function that must return TRUE for an attribute value vector to be transformed. The `transnonmod.all` function is used to generate all-attribute wrappers around two simple transformations of continuous attributes: centering (this time treated as a nonmodeling transformation) and integer division by 10. The latter yields two new attributes, representing the quotient and the remainder. They are both subsequently applied to the *weatherc* data.

```
transnonmod.all <- function(transnm, condf=function(v) TRUE)
{
  function(formula, data, ...)
  {
    attributes <- x.vars(formula, data)
    as.data.frame(sapply(names(data),
                    function(a)
                        if (a %in% attributes && condf(data[[a]]))
                            transnm(data[[a]], ...)
                        else data[[a]],
                    simplify=FALSE))
  }
}

# simple centering (mean subtraction) transformation
center.nm <- transnonmod.all(function(v) v-mean(v), is.numeric)
# performed on the weatherc data
center.nm(play~., weatherc)

# simple round to a multiple of 10 transformation
divmod10 <- transnonmod.all(function(v) cbind(v %/% 10, v %% 10), is.numeric)
# performed on the weatherc data
divmod10(play~., weatherc)
```

Not surprisingly, the first transformed dataset is the same as obtained in the previous example. Note, however, that while the latter was obtained by applying the previously created transformation model (which could also be applied to any other instances from the same domain), the one presented here was obtained without a model. Integer division is actually a better example of nonmodeling transformations, though, since the transformed attribute values for each instance do not indeed depend on any other instances.

17.3   Simple transformations

The most commonly used modeling and nonmodeling attribute transformations perform simple arithmetic or logical operations, needed to adjust the data to the requirements or capabilities of some classification, regression, or clustering algorithms. This section reviews those that are particularly popular and routinely applied.

17.3.1   Standardization

Standardization is used to compensate for differences in continuous attribute distributions that could mislead some modeling algorithms. This applies, in particular, to dissimilarity-based clustering algorithms, presented in Chapters 12 and 13. They all rely on instance dissimilarity measures, such as described in Chapter 11. Many of those may be misled by substantial differences in attribute ranges and distributions. Standardization may help by making all attributes have the same mean of 0 and standard deviation of 1, which is accomplished by replacing each continuous attribute a by a new attribute a' defined as follows:

$$a'(x) = \frac{a(x) - m_T(a)}{s_T(a)} \tag{17.1}$$

where $m_T(a)$ and $s_T(a)$ are the mean and standard deviation of the original attribute, respectively, on the training set T.

Notice that using training set statistics to calculate new attribute values makes standardization a modeling transformation. The transformation model determined using the training set, represented by the mean and standard deviation values, can then be applied to an arbitrary new dataset from the same domain. However, it is not entirely uncommon to see standardization being performed in a – severely flawed – nonmodeling way, either on the complete data or separately on particular data subsets.

Example 17.3.1 The following R code uses the `transmod.all` and `predict.transmod` wrapper generators from Example 17.2.1 to implement standardization as a modeling transformation. The resulting `std.all` and `predict.std` functions, for transformation model creation and application, are then used to standardize all continuous attributes in the *weatherc* and *Glass* datasets. For the latter, the standardization model created on the training set is applied to both the training and test set.

```
## single-attribute standardization transformation
std <- function(v)
{
  list(mean=mean(v, na.rm=TRUE), sd=sd(v, na.rm=TRUE))
}

## standardization of all continuous attributes
std.all <- transmod.all(std, is.numeric)

## standardization model prediction
predict.std <- predict.transmod(function(m, v)  (v-m$mean)/m$sd)

  # standardization model for the weatherc data
w.stdm <- std.all(play~., weatherc)
```

```
  # applied to the weatherc data
w.std <- predict.std(w.stdm, weatherc)

  # standardization model for the Glass data
g.stdm <- std.all(Type~., g.train)
  # applied to the training and test sets
g.train.std <- predict.std(g.stdm, g.train)
g.test.std <- predict.std(g.stdm, g.test)
```

17.3.2 Normalization

While standardization unifies the distributions of continuous attributes, normalization unifies their ranges. It is similarly motivated and has a similar scope of applications. Being less popular than standardization, it may be preferred in some situations. It forces all continuous attributes to have the same range, usually [0, 1] or [−1, 1], by appropriate scaling. Assuming the former, the new attribute replacing attribute a is defined as

$$a'(x) = \frac{a(x) - \min_{x' \in T} a(x)}{\max_{x' \in T} a(x) - \min_{x' \in T} a(x)} \tag{17.2}$$

This view of normalization as attribute scaling is the most common in data mining, but the term "normalization" tends to be used with other meanings as well, particularly in statistics.

Similarly to standardization, normalization is clearly a modeling transformation (even if not always used as such in practice), as it requires training set minimum and maximum attribute values. When applied to new data, they may yield some transformed values out of the [0, 1] interval, if the original values fall beyond the range observed on the training data, but this is usually acceptable. For obvious reasons, normalization is extremely sensitive in outliers and should be always used with particular care.

Example 17.3.2 Strictly following the pattern of the previous example, the R code presented below implements and demonstrates continuous attribute normalization.

```
## single-attribute normalization transformation
nrm <- function(v)
{
  list(min=min(v, na.rm=TRUE), max=max(v, na.rm=TRUE))
}

## normalization of all continuous attributes
nrm.all <- transmod.all(nrm, is.numeric)

## standardization model prediction
predict.nrm <- predict.transmod(function(m, v) (v-m$min)/(m$max-m$min))

  # normalization model for the weatherc data
w.nrmm <- nrm.all(play~., weatherc)
  # applied to the weatherc data
w.nrm <- predict.nrm(w.nrmm, weatherc)

  # normalization model for the Glass data
g.nrmm <- nrm.all(Type~., g.train)
```

```
# applied to the training and test sets
g.train.nrm <- predict.nrm(g.nrmm, g.train)
g.test.nrm <- predict.nrm(g.nrmm, g.test)
```

17.3.3   Aggregation

A large number of discrete attribute values may be problematic for some implementations of modeling algorithms, such as decision trees or the naïve Bayes classifier, increasing the computation time at best or sometimes making them refuse to work altogether. The former is due to the obvious dependence of their computational expense on the number of attribute values and the latter may be due to the limited capacity of some adopted internal data structures. Even if a model can be created in acceptable time, its comprehensibility may be considerably reduced. This provides motivation for discrete attribute value aggregation.

In principle, aggregation can be addressed using quite refined algorithms, similar to those used for discretizing continuous attributes, extensively discussed in Chapter 18. It is much more common and often sufficient to stick with a much simpler approach of just combining a number of the least frequently occurring values. It may be unobvious at first, but even in this basic form this should be viewed again a modeling transformation, with the training set used to obtain attribute value frequencies and decide which values will be retained and which combined.

Example 17.3.3 A simple frequency-based version of discrete attribute aggregation is implemented as a modeling transformation by the R code presented below. The single-attribute aggregation function `agg` creates an aggregation model containing the list of the specified number of the most frequent original values to retain and the new value to replace the remaining original values. Multiattribute functions for aggregation model creation and application are then generated using the `transmod.all` and `predict.transmod` functions from Example 17.2.1 and demonstrated on the *weatherc* data. Its tiny size does not provide much opportunities for aggregation, with only the `outlook` attribute taking more than two values.

```
## single-attribute aggregation transformation with m most frequent values retained
## and others replaced by comb.val
agg <- function(v, m, comb.val="other")
{
  list(retained=names(sort(table(v), decreasing=TRUE))[1:min(m, nlevels(v))],
      combined=comb.val)
}

## normalization of all discrete attributes
agg.all <- transmod.all(agg, is.factor)

## aggregation model prediction
predict.agg <- predict.transmod(function(m, v)
                                  factor(ifelse(v %in% m$retained,
                                                as.character(v),
```

```
                                    ifelse(is.na(v), NA, m$combined)),
                       levels=c(m$retained, m$combined)))

  # aggregation model for the weatherc data
w.aggm <- agg.all(play~., weatherc, 1)
  # applied to the weatherc data
w.agg <- predict.agg(w.aggm, weatherc)
```

17.3.4   Imputation

Some modeling algorithms include internal techniques for handling missing attribute values
with a possibly minimal loss of predictive performance. Such techniques were presented for
decision trees in Section 3.7 and for regression trees in Section 9.7. For some other modeling algorithms simply ignoring missing values may be a perfectly acceptable solution, as
discussed for the naïve Bayes classifier in Section 4.4.4. It is often necessary, however, to take
care of missing values outside a modeling algorithm, to make it possible to create models and
generate predictions using incomplete data. Filtering out incomplete instances from the data
if they are relatively sparse may work for model creation, but is of course unacceptable for
prediction. With a higher rate of missingness all or nearly all instances may be incomplete and
all or nearly all attributes may have missing values, making any simple omitting techniques
useless. This is where the imputation transformation is needed.

Imputation is basically filling in missing values with reasonable guesses. In the simplest
and most common case, this is performed using means or medians for continuous attributes
and modes for discrete attributes. A more refined form of imputation is also possible where a
classification or regression model is created to predict missing values of a particular attribute
based on the available values of other attributes, but its additional complexity does not necessarily always pay off with predictive performance improvement. What is essential, in any
case, is to consider imputation a modeling attribute transformation, with a transformation
model determined using the training set also applicable to any other datasets from the same
domain. In the simplest but most common case of mean, median, or mode imputation, it is
the means, medians, or modes of particular attributes that have to be determined using the
training set and reapplied whenever transforming new data prior to prediction.

Example 17.3.4 The R code presented below provides a simple implementation of missing
value imputation as a modeling transformation. The imputed values are stored in an imputation model that can be then applied to an arbitrary dataset from the same domain. This
is demonstrated using copies of the *weatherc* and *Glass* datasets with some attribute values
removed. For the latter the imputation model determined on the training set is applied both to
the training and test sets.

```
## single-attribute mean/median/mode imputation transformation
imp <- function(v, med=FALSE)
{
  if (!is.numeric(v))
```

```
    modal(v)
  else if (med)
    median(v, na.rm=TRUE)
  else
    mean(v, na.rm=TRUE)
}

## imputation for all attributes
imp.all <- transmod.all(imp)

## imputation model prediction
predict.imp <- predict.transmod(function(m, v) { v[is.na(v)] <- m; v } )

weathercm <- weatherc
weathercm$outlook[c(1, 3)] <- NA
weathercm$temperature[c(2, 4)] <- NA
weathercm$humidity[c(3, 5)] <- NA
weathercm$wind[c(4, 6)] <- NA

gm.train <- g.train
gm.train[sample.int(nrow(gm.train), 0.1*nrow(gm.train)),
        sample.int(ncol(gm.train)-1, 3)] <- NA
gm.test <- g.test
gm.test[sample.int(nrow(gm.test), 0.1*nrow(gm.test)),
        sample.int(ncol(gm.test)-1, 3)] <- NA

  # imputation model for the weatherc data
wm.impm <- imp.all(play~., weathercm, med=TRUE)
  # applied to the weatherc data
wm.imp <- predict.imp(wm.impm, weathercm)

  # imputation model for the Glass data
gm.impm <- imp.all(Type~., gm.train)
  # applied to the training and test sets
gm.train.imp <- predict.imp(gm.impm, gm.train)
gm.test.imp <- predict.imp(gm.impm, gm.test)
```

17.3.5   Binary encoding

Not only the large number of discrete attribute values may cause problems for modeling algorithms. Some of them are inherently limited to continuous attributes only and cannot directly process discrete attributes at all. This is the case, in particular, for modeling algorithms using instance dissimilarity measures without discrete attribute handling capability or using parametric model representation. Specifically, the issue always arises for linear classification and regression algorithms and also often for dissimilarity-based clustering algorithms. This is when discrete attributes may have to be transformed to a numerical representation.

A naïve approach of just numbering discrete values consecutively and using such ordinal numbers as new values may be only acceptable for ordered attributes, which are often treated as continuous anyway. Otherwise it would introduce an objectively nonexisting and meaningless ordering into attribute values, likely to mislead the model creation process.

A much more reasonable approach is based on simple 1-of-k coding. The idea is to replace a discrete k-valued attribute $a : X \rightarrow \{v_1, v_2, \ldots, v_k\}$ with k binary attributes $a_j : X \rightarrow \{v_+, v_-\}$, defined for $j = 1, 2 \ldots, k$ as

$$a_j(x) = \begin{cases} v_+ & \text{if} \quad a(x) = v_j \\ v_- & \text{otherwise} \end{cases} \tag{17.3}$$

where v_+ and v_- are different numbers (typically 1 and 0 or 1 and -1, respectively). These new attributes can be then treated as continuous for any further calculations. Sometimes this transformation is performed internally by modeling algorithms when processing discrete attributes, without explicitly creating a transformed dataset copy.

Notice that the number of binary attributes replacing each discrete attributes can be actually decreased by 1 without reducing the representation power. These binary attributes, defines as presented above, would correspond to arbitrarily selected $k - 1$ out of k values of the original attributes. Whenever all of them take a value of v_-, this indicates that the original attribute takes the remaining kth value. More precisely, a discrete k-valued attribute $a : X \rightarrow \{v_1, v_2, \ldots, v_k\}$ would be replaced with $k - 1$ binary attributes $a_j : X \rightarrow \{v_+, v_-\}$, defined for $j = 1, 2 \ldots, k - 1$ in the same way as before. The value $a(x) = v_j$ for any $j = 1, 2, \ldots, k - 1$ is represented by $a_j(x) = v_+$ and $a_{j'}(x) = v_-$ for $j' \neq j$, and the value $a(x) = v_k$ is represented by $a_1(x) = a_2(x) = \cdots = a_{k-1}(x) = v_-$. This modification is usually desirable, since it simplifies the representation and reduces the number of parameters used, which is likely to simplify the model creation process and reduce the risk of overfitting. In the basic form with as many binary attributes as discrete attribute values the encoding is redundant.

Unlike the other simple transformations discussed above, binary encoding is a nonmodeling transformation, since the transformed attribute values for each instance are determined based on the original values and the encoding scheme only. The latter is predetermined rather than derived from data. Binary attributes representing an originally continuous attribute by 1-of-k encoding are referred to as *dummy coding* or *treatment contrasts* in statistical terminology and represent a specific instantiation of a more general methodology of introducing continuous attributes representing discrete attribute values or their combinations to parametric classification and regression models. The binary attributes introduced by dummy coding are referred to as *dummy variables*.

Example 17.3.5 The discrete attribute encoding transformation as described above is implemented and illustrated by the following R code. The `discode1` function transforms a single discrete attribute value by returning a vector of binary values. The `discode.a` function applies this transformation to all values of a single attribute in a dataset (a complete dataset column). The `transnonmod.all` nonmodeling transformation wrapper generator from Example 17.2.2 applied to the latter yields the `discode` function that performs discrete value encoding for all attributes in a dataset. All these functions accept three optional arguments: `b` that specifies the set of binary values to use, `red` that can be used to request the redundant version of the transformation, and `na.all` that instructs them to encode missing attribute values as if the attribute took all of its possible values simultaneously rather than return a vector of missing values. The latter may turn out convenient when dealing with datasets containing missing values.

```
## binary encoding of a single discrete attribute value
discode1 <- function(v, b=c(0,1), red=FALSE, na.all=FALSE)
{
  r <- 1-as.integer(red)
  if (is.factor(v) && ! (is.na(v) && na.all))
    b[1+as.integer(v==levels(v)[1:(nlevels(v)-r)])]
  else if (is.factor(v))
    rep(b[2], nlevels(v)-r)
  else
    v
}

## binary encoding of an discrete attribute
discode.a <- function(a, b=c(0,1), red=FALSE, na.all=FALSE)
{
  do.call(rbind, lapply(a, discode1, b, red, na.all))
}

## binary encoding of all discrete attributes in a given dataset
discode <- transnonmod.all(discode.a)

  # encoding a single attribute value
discode1(weatherc$outlook[1])
discode1(weatherc$outlook[1], red=TRUE)
discode1(factor(NA, levels=levels(weatherc$outlook)))
discode1(factor(NA, levels=levels(weatherc$outlook)), na.all=TRUE)

  # encoding single attributes of the weatherc data
discode.a(weatherc$outlook)
discode.a(weatherc$temperature)
discode.a(weatherc$outlook, b=c(-1,1), red=TRUE)
discode.a(weatherc$wind, b=c(-1,1))

  # encoding single instances of the weatherc data
discode(~., weatherc[1,])
discode(~., weatherc[1,], red=TRUE)

  # encoding the complete weatherc data
discode(~., weatherc)
  # leave the target attribute unchanged
discode(play~., weatherc)
```

17.4 Multiclass encoding

While most popular classification algorithms, such as decision trees and the naïve Bayes classifier, can handle an arbitrary number of classes (within reasonable practical limits), there are also algorithms that heavily rely on the assumption of dealing with two-class classification tasks only. This is the case, in particular, for linear classification algorithms presented in Chapter 5, for the highly successful AdaBoost ensemble modeling algorithm presented in Section 15.5.3, and the increasingly popular SVM algorithm presented in Section 16.2. There also performance measures, in particular those derived from the confusion matrix and the related ROC analysis, presented in Sections 7.2.4 and 7.2.5, specifically designed for the two-class case only. While many practical classification task instantiations may actually address binary classification only – with diagnostics and anomaly (in particular, fraud) detection being the most typical examples – multiclass tasks are also common in data mining practice. Modeling algorithms and model evaluation techniques designed for the former may be applied to the latter by means of binary multiclass encoding, based on decomposing the original multiclass task into a number of two-class tasks.

17.4.1 Encoding and decoding functions

Multiclass encoding is based on the decomposition of the original classification task with a multiclass target concept into a number of tasks with two-class target concepts. Consider replacing the concept $c : X \rightarrow C$, where $|C| > 2$, with a set of m concepts $c_1, c_2, \ldots, c_m$, $c_i : X \rightarrow \{0, 1\}$. This can be represented by a encoding function $\kappa : C \rightarrow \{0, 1\}^m$ that assigns a binary vector of length m to each class label from C. Then the binary concepts are defined as follows:

$$c_i(x) = \kappa(c(x))[i] \tag{17.4}$$

for $i = 1, 2, \ldots, m$, where the square bracket indexing notation is used to refer to binary vector elements. A separate classification model h_i can then be created for each concept c_i. For these models – referred to as *binary models* thereafter – to be useful, their predictions have to be combined appropriately, to obtain the original class labels rather that 0/1 vectors. This can be accomplished using a decoding function $\kappa^{-1} : \{0, 1\}^m \rightarrow C$, which may be considered the inverse of the encoding function:

$$h_*(x) = \kappa^{-1}(\mathbf{h}(x)) \tag{17.5}$$

where $\mathbf{h}(x)$ denotes the vector of binary model predictions $h_1(x), h_2(x), \ldots, h_m(x)$ for instance x.

In principle, an arbitrary binary code can be adopted for multiclass task decomposition, including even the simplest natural binary or Grey code, with codeword length equal $\lceil \log_2 |C| \rceil$. These can be found not to be particularly well suited to this application, though. This is because, due to the lack of code redundancy, even the misclassification of an instance by a single binary model will usually cause the decoded class to be incorrect. Selected codes that are more appropriate for multiclass task decomposition are presented in the subsequent sections.

It is worthwhile to explicitly mention one difficulty that may arise during the prediction decoding process. For most commonly used encoding functions, due to their redundancy, only a subset of the set of all length-m binary vectors $\{0, 1\}^m$ correspond to class labels from C:

$$C_\kappa = \{\kappa(d) \mid d \in C\} \subset \{0, 1\}^m \tag{17.6}$$

which are *valid codewords* for the encoding function κ. This leaves the inverse function κ^{-1} undefined for binary vectors outside this subset, which may occur when applying models $h_1, h_2, \ldots, h_m$. Special precautions may be required to make sure that predicted class decoding is always possible. One possibility is to – explicitly or implicitly – use a dissimilarity measure over binary strings $\delta : \{0, 1\}^m \times \{0, 1\}^m \rightarrow \mathcal{R}$ to identify the least dissimilar valid codeword to that produced by binary model application:

$$\kappa^{-1}(\mathbf{h}(x)) = \arg \min_{d \in C} \delta(\mathbf{h}(x), \kappa(d)) \tag{17.7}$$

This effectively extends the domain of κ^{-1} from C_κ (the codomain of κ) to $\{0, 1\}^m$. The κ^{-1} notation will be used to refer to the decoding function regardless of whether it is strictly the inverse of the encoding function or an extension thereof.

Multiclass decomposition by using encoding and decoding functions strongly resembles ensemble modeling discussed in Chapter 15, and may be actually considered a specific form thereof. Multiple models are created for the same task and then combined. What is specific to multiclass decomposition are the mechanisms of base model creation and aggregation.

Example 17.4.1 To illustrate the idea of multiclass task decomposition, the following R code defines the `multi.class` function which is a wrapper generator. It takes a modeling algorithm and a prediction function on input and returns the list of two functions, for multiclass model creation and prediction. The former uses the supplied encoding function to generate multiple binary concepts and calls the modeling algorithm for each of them, to create the corresponding model. It returns the list of binary models obtained that way along with the set of original class labels that will be needed for prediction. The latter applies all the binary models to the data and then uses the supplied decoding function to generate multiclass predictions based on their two-class predictions. The encoding function takes three arguments, the vector of class labels for all instances, the target concept attribute name, and the vector of all possible class labels, and returns a matrix with rows corresponding to instances and columns corresponding to the new binary concepts. The decoding function takes two arguments, a matrix of binary model predictions, with rows corresponding to instances and columns corresponding to the binary models, and the vector of possible classes of the original multiclass concept (more precisely, a factor with both the contents and levels equal to the set of original class labels), and returns the vector of decoded predictions. Trivial encoding and decoding functions, using the natural binary code, are defined and used for this illustration, as more practically useful encoding schemes will be discussed later and illustrated by subsequent examples. They use the `int2binvec` and `binvec2int` functions. The implementation of decoding is particularly simplistic, as instead of matching the codeword against a list of valid codewords to find the closest match it simply converts the codeword to an integer and forces it to be between 1 and the number of classes. This is performed using the `clip.val` function. Some other utility functions are also used: `x.vars`, `y.var`, and `flevels`. The resulting multiclass wrappers around `rpart` and `naiveBayes` are applied to the *Vehicle Silhouettes* and *Glass* datasets.

> dmr.util

> dmr.util

> dmr.util

```
## natural binary multiclass encoding function
multi.enc.nbc <- function(d, class, clabs=levels(d))
{
  `colnames<-`(ed <- t(sapply(as.integer(d), int2binvec, length(clabs))),
          paste(class, 1:ncol(ed), sep="."))
}
```

```
## natural binary multiclass decoding function
multi.dec.nbc <- function(pred, clabs)
{
  clabs[clip.val(apply(pred, 1, binvec2int), 1, length(clabs))]
}

## generate a multiclass wrapper around alg using predf for prediction
## and the specified encoding and decoding functions
multi.class<-function(alg, predf=predict,
                       encode=multi.enc.nbc, decode=multi.dec.nbc)
{
  mc.alg <- function(formula, data, ...)
  {
    attributes <- x.vars(formula, data)
    class <- y.var(formula)
    aind <- names(data) %in% attributes
    clabs <- flevels(data[[class]])

    d.enc <- encode(data[[class]], class, clabs)
    binmodels <- lapply(1:ncol(d.enc),
                        function(i)
                        alg(make.formula(colnames(d.enc)[i], attributes),
                           `names<-`(cbind(data[,aind],
                                           factor(d.enc[,i], levels=0:1)),
                                     c(names(data[,aind]), colnames(d.enc)[i])),
                           ...))
    list(binmodels=binmodels, clabs=clabs)
  }

  mc.predict <- function(model, data, ...)
  {
    decode(sapply(model$binmodels, predf, data, ...), model$clabs)
  }

  list(alg=mc.alg, predict=mc.predict)
}

  # encoding class labels and decoding verification
v.nbc <- multi.enc.nbc(Vehicle$Class, "Class")
err(multi.dec.nbc(v.nbc, levels(Vehicle$Class)), Vehicle$Class)

  # basic encoding applied to rpart
rp.n <- multi.class(rpart, predf=function(...) predict(..., type="c"))
v.tree.n <- rp.n$alg(Class~., v.train)
v.tree.n.pred <- rp.n$predict(v.tree.n, v.test)
g.tree.n <- rp.n$alg(Type~., g.train)
g.tree.n.pred <- rp.n$predict(g.tree.n, g.test)

  # basic encoding applied to naive Bayes
nb.n <- multi.class(naiveBayes)
v.nb.n <- nb.n$alg(Class~., v.train)
v.nb.n.pred <- nb.n$predict(v.nb.n, v.test)
g.nb.n <- nb.n$alg(Type~., g.train)
g.nb.n.pred <- nb.n$predict(g.nb.n, g.test)
```

17.4.2   1-ok-k encoding

The simplest and most commonly used approach to multiclass encoding is based on the same 1-of-k code that is typically used for the binary encoding of discrete attributes. This technique uses the encoding function defined as follows:

$$\kappa(d)[i] = \begin{cases} 1 & \text{if} \quad d = d_i \\ 0 & \text{otherwise} \end{cases} \tag{17.8}$$

for $i = 1, 2, \ldots , |C|$, where $C = \{d_1, d_2, \ldots , d_{|C|}\}$ is the original set of class labels. Each class label is therefore encoded by a binary vector of length $|C|$ with a single 1 on the corresponding position.

Basic 1-of-k encoding is unfortunately prone to the above-mentioned issue of the decoding function not being properly defined as the inverse of the encoding function in its particularly severe form. While the codomain of the encoding function contains exactly $|C|$ different binary vectors, each with a single 1, a binary string with two or more 1's may be obtained during prediction, if several binary models return class 1 for the same instance. Any resolution scheme for such prediction conflicts based on codeword dissimilarity would be totally arbitrary and possibly leading to degraded prediction quality, since a codeword with 1's on several positions is equally dissimilar to each valid codeword with a single 1 on one of these positions.

This is why 1-of-k multiclass encoding is applied with scoring classification models only, which are, as discussed in Section 1.3.3, capable of not only producing 0/1 class labels, but also real-valued scores $\pi(x)$ for each instance x. This makes it possible to define the decoding function as a function of scores, $\kappa^{-1} : \mathcal{R}^m \to C$, rather than class labels:

$$h_*(x) = \kappa^{-1}(\pi(x)) = \arg\max_i \pi_i(x) \tag{17.9}$$

where $\pi(x)$ denotes the vector of scores $\pi_1(x), \pi_2(x), \ldots , \pi_{|C|}(x)$ for instance x and π_i is the scoring function of model h_i. This simply follows the recommendation of the model that returns the highest score for the instance being classified. While this is not, strictly speaking, the inverse of the encoding function, we continue to use the κ^{-1} notation.

Example 17.4.2 To illustrate the 1-of-k multiclass encoding technique, the following R code defines the encoding and decoding functions based on the 1-of-k code and uses them as arguments to the `multi.class` function from the previous example. This generates 1-of-k encoding wrappers around the `rpart` and `naiveBayes` functions which are then demonstrated on the *Vehicle Silhouettes* and *Glass* datasets. The encoding functions assign one binary concept to each of the multiple original classes. The decoding function assumes that the predictions of the binary models created for these concepts are real-valued scores and returns the class labels corresponding to the maximum scores. This is achieved by using the second column (corresponding to class 1) of probabilistic predictions for `rpart` and `naiveBayes` models.

```
## 1-of-k multiclass encoding function
multi.enc.1ofk <- function(d, class, clabs=levels(d))
{
  `colnames<-`(sapply(clabs, function(cl) as.integer(d==cl)),
               paste(class, clabs, sep="."))
}


## 1-of-k multiclass encoding function
multi.dec.1ofk <- function(pred, clabs)
{
  clabs[max.col(pred)]
}


  # encoding class labels and decoding verification
v.1ofk <- multi.enc.1ofk(Vehicle$Class, "Class")
err(multi.dec.1ofk(v.1ofk, levels(Vehicle$Class)), Vehicle$Class)


  # 1-of-k encoding applied to rpart
rp.1 <- multi.class(rpart, predf=function(...) predict(...)[,2],
                    encode=multi.enc.1ofk, decode=multi.dec.1ofk)
v.tree.1 <- rp.1$alg(Class~., v.train)
v.tree.1.pred <- rp.1$predict(v.tree.1, v.test)
g.tree.1 <- rp.1$alg(Type~., g.train)
g.tree.1.pred <- rp.1$predict(g.tree.1, g.test)


  # 1-of-k encoding applied to naive Bayes
nb.1 <- multi.class(naiveBayes, predf=function(...) predict(..., type="r")[,2],
                    encode=multi.enc.1ofk, decode=multi.dec.1ofk)
v.nb.1 <- nb.1$alg(Class~., v.train)
v.nb.1.pred <- nb.1$predict(v.nb.1, v.test)
g.nb.1 <- nb.1$alg(Type~., g.train)
g.nb.1.pred <- nb.1$predict(g.nb.1, g.test)
```

17.4.3   Error-correcting encoding

Multiclass encoding based on the 1-of-k code is only applicable with scoring binary models and heavily depends on their ability to deliver reliable scores. While they happen to be the type of models for which multiclass encoding is most often needed and used, a different approach is needed for nonscoring models. A universal technique that works well for arbitrary binary models is based on *error-correcting* codes. Such codes, commonly used for data transfer through unreliable communication channels or data storage on unreliable media, are designed with purposeful redundancy to minimize the risk of incorrect codeword decoding even if some of its bits are flipped. This perfectly suits the requirements of multiclass task decomposition, by reducing the risk of misclassification of an instance even if it is misclassified by some binary models.

The following two desirable properties of error-correcting codes are crucial for their application to multiclass encoding:

Codeword separation. All pairs of valid codewords should differ on possibly many bits.

Bit independence. There should be no correlation between individual bits.

The former makes correct decoding possible with a small number of bits flipped. The latter avoids using overly many useless bits. There are several possible approaches to designing codes with these properties. If the required number of different codewords is relatively small, as in most classification tasks, the following code design scheme can be, applied with k denoting the number of codewords:

Codeword 1. $2^{k-1} - 1$ ones.

Codeword 2. 2^{k-2} zeroes followed by $2^{k-2} - 1$ ones.

Codeword 3. 2^{k-3} zeroes followed by 2^{k-3} ones followed by 2^{k-3} zeroes followed by $2^{k-3} - 1$ ones.

...

Codeword i. Alternating sequences of 2^{k-i} zeros and ones (with one bit dropped from the last sequence of ones).

This code generation scheme can be verified to be equivalent to the following recursive definition:

$$v_k(i) = \text{rev}(\overline{v_{k-1}(i-1) \bullet 1}) \bullet v_{k-1}(i-1) \tag{17.10}$$

where

- $v_k(i)$ denotes the ith codeword of the code for a total of k codewords,

- neg denotes the bit negation operation,

- rev denotes the reversion operation,

- "$\bullet$" denotes the concatenation operation.

This produces $2^{(k-1)} - 1$ output bits, making the code hardly applicable to more than a few codewords, but it remains perfectly sufficient for classification tasks with up to 7 or 8 classes. With more classes, selected bits may have to dropped, ideally without significantly degrading the code's properties. In the simplest case, a random sample of encoding bits can be retained, but more refined approaches employ optimization techniques for bit selection.

Example 17.4.3 The following R code defines the `ecc` function for generating codewords of the error-correcting code discussed above. It is then used to generate code tables for 3, 4, and 5 codewords, illustrating code properties. The rows of the resulting matrices correspond to codewords and their columns correspond to bits.

```
## error correcting codeword for i out of k
ecc <- function(i, k)
{
  if (i>1)
    c(rev(1-c(ecc(i-1, k-1), 1)), ecc(i-1,k-1))
  else
    rep(1, 2^(k-1)-1)
}

  # error-correcting code for 3 codewords
t(sapply(1:3, ecc, 3))
```

```
   # error-correcting code for 4 codewords
t(sapply(1:4, ecc, 4))
   # error-correcting code for 5 codewords
t(sapply(1:5, ecc, 5))
```

The resulting codes are presented in Table 17.1.

A multiclass encoding function based on the error-correcting code presented above can be defined as follows:

$$\kappa(d) = v_{|C|}(\iota_C(d)) \tag{17.11}$$

where ι_C assigns unique consecutive numbers between 1 and $|C|$ to each class from C in an arbitrary way (e.g., based on the ordering implied by their internal representation). Decoding for a vector of binary model predictions requires finding the closest (least dissimilar) codeword:

$$\kappa^{-1}(\mathbf{h}(x)) = \arg\min_{d \in C} \delta_{\mathrm{ham}}(\mathbf{h}(x), \kappa(d)) \tag{17.12}$$

where δ_{ham} is the Hamming distance, presented in Section 11.3.6 in the context of dissimilarity-based clustering (the number of different bits).

Error-correcting multiclass encoding can be expected to deliver superior results in comparison to the naïve approaches using the natural binary or Grey code due to its resistance to (a small number of) binary model mistakes. When applied to scoring classifiers, by using their class label predictions rather than scores, it is likely to outperform the 1-of-k encoding.

Table 17.1 Error-correcting codes for multiclass encoding.

3 classes		
1	1	1
0	0	1
0	1	0

4 classes						
1	1	1	1	1	1	1
0	0	0	0	1	1	1
0	0	1	1	0	0	1
0	1	0	1	0	1	0

5 classes														
1	1	1	1	1	1	1	1	1	1	1	1	1	1	1
0	0	0	0	0	0	0	0	1	1	1	1	1	1	1
0	0	0	0	1	1	1	1	0	0	0	0	1	1	1
0	0	1	1	0	0	1	1	0	0	1	1	0	0	1
0	1	0	1	0	1	0	1	0	1	0	1	0	1	0

Actually, it makes sense and may be beneficial to apply error-correcting multiclass encoding, as another ensemble modeling technique, even with algorithms that are capable of handling multiple classes directly.

Example 17.4.4 Error-correcting multiclass encoding is illustrated by the R code presented below. It defines the `multi.ecc` function that can be used to generate the encoding and decoding functions for the specified number of classes. The optional `maxbits` argument may be used to limit the number of codeword bits by simple sampling. The function internally generates the code table using the `ecc` function from the previous example and stores it for the encoding and decoding functions. The latter are passed as the `enc` and `dec` components of the returned list. Notice that, while the decoding function does not explicitly call a function for Hamming distance calculation, it does compare the supplied codeword to all code table entries looking for the one with the least number of differing bits. Error-correcting wrappers for 4 and 6 classes (as appropriate for the *Vehicle Silhouettes* and *Glass* data, respectively) are then created around `rpart` and `naiveBayes` and demonstrated exactly as in the previous two examples.

```
## error-correcting multiclass encoding and decoding function generator
## for k classes with up to maxbits bits
multi.ecc <- function(k, maxbits=Inf)
{
  code <- sapply(1:k, ecc, k)
  if (nrow(code)>maxbits)
    code <- code[sample.int(nrow(code), maxbits),]

  enc <- function(d, class, clabs=levels(d))
  {
    `colnames<-`(t(code[,d]), paste(class, 1:nrow(code), sep = "."))
  }

  dec <- function(pred, clabs)
  {
    clabs[apply(pred, 1, function(p) which.min(colSums(p!=code)))]
  }
  list(enc=enc, dec=dec)
}

  # error correcting encoding/decoding for 4 and 6 classes
multi.ecc4 <- multi.ecc(4)
multi.ecc6 <- multi.ecc(6)

  # error-correcding encoding applied to rpart
rp.e4 <- multi.class(rpart, predf=function(...) predict(..., type="c"),
                     encode=multi.ecc4$enc, decode=multi.ecc4$dec)
rp.e6 <- multi.class(rpart, predf=function(...) predict(..., type="c"),
                     encode=multi.ecc6$enc, decode=multi.ecc6$dec)
v.tree.e <- rp.e4$alg(Class~., v.train)
v.tree.e.pred <- rp.e4$predict(v.tree.e, v.test)
g.tree.e <- rp.e6$alg(Type~., g.train)
g.tree.e.pred <- rp.e6$predict(g.tree.e, g.test)

  # error-correcting encoding applied to naive Bayes
```

```
nb.e4 <- multi.class(naiveBayes, encode=multi.ecc4$enc, decode=multi.ecc4$dec)
nb.e6 <- multi.class(naiveBayes, encode=multi.ecc6$enc, decode=multi.ecc6$dec)

v.nb.e <- nb.e4$alg(Class~., v.train)
v.nb.e.pred <- nb.e4$predict(v.nb.e, v.test)
g.nb.e <- nb.e6$alg(Type~., g.train)
g.nb.e.pred <- nb.e6$predict(g.nb.e, g.test)
```

17.4.4 Effects of multiclass encoding

In most cases using multiclass encoding techniques is a matter of necessity rather than choice. They are needed to bypass the limitations of some classification algorithms that are unable to handle more than two classes, but have other desirable properties. In such cases, different encoding techniques can be compared to one another, but not to the original algorithm. Nothing prevents, however, applying multiclass encoding in combination with algorithms that can handle multiple classes directly. This makes it possible to evaluate their effects in comparison to the common baseline, corresponding to no encoding used. They are not unlikely to give prediction quality improvement, at least with the most refined encoding schemes.

Example 17.4.5 The following R code collects misclassification error values for the directly multiclass and encoded multiclass `rpart` models created in the previous examples, applying the `err` function. The results are presented using barplots.

> Ex. 7.2.1
> dmr.claseval

```
v.tree.err <- c(direct=err(v.tree.pred, v.test$Class),
                nbc=err(v.tree.n.pred, v.test$Class),
                'lofk'=err(v.tree.1.pred, v.test$Class),
                ecc=err(v.tree.e.pred, v.test$Class))

v.nb.err <- c(direct=err(v.nb.pred, v.test$Class),
              nbc=err(v.nb.n.pred, v.test$Class),
              'lofk'=err(v.nb.1.pred, v.test$Class),
              ecc=err(v.nb.e.pred, v.test$Class))

g.tree.err <- c(direct=err(g.tree.pred, g.test$Type),
                nbc=err(g.tree.n.pred, g.test$Type),
                'lofk'=err(g.tree.1.pred, g.test$Type),
                ecc=err(g.tree.e.pred, g.test$Type))

g.nb.err <- c(direct=err(g.nb.pred, g.test$Type),
              nbc=err(g.nb.n.pred, g.test$Type),
              'lofk'=err(g.nb.1.pred, g.test$Type),
              ecc=err(g.nb.e.pred, g.test$Type))

barplot(v.tree.err, main="Vehicle Silhouettes, rpart", ylab="Error", las=2)
lines(c(0, 5), rep(v.tree.err[1], 2), lty=2)

barplot(v.nb.err, main="Vehicle Silhouettes, naiveBayes", ylab="Error", las=2)
lines(c(0, 5), rep(v.nb.err[1], 2), lty=2)
```

```
barplot(g.tree.err, main="Glass, rpart", ylab="Error", las=2)
lines(c(0, 5), rep(g.tree.err[1], 2), lty=2)

barplot(g.nb.err, main="Glass, naiveBayes", ylab="Error", las=2)
lines(c(0, 5), rep(g.nb.err[1], 2), lty=2)
```

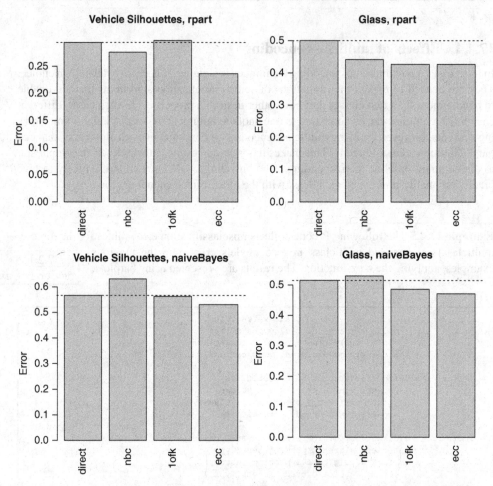

Figure 17.1   The barplots of misclassification error for models using multiclass encoding.

The obtained barplots, presented in Figure 17.1, confirm predictive performance improve-ment due to multiclass encoding is possible indeed. Not surprisingly, the error-correcting encoding technique turns out superior to the other two less refined approaches.

17.5   Conclusion

Attribute transformation is apparently one of the simplest topics in data mining that does not receive very much attention. Most commonly used general-purpose transformations have clear motivation and very simple definitions that do not need extensive discussion. More refined transformations, whenever necessary, are usually domain specific, possibly incorporating background knowledge, and fall beyond the scope of this chapter anyway. What tends to be overlooked and therefore has been highlighted above with particular emphasis is the distinction between modeling and nonmodeling transformations. It is important to understand how the dependence on dataset statistics of even the simplest and most popular standardization and normalization transformations justifies treating them as modeling, with transformation models determined on the training set and applicable to arbitrary other datasets from the same domain. The extremely simple representation of such transformation models is completely irrelevant for this modeling perspective that is, unfortunately, not always adopted. Alternative approaches, that calculate statistics needed to perform transformations separately on each dataset to be transformed or combine all such datasets for the purpose of transformation, suffer from important deficiencies.

Multiclass encoding techniques occupy a distinct place in the data transformation field. What makes them special is that, by transforming the target concept, they effectively decompose the original classification task into subtasks. By such decomposition they enhance the capabilities of classification algorithms originally limited to two classes, making them applicable to arbitrary classification tasks. This is an important achievement, given the fact that some of those algorithms, such as AdaBoost presented in Section 15.5.3 and SVM presented in Section 16.2, often deliver high-quality predictions. Multiclass encoding techniques can also be combined with algorithms that have direct multiclass capabilities, and – with an appropriate encoding scheme – possibly improve their predictions. In this application scenario, multiclass encoding can be considered another approach to ensemble modeling. An associated disadvantage is the unavoidable loss of model human readability, which may cause direct multiple class handling to be preferred for algorithms that produce human-readable model representations.

17.6   Further readings

While most books on data mining include some discussion of data transformation, the scope of techniques covered differ significantly. Unlike this chapter, limited to simple single attribute transformations, they often discuss more complex mathematical transformations of both single and multiple attributes, including, e.g., PCA and SVD as well as the Fourier, Karhunen–Loève, and wavelet transformations. This is the case, in particular, for Theodoridis and Koutroumbas (2008) and, to some extent, for Han *et al.* (2011). Such complex transformations may be viewed as generating new attributes, which is often referred to as *feature extraction*.

Considering only simple transformations, Witten *et al.* (2011) best match the scope and perspective adopted in this chapter, briefly discussing normalization, standardization, binary encoding, and multiclass encoding. Weiss and Indurkhya (1997) presented other similarly simple transformations that may be useful in specific applications, such as continuous attribute smoothing and basic time series attribute transformations (value differences or ratios). Pyle (1999) gives a broad discussion of various data preparation tasks, including attribute transformation. The distinguishing between modeling and nonmodeling transformations emphasized in this chapter tends not to be explicitly noticed and discussed in these presentations.

The binary encoding of discrete attributes, or dummy coding, is a widely used standard approach to using them for linear classification and regression models, dating back at least to Suits (1957). Linear regression and statistical modeling books usually discuss this technique, as well as other types of contrasts (e.g., Draper and Smith 1998; Freedman *et al*. 2007; Glantz and Slinker 2000).

While the standardization transformation has a well-established meaning and is applied in the same way, there is some ambiguity about the meaning of normalization, as can be easily seen by consulting dictionaries of statistical terms (e.g., Dodge 2006; Everitt and Skrondal 2010). The scaling view of normalization adopted in this chapter appears to be the most common in data mining practice.

Most of early work on multiclass encoding occurred in the context of neural networks, which are only capable of producing binary class predictions or numerical scores. Nilsson (1965) may have been the first to describe the basic and self-suggesting 1-of-k multiclass encoding scheme, applied numerous times since then. Zadrozny and Elkan (2002) used the 1-of-k encoding for multiclass probability prediction. Sejnowski and Rosenberg (1987) presented a more refined approach for their *NETtalk* system, using a distributed output code, with each class represented by a unique binary string. Subsequent successful two-class classification algorithms, such as AdaBoost (presented in Chapter 15) and SVM (presented in Chapter 16), stimulated more intense interest in this issue.

Using error-correcting codes for multiclass classification was first proposed by Dieterich and Bakiri (1995), who also discussed the desirable code properties and designed the particular coding scheme presented in Section 17.4.3, as well as possible approaches to handling large numbers of classes when the resulting codewords are unacceptably long. Crammer and Singer (2000) further studied the issue of good code design. A similar idea was subsequently adapted by Ferng and Lin (2011) to multilabel classification, in which each instance is assigned several class labels. This variation of the classification task has been studied in the machine learning literature (Tsoumakas and Katakis 2007), but is rarely encountered in the practice of data mining and therefore not covered in this book. A systematic framework for multiclass encoding and its theoretical analysis were presented by Allwein *et al*. (2000).

References

Allwein EL, Schapire RE and Singer Y 2000 Reducing multiclass to binary: A unifying approach for margin classifiers. *Journal of Machine Learning Research* **1**, 113–141.

Crammer K and Singer Y 2000 On the learnability and design of output codes for multiclass problems *Proceedings of the Thirteenth Annual Conference on Computational Learning Theory (COLT-2000)*. Morgan Kaufmann.

Dieterich TG and Bakiri G 1995 Solving multiclass learning problems via error-correcting output codes. *Journal of Artificial Intelligence Research* **2**, 263–286.

Draper NR and Smith H 1998 *Applied Regression Analysis* 3rd edn. Wiley.

(ed. Dodge Y) 2006 *The Oxford Dictionary of Statistical Terms* 6th edn. Oxford University Press.

Everitt BS and Skrondal A 2010 *The Cambridge Dictionary of Statistics* 4th edn. Cambridge University Press.

Ferng CS and Lin HT 2011 Multi-label classification with error-correcting codes. *Journal of Machine Learning Research* **20**, 281–295.

Freedman DA, Pisani R and Purves R 2007 *Statistics* 4th edn. Norton.

Glantz SA and Slinker BS 2000 *Primer of Applied Regression and Analysis of Variance* 2nd edn. McGraw-Hill.

Han J, Kamber M and Pei J 2011 *Data Mining: Concepts and Techniques* 3rd edn. Morgan Kaufmann.

Nilsson NJ 1965 *Learning Machines: Foundations of Trainable Pattern-Classifying Systems*. McGraw-Hill.

Pyle D 1999 *Data Preparation for Data Mining*. Morgan Kaufmann.

Sejnowski TJ and Rosenberg CR 1987 Parallel networks that learn to pronounce English text. *Journal of Complex Science* **1**, 145–168.

Suits DB 1957 Use of dummy variables in regression equations. *Journal of the American Statistical Association* **52**, 548–551.

Theodoridis S and Koutroumbas K 2008 *Pattern Recognition* 4th edn. Academic Press.

Tsoumakas G and Katakis I 2007 Multi-label classification: An overview. *International Journal of Data Warehousing and Mining* **3**, 1–13.

Weiss SM and Indurkhya N 1997 *Predictive Data Mining: A Practical Guide*. Morgan Kaufmann.

Witten IH, Frank E and Hall MA 2011 *Data Mining: Practical Machine Learning Tools and Techniques* 3rd edn. Morgan Kaufmann.

Zadrozny B and Elkan C 2002 Transforming classifier scores into accurate multiclass probability estimates *Proceedings of the Eighth ACM SIGKDD International Conference on Knowledge Discovery and Data Mining*. ACM Press.

18

Discretization

18.1 Introduction

Continuous attribute discretization – which basically consists in creating discrete attributes to replace originally continuous attributes – belongs to the most frequently used forms of attribute transformation. Unlike most of others, it is sufficiently complex to give space for a variety of algorithms of the varying level of refinement and computational expense. Some of them are actually much closer to classification algorithms than to simple arithmetic transformations. This, along with the possible impact of discretization on the process and results of subsequent modeling, justifies presenting them separately in this chapter.

Discretized attribute values correspond to intervals of the original continuous attribute's values, to which its range is divided. Determining the number and boundaries of these intervals in order to preserve the original attribute's predictive utility is the major challenge addressed by discretization algorithms. The most successful of them are those that take into account the purpose the discretized attribute is supposed to be used for, which is usually creating a classification model. Such discretization algorithms receive most attention in this chapter.

Example 18.1.1 Discretization algorithms presented in this chapter will be illustrated by R code examples, containing simple implementations and usage demonstrations thereof. The latter will be using the tiny *weatherc* data, as well as more realistically sized *Vehicle Silhouettes* and *Glass* datasets from the `mlbench` package. To observe the effects of discretization, classification models will be created based on the original and discretized datasets using the R implementations of the decision tree and naïve Bayes algorithms (provided by the `rpart` and `e1071` packages). Several DMR packages provide auxiliary functions for the illustrative implementations and example calls. The following R code sets up the environment for this chapter's demonstrations by loading the packages and datasets, and partitioning the larger two of those into training, and also tests subsets for simple holdout model evaluation.

> Ex. 1.3.2
> dmr.data

Data Mining Algorithms: Explained Using R, First Edition. Paweł Cichosz.
© 2015 John Wiley & Sons, Ltd. Published 2015 by John Wiley & Sons, Ltd.

```
library(dmr.claseval)
library(dmr.stats)
library(dmr.trans)
library(dmr.util)

library(rpart)
library(e1071)

data(weatherc, package="dmr.data")
data(Vehicle, package="mlbench")
data(Glass, package="mlbench")

set.seed(12)

rv <- runif(nrow(Vehicle))
v.train <- Vehicle[rv>=0.33,]
v.test <- Vehicle[rv<0.33,]

rg <- runif(nrow(Glass))
g.train <- Glass[rg>=0.33,]
g.test <- Glass[rg<0.33,]
```

18.2   Discretization task

Discretization consists in replacing an originally continuous attribute by a discrete attribute, with different values assigned to particular intervals of the original attribute's range. This section outlines the motivation behind discretization and discusses the definition of the task in more detail.

18.2.1   Motivation

Replacing continuous attributes with their newly created discrete counterparts – ideally without loosing too much of the precious information contained therein – is an obvious necessity if one is willing to apply a data mining algorithm that is incapable of handling continuous attributes on its own. This may be the case, e.g., for simple implementations of decision trees or the naïve Bayes classifier. This is more common in classification algorithms than in regression or clustering algorithms, since the latter are usually much more friendly to continuous attributes and more likely to require a semi-inverse numerical encoding transformation, as described in Section 17.3.5. For some attribute selection filters presented in Section 19.4 it may also be necessary or convenient to have a uniform attribute relationship measure applied to all attributes regardless of their type. One way to achieve this is to use a relationship measure for discrete attributes, such as the mutual information or the symmetric uncertainty presented in Section 2.5.3, after discretizing continuous attributes.

Practical necessity is not the only motivation for discretization, though. It may be sometimes a good idea to discretize continuous attributes even if algorithms to be subsequently applied could cope with them directly. This is because there may be two undesirable effects of in-algorithm continuous attribute processing:

- increased computational expense,

- more opportunities to overfit.

Again, these are much more likely for classification algorithms than any other data mining algorithms, because many of them have been developed with the primary focus on discrete attributes and tend to be applied to datasets with the majority of attributes being discrete. This is when continuous attributes, particularly if they take many different values (perhaps approaching the number of instances in the training set), may incur a substantial amount of extra computation and increase the risk of overfitting. The latter results from the possibility of making very fine-grained separations between instances even with a single continuous attribute. It makes it indeed easier to create a near-perfectly fitted model with poor generalization performance.

Properly performed discretization may help us to fight the two problems described above. Creating models with continuous attributes used in their original form may be computationally more economic, particularly when multiple models are created using the same discretized dataset. It may also, by reducing the useless excessive information contained in the original continuous attributes and preserving what is most likely to be predictively useful, increase the chance of arriving at properly generalizing models. These are just possibilities, not guarantees, but they are sufficient to consider discretization as a potentially useful attribute transformation even when it is not strictly necessary.

18.2.2   Task definition

As a type of attribute transformation, discretization clearly inherits the general attribute transformation task definition from Section 17.2, although there are some noteworthy discretization-specific issues.

18.2.2.1   Target task

The *target task* for discretization is usually one of the three major inductive learning tasks introduced in Section 1.2, classification, regression, or clustering, with the latter two being much less common than the first one, for reasons already discussed above. The description of discretization algorithms presented in this section does not explicitly consider target tasks other than classification, although some of them are target task independent and others could be modified at least for the regression task easily, in the unlikely case discretization for the latter would be needed at all.

18.2.2.2   Target attribute

If the target task is classification (or regression, even if this is quite untypical, to say the least), a *target attribute* is also specified. It may be used during discretization to explicitly or implicitly observe the predictive utility of the original attribute and its possible loss due to discretization, as well as to make efforts to minimize the latter. Discretization algorithms that exploit this possibility, i.e., take the target attribute into account, are referred to as *supervised* whereas those which do not use the target attribute, even if it is available, are called *unsupervised*. The former, which are usually preferred, differs from the most simple data transformations, for which the target attribute is completely irrelevant.

18.2.2.3   Discretized attribute

While practical implementations of discretization algorithms can be usually applied to all continuous attributes in a dataset at the same time, we will consider discretizing a single attribute as a single instance of the discretization task. This also preserves consistency with the general attribute transformation task definition presented in Section 17.2, assuming one attribute being transformed. Discretizing multiple attributes can then be viewed as multiple independent instances of the task. Some discretization algorithms may not adhere to this view, if they exploit relationships among attributes. This is not the case for algorithms described in this chapter, though.

The continuous attribute to discretize a is supposed to be replaced by a newly created discrete attribute a' with unique values corresponding to nonoverlapping intervals (bins) completely covering the co-domain of the original attribute. The definition of such a new attribute can be written as follows:

$$a'(x) = \begin{cases} v'_1 & \text{if } a(x) \in (-\infty, b_1] \\ v'_2 & \text{if } a(x) \in (b_1, b_2] \\ \dots \\ v'_k & \text{if } a(x) \in (b_{k-1}, \infty) \end{cases}$$

where $v'_1, v'_2, \dots, v'_k$ designate new discrete values corresponding to k intervals to which the original attribute is discretized and $b_1, b_2, \dots, b_{k-1}$ are the cut points or *breaks* from its co-domain, representing interval bounds.

When discussing discretization algorithms, it is often necessary to refer to the subsets of the training set corresponding to particular intervals. The following notational convention will be used for this purpose:

$$T_1 = \{x \in T \mid a(x) \in (-\infty, b_1]\} \tag{18.1}$$

$$T_2 = \{x \in T \mid a(x) \in (b_1, b_2]\} \tag{18.2}$$

$$\dots$$

$$T_k = \{x \in T \mid a(x) \in (b_{k-1}, \infty)\} \tag{18.3}$$

assuming the set of breaks is $\{b_1, b_2, \dots, b_{k-1}\}$.

18.2.2.4   Training set

Discretization is usually performed on the same dataset which serves as the target task's training set, but – whenever convenient or necessary – another subset of the available data can be used. The precautions mentioned in Section 17.2.4 for the general data transformation task apply here as well. Since discretization has, possibly substantial, impact on models created using the discretized data, it definitely belongs to the modeling process and its training set becomes a part of the data used for model creation. This is true even if the actual training set passed to the modeling algorithm is completely independent.

18.2.3   Discretization as modeling

What is important to underline and sometimes gets overlooked is that replacing the original continuous attribute with a newly created discrete attribute cannot be simply reduced to replacing the original continuous values with new discrete values in the training set. What discretization should produce is not (or at least not just) a modified copy of the data, but the definition of a new attribute that can be subsequently applied to transform any dataset from the same domain from which the training set was originated. It is therefore a modeling attribute transformation, as discussed in Section 17.2.5, with the definition of the new discrete attribute playing the role of a model created based on the training set and then available to be applied to any other dataset from the same domain.

Adopting the modeling view of discretization is necessary to properly integrate it with the target modeling task. To have model produced using the discretized training set applicable to new data (possibly a single instance), *the very same* discretization breaks should be applied to the latter. This is the case even if the new dataset was actually available at the time of modeling and could be merged with the training set for discretization or if it is sufficiently large to be discretized independently (for supervised discretization algorithms it would also have to labeled). The former would effectively use the new data for training and the latter would usually yield inconsistencies between discretization intervals assumed by the model creation and appearing in the new data. Both are clearly unacceptable. The modeling view of discretization makes it possible to consider the target task's model tied to the discretization model that were applied to the training set beforehands (i.e., the particular sets of interval breaks) . It is the very same discretization model that needs to be subsequently applied to any datasets for which the target task's model would be used to predict.

Example 18.2.1 Discretization algorithm implementations that will be presented in subsequent examples will follow the assumption of working with one discretized attribute at a time. To make them applicable to all continuous attributes in a dataset in a convenient way, the following R code defines the `disc.all` function which creates, when applied to a single-attribute discretization function, a wrapper around it that applies it sequentially to multiple attributes specified using the standard formula interface. The target (left-hand side) attribute specified by the formula is also retrieved and passed to the discretization function if it is found to expect the `target` argument. The `x.vars` and `y.var` functions are used to extract input and target attribute names from the supplied formula. `dmr.util`
Note that the `class` attribute of the object representing discretization breaks for all continuous attributes is set to `disc`, to enable appropriate prediction method dispatching. To demonstrate the wrapper generator, it is applied to a simple random discretization function, serving only the illustration purpose. The multiattribute wrapper function returned by `disc.all` is then applied to the *weatherc* data.

```
## create a wrapper for all attributes discretization
disc.all <- function(disc)
{
  disc1 <- function(v, k, class, ...)
  {
    if (is.numeric(v))
    {
      if (is.null(formals(disc)$class))
        disc(v, k=k, ...)  # unsupervised
      else
        disc(v, k=k, class=class, ...)  # supervised
    }
  }
```

```
    function(formula, data, k=5, ...)
    {
        attributes <- x.vars(formula, data)
        class <- y.var(formula)
        if (length(k)==1)
          k <- sapply(attributes, function(a) k)
        km <- match(attributes, names(k))
        'class<-'(mapply(function(a, k1) disc1(data[[a]], k=k1, class=data[[class]], ...),
                         attributes, k[km], SIMPLIFY=FALSE),
                  "disc")
    }
}
## apply discretization breaks to a dataset
predict.disc <- predict.transmod(function(m, v) cut(v, c(-Inf, m, Inf)))

    # random all-attributes discretization
disc.rand <- disc.all(function(v, k=3) sort(runif(k-1, min=min(v), max=max(v))))

    # random discretization for the weatherc data
w.dr3m <- disc.rand(play~., weatherc, 3)
w.dr43m <- disc.rand(play~., weatherc, list(temperature=4, humidity=3))

    # apply discretization breaks to the weatherc data
w.dr3 <- predict(w.dr3m, weatherc)
w.dr43 <- predict(w.dr43m, weatherc)
```

Notice that the `predict.disc` function, which applies discretization breaks to actually replace continuous attribute values with the corresponding interval labels, is obtained using the wrapper generator for modeling transformation prediction, `predict.transmod`. As a matter of fact, it would be straightforward to use the corresponding modeling transformation wrapper generator, `transmod.all`, instead of the `disc.all` function defined above. The latter adds some discretization-specific capabilities, however: specifying the number of intervals on per-attribute basis and detecting whether the supplied single-attribute discretization function is supervised or not.

> Ex. 17.2.1
> `dmr.trans`

18.2.4   Discretization quality

As discussed above, the purpose of discretization is to make subsequent modeling – performing the target task – easier (i.e., to make it possible to use algorithms that do not handle continuous attributes or reducing the computational expense of those that do without substantial model quality loss) or more effective (i.e., to make it possible to obtain models with better generalization capabilities). Hence, there are no other requirements for the new discrete attribute attribute than its utility in performing the target task. This can be considered roughly equivalent to the following postulates:

Arity reduction. The number of values of the new discrete attribute should be considerably less than the number of distinct values of the original continuous attribute observed in the training set.

Predictiveness preservation. The new discrete attribute should retain most of the predictive power of the original continuous attribute.

There is an obvious tradeoff between these two which has to be either resolved by discretization algorithms or by their user, who may be required to set up parameters, typically including the desired number of intervals. The ultimate verification of the effects of discretization is always provided by the evaluation of models obtained when performing the target task using discretized data.

18.3   Unsupervised discretization

Unsupervised discretization algorithms, which do not take any target attribute into account, are based entirely on the observed distribution of the continuous attribute to be discretized in the training set. It is no surprising therefore that they tend to be less refined than supervised algorithms. The set of breaks for attribute a determined on the training set T using an unsupervised discretization algorithm will be designated by $B_T(a)$.

18.3.1   Equal-width intervals

The simplest discretization algorithm that can be considered useful is based on the principle of equal interval width. After determining the range of the attribute to be discretized in the training set

$$\min_T(a) = \min_{x \in T} a(x) \tag{18.4}$$

$$\max_T(a) = \max_{x \in T} a(x) \tag{18.5}$$

it just needs to be partitioned into k bins of equal width

$$\Delta_T^k(a) = \frac{\max_T(a) - \min_T(a)}{k} \tag{18.6}$$

using the following set of $k - 1$ breaks:

$$b_1 = \min_T(a) + \Delta_T^k(a) \tag{18.7}$$

$$b_2 = \min_T(a) + 2\Delta_T^k(a) \tag{18.8}$$

$$\ldots$$

$$b_{k-1} = \min_T(a) + (k - 1)\Delta_T^k(a) = \max_T(a) - \Delta_T^k(a) \tag{18.9}$$

or shortly

$$B_T(a) = \{\min_T(a) + i\Delta_T^k(a) \mid i = 1, 2, \ldots, k - 1\} \tag{18.10}$$

What gives this algorithm its unrivaled simplicity (conceptual, implementational, and computational) also makes it unfortunately prone to problems with the predictive utility of the created discrete attribute. By looking at the range only, it ignores all other possibly important aspects of the original attribute's distribution, which may be nonuniform, asymmetric, and suffering from outliers. The latter are particularly dangerous, as they are likely to mislead the algorithm completely, leading it to creating several empty or nearly empty intervals. If using this simple algorithm in emergency situations (lack of time and tools needed for more refined and reliable algorithms), one should at least take care of checking the data for the presence of outliers and filtering them out if necessary.

Example 18.3.1 Equal-width discretization is implemented and demonstrated by the following R code. It defines the `disc.eqwidth1` function that performs the equal-width discretization of a single-attribute and creates a multiattribute wrapper around it using the previously presented `disc.all` function.

```
## equal-width discretization for a single attribute
disc.eqwidth1 <- function(v, k=5)
{
  w <- diff(r <- range(v))/k
  seq(r[1]+w, r[2]-w, w)
}

## equal-width discretization for a dataset
disc.eqwidth <- disc.all(disc.eqwidth1)

  # equal-width discretization of the temperature attribute in the weatherc data
disc.eqwidth1(weatherc$temperature, 4)

  # equal-width discretization for the weatherc data
disc.eqwidth(play~., weatherc, 3)
disc.eqwidth(play~., weatherc, list(temperature=4, humidity=3))

  # equal-width discretization for the Vehicle Silhouettes data
v.disc.ew <- disc.eqwidth(Class~., v.train, 7)
summary(predict(v.disc.ew, v.train))

  # equal-width discretization for the Glass data
g.disc.ew <- disc.eqwidth(Type~., g.train, 7)
summary(predict(g.disc.ew, g.train))
```

For the two larger datasets used by the demonstrations the obtained intervals are applied to the data and the distributions of the resulting discrete attributes are summarized. Note extreme differences between instance counts for different intervals that can be observed for many attributes.

18.3.2 Equal-frequency intervals

It takes relatively little additional effort to overcome the most striking deficiency of equal-width discretization. The equal-frequency algorithm is outlier-resistant and adapts itself to the actual value distribution of the attribute being discretized, yielding increased interval resolution wherever higher value density is observed. Its operation principle reflected by the name is to select interval breaks so as to achieve the same number of training set instances corresponding to each bin, or – since this cannot be usually achieved exactly – to minimize the differences among bin frequencies.

To perform the equal-frequency discretization of attribute a to k intervals, it is necessary to partition the training set T using a into k subsets $T_1, T_2, \ldots, T_k$ satisfying the conditions given by Equations 18.1 -18.3 such that $|T_1| \approx |T_2| \approx \cdots \approx |T_k| \approx \frac{|T|}{k}$. Then $B_T(a) = \{b_1, b_2, \ldots, b_{k-1}\}$ is the set of equal-frequency discretization breaks.

Perfect interval frequency equality may be possible to obtain only if the training set size is a multiple of k and finding appropriate breaks is not prevented by two or more instances having the same value of a. Otherwise an approximation to equal-frequency partitioning needs to be accepted. There are several possible approaches that can be used to obtain such approximations. They all work similarly well if the attribute being discretized exhibits sufficient variability in the training set and may differ more noticeably otherwise. In extreme cases, if the number of training instances with the same value of a approaches or exceeds $\frac{|T|}{k}$, they may produce intervals with clearly nonequal frequencies or end up with less intervals than requested. One simple approach that should be usually sufficient is to use quantiles of order $\frac{i}{k}$ for $i = 1, 2, \ldots, k - 1$ as breaks, after removing any duplicates they should appear.

Example 18.3.2 The R code presents and demonstrates a simple implementation of equal-frequency discretization, using the standard `quantile` function.

```
## equal-frequency discretization for a single attribute
disc.eqfreq1 <- function(v, k=5) { unique(quantile(v, seq(1/k, 1-1/k, 1/k))) }

## equal-frequency discretization for a dataset
disc.eqfreq <- disc.all(disc.eqfreq1)

  # equal-width discretization of the temperature attribute in the weatherc data
disc.eqfreq1(weatherc$temperature, 4)

  # equal-frequency discretization for the weatherc data
disc.eqfreq(play~., weatherc, 3)
disc.eqfreq(play~., weatherc, list(temperature=4, humidity=3))

  # equal-frequency discretization for the Vehicle Silhouettes data
v.disc.ef <- disc.eqfreq(Class~., v.train, 7)
summary(predict(v.disc.ef, v.train))

  # equal-frequency discretization for the Glass data
g.disc.ef <- disc.eqfreq(Type~., g.train, 7)
summary(predict(g.disc.ef, g.train))
```

Notice that the number of created discretization intervals is less than requested for some attributes. The corresponding interval counts, while not perfectly equal, are usually much more uniform than observed previously for equal-width discretization.

18.3.3   Nonmodeling discretization

In the vast majority of discretization usage scenarios, discretized datasets are used for the classification model creation. As discussed in Section 18.2.3, this is when discretization has to be performed as a modeling transformation, to make it possible to apply the same interval breaks to whatever new data the created classification model will be used for prediction. However, discretization may sometimes be a useful data preprocessing operation even if subsequently applied modeling algorithms receive and process continuous attributes in their original form. This is the case when a discretized dataset version is used "temporarily" for performing some data analysis tasks. Even if the results of these tasks have impact on subsequent model creation, there is no need to retain discretization interval breaks as long as the model to be created only uses the original continuous attributes. In such situations, discretization could be performed in a *nonmodeling* way, without retaining any reusable discretization model. This is possible for both unsupervised and supervised discretization, although it s more common and may appear more natural with the former.

One noteworthy potential application of nonmodeling discretization is the attribute relationship detection using statistical techniques presented in Section 2.5 for domains with both discrete and continuous attributes. It is often desirable not only to detect statistically strong relationships between attributes of different types (discrete–discrete, discrete–continuous, continuous–continuous), but also rank attribute pairs with respect to the strength of their relationship. Since relationship measures applicable to different attribute types can hardly be comparable, one simple solution would be to temporarily replace all continuous attributes with their discretized counterparts, to enable applying the same relationship measure to all attribute pairs, regardless of their original type. This may be particularly convenient for filtering attribute selection techniques, presented in Section 19.4.

Example 18.3.3 The `disc.nm` function defined by the following R code returns a simple nonmodeling wrapper around the supplied discretization function that discards the obtained discretization model after applying it to the same data on which it has been created. It is then used to generate nonmodeling wrappers around the implementations of the equal-width and equal-frequency discretization algorithms from Examples 18.3.1 and 18.3.2. The latter are demonstrated on the *weatherc* data.

```
## create a nonmodeling discretization wrapper disc.nm <-
function(disc) { function(formula, data, k=5, ...)
predict(disc(formula, data, k, ...), data) }

## nonmodeling equal-width discretization discnm.eqwidth <-
disc.nm(disc.eqwidth)

## nonmodeling equal-frequency discretization discnm.eqfreq <-
disc.nm(disc.eqfreq)

  # nonmodeling discretization for the weatherc data
discnm.eqwidth(play~., weatherc, 4)
discnm.eqfreq(play~., weatherc, 3)
discnm.eqfreq(play~., weatherc, list(temperature=4, humidity=3))
```

18.4 Supervised discretization

Supervised discretization requires considerably more effort than the simple unsupervised algorithms discussed above. The effort is usually well paid off, though, by the possibility to adjust interval breaks so as to preserve the discretized attribute's predictive utility with respect to the target attribute. This section will focus entirely on the most common case where the target task is classification and the target attribute is the target concept, assigning class labels to instances. The set of breaks for attribute a with respect to the target concept c, determined on the training set using a supervised discretization algorithm, will be designated by $B_{T,c}(a)$.

18.4.1 Pure-class discretization

One particularly simple approach to supervised discretization that is not directly useful by itself, but may serve as the starting point for discussing more refined algorithms, consists in keeping attribute values corresponding to different classes in separate intervals, as far as possible, and with as many intervals used as necessary. Basically, each interval should only contain attribute values for instances of the same class, unless some instances of different classes have the values of the attribute being discretized equal. This guarantees that the newly created discrete attribute will fully preserve the predictive utility of the original continuous attribute, at least on the training set, but – in the extreme case – may use as many intervals as unique attribute values occurring therein.

Technically, pure-class discretization may be performed by ordering the training set T with respect to the continuous attribute a to be discretized and then adding a break point in the middle between each pair of consecutive different values that correspond to instances of different classes. If $x_1, x_2, \ldots, x_{|T|}$ is the sequence of all training instances ordered such that $a(x_1) \leq a(x_2) \leq \cdots \leq a(x_{|T|})$, then the set of interval breaks can be determined as

$$B_{T,c}(a) = \left\{ \frac{a(x_{i-1}) + a(x_i)}{2} \middle| \begin{array}{l} i = 2, 3, \ldots, |T|, \\ \\ c(x_i) \neq c(x_{i-1}), a(x_i) \neq a(x_{i-1}) \end{array} \right\} \tag{18.11}$$

Example 18.4.1 Pure-class discretization is implemented by the `disc.pure1` function defined in the following R code. It accepts the `k` argument for compatibility with the wrapper-generation function `disc.all` from Example 18.2.1, but its value is ignored. The application of the generated `disc.pure` wrapper to the *weatherc*, *Vehicle Silhouettes*, and *Glass* datasets is demonstrated.

```
## pure-class discretization for a single attribute
disc.pure1 <- function(v, class, k=NULL)
{
  ord <- order(v)
  class <- class[ord]
  v <- v[ord]
  b <- diff(as.integer(class))!=0 & diff(v)!=0
  (v[1:(length(v)-1)][b]+v[2:length(v)][b])/2
}

## pure-class discretization for a dataset
disc.pure <- disc.all(disc.pure1)

  # pure-class discretization for the weatherc data
disc.pure(play~., weatherc)

  # pure-class discretization for the Vehicle Silhouettes data
v.disc.p <- disc.pure(Class~., v.train)
summary(predict(v.disc.p, v.train), maxsum=100)

  # pure-class discretization for the Glass data
g.disc.p <- disc.pure(Type~., g.train)
summary(predict(g.disc.p, g.train), maxsum=100)
```

The number of intervals obtained for the two larger datasets reaches a few dozen for several attributes, which would be usually much more than desired.

18.4.2   Bottom-up discretization

While pure-class discretization is of rather little practical use by itself, as it fails to achieve the expected benefits of discretization due to an overly large number of intervals usually created, several more refined algorithms may be considered that start from pure-class interval breaks and then proceed by eliminating some (or, not untypically, most) of them according to some specific utility measures. Repeatedly merging selected adjacent intervals by removing breaks between them is referred to as *bottom-up discretization* and the simple pure-class algorithm provides a perfect initialization for it.

18.4.2.1   Algorithm scheme

The bottom-up discretization algorithm scheme is presented below. It can be instantiated by specifying an initialization operation used to set the initial breaks, stop criteria used to determine whether a satisfactory set of breaks has been reached, and an evaluation function that measures the utility of each currently existing break and can be used to determine the least useful one to drop. The latter, for break b, attribute a, target concept c, and training set T, is designated by $\varrho_{T,c}(b, a)$ and assumed to assign less values to less useful breaks. On each iteration the two intervals separated by the least useful break are merged.

1: find the set of initial breaks $B_{T,c}(a)$;
2: **while** stop criteria are not satisfied **do**
3:     $B_{T,c}(a) := B_{T,c}(a) - \{\arg min_{b \in B_{T,c}(a)}\, \varrho_{T,c}(b, a)\}$;
4: **end while**

While it is not reflected by the above algorithm scheme to keep it simple, it is worthwhile to notice that after each merge operation (break removal) only the utilities of the remaining left and right neighbor breaks may change (as long as the evaluation function takes into account only the two intervals separated by the break being evaluated, which is a perfectly reasonable assumption). Only these two therefore need to be re-evaluated for the next iteration.

Example 18.4.2 The bottom-up discretization scheme is implemented by the following R code. The `disc.bottomup1` function that discretizes a single attribute receives two functions as arguments: one used for initialization (defaulting to pure-class discretization) and the other used for break evaluation. The latter has to be a function that takes three breaks as arguments (the one to be evaluated along with its left and right neighbors) as well as the values of the attribute to discretize and the corresponding classes for the training set. The evaluation function is initially applied to all breaks, but on subsequent iterations it is only used for breaks that need to re-evaluated after dropping a neighbor of theirs. The stop criteria are specified either as the maximum number of intervals or as the maximum utility a dropped break may have. The `shift.left` and `shift.right`   `dmr.util` functions are used to shift the breaks vector left and right, to identify right and left neighbors for each break, respectively. A multiattribute wrapper around the `disc.bottomup1` function is created as usual and demonstrated on the *weatherc* data, with a trivial identity break evaluation function, serving the illustration purpose only. With each break's utility being estimated as its value, the bottom-up discretization process simply removes as many smallest breaks as necessary to meet the stop criteria.

```
## bottom-up discretization for a single attribute
disc.bottomup1 <- function(v, class, k=5, initf=disc.pure1, evalf, maxev=Inf)
{
  breaks <- initf(v, class)
  utils <- mapply(evalf, breaks,
                  shift.right(breaks, first=-Inf), shift.left(breaks, last=Inf),
                  MoreArgs=list(v, class))
  b <- which.min(utils)

  while (length(breaks)+1>k && utils[b]<=maxev)
  {
    breaks <- breaks[-b]
    utils <- utils[-b]
    if (b>1)
      utils[b-1] <- evalf(breaks[b-1],
                          ifelse(b>2, breaks[b-2], -Inf),
                          ifelse(b<=length(breaks), breaks[b], Inf),
                          v, class)
    if (b<=length(breaks))
      utils[b] <- evalf(breaks[b],
                        ifelse(b>1, breaks[b-1], -Inf),
                        ifelse(b<length(breaks), breaks[b+1], Inf),
                        v, class)

    b <- which.min(utils)
  }
  breaks
}

## bottom-up discretization for a dataset
disc.bottomup <- disc.all(disc.bottomup1)

  # bottom-up discretization of the temperature attribute in the weatherc data
disc.bottomup1(weatherc$temperature, weatherc$play, 3,
               evalf=function(b, bl, br, v, class) b)

  # bottom-up discretization for the weatherc data
disc.bottomup(play~., weatherc, 3, evalf=function(b, bl, br, v, class) b)
```

18.4.2.2 Initialization

As mentioned above, the results of pure-class discretization can serve as the ideal initial set
of breaks for bottom-up algorithms. This is because any reasonable discretization may be
obtained by merging some of pure-class intervals and there would never be need for any
smaller (more fine-grained) intervals than those. With that being said, the bottom-up discretization process can actually start with any set of initial intervals, both more fine-grained
and more coarse-grained than provided by pure-class discretization. While in the latter case
the final discretization quality may be degraded, in the former case most algorithms (using reasonable interval merge criteria) may just take more merging iterations to arrive at a satisfactory
result. For such algorithms, it is safe (albeit inefficient) to start bottom-up discretization even
from singleton intervals, each containing just a single value of the attribute being discretized.
They can be obtained by using mid-points between each pair of the attribute's consecutive
values as initial discretization breaks.

18.4.2.3 Merge criterion

The criterion used to evaluate the utility of breaks, i.e., determine the pair of intervals to merge, is the central component of bottom-up discretization algorithms. A variety of approaches are possible, with different level of refinement and complexity, including:

Instance count. Drop breaks that separate intervals with the least (individual or total) instance count (to eliminate intervals with too little instances).

Dominating class count. Drop breaks that separate intervals with the least (individual or total) dominating class count (to eliminate intervals with too little dominating class representatives).

Misclassification count. Drop breaks that separate intervals with the least misclassification count increase after merging (to minimize the misclassification error possible to obtain).

Class impurity. Drop breaks that separate intervals with the least class impurity increase after merging (to prevent the loss of predictive utility).

Class distribution dissimilarity. Drop breaks that separate intervals with the least dissimilar class distribution (to prevent the loss of predictive utility).

Break utility functions representing these approaches will be discussed in the corresponding subsections below. When considering a break $b \in B_{T,c}(a)$, we will refer to the subsets of the training instances corresponding to the two intervals they separate as follows:

$$T_{b-} = \{x \in T \mid a(x) \in (b_-, b]\} \tag{18.12}$$

$$T_{b+} = \{x \in T \mid a(x) \in [b, b_+]\} \tag{18.13}$$

where

- b_- is the left neighbor break of b in $B_{T,c}(a)$ or $-\infty$ if $b = \min B_{T,c}(a)$,
- b_+ is the right neighbor break of b in $B_{T,c}(a)$ or ∞ if $b = \max B_{T,c}(a)$.

Instance count One of the simplest possible merge criteria for bottom-up discretization is based on instance count, i.e., the number of training instances corresponding to each interval. It is motivated by the belief that too small intervals (corresponding to small data subsets) are not useful, may be misleading, and therefore should be merged, even at the cost of introducing considerable class impurity. It makes sense, in particular, to consider the total instance count for a pair of adjacent intervals, which – if minimized – would identify intervals that have small instance count on the average. The corresponding break utility function may be defined as simply as

$$\varrho_{T,c}(b, a) = |T_{b-}| + |T_{b+}| \tag{18.14}$$

i.e., as the sum of instance counts for the two intervals separated by the break being evaluated. Such break evaluation results in the two intervals with the least total number of corresponding training instances being merged on each iteration. This would have the disadvantage of not being able to merge a single small-count interval with any of its neighbors if they are both large count. One alternative would therefore be to always identify a single least-count interval and then merge it with one of its neighbors (obviously, the less-populated one). This could be accomplished with the following, somewhat more complex, break utility function:

$$\varrho_{T,c}(b, a) = \min\{|T_{b-}|, |T_{b+}|\} + \gamma \max\{|T_{b-}|, |T_{b+}|\} \qquad (18.15)$$

The first term is responsible for identifying the least-count interval, for which the evaluated break is either the left or right bound. The second term is needed to decide whether the least-count interval should be merged with its left or right neighbor. If used with sufficiently small γ (in particular, $\gamma < \frac{1}{|T|}$), it will only break ties occurring for the first term, which is what we want. To see why this formula works, assume $|T_i| = \min_{j=1,\ldots,k} |T_j|$ to be the least per-interval instance count. Under the notational convention introduced above, the left bound of the corresponding interval is then b_{i-1} and the right bound is b_i. The utility of breaks b_{i-1} and b_i will be calculated by the above formula as

$$\varrho_{T,c}(b_{i-1}, a) = |T_i| + \gamma|T_{i-1}| \qquad (18.16)$$

$$\varrho_{T,c}(b_i, a) = |T_i| + \gamma|T_{i+1}| \qquad (18.17)$$

since both $|T_{i-1}| \geq |T_i|$ and $|T_{i+1}| \geq |T_i|$. The less of these evaluations will therefore indicate whether the least-count interval should be merged with its left or right neighbor.

Notice that this merge criterion (in any version), making no use of class labels, is clearly unsupervised. If combined with pure-class interval initialization, it will therefore yield, in a sense, a hybrid supervised–unsupervised discretization algorithm. If used with singleton interval initialization, it would become just an overly complex and inefficient way of performing equal-frequency discretization. As already mentioned before, though, these two initialization methods may actually become equivalent in the extreme case where every two consecutive values of the continuous attribute being discretized correspond to instances of different classes. Instance count-based merge criterion will therefore be useful only if the pattern of attribute values and classes is substantially different from this extreme. Pure-class discretization will then yield a reasonable initial set of interval breaks that may only need relatively minor corrections, accomplished within a small number of merging iterations. Such corrections can be handled by the instance count approach easily and quickly. Whenever the number of necessary merging iterations becomes large, the unsupervised nature of the merge criterion dominates the algorithm, potentially leading to the loss of the original attribute's predictive utility. It makes therefore most sense to think of instance count-based bottom-up discretization as of a straightforward enhancement of pure-class discretization than as of a stand-alone algorithm useful by its own.

Example 18.4.3 The following R code implements and demonstrates the instance count break utility function in the two versions discussed above: based on the total count for the two intervals separated by the break to be evaluated and based on their minimum individual count.

```
## instance count break evaluation
## using the total instance count after merging
evdisc.incount1 <- function(b, bl, br, v, class) { sum(v>bl & v<=br) }
```

```
## instance count break evaluation
## using the minimum individual instance count before merging
evdisc.incount2 <- function(b, bl, br, v, class, gamma=1/length(v))
{
  min(sum(v>bl & v<=b), sum(v>b & v<=br)) +
    gamma*max(sum(v>bl & v<=b), sum(v>b & v<=br))
}

  # instance count bottom-up discretization for the weatherc data
disc.bottomup(play~., weatherc, 3, evalf=evdisc.incount1)
disc.bottomup(play~., weatherc, 3, evalf=evdisc.incount2)

  # instance count bottom-up discretization for the Vehicle Silhouettes data
v.disc.bu.ic1 <- disc.bottomup(Class~., v.train, 7, evalf=evdisc.incount1)
v.disc.bu.ic2 <- disc.bottomup(Class~., v.train, 7, evalf=evdisc.incount2)
summary(predict(v.disc.bu.ic1, v.train))
summary(predict(v.disc.bu.ic2, v.train))

  # instance count bottom-up discretization for the Glass data
g.disc.bu.ic1 <- disc.bottomup(Type~., g.train, 7, evalf=evdisc.incount1)
g.disc.bu.ic2 <- disc.bottomup(Type~., g.train, 7, evalf=evdisc.incount2)
summary(predict(g.disc.bu.ic1, g.train))
summary(predict(g.disc.bu.ic2, g.train))
```

Dominating class count The dominating class count-based merge criterion is nearly as simple as the instance count-based one, but – unlike the latter – it remains supervised. It is based on the belief that useful discretization intervals should contain values corresponding to sufficiently many instances of the dominating class. The number of such instances for an interval is referred to as the interval's dominating class count. It therefore merges intervals with the least dominating class counts. There are at least the following three different exact interpretations of this general principle.

- Merge intervals with the least dominating class count after merging. The corresponding break utility function is defined as follows:

$$\varrho_{T,c}(b, a) = \max_{d \in C} |T_{b-}^d \cup T_{b+}^d| \tag{18.18}$$

This will drop breaks between intervals that have small combined dominating class counts.

- Merge intervals with the least sum of dominating class counts before merging, i.e., minimizing the following utility function:

$$\varrho_{T,c}(b, a) = \max_{d \in C} |T_{b-}^d| + \max_{d \in C} |T_{b+}^d| \tag{18.19}$$

This is equivalent to the previous interpretation if the two intervals separated by the evaluated break have the same dominating class. It will yield higher values for pairs of intervals with different dominating classes and therefore is less likely to pick them for merging, which is usually preferred.

- Merge the interval with the least dominating class count with the one of its neighbors that has more instances of the same class. This can be achieved by eliminating the break that minimizes the following utility function:

$$\varrho_{T,c}(b, a) = \min\{\max_{d \in C} |T^d_{b-}|, \max_{d \in C} |T^d_{b+}|\} - \gamma \max\{|T^{d_*}_{b-}|, |T^{d_*}_{b+}|\} \qquad (18.20)$$

where

$$d_* = \begin{cases} \arg\max_{d \in C} |T^d_{b-}| & \text{if } \max_{d \in C} |T^d_{b-}| \le \max_{d \in C} |T^d_{b+}| \\ \arg\max_{d \in C} |T^d_{b+}| & \text{otherwise} \end{cases} \qquad (18.21)$$

is the dominating class in whichever of the two sets T_{b-}, T_{b+} has less dominating class count. The first term is the minimum dominating class count for the two intervals separated by the evaluated break, and it identifies which interval needs to be merged with its left or right neighbor, by dropping its left or right bound from the set of breaks. The second term is the maximum number of occurrences of the minimum-count dominating class identified by the first term in the two intervals separated by the evaluated break, multiplied by a γ coefficient. If the latter is sufficiently small (e.g., $\gamma \le \frac{1}{|T|}$), this will make the second term matter only for breaking ties occurring when comparing breaks with respect to the first term. In effect, the minimization of the above break utility function will first identify the interval with the least dominating class count and then choose either its left or right bound, depending on whether the left or right neighbor has more instances of its dominating class.

The above two-term break utility formula is constructed and can be explained similarly as presented above for the instance count criterion. Assuming $|T^{d_*}_i| = \min_{j=1,\dots,k} \max_{d \in C} |T^d_j|$ is the least per-interval dominating class count, the corresponding interval's left and right bounds are breaks b_{i-1} and b_i (unless $i = 1$ or $i = k$). These breaks would then be evaluated as

$$\varrho_{T,c}(b_{i-1}, a) = |T^{d_*}_i| - \gamma |T^{d_*}_{i-1}| \qquad (18.22)$$

$$\varrho_{T,c}(b_i, a) = |T^{d_*}_i| - \gamma |T^{d_*}_{i+1}| \qquad (18.23)$$

Choosing the less of these two corresponds to choosing the neighbor interval with more instances of the dominating class d_*.

The last interpretation of the dominating class count criterion is the most refined and appears to make most sense. It attempts to always eliminate the single interval with the least dominating class count by merging it with the one of its neighbors for which it is a better match. While it requires somewhat more implementational and computational effort than the interval-count criterion, it is more likely to deliver good quality discretization breaks. Still, it remains useful mostly as a way to correct an overly fine-grained initial pure-class discretization.

Example 18.4.4 The following R code implements the three versions of the dominating class count break utility function discussed above and demonstrates their application.

```
## dominating class count break evaluation
## using the total dominating class count after merging
evdisc.dccount1 <- function(b, bl, br, v, class) { max(table(class[v>bl & v<=br])) }

## dominating class count break evaluation
## using the sum of individual dominating class counts before merging
```

```
evdisc.dccount2 <- function(b, bl, br, v, class)
{
  max(table(class[v>bl & v<=b])) + max(table(class[v>b & v<=br]))
}

## dominating class count break evaluation
## using the minimum individual dominating class count before merging
evdisc.dccount3 <- function(b, bl, br, v, class, gamma=1/length(v))
{
  ccl <- table(class[v>bl & v<=b])  # class counts: left
  ccr <- table(class[v>b & v<=br])  # and right

  dcl <- which.max(ccl)  # dominating classes: left
  dcr <- which.max(ccr)  # and right

  dcmin <- ifelse(ccl[dcl]<=ccr[dcr], dcl, dcr)
  min(ccl[dcl], ccr[dcr]) - gamma*max(ccl[dcmin], ccr[dcmin])
}

  # dominating class count bottom-up discretization for the weatherc data
disc.bottomup(play~., weatherc, 3, evalf=evdisc.dccount1)
disc.bottomup(play~., weatherc, 3, evalf=evdisc.dccount2)
disc.bottomup(play~., weatherc, 3, evalf=evdisc.dccount3)

  # dominating class count bottom-up discretization for the Vehicle Silhouettes data
v.disc.bu.dc1 <- disc.bottomup(Class~., v.train, 7, evalf=evdisc.dccount1)
v.disc.bu.dc2 <- disc.bottomup(Class~., v.train, 7, evalf=evdisc.dccount2)
v.disc.bu.dc3 <- disc.bottomup(Class~., v.train, 7, evalf=evdisc.dccount3)
summary(predict(v.disc.bu.dc1, v.train))
summary(predict(v.disc.bu.dc2, v.train))
summary(predict(v.disc.bu.dc3, v.train))

  # dominating class count bottom-up discretization for the Glass data
g.disc.bu.dc1 <- disc.bottomup(Type~., g.train, 7, evalf=evdisc.dccount1)
g.disc.bu.dc2 <- disc.bottomup(Type~., g.train, 7, evalf=evdisc.dccount2)
g.disc.bu.dc3 <- disc.bottomup(Type~., g.train, 7, evalf=evdisc.dccount3)
summary(predict(v.disc.bu.dc1, g.train))
summary(predict(v.disc.bu.dc2, g.train))
summary(predict(v.disc.bu.dc3, g.train))
```

Misclassification count The dominating class count criterion may merge intervals with different dominating classes, resulting in considerable class minorities, i.e., the new interval containing several attribute values corresponding to instances of classes different than its dominating class. No classification model using the discretized attribute alone would be able to classify those instances correctly. They will be therefore referred to as *misclassified instances* for a given set of discretization breaks. The misclassification count merge criterion is an attempt to keep those minorities as small as possible by always merging intervals with the least total number of instances *additionally* misclassified due to merging. It is therefore not the misclassification count before or after merging for the two candidate intervals, separated by the break being evaluated, but the raise thereof due to merging, which translates to the following break utility function:

$$
\begin{aligned}
\varrho_{T,c}(b,a) = {} & |T_{b-} \cup T_{b+}| - \max_{d \in C} |T_{b-}^d \cup T_{b+}^d| \\
& - (|T_{b-}| - \max_{d \in C} |T_{b-}^d| + |T_{b+}| - \max_{d \in C} |T_{b+}^d|)
\end{aligned}
$$

(18.24)

It calculates the misclassification count (for the interval that would be obtained after merging and for the two candidates for merging) as the difference between the total instance count and its dominating class count. Using this difference for break utility evaluation minimizes the misclassification error of a basic single-attribute model that for each instance predicts the dominating class of the corresponding interval.

If starting from pure-class initial intervals, with the minimum possible number of misclassified instances, the misclassification count merge criterion will always attempt to add as few misclassified instances as possible. If starting from singleton intervals, it will do the same, ultimately arriving at pure-class intervals and continuing from that point, until the stop criteria are satisfied.

Notice that, unlike the merge criteria presented before, the misclassification count criterion *will always* merge any two adjacent intervals with the same dominating class, as merging them never adds new misclassified instances. While it may appear perfectly reasonable, it is not necessarily always the best way to maximize the predictive power of the new discrete attribute, since identically labeled intervals may still differ substantially in their class distribution, both for the dominating class and minority classes.

Example 18.4.5 The following R code implements and demonstrates the misclassification count-based break utility function.

```
## misclassification count break evaluation
evdisc.mcount <- function(b, bl, br, v, class)
{
  mcount <- function(cond) { sum(cond) - max(table(class[cond])) }

  mcount(v>bl & v<=br) - (mcount(v>bl & v<=b) + mcount(v>b & v<=br))
}

  # misclassification count bottom-up discretization for the weatherc data
disc.bottomup(play~., weatherc, 3, evalf=evdisc.mcount)

  # misclassification count bottom-up discretization for the Vehicle Silhouettes data
v.disc.bu.mc <- disc.bottomup(Class~., v.train, 7, evalf=evdisc.mcount)
summary(predict(v.disc.bu.mc, v.train))

  # misclassification count bottom-up discretization for the Glass data
g.disc.bu.mc <- disc.bottomup(Type~., g.train, 7, evalf=evdisc.mcount)
summary(predict(g.disc.bu.mc, g.train))
```

Class impurity One deficiency of misclassification-count merging is that it is always concerned about every single misclassified instance to the same extent regardless of whether merging small- or large-count intervals. While for the former a single instance does indeed make a substantial difference, it should not matter that much for the latter. One self-suggesting modification that could address this issue would be using relative misclassification rates (i.e., per-interval misclassification error values) rather than absolute counts, but employing an impurity measure is usually a better approach. This is because, while either misclassification counts or rates are completely blind to the distribution of nonmajority classes, impurity measures take it into account. Even for two-class classification tasks, with exactly one nonmajority class, impurity measures, proven successful for decision tree growing, may turn out to be superior to simple misclassification rates, being able not to necessarily always merge identically labeled adjacent intervals. This may help us to retain more predictive utility of the discretized attribute.

Just like for the misclassification count criterion, it makes sense to use the change of impurity due to merging rather than the actual impurity after merging for break utility evaluation. By comparing the impurity after merging with the impurity before merging we will be able to assess the impurity raise due to merging, and then select the break that yields the minimum raise. Using the impurity alone would lead us into merging intervals that have rather low impurity after merging, which is not necessarily bad, but merging intervals for which the impurity does not raise much due to merging is an arguably better idea. The impurity difference before and after merging could be considered a measure of how much predictively useful information gets lost rather than how much is retained. The former is what one should be primarily concerned with when selecting intervals to merge.

It is easy to notice, however, that both the impurity and (even more evidently) the impurity increase applied as a merge criterion directly will strongly resist against merging even small pure-class intervals with different classes if they have the same or similar instance count. This is because such an operation would yield the maximum possible impurity increase (and the maximum possible impurity after merging): from perfect purity before merging to perfect impurity thereafter. More generally, it is likely to delay or entirely avoid merging small intervals with different dominating classes. While this behavior may at first appear reasonable, it actually leads to substantial problems. In many cases one or few huge-count intervals would be created with the other left as small-count pure-class intervals, since the utility function would often recommend dropping breaks that separate large-count intervals from small-count ones. This is clearly undesirable and can be avoided by weighting the impurity-based break utility with the total instance count after merging. This makes the discretization process more sensitive to impurity increases that correspond to many instances. Using the entropy to measure impurity, which is by far the most popular choice, the proposed break evaluation function may be defined as follows:

$$\varrho_{T,c}(b,a) = |T_{b-} \cup T_{b+}|(E_{T_{b-} \cup T_{b+}}(c) - E_{T_{b-} \cup T_{b+}}(c|b)) \tag{18.25}$$

where

$$E_{T_{b-} \cup T_{b+}}(c|b) = \frac{|T_{b-}|}{|T_{b-} \cup T_{b+}|} E_{T_{b-}}(c) + \frac{|T_{b+}|}{|T_{b-} \cup T_{b+}|} E_{T_{b+}}(c) \tag{18.26}$$

is, according to the definition presented in Section 2.5.3, the weighted entropy before merging or, equivalently, the conditional entropy of the target concept in the merged interval given the partitioning into subintervals by the break being evaluated.

Notice, by the way, that the impurity increase is no different from the mutual information between the target concept and interval membership, calculated on the $T_{b-} \cup T_{b+}$ set, and instance count weighting makes it proportional to the corresponding log-likelihood ratio statistic.

The class impurity criterion is an attempt to overcome the weaknesses of the misclassification count criterion and its additional complexity is likely to pay off by better discretization quality. The improvement is more likely to be observed when many intervals have to be merged, i.e., when the final number of intervals is much less than the initial one obtained by pure-class discretization (if started from singleton intervals, class impurity merging will clearly arrive at pure-class intervals at some point anyway, as it will merge intervals with instances of the same class only in the first place).

Example 18.4.6 The following R code defines and demonstrates the `evdisc.entropy` function that implements class impurity-based break evaluation, using the entropy as the impurity measure, using the `entropy` and `entropy.cond` functions.

> Ex. 2.4.26
> dmr.stats

```
## entropy break evaluation
evdisc.entropy <- function(b, bl, br, v, class)
{
  sum(v>bl & v<=br)*(entropy(class[v>bl & v<=br])-
                  entropy.cond(class[v>bl & v<=br], v[v>bl & v<=br]<=b))
}

# entropy bottom-up discretization for the weatherc data
disc.bottomup(play~., weatherc, 3, evalf=evdisc.entropy)

# entropy bottom-up discretization for the Vehicle Silhouettes data
v.disc.bu.e <- disc.bottomup(Class~., v.train, 7, evalf=evdisc.entropy)
summary(predict(v.disc.bu.e, v.train))

# entropy bottom-up discretization for the Glass data
g.disc.bu.e <- disc.bottomup(Type~., g.train, 7, evalf=evdisc.entropy)
summary(predict(g.disc.bu.e, g.train))
```

Class distribution dissimilarity   It can easily be seen that the raise of class impurity due to interval merging will be small if the class distribution for the corresponding subsets of training instances is similar. This suggests that explicitly measuring and minimizing class distribution dissimilarity (if it is supposed to represent break utility) for the two intervals separated by the break being evaluated is another related and possibly useful merge criterion. Measuring class distribution dissimilarity is equivalent to measuring the relationship between the target concept (which is a discrete attribute) and interval membership (which can be considered a binary attribute, indicating either the left or right of the two intervals separated by the break being evaluated). This makes it possible to employ any relationship detection statistics presented in Section 2.5.3, with the χ^2 statistic being the most popular choice. The corresponding break utility function can be then defined as

$$\varrho_{T,c}(b, a) = \chi^2_{T_{b-} \cup T_{b+}}(c, b) \qquad (18.27)$$

where

$$\chi^2_{T_{b-} \cup T_{b+}}(c, b) = \sum_{d \in C} \frac{(|T^d_{b-}| - e^d_{b-})^2}{e^d_{b-}} + \sum_{d \in C} \frac{(|T^d_{b+}| - e^d_{b+})^2}{e^d_{b+}} \qquad (18.28)$$

is the χ^2 statistic for target concept c and interval break b, and for $s = -, +$

$$e^d_{bs} = \frac{|T_{bs}| \cdot |T^d_{b-} \cup T^d_{b+}|}{|T_{b-} \cup T_{b+}|} \qquad (18.29)$$

is the expected number of training instances of class d in the interval on the left (for $s = -$) or right (for $s = +$) side of b, assuming no relationship between the target concept and interval membership.

Depending on the particular choice of the relationship detection measure employed, break evaluation based on class distribution dissimilarity may or may not be prone to the same effect of overly strong resistance against merging small-count intervals with different dominating classes, possibly asking for instance count-based weighting as discussed above. For the most popular χ^2 statistic it is not the case, as it implicitly includes instance count weighting internally (with the relative distribution kept constant, the statistic's value is proportional to the data size).

As the class impurity criterion, the class distribution dissimilarity criterion belongs to the most refined merge criteria for bottom-up discretization and may be expected to often yield good results. It is in fact used by the best-known bottom-up discretization algorithm called *ChiMerge*. As a matter of fact, the impurity-based and distribution dissimilarity-based break utility functions can be actually considered variations of the same approach, which evaluates breaks according to the strength of relationship between the target concept and interval membership, with just different approaches to measuring this relationship. In particular, as mentioned above, the impurity increase can be verified to be the same as the mutual information, and with instance count weighting applied it becomes proportional to the log-likelihood ratio statistic.

Example 18.4.7 The R code presented below implements and demonstrates class distribution dissimilarity-based break evaluation, using the χ^2 statistic, calculated by the standard `chisq.test` function. Notice the use of the `correct=FALSE` argument which prevents applying a small-data correction. While normally useful, it would ruin the capability of merging very small-count intervals reasonably, by making them always appear to have similar class distributions.

```
## chi-square break evaluation
evdisc.chisq <- function(b, bl, br, v, class)
{
  chisq.test(class[v>bl & v<=br], v[v>bl & v<=br]<=b, correct=FALSE)$statistic
}


  # chi-square bottom-up discretization for the weatherc data
disc.bottomup(play~., weatherc, 3,evalf=evdisc.chisq)

  # chi-square bottom-up discretization for the Vehicle Silhouettes data
v.disc.bu.chi <- disc.bottomup(Class~., v.train, 7, evalf=evdisc.chisq)
summary(predict(v.disc.bu.chi, v.train))

  # chi-square bottom-up discretization for the Glass data
g.disc.bu.chi <- disc.bottomup(Type~., g.train, 7, evalf=evdisc.chisq)
summary(predict(g.disc.bu.chi, g.train))
```

18.4.2.4 Stop criteria

One stop criterion that is likely to be useful for all bottom-up discretization algorithms (and, as we will see, for top-down algorithms as well) is the number of intervals to be created. If used in combination with other stop criteria, it would be the *minimum number of intervals allowed* (with no further merging permitted after reaching it), as those other stop criteria could terminate the discretization process earlier). The latter are usually closely tied to the merge criteria discussed above. For each of the merge criteria it is straightforward to specify the corresponding stop criterion with which it can be coupled:

Instance count. Stop if the minimum per-interval instance count is sufficiently large.

Dominating class count. Stop if the minimum per-interval dominating class count is sufficiently large.

Misclassification count. Stop if the minimum per-interval misclassification count is sufficiently large.

Class impurity. Stop if the minimum per-interval class impurity is sufficiently large.

Class distribution dissimilarity. Stop if the minimum class distribution similarity for a pair of adjacent intervals is sufficiently large.

An alternative and nonequivalent, but closely related universal stop criterion could consist in comparing the minimum value of the break utility function (indicating the best possible break to be dropped) against a threshold, specified as a parameter:

$$\min_{b \in B_{T,c}(a)} \varrho_{T,c}(b,a) > \theta_\varrho \qquad (18.30)$$

This is actually adopted in the illustrative implementation of bottom-up discretization and used in the examples presented above (more precisely, the negated version of this inequality is used as a continuation condition).

18.4.3   Top-down discretization

The top-down discretization process is essentially a reverse of the bottom-up one. It starts with one interval covering the whole co-domain of the original continuous attribute (i.e., with no breaks) and then continues to add breaks until a satisfactory set of intervals is reached.

18.4.3.1   Algorithm scheme

Top-down discretization is described by the algorithm scheme presented below. It assumes that a set of candidate breaks is maintained from which one new break is selected in each iteration, corresponding to one existing interval being cut into two subintervals. Its exact operation is controlled by a cut criterion used to select new breaks and stop criteria used to terminate the process. The cut criterion is assumed to be represented by a break utility function $\varrho_{T,c}$, as before for bottom-up discretization, but this time it is to be maximized (i.e., the most useful new break is added, whereas the least useful break was dropped in the bottom-up approach).

1: $B_{T,c}(a) := \emptyset$;
2: initialize the set of candidate breaks $B'_{T,c}(a)$;
3: **while** $B'_{T,c}(a) \neq \emptyset$ and stop criteria are not satisfied **do**
4:      $b_* := \arg\max_{b \in B'_{T,c}(a)} \varrho_{T,c}(b,a)$;
5:      $B_{T,c}(a) := B_{T,c}(a) \cup \{b_*\}$;
6:      $B'_{T,c}(a) := B'_{T,c}(a) - \{b_*\}$;
7: **end while**

Since it does indeed look as a straightforward reverse of bottom-up discretization (with just a separate candidate break set used), it might appear just like a rearranged version of the former, with only minor implementational differences and without substantial capability differences. While indeed the two approaches to supervised discretization may be combined with similar cut and stop criteria and may yield similar results, top-down discretization is better suited to a much more common discretization scenario where the desired final number of intervals is much less than the number of singleton or pure-class intervals. In such situations it just needs to add a small number of breaks, avoiding a lot of work that is required for bottom-up discretization to deal with many small-count intervals. This may not only mean computational savings, but also different – and possibly better – results due to not having to deal with small data subsets at any stage.

While the above formulation of the top-down discretization scheme is quite natural and straightforward to implement, an alternative recursive formulation appears to be more popular, where the algorithm receives an interval to discretize as its argument. It first uses a single break to cut interval into two subintervals and then invokes itself for the latter. It is noteworthy that whatever cut criterion is used by such recursive top-down discretization, it will be only applied to select the best new break points *within intervals*, and not to determine the order in which intervals should be divided. The latter will be clearly the depth-first order implied by the recursive algorithm invocations. This will not have any impact on the final results for stop criteria that only depend on the properties of individual intervals, but will clearly matter a lot for the basic and common stop criterion of the (maximum) number of intervals. The depth-first recursive formulation of top-down discretization is not applicable with this stop criterion and the above best-first iterative formulation should be used instead.

One possible advantage of the recursive version of top-down discretization is that it could employ simpler cut criteria, as their responsibility is limited to selecting break points within a single interval at a time. This may be indeed much easier than picking both an interval to cut and a break to use. Still, with the best-first version being generally more useful and better controllable, the remainder of this section will focus entirely on the latter.

It is worthwhile to notice that top-down discretization can be viewed as a special simple case of decision tree growing, with only one continuous attribute available. Splits selected at decision tree nodes correspond to discretization breaks and leaves to finally obtained discretization intervals. It may be even possible to employ a decision tree growing implementation to perform top-down discretization as long as it permits specifying stop criteria appropriate for the latter.

Example 18.4.8 The top-down discretization scheme is implemented and demonstrated by the following R code. It uses the `closest.below`, `closest.above`, and `insert.ord` auxiliary functions for finding the left and right existing neighbor `dmr.util` breaks for a given candidate break as well as for inserting the selected candidate break to the ordered vector of breaks, respectively. Similarly as before for bottom-up discretization, the `disc.topdown1` function has to be supplied with a break evaluation function and, optionally, with a candidate-generation function that returns a set of candidate breaks to be considered, defaulting to pure-class discretization. Notice that the supplied break evaluation function is internally wrapped to add checking for empty intervals (with no instances) and prevent trying to break such intervals. For the purpose of demonstration, a meaningless trivial break evaluation function is used that returns a candidate break's value as its utility, just like in Example 18.4.2. This has the effect of adding breaks in the greatest first order.

```
## top-down discretization for a single attribute
disc.topdown1 <- function(v, class, k=5, candf=disc.pure1, evalf, minev=0)
{
  evalf.td <- function(b, bl, br, v, class)
  {
    if (any(v>bl & v<=br))
     evalf(b, bl, br, v, class)
    else
     -Inf
  }

  breaks <- NULL
  candidates <- candf(v, class)

  utils <- mapply(evalf.td, candidates,
                  closest.below(candidates, breaks),
                  closest.above(candidates, breaks),
                  MoreArgs=list(v, class))
  b <- which.max(utils)

  while (length(candidates)>0 && length(breaks)+1<k && utils[b]>=minev)
  {
    breaks <- insert.ord(breaks, candidates[b])
    candidates <- candidates[-b]
    utils <- utils[-b]
    if (b>1)
      utils[b-1] <- evalf.td(candidates[b-1],
                             closest.below(candidates[b-1], breaks),
                             closest.above(candidates[b-1], breaks),
                             v, class)
    if (b<=length(candidates))
      utils[b] <- evalf.td(candidates[b],
                           closest.below(candidates[b], breaks),
                           closest.above(candidates[b], breaks),
                           v, class)
    b <- which.max(utils)
  }
  breaks
}

## top-down discretization for a dataset
disc.topdown <- disc.all(disc.topdown1)

  # top-down discretization of the temperature attribute in the weatherc data
disc.topdown1(weatherc$temperature, weatherc$play, 3,
              evalf=function(b, bl, br, v, class) b)

  # top-down discretization for the weatherc data
disc.topdown(play~., weatherc, 3, evalf=function(b, bl, br, v, class) b)
```

18.4.3.2 Initialization

Literally speaking, the initialization of top-down discretization is trivial, as it simply starts with the empty set of breaks, $B_{T,c}(a) = \emptyset$. It is however convenient conceptually – although not necessary implementationally – to assume that a set of candidate breaks that are considered by the cut criterion on each iteration, designated by $B'_{T,c}(a)$, is also explicitly maintained. It is usually assumed to initially contain exactly one value between each pair of consecutive values of the attribute being discretized occurring in the training set, which is clearly the same as the set of breaks for singleton intervals. As discussed above, it is most common to select mid-points between every two values to separate as such breaks. One may notice, though, that under any reasonable cut criterion no break between attribute values corresponding to the instances of the same class would ever be added. This observation makes it possible to exclude such breaks from the set of candidates, which is clearly equivalent to using pure-class discretization breaks to initialize the set of candidate breaks for top-down discretization.

18.4.3.3 Cut criterion

To discuss different cut criteria that may be used to instantiate the generic top-down discretization scheme, we will adopt a similar notation as used before during the discussion of the bottom-up approach. When considering a candidate break $b \in B'_{T,c}(a)$, we will designate the closest less and greater previously added breaks in $B_{T,c}(a)$ as b_- and b_+, and refer to the following subsets of training instances:

$$T_{b-} = \{x \in T \mid a(x) \in (b_-, b]\} \tag{18.31}$$

$$T_{b+} = \{x \in T \mid a(x) \in (b, b_+]\} \tag{18.32}$$

corresponding to the intervals between candidate break b and the closest existing breaks b_- and b_+.

The cut criteria for top-down discretization may be direct counterparts of the bottom-up merge criteria discussed above. In particular, useful cut criteria include:

Misclassification count. Add breaks that yield the greatest misclassification count reduction.

Class impurity. Add breaks that yield the greatest class impurity increase.

Class distribution dissimilarity. Add breaks that yield intervals with the greatest class distribution dissimilarity.

Whilst the two simplest merge criteria, based on instance count and dominating class count, could also be adapted to top-down discretization, they have no counterparts that would be truly useful as cut criteria and will not be considered. This is because what they could be most reasonably used for is only determining which interval to divide rather than which exactly possible break to use.

It is easy to verify that the very same formulae as presented before for merge criteria can also serve as break utility function definitions corresponding to their cut criteria counterparts. This perfect break utility function reusability is not at all surprising. While a merge criterion selects a break to drop and a cut criterion selects a break to add, they may share exactly the same measure of break utility, with the former minimizing it and the latter maximizing it for break selection. As before for bottom-up discretization, one may expect the class impurity and class distribution dissimilarity criteria to usually yield best discretization results. It is the former that is actually the most popular for top-down discretization, with the entropy being used as the impurity measure. The resulting algorithm is commonly referred to as entropy-based discretization.

As mentioned above, adopting the recursive formulation of bottom-up discretization reduces the responsibility of the cut criterion to selecting breaks within a single interval at a time. In that case, simplified versions of some of the above cut criteria would be sufficient to achieve the same effects. In particular, for the most popular class impurity criterion, it would be sufficient just to use the weighted impurity after cutting, i.e., the conditional entropy after cutting (if the entropy is used to measure the impurity):

$$E_{T_{b-} \cup T_{b+}}(c|b) = \frac{|T_{b-}|}{|T_{b-} \cup T_{b+}|} E_{T_{b-}}(c) + \frac{|T_{b+}|}{|T_{b-} \cup T_{b+}|} E_{T_{b+}}(c) \qquad (18.33)$$

Neither subtracting it from the impurity before cutting $E_{T_{b-} \cup T_{b+}}(c)$, nor applying instance count weighting, would have any effect on the selected break, if $T_{b-} \cup T_{b+}$ remains constant due to dealing with a single interval only. Focusing on the preferred best-first version of bottom-up discretization, though, we do not adopt any such simplifications of break utility functions.

Example 18.4.9 The R code presented below demonstrates top-down discretization with the misclassification count, class impurity (using the entropy), and class distribution dissimilarity (using the χ^2 statistic) break evaluation functions, previously defined and used in merge criteria examples for bottom-up discretization.

```
    # misclassification count top-down discretization for the weatherc data
disc.topdown(play~., weatherc, 3, evalf=evdisc.mcount)
    # misclassification count top-down discretization for the Vehicle Silhouettes data
v.disc.td.mc <- disc.topdown(Class~., v.train, 7, evalf=evdisc.mcount)
summary(predict.disc(v.disc.td.mc, v.train))
    # misclassification count top-down discretization for the Glass data
g.disc.td.mc <- disc.topdown(Type~., g.train, 7, evalf=evdisc.mcount)
summary(predict.disc(g.disc.td.mc, g.train))

    # entropy top-down discretization for the weatherc data
disc.topdown(play~., weatherc, 3, evalf=evdisc.entropy)
    # entropy top-down discretization for the Vehicle Silhouettes data
v.disc.td.e <- disc.topdown(Class~., v.train, 7, evalf=evdisc.entropy)
summary(predict.disc(v.disc.td.e, v.train))
    # entropy top-down discretization for the Glass data
g.disc.td.e <- disc.topdown(Type~., g.train, 7, evalf=evdisc.entropy)
summary(predict.disc(g.disc.td.e, g.train))

    # chi-square top-down discretization for the weatherc data
disc.topdown(play~., weatherc, 3, evalf=evdisc.chisq)
    # chi-square top-down discretization for the Vehicle Silhouettes data
v.disc.td.chi <- disc.topdown(Class~., v.train, 7, evalf=evdisc.chisq)
summary(predict.disc(v.disc.td.chi, v.train))
    # chi-square top-down discretization for the Glass data
g.disc.td.chi <- disc.topdown(Type~., g.train, 7, evalf=evdisc.chisq)
summary(predict.disc(g.disc.td.chi, g.train))
```

18.4.3.4   Stop criteria

Similarly as for bottom-up discretization, the simplest useful stop criterion is based on the number of discretization intervals. For the top-down approach, if used in combination with other stop criteria, it would be the *maximum number of intervals* allowed (with no further cutting permitted after reaching it, as those other stop criteria could terminate the discretization process earlier). As explained above, this criterion should not be used with the recursive version of the top-down algorithm scheme, which divides intervals in a depth-first order.

Other reasonable stop criteria include:

Instance count. Stop if the maximum per-interval instance count is sufficiently small.

Dominating class count. Stop if the maximum per-interval dominating class count is sufficiently small.

Misclassification count. Stop if the maximum per-interval misclassification count is sufficiently small.

Class impurity. Stop if the maximum per-interval class impurity is sufficiently small.

Class distribution dissimilarity. Stop if the maximum class distribution dissimilarity for a pair of adjacent intervals is sufficiently small.

The instance count and dominating class count criteria, although not found useful for candidate break selection, are worth listing here, since they are simple to implement and use while capable of delivering satisfactory results in many cases. They can be used in combination with any cut criteria. The remaining criteria are more refined and may sometimes perform better, but setting the threshold parameters (to define "sufficiently small") for them may be much more difficult and they should be rather considered tied to the corresponding cut criteria. A closely related and usually similarly useful universal stop criterion can be obtained, however, by simply comparing the maximum value of the adopted break utility function (indicating the best possible break to be added) against a threshold, specified as a parameter:

$$\max_{b \in B'_{T,c}(a)} \varrho_{T,c}(b, a) < \theta_o \qquad (18.34)$$

which stops adding breaks if the best possible break is not sufficiently good. The negated version of this inequality is used as a continuation condition in the illustrative implementation of top-down discretization presented and used in the previous example. It suffers from the same inconvenience, related to parameter setup, although a reasonable heuristic for adjusting the required minimum break utility can be adopted at least for the entropy-based cut criterion. It is derived using the minimum description length principle and falls beyond the scope of this book.

18.5   Effects of discretization

Since discretization is a part of the modeling process, its effects have to be therefore evaluated using an independent data subset. This can be accomplished by employing a selected evaluation procedure, such as those presented in Section 7.3. For this evaluation procedure the model creation process includes both discretization and model building using the discretized dataset. It can be as simple as the hold-out procedure (i.e., holding out a separate data subset for the evaluation purpose), although more refined and reliable procedures, such as k-fold cross-validation, should be preferred when the data size and computational resources permit.

Example 18.5.1 A simple experiment that evaluates the effects of discretization is performed by the following R code. It uses the datasets discretized using discretization models from the previous examples to create and evaluate decision tree and naïve Bayes classification models. Their test set performance is evaluated using the misclassification error, implemented by the `err` function. Previously created models for original dataset versions are included for comparison. Notice that pure-class discretization models are not used. This is because the resulting discrete attributes have large numbers of values, which makes model creation computationally expensive and overly both time- and memory consuming.

Ex. 7.2.1
`dmr.claseval`

```
  # discretization models for the Vehicle Silhouettes data
v.disc <- list(nodisc='class<-'(list(), "disc"),
               ew=v.disc.ew, ef=v.disc.ef,
               bu.ic1=v.disc.bu.ic1, bu.ic2=v.disc.bu.ic2,
               bu.dc1=v.disc.bu.dc1, bu.dc2=v.disc.bu.dc2, bu.dc3=v.disc.bu.dc3,
               bu.mc=v.disc.bu.mc, bu.e=v.disc.bu.e, bu.chi=v.disc.bu.chi,
               td.mc=v.disc.td.mc, td.e=v.disc.td.e, td.chi=v.disc.td.chi)

  # discretization models for the Glass data
g.disc <- list(nodisc='class<-'(list(), "disc"),
               ew=g.disc.ew, ef=g.disc.ef,
               bu.ic1=g.disc.bu.ic1, bu.ic2=g.disc.bu.ic2,
               bu.dc1=g.disc.bu.dc1, bu.dc2=g.disc.bu.dc2, bu.dc3=g.disc.bu.dc3,
               bu.mc=g.disc.bu.mc, bu.e=g.disc.bu.e, bu.chi=g.disc.bu.chi,
               td.mc=g.disc.td.mc, td.e=g.disc.td.e, td.chi=g.disc.td.chi)

  # misclassification error values for the Vehicle Silhouettes data
v.err <- lapply(v.disc,
               function(dm)
               {
                 v.train.d <- predict(dm, v.train)
                 v.test.d <- predict(dm, v.test)
                 v.tree.d <- rpart(Class~., v.train.d)
                 v.nb.d <- naiveBayes(Class~., v.train.d)
                 list(tree=err(predict(v.tree.d, v.test.d, type="c"),
                               v.test.d$Class),
                      nb=err(predict(v.nb.d, v.test.d), v.test.d$Class))
               })

  # misclassification error values for the Glass data
g.err <- lapply(g.disc,
               function(dm)
               {
                 g.train.d <- predict(dm, g.train)
                 g.test.d <- predict(dm, g.test)
                 g.tree.d <- rpart(Type~., g.train.d)
                 g.nb.d <- naiveBayes(Type~., g.train.d)
                 list(tree=err(predict(g.tree.d, g.test.d, type="c"),
                               g.test.d$Type),
                      nb=err(predict(g.nb.d, g.test.d), g.test.d$Type))
               })

  # error comparison
v.tree.err <- sapply(v.err, function(e) e$tree)
g.tree.err <- sapply(g.err, function(e) e$tree)
v.nb.err <- sapply(v.err, function(e) e$nb)
g.nb.err <- sapply(g.err, function(e) e$nb)

barplot(v.tree.err, main="Vehicle Silhouettes, rpart", ylab="Error", las=2)
lines(c(0, 17), rep(v.tree.err[1], 2), lty=2)
```

```
barplot(g.tree.err, main="Glass, rpart", ylab="Error", las=2)
lines(c(0, 17), rep(g.tree.err[1], 2), lty=2)

barplot(v.nb.err, main="Vehicle Silhouettes, naiveBayes", ylab="Error", las=2)
lines(c(0, 17), rep(v.nb.err[1], 2), lty=2)

barplot(g.nb.err, main="Glass, naiveBayes", ylab="Error", las=2)
lines(c(0, 17), rep(g.nb.err[1], 2), lty=2)
```

Barplots that compare the misclassification error for the two classification algorithms used, with and without discretization, produced by the above code, are presented in Figures 18.1 and 18.2. As it could be expected, naïve Bayes models benefit from being supplied discretized data more than decision trees. Discretization appears to always yield worse decision tree models for the *Vehicle Silhouettes* data, sometimes substantially. Surprisingly, in this case supervised discretization algorithms do not turn out superior to unsupervised ones. There is a noticeable error reduction due to discretization in all the remaining cases, though, with supervised discretization sometimes – but not always – giving the best results. This experimental study, serving the illustration purpose only, is simplified in many ways (the most important of which are using two datasets only and no tuning of the number of intervals or the maximum/minimum break utility stop criteria), so it hardly provides any basis for generally reliable conclusions.

18.6   Conclusion

Several traditional machine learning algorithms or their simple implementations assumed discrete attributes, making discretization a necessity during the early period when data mining emerged as a kind of "applied machine learning." This has changed substantially since then and most currently used practical algorithms and their implementations support both discrete and continuous attributes. Even then there is sufficient motivation to at least consider discretization as a usually optional, but potentially useful attribute transformation. And if discretization is often not a necessity, but a deliberate choice for model readability improvement, computational savings, or overfitting prevention, its quality becomes even more important.

The two primitive unsupervised algorithms and the two more refined supervised ones (with multiple instantiations using different break utility measures) presented in this chapter are far from being sufficient representatives of the variety of existing discretization techniques, particularly those more recently developed. They should be sufficient, though, to highlight the utility of discretization and its usage scenarios. The modeling perspective, according to which discretization algorithms use the training set to create discretization models (represented as simply as by sets of interval breaks) applicable to new data from the same domain is essential for the latter, although not always properly recognized and adhered to.

18.7   Further readings

The utility of discretization and the importance of its quality has gained appreciation over the last two decades. Most work of discretization occured in the area of machine learning, in which

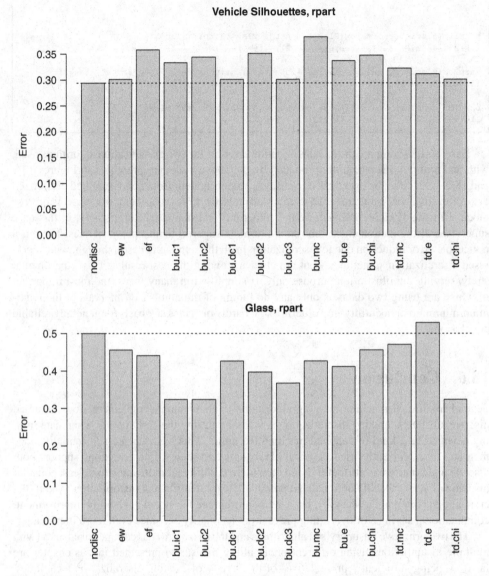

Figure 18.1   The barplots of the test set misclassification error for the decision tree models using discretized attributes.

classification algorithms that "prefer" discrete attributes, such as decision trees, rule induction, or the naïve Bayes classifier, are much more often studied than in statistics. Contemporary data mining books that cover such algorithms usually include an adequate representation of discretization techniques as well (e.g., Cios *et al.* 2007; Han *et al.* 2011, Tan *et al.* 2013, Witten *et al.* 2011). There are also numerous review articles written and experimental comparisons performed over the years that can be referred to for description of a greater variety of algorithms and information on their performance (Aguilar *et al.* 2004; Dougherty *et al.* 1995; Garcia *et al.* 2013; Kotsiantis and Kanellopoulos 2006; Liu *et al.* 2002). They present

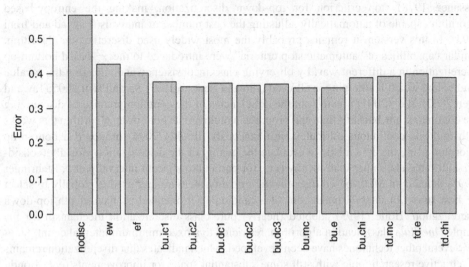

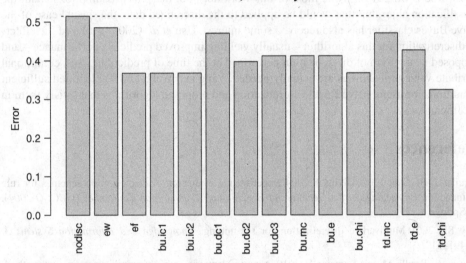

Figure 18.2   The barplots of the test set misclassification error for the naïve Bayes models using discretized attributes.

a much broader spectrum of discretization techniques than this chapter, but do not put similar emphasis on the modeling view of discretization.

The increase of interest in discretization started from the works of Catlett (1991) and Kerber (1992). The former contributed a basic top-down entropy-based discretization algorithm and the latter the bottom-up *ChiMerge* algorithm using the χ^2 statistic. Both these algorithms were subsequently investigated and refined by other authors. Fayyad and Irani (1992) examined and confirmed the suitability of the entropy for interval break evaluation.

Shortly thereafter they introduced a minimum description length (MDL, Grünwald 2007; Rissanen 1978) stop criterion for top-down discretization, making the entropy-based algorithm capable of automatically adjusting the best number of intervals (Fayyad and Irani 1993). In this version, it remains probably the most widely used discretization algorithm. Similar capabilities of "automatic stop criteria" were introduced to the χ^2-based bottom-up discretization in a different way, by observing class inconsistency and adjusting the p-value level below which no merging is allowed (Liu and Setiono 1997; Su and Hsu 2005; Tay and Shen 2002). Bay (2001) pointed out the deficiencies of the common practice of discretizing one continuous attribute at a time and proposed a multivariate bottom-up algorithm that works with multiple continuous attributes simultaneously. Boulle (2004) proposed a bottom-up algorithm using the χ^2 statistic to evaluate the quality of the discretization globally (considering all intervals) rather than locally (i.e., for particular adjacent interval pairs). Pfahringer (1995) defined an MDL-based discretization quality measure and used it globally to select the best subset of interval breaks out of a candidate set obtained by fixed-depth top-down discretization. Holte (1993) adopted (nearly) pure-class discretization mechanism for his simple *One-Rule* classification algorithm, restricted to a fixed minimum interval count.

Several other techniques have been developed, which indicates that discretization remains an attractive research topic with still some substantial space for improvements (e.g., Bondu *et al.* 2010; Ching *et al.* 1995; Kurgan and Cios 2004; Liu and Wang 2005; Zighed *et al.* 1998; Zou *et al.* 2013). While most of this work does not make any assumptions about the classification algorithm that would be applied to the discretized data, the special case of the naïve Bayes classifier has also received some interest. Hsu *et al.* (2003) analyzed the effects of discretization for this algorithm – usually yielding improved predictive performance – and proposed a lazy form of discretization performed at the time of prediction, when conditional attribute value probabilities are actually needed. Yang and Webb (2009) identified sufficient conditions for useful naïve Bayes discretization and proposed algorithms that satisfy them in an efficient way.

References

Aguilar J, Bacardit J and Divina F 2004 Experimental evaluation of discretization schemes for rule induction *Proceedings of the Genetic and Evolutionary Computation Conference (GECCO-2004)*. Springer.

Bay SD 2001 Multivariate discretization for set mining. *Knowledge and Information Systems* **3**, 491–512.

Bondu A, Boulle M and Lemaire V 2010 A non-parametric semisupervised discretization method. *Knowledge and Information Systems* **24**, 35–57.

Boulle M 2004 Khiops: A statistical discretization method of continuous attributes. *Machine Learning* **55**, 53–69.

Catlett J 1991 On changing continuous attributes into ordered discrete attributes *Proceedings of the European Working Session on Learning*. Springer.

Ching JY, Wong AKC and Chan KCC 1995 Class-dependent discretization for inductive learning from continuous and mixed-mode data. *IEEE Transactions on Pattern Analysis and Machine Intelligence* **17**, 641–651.

Cios KJ, Pedrycz W, Swiniarski RW and Kurgan L 2007 *Data Mining: A Knowledge Discovery Approach*. Springer.

Dougherty J, Kohavi R and Sahami M1995 Supervised and unsupervised discretization of continuous features *Proceedings of the Twelfth International Conference on Machine Learning (ICML-95)*. Morgan Kaufmann.

Fayyad UM and Irani KB 1992 On the handling of continuous-valued attributes in decision tree generation. *Machine Learning* pp. 87–102.

Fayyad UM and Irani KB1993 Multi-interval discretization of continuous-valued attributes for classification learning *Proceedings of the Thirteenth International Joint Conference on Artificial Intelligence (IJCAI-93)*. Morgan Kaufmann.

Garcia S, Luengo J, Sáez JA, López V and Herrera F 2013 A survey of discretization techniques: Taxonomy and empirical analysis in supervised learning. *IEEE Transactions on Knowledge and Data Engineering* **25**, 734–750.

Grünwald P 2007 *The Minimum Description Length Principle*. MIT Press.

Han J, Kamber M and Pei J 2011 *Data Mining: Concepts and Techniques* 3rd edn. Morgan Kaufmann.

Holte RC 1993 Very simple classification rules perform well on most commonly used datasets. *Machine Learning* **11**, 63–90.

Hsu CN, Huang HJ and Wong TT 2003 Implications of the Dirichlet assumption for discretization of continuous variables in naive Bayesian classifiers. *Machine Learning* **53**, 235–263.

Kerber R 1992 ChiMerge: Discretization of numeric attributes *Proceedings of the Tenth National Conference on Artificial Intelligence (AAAI-92)*. AAAI Press.

Kotsiantis S and Kanellopoulos D 2006 Discretization techniques: A recent survey. *GESTS International Transactions on Computer Science and Engineering* **32**, 47–58.

Kurgan LA and Cios KJ 2004 CAIM discretization algorithm. *IEEE Transactions on Knowledge and Data Engineering* **16**, 145–153.

Liu H, Hussain F, Tan CL and Dash M 2002 Discretization: An enabling technique. *Data Mining and Knowledge Discovery* **6**, 393–423.

Liu H and Setiono R 1997 Feature selection via discretization. *IEEE Transactios on Knowledge and Data Engineering* **9**, 642–645.

Liu X and Wang H 2005 A discretization algorithm based on a heterogeneity criterion. *IEEE Transactions on Knowledge and Data Engineering* **17**, 1166–1173.

Pfahringer B 1995 Compression-based discretization of continuous attributes *Proceedings of the Twelfth International Conference on Machine Learning (ICML-95)*. Morgan Kaufmann.

Rissanen J 1978 Modeling by shortest data description. *Automatica* **14**, 465–658.

Su CT and Hsu JH 2005 An extended Chi2 algorithm for discretization of real value attributes. *IEEE Transactions on Knowledge and Data Engineering* **17**, 437–441.

Tan PN, Steinbach M and Kumar V 2013 *Introduction to Data Mining* 2nd edn. Addison-Wesley.

Tay FEH and Shen L 2002 A modified Chi2 algorithm for discretization. *IEEE Transactions on Knowledge and Data Engineering* **14**, 666–670.

Witten IH, Frank E and Hall MA 2011 *Data Mining: Practical Machine Learning Tools and Techniques* 3rd edn. Morgan Kaufmann.

Yang Y and Webb GI 2009 Discretization for naive-Bayes learning: Managing discretization bias and variance. *Machine Learning* **74**, 39–74.

Zighed DA, Rabaseda S and Rakotomalala R 1998 Fusinter: A method for discretization of continuous attributes for supervised learning. *International Journal of Uncertainty, Fuzziness and Knowledge-Based Systems* **6**, 307–326.

Zou L, Yan D, Karimi HR and Shi P 2013 An algorithm for discretization of real value attributes based on interval similarity. *Journal of Applied Mathematics* **2013**, 8.

19

Attribute selection

19.1   Introduction

Unlike the classification, regression, or clustering tasks, the attribute selection task is not a data mining task *per se*, as – contrary to the former – it is not concerned with delivering models representing generalizations of predictively useful relationships discovered in the data. While the results of attribute selection – taking the form of a subset of attributes – can be interesting and insightful on their own – there is always another task (usually classification, regression, or clustering) for which attribute selection serves as preprocessing, and which provides the motivation, context, and quality criteria therefore. This puts attribute selection in the same category as data transformation techniques discussed in Chapter 17. What distinguishes attribute selection from the latter is that it often requires much more refined and computationally demanding algorithms to be adequately performed, and its impact on the final model quality may be even more substantial. It cannot be therefore reduced to something purely technical and trivial by any means, and definitely deserves more interest than it usually receives.

Example 19.1.1 Attribute selection algorithms presented in this chapter will be illustrated by R code examples, using the toy *weather*, *weatherc*, and *weatherr* datasets from Examples 1.3.1, 1.3.2, and 1.4.1, as well as the more realistic *Vehicle Silhouettes*, `dmr.data` *Soybean*, and *Boston Housing* datasets from the `mlbench` package. The following R code sets up the environment for these demonstrations by loading the datasets, as well as several DMR packages providing auxiliary functions and the `rpart` package for creating decision and regression tree models. The larger datasets are split into train and test subsets for subsequent demonstrations of the effects of attribute selection.

Data Mining Algorithms: Explained Using R, First Edition. Paweł Cichosz.
© 2015 John Wiley & Sons, Ltd. Published 2015 by John Wiley & Sons, Ltd.

```
library(dmr.claseval)
library(dmr.disc)
library(dmr.regeval)
library(dmr.stats)
library(dmr.trans)
library(dmr.util)

library(rpart)

data(weather, package="dmr.data")
data(weatherc, package="dmr.data")
data(weatherr, package="dmr.data")

data(Vehicle, package="mlbench")
data(Soybean, package="mlbench")
data(BostonHousing, package="mlbench")

set.seed(12)

rv <- runif(nrow(Vehicle))
v.train <- Vehicle[rv>=0.33,]
v.test <- Vehicle[rv<0.33,]

rs <- runif(nrow(Soybean))
s.train <- Soybean[rs>=0.33,]
s.test <- Soybean[rs<0.33,]

rbh <- runif(nrow(BostonHousing))
bh.train <- BostonHousing[rbh>=0.33,]
bh.test <- BostonHousing[rbh<0.33,]
```

19.2   Attribute selection task

The attribute selection task essentially consists in selecting a subset of originally available attributes to be subsequently used for model creation. This section outlines the motivation behind attribute selection and discusses the definition of the task in more detail.

19.2.1   Motivation

As mentioned above, identifying a subset of predictively useful attributes provides possibly interesting insights about the domain and can impact the data mining process, but the primary motivation is the hope for better models. Although reducing the informational content of the data might appear harmful, using a smaller subset of available attributes can actually improve the true performance of created models by reducing the risk of overfitting. Less attributes means a smaller space to search during model construction and less opportunities to make wrong decisions and arrive at misleading, insufficiently justified generalizations.

Contrary to a popular misconception, attribute selection is not usually motivated by computational savings, possible when creating and applying models using a smaller attribute subset. This is because the attribute selection process is often much more demanding computationally than actual model building, and would pay off in computational terms only if a particularly simple attribute selection algorithm is applied or the selected attribute subset is subsequently used for creating a large number of models (e.g., based on different data samples, using different modeling algorithms with different parameter settings, etc.).

With that being said, reducing the number of attributes is usually desirable even if it does not improve but only retains the predictive performance. This is because models using smaller attribute subsets are easier to inspect and understand as well as more efficient and safer to apply. In particular, they are less likely to be affected by possible data quality issues at the time of prediction.

19.2.2   Task definition

Given the nature of the attribute selection task as preprocessing for inductive learning tasks, it is not surprising that it inherits the major parts of its definition from the latter. In particular, it adopts exactly the same view of the data as a set of instances from a domain, described by a number of attributes, more extensively discussed in Section 1.2.

19.2.2.1   Target task

The modeling task for which attribute selection is to be performed is the *target task*. It is usually one of the three major inductive learning tasks, classification, regression, or clustering.

19.2.2.2   Target algorithm

For some attribute selection methods it is necessary to specify not only the target modeling task, but also the *target algorithm* that will be used for model creation.

19.2.2.3   Training set

Attribute selection is often performed on the same dataset which serves as the training set for the target task. It is not uncommon to use a smaller subset thereof when computational savings are necessary, given the computational expense associated with most attribute selection algorithms. For sufficiently large datasets, entirely separate subsets can be used for attribute selection and for subsequent model creation, making the attribute selection training set disjoint with the modeling training set.

The exact usage of the training set heavily depends on a particular attribute selection algorithm. In particular, attribute selection algorithms may perform some data partitioning internally and use different subsets for different operations (e.g., create a model on one subset and evaluate it on another subset). Regardless of such internal data management issues, from the external perspective the attribute selection training set should be viewed as the set of all instances that are used in any way and in any part of the attribute selection process and may have any impact on its final outcome.

As underlined in Section 7.3.2, reliable model performance estimates can be only obtained on data strictly separate from that used in the modeling process. Otherwise there is a significant risk of evaluation overfitting, i.e., arriving at unreliably optimistic model performance estimates. Whenever attribute selection becomes a part of the modeling process, the attribute selection training set clearly becomes a part of the data that impacts the finally obtained model, even if it is entirely disjoint with the training set supplied to the applied modeling algorithm.

It is therefore important to strictly separate the data used for attribute selection from that subsequently used for the evaluation of models created when performing the target task after attribute selection.

19.2.2.4   Attribute subset

The attribute selection task consists in finding a subset of the set of all attributes defined on the domain and available in the training set to be subsequently used for the target task. Roughly speaking, this subset should be substantially smaller than the original full set of attributes yet sufficient to obtain good quality models. The exact criteria to be satisfied by the selected subset usually take one of the following forms:

- the smallest subset that makes it possible to obtain a model of a specified minimum estimated quality (e.g., the same as with the full original set of attributes),
- a subset of a specified maximum size that makes it possible to obtain a model of the best estimated quality,
- a subset that makes it possible to obtain a model of the best estimated quality, regardless of the size,

or some mixture of these three. Most typically, maximizing the estimated model quality is the primary criterion, with the preference for smaller attribute subsets playing a supplementary role.

19.2.3   Algorithms

Any algorithm that can use a provided training set to select a subset of attributes for a specified target task, meeting criteria like those presented above, can be considered an attribute selection algorithm. Some attribute selection algorithms may be dedicated only to particular target tasks and particular target algorithms. General-purpose attribute selection algorithms can be applied to select attributes for arbitrary target algorithms, and – sometimes – also for different target tasks. They fall into two main categories:

Attribute selection filters. Algorithms that rank single attributes or attribute subsets with respect to some utility measures, dependent on the target task, but independent of the target algorithm.

Attribute selection wrappers. Algorithms that rank attribute subsets with respect to the estimated true performance of models obtained by applying the target algorithm to the training set using these attribute subsets.

The primary difference between these two is in the criteria used to estimate the utility of candidate attribute subsets. For the latter, it involves actually building a model using the attribute subset to be evaluated by applying the target algorithm, while for the former some other (usually, but not necessarily simpler) utility criteria are used. Some popular filtering attribute selection algorithms actually estimate the utility of single attributes only rather than of attribute subsets, generating an attribute ranking that can be used to select a number of top-ranked attributes.

19.3   Attribute subset search

A search strategy is needed for any attribute selection technique that is based on evaluating attribute subsets rather than single attributes. This includes primarily attribute selection wrappers presented in Section 19.5, but also some of attribute selection filters presented in Section 19.4. An attribute selection search strategy is responsible for generating candidate attribute subsets in a systematic, but nonexhaustive way. Ideally, it should make it possible to identify a near-optimum attribute subset after examining only a fraction of the whole attribute subset space. It becomes particularly important if the space is huge, which is usually the case, unless the number of attributes is small (which would make attribute selection questionable, though).

19.3.1   Search task

Attribute selection search is an instantiation of the generic search task which casts problem solving as traversing the space of problem *states*, searching for those that represent good solutions. Problem solving by search belongs to the most thoroughly studied topics in artificial intelligence and for the most part it lies beyond the scope of our interest in this chapter; hence, it will be only briefly and superficially discussed here, to the extent necessary to present the most common attribute selection search strategies.

A search task is usually defined by specifying:

State space. The set of all possible states of the problem to be solved.

Operators. Actions that generate one or more new states based on the current state.

Initial state. The state at which the problem solving process starts.

Final states. The subset of states corresponding to acceptable problem solutions.

Evaluation function. (Optionally) a function that evaluates the quality of problem solutions represented by final states.

Cost function. (Optionally) a function that assigns a cost value to each operator application.

The task then consists in finding a minimum-cost sequence of operators that leads from the initial state to a possibly best-evaluated final state. An algorithm capable of performing this task is a *search algorithm*. The major component of any search algorithm is its *search strategy*, responsible for selecting next states out of all those that can be generated by applying the available operators to the current state, as well as for determining when the search for better solutions should be terminated, after one or more acceptable solutions have been already found. There are two major genres of those:

Blind (uninformed). Not using any information on the properties of particular states.

Heuristic (informed). Using information on the properties of particular states, usually represented by the numerical values of a *heuristic function* which indicates which states should be preferred.

The purpose of the latter is to improve the efficiency of the search process by making it possible to arrive at a good solution without visiting an excessive number of states. In general, the heuristic function should not be confused with the evaluation function: the former is applicable to any state and indicates how promising it is as the next search direction whereas the latter is applicable to final states only and measures the quality of problem solutions they represent.

There are several variations of the search task definition depending on the particular type of problem to be solved. For some problems there is no evaluation function, since there is only one final state or all final states represent equally good solutions. For some problems there is no cost function, since it is only the quality of the solution that matters and not the number of steps required to arrive at it. In fact, for most search task instantiations one would either have the evaluation function or the cost function specified, but rarely both. When the cost function is specified, final states are often perfectly known *a priori* and the purpose of the search process is to find the sequence of operator applications that transform the initial state to one of them. When the evaluation function is specified, there are usually many final states (or, sometimes, even all states are final) and the purpose of the search process is to just find one that is not far from the best, regardless of the sequence of operators that is used to reach it. The latter is actually the case for attribute selection search, where:

- the state space consists of all possible nonempty attribute subsets,

- all states are final since all attribute subsets can be accepted as selection results,

- a possibly best subset needs to be found,

- the initial state and operators can be defined arbitrarily as long as they serve the above purpose, which makes it possible to consider them elements of the search strategy rather than of the task.

Since all states are final, the same function can be used as the evaluation function and as the heuristic function. Although this is not necessarily a universally good idea for all search tasks with this property, it tends to work well for attribute selection search. This assumption, with the evaluation function being already preselected, reduces the design of an attribute selection search strategy to determining the initial state, the available search operators, the next states selection criteria based on the evaluation function, and the stop criteria that control the termination of the search process.

19.3.2   Initial state

There are two natural possible initial states for attribute selection search:

- the empty attribute set,

- the complete original attribute set,

and one of those two is adopted in most situations. Sometimes it may make sense to start from a random attribute set (particularly if attribute selection is to be performed several times, with the ultimately best subset selected from the outcomes of these multiple runs) or from a particular subset preselected based on the available domain knowledge or using a simple attribute selection algorithm that does not require subset space search (usually a simple statistical filter).

Example 19.3.1 The following R code defines two one-line functions that initialize attribute selection search either to the subset containing all or none of the original attributes. The functions are demonstrated using attribute names for the *weather* data.

```
## attribute selection search initialization
asel.init.none <- function(attributes) { character(0) }
asel.init.all <- function(attributes) { attributes }

  # attribute selection search initialization for the weather data
asel.init.none(names(weather)[-5])
asel.init.all(names(weather)[-5])
```

19.3.3   Search operators

With the state space consisting of all possible attribute subsets the role of search operators is to generate new subsets based on the current subset. The two simplest and most natural ways of doing this are:

Attribute addition. Adding a single attribute to the current subset (unless it already contains all attributes) – applied to an attribute subset $\mathbf{A}$, it generates $|\mathbf{A}'|$ new subsets, one for each $a \in \mathbf{A}'$, where $\mathbf{A}'$ denotes the set of attributes from the complete original attribute set that are not in $\mathbf{A}$.

Attribute removal. Removing a single attribute from the current subset (unless it already contains one attribute only) – applied to an attribute subset $\mathbf{A}$, it generates $|\mathbf{A}|$ new subsets, one for each $a \in \mathbf{A}$.

Note that it makes little sense to consider any more complex operators that add or remove multiple attributes at a time (or add some attributes while removing some other attributes) as these two are sufficient to make a transition between two arbitrary attribute subsets possible in a number of steps that is bound by the sum of their sizes. Adding or removing even a single attribute in an attribute subset may have huge impact on its predictive utility, which makes it hard to justify changing more than one attribute at a time.

19.3.4   State selection

With the two possible initial states and the two search operators presented above there are several possibilities of organizing the search process. The following two basic approaches are obtained when only one search operator is used:

Forward selection. Start from an empty set as the initial state and use the attribute addition operator only.

Backward elimination. Start from the complete original attribute set as the initial state and use the attribute removal operator only.

They keep the search process one-directional and monotonic with respect to attribute subset size, which prevents excessive computational expense, but may lead to suboptimal results, since attribute subset utility does not necessarily change monotonically with attribute subset size.

By permitting the use of the other operator in each of these two simple search strategies one can increase the chance of arriving at a good finally selected subset. The downside of such extensions is the increased search algorithm complexity and computational cost. The latter results not only from the usually greater number of evaluated attribute subsets, but also from the necessity of preventing search loops. While it is straightforward to forbid directly going back to the previous state, in general loop prevention would require memorizing all states (attribute subsets) visited so far to avoid considering them again.

Example 19.3.2 Functions that implement forward and backward next state generation for attribute selection search are defined and demonstrated by the R code presented below.

```
## forward attribute selection search next state generation
asel.next.forward <- function(subset, attributes)
{
  lapply(setdiff(attributes, subset), function(a) c(subset, a))
}

## backward attribute selection search next state generation
asel.next.backward <- function(subset, attributes)
{
  if (length(subset)>1)
    lapply(1:length(subset), function(i) subset[-i])
  else
    list()
}

  # attribute selection next state generation for the weather data
asel.next.forward(c("outlook", "humidity"), names(weather)[-5])
asel.next.backward(c("outlook", "humidity"), names(weather)[-5])
```

19.3.5   Stop criteria

With pure forward selection and backward elimination there is a natural ultimate stop criterion, which terminates search when no next states can be generated by the corresponding attribute addition or attribute removal operator. In practice, such ultimate stop criteria are usually supplemented with some mechanism that prevents continuing search when there is little hope for further improvement. This is usually achieved by specifying the maximum number of search steps without the improvement of the evaluation function value. This is not the case for hybrid strategies that use both the attribute addition and removal operators, which can always generate next states unless they are equipped with some form of state memory to avoid re-vising past states.

19.3.5.1   Additional preference criteria

The subset evaluation function, guiding the attribute subset space search process, is primarily based on the predictive utility of candidate subsets. It may be measured in several ways, specific to particular search-based attribute selection algorithms. If any other criteria, apart from subset predictive utility, are adopted for a particular attribute selection task, they can be incorporated into the search evaluation function as well. This may include, in particular, preference for smaller attribute subsets or domain-specific attribute costs. Both of these can be represented by appropriate penalty terms added to the predictive utility estimate. In particular, one simple approach to introduce the preference for smaller subsets when evaluating attribute subset $\mathbf{A}$ is to add a penalty term $\gamma|\mathbf{A}|$, where γ is a penalty coefficient that controls the tradeoff between subset size and utility.

19.3.5.2   Greedy search

Regardless of whether using forward selection, backward elimination, or a mix of these two, at each search step a number of next states can be generated. The evaluation function (which plays the role of the heuristic function for attribute selection search) applied to all of them indicates which one will be selected for the next step. In the simplest and most common case all the other candidate subsets are discarded. This approach can be referred to as *greedy search* and is described by the algorithm presented below. Depending on the particular initial state, search operators, and stop criteria choices, it can be instantiated into a several different search strategies, including greedy forward selection and greedy backward elimination as well as hybrids of these two.

```
 1: initialize A; A_* := A;
 2: repeat
 3:     if stop criteria are satisfied then return A_*;
 4:     end if
 5:     use search operators to generate the set of candidate next attribute subsets S;
 6:     select the best-evaluated subset A ∈ S;
 7:     if A is better than A_* then
 8:         A_* := A;
 9:     end if
10: until true;
```

More refined search strategies are also possible that preserve more than one best-evaluated subset, either to use them all for candidate generation in the next step, or to possibly consider them later when the search path starting from the best subset no longer appears promising. It may be also reasonable to consider for candidate generation one or more of the best subsets generated on any iteration so far rather than just on the single last iteration. In any case, it is important to keep track of the best subset found so far, returned when the stop criteria for the search process are satisfied.

Of the two basic variants of greedy search, forward selection and backward elimination, the former may be usually expected to yield smaller subsets (as it starts from the empty subset). On the other hand, the latter is likely to be faster (as long as the cost of the modeling algorithm does not heavily depend on the number of attributes), since – starting from the full original attribute set – it may encounter the no further improvement stop condition sooner.

Example 19.3.3 The R code presented below implements a generic greedy search strategy for attribute selection. The `asel.search.greedy` function iteratively generates and evaluates candidate subsets, keeping track of the best subset found so far, until a specified maximum number of iterations lead to no improvement. The evaluation function specified via the `evalf` argument is assumed to assign higher values to better attribute subsets, which means it will be maximized. The `penalty` parameter specifies the penalty coefficient used to incorporate preference for smaller attribute subsets. The exact search strategy used is specified by providing the `initf` and `nextf` arguments, which are functions used to generate the initial attribute subset and the candidate next subsets. The `asel.init.all` and `asel.init.none` functions defined in Example 19.3.1 (returning the empty set of attributes, to initialize forward selection, and returning the set of all attributes, to initialize backward elimination, respectively) can be used for the former and the `asel.next.forward` and `asel.next.backward` functions from Example 19.3.2 (generating candidate next subsets by forward selection and by backward elimination, respectively) for the latter (with backward elimination settings used by default). The search process terminates when no new candidate subsets can be generated or the number of search steps without evaluation improvement exceeds the value specified by the `max.noimp` parameter. The `asel.search.greedy` function is demonstrated using the *weather* data with a trivial and useless evaluation function returning a constant value of 1, serving the illustration purpose only. There will be several opportunities to demonstrate attribute subset space search with more meaningful evaluation functions when illustrating particular search-based attribute selection algorithms.

```
## greedy attribute selection search
asel.search.greedy <- function(attributes, target, evalf,
                               initf=asel.init.all, nextf=asel.next.backward,
                               max.noimp=3, penalty=0.01)
{
  ev <- function(subset)
  {
    ifelse(is.finite(v<-evalf(subset, target)), v-penalty*length(subset)*abs(v), v)
  }

  best.subset <- subset <- initf(attributes)
  best.eval <- eval <- ev(subset)
  noimp <- 0

  while (noimp < max.noimp)
  {
    candidates <- nextf(subset, attributes)
    if (length(candidates)>0)
    {
      cand.eval <- sapply(candidates, ev)
      cand.best <- which.max(cand.eval)
      noimp <- ifelse(cand.eval[cand.best]>eval, 0, noimp+1)
      subset <- candidates[[cand.best]]
      eval <- cand.eval[cand.best]
      if (eval>best.eval)
      {
        best.subset <- subset
        best.eval <- eval
      }
    }
    else
      break
  }

  list(subset=best.subset, eval=best.eval)
}

  # greedy attribute selection search for the weather data
asel.search.greedy(names(weather)[-5], "play", evalf=function(subset, target) 1)
asel.search.greedy(names(weather)[-5], "play", evalf=function(subset, target) 1,
                   penalty=0)
asel.search.greedy(names(weather)[-5], "play", evalf=function(subset, target) 1,
                   initf=asel.init.none, nextf=asel.next.forward)
asel.search.greedy(names(weather)[-5], "play", evalf=function(subset, target) 1,
                   initf=asel.init.none, nextf=asel.next.forward, penalty=0)
```

Not surprisingly, the empty attribute subset is returned in all cases except when running backward elimination with no subset size penalty, when the full attribute set is returned.

19.4   Attribute selection filters

The filtering approach to the attribute selection task is based on evaluating individual attributes or candidate attribute subsets with respect to their predictive utility. Utility measures used for this purpose are task dependent (i.e., they depend on the target task), but algorithm independent (i.e., they do not depend on the target algorithm). With these basic features being common, there is a variety of attribute selection filters, some representatives of which are presented in this section.

Two major flavors of filtering attribute selection algorithms can be distinguished depending on whether they evaluate individual attributes or candidate attribute subsets. The former generate attribute utility ranks, which can be used to select a number of top-ranked attributes by cutting off a portion of the obtained ranking. Such algorithms can be fully specified by the adopted utility measure and cutoff criteria. The latter consider a number of candidate attribute subsets and select the one with the highest estimated utility. They can be specified by the adopted utility measure and search strategy for searching the space of candidate subsets.

19.4.1   Simple statistical filters

A particularly popular approach to attribute selection is based on applying statistical measures of relationship as attribute utility measures. The idea is to evaluate attributes with respect to the strength of their relationship to the target attribute of the target task, i.e., the target concept for the classification task or the target function for the regression task. For the clustering task, an attribute representing an external labeling of instances or cluster membership with respect to a previously created clustering model, as used by external clustering quality measures discussed in Section 14.4, could serve sometimes the same purpose.

The exact kind of statistical relationship measures that can be used depends on attribute types and the target task. Depending on whether the attribute to be evaluated and the target attribute are discrete or continuous, we may need statistics for measuring the relationship:

Between two discrete attributes. Like the symmetric uncertainty, the mutual information, or the χ^2 statistic presented in Section 2.5.3 (with the first of those being often preferred due to its built-in normalization).

Between two continuous attributes. Like the linear correlation or the rank correlation presented in Section 2.5.2.

Between one continuous and one discrete attribute. Like the t statistic for two means, one-way ANOVA, or the Kruskal–Wallis statistic presented in Section 2.5.4.

One practical problem that arises when simple filters are applied to mixed attribute sets (containing both discrete and continuous attributes) is the questionable (to say the least) comparability of utility values obtained using different relationship measures, appropriate for particular attribute types. The same range and distribution of the adopted relationship measures would be necessary to rank attributes of different types with respect to their utility in a truly meaningful way. This would be possible in some cases by using the appropriately transformed statistics of the statistical tests based on the adopted relationship measure. To keep simple filters simple, though, this is usually not addressed in any other way than simply:

- filtering discrete and continuous attributes separately (and then selecting a number of top-ranked attributes from the two groups),

- using the p-value of corresponding statistical test (or its 1's complement) rather than the underlying relationship measure (i.e., measuring the significance of attribute relationships rather than their strength),

- discretizing continuous attributes and using a relationship measure for discrete attributes only.

Whereas the first approach is inconvenient and may lead to suboptimal results, it is sometimes a sufficient solution in practice. The second approach is apparently convenient and elegant, but may not work as expected for large datasets, where most relationships are likely to be maximally statistically significant (with p-values indistinguishable within the limits of arithmetic precision), resulting in all or nearly all attributes being selected. The deficiency of the last approach is that by using the discretized versions of continuous attributes it may underestimate their predictive utility. Making it possible to apply the same relationship measure for all attribute pairs is a very significant advantage, though, that usually outweighs this risk. It is worthwhile to note that a simple nonmodeling form of discretization can be applied here, as discussed in Section 18.3.3, since there is no need to retain and re-apply discretization interval breaks to any other datasets.

Example 19.4.1 The following R code defines a function that automates attribute filtering using simple statistical attribute utility measures. The function accepts three relationship measures on input, for the discrete–discrete, discrete–continuous, and continuous–continuous attribute type combinations (passed via the `dd`, `cd`, and `cc` arguments, respectively). For the purpose of illustration, default relationship measures are defined as 1's complements of the p-values obtained using the χ^2 test, the Kruskal–Wallis test, and the rank correlation test, despite the reservations against this approach discussed above. Attributes to consider are specified using the formula interface. The `x.vars` and `y.var` functions are used to extract attribute names from the supplied formula, and the `attr.type` function to determine the attribute type.

| dmr.util |

The function is applied to the *weather*, *weatherc*, *weatherr*, *Vehicle Silhouettes*, *Soybean*, and *Boston Housing* datasets. For the three small datasets, apart from filtering all attributes with the above-mentioned default relationship measures, the separate filtering of discrete and continuous attributes is also demonstrated with more appropriate relationship measures: the symmetric uncertainty for the discrete–discrete case, implemented by the `symunc` function, the Kruskal–Wallis test statistic for the continuous–discrete case, and the squared rank correlation for the continuous-continuous case. For the three larger datasets, attribute utility is measured using the symmetric uncertainty only, after transforming continuous attributes by equal-frequency discretization implemented by the `discnm.eqfreq` function. Note that attribute filtering for the *Vehicle Silhouettes*, *Soybean*, and *Boston Housing* data is performed on their training subsets, and the obtained attribute utilities are stored in the `v.utl.simple`, `s.utl.simple`, and `bh.utl.simple` variables. The same practice will be followed by subsequent examples, and Example 19.6.1 will use the corresponding test subsets to evaluate the performance of models created after attribute selection.

| Ex. 2.5.7 |
| dmr.stats |

| Ex. 18.3.3 |
| dmr.disc |

```
dd.chi2 <- function(a1, a2) 1-chisq.test(a1, a2)$p.value
cd.kruskal <- function(a1, a2) 1-kruskal.test(a1, a2)$p.value
cc.spearman <- function(a1, a2) 1-cor.test(a1, a2, method="spearman")$p.value

## simple statistical attribute selection filter
simple.filter <- function(formula, data, dd=dd.chi2, cd=cd.kruskal, cc=cc.spearman)
{
  attributes <- x.vars(formula, data)
  target <- y.var(formula)

  utility <- function(a)
  {
    unname(switch(attr.type(data[[a]], data[[target]]),
               dd = dd(data[[a]], data[[target]]),
               cd = cd(data[[a]], data[[target]]),
               dc = cd(data[[target]], data[[a]]),
               cc = cc(data[[a]], data[[target]])))
  }

  sort(sapply(attributes, utility), decreasing=TRUE)
}

  # simple filter for the weather data
simple.filter(play~., weather)
simple.filter(play~., weather, dd=symunc)

  # simple filter for the weatherc data
simple.filter(play~., weatherc)
simple.filter(play~outlook+wind, weatherc, dd=symunc)
simple.filter(play~temperature+humidity, weatherc,
            cd=function(a1, a2) kruskal.test(a1, a2)$statistic)

  # simple filter for the weatherr data
simple.filter(playability~., weatherr)
simple.filter(playability~outlook+wind, weatherr,
            cd=function(a1, a2) kruskal.test(a1, a2)$statistic)
simple.filter(playability~temperature+humidity, weatherr,
            cc=function(a1, a2) cor(a1, a2, method="spearman")^2)

  # simple filter for the Vehicle Silhouettes data
v.utl.simple <- simple.filter(Class~., discnm.eqfreq(~., v.train, 7), dd=symunc)
  # simple filter for the Soybean data
s.utl.simple <- simple.filter(Class~., Soybean, dd=symunc)
  # simple filter for the BostonHousing data
bh.utl.simple <- simple.filter(medv~., discnm.eqfreq(~., bh.train, 7), dd=symunc)
```

Simple statistical filters are computationally efficient and easy to apply. They are also widely available, since the underlying utility measures can be calculated using virtually all analytical software kits, even those that do not offer explicit attribute selection functionality. Their single, but severe drawback is that they evaluate attribute predictive utility in a context-insensitive way, i.e., the utility of each attribute does not depend on the presence or absence of other attributes in any way. This makes it possible for several closely related attributes that exhibit strong relationship to the target attribute to be considered similarly highly useful, even if selecting one of them would make the others unnecessary. This issue will be referred to as *apparent utility*. Similarly, this makes it possible for some attributes that individually appear unrelated to the target attribute to be considered useless, even if they are essential when used in combination, which can be called *apparent disutility* and is not as frequent as apparent utility, but also harder to fight. The remaining attribute selection filters presented in this section make various attempts to achieve some context-sensitivity of their attribute utility measures.

19.4.2   Correlation-based filters

The idea of correlation-based attribute selection (originally called correlation-based feature selection, or CFS) is to use statistical relationship measures – such as those on which simple statistical filters are based – in a more refined, context-sensitive way. This is achieved by considering a number of candidate attribute subsets and evaluating them both with respect to their relationship to the target attribute and with respect to their mutual relationships. A specific correlation-based filter algorithm can therefore be fully described by

- a set of relationship measures appropriate for different attribute types (discrete–discrete, discrete–continuous, continuous–continuous),

- a search strategy to systematically (but usually nonexhaustively) consider candidate subsets.

The latter has been discussed in Section 19.3 and now we may focus on the former.

19.4.2.1   Relationship measures

Despite the term "correlation" used in the name under which this approach is known, the employed relationship measure does not have to be the linear or rank correlation, unless both the target attributes and all other attributes are continuous. With discrete or mixed attribute spaces, appropriate relationship measures are needed for different attribute types, but they should ideally have the same range and distribution so that they can be aggregated additively in a meaningful way. A recommended approach is therefore to choose a single measure that can handle different attribute types (or can be adapted accordingly). This can be easily achieved in one of the following two ways:

- using a normalized (i.e., insensitive to the number of attribute values) discrete-attribute relationship measure such as the symmetric uncertainty, after discretizing any continuous attributes (including the target attribute, if the target task is regression),

- using a continuous-attribute relationship measure such as (typically, the absolute value or square of) the linear or rank correlation, after encoding any discrete attributes numerically.

The former has already been presented before as an approach to overcoming the difficulties with applying simple statistical filters to mixed attribute spaces. The latter – based on a roughly inverse attribute transformation – employs the same discrete attribute binary encoding technique presented in Section 17.3.5 that is commonly applied for parametric classification and regression or dissimilarity calculation with discrete attributes. It is based on simple 1-of-k coding, which replaces a discrete k-valued attribute with k binary attributes. Then the correlation between a_1 and a_2, if a_1 is discrete and a_2 is continuous, is calculated as the probability-weighted average of the correlations between a_2 and the binary attributes replacing a_1:

$$\rho_T(a_1, a_2) = \sum_{v \in A_1} P(a_1 = v)\rho_T(a_{1v}, a_2) \tag{19.1}$$

where a_{1v} is the binary attribute corresponding to $v \in A_1$ and $P(a_1 = v)$ is the probability of a_1 taking the value of v, estimated on the training set in the usual way. When both a_1 and a_2 are discrete, all pairwise binary attribute correlations are considered, weighted by the corresponding probabilities:

$$\rho_T(a_1, a_2) = \sum_{v_1 \in A_1} \sum_{v_2 \in A_2} P(a_1 = v_1, a_2 = v_2) \rho_T(a_{1v_1}, a_{2v_2}) \tag{19.2}$$

Example 19.4.2 The R code presented below implements and demonstrates correlation calculation for continuous or discrete attributes, with the latter processed by binary encoding and the resulting multiple correlations averaged as described above. The absolute value of the rank correlation serves as a base correlation measure, passed as the default value of the `corf` argument. Discrete attribute encoding is performed using the `discode.a` function, called with the `na.all=TRUE` argument to reasonably handle missing values of discrete attributes. Similarly, the default `corf` function is a wrapper around standard `cor` that specifies the `use="complete.obs"` argument to omit missing values for continuous attributes. The `pdisc` function is used to calculate discrete attribute probability distributions.

> Ex. 17.3.5
> dmr.trans

> Ex. 2.4.22
> dmr.stats

```
## discrete attribute correlation using binary encoding
discor <- function(a1, a2,
                   corf=function(a1, a2)
                        abs(cor(a1, a2, method="spearman", use="complete.obs")))
{
  switch(attr.type(a1, a2),
         cc=p12<-1,
         dc=p12<-as.matrix(pdisc(a1)),
         cd=p12<-t(as.matrix(pdisc(a2))),
         dd=p12<-pdisc(a1, a2))

  a1dc <- discode.a(a1, red=TRUE, na.all=TRUE)
  a2dc <- discode.a(a2, red=TRUE, na.all=TRUE)
  cor12 <- outer(1:ncol(a1dc), 1:ncol(a2dc),
                 Vectorize(function(i, j) corf(a1dc[,i], a2dc[,j])))
  weighted.mean(cor12, p12)
}

  # two continuous attributes
discor(weatherc$temperature, weatherc$humidity)
  # one discrete and one continuous attribute
discor(weatherc$outlook, weatherc$temperature)
discor(weatherc$temperature, weatherc$play)
  # two discrete attributes
discor(weatherc$outlook, weatherc$play)
  # attributes with missing values
discor(Soybean$seed, Soybean$roots)
```

19.4.2.2   Subset evaluation

The utility of an attribute subset **A** with respect to the target attribute a_0 is calculated by correlation-based filters in the following way:

$$u_T(\mathbf{A}) = \frac{|\mathbf{A}| R_T(\mathbf{A}, a_0)}{\sqrt{|\mathbf{A}| + |\mathbf{A}|(|\mathbf{A}| - 1) R_T(\mathbf{A})}} \tag{19.3}$$

where $R_T(\mathbf{A}, a_0)$ and $R_T(\mathbf{A})$ are the mean strength of relationship between attributes from $\mathbf{A}$ and the target attribute, and the mean strength of pairwise relationship between attributes from $\mathbf{A}$, respectively, calculated on the training set T using the adopted relationship measure ρ:

$$R_T(\mathbf{A}, a_0) = \frac{1}{|\mathbf{A}|} \sum_{a \in \mathbf{A}} \rho_T(a, a_0) \tag{19.4}$$

$$R_T(\mathbf{A}) = \frac{1}{|\mathbf{A}|(|\mathbf{A}| - 1)} \sum_{a_1 \in \mathbf{A}} \sum_{a_2 \in \mathbf{A} - \{a_1\}} \rho_T(a_1, a_2) \tag{19.5}$$

It is easy to see that the definition of $u_T(\mathbf{A})$ can be rewritten as

$$u_T(\mathbf{A}) = \frac{\sum_{a \in \mathbf{A}} \rho_T(a, a_0)}{\sqrt{|\mathbf{A}| + \sum_{a_1 \in \mathbf{A}} \sum_{a_2 \in \mathbf{A} - \{a_1\}} \rho_T(a_1, a_2)}} \tag{19.6}$$

which is actually a simpler form than the previous one, appearing in the original description of CFS filters.

The subset utility measure of correlation-based filters favors attribute subsets that combine strong relationships with the target attribute and weak or sparse mutual relationships, which is intuitively clear and perfectly reasonable although the particular formula used to tradeoff between these types of relationships is not necessarily obvious. It is borrowed from classical psychometric test theory and has been found to represent a good estimate of the relationship between the composite attribute that could be obtained by combining all attributes from the subset and the target attribute. The numerator provides a measure of the combined predictive power of all attributes in the subset and the denominator applies a punishment for redundancy that manifests itself by their mutual relationships.

Example 19.4.3 Correlation-based attribute selection is implemented and demonstrated by the R code presented below. It uses the search strategy implementation provided by the `asel.search.greedy`, `asel.init.none`, `asel.init.all`, `asel.next.forward`, and `asel.next.backward` functions, defined in Examples 19.3.1, 19.3.2, and 19.3.3, with backward elimination performed by default. They are combined with correlation-based attribute subset evaluation implemented by the `cfs.eval` function, which accepts a matrix of pairwise attribute correlations on input. The `cfs.filter` function, which organizes the correlation-based filtering process, takes care of calculating the matrix, using the relationship measure specified by the `corf` argument. It defaults to the symmetric uncertainty, implemented by the `symunc` function. To make it applicable to datasets with continuous attributes, they are discretized in example calls in the same way as in Example 19.4.1. The alternative approach of using the absolute value of the rank correlation, with discrete attribute support via binary encoding, implemented by the `discor` function from the previous example, is also demonstrated.

> Ex. 2.5.7
> dmr.stats

```
## subset evaluation for correlation-based filters
cfs.eval <- function(subset, target, data, cormat)
{
  if (length(subset <- unique(subset))>0)
  {
```

```
      cor.at <- mean(sapply(subset, function(a) cormat[a,target]))
      cor.aa <- mean(outer(subset, subset,
                          Vectorize(function(a1, a2)
                                      ifelse(a1!=a2, cormat[a1,a2], NA))),
                    na.rm=TRUE)
      cor.aa <- ifelse(is.finite(cor.aa), cor.aa, 0)
      length(subset)*cor.at/
        sqrt(length(subset)+length(subset)*(length(subset)-1)*cor.aa)
  }
  else
    -Inf
}

## correlation-based filter
cfs.filter <- function(formula, data, corf=symunc,
                        searchf=asel.search.greedy,
                        initf=asel.init.all, nextf=asel.next.backward)
{
  target <- y.var(formula)
  attributes <- x.vars(formula, data)
  atnames <- c(attributes, target)

  cormat <- outer(1:length(atnames), 1:length(atnames),
                  Vectorize(function(i, j)
                              ifelse(j<i, corf(data[[atnames[i]]], data[[atnames[j]]]),
                                  NA)))
  cormat[upper.tri(cormat)] <- t(cormat)[t(lower.tri(cormat))]
  dimnames(cormat) <- list(atnames, atnames)

  searchf(attributes, target,
          evalf=function(subset, target) cfs.eval(subset, target, data, cormat),
          initf=initf, nextf=nextf)
}

  # correlation-based filter for the weather data
cfs.filter(play~., weather)
cfs.filter(play~., weather, initf=asel.init.none, nextf=asel.next.forward)
cfs.filter(play~., weather, corf=discor)
cfs.filter(play~., weather, corf=discor,
          initf=asel.init.none, nextf=asel.next.forward)

  # correlation-based filter for the weatherc data
cfs.filter(play~., discnm.eqfreq(~., weatherc, 4))
cfs.filter(play~., weatherc, corf=discor)
cfs.filter(play~., weatherc, corf=discor,
          initf=asel.init.none, nextf=asel.next.forward)

  # correlation-based filter for the weatherr data
cfs.filter(playability~., weatherr, corf=discor)
cfs.filter(playability~., weatherr, corf=discor,
          initf=asel.init.none, nextf=asel.next.forward)

  # correlation-based filter for the Vehicle Silhouettes data
v.sel.cfs <- cfs.filter(Class~., discnm.eqfreq(~., v.train, 7))$subset
  # correlation-based filter for the Soybean data
s.sel.cfs <- cfs.filter(Class~., s.train)$subset
  # correlation-based filter for the Boston Housing data
bh.sel.cfs <- cfs.filter(medv~., discnm.eqfreq(~., bh.train, 7))$subset
```

19.4.3   Consistency-based filters

The consistency-based approach to attribute selection is directly applicable to the classification target task only. It can be roughly described as an attempt to identify the smallest attribute subset sufficient to perfectly discriminate among classes. Each value combination of attributes from such a subset corresponds to a *consistent* subset of training instances, i.e., containing instances of a single class only. It is often impossible and usually unreasonable to achieve this goal exactly, since

- it may require searching the attribute subset space exhaustively – which is computationally prohibitive for all but the smallest attribute sets,

- perfect consistency may be impossible to achieve for real-world data (particularly affected by noise), and actually may be in conflict with one of the goals of attribute selection, which is overfitting prevention.

This is why practical consistency-based filters try to approach the consistency in an approximate way, by using inexhaustive search strategies, such as those presented in Section 19.3, and relaxing the notion of consistency (accepting minor inconsistencies). The latter requires an attribute subset evaluation function that measures the degree of its consistency.

There are number of consistency-based attribute selection algorithms that use different evaluation functions. The one that most directly represents the notion of consistency is based on calculating the *inconsistency rate* for an attribute subset, defined as follows:

$$v_T(\mathbf{A}) = \frac{1}{|T/\mathbf{A}|} \sum_{T' \in T/\mathbf{A}} v_{T'}(c) \qquad (19.7)$$

where $T/\mathbf{A}$ denotes all subsets of T induced by the attribute subset $\mathbf{A}$ (corresponding to all distinct value combinations of attributes from $\mathbf{A}$ occurring in T) and $v_{T'}(c)$ is the *inconsistency count* of the target concept on data subset T' calculated as the number of instances from the subset with a nonmajority class label:

$$v_{T'}(c) = \left| \{x \in T' \mid c(x) \neq \arg \max_{d \in C} |T'^d|\} \right| \qquad (19.8)$$

It is easy and worthwhile to note that the inconsistency rate is monotonic. By adding one or more attributes to an attribute subset, we cannot increase its inconsistency rate. It makes it possible to simplify the attribute subset space search process.

Another idea is to relax the notion of consistency by adopting an impurity measure for this purpose, such as presented in Section 2.4.2, applying it over all data subsets induced by the attribute subset being evaluated, and aggregating the results appropriately. With the entropy serving as the impurity measure, the corresponding formula can be written as follows:

$$E_T(\mathbf{A}) = \sum_{T' \in T/\mathbf{A}} \frac{|T'|}{|T|} E_{T'}(c) \qquad (19.9)$$

where $E_{T'}(c)$ is the entropy of the target concept on data subset T'. This is the weighted average entropy over all data subsets induced by the attribute subset.

Consistency-based filters can be safely applied only when all attributes are discrete, since continuous attributes usually exhibit sufficient diversity of values to yield an overwhelming number of distinct value combinations. This is why it makes sense to discretize continuous attributes prior to consistency-based filtering. In particular, by discretizing the target attribute, consistency-based attribute filtering can be applied even if the target task is regression.

Example 19.4.4 A simple version of consistency-based attribute filtering is implemented and demonstrated by the R code presented below. It defines the `cons.eval` function that evaluates an attribute subset by the weighted average entropy over the induced data subsets, and the `cons.filter` function that uses it as an evaluation function to guide the attribute subset space search process. The former uses the `entropy` function, and the `digest` function from the `digest` package (optionally, if available) to generate MD5 hash strings for attribute value vectors, which makes it possible to efficiently identify data subsets induced by the attribute subset being evaluated. As in Example 19.4.3, the search strategy implementation provided by the `asel.search.greedy`, `asel.init.none`, `asel.init.all`, `asel.next.forward`, and `asel.next.backward` functions, is employed, and continuous attributes are discretized using the `discnm.eqfreq` function. For the *weatherc* data consistency-based filtering is demonstrated both with and without discretization.

> Ex. 2.4.26
> dmr.stats

> Ex. 18.3.3
> dmr.disc

```
## subset evaluation for consistency-based filters
cons.eval <- function(subset, target, data)
{
  if (require(digest, quietly=TRUE))
    hashfun <- function(x) digest(as.numeric(x))
  else
    hashfun <- function(x) paste(as.numeric(x), collapse="")
  aind <- names(data) %in% subset
  datahash <- sapply(1:nrow(data), function(j) hashfun(data[j,aind]))
  -sum(sapply(unique(datahash),
              function(xh)
              sum(datahash==xh)/nrow(data)*entropy(data[datahash==xh,target])))
}

## consistency-based filter
cons.filter <- function(formula, data,
                        searchf=asel.search.greedy,
                        initf=asel.init.all, nextf=asel.next.backward)
{
  target <- y.var(formula)
  attributes <- x.vars(formula, data)
  searchf(attributes, target,
          evalf=function(subset, target) cons.eval(subset, target, data),
          initf=initf, nextf=nextf)
}

  # consistency-based filter for the weather data
cons.filter(play~., weather)
cons.filter(play~., weather, initf=asel.init.none, nextf=asel.next.forward)

  # consistency-based filter for the weatherc data
cons.filter(play~., discnm.eqfreq(~., weatherc, 4))
cons.filter(play~., discnm.eqfreq(~., weatherc, 4),
            initf=asel.init.none, nextf=asel.next.forward)
```

```
cons.filter(play~., weatherc)
cons.filter(play~., weatherc, initf=asel.init.none, nextf=asel.next.forward)

  # correlation-based filter for the weatherr data
cons.filter(playability~., discnm.eqfreq(~., weatherr, 4))
cons.filter(playability~., discnm.eqfreq(~., weatherr, 4),
          initf=asel.init.none, nextf=asel.next.forward)

  # consistency-based for the Vehicle Silhouettes data
v.sel.cons <- cons.filter(Class~., discnm.eqfreq(~., v.train, 7))$subset
  # consistency-based for the Soybean data
s.sel.cons <- cons.filter(Class~., s.train)$subset
  # consistency-based for the Boston Housing data
bh.sel.cons <- cons.filter(medv~., discnm.eqfreq(~., bh.train, 7))$subset
```

Notice that the forward-search consistency-based filter applied to the *weatherc* data without discretization selects a single continuous attribute. The example calls for the larger datasets make it possible to observe the computational expense of consistency-based filtering, since they take a considerable amount of time (particularly for the *Soybean* data, with 35 attributes).

19.4.4 RELIEF

The RELIEF algorithm is an attribute selection filter originally designed for the classification task that estimates attribute utility based on the observed difference of their values for similar instances. Basically, an attribute is considered more useful if it appears to help distinguish between similar instances of different classes and less useful if it contributes to distinguishing between similar instances from the same class.

19.4.4.1 Basic algorithm

The estimation of attribute utility is performed by going through a number of randomly selected training instances and for each of them identifying the most similar instances of the same class and of a different class. Attribute value differences between the currently processed instance and these neighbors contribute to the utility of each attribute. The algorithm can be written as follows:

```
1: for i = 1, 2, ..., n do
2:     u_i := 0;
3: end for
4: for j = 1, 2, ..., K do
5:     randomly select x ∈ T;
6:     T_x^1 := arg^k min_{x'∈T^{c(x)}} δ(x, x');
7:     T_x^0 := arg^k min_{x'∈T^{c(x)}} δ(x, x');
8:     for i = 1,2,...,n do
9:         u_i := u_i + 1/K ( 1/|T_x^0| Σ_{x_0∈T_x^0} δ_i (x, x_0) − 1/|T_x^1| Σ_{x_1∈T_x^1} δ_i (x, x_1) );
10.    end for
11. end for
```

The algorithm processes K randomly selected training instances, where K is a parameter that controls the tradeoff between the reliability of attribute utility estimation and the computational expense. For each selected instance x, it identifies the set of k most similar instances of the same class, designated by T_x^1, and the set of k most similar instances of other classes, designated by T_x^0, where $k \geq 1$ is another parameter. These instances, called the *nearest hits* and the *nearest misses*, are determined based on a dissimilarity measure δ. The arg ^{k} min notation is used to refer to the set of k instances with the least dissimilarity to x.

In principle, any reasonable dissimilarity measure could be adopted for finding the nearest hits and the nearest misses, including in particular those presented in Chapter 11, but the RELIEF algorithm assumes it is defined simply as the sum of per-attribute differences:

$$\delta(x_1, x_2) = \sum_{i=1}^{n} \delta_i(x_1, x_2) \tag{19.10}$$

where δ_i calculates the difference of values of attribute a_i, depending on its type:

For continuous attributes:

$$\delta_i(x_1, x_2) = \frac{|a_i(x_1) - a_i(x_2)|}{\max_{x \in T} a_i(x) - \min_{x \in T} a_i(x)} \tag{19.11}$$

For discrete attributes:

$$\delta_i(a_i(x_1), a_i(x_2)) = \begin{cases} 0 & \text{if } a_i(x_1) = a_i(x_2) \\ 1 & \text{if } a_i(x_1) \neq a_i(x_2) \end{cases} \tag{19.12}$$

which is actually the same as the contribution of individual attributes to Gower's dissimilarity, defined in Section 11.3.7.

Attribute value differences calculated that way are also used to update attribute utility estimates, initialized to 0. An attribute's utility is decreased proportionally to the averaged difference between the currently processed instance and its nearest hits, and increased proportionally to the averaged difference between the currently processed instance and its nearest misses. This rewards attributes for contributions to distinguishing between similar instances of different classes, and punishes for contributions to distinguishing between similar instances of the same class. With attribute value differences defined to fall in the [0, 1] interval and with K updates scaled down by the $\frac{1}{K}$ coefficient, the estimated attribute utilities remain between -1 and 1.

The RELIEF algorithm achieves the desired context sensitivity by focusing on how particular attributes help to distinguish between instances that are similar, i.e., have the same or close values of most other attributes. This provides the context: the differences observed for an attribute contribute to its estimated utility only if most other attributes do not differ substantially, and are ignored otherwise. This is why the algorithm is not fooled by several closely related attributes being apparently highly useful, i.e., it is not susceptible to the apparent utility issue. Although it is not similarly secured against apparent disutility, it at least captures the predictive utility much better than simple statistical filters when there are dependences among attributes. The original formulation of the algorithm assumed using just one nearest hit and nearest miss, but $k > 1$ was found to improve results, particularly for noisy data.

Example 19.4.5 The R code presented below implements the RELIEF algorithm in a straightforward way that closely matches the previously presented pseudocode. The auxiliary `relief.diss` function calculates the dissimilarity measure, returned as the vector of per-attribute value differences that can be either summed up (when searching for the nearest hits and misses) or used individually (when updating attribute utilities). It uses the `ranges` function to determine attribute ranges and the `arg.min` function to perform the arg min operation. The algorithm is demonstrated using the *weather*, *weatherc*, *Vehicle Silhouettes*, and *Soybean* datasets. The K argument – the number of randomly drawn instances – defaults to 10% of the dataset size, but for these small datasets it is explicitly set to 100 to increase the stability of results. The number of nearest hits and misses defaults to 1, but the effect of $k = 3$ is also demonstrated.

`dmr.util`

```
## RELIEF dissimilarity
relief.diss <- function(x1, x2, rngs)
{
  ifelse(is.na(rd <- mapply(function(v1, v2, r)
                            ifelse(is.numeric(v1), abs(v1-v2)/r, v1!=v2),
                            x1, x2, rngs)),
         0, rd)
}

## RELIEF filter
relief.filter <- function(formula, data, k=1, K=floor(0.1*nrow(data)))
{
  attributes <- x.vars(formula, data)
  class <- y.var(formula)
  aind <- names(data) %in% attributes
  rngs <- ranges(data)[aind]

  util <- sapply(attributes, function(a) 0)

  for (i in 1:K)
  {
    xi <- sample(nrow(data), 1)
    x <- data[xi, ]
    data.x <- data[-xi,]
    hits <- arg.min((1:nrow(data.x))[data.x[[class]]==x[[class]]],
                    function(j) sum(relief.diss(data.x[j,aind], x[aind], rngs)),
                    k=k)
    misses <- arg.min((1:nrow(data.x))[data.x[[class]]!=x[[class]]],
                      function(j) sum(relief.diss(data.x[j,aind], x[aind], rngs)),
                      k=k)
    util <- util +
      rowSums(sapply(misses,
                     function(j) relief.diss(data.x[j,aind], x[aind], rngs)))/K-
        rowSums(sapply(hits,
                       function(j) relief.diss(data.x[j,aind], x[aind], rngs)))/K
  }
  sort(util, decreasing=TRUE)
}

  # RELIEF for the weather data
relief.filter(play~., weather, K=100)
relief.filter(play~., weather, k=3, K=100)

  # RELIEF for the weatherc data
relief.filter(play~., weatherc, K=100)
relief.filter(play~., weatherc, k=3, K=100)

  # RELIEF for the Vehicle Silhouettes data
```

```
v.utl.rel <- relief.filter(Class~., v.train, k=3, K=200)
  # RELIEF for the Soybean data
s.utl.rel <- relief.filter(Class~., s.train, k=3, K=200)
```

For the two smallest datasets used in this example, the obtained attribute rankings do not substantially differ from those produced by the simple statistical filter presented in Example 19.4.1. The only difference is the lower position of the humidity attribute. It disappears for $k = 3$, although using more than single nearest hits and misses for datasets that small is definitely objectionable. For the *Vehicle Silhouettes* and *Soybean* dataset, the results of the RELIEF algorithm differ from those obtained using the simple statistical filter more substantially.

19.4.4.2 Enhancements

The RELIEF algorithm as presented above handles both discrete and continuous attributes, and already incorporates one enhancement that was not present in its original formulation – using more than one nearest hit and nearest miss. It still leaves space for other enhancements, though, which make it more widely applicable to real-world datasets. These, contributed by further work on the algorithm, include missing attribute value handling, improved multiple class processing for target classification tasks with more than two classes, and support for the regression target task.

Missing value handling   Attribute values are directly used in the RELIEF algorithm only for determining attribute value differences, which are needed for two purposes:

- dissimilarity calculation to identify the nearest hits and the nearest misses,
- utility update.

These are the two operations of RELIEF that are affected by missing values. Whereas the simplistic approach of entirely ignoring them when performing these two operations might work sufficiently well when relatively few values are missing, in general one can do somewhat better than that.

One reasonable approach is to redefine $\delta_i(x_1, x_2)$ if one or both of $a_i(x_1)$, $a_i(x_2)$ are missing as the *expected* difference between $a_i(x_1)$ and $a_i(x_2)$, given $c(x_1)$ and $c(x_2)$. For continuous attributes, it reduces simply to replacing $a_i(x)$, if it is missing, by the mean value of a_i in $T^{c(x)}$, i.e., $m_{T^{c(x)}}(a_i)$. It becomes more complicated for discrete attributes, where the expected difference is the probability of two attribute values (one or both of which are missing) being different, estimated based on the training set. If the value of a_i is missing for x_1 and available for x_2, this translates to

$$\delta_i(x_1, x_2) = 1 - P_T(a_i = a_i(x_2)|c = c(x_1)) \tag{19.13}$$

i.e., 1's complement of the probability that an instance of class $c(x_1)$ has the value of a_i equal $a_i(x_2)$. If both $a_i(x_1)$ and $a_i(x_2)$ are missing, this requires considering all possible values of a_i as follows:

$$\delta_i(x_1, x_2) = 1 - \sum_{v \in A_i} P_T(a_i = v | c = c(x_1)) P_T(a_i = v | c = c(x_2)) \qquad (19.14)$$

This is a direct instantiation of the general internal missing value handling approach for dissimilarity measures outlined in Section 11.5.

Multiclass target tasks The RELIEF algorithm was originally developed as dealing with two-class classification tasks only, assuming the nearest misses are selected from the single class other than the class of the currently processed instance. The version of the algorithm presented above drops this assumption simply by permitting that the nearest misses are selected from any classes other than the class of the currently processes instance. This straightforward approach to handling multiple classes is not necessarily sufficient, though, since it does not guarantee that all classes would be actually taken into account when rewarding attributes for their discrimination capabilities. It is not unlikely that due to the similarity patterns exhibited by the data the nearest misses might come entirely or mostly from one or few out of several other classes, which makes the algorithm ignore attribute value differences for the remaining classes. In effect, the final utility estimates could favor attributes that help us to distinguish between similar instances from some but not necessarily all classes.

A slightly more complex, but arguably superior approach consists in finding a specified number of nearest misses from each class other than the class of the currently processed instance, and averaging their contributions, weighted by the corresponding estimated class probabilities. When processing instance x, instead of the two T_x^1 and T_x^0 subsets, we would therefore consider multiple T_x^d nearest neighbor subsets for each $d \in C$:

$$T_x^d := \arg^k \min_{x' \in T^d} \delta(x, x') \qquad (19.15)$$

with $T_x^{c(x)}$ containing the nearest hits and all of T_x^d for $d \neq c(x)$ containing the nearest misses. The modified utility update rule can be written as follows:

$$u_i := u_i + \frac{1}{K} \left(\sum_{d \in C - \{c(x)\}} P_T(d) \frac{1}{|T_x^d|} \sum_{x_0 \in T_x^d} \delta_i(x, x_0) - \frac{1}{|T_x^{c(x)}|} \sum_{x_1 \in T_x^{c(x)}} \delta_i(x, x_1) \right) \qquad (19.16)$$

where $P_T(d)$ is the probability of class d estimated from the training set. It has been found indeed to improve the reliability of RELIEF attribute utility estimates for target classification tasks with more than two classes, particularly for noisy data.

A similar effect could be obtained by decomposing the original multiclass classification task into a number of two-class tasks using one of encoding techniques presented in Section 17.4, applying the original RELIEF algorithm to each of them separately, and averaging the results.

Continuous target attribute   The assumption of dealing with the classification task is only at heart of the RELIEF algorithm, since it makes it possible to update attribute utility estimates based on the nearest hits and the nearest misses. This idea is clearly not applicable to continuous target attributes. The algorithm can be adapted to handle the regression target task, though, in a way that leaves at least other of its essential design principles unchanged.

The original RELIEF utility update rule rewards attributes for contributions to distinguishing between similar instances from different classes (based on the nearest misses) and punishes them for contributions to distinguishing between similar instances from the same class (based on the nearest hits). When the target attribute is continuous, the set of nearest neighbors of the currently processed instance x, that will be referred to as T_x, is not partitioned into the nearest hits and the nearest misses. The absolute difference between their target function values can be used instead of this partitioning to determine their contribution to attribute utility update. Attributes that have substantially different values for instances with substantially different target function values would then have their utility estimates increased, and those that have substantially different values for instances with close target function values would have their utility estimates decreased.

It also makes sense to take into account the dissimilarity between the currently processed instance and its nearest neighbors when calculating attribute utility updates. The most similar neighbors should contribute to attribute utility estimates to a greater extent than those not so similar. One way to achieve this is to use an exponentially decaying function of the rank with respect to dissimilarity, normalized to sum up to one 1 on T_x:

$$W_{T_x}(x, x') = \frac{\exp\left(-\frac{r_{x,x'}^2}{\sigma^2}\right)}{\sum_{x'' \in T_x} \exp\left(-\frac{r_{x,x''}^2}{\sigma^2}\right)} \tag{19.17}$$

where $r_{x,x'}$ is the rank of x' in the set of nearest neighbors T_x with respect to the $\delta(x', x)$ dissimilarity and σ is a parameter that controls the exponential decay. This quantity is used to weight attribute value differences, calculated as in the original RELIEF algorithm. The modified algorithm is presented below.

1: $\Delta^f := 0$
2: **for** $i = 1, 2, \dots, n$ **do**
3:    $\Delta_i := 0; \Delta_i^f := 0;$
4: **end for**
5: **repeat**
6:    randomly select $x \in T$;
7:    $T_x := \arg^k \min_{x' \in T} \delta(x, x');$
8:    $\Delta^f := \Delta^f + \sum_{x' \in T_x} |f(x') - f(x)| W_{T_x}(x, x');$
9:    **for** $i = 1, 2, \dots, n$ **do**
10:     $\Delta_i := \Delta_i + \sum_{x' \in T_x} \delta_i(x, x') W_{T_x}(x, x');$
11:     $\Delta_i^f := \Delta^f + \sum_{x' \in T_x} |f(x') - f(x)| \delta_i(x, x') W_{T_x}(x, x');$
12:    **end for**
13: **until** K instances are processed;
14: **for** $i = 1, 2, \dots, n$ **do**
15:    $u_i := \Delta_i^f / \Delta^f - (\Delta_i - \Delta_i^f)/(K - \Delta^f);$
16: **end for**

For each instance x, the algorithm identifies the set of its k nearest neighbors T_x and uses them to incrementally update the following quantities:

- Δ^f – the weighted sum of absolute target function value differences,

- Δ_i – the weighted sum of value differences for attribute a_i, for $i = 1, 2, \ldots, n$,

- Δ_i^f – the weighted sum of value differences for attribute a_i multiplied by absolute target function value differences.

All these differences are measured between x and all its neighbors $x' \in T_x$ and summed up after weighting by $W_{T_x}(x, x')$. When the main loop of the algorithm is completed, they are used to obtain attribute utility estimates, with Δ_i^f / Δ^f representing the "reward" assigned to attribute a_i for its contribution to (desirable) distinguishing between similar instances with substantially different target function values, and $(\Delta_i - \Delta_i^f)/(K - \Delta^f)$ representing the "punishment" assigned to attribute a_i for its contribution to (undesirable) distinguishing between similar instances with close target function values. The $K - \Delta^f$ denominator only makes sense for $\Delta^f < K$, which can be easily achieved by normalizing target function values to fit in the $[0, 1]$ interval. This will make the absolute differences between target function values for any instance pairs to fall in the same interval.

Example 19.4.6 The regression version of the RELIEF algorithm is implemented by the following R code. Its application to the *weatherr* and *Boston Housing* datasets is demonstrated. Target function value normalization is internally performed using the `nrm.all` and `predict.nrm` functions. The `make.formula` function is used to construct the formula argument for the former.

> Ex. 17.3.2
> dmr.trans

> dmr.util

```
## regression RELIEF filter
rrelief.filter <- function(formula, data, k=1, K=floor(0.1*nrow(data)), sigma=10)
{
  attributes <- x.vars(formula, data)
  target <- y.var(formula)
  aind <- names(data) %in% attributes
  rngs <- ranges(data)[aind]
  data <- predict.nrm(nrm.all(make.formula("", target), data), data)

  delta.f <- 0
  delta.i <- delta.fi <- sapply(attributes, function(a) 0)

  for (i in 1:K)
  {
    xi <- sample(nrow(data), 1)
    x <- data[xi, ]
    data.x <- data[-xi,]
    diss <- sapply(1:nrow(data.x),
                   function(j) sum(relief.diss(data.x[j,aind], x[aind], rngs)))
    neighbors <- arg.min(1:nrow(data.x), function(j) diss[j], k=k)
    nbranks <- rank(diss[neighbors])
    weights <- (weights <- exp(-(nbranks/sigma)^2))/sum(weights)
    adiff <- sapply(neighbors, function(j)
                    relief.diss(data.x[j,aind], x[aind], rngs))
    fdiff <- abs(data.x[neighbors,target]-x[,target])
    delta.f <- delta.f + sum(fdiff*weights)
```

```
    delta.i <- delta.i + rowSums(adiff %*% diag(weights, nrow=length(weights)))
    delta.fi <- delta.fi +
                    rowSums(adiff %*% diag(fdiff*weights, nrow=length(weights)))
  }
  sort(util<-delta.fi/delta.f-(delta.i-delta.fi)/(K-delta.f), decreasing=TRUE)
}

  # RELIEF for the weatherr data
rrelief.filter(playability~., weatherr, K=100)
rrelief.filter(playability~., weatherr, k=3, K=100)

  # RELIEF for the Boston Housing data
bh.utl.rel <- rrelief.filter(medv~., bh.train, k=3, K=200)
```

19.4.5   Random forest

Random forests, as described in Section 15.5.4, are a successful example of model ensembles, consisting of multiple decision or regression trees, grown from bootstrap data samples using randomized algorithms for greater diversity. A useful side effect of creating random forest models is the evaluation of attribute predictive utility – or importance, in standard random forest terminology. The technique used for this purpose, based on estimating the effect of attribute perturbation by evaluating individual trees on their out-of-bag instances, is described in Section 15.5.4.

Random forest-based attribute importance belongs to the most reliable and widely used measures of attribute predictive utility. It is context sensitive, since each attribute is evaluated with other attributes being present. An attribute that is rarely needed to grow a tree or usually appears on bottom levels, is likely to have little effect on the overall accuracy and to be considered rather unimportant. An attribute that appears in most trees, usually on top levels, is likely to have substantial effect on the overall performance and to be considered highly important. In both cases, the appearance and position of an attribute in particular trees, as well as the impact of its perturbation on the accuracy, clearly depends on other attributes available during tree growing.

Random forest attribute filters involve building multiple models, which makes them superficially similar to attribute selection wrappers. They remain substantially different, though, in the purpose of building the models and the way of using them. And of course there is no reason to consider decision trees and regression trees as the only possible target algorithms for random forest filters. What these approaches have in common, though, is the computational expense of building hundreds or thousands of models, that may sometimes enforce aggressive sampling to reduce the training set size.

Example 19.4.7 Random forest-based attribute filtering is illustrated by the following R code, which defines a convenience wrapper around the appropriate functions from the random-Forest package and demonstrates its application.

```
## random forest filter
rf.filter <- function(formula, data, ...)
{
  if (require(randomForest, quietly=TRUE))
```

```
  {
    rf <- randomForest(formula, na.roughfix(data), importance=TRUE, ...)
    sort(importance(rf, type=1)[,1], decreasing=TRUE)

  }
.else
  {
    attributes <- x.vars(formula, data)
    names<-(rep(1, length(attributes), attributes))
  }
}

  # random forest filter for the weather data
rf.filter(play~., weather)
  # random forest filter for the weatherc data
rf.filter(play~., weatherc)
  # random forest filter for the weatherr data
rf.filter(playability~., weatherr)

  # random forest filter for the Vehicle Silhouettes data
v.utl.rf <- rf.filter(Class~., v.train)
  # random forest filter for the Soybean data
s.utl.rf <- rf.filter(Class~., s.train)
  # random forest filter for the BostonHousing data
bh.utl.rf <- rf.filter(medv~., bh.train)
```

19.4.6   Cutoff criteria

Attribute selection filters that evaluate single attributes require some cutoff criteria to select a number of top-ranked attributes. These can be simple rules, such as

1. select top k attributes for some $0 < k < n$, where n is the number of all attributes,

2. select a fraction of p top attributes for some $0 < p < 1$.

The k or p parameters cannot be reasonably set up other than by tuning, i.e., trying a number of values, and building and evaluating models obtained using the selected attributes. Alternatively, the attribute ranking can be analyzed searching for the biggest predictive utility gaps between two consecutive attributes. These identify the most reasonable cutoff points, all of which can be tried or one of which can be chosen based on other task-specific preferences.

Example 19.4.8 The following R code implements simple cutoff criteria for attribute selection, based on a named vector of attribute utilities sorted in a decreasing order. The number of attributes to select can be specified directly via the k parameter or as a fraction of all attributes via the p parameter. If neither of these is specified, the cutoff point is identified as the one corresponding to the biggest utility gap. The application of these cutoff criteria is demonstrated for the *weather*, *weatherc*, and *weatherr* data using random forest-based attribute utilities. For the larger datasets, the top halves of their available attribute sets with respect to the simple statistical, RELIEF, and random forest filters are selected.

```
## cutoff based on decreasingly sorted named attribute utilities
cutoff <- function(utils, k=NA, p=NA)
{
  k <- ifelse(is.na(k), round(p*length(utils)), k)
  k <- ifelse(is.na(k), which.max(-diff(utils)), k)
  k <- ifelse(is.na(k), 1, k)

  names(utils)[1:min(k, length(utils))]
}

  # cutoff based on the random forest filter for the weather data
cutoff(rf.filter(play~., weather), k=3)
  # cutoff based on the random forest filter for the weatherc data
cutoff(rf.filter(play~., weatherc), p=0.5)
  # cutoff based on the random forest filter for the weatherr data
cutoff(rf.filter(playability~., weatherr))

  # cutoff for the Vehicle Silhouettes data
v.sel.simple <- cutoff(v.utl.simple, p=0.5)
v.sel.rf <- cutoff(v.utl.rf, p=0.5)
v.sel.rel <- cutoff(v.utl.rel, p=0.5)

  # cutoff for the Soybean data
s.sel.simple <- cutoff(s.utl.simple, p=0.5)
s.sel.rf <- cutoff(s.utl.rf, p=0.5)
s.sel.rel <- cutoff(s.utl.rel, p=0.5)

  # cutoff for the Boston Housing data
bh.sel.simple <- cutoff(bh.utl.simple, p=0.5)
bh.sel.rf <- cutoff(bh.utl.rf, p=0.5)
bh.sel.rel <- cutoff(bh.utl.rel, p=0.5)
```

Not surprisingly, the attribute subsets obtained using different attribute selection filters contain some common attributes, the utility of which may therefore be considered double- (or, actually, triple-) confirmed. Nevertheless, the discrepancies between them may be perfectly sufficient to yield models of vastly different quality.

19.4.7   Filter-driven search

When computational resources permit and the target algorithm is predetermined, the best cutoff level can be determined by applying a subset evaluation function, such as used by correlation-based filters, consistency-based filters, or attribute selection wrappers. The resulting hybrid technique can be viewed as combining single attribute evaluation for for ranking attributes and subset evaluation for selecting the best subset of top-level attributes.

As discussed above, attribute selection filters that evaluate candidate attribute subsets rather than single attributes use attribute subset search strategies. Filtering techniques that generate per-attribute utility estimates, on the other hand, can become helpful for organizing the search process in an alternative way that radically cuts down the number of candidate subsets to consider. This is possible by following either the forward selection or backward elimination pattern, but instead of greedily selecting the best-evaluated subset out of all candidates obtained by applying the addition or removal operator, respectively, to always generate only one candidate, obtained by:

- adding the next highest utility attribute not yet in the subset,

- or removing the next lowest utility attribute from the subset.

Both these approaches are equivalent, as they generate and evaluate subsets of top k attributes for each $k = 1, 2, \dots, n$, where n is the number of original attributes. The number of different attribute subsets considered using this approach is therefore equal to the size of the original attribute set, which is usually much less than considered using forward selection or backward elimination.

Clearly, the results obtained using filter-driven search severely depend on the quality of the utility estimates used. Simple statistical filters may not guide the search process properly. Good utility estimates, such as produced by the RELIEF or random forest filters, are costly to obtain, making the computational savings due to reducing the number of candidate subsets not so obvious.

Example 19.4.9 The filter-driven search strategy is implemented by the following R code. The `asel.search.filter` function takes a vector of named attribute utilities on input, generates all subsets of top k attributes with respect to these utilities for k between 1 and the number of original attributes, and returns the best subset according to the specified evaluation function. For the demonstrations using the *weather* data, the meaningless evaluation function returning a fixed value of 1 is used. A simple symmetric uncertainty-based statistical filter is employed to guide the search, using the `simple.filter` function from Example 19.4.1 and the `mutinfo` function for mutual information calculation.

> Ex. 2.5.6
> dmr.stats

```
## filter-driven attribute selection search
asel.search.filter <- function(attributes, target, utils, evalf, penalty=0.01)
{
  ev <- function(subset)
  {
    ifelse(is.finite(v<-evalf(subset, target)), v-penalty*length(subset)*abs(v), v)
  }

  subsets <- unname(lapply(utils, function(u) names(utils)[utils>=u]))
  subsets.eval <- sapply(subsets, ev)
  s.best <- which.max(subsets.eval)

  list(subset=subsets[[s.best]], eval=subsets.eval[s.best])
}

  # filter attribute selection search for the weather data
  # using mutual information-based utility estimates
asel.search.filter(names(weather)[-5], "play",
                   simple.filter(play~., weather, dd=symunc),
                   evalf=function(subset, target) 1)
asel.search.filter(names(weather)[-5], "play",
                   simple.filter(play~., weather, dd=symunc),
                   evalf=function(subset, target) 1, penalty=0)
asel.search.filter(names(weather)[-5], "play",
```

```
                    simple.filter(play~., weather, dd=symunc),
                    evalf=function(subset, target) 1)
asel.search.filter(names(weather)[-5], "play",
                    simple.filter(play~., weather, dd=symunc),
                    evalf=function(subset, target) 1, penalty=0)
```

19.5   Attribute selection wrappers

The motivation behind attribute selection wrappers is based on the belief that there is no single best (most predictively useful) attribute subset for any given domain and target task, since different target algorithms are likely to reach their maximum model quality for different attribute subsets. A reasonable approach to attribute selection would therefore be *wrapping* the selected target algorithm within an attribute subset evaluation function, that builds a model using the algorithm and evaluates its quality, usually by calculating an estimate of its true performance. The resulting subset evaluation function, combined with an attribute subset space search method that generates a number of candidate subsets to evaluate, yields an attribute selection wrapper.

Unlike attribute selection filters, the wrapper approach explicitly uses the estimated model quality as the subset utility measure when searching for the best attribute subset. Whereas the limited reliability of the estimation, inherent to all model evaluation procedures, as well as the inexhaustiveness of the search, necessary to make it computationally feasible, may often prevent attribute selection wrappers from identifying the truly best possible attribute subset, they attempt to directly (albeit approximately) achieve this natural attribute selection objective. Attribute selection filters try to do the same not only approximately, but – more importantly – indirectly and implicitly. While this does not prevent some of them from delivering excellent results, attribute selection wrappers are definitely more predictable.

19.5.1   Subset evaluation

Unlike attribute selection filters, the wrapper approach to attribute selection is inherently tied to evaluating complete attribute subsets rather than single attributes. An attribute subset is evaluated by giving it a try – using the target algorithm to create a model (or multiple models) based on the training set and then evaluating its (or their) quality. Since the main purpose of attribute selection is to improve the true performance of the model that can be created from the training set, the evaluation should provide a reliable estimation thereof, with the misclassification error defined in Section 7.2.1 and the mean square error defined in Section 10.2.3 being the most popular (but not necessarily always the most appropriate) performance measures for the classification and regression target tasks, respectively. This is accomplished using an appropriate evaluation procedure, and k-fold cross-validation presented in Section 7.3.4 is the most natural choice, unless working with a large dataset for which the simple holdout procedure may be sufficiently good and less computationally demanding alternative. Of the two evaluation characteristics discussed in Section 7.3.2, bias and variance, that usually need to properly balanced, the variance is particularly crucial for attribute selection. Since candidate subset evaluation is used to guide the attribute subset space, high evaluation variance may lead to misleading search space directions being taken. The bias is more tolerable as long as it remains systematic and does not favor some attribute subsets against the others.

If cross-validation with k-folds is employed for attribute subset evaluation, actually k models are created end evaluated for each candidate attribute subset, to produce the aggregated evaluation. To keep the evaluation variance low, this may be repeated n times with the performance estimates averaged (as per $n \times k$-fold cross-validation). To allocate computational resources most economically, the value of n can be adjusted dynamically depending on the actually observed variance. One simple reasonable approach is to start with $n = 2$ and then allow additional iterations, as long as the dispersion of the performance estimates produced by the previous ones is above an acceptance threshold and a maximum number of iterations has not been reached. The latter is necessary to prevent excessive computational expense. The former may be measured using the variance or the standard deviation, although relative measures such as the coefficient of variation or the relative standard deviation may be more convenient, since they make it possible to set the acceptance threshold in a dataset- and algorithm-independent way. These and other dispersion measures are described in Section 2.4.1.

A similar on the fly adjustment procedure is obviously not possible for k, as the number of cross-validation folds needs to be predetermined. The value of $k = 5$ is usually considered a reasonable compromise between the computational expense and reliability, with the above-mentioned possibility to switch to holdout if the dataset is sufficiently large.

The model evaluation process internally employed by attribute selection wrappers should never be used as a source of reliable quality assessment of the final model created using the selected attribute subset. This is because, regardless of the evaluation procedure employed, attribute selection wrappers naturally fit to their training set by evaluating multiple candidate subsets and using these evaluations to guide the search for better subsets. As more extensively discussed in Section 7.3.2, no part of the data used for final model creation (in any way, including in particular attribute selection) may provide reliable true performance estimates, and failing to perform independent final model evaluation may lead to evaluation overfitting.

Example 19.5.1 The following R code implements the wrapper evaluation of candidate attribute subsets and demonstrates it using the *weather* and *weatherr* datasets. The `wrapper.eval` function makes it possible to specify the modeling algorithm and its arguments as well as the evaluation procedure and its arguments, defaulting to 5-fold cross-validation. The performance measure to be calculated defaults to the misclassification error for the classification target task and to the mean square error for the regression target task. Since these most common performance measures take lower values for better models, the finally returned evaluation value is negated, to adhere to the more intuitive convention that higher evaluation means better subsets as well as to the expectations of the `asel.search.greedy` function from Example 19.3.3. The specified evaluation procedure is called multiple times until a sufficiently small relative standard deviation of the calculated performance measure is obtained (below `minrelsd`) or the maximum number of `maxn` iterations is reached. The `make.formula` function is used to construct the formula argument to the evaluation procedure on the fly. The default cross-validation evaluation procedure is implemented by the `crossval` function.

> `dmr.util`

> `Ex. 7.3.2`
> `dmr.claseval`

```
## subset evaluation for attribute selection wrappers
wrapper.eval <- function(subset, target, data, alg, args=NULL, predf=predict,
                         perf=if (is.numeric(data[[target]])) mse else err,
                         evproc=crossval, evargs=list(k=5),
                         minrelsd=0.01, maxn=5)
{
  if (length(subset <- unique(subset))>0)
  {
    aind <- names(data) %in% subset
    ev <- do.call(evproc, c(list(alg, make.formula(target, subset), data,
                            args=args, predf=predf),
                        evargs))
    p <- perf(ev$pred, ev$true)
    repeat
    {
      ev <- do.call(evproc, c(list(alg, make.formula(target, subset), data,
                              args=args, predf=predf),
                          evargs))
      p <- c(p, perf(ev$pred, ev$true))
      if (length(p)>=maxn || sd(p)/abs(mean(p))<=minrelsd)
        break;
    }
    -mean(p)
  }
  else
    -Inf
}

# wrapper evaluation for the weather data
wrapper.eval(c("outlook", "temperature"), "play", weather,
             rpart, args=list(minsplit=2),
             predf=function(...) predict(..., type="c"))
wrapper.eval(c("outlook", "temperature", "humidity"), "play", weather,
             rpart, args=list(minsplit=2),
             predf=function(...) predict(..., type="c"))
wrapper.eval(names(weather)[-5], "play", weather,
             rpart, args=list(minsplit=2),
             predf=function(...) predict(..., type="c"))

# wrapper evaluation for the weatherc data
wrapper.eval(c("outlook", "temperature"), "play", weatherc,
             rpart, args=list(minsplit=2),
             predf=function(...) predict(..., type="c"))
wrapper.eval(c("outlook", "temperature", "humidity"), "play", weatherc,
             rpart, args=list(minsplit=2),
             predf=function(...) predict(..., type="c"))
wrapper.eval(names(weatherc)[-5], "play", weatherc,
             rpart, args=list(minsplit=2),
             predf=function(...) predict(..., type="c"))

# wrapper evaluation for the weatherr data
wrapper.eval(c("outlook", "temperature"), "playability", weatherr,
             rpart, args=list(minsplit=2))
wrapper.eval(c("outlook", "temperature", "humidity"), "playability", weatherr,
             rpart, args=list(minsplit=2))
wrapper.eval(names(weatherr)[-5], "playability", weatherr,
             rpart, args=list(minsplit=2))
```

Notice the use of the `minsplit=2` argument for `rpart`, which would create a single-leaf tree for the small datasets used in the above calls if left at default settings.

19.5.2   Wrapper attribute selection

An attribute selection wrapper is obtained by combining a wrapper evaluation function with an attribute subset search strategy. For the former, this requires selecting a performance measure and an evaluation procedure to be employed for assessing the quality of models created using the attribute subset being evaluated. For the latter, this requires selecting the initialization and next state selection methods as well as stop criteria. Many different combinations are possible that lead to a variety of particular attribute selection wrappers. The most common choices are those already mentioned above:

- misclassification error or mean square error as the performance measure (depending on the target task),

- k-fold cross-validation or holdout as the evaluation procedure, depending on the data size,

- forward selection, backward selection, a mix of these two, or a filter-driven search strategy.

Example 19.5.2 The R code presented below demonstrates how to combine wrapper subset evaluation with state space search to create an attribute selection wrapper. The `wrapper.select` function applies the function specified via the `searchf` argument to search the attribute subset space using the `wrapper.eval` function for subset evaluation. Its application to the *weather, weatherc, weatherr, Vehicle Silhouettes, Soybean,* and *Boston Housing* datasets is presented. Not surprisingly, the calls for the three larger datasets require substantial time to complete. As in the previously presented examples of attribute selection filters, the attribute subsets obtained for them are retained, to be subsequently used for model creation in Example 19.6.2.

```
## wrapper attribute selection
wrapper.select <- function(formula, data, alg, args=NULL, predf=predict,
                        searchf=asel.search.greedy,
                        initf=asel.init.all, nextf=asel.next.backward,
                        perf=if (is.numeric(data[[target]])) mse else err, ...)
{
  target <- y.var(formula)
  attributes <- x.vars(formula, data)
  searchf(attributes, target,
        evalf=function(subset, target)
              wrapper.eval(subset, target, data, alg, args, predf, perf, ...),
        initf=initf, nextf=nextf)
}


  # wrapper selection for the weather data
wrapper.select(play~., weather, rpart, args=list(minsplit=2),
            predf=function(...) predict(..., type="c"))
wrapper.select(play~., weather, rpart, args=list(minsplit=2),
            predf=function(...) predict(..., type="c"),
            initf=asel.init.none, nextf=asel.next.forward)

  # wrapper selection for the weatherc data
wrapper.select(play~., weatherc, rpart, args=list(minsplit=2),
            predf=function(...) predict(..., type="c"))
```

```
wrapper.select(play~., weatherc, rpart, args=list(minsplit=2),
               predf=function(...) predict(..., type="c"),
               initf=asel.init.none, nextf=asel.next.forward)

# wrapper selection for the weatherr data
wrapper.select(playability~., weatherr, rpart, args=list(minsplit=2))
wrapper.select(playability~., weatherr, rpart, args=list(minsplit=2),
               initf=asel.init.none, nextf=asel.next.forward)

# wrapper selection for the Vehicle Silhouettes data
v.sel.fwd <- wrapper.select(Class~., v.train, rpart,
                            predf=function(...) predict(..., type="c"),
                            initf=asel.init.none, nextf=asel.next.forward)
v.sel.bwd <- wrapper.select(Class~., v.train, rpart,
                            predf=function(...) predict(..., type="c"))

# wrapper selection for the Soybean data
s.sel.fwd <- wrapper.select(Class~., s.train, rpart,
                            predf=function(...) predict(..., type="c"),
                            initf=asel.init.none, nextf=asel.next.forward)
s.sel.bwd <- wrapper.select(Class~., s.train, rpart,
                            predf=function(...) predict(..., type="c"))

# wrapper selection for the Boston Housing data
bh.sel.fwd <- wrapper.select(medv~., bh.train, rpart,
                             initf=asel.init.none, nextf=asel.next.forward)
bh.sel.bwd <- wrapper.select(medv~., bh.train, rpart)
```

Example 19.5.3 The following R code demonstrates wrapper attribute selection using filter-driven search by defining the `wrapper.filter.select` function and applying it to the same datasets as used in the previous example. For the *weather*, *weatherc*, and *weatherr* datasets, attribute utilities generated by the simple statistical filter from Example 19.4.1 are used. For the three larger datasets the search process is guided by the random forest attribute utilities from Example 19.4.7.

```
## filter-driven wrapper attribute selection
wrapper.filter.select <- function(formula, data, utils, alg, args=NULL,
                                  predf=predict, searchf=asel.search.filter,
                                  perf=if (is.numeric(data[[target]])) mse else err,
                                  ...)
{
  target <- y.var(formula)
  attributes <- x.vars(formula, data)
  searchf(attributes, target, utils,
          evalf=function(subset, target)
                wrapper.eval(subset, target, data, alg, args, predf, perf, ...))
}

# simple filter-driven wrapper selection for the weather data
wrapper.filter.select(play~., weather, simple.filter(play~., weather),
                      rpart, args=list(minsplit=2),
                      predf=function(...) predict(..., type="c"))
```

```
  # simple filter-driven wrapper selection for the weatherc data
wrapper.filter.select(play~., weatherc, simple.filter(play~., weatherc),
                      rpart, args=list(minsplit=2),
                      predf=function(...) predict(..., type="c"))

  # simple filter-driven wrapper selection for the weatherr data
wrapper.filter.select(playability~., weatherr,
                      simple.filter(playability~., weatherr),
                      rpart, args=list(minsplit=2))

  # RF filter-driven wrapper selection for the Vehicle Silhouettes data
v.sel.flt <- wrapper.filter.select(Class~., Vehicle, rf.filter(Class~., v.train),
                                   rpart, predf=function(...)
                                   predict(..., type="c"))

  # RF filter-driven wrapper selection for the Soybean data
s.sel.flt <- wrapper.filter.select(Class~., Soybean, rf.filter(Class~., s.train),
                                   rpart, predf=function(...)
                                   predict(..., type="c"))

  # RF filter-driven wrapper selection for the Boston Housing data
bh.sel.flt <- wrapper.filter.select(medv~., BostonHousing,
                                    rf.filter(medv~., bh.train), rpart)
```

The evaluations of the returned attribute subsets are worse than those previously received using greedy search, which is not at all surprising given the vastly less number of generated candidate subsets. This does not necessarily mean that the subsets would indeed lead to substantially worse models, though, since only independent data subset evaluation could verify this reliably.

19.6   Effects of attribute selection

Attribute selection has to be always considered as a part of the modeling process in a broader sense. Its effects have to be therefore evaluated using an independent data subset. This can be accomplished by employing a selected evaluation procedure *outside* attribute selection. For this evaluation procedure the model creation process includes both attribute selection and model building using the selected attribute subset. It can be as simple as using the holdout procedure (i.e., holding out a separate data subset for the evaluation purpose), although more refined and reliable procedures, such as *k*-fold cross-validation, should be preferred when the data size and computational resources permit.

Example 19.6.1 A simple experiment that evaluates the effects of filter attribute selection is performed by the following R code. A manual holdout procedure is used to estimate the true performance of decision and regression trees for the *Vehicle Silhouettes*, *Soybean*, and *Boston Housing* datasets. They will be created using the `rpart` package, both without attribute selection and with filtering attribute selection performed in the previously presented examples.

```
  # selected attribute subsets
v.self <- list(nosel=setdiff(names(v.train), "Class"),
```

```
               simple=v.sel.simple, cfs=v.sel.cfs, cons=v.sel.cons,
               rel=v.sel.rel, rf=v.sel.rf)
s.self <- list(nosel=setdiff(names(s.train), "Class"),
               simple=s.sel.simple, cfs=s.sel.cfs, cons=s.sel.cons,
               rel=s.sel.rel, rf=s.sel.rf)
bh.self <- list(nosel=setdiff(names(bh.train), "medv"),
                simple=bh.sel.simple, cfs=bh.sel.cfs, cons=bh.sel.cons,
                rel=bh.sel.rel, rf=bh.sel.rf)


  # misclassification error
v.errf <- sapply(v.self, function(sel)
                        {
                           tree <- rpart(make.formula("Class", sel), v.train)
                           err(predict(tree, v.test, type="c"), v.test$Class)
                        })
s.errf <- sapply(s.self, function(sel)
                        {
                           tree <- rpart(make.formula("Class", sel), s.train)
                           err(predict(tree, s.test, type="c"), s.test$Class)
                        })
bh.errf <- sapply(bh.self, function(sel)
                         {
                            tree <- rpart(make.formula("medv", sel), bh.train)
                            mse(predict(tree, bh.test), bh.test$medv)
                         })

  # attribute subset size
v.sizef <- sapply(v.self, length)
s.sizef <- sapply(s.self, length)
bh.sizef <- sapply(bh.self, length)

barplot(v.errf, main="Vehicle Silhouettes", ylab="Error", las=2)
lines(c(0, 8), rep(v.errf[1], 2), lty=2)
barplot(v.sizef, main="Vehicle Silhouettes", ylab="Size", las=2)

barplot(s.errf, main="Soybean", ylab="Error", las=2)
lines(c(0, 8), rep(s.errf[1], 2), lty=2)
barplot(s.sizef, main="Soybean", ylab="Size", las=2)

barplot(bh.errf, main="Boston Housing", ylab="Error", las=2)
lines(c(0, 8), rep(bh.errf[1], 2), lty=2)
barplot(bh.sizef, main="Boston Housing", ylab="Size", las=2)
```

Barplots that compare the performance (misclassification error or mean square error) and attribute set size before and after the application of attribute selection filters, produced by the above code, are in Figures 19.1, 19.2, and 19.3. Attribute selection appears to improve model quality for the *Boston Housing* data only, with no substantial effect or some predictive performance degradation observed for the other two datasets. In any case, it at least makes it possible to create models as good or nearly as good as without attribute selection using a fraction of the original number of attributes. Not all attribute selection filters turn out equally successful, though. The correlation-based filter yields small attribute subsets which may however degrade model performance. The consistency-based filter selects all available attributes for the *Vehicle Silhouettes* data, a small subset of attributes resulting in a poor model for the *Soybean* data, and about a half of attributes preserving model quality for the *Boston Housing* data. The simple statistical filter turns out surprisingly successful compared

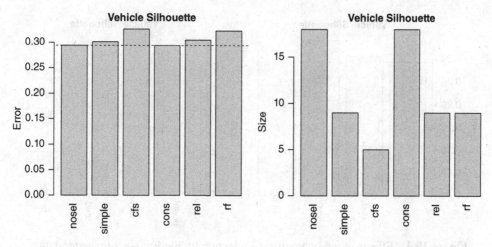

Figure 19.1   Effects of filtering attribute selection for the *Vehicle Silhouettes* data.

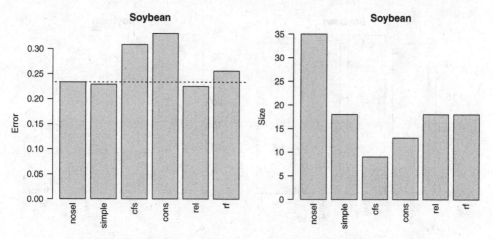

Figure 19.2   Effects of filtering attribute selection for the *Soybean* data.

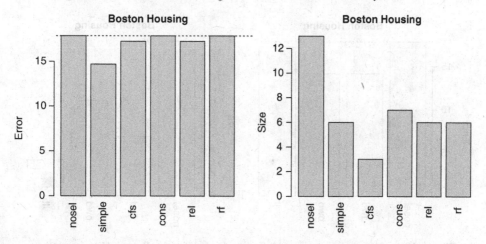

Figure 19.3   Effects of filtering attribute selection for the *Boston Housing* data.

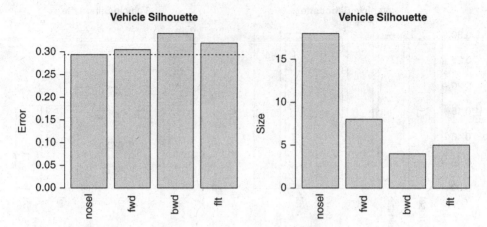

Figure 19.4   Effects of wrapper attribute selection for the *Vehicle Silhouettes* data.

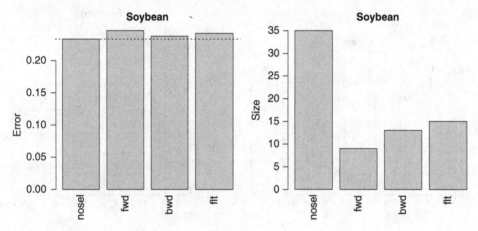

Figure 19.5   Effects of wrapper attribute selection for the *Soybean* data.

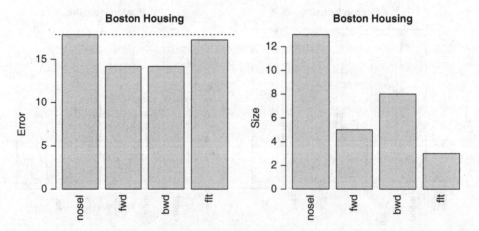

Figure 19.6   Effects of wrapper attribute selection for the *Boston Housing* data.

to the more refined and theoretically superior RELIEF and random forest filters, at least under the arbitrary 50% cutoff applied.

Example 19.6.2 The R code presented below demonstrates a simple experiment that evaluates the effects of wrapper attribute selection, using the same design and datasets as in the previous example for attribute selection filters.

```
 # selected attribute subsets
v.selw <- list(nosel=setdiff(names(v.train), "Class"),
               fwd=v.sel.fwd$subset, bwd=v.sel.bwd$subset, flt=v.sel.flt$subset)
s.selw <- list(nosel=setdiff(names(s.train), "Class"),
               fwd=s.sel.fwd$subset, bwd=s.sel.bwd$subset, flt=s.sel.flt$subset)
bh.selw <- list(nosel=setdiff(names(bh.train), "medv"),
               fwd=bh.sel.fwd$subset, bwd=bh.sel.bwd$subset, flt=bh.sel.flt$subset)

 # misclassification error
v.errw <- sapply(v.selw, function(sel)
                   {
                     tree <- rpart(make.formula("Class", sel), v.train)
                     err(predict(tree, v.test, type="c"), v.test$Class)
                   })
s.errw <- sapply(s.selw, function(sel)
                   {
                     tree <- rpart(make.formula("Class", sel), s.train)
                     err(predict(tree, s.test, type="c"), s.test$Class)
                   })
bh.errw <- sapply(bh.selw, function(sel)
                   {
                     tree <- rpart(make.formula("medv", sel), bh.train)
                     mse(predict(tree, bh.test), bh.test$medv)
                   })

 # attribute subset size
v.sizew <- sapply(v.selw, length)
s.sizew <- sapply(s.selw, length)
bh.sizew <- sapply(bh.selw, length)

barplot(v.errw, main="Vehicle Silhouettes", ylab="Error", las=2)
lines(c(0, 5), rep(v.errw[1], 2), lty=2)
barplot(v.sizew, main="Vehicle Silhouettes", ylab="Size", las=2)

barplot(s.errw, main="Soybean", ylab="Error", las=2)
lines(c(0, 5), rep(s.errw[1], 2), lty=2)
barplot(s.sizew, main="Soybean", ylab="Size", las=2)

barplot(bh.errw, main="Boston Housing", ylab="Error", las=2)
lines(c(0, 5), rep(bh.errw[1], 2), lty=2)
barplot(bh.sizew, main="Boston Housing", ylab="Size", las=2)
```

Barplots that compare the performance (misclassification error or mean square error) and attribute set size without selection, with forward selection, and with backward elimination, produced by the above code, are presented in Figures 19.4, 19.5, and 19.6.

Similarly as in the previous example for attribute selection filters, wrapper attribute selection improves model quality for the *Boston Housing* data only, with forward selection and backward elimination yielding the same performance (although different attribute subsets) and filtering selection clearly less effective. For the remaining datasets it only makes it possible to reduce the number of attributes. The sizes of the attribute subsets obtained using the forward, backward, and filter-driven search strategies do not differ substantially. The results of wrapper attribute selection might be better with other setups of the `minrelsd` and `maxn` parameters controlling the internal evaluation procedure.

19.7   Conclusion

The attribute selection task plays a special role in data mining. While rarely exposed as the primary objective, it may be often a required step, essential for obtaining high-quality models. This is particularly true for domains where the number of attributes can not only be as large as hundreds or thousands, but also for moderate attribute sets of several dozens or less it would not harm to try attribute selection. It is by no means guaranteed to always give model quality improvements, but whenever time and resources permit, it is worthwhile to investigate. Even if no improvement is obtained, the resulting insight on the domain may be a useful side effect.

The distinctive feature of attribute selection filters is their algorithm independence – they are not untied to a particular target algorithm, but do not even require one to be specified at all. The selected attribute subset can be used with arbitrary algorithms appropriate for the target modeling task. While it can be considered an important advantage, making it possible to perform attribute selection once and then build multiple models using a variety of algorithms, it may also turn out to be a deficiency. Since different algorithms represent and create their models in a substantially different way, they are not unlikely to require different attribute subsets to achieve the best possible model performance. In particular, filter methods are naturally focused on global predictive utility (in the whole domain or in major parts thereof) and may fail to include attributes that turn out locally useful (in selected small domain regions), which is likely to sometimes affect the performance of models created by algorithms capable of exploiting such attributes (in particular, decision trees and regression trees). This is the major argument behind the alternative wrapper approach to attribute selection.

With that reservation worthwhile to keep in mind, filtering attribute selection algorithms remain useful and can be safely recommended whenever wrappers are impossible or inconvenient to apply. Clearly, more refined context-sensitive filters should be preferred whenever possible, since simple context-insensitive statistical filters can be often misleading and should only be used when enforced by computational economics reasons.

The general assumption of attribute selection wrappers – that different target algorithms may require different attribute subsets to reach the best model quality – is in general hardly questionable, given the diversity of model representation and identification methods employed by modeling algorithms. It is not always desirable and acceptable, however, to perform attribute selection independently for each target algorithm considered for a given application. While some attribute selection filters employ essentially the same subset space search schemes, their subset evaluation functions are usually much easier and faster to calculate than the full model building and evaluation cycle needed by attribute selection wrappers.

When computational savings are not necessary (which may be the case, e.g., when working with small data or the repertoire of modeling algorithms to consider has been narrowed down by prior analysis) and the ultimate model quality has top priority (which is usually the case), the wrapper approach to attribute selection should be usually the best way to go. Attribute selection filters remain a useful alternative when a single attribute subset for multiple target algorithms is to be selected or when a faster and more computationally efficient attribute selection process is needed, particularly for large datasets.

One potential caveat when using attribute selection wrappers is failing to spare a separate dataset for evaluating models created using the attribute subset obtained after attribute selection. It may be indeed tempting to treat the performance estimate produced by wrapper evaluation for the best attribute subset as representing the expected model quality on new data. As mentioned above and more extensively discussed in Section 7.3.2, this is likely to lead to evaluation overfitting, i.e., overoptimistic model quality assessment.

19.8   Further readings

During recent years, attribute selection has evolved from a relatively small research niche to one of mainstream machine learning and data mining directions. Its practical utility has become recognized and the increased computational power makes it more common to see attribute selection performed as part of practical data mining projects. It has not yet attracted a similar level of interest and appreciation as the most popular modeling algorithms, but has already found its way into recent data mining books (e.g., Cios et al. 2007; Tan et al. 2013; Witten et al. 2011). Some of them present attribute selection as one of possible approaches to data reduction (Han et al. 2011; Theodoridis and Koutroumbas 2008), with other possibilities including mathematical transformations that replace the original set of attributes with a different smaller subset, e.g., based on matrix factorization.

A somewhat dated, but thorough review of attribute selection algorithms is given by Liu and Motoda (1998). The more recent collection of articles (Liu and Motoda 2007) covers important directions of attribute selection research. Guyon and Elisseeff (2003) reviewed major filter and wrapper attribute selection algorithms. Guyon et al. (2004) summarized the results of an attribute selection competition that provided useful insights about the utility of different algorithms. Blum and Langley (1997) discussed both attribute and training instance selection in a unified perspective. Forman (2003) experimentally evaluated a number of attribute utility measures in the speficic application domain of text classification. Liu et al. (2010) presented a unified view of different attribute selection algorithms and a review of ongoing research.

The RELIEF algorithm introduced by Kira and Rendell (1992a,b) belongs to the first refined attribute selection filters that attempt to achieve context sensitivity when evaluating the predictive utility of attributes. Subsequent extensions added more capabilities, including handling missing attribute values and multiple classes (Kononenko 1998) as well as continuous target attributes (Robnik-Šikonja and Kononenko 1997). Robnik-Šikonja and Kononenko (2003) presented an in-depth analysis of the extended versions of RELIEF. The random forest measure of attribute utility, originally called variable importance, was proposed by Breiman (2001) and then used or modified by other authors (e.g., Genuer et al. 2010; Hapfelmeier and Ulm 2013; Sandri and Zuccolotto 2006).

The attribute subset space search problem is an instantiation of the general search problems in artificial intelligence for which several algorithms are available (e.g., Russell and Norvig 2011). The work of Marill and Green (1963) is an early example of backward elimination. Forward selection and backward elimination strategies were discussed by Kittler (1978). Siedlecki and Sklansky (1988) used more refined beam and bi-directional search strategies. Xu et al. (1989) presented a best-first search strategy for attribute selection. Evolutionary search was also considered as an alternative approach by several authors (e.g. Tan et al. 2009; Vafaie and De Jong 1992). The search view of attribute selection was emphasized by Langley (1994).

The correlation-based filter was proposed by Hall (1999, 2000). Dash et al. (2000) introduced the consitency-based filter, later more extensively described and evaluated by Dash and Liu (2003), although a similar idea of searching for a minimum attribute subset sufficient for making consistent predictions appeared before (Almuallim and Dietterich 1992). Shin et al. (2011) presented several consistency measures and a greedy backward elimination consistency-based selection algorithm.

The seminal article by Kohavi and John (1997) argued for the advantages of wrapper attribute subset evaluation and discussed how it can be combined with different search strategies. Strazucci and Utgoff (2004) demonstrated how the number of subsets to evaluate can be reduced by an appropriate randomization technique. Dy and Brodley (2004) adopted the idea of wrapper selection to the clustering target task, when working with unlabeled data. A survey and a unified view of attribute selection algorithms for classification and clustering was presented by Liu and Yu (2005).

References

Almuallim H and Dietterich TG 1992 Efficient algorithms for identifying relevant features *Proceedings of the Ninth Canadian Conference on Artificial Intelligence*. Morgan Kaufmann.

Blum AL and Langley P 1997 Selection of relevant features and examples in machine learning. *Artificial Intelligence* **97**, 245–271.

Breiman L 2001 Random forests. *Machine Learning* **45**, 5–32.

Cios KJ, Pedrycz W, Swiniarski RW and Kurgan L 2007 *Data Mining: A Knowledge Discovery Approach*. Springer.

Dash M and Liu H 2003 Consistency-based search in feature selection. *Artificial Intelligence* **151**, 155–176.

Dash M, Liu H and Motoda H 2000 Consistency-based feature selection *Proceedings of the Fourth Asia-Pacific Conference on Knowledge Discovery and Data Mining (PAKDD-2000)*. Springer.

Dy JG and Brodley CE 2004 Feature selection for unsupervised learning. *Journal of Machine Learning Research* **5**, 845–889.

Forman G 2003 An extensive empirical study of feature selection measures for text classification. *Journal of Machine Learning Research* **3**, 1289–1305.

Genuer R, Poggi JM and Tuleau-Malot C 2010 Variable selection using random forests. *Pattern Recognition Letters* **31**, 2225–2236.

Guyon IM, Ben-Hur A, Gunn S and Dror G 2004 Result analysis of the NIPS 2003 feature selection challenge *Advances in Neural Information Processing Systems 17*. MIT Press.

Guyon IM and Elisseeff A 2003 An introduction to variable and feature selection. *Journal of Machine Learning Research* **3**, 1157–1182.

Hall MA 1999 *Correlation-Based Feature Selection for Machine Learning* PhD thesis University of Waikato, Department of Computer Science.

Hall MA 2000 Correlation-based feature selection for discrete and numeric class machine learning *Proceedings of the Seventeenth International Conference on Maching Learning (ICML=2000)*. Morgan Kaufmann.

Han J, Kamber M and Pei J 2011 *Data Mining: Concepts and Techniques* 3rd edn. Morgan Kaufmann.

Hapfelmeier A and Ulm K 2013 A new variable selection approach using random forests. *Computational Statistics and Data Analysis* **60**, 50–69.

Kira K and Rendell L 1992a The feature selection problem: Traditional methods and a new algorithm *Proceedings of the Tenth National Conference on Artificial Intelligence (AAAI-92)*. AAAI Press.

Kira K and Rendell L 1992b A practical approach to feature selection *Proceedings of the Ninth International Workshop on Machine Learning (ICML-92)*. Morgan Kaufmann.

Kittler J 1978 Feature set search algorithms In *Pattern Recognition and Signal Processing* (ed. Chen CH) Sijthoff and Noordhoff.

Kohavi R and John GH 1997 Wrappers for feature subset selection. *Artificial Intelligence* **97**, 273–324.

Kononenko I 1998 Estimating attributes: Analysis and extensions of RELIEF *Proceedings of the Seventh European Conference on Machine Learning (ECML-94)*. Springer.

Langley P 1994 Selection of relevant features in machine learning *Proceedings of the AAAI Fall Symposium on Relevance*. AAAI Press.

Liu H and Motoda H 1998 *Feature Selection for Knowledge Discovery and Data Mining*. Springer.

Liu H and Motoda H 2007 *Computational Methods of Feature Selection*. Chapman and Hall.

Liu H and Yu L 2005 Toward integrating feature selection algorithms for classification and clustering. *IEEE Transactions on Knowledge and Data Engineering* **17**, 1–12.

Liu H, Motoda H, Setiono R and Zhao Z 2010 Feature selection: An evel-evolving frontier in data mining *Proeedings of the Fourth Workshop on Feature Selection in Data Mining (FSDM-10)*. JMLR Workshop and Conference Proceedings.

Marill T and Green DM 1963 On the effectiveness of receptors in recognition systems. *IEEE Transactions on Information Theory* **9**, 11–17.

Robnik-Šikonja M and Kononenko I 1997 An adaptation of Relief for attribute estimation in regression *Proceedings of the Fourteenth International Conference on Machine Learning (ICML-97)*. Morgan Kaufmann.

Robnik-Šikonja M and Kononenko I 2003 Theoretical and empirical analysis of ReliefF and RReliefF. *Machine Learning* **53**, 23–69.

Russell S and Norvig P 2011 *Artificial Intelligence: A Modern Approach* 3rd edn. Prentice Hall.

Sandri M and Zuccolotto P 2006 Variable selection using random forests *Data Analysis, Classification and the Forward Search: Proceedings of the Meeting of the Classification and Data Analysis Group (CLADAG) of the Italian Statistical Society*. Springer.

Shin K, Fernandes D and Miyazaki S 2011 Consistency measures for feature selection: A formal definition, relative sensitivity comparison and a fast algorithm *Proceedings of the Twenty-Second International Joint Conference on Artificial Intelligence (IJCAI-2011)*. AAAI Press.

Siedlecki W and Sklansky J 1988 On automatic feature selection. *International Journal of Pattern Recognition and Artificial Intelligence* **2**, 197–220.

Strazucci DJ and Utgoff PE 2004 Randomized variable elimination. *Journal of Machine Learning Research* **5**, 1331–1362.

Tan KC, Teoh EJ, Yu Q and Goh KC 2009 A hybrid evolutionary algorithm for attribute selection in data mining. *Expert Systems with Applications* **36**, 8616–8630.

Tan PN, Steinbach M and Kumar V 2013 *Introduction to Data Mining* 2nd edn. Addison-Wesley.

Theodoridis S and Koutroumbas K 2008 *Pattern Recognition* 4th edn. Academic Press.

Vafaie H and De Jong K 1992 Genetic algorithms as a tool for feature selection in machine learning *Proceedings of the Fourth International Conference on Tools with Artificial Intelligence (ICTAI-92)*. IEEE Computer Society Press.

Witten IH, Frank E and Hall MA 2011 *Data Mining: Practical Machine Learning Tools and Techniques* 3rd edn. Morgan Kaufmann.

Xu L, Yan P and Chang T 1989 Best-first strategy for feature selection *Proceedings of the Ninth International Conference on Pattern Recognition (ICPR=89)*. IEEE Computer Society Press.

20

Case studies

20.1 Introduction

The previous chapters of this book contain R code examples that illustrate the operation mechanics of data mining algorithms. While some of the datasets used for these illustrations and demonstrations can be considered realistic (even if small), their scope is by far too limited to adequately represent the typical data mining process. This is supposed to be partially compensated by this chapter, containing case studies that somewhat better portray the path from data to models that has to be traversed in a real-world data mining project.

With the above being said, the case studies remain limited with respect to the scope and depth in comparison to what would be usually done in reality. To make them easily reproducible, they all use publicly available datasets that can be loaded to R with single function calls, are relatively clean, and require very limited preprocessing. To keep the computational requirements within the reach of even aged and low-performant personal computers, computationally intensive operations are avoided. No more than two or three modeling algorithms are used in each study with none or limited parameter tuning, hopefully providing an encouragement for the reader to continue with other algorithms and parameter setups. Little or no statistical exploration of attribute distribution and relationships is included. Some methods of analysis that could be applied to all the datasets are only demonstrated for one of them to avoid uninstructive repetitions. All this helps us to keep the amount of code to be presented within reasonable limits and its contents easy to follow, but departs from the full realism of data mining practice. On the other hand, it leaves space for the reader's own analytic initiative.

The best possible use of the case studies is to repeat them completely in R, taking care to understand all function calls and inspect their output. Introducing modifications or trying to similarly proceed with other datasets would be a natural and useful follow-up. It also makes

Data Mining Algorithms: Explained Using R, First Edition. Paweł Cichosz.
© 2015 John Wiley & Sons, Ltd. Published 2015 by John Wiley & Sons, Ltd.

sense to adhere to the order of case study presentation in this chapter, since the level of detail of the provided explanations and result discussion gradually decreases.

20.1.1 Datasets

The following datasets used for the case studies all come from the UCI Machine Learning Repository:

Census income. Dataset from the 1994–1995 US Census population surveys, with demographic and employment attributes.

Communities and crime. Dataset combining the socioeconomic attributes from 1990 US Census community survey, local police department data from the 1990 US LEMAS survey,[1] and the with the 1995 FBI UCR crime statistics.[2]

Cover type. Dataset containing cartographic attributes with the forest cover type from the US Forest Service.

Appendix C can be referred to for information how to locate the corresponding UCI repository pages, providing more details about the origin and past usage of these datasets, as well as attribute characteristics. Each of these datasets has a designated target attribute for classification or regression, but of course they can also be used to perform clustering.

All the datasets are loaded from the corresponding files downloaded from the repository using the `read.table` function. The presented data loading calls assume the files are placed in the `Data` directory, located one level above R's current working directory, and have to be modified for other settings.

20.1.2 Packages

The case studies will use data mining algorithm implementations from CRAN packages. While the same or similar algorithms, sometimes in more refined versions, are also provided by the popular *Weka* library and available from R via the `RWeka` interface package, only native R packages are used.

Selected functions from the DMR family of packages accompanying this book will be also applied when convenient, particularly for the purpose of model evaluation. They will be used even if the same or similar functionality is available from CRAN packages, to make it easier for the reader to fully understand their operation by referring to the examples where they were defined. All the packages are listed in Appendix B and will be explicitly loaded when necessary.

20.1.3 Auxiliary functions

Besides CRAN and DMR packages, a few R functions specifically written for this chapter will come handy. They are all very simple utility functions that automate certain operations related to `rpart` decision and regression trees.

---

[1] Law Enforcement Management and Administrative Statistics.
[2] Uniform Crime Reporting.

The following three functions will be used for accessing cost-complexity tables generated by the `rpart` decision and regression tree implementations, available in the `cptable` component of `rpart` objects:

```
## complexity parameter corresponding to the minimum cross-validated error
cpmin <- function(cptab) { cptab[which.min(cptab[,4])[1], 1] }

## complexity parameter corresponding to the smallest tree within 1-SD
## from the minimum cross-validated error
cp1sd <- function(cptab)
{ cptab[which(cptab[,4]<min(cptab[,4]) + cptab[which.min(cptab[,4]),5])[1], 1] }

## sequence of complexity parameter values corresponding to the minimum
## cross-validated error, s next smaller trees, and l next larger trees
cpminrange <- function(cptab, s=5, l=5)
{
  m <- which.min(cptab[,4])[1]
  cptab[max(m-s, 1):min(m+l, nrow(cptab)), 1]
}
```

The `cpmin` function selects a complexity parameter value for cost-complexity pruning corresponding to the minimum cross-validated error. The `cp1sd` function selects a complexity parameter value for cost-complexity pruning using the one standard deviation rule. The `cpminrange` function identifies a sequence of the most promising cost-complexity values around the minimum-error one. The actual usage demonstrated later will clarify the purpose and the correct way of applying these functions. Section 3.4.2 can be referred to for background on cost-complexity pruning. The fourth column of the cost-complexity table, used by the functions, contains the cross-validated error corresponding to particular complexity parameter values.

The `rpart.pmin` and `rpart.p1sd` functions defined below wrap `rpart` tree growing and pruning in a single function. It may be sometimes convenient to have one call performing the whole model creation process.

```
## grow and prune an rpart tree using minimum-error cost-complexity pruning
rpart.pmin <- function(formula, data, ...)
{
  tree.f <- rpart(formula, data, minsplit=2, cp=0, ...)
  prune(tree.f, cpmin(tree.f$cptable))
}

## grow and prune an rpart tree using 1-SD cost-complexity pruning
rpart.p1sd <- function(formula, data, ...)
{
  tree.f <- rpart(formula, data, minsplit=2, cp=0, ...)
  prune(tree.f, cp1sd(tree.f$cptable))
}
```

The two more refined auxiliary functions defined below automate piecewise linear model creation and prediction in R, with a regression tree used to decompose the domain into regions and linear models assigned to particular regions. This can be thought of as a simple form of model trees, discussed in a more general setting in Section 9.8, in which the tree structure is created using the `rpart` function, linear models for leaves are created using the `lm` function, and some simple additional code takes care of matching the latter with the former. Such models are created by the `lmrpart` function and applied by the `predict.lmrpart` prediction method. The optional `skip.attr` argument for the former may be used to request skipping attributes used for splitting in the tree structure during linear model creation. The `make.formula` function is used to construct the formula argument for `lm`.

It would be actually a better idea to independently adjust the set of attributes for each leaf's linear model, skipping only those occurring on the corresponding path, but using the common set of attributes keeps the implementation simple.

> dmr.util

```
## create a piecewise linear model represented by an rpart regression tree
## with linear models corresponding to leaves
lmrpart <- function(formula, data, skip.attr=FALSE, ...)
{
  m.tree <- rpart(formula, data, ...)
  m.leaves <- sort(unique(predict(m.tree, data)))
  lmattr <- if (skip.attr)
               setdiff(x.vars(formula, data), setdiff(m.tree$frame$var, "<leaf>"))
            else "."
  m.lm <- `names<-`(lapply(m.leaves, function(l)
                                 lm(make.formula(y.var(formula), lmattr),
                                     data[predict(m.tree, data)==l,])),
                   m.leaves)
  `class<-`(list(tree=m.tree, lm=m.lm), "lmrpart")
}

## prediction method for lmrpart
predict.lmrpart <- function(model, data)
{
  leaves <- as.character(predict(model$tree, data))
  sapply(1:nrow(data),
         function(i) predict(model$lm[[leaves[i]]], data[i,]))
}
```

No usage examples for these functions are provided here, as they will appear naturally when needed in the case studies.

20.2   Census income

This case study is devoted to the classification of individuals described by socioeconomic Census attributes into income categories. The R packages used by this case study can be loaded as follows:

```
library(dmr.claseval)
library(dmr.util)
library(dmr.trans)

library(rpart)
library(rpart.plot)
library(randomForest)
```

20.2.1   Data loading and preprocessing

The following code loads the *Census Income* data and sets attribute names based on the data
description available in its repository page.

```
census <- read.table("../Data/census-income.data",
                     sep=",", na.strings="?", strip.white=TRUE)
census.test <- read.table("../Data/census-income.test",
                          sep=",", na.strings="?", strip.white=TRUE)
names(census) <- c("age",
                   "class.of.worker",
                   "detailed.industry.recode",
                   "detailed.occupation.recode",
                   "education",
                   "wage.per.hour",
                   "enroll.in.edu.inst.last.wk",
                   "marital.stat",
                   "major.industry.code",
                   "major.occupation.code",
                   "race",
                   "hispanic.origin",
                   "sex",
                   "member.of.a.labor.union",
                   "reason.for.unemployment",
                   "full.or.part.time.employment.stat",
                   "capital.gains",
                   "capital.losses",
                   "dividends.from.stocks",
                   "tax.filer.stat",
                   "region.of.previous.residence",
                   "state.of.previous.residence",
                   "detailed.household.and.family.stat",
                   "detailed.household.summary.in.household",
                   "instance.weight",
                   "migration.code.change.in.msa",
                   "migration.code.change.in.reg",
                   "migration.code.move.within.reg",
                   "live.in.this.house.1.year.ago",
                   "migration.prev.res.in.sunbelt",
                   "num.persons.worked.for.employer",
                   "family.members.under.18",
                   "country.of.birth.father",
                   "country.of.birth.mother",
                   "country.of.birth.self",
                   "citizenship",
                   "own.business.or.self.employed",
                   "fill.inc.questionnaire.for.veterans.admin",
                   "veterans.benefits",
                   "weeks.worked.in.year",
                   "year",
                   "income")
names(census.test) <- names(census)
```

There are actually two datasets loaded, the training set and the accompanying test set. All
subsequent preprocessing applies to both of them.

We need to make sure that discrete attributes that are just represented numerically in
the dataset are not treated as numbers. The following code achieves this by applying the

`as.factor` function to the numerically represented discrete attributes. Notice that the `year` is also treated as discrete, since it only distinguishes between instances from the 1994 and 1995 surveys.

```
ci.discrete <- c("detailed.industry.recode", "detailed.occupation.recode",
                 "own.business.or.self.employed", "veterans.benefits", "year")
for (a in ci.discrete)
{
  census[[a]] <- as.factor(census[[a]])
  census.test[[a]] <- as.factor(census.test[[a]])
}
```

As we can find out from the description of the dataset, the `instance.weight` attribute, related to Census data gathering procedures, is not meaningful for classification. The following code removes it:

```
census$instance.weight <- NULL
census.test$instance.weight <- NULL
```

It will be convenient to modify the class labels stored in the `income` factor column, which may be prone to typing errors in their original form:

```
ci.labels <- c("low", "high")
census$income <- factor(ifelse(census$income=="50000+.", "high", "low"),
                        levels=ci.labels)
census.test$income <- factor(ifelse(census.test$income=="50000+.", "high", "low"),
                             levels=ci.labels)
```

By explicitly specifying factor levels, we override the default lexicographic ordering and make sure that the `low` class comes before the `high` class, which will also turn out convenient during model evaluation. This is because we will consider the latter positive when calculating confusion matrix-based performance measures.

The test set will be reserved solely for the final evaluation of created models, but several decisions will have to be made before the ultimate models are achieved. These decisions are based on building and evaluating several trial models. Therefore, the original training set is randomly decomposed into a 67% training and 33% validation subsets as follows, with the random generator seed explicitly initialized for reproducibility:

```
set.seed(12)

rci <- runif(nrow(census))
ci.train <- census[rci>=0.33,]
ci.val <- census[rci<0.33,]
ci.train.small <- census[rci>=0.9,]
```

The last `ci.train.small` subset is a small sample of the training subset, intended to be used for the most computationally intensive operation of random forest growing.

20.2.2   Default model

The following line of code builds a basic decision tree model on the training subset, with all parameters taking default values, no misclassification costs specified, no attribute selection, and no pruning.

```
ci.tree.d <- rpart(income~., ci.train)
```

To evaluate this default tree on the validation subset, we will calculate the misclassification error using the `err` function:

> Ex. 7.2.1
> `dmr.claseval`

```
ci.tree.d.pred <- predict(ci.tree.d, ci.val, type="c")
err(ci.tree.d.pred, ci.val$income)
```

Although the error value of less than 0.05 may appear quite promising, inspecting the confusion matrix calculated using the `confmat` function allows one to better verify the actual utility of this model.

> Ex. 7.2.4
> `dmr.claseval`

```
ci.tree.d.cm <- confmat(ci.tree.d.pred, ci.val$income)
```

The rows of the produced matrix correspond to predicted class labels and the columns to the true ones. Not surprisingly, the `high` class, which has considerably less representation in the training set, is much harder to predict. With this more interesting class considered "positive," the true positive rate and the false positive rate, two convenient indicators of model quality based on the confusion matrix defined in Section 7.2.4, can be calculated using the `tpr` and `fpr` functions:

> Ex. 7.2.6
> `dmr.claseval`

```
ci.tree.d.tpr <- tpr(ci.tree.d.cm)
ci.tree.d.fpr <- fpr(ci.tree.d.cm)
```

While the false positive rate of just slightly above 0.01 is more than satisfactory, the true positive rate below 0.4 shows that the majority of the positive class (high income) remains undetected by the model.

Finally, the following code shows how to produce the ROC curve and calculate the area under the curve, as discussed in Section 7.2.5, using the `roc` and `auc` functions:

> Ex. 7.2.11, 7.2.14
> `dmr.claseval`

```
ci.tree.d.roc <- roc(predict(ci.tree.d, ci.val)[,2], ci.val$income)
plot(ci.tree.d.roc$fpr, ci.tree.d.roc$tpr, type="l", xlab="FP rate", ylab="TP rate")
points(ci.tree.d.fpr, ci.tree.d.tpr, pch=8)
auc(ci.tree.d.roc)
```

Calling the `predict` function without the `type="c"` argument makes it return class probabilities for each classified instance, in the order corresponding to the ordering of class labels. Given our ordering ("negative" class `low` first, "positive" class `high` second), we need to take the second column of the results returned by the `predict` functions to get the "positive" class probability.

The obtained plot is presented in Figure 20.1. The asterisk mark represents the model's default operating point, obtained when using the `predict` function with `type="c"`.

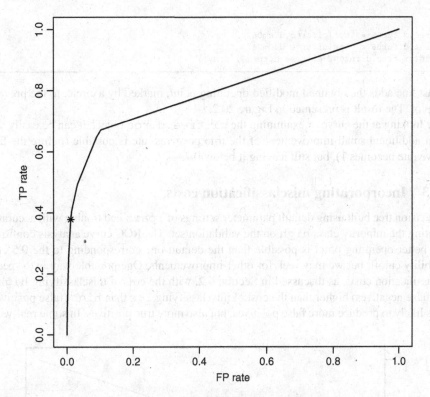

Figure 20.1   The ROC curve of the default decision tree for the *Census Income* data.

It corresponds to classifying each instance to the more likely of the two classes, with respect to the probability estimates from decision tree leaves. The ROC curve shows that the very same decision tree could produce considerably more true positives, without adding too many false positives, just by changing the cutoff value used to assign class labels based on estimated class probabilities. To exploit this possibility, we need to replace the default classification rule applied by predict(..., type="c"), selecting the most likely class, which corresponds to a cutoff value of 0.5, with an alternative rule using an appropriately selected probability cutoff.

The following code identifies cutoff values corresponding to the true positive rate above 0.6:

```
ci.tree.d.cut06 <- ci.tree.d.roc$cutoff[ci.tree.d.roc$tpr>0.6]
```

The first element of the resulting vector is the least cutoff value with the true positive rate greater than 0.6. To generate predicted class labels with this cutoff value, we will use the cutclass function as follows:

Ex. 7.2.12
dmr.claseval

```
ci.tree.d.cm06 <- confmat(cutclass(ci.tree.d.prob, ci.tree.d.cut06[1], ci.labels),
                          ci.val$income)
```

```
ci.tree.d.tpr06 <- tpr(ci.tree.d.cm06)
ci.tree.d.fpr06 <- fpr(ci.tree.d.cm06)
points(ci.tree.d.fpr06, ci.tree.d.tpr06, pch=1)
```

The last line adds the obtained modified operating point, marked by a circle, to the previous ROC plot. The result is presented in Figure 20.2.

By looking at the curve or examining the `ci.tree.d.roc` object, it can be easily seen that an additional small improvement of the true positives rate is possible (before the false positive rate becomes 1), but still leaving it below 0.7.

20.2.3 Incorporating misclassification costs

The decision tree built using default parameter settings of `rpart` had trouble with accurately predicting the minority class `high` on the validation set. The ROC curve analysis confirmed that a better operating point is possible than the default one corresponding to the 0.5 class probability cutoff, but we may seek for other improvements. One possible way is to specify misclassification costs, as discussed in Section 6.2, with the cost of misclassifying `high` as `low` (false negatives) higher than the cost of misclassifying `low` than `high` (false positives). This is likely to produce more false positives, but also more true positives. In some real-world

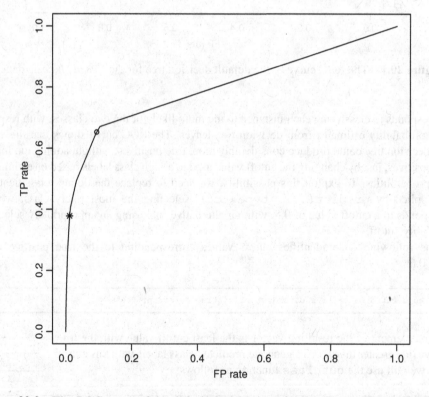

Figure 20.2 The ROC curve of the default decision tree for the *Census Income* data with the operating point shifted to reach a TP rate above 0.6.

applications the "true" misclassification costs may be known based on the underlying "business" model, but in our case we will experiment with a few arbitrarily selected settings to examine their effects.

Misclassification costs can be specified for `rpart` using the `loss` slot of the parameter list passed via the `parms` argument. This is easier than using general-purpose cost-sensitive wrappers defined in the examples presented in Section 6.3 and sufficient for this study. The following `rpart` call builds a decision tree with a 2:1 cost matrix, i.e., with the cost of false negatives twice greater than the cost of false positives:

```
ci.cost2 <- matrix(c(0, 1, 2, 0), nrow=2, byrow=TRUE)
ci.tree.c2 <- rpart(income~., ci.train, parms=list(loss=ct.cost2))
```

The rows of the misclassification cost matrix correspond to true class labels and the columns to predictions, with the ordering of labels corresponding to the ordering of levels of the class column in the training dataset, which is `low` (negative) first and `high` (positive) second in our case. The matrix used above specifies that the cost of predicting `low` instead of `high` is 2, whereas the cost of predicting `high` instead of `low` is 1. The effect of such a cost specification is that `rpart` puts twice as much weight to the instances of the `high` class as to the instances of the `low` class, which affects both split selection and stop criteria used when growing the tree, as well as class label assignment.

The resulting decision tree can be evaluated on the validation set in the exactly same way as before:

```
ci.tree.c2.pred <- predict(ci.tree.c2, ci.val, type="c")
err(ci.tree.c2.pred, ci.val$income)
  # confusion matrix
ci.tree.c2.cm <- confmat(ci.tree.c2.pred, ci.val$income)
  # true positives/false negative rates
ci.tree.c2.tpr <- tpr(ci.tree.c2.cm)
ci.tree.c2.fpr <- fpr(ci.tree.c2.cm)
```

Note the improvement of the true positive rate, which now approaches 0.45 (at the cost of an increased increased false positive rate above 0.02). We reach an arguably more preferred performance level than for the default model, but without even coming close to its performance at the shifted operating point with the modified probability cutoff.

The following code produces the ROC curve (adding the curve previously obtained for the default model plotted with a dashed line for comparison) and calculates the area under the curve for the decision tree incorporating the 2:1 cost matrix:

```
ci.tree.c2.prob <- predict(ci.tree.c2, ci.val)[,2]
ci.tree.c2.roc <- roc(ci.tree.c2.prob, ci.val$income)
plot(ci.tree.c2.roc$fpr, ci.tree.c2.roc$tpr, type="l", xlab="FP rate", ylab="TP rate")
points(ci.tree.c2.fpr, ci.tree.c2.tpr, pch=8)
lines(ci.tree.d.roc$fpr, ci.tree.d.roc$tpr, lty=2)
points(ci.tree.d.fpr, ci.tree.d.tpr, pch=4)
auc(ci.tree.c2.roc)
```

The obtained plot is presented in Figure 20.3. As shown in the figure, the default operating point (i.e., the one corresponding to the default classification cutoff used by

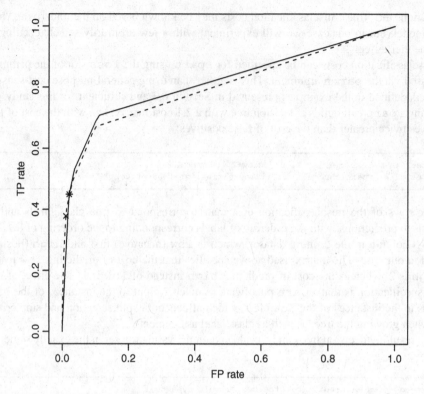

Figure 20.3 The ROC curve of the cost-sensitive decision tree for the *Census Income* data using the 2:1 misclassification cost matrix.

`predict(..., type="c"))` is marked by an asterisk (with the default operating point for the previously created default decision tree marked by an "X"). The area under the ROC curve is slightly greater than for the default model, suggesting that a more favorable operating point could be possible. It can also be immediately seen by comparing the ROC curves. Now we can set the goal of achieving a true positive rate of more than 0.7, more aggressively than before for the default model. The shifted operating point – added to the ROC curve presented in Figure 20.4 – is identified and used as follows.

```
ci.tree.c2.cut07 <- ci.tree.c2.roc$cutoff[ci.tree.c2.roc$tpr>0.7]
ci.tree.c2.cm07 <- confmat(cutclass(ci.tree.c2.prob, ci.tree.c2.cut07[1], ci.labels),
                    ci.val$income)
ci.tree.c2.tpr07 <- tpr(ci.tree.c2.cm07)
ci.tree.c2.fpr07 <- fpr(ci.tree.c2.cm07)
points(ci.tree.c2.fpr07, ci.tree.c2.tpr07, pch=1)
```

A true positive rate of more than 0.7 was not possible with the default model. Even given an increased false positive rate of more than 0.1, this would be usually considered a better performance level, which shows that incorporating a 2:1 misclassification cost matrix brought an improvement. Comparing the ROC curves clearly confirms that the current model is capable

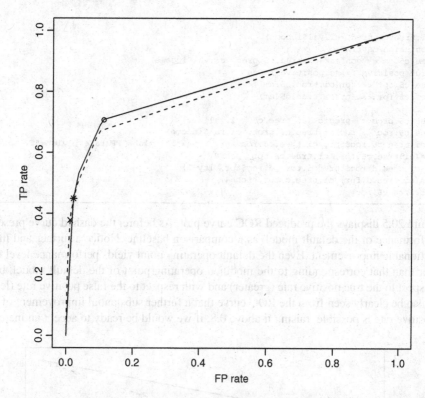

Figure 20.4 The ROC curve of the cost-sensitive decision tree for the *Census Income* data using the 2:1 misclassification cost matrix with the operating point shifted to reach a TP rate above 0.7.

of achieving more true positives with more (but still acceptably little) false positives. What it is unfortunately not capable of is reducing the false positive rate with keeping the same level of the true positive rate.

It is worthwhile to notice, by the way, that the same effect as obtained using the 2:1 cost matrix can be achieved in the following alternative way, by specifying explicit instance weights:

```
ci.tree.w2 <- rpart(income~., census.train,
                    weights=ifelse(census.train$income=="high", 2, 1))
```

It can be easily verified that the resulting decision tree has the same structure and achieves exactly the same performance as the one built using the cost matrix.

To examine the effects of putting even more weight to the positive class, consider using a 5:1 cost matrix, as in the code below.

```
ci.cost5 <- matrix(c(0, 1, 5, 0), nrow=2, byrow=TRUE)
ci.tree.c5 <- rpart(income~., ci.train, parms=list(loss=ci.cost5))
  # error
ci.tree.c5.pred <- predict(ci.tree.c5, ci.val, type="c")
```

```
err(ci.tree.c5.pred, ci.val$income)
  # confusion matrix
ci.tree.c5.cm <- confmat(ci.tree.c5.pred, ci.val$income)
  # true positive/false positive rates
ci.tree.c5.tpr <- tpr(ci.tree.c5.cm)
ci.tree.c5.fpr <- fpr(ci.tree.c5.cm)
  # ROC
ci.tree.c5.prob <- predict(ci.tree.c5, ci.val)[,2]
ci.tree.c5.roc <- roc(ci.tree.c5.prob, ci.val$income)
plot(ci.tree.c5.roc$fpr, ci.tree.c5.roc$tpr, type="l", xlab="FP rate", ylab="TP rate")
points(ci.tree.c5.fpr, ci.tree.c5.tpr, pch=8)
lines(ci.tree.d.roc$fpr, ci.tree.d.roc$tpr, lty=2)
points(ci.tree.d.fpr, ci.tree.d.tpr, pch=4)
auc(ci.tree.c5.roc)
```

Figure 20.5 displays the produced ROC curve plot. As before, the dashed curve presents the performance of the default model, as a comparison baseline. Notice a further and more unquestionable improvement. Even the default operating point yields performance level that is better than that corresponding to the modified operating point of the default model, both with respect to the true positive rate (greater) and with respect to the false positive rate (less). It can also be clearly seen from the ROC curve that a further substantial improvement of the true positive rate is possible, raising it above 0.8, if we would be ready to accept an increase

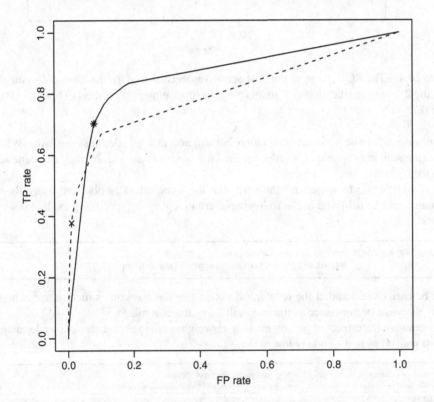

Figure 20.5 The ROC curve of the cost-sensitive decision tree for the *Census Income* data using the 5:1 misclassification cost matrix.

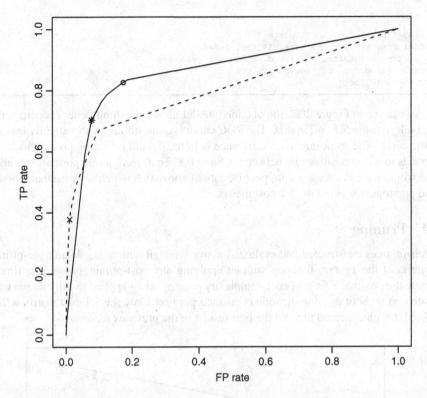

Figure 20.6   The ROC curve of the cost-sensitive decision tree for the *Census Income* data using the 5:1 misclassification cost matrix with the operating point shifted to reach a TP rate above 0.8.

of the false positive rate (still leaving it below 0.2, though). The following code exploits this possibility, adding a modified operating point marked by a circle to the plot, the resulting modified form of which is presented in Figure 20.6.

```
ci.tree.c5.cut08 <- ci.tree.c5.roc$cutoff[ci.tree.c5.roc$tpr>0.8]
ci.tree.c5.cm08 <- confmat(cutclass(ci.tree.c5.prob, ci.tree.c5.cut08[1], ci.labels),
                    ci.val$income)
ci.tree.c5.tpr08 <- tpr(ci.tree.c5.cm08)
ci.tree.c5.fpr08 <- fpr(ci.tree.c5.cm08)
points(ci.tree.c5.fpr08, ci.tree.c5.tpr08, pch=1)
```

Now we might be tempted to see if using an even higher ratio of misclassification costs, say 10:1, brings some additional improvement. The following code builds and evaluates decision tree using a 10:1 cost matrix. Again, the ROC plot includes the curve for the default model using a dashed line for comparison.

```
ci.tree.c10.prob <- predict(ci.tree.c10, ci.val)[,2]
ci.tree.c10.roc <- roc(ci.tree.c10.prob, ci.val$income)
plot(ci.tree.c10.roc$fpr, ci.tree.c10.roc$tpr, type="l",
```

```
    xlab="FP rate", ylab="TP rate")
points(ci.tree.c10.fpr, ci.tree.c10.tpr, pch=8)
lines(ci.tree.d.roc$fpr, ci.tree.d.roc$tpr, lty=2)
points(ci.tree.d.fpr, ci.tree.d.tpr, pch=4)
auc(ci.tree.c10.roc)
```

As we can see in Figure 20.7, the obtained model does not exhibit better properties than the one built with the 5:1 cost matrix. The ROC curve is quite similar, with a slightly less area under the curve. The most important difference is in the default operating point, which now corresponds to a true positive rate between 0.7 and 0.8. To achieve a true positive rate above 0.8, one would have to accept a false positive rate of above 0.2, which is worse than possible with the previous tree using the 5:1 cost matrix.

20.2.4 Pruning

The decision trees constructed and evaluated above were all built using default pre-pruning stop criteria of the `rpart` function without applying any post-pruning. Now it is time to investigate the possible effects of cost-complexity pruning when applied to fully grown trees, both with and without misclassification costs incorporated. Only the 5:1 cost matrix will be considered, which appeared to yield the best results in the previous section.

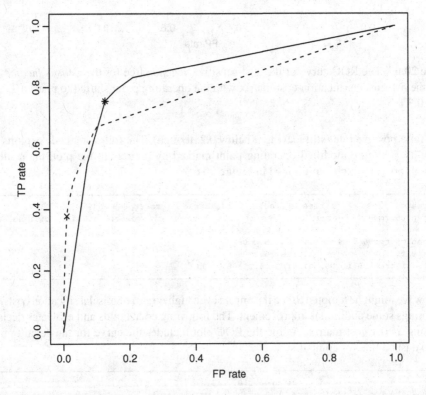

Figure 20.7 The ROC curve of the cost-sensitive decision tree for the *Census Income* data using the 10:1 misclassification cost matrix.

One can request a fully grown tree (with no pre-pruning) by specifying the cp and minsplit parameters of rpart as in the following call:

```
ci.tree.f <- rpart(income~., ci.train, minsplit=2, cp=0)
```

Unfortunately, it takes considerably longer than with default stop criteria, and – depending on the available computational power and patience – one may prefer to choose larger values of minsplit and/or cp, accepting the theoretical risk that the resulting tree might be pre-pruned too much (i.e., some nodes that would not be pruned by cost-sensitivity pruning using a subsequently selected complexity parameter value, would get pre-pruned). Setting minsplit=10 and cp=0.0001 should be safe in our case.

When the fully (or almost fully) grown tree is ready, one can either select the value of the complexity parameter yielding the minimum cross-validated error, or apply the one-standard-deviation rule (1-SD for short), which recommends the cp value yielding the smallest tree with cross-validated error within one standard deviation of the minimum cross-validated error. This selection is performed by the following R code, using the cpmin and cp1sd functions defined in Section 20.1.3:

```
  # minimum-error cost-complexity pruning
ci.tree.pmin <- prune(ci.tree.f, cpmin(ci.tree.f$cptable))
  # 1-sd cost-complexity pruning
ci.tree.p1sd <- prune(ci.tree.f, cp1sd(ci.tree.f$cptable))
```

Since the selection of the complexity parameter is based on internally cross-validated errors, which are obtained by randomly partitioning the training set into cross-validation folds, several independent runs of tree growing may yield slightly different complexity tables. This may cause different cp value selections, yielding different pruned trees.

Note the difference in the sizes of the default (pre-pruned), fully grown, and pruned decision trees, which can be compared as follows:

```
c(default=nrow(ci.tree.d$frame), full=nrow(ci.tree.f$frame),
  pruned.min=nrow(ci.tree.pmin$frame), pruned.1sd=nrow(ci.tree.p1sd$frame))
```

The frame component of the rpart object represents the structure of the decision tree, as has as many rows as the total number of nodes and leaves. The pruned tree is much smaller than the fully grown tree, but noticeably larger than the default (pre-pruned) one, which suggests that the default level of pre-pruning might have been too aggressive.

Now the fully grown and pruned trees can be evaluated in exactly the same way as before, by calculating error values, the true positive and false positive rates, and plotting the ROC curves. The following code evaluates the fully grown tree and plots its ROC curve, with the ROC curve for the default model shown using a dashed line for comparison.

```
ci.tree.f.pred <- predict(ci.tree.f, ci.val, type="c")
err(ci.tree.f.pred, ci.val$income)

ci.tree.f.cm <- confmat(ci.tree.f.pred, ci.val$income)
ci.tree.f.tpr <- tpr(ci.tree.f.cm)
ci.tree.f.fpr <- fpr(ci.tree.f.cm)

ci.tree.f.prob <- predict(ci.tree.f, ci.val)[,2]
```

```
ci.tree.f.roc <- roc(ci.tree.f.prob, ci.val$income)
plot(ci.tree.f.roc$fpr, ci.tree.f.roc$tpr, type="l", xlab="FP rate", ylab="TP rate")
points(ci.tree.f.fpr, ci.tree.f.tpr, pch=8)
lines(ci.tree.d.roc$fpr, ci.tree.d.roc$tpr, lty=2)
points(ci.tree.d.fpr, ci.tree.d.tpr, pch=4)
auc(ci.tree.f.roc)
```

The produced ROC curve plot is presented in Figure 20.8. The fully grown tree definitely represents a poor model, although it becomes evident only after looking at the ROC curve. The default operating point is no worse than for the default model, it just represents a different (and arguably better) compromise between true and false positives. The error value is only marginally greater. However, whereas for the default model one could shift the operating point to achieve a much more satisfactory performance level, this is impossible for the fully grown tree. In particular, the highest possible true positive rate, without reaching a false positive rate value of 1, is about 0.52. The area under the ROC curve is substantially less as well. Notice, by the way, that the ROC curve of the fully grown tree is much more smooth. This is a side effect of its big size – with so many leaves there are many distinct class probability values, and even small changes of the cutoff level result in some instances being

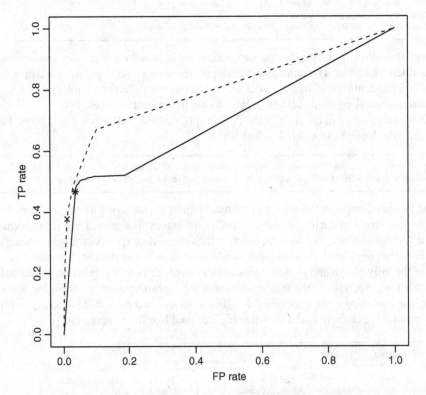

Figure 20.8 The ROC curve of the fully grown decision tree for the *Census Income* data.

classified differently, which makes the number of points of which the ROC curve is created much larger.

The pruned trees can be evaluated and their ROC curves plotted (again, with the ROC curve of the default model shown for comparison) as follows.

```
ci.tree.pmin.pred <- predict(ci.tree.pmin, ci.val, type="c")
err(ci.tree.pmin.pred, ci.val$income)

ci.tree.pmin.cm <- confmat(ci.tree.pmin.pred, ci.val$income)
ci.tree.pmin.tpr <- tpr(ci.tree.pmin.cm)
ci.tree.pmin.fpr <- fpr(ci.tree.pmin.cm)

ci.tree.pmin.prob <- predict(ci.tree.pmin, ci.val)[,2]
ci.tree.pmin.roc <- roc(ci.tree.pmin.prob, ci.val$income)
plot(ci.tree.pmin.roc$fpr, ci.tree.pmin.roc$tpr, type="l",
    xlab="FP rate", ylab="TP rate", main="Minimum error CCP")
points(ci.tree.pmin.fpr, ci.tree.pmin.tpr, pch=8)
lines(ci.tree.d.roc$fpr, ci.tree.d.roc$tpr, lty=2)
points(ci.tree.d.fpr, ci.tree.d.tpr, pch=4)
auc(ci.tree.pmin.roc)

ci.tree.p1sd.pred <- predict(ci.tree.p1sd, ci.val, type="c")
err(ci.tree.p1sd.pred, ci.val$income)

ci.tree.p1sd.cm <- confmat(ci.tree.p1sd.pred, ci.val$income)
ci.tree.p1sd.tpr <- tpr(ci.tree.p1sd.cm)
ci.tree.p1sd.fpr <- fpr(ci.tree.p1sd.cm)

ci.tree.p1sd.prob <- predict(ci.tree.p1sd, ci.val)[,2]
ci.tree.p1sd.roc <- roc(ci.tree.p1sd.prob, ci.val$income)
plot(ci.tree.p1sd.roc$fpr, ci.tree.p1sd.roc$tpr, type="l",
    xlab="FP rate", ylab="TP rate", main="1-SD CCP")
points(ci.tree.p1sd.fpr, ci.tree.p1sd.tpr, pch=8)
lines(ci.tree.d.roc$fpr, ci.tree.d.roc$tpr, lty=2)
points(ci.tree.d.fpr, ci.tree.d.tpr, pch=4)
auc(ci.tree.p1sd.roc)
```

Figure 20.9 shows the plotted ROC curves. It can be immediately seen that the pruned trees clearly outperform both the fully grown tree and the default tree, although – again – it takes looking at the ROC curves to fully appreciate the improvement. Those for the pruned trees are consistently above the one for the default model, which shows that the improvement is unquestionable. Whatever true positive rate is required, we can get it with less false positives. The ROC curves of the two pruned trees are very similar to each other and it makes sense to prefer the smaller tree in this situation.

The default operating points of the pruned trees are actually quite similar to that of the default model. Using a different probability cutoff value makes it possible, however, to raise the true positive rate above 0.8 while keeping the false positive rate at about 0.2 – something that was not possible with the default model. The following R code exploits this possibility, adding a shifted operating point marked by a circle to the plot, for the tree pruned using the one-standard-deviation rule. The resulting modified plot is presented in Figure 20.10.

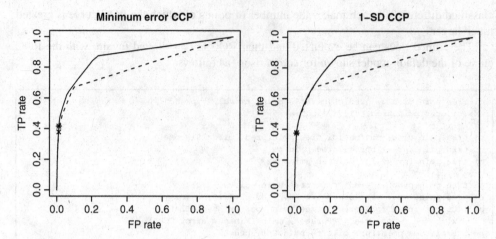

Figure 20.9   The ROC curves of the pruned decision trees for the *Census Income* data.

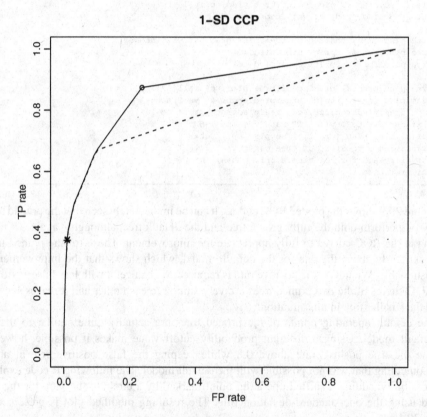

Figure 20.10   The ROC curve of the 1-SD pruned decision tree for the *Census Income* data with the operating point shifted to reach a TP rate above 0.8.

```
ci.tree.p1sd.cut08 <- ci.tree.p1sd.roc$cutoff[ci.tree.p1sd.roc$tpr>0.8]
ci.tree.p1sd.cm08 <- confmat(cutclass(ci.tree.p1sd.prob, ci.tree.p1sd.cut08[1],
                                       ci.labels),
                             ci.val$income)
ci.tree.p1sd.tpr08 <- tpr(ci.tree.p1sd.cm08)
ci.tree.p1sd.fpr08 <- fpr(ci.tree.p1sd.cm08)
points(ci.tree.p1sd.fpr08, ci.tree.p1sd.tpr08, pch=1)
```

The process of building a fully grown tree and cost-complexity pruning can be repeated with misclassification costs incorporated. The following call uses the 5:1 cost matrix for growing:

```
ci.tree.c5f <- rpart(income~., ci.train, minsplit=2, cp=0, parms=list(loss=ci.cost5))
```

The obtained fully grown tree can be evaluated by calculating its error value, true positives and false positive rates, and plotting the ROC curve, as presented below. This time the dashed curve shown for comparison represents the performance of the tree built with default stop criteria (pre-pruning) and the same 5:1 cost matrix.

```
ci.tree.c5f.pred <- predict(ci.tree.c5f, ci.val, type="c")
err(ci.tree.c5f.pred, ci.val$income)

ci.tree.c5f.cm <- confmat(ci.tree.c5f.pred, ci.val$income)
ci.tree.c5f.tpr <- tpr(ci.tree.c5f.cm)
ci.tree.c5f.fpr <- fpr(ci.tree.c5f.cm)

ci.tree.c5f.prob <- predict(ci.tree.c5f, ci.val)[,2]
ci.tree.c5f.roc <- roc(ci.tree.c5f.prob, ci.val$income)
plot(ci.tree.c5f.roc$fpr, ci.tree.c5f.roc$tpr, type="l",
     xlab="FP rate", ylab="TP rate")
points(ci.tree.c5f.fpr, ci.tree.c5f.tpr, pch=8)
lines(ci.tree.c5.roc$fpr, ci.tree.c5.roc$tpr, lty=2)
points(ci.tree.c5.fpr, ci.tree.c5.tpr, pch=4)
auc(ci.tree.c5f.roc)
```

As can be seen in Figure 20.11, the fully grown tree is clearly inferior to the pre-pruned one, both with respect to its default operating point and its – virtually nonexistent – improvement potential. It cannot deliver much more than 0.5 of true positive rate, despite incorporating misclassification costs. It is no better (slightly worse, actually) than the fully-grow tree built without specifying nonuniform misclassification costs. This may appear surprising at first, but some more thoughts are enough to realize that misclassification costs can only affect predictions generated probabilistically (i.e., by inaccurate leaves), where they may justify selecting a less probable class. A fully grown tree may have several perfectly accurate leaves (fitting the training set exactly), and these leaves will always predict their assigned class labels regardless of misclassification costs.

To prune the fully grown cost-sensitive decision tree, we will apply the complexity parameter values previously determined for its cost-insensitive counterpart:

```
# minimum-error cost-complexity pruning (with cp determined based on ci.tree.f)
ci.tree.c5pmin <- prune(ci.tree.c5f, cpmin(ci.tree.f$cptable))
# 1-sd cost-complexity pruning  (with cp terming based on ci.tree.f)
ci.tree.c5p1sd <- prune(ci.tree.c5f, cp1sd(ci.tree.f$cptable))
```

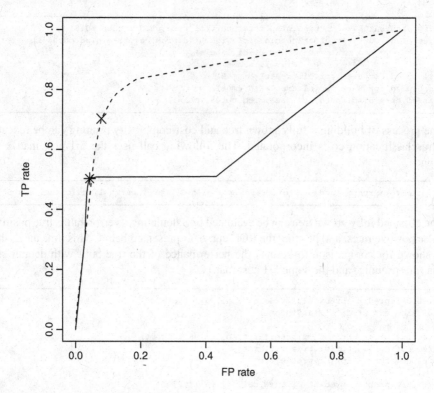

Figure 20.11 The ROC curve of the fully grown cost-sensitive decision tree for the *Census Income* data using the 5:1 cost matrix.

This may appear strange and questionable, but turns out useful in this case, since the more obvious approach of determining the `cp` values for pruning based on the cost-complexity table for the cost-sensitive fully grown tree fails, overpruning to a single leaf. This suggests that the misclassification error-based pruning criterion adopted by cost-complexity pruning is not necessarily a good choice for cost-sensitive models. We may still see whether the fully grown cost-sensitive tree – pruned to cost-complexity levels identified for the fully grown cost-insensitive tree – can give any improvement over the pre-pruned cost-sensitive models created before. The usual evaluation process is performed by the following code:

```
ci.tree.c5pmin.pred <- predict(ci.tree.c5pmin, ci.val, type="c")
err(ci.tree.c5pmin.pred, ci.val$income)

ci.tree.c5pmin.cm <- confmat(ci.tree.c5pmin.pred, ci.val$income)
ci.tree.c5pmin.tpr <- tpr(ci.tree.c5pmin.cm)
ci.tree.c5pmin.fpr <- fpr(ci.tree.c5pmin.cm)

ci.tree.c5pmin.prob <- predict(ci.tree.c5pmin, ci.val)[,2]
ci.tree.c5pmin.roc <- roc(ci.tree.c5pmin.prob, ci.val$income)
plot(ci.tree.c5pmin.roc$fpr, ci.tree.c5pmin.roc$tpr, type="l",
    xlab="FP rate", ylab="TP rate", main="Minimum error CCP")
points(ci.tree.c5pmin.fpr, ci.tree.c5pmin.tpr, pch=8)
lines(ci.tree.c5.roc$fpr, ci.tree.c5.roc$tpr, lty=2)
points(ci.tree.c5.fpr, ci.tree.c5.tpr, pch=4)
auc(ci.tree.c5pmin.roc)
```

```
ci.tree.c5p1sd.pred <- predict(ci.tree.c5p1sd, ci.val, type="c")
err(ci.tree.c5p1sd.pred, ci.val$income)

ci.tree.c5p1sd.cm <- confmat(ci.tree.c5p1sd.pred, ci.val$income)
ci.tree.c5p1sd.tpr <- tpr(ci.tree.c5p1sd.cm)
ci.tree.c5p1sd.fpr <- fpr(ci.tree.c5p1sd.cm)

ci.tree.c5p1sd.prob <- predict(ci.tree.c5p1sd, ci.val)[,2]
ci.tree.c5p1sd.roc <- roc(ci.tree.c5p1sd.prob, ci.val$income)
plot(ci.tree.c5p1sd.roc$fpr, ci.tree.c5p1sd.roc$tpr, type="l",
    xlab="FP rate", ylab="TP rate", main="1-SD CCP")
points(ci.tree.c5p1sd.fpr, ci.tree.c5p1sd.tpr, pch=8)
lines(ci.tree.c5.roc$fpr, ci.tree.c5.roc$tpr, lty=2)
points(ci.tree.c5.fpr, ci.tree.c5.tpr, pch=4)
auc(ci.tree.c5p1sd.roc)
```

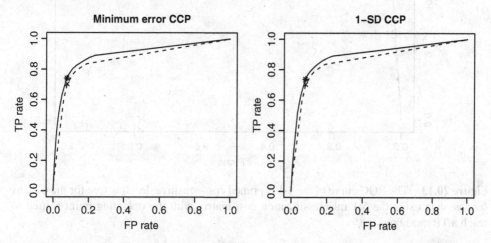

Figure 20.12 The ROC curves of the pruned cost-sensitive decision tree for the *Census Income* data using the 5:1 misclassification cost matrix.

The obtained ROC plots, with the dashed line representing the ROC curve for the pre-pruned 5:1 cost-sensitive tree, are displayed in Figure 20.12. The two pruned trees outperform their pre-pruned counterpart and make it possible to raise the true positive rate above 0.85 while keeping the false positive rate below 0.2. This is demonstrated below for the smaller of the pruned trees:

```
ci.tree.c5p1sd.cut085 <- ci.tree.c5p1sd.roc$cutoff[ci.tree.c5p1sd.roc$tpr>0.85]
ci.tree.c5p1sd.cm085 <- confmat(cutclass(ci.tree.c5p1sd.prob,
                                ci.tree.c5p1sd.cut085[1], ci.labels),
                         ci.val$income)
ci.tree.c5p1sd.tpr085 <- tpr(ci.tree.c5p1sd.cm085)
ci.tree.c5p1sd.fpr085 <- fpr(ci.tree.c5p1sd.cm085)
points(ci.tree.c5p1sd.fpr085, ci.tree.c5p1sd.tpr085, pch=1)
```

Figure 20.13 presents the ROC curve with a new operating point marked by a circle. This is arguably the most preferable of all operating points we have managed to achieve so far. It

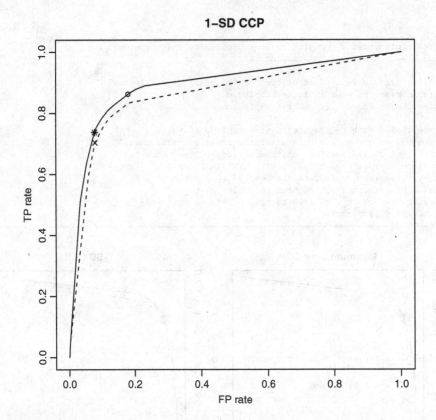

Figure 20.13 The ROC curve of the 1-SD pruned cost-sensitive decision tree for the *Census Income* data using the 5:1 misclassification cost matrix with the operating point shifted to reach a TP rate above 0.85.

represents indeed a striking improvement in comparison to the default operating point for the default cost-insensitive decision tree we started from.

20.2.5 Attribute selection

To see whether a smaller attribute subset, by constraining the model search space, would improve the predictive power of decision trees for the *Census Income* data, one may employ the random forest algorithm for attribute utility estimation, as discussed in Section 19.4.5. Some more data preprocessing is needed to meet the requirements of the `randomForest` package, which does not permit more than 32 discrete attribute values (factor levels) and any missing values. The former can be ensured by applying the aggregation transformation, as described in Section 17.3.3, using the `agg.all` and `predict.agg` functions, and the latter – by applying the imputation transformation, as described in Section 17.3.4, using the `imp.all` and `predict.imp` functions. There are other functions that could be used to perform aggregation and imputation in R (e.g., the `combine.levels` function

Ex. 17.3.3
`dmr.trans`

Ex. 17.3.4
`dmr.trans`

from the `Hmisc` package and the `na.roughfix` and `rfImpute` functions from the `randomForest` package), but they do not support the modeling view of these transformations, discussed in Section 17.2.5, making it problematic to apply a predictive model created on the transformed training data to new instances. It would not matter if the random forest algorithm were used for attribute selection only, but once the forest is grown, it would be a waste not to consider it as a candidate classification model as well.

Aggregation and imputation is performed by the code presented below. Notice that the small sample of the training set is used to reduce the computational expense of random forest growing.

```
 # aggregation (ensure no more than 32 discrete attribute values)
ci.aggm <- agg.all(income~., ci.train.small, 31)
cirf.train <- predict.agg(ci.aggm, ci.train.small)
cirf.val <- predict.agg(ci.aggm, ci.val)
 # imputation (ensure no missing values)
cirf.impm <- imp.all(income~., cirf.train)
cirf.train <- predict.imp(cirf.impm, cirf.train)
cirf.val <- predict.imp(cirf.impm, cirf.val)
```

Now the random forest model can be created as follows:

```
ci.rf <- randomForest(income~., cirf.train, importance=TRUE)
```

with the `importance=TRUE` argument used to request calculating attribute utility estimates. Before using them for attribute selection, let us evaluate the random forest model on the validation subset, as with all the decision tree models created so far.

```
ci.rf.pred <- predict(ci.rf, cirf.val)
err(ci.rf.pred, cirf.val$income)
ci.rf.cm <- confmat(ci.rf.pred, cirf.val$income)

ci.rf.tpr <- tpr(ci.rf.cm)
ci.rf.fpr <- fpr(ci.rf.cm)

ci.rf.prob <- predict(ci.rf, cirf.val, type="p")[,2]
ci.rf.roc <- roc(ci.rf.prob, cirf.val$income)
plot(ci.rf.roc$fpr, ci.rf.roc$tpr, type="l", xlab="FP rate", ylab="TP rate")
points(ci.rf.fpr, ci.rf.tpr, pch=8)
auc(ci.rf.roc)
```

The ROC curve is presented in Figure 20.14. While the default operating point, marked by an asterisk, is rather unimpressive, the curve indicates a very good predictive performance potential, clearly superior to even the best decision tree created so far. The area under the ROC curve exceeds 0.9 and an operating point with a true positive rate above 0.9 is possible with no more than 0.2 false positive rate. The following code identifies the correspondingly shifted operating point and adds it to the plot presented in Figure 20.15.

```
ci.rf.cut09 <- ci.rf.roc$cutoff[ci.rf.roc$tpr>0.9]
ci.rf.cm09 <- confmat(cutclass(ci.rf.prob, ci.rf.cut09[1], ci.labels), ci.val$income)
```

```
ci.rf.tpr09 <- tpr(ci.rf.cm09)
ci.rf.fpr09 <- fpr(ci.rf.cm09)
points(ci.rf.fpr09, ci.rf.tpr09, pch=1)
```

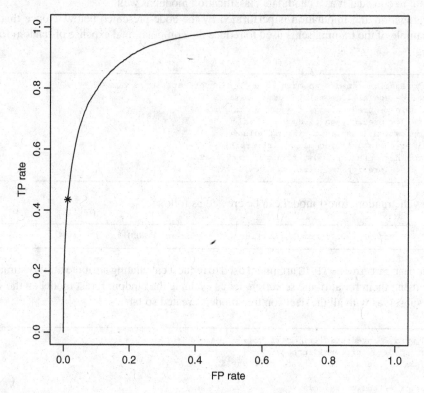

Figure 20.14   The ROC curve of the random forest model for the *Census Income* data.

It is worthwhile to notice that the random forest model outperforms the previous decision tree models despite using a considerably smaller training sample and a single default parameter setup. The capability to deliver good models without extensive parameter tuning is definitely an important practical advantage. What may be disappointing is the default operating point, but we have seen that shifting it at the time of prediction is straightforward. This could become unnecessary by controlling the class balance of bootstrap samples using the `sampsize` parameter, which usually gives better results than specifying class weights via the `classwt` parameter. The reader may want to explore both these possibilities.

Finally getting to attribute selection, the following R code produces a plot of attribute utility estimates and, somewhat arbitrarily, takes top 10%, 25%, and 50% of them as candidate attribute subsets, adding also the full attribute subset for easy direct comparison. The plot is presented in Figure 20.16. It is certainly reassuring to see `education` at the top of the attribute utility ranking for income prediction.

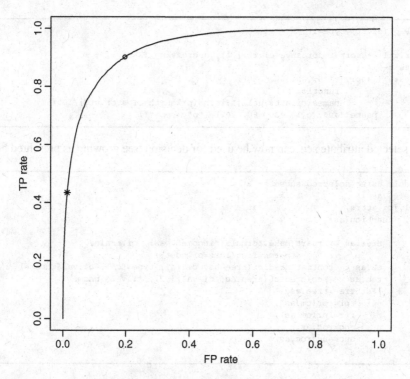

Figure 20.15   The ROC curve of the random forest model for the *Census Income* data with the operating point shifted to reach a TP rate above 0.9.

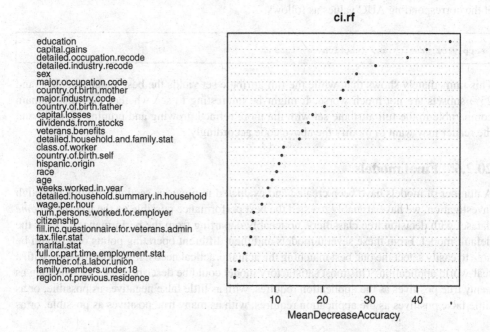

Figure 20.16   The random forest attribute utilities for the *Census Income* data.

```
varImpPlot(ci.rf, type=1)

ci.attr.utl <- sort(importance(ci.rf)[,1], decreasing=TRUE)
ci.asets <-
  'names<-'(lapply(c(10, 25, 50, 100),
                   function(p)
                   names(ci.attr.utl)[1:round(p*length(ci.attr.utl)/100)]),
            paste("as", c(10, 25, 50, 100), "p", sep=""))
```

The selected attribute sets can now be used for decision tree growing as presented below:

```
# models using selected subsets
ci.tree.c5.as <-
  lapply(ci.attrs,
         function(as)
         {
           tree.as <- rpart(make.formula("income", as), ci.train,
                            parms=list(loss=ci.cost5))
           cm.as <- confmat(predict(tree.as, ci.val, type="c"), ci.val$income)
           roc.as <- roc(predict(tree.as, ci.val)[,2], ci.val$income)
           list(tree=tree.as,
                tpr=tpr(cm.as),
                fpr=fpr(cm.as),
                roc=roc.as,
                auc=auc(roc.as))
         })
```

This creates and evaluates 5:1 cost-sensitive decision trees (with default stop criteria and no pruning) using each selected attribute subset, with the full set of attributes also included for easier comparison. To quickly assess the predictive power of the obtained trees one can look at the corresponding AUC values as follows:

```
sapply(ci.tree.c5.as, function(ta) ta$auc)
```

This immediately shows that, while the full attribute set yields the best model, the 50% and 25% subsets are not much worse. It might be interesting to see whether they also remain comparable to the full attribute set with the more refined growing and pruning scenario and the reader may want to modify the above code accordingly.

20.2.6   Final models

A number of models have been created and evaluated in previous sections. By this thorough investigation, we have arguably reached the best performance possible on the *Census Income* dataset with decision tree classifiers, noticeably departing from the performance level of the default model. From these several models and their different operating points one should be able to easily select the one best matching the actual practical needs, if we were dealing with a real-world application. This most satisfactory model could be described, e.g., as achieving as many true positives as the application requires, with as little false negatives as possible, or as little false positives as the application requires, with as many true positives as possible, or as maximizing some combined performance metric (such as the *F*-measure), or as minimizing the underlying application-specific misclassification costs, etc.

Once the most promising models have been identified using the validation set, the same modeling procedures could be applied to the combined training and validation sets to produce the final models ready for deployment. However, since the best models created in this section use shifted operating points determined based on their validation set performance, the validation set has already been used within the modeling procedure. Therefore, we will accept the pruned cost-sensitive decision tree and the random forest, with operating points shifted appropriately using the validation set, as two potential final models and evaluate their performance on the test set, which contains instances not used for any purpose so far. The reason why the operating point adjustment is not performed using the test set – which would make it possible to create models on the combined training and validation sets – is that it is used in this study to simulate the "new data" to which any final prediction model is eventually applied. It can therefore be used to see whether the performance expectations based on the previous validation set evaluation are confirmed, but not for model adjustments. In real application conditions, the true class labels for new data would be clearly unknown at the time of prediction, making any such adjustments impossible.

The test set evaluation of the pruned decision tree using the 5:1 misclassification cost matrix is performed by the following code, which identifies the default operating point, the ROC curve, and the shifted operating point. The latter uses the probability cutoff value previously determined using the validation set.

```
 # default operating point
ci.tree.c5p1sd.test.pred <- predict(ci.tree.c5p1sd, census.test, type="c")
ci.tree.c5p1sd.test.cm <- confmat(ci.tree.c5p1sd.test.pred, census.test$income)
ci.tree.c5p1sd.test.tpr <- tpr(ci.tree.c5p1sd.test.cm)
ci.tree.c5p1sd.test.fpr <- fpr(ci.tree.c5p1sd.test.cm)
 # ROC
ci.tree.c5p1sd.test.prob <- predict(ci.tree.c5p1sd, census.test)[,2]
ci.tree.c5p1sd.test.roc <- roc(ci.tree.c5p1sd.test.prob, census.test$income)
plot(ci.tree.c5p1sd.test.roc$fpr, ci.tree.c5p1sd.test.roc$tpr, type="l",
     xlab="FP rate", ylab="TP rate", main="Decision tree")
points(ci.tree.c5p1sd.test.fpr, ci.tree.c5p1sd.test.tpr, pch=8)
auc(ci.tree.c5p1sd.roc)
 # operating point shifted based on the validation set
ci.tree.c5p1sd.test.cm085 <- confmat(cutclass(ci.tree.c5p1sd.test.prob,
                                       ci.tree.c5p1sd.cut085[1], ci.labels),
                                 census.test$income)
ci.tree.c5p1sd.test.tpr085 <- tpr(ci.tree.c5p1sd.test.cm085)
ci.tree.c5p1sd.test.fpr085 <- fpr(ci.tree.c5p1sd.test.cm085)
points(ci.tree.c5p1sd.test.fpr085, ci.tree.c5p1sd.test.tpr085, pch=1)
```

The same is repeated below for the random forest model. Notice that this requires preprocessing the test set by aggregation and imputation, as before for random forest model creation. This is performed by applying the aggregation and imputation models previously created on the training set.

```
 # test set preprocessing
cirf.test <- predict.agg(ci.aggm, census.test)
cirf.test <- predict.imp(cirf.impm, cirf.test)
```

```
    # default operating point
ci.rf.test.pred <- predict(ci.rf, cirf.test)
ci.rf.test.cm <- confmat(ci.rf.test.pred, cirf.test$income)
ci.rf.test.tpr <- tpr(ci.rf.test.cm)
ci.rf.test.fpr <- fpr(ci.rf.test.cm)
    # ROC
ci.rf.test.prob <- predict(ci.rf, cirf.test, type="p")[,2]
ci.rf.test.roc <- roc(ci.rf.test.prob, cirf.test$income)
plot(ci.rf.test.roc$fpr, ci.rf.test.roc$tpr, type="l",
     xlab="FP rate", ylab="TP rate", main="Random forest")
points(ci.rf.test.fpr, ci.rf.test.tpr, pch=8)
auc(ci.rf.test.roc)
    # operating point shifted based on the validation set
ci.rf.test.cm09 <- confmat(cutclass(ci.rf.test.prob, ci.rf.cut09[1], ci.labels),
                           cirf.test$income)
ci.rf.test.tpr09 <- tpr(ci.rf.test.cm09)
ci.rf.test.fpr09 <- fpr(ci.rf.test.cm09)
points(ci.rf.test.fpr09, ci.rf.test.tpr09, pch=1)
```

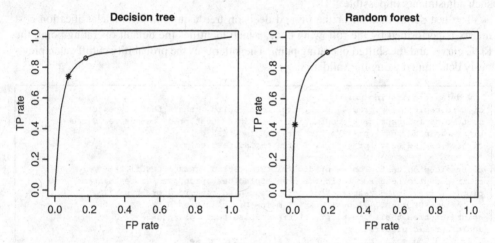

Figure 20.17 The test set ROC curves of the 1-SD pruned cost-sensitive decision tree using the 5:1 misclassification cost matrix and the random forest for the *Census Income* data.

The obtained test set ROC curves, with the default operating points marked by asterisks and the shifted operating points marked by circles, are presented in Figure 20.17. It is easy to see that the test set performance of the two models is nearly the same as previously observed on the validation set. If we were deploying these models for a real application, we would not therefore be disappointed by the results. While the random forest is noticeably better, the interpretability of the decision tree remains its important advantage that in some applications could outweigh the performance difference. The tree structure can be verified to be indeed quite comprehensible and makes it possibly to explain the predictions quite well, although slightly too large to fit on a book page. The following code generates a plot of a top part of the tree presented in Figure 20.18, obtained by more aggressive pruning:

```
prp(prune(ci.tree.c5p1sd, 0.01), varlen=8, faclen=2)
```

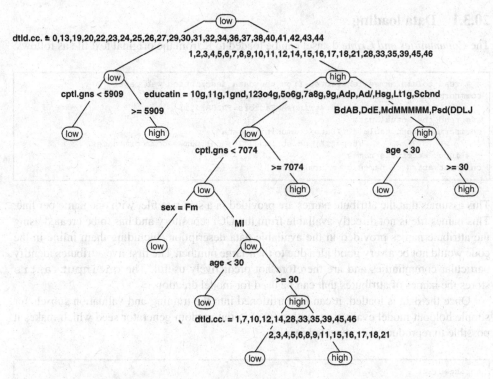

Figure 20.18 The 1-SD pruned cost-sensitive decision tree using the 5:1 misclassification cost matrix for the *Census Income* data.

While the comprehensibility of the tree is certainly limited due to the large number of values of the discrete attributes used for splitting (and abbreviations necessary to put attribute name and value labels in the plot), it can be definitely inspected and understood, using an attribute value dictionary.

20.3 Communities and crime

The primary objective of this study is to predict the number of violent crimes (per population) in US communities based on attributes describing their sociodemographic and economic profiles as well as local law enforcement agencies. Regression trees and linear models, combined with random forest attribute selection, will be used for modeling. The set of packages used by this case study is loaded by the following code:

```
library(dmr.regeval)
library(dmr.util)
library(dmr.trans)

library(rpart)
library(rpart.plot)
library(randomForest)
```

20.3.1   Data loading

The *Communities and Crime* dataset can be loaded to R from the original text file as follows:

```
  # read column names (extracted from the data description web page)
commnorm.names <- read.table("../Data/commnorm.names",
                             stringsAsFactors=FALSE)[,1]
  # read the actual data
commnorm <- read.table("../Data/communities.data",
                       sep=",", na.strings="?", col.names=commnorm.names)
  # input attribute names
cn.input.attrs <- names(commnorm)[6:127]
```

This assumes that the attribute names are provided in a separate file, with one name per line. This names file is not directly available from the UCI repository and has to be created using the attribute names provided in the available data description. Including them inline in the code would not be a very good idea due to their large number. The first five attributes identify particular communities and are therefore not predictively useful. The `cn.input.attrs` stores the names of attributes that can be used for model creation.

Once the data is loaded, it can be partitioned into the training and validation subsets for simple holdout model evaluation, with a fixed initial random generator seed which makes it possible to reproduce the presented results:

```
set.seed(12)

rcn <- runif(nrow(commnorm))
cn.train <- commnorm[rcn>=0.33,]
cn.val <- commnorm[rcn<0.33,]
```

Unlike in the previous study, there is no designated separate test file. Since the number of available instances is small, particularly compared to the number of attributes, such a test set for final model evaluation will not be created. The validation set will be used to evaluate a number of candidate models and identify the most promising ones, but their expected true performance on new data will not be estimated.

20.3.2   Data quality

The *Crime and Communities* data can be easily seen to have some quality issues. In particular, it turns out to suffer from extremely severe attribute value missingness. Only about 5–6% of instances in the training and validation sets have no missing values:

```
sum(complete.cases(cn.train))/nrow(cn.train)
sum(complete.cases(cn.val))/nrow(cn.val)
```

The below code snippet identifies attributes for which more values are missing than available.

```
   # attributes with many (>50%) missing values
cn.input.attrs.miss <-
   names(which(sapply(cn.input.attrs,
                      function(a)
                      sum(is.na(cn.train[a]))/nrow(cn.train))>0.5))
```

There are 22 such attributes, mostly describing local law enforcement agencies. Similarly, attributes that appear to have many outlying values can be identified as follows:

```
   # attributes with many (>10%) outliers
cn.input.attrs.out <-
   names(which(sapply(cn.input.attrs,
                      function(a)
                      length(boxplot(cn.train[a], range=2, plot=FALSE)$out)/
                      nrow(cn.train))>0.1))
```

There are five attributes with more than 10% outlying values, based on the standard quartile-based criterion described in Section 2.4.1 (with the inter-quartile range multiplier set to 2). They may be rather indicative of high diversity of US communities than of data corruption, though.

It is also easy to see that some attribute pairs are very strongly correlated:

```
cn.input.attrs.cor <- cor(cn.train[,cn.input.attrs], use="pairwise.complete.obs")
cn.input.attrs.corind <- which(upper.tri(cn.input.attrs.cor) &
                               abs(cn.input.attrs.cor)>0.98, arr.ind=TRUE)
cn.input.attrs.corpairs <- data.frame(a1=cn.input.attrs[cn.input.attrs.corind[,1]],
                               a2=cn.input.attrs[cn.input.attrs.corind[,2]])
```

This identifies the following 14 attribute pairs exhibiting near-perfect correlation:

	a1	a2
1	population	numbUrban
2	FemalePctDiv	TotalPctDiv
3	PctFam2Par	PctKids2Par
4	PctRecentImmig	PctRecImmig5
5	PctRecImmig5	PctRecImmig8
6	PctRecImmig5	PctRecImmig10
7	PctRecImmig8	PctRecImmig10
8	PctLargHouseFam	PctLargHouseOccup
9	PctPersOwnOccup	PctHousOwnOcc
10	OwnOccLowQuart	OwnOccMedVal
11	OwnOccMedVal	OwnOccHiQuart
12	RentMedian	RentHighQ
13	RentMedian	MedRent
14	LemasSwFTPerPop	PolicPerPop

While this is not a quality issue, strictly speaking, it at least suggests that not all attributes may be actually predictively useful. It would not be unreasonable to consider removing some attributes of particularly questionable quality, clipping outlying values, or leaving only single representatives of each strongly correlated attribute pair. The reader may want to explore these possibilities. In this study, we will address the utility of attributes by random forest attribute selection later and for now leave all attributes in place. The only attribute transformation that is applied below is the one actually required by some of the regression algorithms to be used: missing value imputation. Similarly as in the previous study and as presented in Section 17.3.4, it is performed as a modeling transformation, with the imputation model determined on the training set applied both to the training and validation sets (and potentially applicable to any new data, should such be available).

```
cn.impm <- imp.all(make.formula(NULL, cn.input.attrs), cn.train)
cni.train <- predict.imp(cn.impm, cn.train)
cni.val <- predict.imp(cn.impm, cn.val)
```

It is worthwhile to mention that attribute values in the dataset used for this study, coming from Census, LEMAS, and crime statistics, are already normalized and partially filtered. The original unnormalized dataset, also available from the UCI repository, suffers from even more severe quality issues.

20.3.3   Regression trees

Our modeling attempts start from regression trees, created using the `rpart` function. Since it handles missing values using surrogate splits, and with many attributes this technique can be expected to be quite successful, the training and validation sets are used in their original form, without imputation applied. The reader may find it worthwhile to verify whether and how using the data processed by imputation changes the achieved predictive performance.

The following code creates a regression tree by calling `rpart` with default parameters:

```
cn.tree.d <- rpart(make.formula("ViolentCrimesPerPop", cn.input.attrs), cn.train)
r2(predict(cn.tree.d, cn.val), cn.val$ViolentCrimesPerPop)
```

and then evaluates its validation set performance by calculating the coefficient of determination (R^2) using the `r2` function. It relates the mean square error to the target function variance, making it easier to interpret. The obtained value of about `dmr.regeval` 0.54 indicates some quite limited, but potentially useful predictive power. In the remainder of this section, we will be seeking for possible improvements.

One obvious direction to consider is regression tree pruning. It can be performed in the same way as demonstrated for decision trees in the *Census Income* study, by creating a fully grown tree first and then identifying the most promising complexity cutoff values from the cost-complexity table. This is performed below using two complexity parameter selection rules: the minimum cross-validated error and the minimum tree size within one standard deviation from the minimum cross-validated error.

```
  # fully grown tree
cn.tree.f <- rpart(make.formula("ViolentCrimesPerPop", cn.input.attrs), cn.train,
                  minsplit=2, cp=0)
r2(predict(cn.tree.f, cn.val), cn.val$ViolentCrimesPerPop)
  # minimum-error cost-complexity pruning
cn.tree.pmin <- prune(cn.tree.f, cpmin(cn.tree.f$cptable))
r2(predict(cn.tree.pmin, cn.val), cn.val$ViolentCrimesPerPop)
  # 1-sd cost-complexity pruning
cn.tree.p1sd <- prune(cn.tree.f, cp1sd(cn.tree.f$cptable))
r2(predict(cn.tree.p1sd, cn.val), cn.val$ViolentCrimesPerPop)
```

Not surprisingly, the fully grown tree turns out to yield poor predictions with an R^2 value below 0.3, but – rather surprisingly – none of the pruned trees outperforms the previously created pre-pruned tree (grown with default stop criteria). The pruned trees are both smaller than the latter, and the larger of them is better than the smaller one, which suggests that cost-complexity pruning may have been too aggressive here due to the variance of the cross-validation performed by `rpart` on the relatively small dataset. On the other hand, there is also no reason to overly trust the validation set performance, either.

Tuning the complexity parameter using a reliable external evaluation procedure, such as $n \times k$-fold cross-validation for some $n > 1$, may help us make a more informed decision about the right pruning level. This is performed by the R code presented below, using the `crossval` function. A sequence of candidate `cp` values around the minimum-error value is preselected and then each of them is evaluated by 10×10 cross-validation, which is a low-variance and low-bias evaluation

> Ex. 7.3.2
> dmr.claseval

procedure. The value maximizing the coefficient of determination is then applied to prune the full-grown tree, and the resulting pruned tree is evaluated on the validation subset. Notice the `xval=0` argument specified for `rpart`. It switches off its internal cross validation for computational savings.

```
# 10x10-fold cross-validated R2 values for the most promising cp sequence
cn.cp.cv <-
  sapply(unname(cpminrange(cn.tree.f$cptable, 5, 10)),
         function(cp)
         {
           cv <- crossval(rpart, make.formula("ViolentCrimesPerPop", cn.input.attrs),
                          cn.train, args=list(cp=cp, minsplit=2, xval=0), n=10)
           `names<-`(r2(cv$pred, cv$true), cp)
         })
cn.tree.pcv <- prune(cn.tree.f, as.numeric(names(cn.cp.cv)[which.max(cn.cp.cv)]))
r2(predict(cn.tree.pcv, cn.val), cn.val$ViolentCrimesPerPop)
```

Several cost-complexity values appear to yield very similar performance, making the final choice of the best `cp` level somewhat arbitrary. The obtained tree turns out to be identical to the default pre-pruned one and slightly better than the previously pruned trees. We now have a stronger reason to believe it might be also good on new data.

20.3.4 Linear models

Having had limited success with regression tree models, let us now see how good predictions can be obtained using linear regression. With the extreme level of missingness in the *Communities and Crime* data that we have observed before and no internal missing value handling capabilities of least-squares linear regression (other than skipping incomplete instances), it is necessary to used transformed datasets for model creation and evaluation, with missing value imputed:

```
cn.lm <- lm(make.formula("ViolentCrimesPerPop", cn.input.attrs), cni.train)
r2(predict(cn.lm, cni.val), cni.val$ViolentCrimesPerPop)
```

The obtained linear model with the coefficient of determination approaching 0.59 outperforms the best regression trees created before.

With so many attributes it makes sense to examine their statistical significance levels reported by the summary function. The following code identifies a subset attributes with *p*-values below 0.05 and repeats model creation using these significant attributes only.

```
signif.attrs <- cn.input.attrs[(summary(cn.lm)$coefficients)[-1,4]<0.05]
cn.lm.s <- lm(make.formula("ViolentCrimesPerPop", signif.attrs), cni.train)
r2(predict(cn.lm.s, cni.val), cni.val$ViolentCrimesPerPop)
```

The simple attribute selection of the subset of significant attributes turned out to reduce rather than improve model quality. More refined attribute selection will be attempted in the next subsection.

20.3.5   Attribute selection

As in the previous study, the random forest algorithm will be employed to produce attribute utility estimations. Similarly as linear regression, the random forest algorithm (in the version implemented by the randomForest package) does not handle missing values and therefore we use transformed data with missing values imputed for model creation and evaluation:

```
cn.rf <- randomForest(make.formula("ViolentCrimesPerPop", cn.input.attrs), cni.train,
                      importance=TRUE)
r2(predict(cn.rf, cni.val[,cn.input.attrs]), cni.val$ViolentCrimesPerPop)
```

The predictive performance of the random forest model is clearly superior to that of both regression trees and linear models. It achieves an R^2 value of nearly 0.65 and this is likely near the top performance level possible for the *Communities and Crime* data, at least with the original set of attributes.

The following code produces the attribute utility plot, presented in Figure 20.19, and picks three subsets, consisting of 10%, 25%, and 50% most useful attributes, as well as the full 100% subset for easy comparison:

```
varImpPlot(cn.rf, type=1)

cn.attr.utl <- sort(importance(cn.rf)[,1], decreasing=TRUE)
cn.asets <-
  'names<-'(lapply(c(10, 25, 50, 100),
                   function(p)
                   names(cn.attr.utl)[1:round(p*length(cn.attr.utl)/100)]),
            paste("as", c(10, 25, 50, 100), "p", sep=""))
```

With all attributes being continuous, it is also particularly straightforward to apply a simple statistical filter using the linear or rank correlation. This is demonstrated below for the rank correlation, and the correspondingly sized subsets of top correlated attributes are identified and appended to the list of considered attribute subsets. The reader may find it interesting to examine how much the random forest and correlation-based rankings of attribute utility differ.

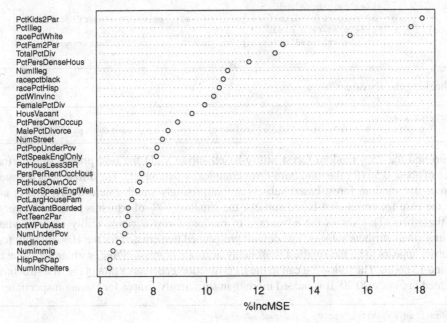

Figure 20.19   The random forest attribute utilities for the *Communities and Crime* data.

```
cn.attr.cor <- sort(abs(cor(cn.train[,cn.input.attrs], cn.train$ViolentCrimesPerPop,
                        method="spearman", use="pairwise.complete.obs")[,1]),
               decreasing=TRUE)
cn.asets <- c(cn.asets,
            'names<-'(lapply(c(10, 25, 50, 100),
                    function(p)
                    names(cn.attr.cor)[1:round(p*length(cn.attr.cor)/100)]),
               paste("as", c(10, 25, 50, 100), "p.cor", sep="")))
```

The following code uses each of candidate attribute subsets for regression tree growing. Both default pre-pruned and pruned trees are created and evaluated, but with no time-consuming cost-complexity tuning demonstrated before. Instead, the minimum-error and one-standard-deviation rules are applied as cp selection heuristics.

```
cn.tree.as <-
  lapply(cn.asets,
      function(as)
      {
          tree.d <- rpart(make.formula("ViolentCrimesPerPop", as), cn.train)
          tree.f <- rpart(make.formula("ViolentCrimesPerPop", as), cn.train,
                      minsplit=2, cp=0)
          tree.pmin <- prune(tree.f, cpmin(tree.f$cptable))
          tree.p1sd <- prune(tree.f, cp1sd(tree.f$cptable))
          list(tree.d=tree.d,
              r2.d=r2(predict(tree.d, cn.val), cn.val$ViolentCrimesPerPop),
```

```
                   tree.pmin=tree.pmin,
                   r2.pmin=r2(predict(tree.pmin, cn.val), cn.val$ViolentCrimesPerPop),
                   tree.p1sd=tree.p1sd,
                   r2.p1sd=r2(predict(tree.p1sd, cn.val), cn.val$ViolentCrimesPerPop))
        })
```

By checking the R^2 values:

```
sapply(cn.tree.as,
       function(ta) c(r2.d=ta$r2.d, r2.pmin=ta$r2.pmin, r2.p1sd=ta$r2.p1sd))
```

we can find the 25% random forest utility-based subset yield the best pruned tree performance of about 0.56 (for the minimum-error complexity parameter). The differences between
different-size random forest-based subsets are surprisingly small. Interestingly, this is not
quite the case for correlation-based subsets, the smallest 10% of which is noticeably worse
than the others. This shows the advantage of random forest attribute utility ranking when
selecting small attribute subsets. The default pre-pruned tree created for the 10% random forest subset appears to be the smallest reasonably accurate regression tree, with an R^2 value of
0.55 and 13 nodes. The following call to the prp function generates a plot of the tree structure,
presented in Figure 20.20. It is indeed more than sufficiently simple for human inspection.

```
prp(cn.tree.as$as10p$tree.d, varlen=0, faclen=0)
```

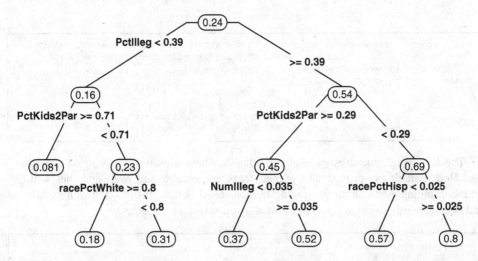

Figure 20.20   The default regression tree for the *Communities and Crime* growing using top
10% attributes.

We can similarly proceed with creating and evaluating linear models for different attribute
subsets:

```
cn.lm.as <-
  lapply(cn.asets,
         function(as)
```

```
            {
                lmod <- lm(make.formula("ViolentCrimesPerPop", as), cni.train)
                list(lm=lmod,
                     r2=r2(predict(lmod, cni.val), cni.val$ViolentCrimesPerPop))
            })

sapply(cn.lm.as, function(ta) ta$r2)
```

The smallest attribute subset selected based on the random forest utility turns out to work best, with an R^2 value of about 0.63 – not far behind the performance level of the random forest model. With so few attributes the model is quite simple and easy to inspect. Its parameter vector is presented below:

```
Coefficients:
      (Intercept)        PctKids2Par           PctIlleg         racePctWhite
          0.18989           -0.33895            0.17937             -0.02939
       PctFam2Par          TotalPctDiv   PctPersDenseHous            NumIlleg
          0.06962            0.33632            0.24192             -0.09450
      racepctblack         racePctHisp          pctWInvInc        FemalePctDiv
          0.16736            0.02117            0.05614             -0.19992
       HousVacant
          0.29761
```

While some more refined nonlinear algorithms, such as support vector regression with a nonlinear kernel, might lead to better performance, this is most probably the simplest good model possible for the *Communities and Crime* data. The reader is encouraged to experiment with other algorithms to examine whether and how much performance advantage they can offer in this case. Another possibility that is worth consideration, but falls beyond the scope of this book, is performing dimensionality reduction of the original attribute set (e.g., via the principal component analysis) instead of attribute selection.

20.3.6   Piecewise-linear models

This section will demonstrate that, with little effort, one can combine regression trees and linear regression to create piecewise linear models. The `lmrpart` function used for this purpose was already defined above and now it remains to review its definition and verify its effects. It performs three basic operations:

- creating a regression tree that serves as the model tree structure by calling `rpart`,

- determining unique predicted target values that serve as leaf identifiers,

- creating a linear model for each leaf by calling `lm`, with the training set restricted to instances from the corresponding leaf and, optionally, the set of attributes restricted to those not used for splitting in the tree structure.

In our case, the function will be called using the top 10% attributes according to the random forest utility measure as follows:

```
cn.mtree <- lmrpart(make.formula("ViolentCrimesPerPop", cn.asets[["as10p"]]),
                    cn.train, cp=0.02, skip.attr=TRUE))
```

where the `cp=0.02` argument is passed to `rpart` to restrict tree complexity. Attributes used for splitting in the tree will be skipped when creating linear models.

The prediction method, also defined above, basically assigns instances to leaves and then applies the corresponding linear models to generate predictions. Their quality can then be evaluated on the validation set:

```
r2(predict(cn.mtree, cni.val), cni.val$ViolentCrimesPerPop)
```

The obtained R^2 value of nearly 0.64 is considerably better than that of the best regression tree created before, but only slightly better than of the best single single linear model. While this may appear disappointing, it is nearly the same performance level as that of the random forest model. Adjusting the complexity of the tree structure or the attribute sets used for tree growing and linear model fitting might lead to some more improvement. This path is not further followed here, since our intention was to demonstrate a very simple automated technique of model tree creation in R than to fully explore its predictive potential.

20.4   Cover type

This study addresses the task of predicting forest cover type based on cartographic attributes. With seven cover types to distinguish, this will provide an opportunity to experience typical challenges encountered in multiclass classification. The study also demonstrates clustering. The R packages used by this case study can be loaded as follows:

```
library(rpart)
library(rpart.plot)
library(randomForest)
library(cluster)

library(dmr.claseval)
library(dmr.cluseval)
library(dmr.trans)
library(dmr.util)
```

20.4.1   Data loading and preprocessing

The *Cover Type* data is loaded to R by the following call to the `read.table` function:

```
covtype <- read.table("../Data/covtype.data", sep=",",
                 col.names=c("Elevation",
                     "Aspect",
                     "Slope",
                     "Horizontal.Distance.To.Hydrology",
                     "Vertical.Distance.To.Hydrology",
                     "Horizontal.Distance.To.Roadways",
                     "Hillshade.9am",
                     "Hillshade.Noon",
                     "Hillshade.3pm",
                     "Horizontal.Distance.To.Fire.Points",
                     paste("Wilderness.Area", 1:4, sep=""),
```

```
                        paste("Soil.Type", 1:40, sep=""),
                        "Cover.Type"))

ct.input.attrs <- setdiff(names(covtype), "Cover.Type")
```

The specified attribute names are based on the data description available in the UCI repository. Notice that there are 4 `Wilderness.Area` attributes and 40 `Soil.Type` attributes,
which – with 10 other attributes – makes the total number of input attributes equal 54. Their
names are stored in the `ct.input.attrs`. The target `Cover.Type` attribute is represented by integer numbers, but has to be treated as a nominal attribute. This is ensured by
converting to a factor:

```
covtype$Cover.Type <- as.factor(covtype$Cover.Type)
```

The original numerical representation of all input attributes is left unchanged, although the
`Wilderness.Area` and `Soil.Type` attributes are actually binary-valued and could be
also converted to factors. This would make no effect for the decision tree and random forest
algorithms used for classification, though.

With more than 580,000 instances, the dataset is sufficiently large to cause some performance issues on slow-CPU or low-memory machines. A half of it will be used as the training
set, with the remaining half equally split into the validation set for intermediate model evaluation and the test set for the evaluation of final most promising models.

```
set.seed(12)

rct <- runif(nrow(covtype))
ct.train <- covtype[rct>=0.5,]
ct.val <- covtype[rct>=0.25 & rct<0.5,]
ct.test <- covtype[rct<0.25,]
```

20.4.2   Class imbalance

It is common for multiclass classification tasks to have unbalanced classes, which is usually
associated with substantially different predictability. This may be the case for the *Cover Type*
data, for which the occurrence counts of particular classes vastly differ. This can be verified
as follows:

```
table(ct.train$Cover.Type)/nrow(ct.train)
```

The first two cover types dominate, accounting for 85% of the training set, and cover type 4
is the least frequent one, with less than 0.5%.

20.4.3   Decision trees

The following code creates the default-setup pre-pruned decision tree and evaluates its validation set predictions. The `xval=0` argument turns off the internal cross-validation to reduce

computation time. It is only needed to calculate error levels for different complexity parameter values used for pruning.

```
ct.tree.d <- rpart(Cover.Type~., ct.train, xval=0)
ct.tree.d.pred <- predict(ct.tree.d, ct.val, type="c")
err(ct.tree.d.pred, ct.val$Cover.Type)
confmat(ct.tree.d.pred, ct.val$Cover.Type)
```

The obtained error level is rather unimpressive and inspecting the confusion matrix immediately reveals that only three out of seven cover types are detected by the model:

```
         true
predicted     1       2      3    4     5     6     7
       1  38212   16958      0    0     0     0  5115
       2  14703   52912   2761    1  2420  1345    14
       3      1     796   6171  686    33  3031     0
       4      0       0      0    0     0     0     0
       5      0       0      0    0     0     0     0
       6      0       0      0    0     0     0     0
       7      0       0      0    0     0     0     0
```

Per-class performance indicators are calculated according to the 1-vs-rest approach presented in Section 7.2.4 using the `confmat01` function as follows:

> Ex. 7.2.8
> `dmr.claseval`

```
ct.tree.d.cm01 <- confmat01(ct.tree.d.pred, ct.val$Cover.Type)
ct.tree.d.tpfp <- sapply(ct.tree.d.cm01,
                  function(cm) c(tpr=tpr(cm), fpr=fpr(cm), fm=f.measure(cm)))
```

which yields the following true positive rate, false positive rate, and *F*-measure values for all the cover types (rounded to two significant digits for greater readability):

```
        1    2     3 4 5 6 7
tpr  0.72 0.75 0.690 0 0 0 0
fpr  0.24 0.29 0.033 0 0 0 0
fm   0.68 0.73 0.630 0 0 0 0
```

Averaging over classes:

```
rowMeans(ct.tree.d.tpfp)
```

we receive an average TP rate of about 0.31 and average *F*-measure of about 0.29. The corresponding averages weighted with class counts:

```
apply(ct.tree.d.tpfp, 1, weighted.mean, table(ct.val$Cover.Type))
```

appear better, since the impact of 0's for the low-frequency classes is minimized.

To see whether pruning can deliver better trees than the default pre-pruned tree created above, we proceed in the usual way by creating a fully grown tree:

```
ct.tree.f <- rpart(Cover.Type~., ct.train, minsplit=2, cp=0)
```

and then selecting appropriate complexity parameter values for cost-complexity pruning. Before that, however, let us evaluate the fully grown tree:

```
ct.tree.f.pred <- predict(ct.tree.f, ct.val, type="c")
ct.tree.f.cm01 <- confmat01(ct.tree.f.pred, ct.val$Cover.Type)
ct.tree.f.tpfp <- sapply(ct.tree.f.cm01,
                    function(cm) c(tpr=tpr(cm), fpr=fpr(cm), fm=f.measure(cm)))
rowMeans(ct.tree.f.tpfp)
```

One can see a huge improvement in all quality indicators of interest. In particular, a much more acceptable average true positive rate of about 0.87 is obtained, with a false positive rate of less than 0.02. The detailed per-class indicators, presented below, after rounding for better readability:

```
        1     2     3      4       5      6      7
tpr 0.920 0.940 0.910 0.79000 0.7700 0.8400 0.9300
fpr 0.045 0.064 0.006 0.00082 0.0036 0.0048 0.0023
fm  0.920 0.930 0.910 0.80000 0.7800 0.8400 0.9300
```

confirm that now all classes are detected with a reasonable accuracy, although clearly cover types 4, 5, and 6 are harder to correctly predict than the others. This is not surprising given their extremely low, 0.5–3% occurrence rate that has been observed before.

Now it remains to see whether pruning can make the performance any better. The following code uses the minimum-error cp value:

```
ct.tree.pmin <- prune(ct.tree.f, cpmin(ct.tree.f$cptable))
ct.tree.pmin.pred <- predict(ct.tree.pmin, ct.val, type="c")
ct.tree.pmin.cm01 <- confmat01(ct.tree.pmin.pred, ct.val$Cover.Type)
ct.tree.pmin.tpfp <- sapply(ct.tree.pmin.cm01,
                    function(cm) c(tpr=tpr(cm), fpr=fpr(cm),
                                   fm=f.measure(cm)))
rowMeans(ct.tree.pmin.tpfp)
```

The per-class performance indicators:

```
        1     2     3      4       5      6      7
tpr 0.920 0.940 0.920 0.78000 0.7600 0.8400 0.9200
fpr 0.043 0.064 0.006 0.00071 0.0031 0.0045 0.0021
fm  0.920 0.940 0.910 0.81000 0.7800 0.8500 0.9300
```

do not substantially differ from those observed for the fully grown tree. Proceeding similarly with the complexity parameter value selected using the one-standard-deviation rule:

```
ct.tree.p1sd <- prune(ct.tree.f, cp1sd(ct.tree.f$cptable))
ct.tree.p1sd.pred <- predict(ct.tree.p1sd, ct.val, type="c")
ct.tree.p1sd.cm01 <- confmat01(ct.tree.p1sd.pred, ct.val$Cover.Type)
ct.tree.p1sd.tpfp <- sapply(ct.tree.p1sd.cm01,
                            function(cm) c(tpr=tpr(cm), fpr=fpr(cm),
                                           fm=f.measure(cm)))
rowMeans(ct.tree.p1sd.tpfp)
```

we observe comparable, but slightly worse performance.

It is interesting to compare the sizes of the default tree, the fully grown tree, and the two pruned trees:

```
c(default=nrow(ct.tree.d$frame), full=nrow(ct.tree.f$frame),
  pruned.min=nrow(ct.tree.pmin$frame), pruned.1sd=nrow(ct.tree.p1sd$frame))
```

The numbers of nodes (including leaves):

```
    default     full pruned.min pruned.1sd
          5    35759      26989      18115
```

speak for themselves. Default stop criteria fail entirely it this case, and much more model complexity is necessary to satisfactorily separate all the seven classes. Full-depth growing gives good results with no signs of overfitting, and pruning only helps to somewhat reduce the tree size.

20.4.4   Class rebalancing

While sufficiently complex trees appear to successfully handle the unbalanced classes of the *Cover Type* data, there is one other self-suggesting approach to increasing the sensitivity of decision trees to less frequent classes. Per-class instance weights can be used to compensate for class frequency differences (similarly as for incorporating misclassification costs). The `rpart` function makes it quite straightforward by its `prior` argument, used as follows:

```
ct.tree.w <- rpart(Cover.Type~., ct.train, xval=0,
                   parms=list(prior=rep(1/nlevels(ct.train$Cover.Type),
                                        nlevels(ct.train$Cover.Type))))
```

which can be easily verified to be exactly equivalent to explicit instance weighting with weights inversely proportional to their class occurrence counts (and is therefore another way of specifying per-class instance weights). Evaluating the obtained model in the usual way:

```
ct.tree.w.pred <- predict(ct.tree.w, ct.val, type="c")
ct.tree.w.cm01 <- confmat01(ct.tree.w.pred, ct.val$Cover.Type)
ct.tree.w.tpfp <- sapply(ct.tree.w.cm01,
                         function(cm) c(tpr=tpr(cm), fpr=fpr(cm), fm=f.measure(cm)))
rowMeans(ct.tree.w.tpfp)
```

we can find that it is indeed quite good at detecting the low-frequency classes, but performs considerably worse for the three dominating classes:

```
        1    2    3     4     5     6     7
tpr  0.42 0.44 0.41 0.860 0.83 0.520 0.95
fpr  0.14 0.15 0.01 0.017 0.17 0.046 0.14
fm   0.51 0.55 0.53 0.310 0.14 0.340 0.33
```

While the class-weighted tree is arguably better than the default tree with no weighting and – with 19 nodes and leaves – much smaller than the pruned trees with no weighting, with so poor dominating class performance it can be hardly considered useful. It makes therefore sense to repeat the pruning process with classes rebalanced by weighting. As usual, this starts from the fully grown tree:

```
ct.tree.w.f <- rpart(Cover.Type~., ct.train, minsplit=2, cp=0,
                     parms=list(prior=rep(1/nlevels(ct.train$Cover.Type),
                                         nlevels(ct.train$Cover.Type))))
```

which can be then evaluated as follows:

```
ct.tree.w.f.pred <- predict(ct.tree.w.f, ct.val, type="c")
ct.tree.w.f.cm01 <- confmat01(ct.tree.w.f.pred, ct.val$Cover.Type)
ct.tree.w.f.tpfp <- sapply(ct.tree.w.f.cm01,
                          function(cm) c(tpr=tpr(cm), fpr=fpr(cm),
                                         fm=f.measure(cm)))
rowMeans(ct.tree.w.f.tpfp)
```

As mentioned earlier, it is the per-class values of the true positive and false positive rate as well as the F-measure that are the most interesting:

```
        1      2      3       4      5      6       7
tpr  0.920  0.920  0.9000  0.76000  0.830  0.8200  0.9200
fpr  0.045  0.064  0.0067  0.00091  0.011  0.0051  0.0028
fm   0.920  0.930  0.9000  0.78000  0.680  0.8200  0.9200
```

They are quite close to those previously observed for the fully grown tree with no class weighting. This is not overly surprising, since for large trees with highly fitted leaves there is actually no space for improvement by introducing instance weights.

It remains to see how much of a change can be obtained by pruning, which should be normally performed as follows:

```
ct.tree.w.p1sd <- prune(ct.tree.w.f, cp1sd(ct.tree.w.f$cptable))
ct.tree.w.pmin <- prune(ct.tree.w.f, cpmin(ct.tree.w.f$cptable))
```

Unfortunately, these calls fail with the rpart version used for preparing this case study due to a bug. As a workaround, we may simply re-grow the trees from scratch using the two complexity parameter values determined above, which is equivalent to pruning (if it worked):

```
ct.tree.w.pmin <- rpart(Cover.Type~., ct.train, xval=0,
                       minsplit=2, cp=cpmin(ct.tree.w.f$cptable),
                       parms=list(prior=rep(1/nlevels(ct.train$Cover.Type),
                                           nlevels(ct.train$Cover.Type))))
```

```
ct.tree.w.plsd <- rpart(Cover.Type~., ct.train, xval=0,
                        minsplit=2, cp=cplsd(ct.tree.w.f$cptable),
                        parms=list(prior=rep(1/nlevels(ct.train$Cover.Type),
                                            nlevels(ct.train$Cover.Type))))
```

The evaluation of the first of these class-weighted pruned trees:

```
ct.tree.w.pmin.pred <- predict(ct.tree.w.pmin, ct.val, type="c")
ct.tree.w.pmin.cm01 <- confmat01(ct.tree.w.pmin.pred, ct.val$Cover.Type)
ct.tree.w.pmin.tpfp <- sapply(ct.tree.w.pmin.cm01,
                        function(cm) c(tpr=tpr(cm), fpr=fpr(cm),
                                       fm=f.measure(cm)))
rowMeans(ct.tree.w.pmin.tpfp)
```

reveals that it performs much better than the class-weighted pre-pruned tree, achieving a TP rate above 0.8 for each class:

```
        1      2      3      4      5      6      7
tpr 0.870  0.850  0.9100  0.8000  0.900  0.870  0.9600
fpr 0.077  0.079  0.0098  0.0013  0.017  0.011  0.0095
fm  0.870  0.880  0.8800  0.7800  0.620  0.790  0.8700
```

In comparison to the fully grown and pruned trees without class weighting, as well as the class-weighted fully grown tree, it is less successful at predicting the two dominating classes, but gives an improvement for most other classes, particularly 4 and 5.

Similarly pruning at the complexity level determined by the one-standard-deviation rule:

```
ct.tree.w.plsd.pred <- predict(ct.tree.w.plsd, ct.val, type="c")
ct.tree.w.plsd.cm01 <- confmat01(ct.tree.w.plsd.pred, ct.val$Cover.Type)
ct.tree.w.plsd.tpfp <- sapply(ct.tree.w.plsd.cm01,
                        function(cm) c(tpr=tpr(cm), fpr=fpr(cm),
                                       fm=f.measure(cm)))
rowMeans(ct.tree.w.plsd.tpfp)
```

may be verified to yield slightly worse average performance and slightly less tree size. Interestingly, the two pruned trees are both considerably smaller than their counterparts created without class weights. This can be verified as follows:

```
c(weighted.pruned.min=nrow(ct.tree.w.pmin$frame),
  weighted.pruned.1sd=nrow(ct.tree.w.plsd$frame))
```

Of all the decision tree models created for the *Cover Type* data those using default stop criteria are clearly useless. The remaining fully grown and pruned trees, either with or without class weighting, can be all considered reasonable and the final choice may depend on the specific requirements of a particular application: the relative importance of correctly predicting for the dominating and rare classes (possibly expressed by misclassification costs) and the level of preference for the reduced tree size (if any). Nevertheless, the class-weighted tree pruned with minimum-error cost-complexity pruning appears to represent a good overall balance of per-class predictive performance and model complexity for most typical requirements.

20.4.5 Multiclass encoding

Although the more complex decision tree models created so far in this study turned out to be quite good multiclass classifiers, the *Cover Type* datasets provides an opportunity to examine the utility of the multiclass encoding technique described in Section 17.4. We will use the most straightforward 1-of-k encoding, in which there is a separate binary scoring model for each class, and the highest score determines the class label at the time of prediction. The corresponding multiclass wrapper around `rpart` can be created using the `multi.enc.1ofk` function as follows:

> Ex. 17.4.2
> dmr.trans

```
rp.1k <- multi.class(rpart, predf=function(...) predict(...)[,2],
                 encode=multi.enc.1ofk, decode=multi.dec.1ofk)
```

It can then be applied to create a set of binary decision trees with default parameter settings:

```
ct.tree.1k <- rp.1k$alg(Cover.Type~., ct.train, xval=0)
```

The resulting model is evaluated by the following code:

```
ct.tree.1k.pred <- rp.1k$predict(ct.tree.1k, ct.val)
ct.tree.1k.cm01 <- confmat01(ct.tree.1k.pred, ct.val$Cover.Type)
ct.tree.1k.tpfp <- sapply(ct.tree.1k.cm01,
                    function(cm) c(tpr=tpr(cm), fpr=fpr(cm), fm=f.measure(cm)))
rowMeans(ct.tree.1k.tpfp)
```

While the observed classification performance is far below the best models created before, it is definitely superior to that delivered by the first decision tree with default parameters. Unlike the latter, which was unable to detect any more than the three most common classes, this one recognizes most of the rare classes (with class 6 as a notable exception):

```
      1    2    3      4       5  6   7
tpr 0.64 0.80 0.850 0.61000 0.13000 0 0.4300
fpr 0.18 0.31 0.031 0.00089 0.00063 0 0.0033
fm  0.66 0.75 0.730 0.68000 0.23000 0 0.5600
```

While this indicates some potential usefulness of the multiclass encoding path for the *Cover Type* data, the overall performance level is still unacceptably poor. The individual trees, grown with default stop criteria, may be too simple to capture the complex patterns needed to distinguish between all classes. Using the following code:

```
sapply(ct.tree.1k$binmodels, function(m) nrow(m$frame))
```

we may find that they have between 5 and 31 nodes, and just above 100 nodes in total (including leaves), which is several times less than was necessary before with directly multiclass trees.

The same directions of search for improvement that were explored for the directly multiclass trees – adjusting tree complexity and rebalancing classes – remain available here. We will focus on the former, but – instead of building per-class fully grown trees and then pruning each of them – we will first try specifying a small complexity parameter value and enabling splitting even small subsets instances:

```
ct.tree.1k.cp <- rp.1k$alg(Cover.Type~., ct.train, xval=0, minsplit=2, cp=1e-5)
```

The particular value of 10^{-5} used in the above call seems to be a good initial guess, as it is about the same order of magnitude as the cp values previously found when pruning the directly multiclass fully grown trees. Evaluating the obtained model:

```
ct.tree.1k.cp.pred <- rp.1k$predict(ct.tree.1k.cp, ct.val)
ct.tree.1k.cp.cm01 <- confmat01(ct.tree.1k.cp.pred, ct.val$Cover.Type)
ct.tree.1k.cp.tpfp <- sapply(ct.tree.1k.cp.cm01,
                             function(cm) c(tpr=tpr(cm), fpr=fpr(cm),
                                            fm=f.measure(cm)))
rowMeans(ct.tree.1k.cp.tpfp)
```

shows unquestionable improvement due to increased tree complexity:

```
tpr 0.920 0.940 0.8900 0.7400 0.7500 0.770 0.9100
fpr 0.044 0.073 0.0055 0.0012 0.0033 0.004 0.0027
fm  0.920 0.930 0.9000 0.7400 0.7700 0.810 0.9200
```

This is in fact quite close to (but slightly worse than) the performance level of the directly multiclass pruned trees (without class rebalancing). It is not unlikely that more carefully adjusted complexity parameter values, possibly different for each class, might make the encoding approach deliver the best model for this study. A simple way to achieve this with the multi.class wrapper generator is to use a single function that performs tree growing and pruning, such as the rpart.pmin function defined above. A multiclass wrapper can then be generated and used similarly as before:

```
rp.1k.pmin <- multi.class(rpart.pmin, predf=function(...) predict(...)[,2],
                          encode=multi.enc.1ofk, decode=multi.dec.1ofk)
ct.tree.1k.pmin <- rp.1k.pmin$alg(Cover.Type~., ct.train)
```

Verifying the validation set performance:

```
ct.tree.1k.pmin.pred <- rp.1k$predict(ct.tree.1k.pmin, ct.val)
ct.tree.1k.pmin.cm01 <- confmat01(ct.tree.1k.pmin.pred, ct.val$Cover.Type)
ct.tree.1k.pmin.tpfp <- sapply(ct.tree.1k.pmin.cm01,
                               function(cm) c(tpr=tpr(cm), fpr=fpr(cm),
                                              fm=f.measure(cm)))
rowMeans(ct.tree.1k.pmin.tpfp)
```

reveals that it is actually very similar to that obtained above with the *ad hoc* chosen cp value:

```
       1      2       3       4       5       6       7
tpr 0.92  0.940  0.9000  0.7800  0.7100  0.8000  0.9200
fpr 0.04  0.064  0.0062  0.0015  0.0044  0.0053  0.0049
fm  0.92  0.940  0.9000  0.7400  0.7200  0.8100  0.8900
```

20.4.6   Final classification models

We arrived at reasonably good cover type prediction models as soon as we departed from the default stop criteria of the rpart function and permitted more complex trees. It turned out to take several thousand nodes to sufficiently well discriminate between the seven different classes. The 1-of-k multiclass encoding technique did not help. The two minimum-error pruned trees, with and without class rebalancing, are the best models we have been able to find, representing somewhat different tradeoffs between the predictive performance for the dominating and rare classes. Now the test set performance of these two can be determined, to see whether and how much it differs from the corresponding estimates obtained on the validation set. In principle, one could re-create the selected models on the combined training and validation test and the reader is encouraged to do so, but for simplicity and computational savings this possibility is not demonstrated here. The same models that were created on the training set and previously evaluated on the validation set are therefore now evaluated on the test set:

```
ct.tree.pmin.test.pred <- predict(ct.tree.pmin, ct.test, type="c")
ct.tree.pmin.test.cm01 <- confmat01(ct.tree.pmin.test.pred, ct.test$Cover.Type)
ct.tree.pmin.test.tpfp <- sapply(ct.tree.pmin.test.cm01,
                      function(cm) c(tpr=tpr(cm), fpr=fpr(cm),
                                     fm=f.measure(cm)))
rowMeans(ct.tree.pmin.test.tpfp)

ct.tree.w.pmin.test.pred <- predict(ct.tree.w.pmin, ct.test, type="c")
ct.tree.w.pmin.test.cm01 <- confmat01(ct.tree.w.pmin.test.pred, ct.test$Cover.Type)
ct.tree.w.pmin.test.tpfp <- sapply(ct.tree.w.pmin.test.cm01,
                      function(cm) c(tpr=tpr(cm), fpr=fpr(cm),
                                     fm=f.measure(cm)))
rowMeans(ct.tree.w.pmin.test.tpfp)
```

The performance is nearly the same as observed on the validation set, and is probably a reliable estimate of the true performance that can be expected on new data.

It is not unlikely that systematic complexity parameter tuning, using a reliable evaluation procedure and a performance measure more appropriate than the misclassification error normally used by cost-complexity pruning, would lead to better results. This would be a computationally demanding process with relatively little performance gain expectations, but even minor increase of predictive power may be worthwhile considerable effort in real-world applications. The complexity parameter tuning process demonstrated in the *Communities and Crime* study provides a pattern to follow for the reader who would like to undertake this challenge. Attribute selection might help find models with better performance or less complexity, as demonstrated in the previous studies. Finally, the more refined multiclass encoding technique with error-correcting codes is another promising – although computationally expensive due to the considerably increased number of binary models – direction to explore.

20.4.7   Clustering

The *Cover Type* data, with its several dozen cartographic attributes, provides an interesting opportunity for terrain clustering. All the attributes can be reasonably considered numeric, which is convenient for dissimilarity calculation and *k*-centers cluster representation. However, the majority of them are actually binary and the remaining ones have very diverse ranges, hugely exceeding the [0, 1] range. This suggests applying the normalization transformation, described in Section 17.3.2, to prevent a small number of large range attributes dominate dissimilarity calculation. It is arguably a better choice here than standardization, since – unlike the latter – it leaves the binary attributes unchanged. The transformation is performed below using the `nrm.all` and `predict.nrm` functions, which create the normalization model on the training set and apply it to all data subsets.

> Ex. 17.3.2
> `dmr.trans`

```
ct.nrmm <- nrm.all(Cover.Type~., ct.train)
ctn.train <- predict.nrm(ct.nrmm, ct.train)
ctn.val <- predict.nrm(ct.nrmm, ct.val)
ctn.test <- predict.nrm(ct.nrmm, ct.test)
```

The *Cover Type* data is relatively large and may cause performance problems for memory- or CPU-intensive clustering algorithms on some machines. The basic *k*-means algorithm from the standard `stats` package should be able to handle the full training set smoothly, but for this study an approximate efficient version of the *k*-medoids algorithm will be adopted, which internally subsamples data, performs partitioning around medoids for each sample, and combines the results appropriately. The algorithm, known as CLARA (*Clustering Large Applications*) is implemented by the `clara` function from the `cluster` package. Its advantages over *k*-means, apart from possibly better robustness, include the immediate availability of training set silhouette width evaluations. This is a highly convenient feature in practice, since – as discussed in Sections 14.2.4 and 14.3.4 – it makes it possible to immediately assess the quality of the obtained clustering models and make a well-informed choice of the number of clusters. The silhouette width values are approximate (based on data subsamples), but calculating exact cluster quality measures for large data may be too computationally expensive to serve this purpose.

The following code creates CLARA models for the sequence of *k* values between 2 and 10, using the normalized training set with input attributes only:

```
ctn.cla <-
  'names<-'(lapply(2:10, function(k)
                      clara(ctn.train[,ct.input.attrs], k,
                            samples=100, sampsize=200, keep.data=FALSE)), 2:10)
```

The `samples` and `sampsize` arguments may be used to adjust the level of tradeoff between the quality of *k*-medoids approximation and the computational expense. Also note the `keep.data=FALSE` argument, used to prevent training set copies from being stored as components of the created `clara` objects.

The average silhouette width for each *k* value can be plotted as follows:

```
plot(2:10, sapply(ctn.cla, function(cm) cm$silinfo$avg.width),
     type="l", xlab="k", ylim=c(0, 0.5))
```

```
lines(2:10, sapply(ctn.cla, function(cm) sd(cm$silinfo$clus.avg.widths)), lty=2)
legend("bottomright", legend=c("average silhouette width",
                               "sd(cluster silhouette widths)"), lty=1:2)
```

The resulting plot, presented in Figure 20.21, also shows the standard deviation of the average per-cluster silhouette width values using a dashed line. The plot suggests 2, 7, or 10 as the best justified numbers of clusters. For $k = 2$ we receive the most uniform quality clusters (a low standard deviation of per-cluster silhouette widths) and for $k = 10$ the best average cluster quality. While $k = 7$ is slightly worse than these two, it is worth considering as a local maximum between the neighbory k values.

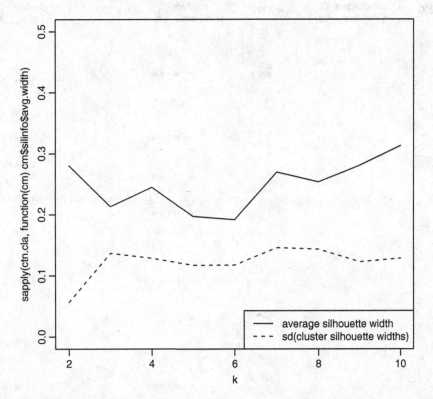

Figure 20.21 The average silhouette width for different k values for the *Cover Type* data.

A more detailed picture of the quality of the two most promising clustering models is provided by their silhouette plots:

```
par(mfrow=c(1, 3))
plot(silhouette(ctn.cla[["2"]]), main="k=2")
plot(silhouette(ctn.cla[["7"]]), main="k=7")
plot(silhouette(ctn.cla[["10"]]), main="k=10")
```

As we can see in Figure 20.22, for $k = 2$ we have indeed two similarly good clusters, whereas for $k = 7$ and $k = 10$ there are two or three clusters considerably worse than the others. All the three clustering models might be useful depending on specific application requirements. The remainder of this study will only use $k = 7$, which provides opportunities for more interesting demonstrations than the basic case of $k = 2$, but is simpler than $k = 10$.

Since there is no prediction method provided in the cluster package, cluster membership for nontraining instances has to be determined by calculating their dissimilarities to cluster medoids and choosing the least dissimilar medoid. This can be performed using the

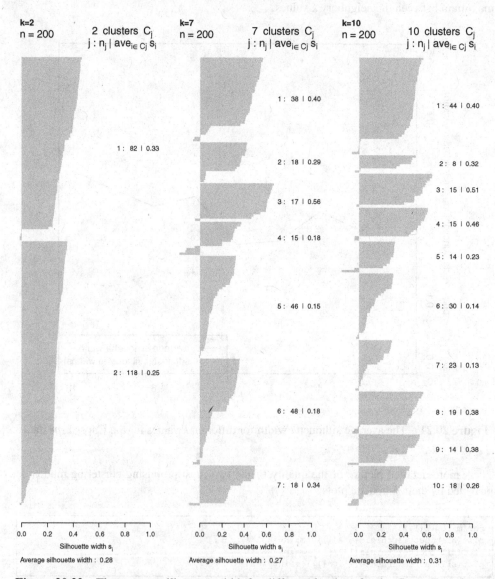

Figure 20.22 The average silhouette width for different k values for the *Cover Type* data.

prediction method for `pam` and `clara` objects defined in Example 14.1.2. Its implementation is very simple, but unfortunately rather inefficient computationally and the following call may take several minutes to complete.

dmr.cluseval

```
ctn.cla7.pred <- predict(ctn.cla[["7"]], ctn.val[,ct.input.attrs])
```

It may be possible to assign new data to clusters more efficiently by creating classifiers serving as cluster membership models. Using a decision tree classifier, grown with default `rpart` settings, this can be performed as follows:

```
ct.cla7.tree <- rpart(make.formula("as.factor(ctn.cla[[\"7\"]]$clustering)",
                                   ct.input.attrs),
                      ct.train, xval=0)
```

Notice that the above `rpart` call uses the original rather than normalized training set. This is because normalization (and standardization) makes no difference for tree structure and quality, and only affects the particular threshold values used for inequality-based splits. Using the original training set makes the tree directly interpretable, with split threshold values coming from the original attribute codomains.

To evaluate the quality of this cluster membership model, we can compare its predictions with the dissimilarity-based cluster assignment on the validation set determined above:

```
ct.cla7.tree.pred <- predict(ct.cla7.tree, ct.val, type="c")
ct.cla7.tree.cm01 <- confmat01(ct.cla7.tree.pred, as.factor(ctn.cla7.pred))
ct.cla7.tree.tpfp <- sapply(ct.cla7.tree.cm01,
                            function(cm) c(tpr=tpr(cm), fpr=fpr(cm),
                                           fm=f.measure(cm)))
rowMeans(ct.cla7.tree.tpfp)
```

This calculates the same per-class performance measures as used above for cover type classification models. Despite using default `rpart` stop criteria that performed so poorly for cover type prediction, this cluster membership model appears quite good:

```
        1     2     3      4      5     6     7
tpr 0.980 0.920 0.79 0.8700 0.940 0.970 0.90
fpr 0.004 0.018 0.00 0.0041 0.013 0.041 0.00
fm  0.980 0.870 0.89 0.9100 0.950 0.920 0.95
```

The true positive rate is above 0.9 on the average, with the false positive rate just above 0.01. This excellent predictive performance – that could be possibly even further improved by adjusting parameters – shows that cluster membership is much easier to represent that the original cover type target attribute. It is also easy to verify that clusters are much more balanced than cover types. The tree can be presented graphically using the following call:

```
prp(ct.cla7.tree, varlen=0, faclen=0)
```

which produces the plot presented in Figure 20.23. The cluster membership model is indeed quite simple and – with sufficient domain knowledge about attribute interpretation – would

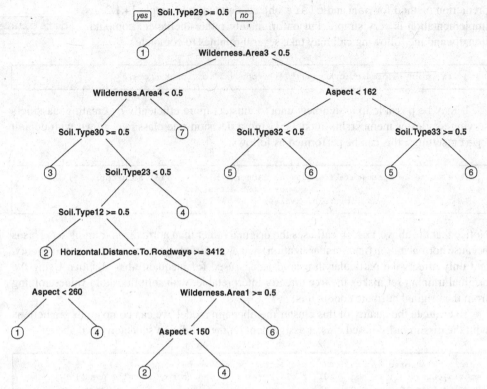

Figure 20.23   The decision tree modeling cluster membership for the *Cover Type* data.

probably permit describing the distinguishing characteristics of particular cluster in a comprehensible way.

20.5   Conclusion

The case studies presented in this chapter – despite their limited scope and adopted simplifications – hopefully give some taste of data mining practice that the other chapters, focused on explaining the mechanics of data mining algorithms, were unable to provide. It not only shows how to put the previously described algorithms at some serious work, but also demonstrates how understanding the algorithms makes their application easier and more effective. This justifies its inclusion in the book.

This chapter is composed of text narration and R code in near equal proportions. While the code is simple, it is not entirely limited to straightforward calls to modeling algorithm implementations. Some basic data preprocessing is performed, multiple models are created and evaluated, different parameter setups and attribute subsets are tried. This should make it possible to appreciate the flexibility of R as an analytic environment, particularly in comparison to point-and-click tools.

A side effect of the case studies is also some evidence of the practical utility of selected R functions defined in this books examples. While standard CRAN implementations of modeling algorithms are usually much more efficient and flexible than that the illustrative implementations presented in these examples, some of them do provide practically useful functions for model evaluation and attribute transformation. This applies, in particular, to the simple aggregation, imputation, and normalization transformations used in this chapter in a way that adheres to the modeling view: with transformation models determined on the training set and applicable to arbitrary other datasets from the same domain. This is not always convenient using their standard R implementations.

20.6   Further readings

Some results obtained while analyzing the datasets used by this case study can be found in the literature. Definitely the most popular of them is the *Census Income* dataset, which was used by Oza and Russell (2001) to experimentally compare incremental and batch versions of bagging and boosting, and by Frank *et al.* (2002) to evaluate an incremental boosting algorithm applicable to large data. Bay (2001) performed experiments with a multivariate discretization algorithm (working with multiple continuous attributes at a time) on this dataset. A smaller version of the dataset (also known as *Adult*) with a subset of attributes was used for several experimental studies (e.g., Chawla *et al.* 2002; Keerthi *et al.* 2005; Kohavi 1996; Platt 1998; Zadrozny and Elkan 2002).

Redmond and Highley (2009) used the *Communities and Crime* dataset to experimentally evaluate a training instance removal technique for memory-based regression. Redmond and Baveja (2002) developed their system for crime similarity pattern identification based on a differently normalized version thereof with less input attributes, but several additional target attributes corresponding to specific crimes. Buczak and Gifford (2010) performed association discovery on a preprocessed version of the data.

The *Cover Type* data was used to perform forest cover type prediction experiments with neural networks and linear discriminant analysis by Blackard and Dean (2000). It was also used as one of the datasets for bagging and boosting experiments reported by Oza and Russell (2001). Gama *et al.* (2003) mentioned the application of their incremental decision tree algorithm to this data. More recently Chandra and Pallath (2007) presented results obtained using a decision tree algorithm designed for large datasets, an enhancement of SLIQ (Mehta *et al.* 1996).

The UCI repository, the source of datasets for this chapter's case studies, as well as examples presented in other chapters, have been serving the machine learning and data mining community for more than a quarter-century (Bache and Lichman 2013). It is impossible to overestimate its impact on research and education in these areas, stimulating progress in algorithms and evaluation methodologies as well as providing students opportunities to get hands-on experience with real-world modeling tasks.

References

Bache K and Lichman M 2013 UCI Machine Learning Repository. School of Information and Computer Science, University of California Irvine.

Bay SD 2001 Multivariate discretization for set mining. *Knowledge and Information Systems* **3**, 491–512.

Blackard JA and Dean DJ 2000 Comparative accuracies of artificial neural networks and discriminant analysis in predicting forest cover types from cartographic variables. *Computers and Electronics in Agriculture* **24**, 131–151.

Buczak AL and Gifford CM 2010 Fuzzy association rule mining for community crime pattern discovery *Proceedings of the ACM SIGKDD Workshop on Intelligence and Security Informatics*. ACM Press.

Chandra B and Pallath PV 2007 Prediction of forest cover using decision trees. *Journal of the Indian Society of Agricultural Statistics* **61**, 192–198.

Chawla NV, Bowyer, K. W. Hall LO and Kegelmeyer WP 2002 SMOTE: Synthetic minority over-sampling technique. *Journal of Artificial Intelligence Research* **16**, 321–357.

Frank, E. Holmes G, Kirkby R and Hall MA 2002 Racing committees for large datasets *Proceedings of the Fifth International Conference on Discovery Science (DS-2002)*. Springer.

Gama J, Rocha R and Medas P 2003 Accurate decision trees for mining high-speed data streams *Proceedings of the Ninth ACM SIGKDD International Conference on Knowledge Discovery and Data Mining*. ACM Press.

Keerthi SS, Duan KB, Shevade SK and Poo AN 2005 A fast dual algorithm for kernel logistic regression. *Machine Learning* **61**, 151–165.

Kohavi R 1996 Scaling up the accuracy of naive-Bayes classifiers: A decision-tree hybrid *Proceedings of the Second International Conference on Knowledge Discovery and Data Mining (KDD-96)*. AAAI Press.

Mehta M, Agrawal R and Rissanen J 1996 SLIQ: A fast scalable classifier for data mining *Proceedings of the Fifth Intl Conference on Extending Database Technology (EDBT-96)*. Springer.

Oza NJ and Russell SJ 2001 Experimental comparisons of online and batch versions of bagging and boosting *Proceedings of the Seventh ACM SIGKDD International Conference on Knowledge Discovery and Data Mining*. ACM Press.

Platt JC 1998 Using analytic QP and sparseness to speed training of support vector machines *Advances in Neural Information Processing Systems 11 (NIPS-98)*. MIT Press.

Redmond MA and Baveja A 2002 A data-driven software tool for enabling cooperative information sharing among police departments. *European Journal of Operational Research* **141**, 660–678.

Redmond MA and Highley T 2009 Empirical analysis of case-editing approaches for numeric prediction *Proceedings of the 2009 International Conference on Systems, Computing Sciences and Software Engineering (SCSS-2009)*. Springer.

Zadrozny B and Elkan C 2002 Transforming classifier scores into accurate multiclass probability estimates *Proceedings of the Eighth ACM SIGKDD International Conference on Knowledge Discovery and Data Mining*. ACM Press.

Closing

To be without some of the things you want is an indispensable part of hapiness.

BERTRAND RUSSELL

Retrospecting

This book has covered a selection of essential data mining algorithms, mostly originating from machine learning and assuming a machine learning perspective. While both algorithm properties and usage principles are discussed, particular attention has been given to their internal operation mechanisms, explained not only by textual descriptions, equations, and occasionally pseudocode, but also by illustrative and simple R code examples. Apart from algorithms directly used for creating predictive models, techniques for model evaluation, data transformation, and attribute selection have been presented that do not always receive as much attention in data mining practice as they deserve. They may all have significant impact on the final model quality. Regrettably, they also provide opportunities to do things wrong, resulting in producing poor models or overoptimistic performance estimates. The adopted modeling view of data mining was extended even to basic statistics and data transformation. This helps not only to achieve consistency, but also highlight possible pitfalls related to the latter. The above-average space occupied in this book by the discussion of model evaluation techniques and – partially related – methods of incorporating misclassification costs to classification models is motivated by their high practical importance, not always sufficiently recognized.

Modeling algorithms presented in the book are deliberately limited to the three major and most general data mining tasks: classification, regression, and clustering. All additionally discussed techniques are also presented in the context provided by these three tasks. A number of other tasks remain undiscussed, including in particular time series forecasting, association rule and temporal pattern discovery, geo-spatial analysis, graph mining and social network analysis, clickstream analysis, and several types of text mining. Even if some of those can be viewed as specific versions of classification, regression, and clustering, they can be much more effectively addressed by dedicated rather than general-purpose algorithms. With these more specific tasks gaining increasingly more attention, a book sticking to the basic general tasks may appear outdated. The core tasks and algorithms still remain the most frequently encountered in practical applications. More importantly, they are essential for understanding data

Data Mining Algorithms: Explained Using R, First Edition. Paweł Cichosz.
© 2015 John Wiley & Sons, Ltd. Published 2015 by John Wiley & Sons, Ltd.

mining and acquiring good data mining practices. After getting intimate with them, extending one's competence to the more specific tasks becomes much easier. It would be definitely great to see a follow-up to this book, covering the most important of those using a similar R-illustrated presentation approach, but including them here by adding several more chapters was not reasonably possible.

Even the coverage of the three basic tasks, classification, regression, and clustering, is far from completeness and mostly limited to the best-known classic algorithms. This is motivated by the naturally limited capacity of a single book on one hand and strong preference for consistence with respect to the presentation style, accessibility, and depth. If the actually reached level of consistency leaves a lot to be desired, striving for completeness would inevitably make it much worse.

The book – as a practical guide to data mining algorithms – does not refer to the underlying machine learning and statistics theory. This is not to question the usefulness of theoretical achievements of these fields by any means. They provide solid grounds for understanding the properties of data mining tasks and the capabilities of data mining algorithms, as well as deeper justifications or refutations for common beliefs and intuitions. While many of those theoretical results do not naturally fit to the adopted scope and style, some of them – in particular, probabilistic and information-theoretic model selection criteria – could become a valuable complement of the presented model performance measures and evaluation procedures. Of all the numerous omissions of this book this may be the hardest to forgive.

Final words

There is obviously no reason to find the adopted presentation method – focusing on algorithmic details, with rather limited maths and usage scenario discussion, but extensive R code illustrations – universally superior to any other possible approaches, e.g., more oriented toward the underlying theory or to possible practical applications. It may appeal to some readers and repel others. Hopefully the latter will not be disappointed since they would not buy or borrow this book in the first place. Unfortunately, even those who find the adopted perspective and presentation interesting and useful may find the book somehow lacking. This is perfectly understandable and I am the last to claim this book is satisfactory in all respects. The selection of algorithms, the scope, depth, and clarity of their presentation, can be all considered justifiably insufficient. The comprehensibility and utility of R code examples can be questioned. Certainly many of them could be rewritten in a more readable form, they could be more consistent with respect to R coding style, more comprehensively explained, and better organized.

With all these reservations, I have found myself early draft versions of the text useful for my teaching and some code snippets useful for my practical data mining work. After making extensive text and code portions available to my students, I have observed a noticeable quality improvement of their project assignments, including more thoughtfully designed experimental procedures for model creation and evaluation and better use of the power of R. This leads me to the belief that there will be readers who find this book a helpful reference, a source of guidance and inspiration, and maybe even enjoyable reading.

A

Notation

This appendix lists notational conventions used throughout this book. While they are usually explained when first used, they are collected here for reference.

A.1 Attribute values

The following notation is used to refer to attribute values of instance x.

Value of attribute a_i for instance x. $a_i(x)$

Co-domain of attribute a_i. A_i

Vector of all attribute values for instance x. $\mathbf{a}(x)$.

A.2 Data subsets

The following notation is used to refer to subsets of dataset S.

Subset of S satisfying condition.

$$S_{condition} = \{x \in S \mid x \text{ satisfies } condition\} \tag{A.1}$$

Subset of S with the value of attribute a equal, not equal v.

$$S_{a=v} = \{x \in S \mid a(x) = v\} \tag{A.2}$$

$$S_{a\neq v} = \{x \in S \mid a(x) \neq v\} \tag{A.3}$$

Subset of S with the value of attribute a less, less or equal, greater, greater, or equal than v.

$$S_{a<v} = \{x \in S \mid a(x) < v\} \tag{A.4}$$

Data Mining Algorithms: Explained Using R, First Edition. Paweł Cichosz.
© 2015 John Wiley & Sons, Ltd. Published 2015 by John Wiley & Sons, Ltd.

$$S_{a \leq v} = \{x \in S \mid a(x) \leq v\} \tag{A.5}$$

$$S_{a > v} = \{x \in S \mid a(x) > v\} \tag{A.6}$$

$$S_{a \geq v} = \{x \in S \mid a(x) \geq v\} \tag{A.7}$$

Subset of S with the value of attribute a_1 equal, not equal the value of attribute a_2.

$$S_{a_1 = a_2} = \{x \in S \mid a_1(x) = a_2(x)\} \tag{A.8}$$

$$S_{a_1 \neq a_2} = \{x \in S \mid a_1(x) \neq a_2(x)\} \tag{A.9}$$

Subset of S with the value of attribute a missing.

$$S_{a=?} = \{x \in S \mid a(x) \text{ is missing}\} \tag{A.10}$$

Subset of S containing instances of class d.

$$S^d = S_{c=d} = \{x \in S \mid c(x) = d\} \tag{A.11}$$

Subset of S containing instances of class d and satisfying condition.

$$S^d_{condition} = \{x \in S \mid c(x) = d \wedge condition\} \tag{A.12}$$

A.3   Probabilities

The following notation is used to refer to probabilities estimated on a dataset S.

Probability of attribute a taking value v. $P_S(a = v)$

Probability of attribute a taking value less, less or equal, greater, greater or equal than v.
  $P_S(a < v)$, $P_S(a \leq v)$, $P_S(a > v)$, $P_S(a \geq v)$

Probability of attribute a having a missing value. $P_S(a = ?)$.

Probability of class d. $P_S(d) = P_S(c = d)$

B

R packages

Despite this book being heavily loaded with R code, it uses existing R packages quite frugally. This is because the example code serves the purpose of illustrating the internal operation mechanics of data mining algorithms rather than demonstrating the usage of their popular R implementations. Even the case studies of Chapter 20 do not employ many of them. Those few that are used in the book provide modeling algorithm implementations and other auxiliary functions, as well as datasets. Apart from the *standard* base, stats, and datasets packages – which are always preinstalled and preloaded – the remaining *contributed* packages have to be explicitly loaded and – depending on the particular R environment, which may come with several popular packages preinstalled – installed from a CRAN repository mirror.

A more important category of R packages related are those that serve as containers for all the reusable R functions defined in this book's examples, as well as some simple utility functions not included in the book due to their lack of didactic value. They all have the dmr. name prefix and are referred to as DMR packages. These can be downloaded from the book's website and then installed from local files.

Whenever a CRAN or DMR package is used by one or more examples presented in a chapter, it is explicitly loaded in the chapter's first example. This appendix contains a full list of packages of both these categories that are required by any code snipped occurring in the book.

B.1 CRAN packages

The Comprehensive R Archive Network (CRAN) is a repository of R sources, binaries, documentation, and contributed packages. The following table lists all the CRAN packages used in this book, with a brief description summarizing what the package is used for. This may sometimes be a fraction of the package's functionality.

CRAN repository links are not provided, since each of them can be constructed using the following template:

`cran.r-project.org/package=`*name*

with *name* replaced by package name.

Package name	What used for
`cluster` (Maechler et al. 2013)	Dissimilarity measures and k-medoids clustering
`digest` (Eddelbuettel 2013)	Hash string generation
`e1071` (Meyer et al. 2014)	Naïve Bayes classifier
`ipred` (Peters and Hothorn 2013)	Bagging model ensembles
`kernlab` (Karatzoglou et al. 2004)	Quadratic programming solver
`lattice` (Sarkar 2008)	Surface and level plots
`Matrix` (Bates and Maechler 2014)	Finding the nearest positive-definite matrix
`mlbench` (Leisch and Dimitriadou 2010)	Example datasets
`randomForest` (Liaw and Wiener 2002)	Random forest model ensembles
`rpart` (Therneau et al. 2014)	Decision and regression trees
`rpart.plot` (Milborrow 2014)	Decision and regression tree plotting
`quadprog` (Turlach 2013)	Quadratic programming solver

B.2 DMR packages

The relationship between DMR packages and this book's chapters is mostly, but not entirely 1:1. Chapter 1 only contributes illustrative datasets for classification, regression, and clustering. The `dmr.util` package does not correspond to any chapter, but is used by most. The definitions of its utility functions do not appear anywhere in the book, but can be seen by inspecting its source code. As a matter of fact, since all functions from all DMR packages are written entirely in R they are all built as source packages, simply typing a function's name in the R command line will list its code.

The main reason for packaging this book's example functions is to make it easier to run the example code, which frequently reuses functions defined in other chapters. These function backward- and forward references are unavoidable and would incur extreme inconveniences of copy-pasting or including source files if they were not organized into a set of packages. Now, whenever an example code snippet calls a function defined in another chapter, it only requires that the corresponding package be loaded. As mentioned above, the first example in each chapter takes care of loading all packages required for the chapter's examples.

The DMR packages with brief descriptions and the corresponding chapter numbers are listed in the table below.

Package name	Description	Chapter
dmr.attrsel	Attribute selection filters and wrappers	19
dmr.bayes	Naïve Bayes classifier	4
dmr.claseval	Classification performance measures, model evaluation procedures	7
dmr.cluseval	Clustering quality measures	14
dmr.data	Illustrative tiny datasets	1
dmr.dectree	Decision trees	3
dmr.disc	Discretization	18
dmr.dissim	Dissimilarity and similarity measures	11
dmr.ensemble	Model ensembles	15
dmr.hierclus	Hierarchical clustering	13
dmr.kcenters	k-centers clustering	12
dmr.kernel	Support vector machines, support vector regression, kernel functions	16
dmr.linclas	Linear classification	5
dmr.linreg	Linear regression	8
dmr.miscost	Incorporating misclassification costs	6
dmr.regeval	Regression performance measures	10
dmr.regtree	Regression trees	9
dmr.rpartutil	Utilities for using rpart decision and regression trees	20
dmr.stats	Descriptive statistics and statistical relationship measures	2
dmr.trans	Attribute transformations	17
dmr.util	Utility functions	

B.3   Installing packages

The standard platform-independent way of installing R packages is by using the install.packages function. For CRAN packages, one just supplies a character string with the package name as an argument, like in the following call:

```
install.packages("randomForest")
```

To request installing multiple packages a vector of package names can be specified:

```
install.packages(c("cluster", "e1071"))
```

When calling `install.packages` for the first time in the current R session, the user will be asked to select a CRAN mirror to use. It usually makes sense to choose the geographically nearest one. With default settings, which should match all standard needs, the function will automatically identify and download the suitable source or binary version of the package (depending on the operating system), compile it if necessary, and install in the default location.

Installing DMR packages from local files can also be performed by the `install.packages` function by providing a package file name instead of a package name and specifying the `repos=NULL` argument:

```
install.packages("dmr.util_1.0.tar.gz", repos=NULL)
```

This assumes that the package file is in the current R working directory, which can be checked using the `getwd` function and modified using the `setwd` function. Otherwise a file path has to be specified, as in this call:

```
install.packages("DMR-packages/dmr.util_1.0.tar.gz", repos=NULL)
```

Alternatively, packages can be installed from local files by running R from the operating system's command line with the `CMD INSTALL` arguments, followed by a package file name, as in the following example:

```
R CMD INSTALL dmr.util_1.0.tar.gz
```

This assumes the package file is in the current directory, otherwise a file path has to be specified:

```
R CMD INSTALL DMR-packages/dmr.util_1.0.tar.gz
```

More precise instructions, covering nonstandard installation options (such as library locations) and including operating system-specific details, are provided in R documentation.

References

Bates D and Maechler M 2014 *Matrix: Sparse and Dense Matrix Classes and Methods*, http://CRAN.R-project.org/package=Matrix.

Eddelbuettel D 2013 *digest: Create Cryptographic Hash Digests of R Objects*, http://CRAN.R-project.org/package=digest.

Karatzoglou A, Smola A, Hornik K and Zeileis A 2004 kernlab – an S4 package for kernel methods in R. *Journal of Statistical Software* **11**, 1–20.

Leisch F and Dimitriadou E 2010 *mlbench: Machine Learning Benchmark Problems*, http://CRAN.R-project.org/package=mlbench.

Liaw A and Wiener M 2002 Classification and regression by randomForest. *R News* **2**, 18–22.

Maechler M, Rousseeuw P, Struyf A, Hubert M and Hornik K 2013 *cluster: Cluster Analysis Basics and Extensions*, http://CRAN.R-project.org/package=cluster.

Meyer D, Dimitriadou E, Hornik K, Weingessel A and Leisch F 2014 *e1071: Misc Functions of the Department of Statistics (e1071)*, TU Wien.

Milborrow S 2014 *rpart.plot: Plot rpart Models*, http://CRAN.R-project.org/package=rpart.plot.

Peters A and Hothorn T 2013 *ipred: Improved Predictors*, http://CRAN.R-project.org/package=ipred.

Sarkar D 2008 *lattice: Multivariate Data Visualization with R*. Springer.

Therneau T, Atkinson B and Ripley B 2014 *rpart: Recursive Partitioning and Regression Trees*, http://CRAN.R-project.org/package=rpart.

Turlach BA 2013 *quadprog: Functions to Solve Quadratic Programming Problems*, http://CRAN.R-project.org/package=quadprog.

C

Datasets

Besides the tiny *weather* family of datasets presented in Chapter 1 and artificially generated datasets in some chapters, the R code examples use a set of real datasets originating from various sources. They are all available for download from the *UCI Machine Learning Repository*. Except for those used by case studies in Chapter 20, the datasets do not actually have to be downloaded from the repository, since they are also available in R packages, `mlbench` and `datasets`. It still makes sense to check the corresponding UCI pages for some basic characteristics of the data as well as information about their origin and past usage. The table presented below lists all the UCI datasets used in this book, providing their original repository names as well R package names, where available. The corresponding links to the UCI pages can be constructed using the following simple template:

```
http://archive.ics.uci.edu/ml/datasets/name
```

with *name* replaced by UCI dataset name.

Dataset	UCI name	R package/name
Census Income	Census-Income+(KDD)	
Communities and Crime	Communities+and+Crime	
Cover Type	Covertype	
Boston Housing	Housing	`mlbench/BostonHousing`
Glass	Glass+Identification	`mlbench/Glass`
HouseVotes84	Congressional+Voting+Records	`mlbench/HouseVotes84`
Iris	Iris	`datasets/iris`
Pima Indians Diabetes	Pima+Indians+Diabetes	`mlbench/PimaIndiansDiabetes`
Soybean	Soybean+(Large)	`mlbench/Soybean`
Vehicle Silhouettes	Statlog+(Vehicle+Silhouettes)	`mlbench/Vehicle`

Data Mining Algorithms: Explained Using R, First Edition. Paweł Cichosz.
© 2015 John Wiley & Sons, Ltd. Published 2015 by John Wiley & Sons, Ltd.

Index

0-1 loss, *see* loss function, 0-1

.632 bootstrap, 225–226, 306

absolute loss, *see* loss function, absolute
accuracy, 191, 196
AdaBoost, 435–437
    instance weighting for, 435
    model weighting for, 435
    multiclass, 438–439
AdaBoost.MH, 439
adaptive boosting, *see* AdaBoost
add-one smoothing, *see* probability,
    Laplace estimate of
agglomerative clustering, *see* hierarchical
    clustering, agglomerative
aggregation, 506
AHC, *see* hierarchical clustering,
    agglomerative
algorithm randomization, 412–413
anomaly detection, 163
    by clustering, 17, 352
ANOVA, *see* F-test
AODE, *see* averaged one-dependence
    estimators
apparent disutility, 570
apparent utility, 570
attribute, xxii, 5
    continuous, 6, 25
        discretization of, *see*
            discretization

discrete, 36
    aggregation of, *see* aggregation
    binary encoding of, *see* attribute
        encoding
hidden, 16
input, *see* input attribute
nominal, 5
numeric, *see* attribute, continuous
observable, 16
ordinal, 6
target, *see* target attribute
attribute encoding, 154, 250, 508–509
attribute sampling, 410
attribute selection
    motivation for, 559–560
    target algorithm for, 560
    target task for, 560
attribute selection filter, 561, 568
    consistency-based, *see*
        consistency-based filter
    correlation-based, *see*
        correlation-based filter
    random forest, *see* random forest, for
        attribute selection
    simple statistical, *see* simple statistical
        filter
attribute selection search, 562, 563
    backward, *see* backward elimination
    filter-driven, 586–587
    forward, *see* forward selection
    greedy, 566

Data Mining Algorithms: Explained Using R, First Edition. Paweł Cichosz.
© 2015 John Wiley & Sons, Ltd. Published 2015 by John Wiley & Sons, Ltd.

Printed in the United States
By Bookmasters